Lecture Notes in Computer Science 1643
Edited by G. Goos, J. Hartmanis and J. van Leeuwen

Springer

Berlin
Heidelberg
New York
Barcelona
Hong Kong
London
Milan
Paris
Singapore
Tokyo

Jaroslav Nešetřil (Ed.)

Algorithms – ESA '99

7th Annual European Symposium
Prague, Czech Republic, July 16-18, 1999
Proceedings

Series Editors

Gerhard Goos, Karlsruhe University, Germany
Juris Hartmanis, Cornell University, NY, USA
Jan van Leeuwen, Utrecht University, The Netherlands

Volume Editor

Jaroslav Nešetřil
Charles University
Department of Applied Mathematics and DIMATIA Centre
Malostranské nám. 25, CZ-11800 Prague 1, Czech Republic
E-mail: nesetril@kam.ms.mff.cuni.cz

Cataloging-in-Publication data applied for

Die Deutsche Bibliothek - CIP-Einheitsaufnahme

Algorithms : 7th annual European symposium ; proceedings / ESA '99, Prague, Czech Republic, July 16 - 18, 1999. Jaroslav Nešetřil (ed.). - Berlin ; Heidelberg ; New York ; Barcelona ; Hong Kong ; London ; Milan ; Paris ; Singapore ; Tokyo : Springer, 1999
(Lecture notes in computer science ; Vol. 1643)
ISBN 3-540-66251-0

CR Subject Classification (1998): F.2, G.1-2, E.1, F.1.3

ISSN 0302-9743
ISBN 3-540-66251-0 Springer-Verlag Berlin Heidelberg New York

Typesetting: Camera-ready by author
SPIN: 10703529 06/3142 – 5 4 3 2 1 0 Printed on acid-free paper

Preface

The 7th Annual European Symposium on Algorithms (ESA '99) was held in Prague, Czech Republic, July 16-18, 1999. This continued the tradition of the meetings which were held in

- 1993 Bad Honnef (Germany)
- 1994 Utrecht (The Netherlands)
- 1995 Corfu (Greece)
- 1996 Barcelona (Spain)
- 1997 Graz (Austria)
- 1998 Venice (Italy)

(The proceedings of previous ESA meetings were published as Springer LNCS volumes 726, 855, 979, 1136, 1284, 1461.)

In the short time of its history ESA (like its sister meeting SODA) has become a popular and respected meeting.

The call for papers stated that the "Symposium covers research in the *use, design, and analysis of efficient algorithms and data structures* as it is carried out in computer science, discrete applied mathematics and mathematical programming. Papers are solicited describing original results in *all areas of algorithmic research*, including but not limited to: Approximation Algorithms; Combinatorial Optimization; Computational Biology; Computational Geometry; Databases and Information Retrieval; Graph and Network Algorithms; Machine Learning; Number Theory and Computer Algebra; On-line Algorithms; Pattern Matching and Data Compression; Symbolic Computation. The algorithms may be sequential, distributed or parallel, and they should be analyzed either mathematically or by rigorous computational experiments. Submissions that report on experimental and applied research are especially encouraged."

A total of 122 papers were received. The program committee thoroughly reviewed the papers (each paper was sent to at least 3 members of the PC) and after an electronic discussion the agreement was reached during the PC meeting in Prague, March 26-28, 1999. (At this meeting, the following members of the PC were physically present: G. Billardi, H. Bodlaender, J. Diaz, A. Goldberg, M. Goemans, M. Kaufmann, B. Monien, J. Matoušek, E. Mayr, J. Nešetřil, P. Widmayer, G. Woeginger, while most of the other PC members cooperated electronically.)

The program committee selected 44 papers, which are presented in this volume.

The program of ESA' 99 included two invited talks by Bernhard Korte (University of Bonn) and Moti Yung (Columbia, New York).

As an experiment, ESA'99 was held at the same time as ICALP'99. Only history will show whether this scheme is suitable to the European context.

The ESA'99 was organized by the Center of Discrete Mathematics, Theoretical Computer Science and Applications (shortly DIMATIA) which is a joint venture of Charles University, the Czech Academy of Science and the Institute for Chemical Technology (all based in Prague) together with its associates (presently in Barcelona, Bielefeld, Bonn, Bordeaux, Budapest, Novosibirsk, Pilsen, and Pisa). Also two similar centers DIMACS (Rutgers University, New Jersey) and PIMS (Vancouver, Canada) are associated members of DIMATIA. We thank all our partners for the support and all the work. We would like to thank all the members of the organizing committee, and particularly Mrs. Hana Čásenská (DIMATIA) and Mrs. Anna Kotěšovcová (CONFORG). The electronic efficiency was made possible by Jiří Fiala, Jiří Sgall, Vít Novák, and Pavel Valtr.

We also thank our industrial supporters Čedok, Telecom, Komerční banka, Mercedes Benz Bohemia, and Conforg (all based in Prague). DIMATIA is also supported by GACR grant 201/99/0242 and MŠMT grant 055 (Kontakt).

We hope that the present volume reflects the manifold spectrum of contemporary algorithmic research and that it convincingly demonstrates that this area is alive and well. We wish the next ESA (to be held in Saarbrücken) success and many excellent contributions.

July 1999 Jaroslav Nešetřil

Organization

ESA'99 was organized by the

Conference Chair

Jaroslav Nešetřil (Charles University)

Organizing Committee

H. Čásenská
J. Fiala
V. Janota
A. Kotěšovcová
J. Kratochvíl
L. Kučera
H. Nešetřilová
V. Novák
J. Sgall
P. Valtr

Program Committee

H. Attiya (Haifa)
G. Ausiello (Rome)
G. Bilardi (Padua)
H. Boadlander (Utrecht)
W. Cook (Houston)
J. Diaz (Barcelona)
A. Frieze (Pittsburgh)
M. Goemans (Louvain)
A. Goldberg (Princeton)
G. Italiano (Venice)
H. Karloff (Atlanta)
M. Kaufmann (Tubingen)
J. K. Lenstra (Eindhoven)
J. Matoušek (Prague)
E. Mayr (Munich)
B. Monien (Paderborn)
B. Shepherd (Murray Hill)
P. Widmayer (Zurich)
J. Wiederman (Prague)
G. Woeginger (Graz)
A. Zvonkin (Bordeaux)

Referees

Karen Aardal
Susanne Albers
Noga Alon
Carme Alvarez
Fabrizio d'Amore
Matthew Andrews
Richard Arratia
Yossi Azar
Ricardo Baeza-Yates
Nicole Bidoit
Vita Bortnikov
Srecko Brlek
Prasad Chalasani
Otfried Cheong
Fabian Chudak
Edith Cohen
Gerard Cornuejols
Bruno Courcelle
Olivier Delgrange
Camil Demetrescu
Stephan Eidenbenz

Jens Ernst
Sandor Fekete
Amos Fiat
Joaquim Gabarro
Giorgio Gambosi
Cyril Gavoille
Leslie Goldberg
Ervin Gyori
Alexander Hall
M. T. Hallett
Johan Hastad
Monika Henzinger
Cor Hurkens
David Janin
Vít Janota
Bala Kalyanasundaram
Claire Kenyon
Bettina Klinz
Ton Kloks
Werner Krandick
Stefano Leonardi
Bertrand Le Saec
Vincenzo Liberatore
Martin Loebl
Flaminia Luccio
Conrado Martínez
Colin McDiarmid
Michael Mitzenmacher
Haiko Müller
Seffi Naor
Gabriele Neyer
Rolf Niedermeier
John Noga
Jordi Petit
Kirk Pruhs
Martin Raab
Yuval Rabani
R. Ravi
Thomas Roos
Guenter Rote
S. Roura
Thomas Schickinger
Christophe Schlick
Konrad Schlude
Petra Schuurman
Maria Serna
Oriol Serra
Ron Shamir
David Sherman
David Shmoys
Alistair Sinclair
Job Smeltink
A. Steger
Cliff Stein
Robert Strandh
Wojtek Szapankowski
Gerard Tel
Prasad Tetali
Roland Ulber
Eli Upfal
Bernard Vauquelin
Marinus Veldhorst
Bram Verweij
Tandy Warnow
Roger Wattenhofer
Gordon Wilfong
Fatos Xhafa
Francis Zane
Wieslaw Zielonka
Lisa Zhang

Table of Contents

ESA'99 Program

Friday

- **9.00 - 10.00:** *Moti Yung:*
  Adaptively Secure Distributed Public-Key Systems (invited talk)
- **10.00:** cofee break
- **10.30:** *R. Ravi and F.S. Salman:*
  Approximation Algorithms for the Traveling Purchaser Problem and its Variants in Network Design
- **10.55:** *K. Diks, E. Kranakis, D. Krizanc and A. Pelc:*
  The impact of knowledge on broadcasting time in radio networks
- **11.20:** *K. Iwama and E. Miyano:*
  Multipacket routing on 2-D meshes and its applications to fault-tolerant routing
- **11.45:** *P. Crescenzi, L. Dardini and R. Grossi:*
  IP Address Lookup Made Fast and Simple
- **12.05:** lunch
- **14.00:** *A. Bar-Noy, A. Freund and S. Naor:*
  On-line load Balancing in a Hierarchical Server Topology
- **14.25:** *F. Meyer auf der Heide, B. Voecking and M. Westermann:*
  Provably good and practical strategies for non-uniform data management in networks
- **14.50:** *S. Phillips and J. Westbrook:*
  Approximation algorithms for restoration capacity planning
- **15.10:** break
- **15.30:** *R. Bar-Yehuda and D. Rawitz:*
  Efficient Algorithms for Integer Programs with Two Variables per Constraint
- **15.55:** *M. Skutella:*
  Convex Quadratic Programming Relaxations for Network Scheduling Problems
- **16.20:** *R. H. Moehring, A. S. Schulz, F. Stork and M. Uetz:*
  Resource Constrained Project Scheduling: Computing Lower Bounds by Solving Minimum Cut Problems
- **16.40:** break
- **17.00:** *L. Epstein and J. Sgall:*
  Approximation Schemes for Scheduling on Uniformly Related and Identical Parallel Machines
- **17.25:** *Y. Azar and O. Regev:*
  Off-line Temporary Tasks Assignment
- **17.50:** *S. Bischof, T. Schickinger and A. Steger:*
  Load Balancing Using Bisectors — A Tight Average-Case Analysis
- after dinner ESA business meeting

Saturday

- **8.30:** *T. Jansen and I. Wegener:*
  On the Analysis of Evolutionary Algorithms — A Proof That Crossover Really Can Help
- **8.55:** *P. Nicodeme, B. Salvy and Ph. Flajolet:*
  Motif statistics
- **9.20:** *V. Heun:*
  Approximate Protein Folding in the HP Side Chain Model on Extended Cubic Lattices
- **9.45:** *A. Crauser and P. Ferragina:*
  On constructing suffix arrays in external memory
- **10.05:** break
- **10.30:** *E. S. Laber, R. Milidiu and A. A. Pessoa:*
  Strategies for Searching with Different Access Costs
- **10.55:** *Ting Chen, Ming-Yang Kao:*
  On the Informational Asymmetry between Upper and Lower Bounds for Ultrametric Evolutionary Trees
- **11.20:** *F. Cicalese and D. Mundici:*
  Optimal binary search with two unreliable tests and minimum adptiveness
- **11.45:** *S. Roura:*
  Improving Mergesort for Linked Lists
- **12.05:** lunch
- **14.00:** *G. Manzini:*
  Efficient algorithms for on-line symbol ranking compression
- **14.25:** *E. Anderson, K. Hildrum, A. R. Karlin, A. Rasala and M. Saks:*
  On List Update and Work Function Algorithms
- **14.50:** *W. Bein, M. Chrobak and L. Larmore:*
  The 3-Server Problem in the Plane
- **15.10:** break
- **15.30:** *V. Berry, Tao Jiang, P. Kearney, Ming Li and T. Wareham:*
  Quartet cleaning: Improved algorithms and simulations
- **15.55:** *B. Gaertner:*
  Fast and Robust Smallest Enclosing Balls
- **16.20:** *E. Nardelli, M. Talamo and P. Vocca:*
  Efficient Searchingfor Multi-Dimensional Data Made Simple
- **16.40:** break
- **17.00:** *B. Chazelle:*
  Geometric Searching over the Rationals
- **17.25:** *D. V. Finocchiaro and M. Pellegrini:*
  On computing the diameter of a point set in high dimensional Eucledian space
- **17.50:** *S. G. Kolliopoulos and S. Rao:*
  A nearly linear-time approximation scheme for the Euclidean k-median problem
- Concert (St. Nicolai Church at Lesser Town)
- Conference dinner (Ledebourgh gardens under Prague Castle)

Sunday

- **8.30 - 9.30:** *B. Korte:*
  How long does a bit live in a computer? (invited talk)
- **9.30:** break
- **10.00:** *A. Bar-Noy, M. M. Halldorsson, G. Kortsarz, R. Salman and H. Shachnai:*
  Sum Multi-Coloring of Graphs
- **10.25:** *P. Krysta and K. Lorys:*
  Efficient Approximation Algorithms for the Achromatic Number
- **10.50:** *T. Ishii, H. Nagamochi, T. Ibaraki:*
  Augmenting an $(l-1)$-Vertex-Connected Multigraph to a k-Edge-Connected and l-Vertex-Connected Multigraph
- **11.25:** *K. Aardal and B. Verweij:*
  An Optimisation Algorithm for Maximum Independent Set with Applications in Map Labelling
- **11.50:** *M.Y. Kao, T.W. Lam, W.K. Sung and H.F. Ting:*
  A Decomposition Theorem for Maximum Weight Bipartite Matchings with Applications in Evolutionary Trees
- **12.10:** lunch
- **14.00:** *D. Peleg and L. Drori:*
  Faster Exact Solutions for Some NP-Hard Problems
- **14.25:** *L. Moura:*
  A polyhedral algorithm for packings and designs
- **14.50:** *A. Akhavi:*
  Threshold phenomena in random lattices and efficient reduction algorithms
- **15.10:** break
- **15.30:** *A. A. Ageev:*
  On Finding the Maximum Number of Disjoint Cuts in Seymour Graphs
- **15.55:** *A. Benczur, Z. Kiraly and J. Forster:*
  Dilworth's Theorem and its application for path systems of a cycle — implementation and analysis
- **16.20:** *J.Cheriyan, T.Jordan and R.Ravi:*
  On 2-coverings and 2-packings of laminar families
- **16.40:** break
- **17.00:** *I. Pak:*
  Random Cayley graphs with O(log G) generators are expanders
- **17.25:** *P. Hell, R. Shamir and R. Sharan:*
  A Fully Dynamic Algorithm for Recognizing and Representing Proper Interval Graphs
- **17.50:** *Xin He, Ming-Yang Kao and Hsueh-I Lu:*
  A Fast General Methodology for Information-Theoretically Optimal Encodings of Graphs

Adaptively-Secure Distributed Public-Key Systems

Yair Frankel[1], Philip MacKenzie[2], and Moti Yung[1]

[1] CertCo, yfrankel@cs.columbia.edu
[2] Information Sciences Research Center, Bell Laboratories, Murray Hill, NJ 07974, philmac@research.bell-labs.com

Abstract. When attacking a distributed protocol, an adaptive adversary is able to decide its actions (e.g., which parties to corrupt) at any time based on its entire view of the protocol including the entire communication history. Proving security of cryptographic protocols against adaptive adversaries is a fundamental problem in cryptography. In this paper we consider "distributed public-key systems" which are secure against an adaptive adversary.

1 Introduction

Distributed public-key systems involve public/secret key pairs where the secret key is distributively held by some number of the servers (using a secret sharing scheme). In these systems, a key is split amongst a set of share-holders and a quorum of servers is needed to act on a common input in order to produce a function value (a signature or a cleartext). As long as an adversary does not corrupt a certain threshold of servers the system remains secure (as opposed to centralized cryptosystems in which the compromise of a single entity breaks the system). Function sharing (Threshold) systems were presented in [16, 17, 15]. Robust function sharing systems, in which the function can be evaluated correctly even if the adversary causes share-holders it controls to misbehave arbitrarily, were presented in [29, 25, 30]. Constructions of these systems are required to be efficient (e.g., they should not involve generic "secure function evaluation" which is assumed impractical [15]). The current trend for specific efficient solutions is reviewed in [31, 28].

A fundamental problem in cryptography is coping with an adaptive adversary who may, while a protocol is running, attack the protocol using actions based on its complete view up to that point in the protocol. This problem was dealt with recently in the context of multi-party computations (secure function evaluation), initially where parties erase some of their information [4], and later even when they do not necessarily do so [6]. An adaptively secure Oblivious Transfer protocol was also given in [3]. In this paper we examine the problem of obtaining adaptive security for distributed public-key systems. None of the known implementations of distributed cryptosystems have been proven secure under the very powerful "adaptive" adversary model. We deal with both discrete-log-based (DL-based) [18] and RSA-based [44] systems.

The major difficulty in proving the security of protocols against adaptive adversaries is being able to efficiently simulate (without actually knowing the secret keys) the view of an adversary which may corrupt parties dynamically, depending on its internal "unknown strategy." The adversary's corruption strategy may be based on values of public ciphertexts, other public cryptographic values in the protocol, and the internal states of previously corrupted parties. For example, the adversary could decide to corrupt next the party whose identity matches a one-way hash function applied to the entire communication history of the protocol so far. Since in an "actual execution," an adaptive adversary *does* obtain such a view, this simulation argument is intuitively necessary to claim that the adversary obtains no useful information, and thus claim that the protocol is secure. Now, when a party is corrupted, its (simulated) internal state must be consistent with the current view of the adversary and publicly available values, which include public ciphertexts and other public cryptographic values (e.g., commitments). Without the secret keys, however, the simulator is not able to determine the true internal states of all parties simultaneously, and thus might have difficulty producing a consistent internal state for an arbitrarily corrupted party. In other words, we may fail to simulate the "on-line" corruption and will need to backtrack and try again to produce a consistent view for the adversary. But after backtracking and proceeding with different ciphertexts or cryptographic values, the adaptive adversary may corrupt different parties (based on its unknown strategy). Since the adversary can corrupt subsets of users it has an exponentially large set of corruption possibilities, and since it applies an unknown strategy, we may not be able to terminate the simulation in expected polynomial time, which is a requirement for claiming security.

In distributed public-key systems, the problem of adaptive security is exacerbated by the fact that there is generally "public function and related publicly-committed robustness information" available to anyone, which as discussed above, needs to be consistent with internal states of parties which get corrupted. This is the main cause of difficulties in the proof of security.

Our Contributions and Techniques: We give a new set of techniques that can be used to construct distributed DL-based and RSA-based public-key systems with adaptive security. Since the simulation-based proofs of the earlier techniques fail against an adaptive adversary, we have to employ new ideas. The driving "meta idea" is to develop techniques that assure, in spite of the "exponential set of behaviors" of the adversary, that the adversary can only "disrupt" the simulation with polynomial probability. This argument will assure simulatability and thus a "proof of security". The basic principle is based on the notion of a "faking server." The simulator exploits the "actions" of this server to assure that the view is simulatable while not knowing the secret key. This server is chosen at random and its public actions are indistinguishable from an honest server to the adversary. We have to backtrack the simulation **only if** the adversary corrupts this special server. Since there is only one faking server, and since regardless of its corruption strategy, the adversary has a polynomial chance (at least $1/(t+1)$) of not corrupting this one server, we will be able to complete the simulation in expected polynomial time.

We employ non-binding encryption and develop the notion of **"detached commitments"**. These commitments are used to ensure correct behavior of servers, yet have no "hard attachment" to the rest of the system, even the secret key itself! We show how to work with these detached commitments, e.g., using "function representation transformations" like "poly-to-sum" and "sum-to-poly" (which we build based on [23]). We also show how to maintain robustness by constructing simulatable "soft attachments" from these detached commitments into the operations of the rest of the system. By using detached commitments, the simulator has the freedom to separate its simulation of the secret key *representation* (which, as it turns out, doesn't even depend on the secret key) and its simulation of secret-key-based *function applications* (which naturally depends on the secret key), thus enabling a proof of security. The soft attachments are constructed using efficient zero-knowledge proofs-of-knowledge. The protocols we give for these are similar to [10], but are novel in the following ways: (1) there is one "setup" protocol for all the "proof" protocols, which allows concurrency in the "proof" protocols, and (2) the setup and proof protocols are not as tightly related (i.e. the setup does not prove knowledge of a commitment based on exactly what the proof protocol is trying to prove) but still achieve statistical zero-knowledge (no reliance on computational assumptions). We believe these ZK-proof techniques will be useful in developing future threshold cryptosystems.

Our techniques maintain "optimal resilience," namely, the protocols can withstand any minority of misbehaving parties (t faults out of l parties while $l \geq 2t+1$ is allowed). Our main results are:

Theorem 1. *There exists an adaptively-secure robust DL-based optimal-resilient (t,l)-threshold public-key system.*

Theorem 2. *There exists an adaptively-secure robust RSA-based optimal-resilient (t,l)-threshold public-key system.*

Beyond robustness of distributed cryptosystem, there is the notion of proactive security, which is a strengthening of the security of a system to cope with mobile adversaries [39]. In proactively-secure systems, the share-holders must maintain the key and re-randomize it periodically. Thus, an adversary which corrupts less than a threshold in every period cannot break the system (even if over time every share-holder is corrupted). Proactive public-key systems were presented in [33, 24, 7, 23, 43] Furthermore, to initiate the above systems without a trusted key generator or "dealer" (whose presence provides a single source of failure) requires a "distributed key generation" procedure. Such protocols were given in [40, 5, 26]. The techniques in this paper for the DL-based systems can be used to construct a proactive DL-based system and also extend to key generation. We present protocols for these, but omit the proofs due to space considerations.

Corollary 3. *There exists an adaptively-secure proactive DL-based optimal-resilient (t,l)-threshold public-key system.*

Corollary 4. *There exists an adaptively-secure DL-based optimal-resilient (t,l)-threshold key-generation system.*

In a companion paper (also [27]) we extend this work to key generation and proactive-maintenance of RSA-based systems, which require additional techniques.

2 Model and Definitions

Our system consists of l servers $\mathcal{S} = \{S_1,\dots,S_l\}$. A server is *corrupted* if it is controlled by the adversary. When a server is corrupted, we assume "for security" that the adversary sees all the information currently on that server. On the other hand, the system should not "open" secrets of unavailable servers. Namely, we separate availability faults from security faults (and do not cause security exposures due to unavailability). We assume that all un-corrupted servers receive all messages that are broadcast, and may retrieve the information from messages encrypted with a public key, if they know the corresponding private key. Our communication model is similar to [34]. All participants communicate via an authenticated bulletin board [8] in a synchronized manner.

The adversary: Our threshold schemes assume **stationary** adversary which stays at the corrupt processor (extensions to mobile adversary as defined in [15] are assumed by the proactive protocol). It is **t-restricted**; namely it can, during the life-time of the system, corrupt at most t servers. The actions of an adversary at any time may include submitting messages to the system to be signed, corrupting servers, and broadcasting arbitrary information on the communication channel. The adversary is **adaptive**; namely it is allowed to base its actions not only on previous function outputs, but on all the information that it has previously obtained during the execution of the protocol.

DL-based and RSA-based systems: For information on the basics of DL-based and RSA-based systems, see Appendix A. It includes, among other information, details of the variants of secret sharing schemes that are used, such as Shamir threshold secret sharing and Pedersen unconditionally-secure threshold verifiable secret sharing ((t,l)-US-VSS) over known groups, along with secret sharing and unconditionally-secure threshold verifiable secret sharing (INT-(t,l)-US-VSS) over the integers, with check shares computed modulo an RSA modulus.

Distributed Public-Key Systems: We will say that the secret key x is shared among the servers, and each server S_i holds share x_i. The public key associated with x will be called y. We say a *(t,l)-threshold system* is a system with l servers that is designed to withstand a t-restricted adaptive adversary. Formal definitions for distributed public-key systems are given in Appendix B.

3 Techniques

The main problem with proving security and robustness against adaptive adversaries is that public values such as ciphertexts and commitments are linked to actual cleartext values in an undeniable fashion. To "detach" ciphertexts from their cleartext values we simply employ semantically-secure non-committing encryption [4]. In fact, our (full) security proofs first assume perfectly secret channels and then add the above (a step we omit here).

A more involved issue concerns the commitments. We know that the collection of techniques needed to underly distributed public-key systems include: distributed representation methods (polynomial sharing, sum (additive) sharing), representation transformers which move between different ways to represent a function (poly-to-sum, sum-to-poly) as well as a set of "elementary" distributed operations (add, multiply, invert). For example, the "poly-to-sum" protocol is executed by $t+1$ servers at a time, and transforms $t+1$-out-of-l polynomial-based sharings to $t+1$-out-of-$t+1$ additive sharings. We need to have such techniques (motivated by [26, 23]) which are secure and robust against adaptive adversaries. We will rely on new zero-knowledge proof techniques (see Appendix C), as well as on shared representation of secrets as explained in Sections A.1 and A.2. The notation "2poly" refers to a polynomial and its companion polynomial shared with (t,l)-US-VSS (which is "unconditionally secure VSS") (or INT-(t,l)-US-VSS (which is the same but over the Integers)). The notation "2sum" refers to two additive sharings, with check shares that contain both additive shares of a server (similar to the check shares in (t,l)-US-VSS). In describing the DL-based protocols, unless otherwise noted we will assume multiplication is performed mod p and addition (of exponents) is performed mod q. In describing the RSA-based protocols, unless otherwise noted we will assume multiplication is performed mod N and addition (of exponents) is performed over the integers (i.e., not "mod" anything).

3.1 2poly-to-2sum

The goal of 2poly-to-2sum is to transform t-degree polynomials $a()$ and $a'()$ used in (t,l)-US-VSS into $t+1$ additive shares for each secret $a(0)$ and $a'(0)$, with corresponding check shares. The idea is to perform interpolation.[1] The DL-based scheme shown in Figure 1 does not actually require any communication, since all check shares can be computed from public information. We note that in Step 2 each s_i and s_i' is a multiple of L, so S_i can actually compute b_i and b_i' over the integers. The RSA-based scheme shown in Figure 2 is similar, but requires S_i to broadcast the check share, since it cannot be computed by every server.

1. **Initial configuration:** (t,l)-US-VSS (parameters: (p,q,g,h)) with t-degree polynomials $a()$ and $a'()$, and a set Λ of $t+1$ server indices. For all $i \in \Lambda$, recall S_i holds shares s_i and s_i' with corresponding check share $A_i = g^{s_i} h^{s_i'}$.
2. S_i computes additive shares $b_i = s_i z_{i,\Lambda}$ and $b_i' = s_i' z_{i,\Lambda}$.
3. Every server computes the check shares $B_i = g^{b_i} h^{b_i'} = A_i^{z_{i,\Lambda}}$ for all $i \in \Lambda$. (Note that there is no communication, since the additive shares can be computed individually by each shareholder, and all check shares can be computed from publicly available verification shares.)

Fig. 1. 2poly-to-2sum: DL-based scheme

3.2 2sum-to-2sum

The goal of 2sum-to-2sum is to randomize additive dual-shares (most likely obtained from a 2poly-to-2sum) and update the corresponding check shares. The DL-based scheme is in Figure 3 and the RSA-based scheme is in Figure 4.

[1] In [23], poly-to-sum also performed a rerandomization of the additive shares. We split that into a separate protocol for efficiency, since sometimes 2poly-to-2sum is used without a rerandomization of additive shares.

1. **Initial configuration:** INT-(t,l)-US-VSS (parameters: (N,g,h)) with t-degree polynomials $a()$ and $a'()$, and a set Λ of $t+1$ server indices. For all $i \in \Lambda$, recall S_i holds shares s_i and s'_i with corresponding check share $A_i = g^{s_i}h^{s'_i}$.
2. For all $i \in \Lambda$, S_i computes the additive shares $b_i = s_i z_{i,\Lambda}$ and $b'_i = s'_i z_{i,\Lambda}$ and publishes $B_i = g^{b_i}h^{b'_i} = A_i^{z_{i,\Lambda}}$.
3. All servers verify B_i for all $i \in \Lambda$ using $(A_i)^{V_{i,\Lambda}} \equiv (B_i)^{V'_{i,\Lambda}}$ where $V_{i,\Lambda} = \prod_{j\in\Lambda\setminus\{i\}}(0-j)$ and $V'_{i,\Lambda} = \prod_{j\in\Lambda\setminus\{i\}}(i-j)$. If the verification for a given B_i fails, each server broadcasts a (Bad,i) message and quits the protocol.

Fig. 2. 2poly-to-2sum: RSA-based scheme

1. **Initial configuration:** There is a set Λ of $t+1$ server indices. For all $i \in \Lambda$, S_i holds additive dual-share (b_i, b'_i), with corresponding check share $B_i = g^{b_i}h^{b'_i}$.
2. For all $i \in \Lambda$, S_i chooses $r_{i,j} \in Z_q$ and $r'_{i,j} \in Z_q$ for $j \in \Lambda\setminus\{i\}$.
3. For all $i \in \Lambda$, S_i sets $r_{i,i} = b_i - \sum_{j\in\Lambda\setminus\{i\}} r_{i,j}$ and $r'_{i,i} = b'_i - \sum_{j\in\Lambda\setminus\{i\}} r'_{i,j}$.
4. For all $i \in \Lambda$, S_i privately transmits $r_{i,j}$ and $r'_{i,j}$ to all S_j for $j \in \Lambda\setminus\{i\}$.
5. For all $i \in \Lambda$, S_i publishes $R_{i,j} = g^{r_{i,j}}h^{r'_{i,j}}$ for $j \in \Lambda\setminus\{i\}$.
6. All servers can compute $R_{i,i} = B_i / \prod_{j\in\Lambda\setminus\{i\}} R_{i,j}$ for all $i \in \Lambda$.
7. For all $j \in \Lambda$, S_j verifies $R_{i,j} \equiv g^{r_{i,j}}h^{r'_{i,j}}$. If the verification fails, S_j broadcasts an (Accuse,i,$R_{i,j}$) message, to which S_i responds by broadcasting $r_{i,j}$ and $r'_{i,j}$. If S_i does not respond, or $R_{i,j} \not\equiv g^{r_{i,j}}h^{r'_{i,j}}$ (which all servers can now test), then each server broadcasts a (Bad,i) message and quits the protocol.
8. For all $j \in \Lambda$, S_j computes $d_j = \sum_{i\in\Lambda} r_{i,j}$, $d'_j = \sum_{i\in\Lambda} r'_{i,j}$, and $D_j = \prod_{i\in\Lambda} R_{i,j}$.

Fig. 3. 2sum-to-2sum: DL-based scheme

3.3 2sum-to-1sum

The goal of 2sum-to-1sum is to reveal check shares corresponding to the first half of additive dual-shares, and prove they are correct. These proofs form the "soft attachments" from the information-theoretically secure verification shares to the computationally secure check shares that must correspond to the actual secret. The DL-based scheme is shown in Figure 5 and the RSA-based scheme is shown in Figure 6.

4 Protocols

We now present protocols for threshold cryptographic function application for both DL-based and RSA-based systems. The security and robustness of these protocols are proven in Appendix E.

4.1 DL-based threshold function application

Here we consider any DL-based (t,l)-threshold function application protocol that works by (1) constructing a verifiable additive representation of the secret x over $t+1$ servers

1. **Initial configuration:** There is a set Λ of $t+1$ server indices. For all $i \in \Lambda$, S_i holds additive dual-share (b_i, b'_i), with corresponding check share $B_i = g^{b_i} h^{b'_i}$.
2. For all $i \in \Lambda$, S_i chooses $r_{i,j} \in Z_N$ and $r'_{i,j} \in Z_N$, for $j \in \Lambda \setminus \{i\}$.
3. For all $i \in \Lambda$, S_i sets $r_{i,i} = b_i - \sum_{j \in \Lambda \setminus \{i\}} r_{i,j}$ and $r'_{i,i} = b'_i - \sum_{j \in \Lambda \setminus \{i\}} r'_{i,j}$.
4. For all $i \in \Lambda$, S_i privately transmits $r_{i,j}$ and $r'_{i,j}$ to all S_j for $j \in \Lambda \setminus \{i\}$.
5. For all $i \in \Lambda$, S_i publishes $R_{i,j} = g^{r_{i,j}} h^{r'_{i,j}}$ for $j \in \Lambda \setminus \{i\}$.
6. All servers can compute $R_{i,i} = B_i / \prod_{j \in \Lambda \setminus \{i\}} R_{i,j}$ for all $i \in \Lambda$.
7. For all $j \in \Lambda$, S_j verifies that each $r_{i,j}$ and $r'_{i,j}$ received is in the correct range and that $R_{i,j} \equiv g^{r_{i,j}} h^{r'_{i,j}}$. If the verification fails, S_j broadcasts an (Accuse,i,$R_{i,j}$) message, to which S_i responds by broadcasting $r_{i,j}$ and $r'_{i,j}$. If S_i does not respond, $r_{i,j}$ or $r'_{i,j}$ is not in the correct range, or $R_{i,j} \not\equiv g^{r_{i,j}} h^{r'_{i,j}}$ (which all servers can now test), then each server broadcasts a (Bad,i) message and quits the protocol.
8. For all $j \in \Lambda$, S_j computes $d_j = \sum_{i \in \Lambda} r_{i,j}$, $d'_j = \sum_{i \in \Lambda} r'_{i,j}$, and $D_j = \prod_{i \in \Lambda} R_{i,j}$.

Fig. 4. 2sum-to-2sum: RSA-based scheme

1. **Initial configuration:** Parameters (p,q,g,h). There is a set Λ of $t+1$ server indices. For all $i \in \Lambda$, S_i holds additive dual-share (d_i, d'_i), with corresponding check share $D_i = g^{d_i} h^{d'_i}$. Also, all servers S_i with $i \in \Lambda$ have performed a ZK-proof-setup protocol ZKSETUP-DL(p,q,g,h) with all other servers.
2. For all $i \in \Lambda$, S_i broadcasts $E_i = g^{d_i}$.
3. For all $i \in \Lambda$, S_i performs a ZK-proof of knowledge ZKPROOF-DL-REP(p,q,g,h,E_i,D_i) with all other servers. Recall that this is performed over a broadcast channel so all servers can check if the ZK-proof was performed correctly.
4. If a server detects that for some $i \in \Lambda$, S_i fails to perform the ZK-proof correctly, that server broadcasts a message (Bad,i) and quits the protocol.

Fig. 5. 2sum-to-1sum: DL-based scheme

with check shares over g, (2) finishing the function application with those $t+1$ servers (we will call this the *additive application* step), if there is no misbehavior, and (3) going back to step (1) if misbehavior is detected, discarding servers which have misbehaved and using $t+1$ remaining servers. We assume that there is a simulator for the additive application step for a message m which can simulate the step with inputs consisting of $t+1$ additive shares with $t+1$ check shares, where at most one additive share does not correspond to its check share, and a signature on m. The simulator fails only if the "faking server" (the one containing the unmatched share and check share) is corrupted, and otherwise provides a view to the adversary which is perfectly indistinguishable from the view the adversary would have in the real protocol. Most robust threshold DL-based protocols against static adversaries, like AMV-Harn [1, 32] and El-Gamal decryption [21] work this way. For an example with AMV-Harn signatures, see [36]. We show how to use this technique for static adversaries to construct a protocol that withstands an adaptive adversary. Specifically, we use a (t,l)-US-VSS representation

1. **Initial configuration:** Parameters (N,e,g,h). There is a set Λ of $t+1$ server indices. For all $i \in \Lambda$, S_i holds additive dual-share (d_i,d'_i), with corresponding check share $D_i = g^{d_i}h^{d'_i}$. Also, all servers S_i with $i \in \Lambda$ have performed a ZK-proof-setup protocol ZKSETUP-RSA(N,e,g) with all other servers.
2. For all $i \in \Lambda$, S_i broadcasts $E_i = m^{d_i}$, where m is the message to be signed, or more generally, the value to which the cryptographic function is being applied.
3. For all $i \in \Lambda$, S_i performs a ZK-proof of knowledge ZKPROOF-IF-REP(N,e,m,g,h,E_i,D_i) with all other servers. Recall that this is performed over a broadcast channel so all servers can check if the ZK-proof was performed correctly.
4. If a server detects that for some $i \in \Lambda$, S_i fails to perform the ZK-proof correctly, that server broadcasts a message (Bad,i) and quits the protocol.

Fig. 6. 2sum-to-1sum: RSA-based scheme

to store the secret, and for function application we use 2poly-to-1sum (shorthand for the concatenation of 2poly-to-2sum, 2sum-to-2sum, and 2sum-to-1sum) to construct the verifiable additive representation of the secret. We call this the *Basic DL-based Threshold Protocol*. The protocol is given in Figure 7.

1. **Initial configuration:** DL-based system parameters: (p,q,g).
2. The dealer generates $h \in_R Z_p^*$, $x,x' \in_R Z_q$, $y = g^x$, and a (t,l)-US-VSS on secrets x,x'.
3. Each (ordered) pair of servers (S_i,S_j) performs ZKSETUP-DL$_{S_i,S_j}(p,q,g,h)$.
4. Each server maintains a list $\mathcal{G}$ of server indices for servers that have not misbehaved (i.e., they are considered good).
5. When a message m needs to be signed, the following DISTAPPLY protocol is run:
   (a) A set $\Lambda \subseteq G$ with $|\Lambda| = t+1$ is chosen in some public way.
   (b) 2poly-to-2sum is run. If there are misbehaving servers, their indices are removed from $\mathcal{G}$ and the protocol loops to Step 5a.
   (c) 2sum-to-2sum is run. If there are misbehaving servers, their indices are removed from $\mathcal{G}$ and the protocol loops to Step 5a.
   (d) 2sum-to-1sum is run. If there are misbehaving servers, their indices are removed from $\mathcal{G}$ and the protocol loops to Step 5a.
   (e) The additive application step of the signature protocol is run. If there are misbehaving servers, their indices are removed from $\mathcal{G}$ and the protocol loops to Step 5a.
   (f) All values created during the signing protocol for m are erased.

Fig. 7. Basic DL-based Threshold Protocol

4.2 RSA-based threshold function application

We define *Basic RSA-based protocols* analogously to Basic DL-based protocols. The main change is that the version of VSS over the integers (INT-(t,l)-US-VSS) should be used. (An example of this type of protocol for RSA signature and decryption functions is given in [23].) This allows the simulator to construct a view for the adversary which is statistically indistinguishable (as opposed to perfectly indistinguishable in the DL-based protocols) from the view the adversary would have in the real protocol. We also shortcut the additive application step, and simply form the partial RSA signatures in the 2sum-to-1sum step. The protocol is given in Figure 8

1. The dealer generates an RSA public/private key (N,e,d), and computes public value x^* and secret value x such that $d \equiv x^* + L^2 x \bmod \phi(N)$, as in [24],[a] Then the dealer generates generators $g,h \in_R Z_N^*$, $x' \in_R Z_{N^3}$, and an INT-(t,l)-US-VSS (with $K = N$, i.e., the range of the secret is assumed to be $[0,N]$) on secrets x,x' with parameters (N,g,h).
2. Each (ordered) pair of servers (S_i,S_j) performs ZKSETUP-RSA$_{S_i,S_j}(N,e,g,h)$.
3. Each server maintains a list $\mathcal{G}$ of server indices for servers that have not misbehaved (i.e., they are considered good).
4. When a message m needs to be signed, the following DISTAPPLY protocol is run:
   (a) A set $\Lambda \subseteq G$ with $|\Lambda| = t+1$ is chosen in some public way.
   (b) 2poly-to-2sum is run. If there are misbehaving servers, their indices are removed from $\mathcal{G}$ and the protocol loops to Step 4a.
   (c) 2sum-to-2sum is run. If there are misbehaving servers, their indices are removed from $\mathcal{G}$ and the protocol loops to Step 4a.
   (d) 2sum-to-1sum is run. If there are misbehaving servers, their indices are removed from $\mathcal{G}$ and the protocol loops to Step 4a. If there is no misbehavior, the signature on m can be computed from the partial signatures generated in this step.
   (e) All values created during the signing protocol for m are erased.

[a] Recall that x^* is computed using only the public values N,e,L.

Fig. 8. Basic RSA-based Threshold Protocol

4.3 2sum-to-2poly

The protocol is given in Figure 9

4.4 DL-based Key Generation

In Figure 10 we give the protocol for DL-based Key Generation. The major issues are: (1) generating h (the element with unknown logarithm) in a distributed way, so as not to rely on a centralized trusted entity to generate h without knowing $DL(h,g)$, yet (2) allowing a reduction in the robustness proof from finding $DL(h',g)$ (for some random h' in a discrete log problem instance) to finding $DL(h,g)$. Finally (3) the participants share their contribution to the public/private key pair (y,x) prior to learning any (information-theoretic) knowledge, so as to avoid "restarts" which may introduce biases regarding the generated key.

1. **Initial configuration:** Parameters (p,q,g,h). There is a set Λ of server indices. For all $i \in \Lambda$, S_i holds additive dual-share (d_i, d'_i), with corresponding check share $D_i = g^{d_i} h^{d'_i}$.
2. For each $i \in \Lambda$, S_i shares d_i and d'_i using (t,l)-US-VSS, say with polynomials $v_i()$ and $v'_i()$.
3. S_j computes the sums $v(j) = \sum_{i \in \Lambda} v_i(j)$ and $v'(j) = \sum_{i \in \Lambda} v'_i(j)$. The verification shares for $v()$ and $v'()$ can be computed from the verification shares for $v_i()$ and $v'_i()$, for $i \in \Lambda$.
4. If a verification fails for the (t,l)-US-VSS from S_i, each server broadcasts (Bad, i). When 2sum-to-2poly is used for key generation, this server is simply removed from Λ, and the protocol proceeds on the smaller set Λ. When 2sum-to-2poly is used for proactive maintenance, the servers quit the protocol.

Fig. 9. 2sum-to-2poly: DL-based scheme

1. **Initial configuration:** DL-based system parameters: (p,q,g).
2. **Generate shares of h:** Each server S_i generates a random $r_i \in Z_q$ and computes $h_i = g^{r_i}$. Then each server S_i broadcasts h_i.
3. Each server S_j tests $h_i^q \equiv 1 \bmod p$, with I the set of indices in which this test passes.
4. Each (ordered) pair of servers (S_i, S_j) from $I \times I$ performs ZKSETUP-DL$_{S_i,S_j}(p,q,g,h_j)$. (This is used as the ZKSETUP-DL protocol for all later ZKPROOF-DL and ZKPROOF-DL-REP protocols.)
5. Each (ordered) pair of servers (S_i, S_j) from $I \times I$ performs ZKPROOF-DL$_{S_i,S_j}(p,q,g,h_i,h_i)$.
6. For all $i \in I$ in which S_i misbehaved in either the ZKSETUP-DL or ZKPROOF-DL protocol, i is removed from I.
7. $h = \prod_{i \in I} h_i$.
8. Each server S_i randomly chooses an additive share of the secret key and a companion $(s_i, s'_i \in_R Z_q)$ and broadcasts the corresponding check share $(g^{s_i} h^{s'_i})$.
9. The servers perform a 2sum-to-2poly to construct a shared representation of the secret $x = \sum_{i=1}^{l} s_i$.
10. The servers perform 2poly-to-1sum, and construct the public key $y = g^x$ from the product of the resulting check shares

Fig. 10. DL-based Distributed Key Generation Protocol

4.5 DL-based proactive maintenance

For DL-based proactive maintenance, we perform an update by running 2poly-to-2sum on the secret polynomials, and then 2sum-to-2poly. After 2sum-to-2poly, each server erases all previous share information, leaving just the new polynomial shares. If there is misbehavior by a server in either protocol, the procedure is restarted with new participants (here restarts do not introduce statistical biases and do not reduce the protocol's security). The protocol is given in Figure 11.

5 Conclusion

To summarize, we have provided protocols for distributed public-key systems that are adaptively secure. Our techniques and protocols are efficient and typically take con-

1. **Initial configuration:** (t,l)-US-VSS (parameters: (p,q,g,h)) with t-degree polynomials $a()$ and $a'()$
2. Each server maintains a list $\mathcal{G}$ of server indices for servers that have not misbehaved (i.e., they are considered good).
3. A set $\Lambda \subseteq G$ with $|\Lambda| = t+1$ is chosen in some public way.
4. 2poly-to-2sum is run. If there are misbehaving servers, their indices are removed from $\mathcal{G}$ and the protocol loops to Step 3.
5. 2sum-to-2poly is run. If there are misbehaving servers (among Λ), their indices are removed from $\mathcal{G}$ and the protocol loops to Step 3.
6. All previous share information is erased.

Fig. 11. DL-based Proactive Maintenance (Key Update) Protocol

stant communication rounds when there are no faults (and a fault may cause a constant delay).

References

[1] G. Agnew, R. C. Mullin, and S. Vanstone. Improved digital signature scheme based on discrete exponentiation. *Electronics Letters*, 26:1024–1025, 1990.

[2] E. Bach. Discrete logarithms and factoring. Technical report, Computer Science Division (EECS), University of California, Berkeley, June 1984.

[3] D. Beaver. Adaptively secure oblivious transfer. In *Advances in Cryptology—ASIACRYPT '98*, Lecture Notes in Computer Science, pages 300–314. Springer-Verlag, Nov. 1998.

[4] D. Beaver and S. Haber. Cryptographic protocols provably secure against dynamic adversaries. In *Advances in Cryptology—EUROCRYPT 92*, volume 658 of *Lecture Notes in Computer Science*, pages 307–323. Springer-Verlag, 24–28 May 1992.

[5] D. Boneh and M. Franklin. Efficient generation of shared RSA keys (extended abstract). In CRYPTO'97 [14], pages 425–439.

[6] R. Canetti, U. Feige, O. Goldreich, and M. Naor. Adaptively secure multi-party computation. In STOC'96 [47], pages 639–648.

[7] R. Canetti, S. Halevi, and A. Herzberg. Maintaining authenticated communication in the presence of break-ins. In PODC'97 [42], pages 15–24.

[8] J. D. Cohen and M. J. Fischer. A robust and verifiable cryptographically secure election scheme (extended abstract). In *26th Annual Symposium on Foundations of Computer Science*, pages 372–382, Portland, Oregon, 21–23 Oct. 1985. IEEE.

[9] R. Cramer. *Modular Design of Secure yet Practical Cryptographic Protocols.* PhD thesis, University of Amsterdam, 1995.

[10] R. Cramer, I. Damgård, and P. MacKenzie. Zk for free: the case of proofs of knowledge. manuscript, 1999.

[11] *Advances in Cryptology—CRYPTO '89*, volume 435 of *Lecture Notes in Computer Science.* Springer-Verlag, 1990, 20–24 Aug. 1989.

[12] *Advances in Cryptology—CRYPTO '91*, volume 576 of *Lecture Notes in Computer Science.* Springer-Verlag, 1992, 11–15 Aug. 1991.

[13] *Advances in Cryptology—CRYPTO '95*, volume 963 of *Lecture Notes in Computer Science.* Springer-Verlag, 27–31 Aug. 1995.

[14] *Advances in Cryptology—CRYPTO '97*, volume 1294 of *Lecture Notes in Computer Science*. Springer-Verlag, 17–21 Aug. 1997.

[15] A. De Santis, Y. Desmedt, Y. Frankel, and M. Yung. How to share a function securely (extended summary). In *Proceedings of the Twenty-Sixth Annual ACM Symposium on the Theory of Computing*, pages 522–533, Montréal, Québec, Canada, 23–25 May 1994.

[16] Y. Desmedt and Y. Frankel. Threshold cryptosystems. In CRYPTO'89 [11], pages 307–315.

[17] Y. Desmedt and Y. Frankel. Shared generation of authenticators and signatures (extended abstract). In CRYPTO'91 [12], pages 457–469.

[18] W. Diffie and M. Hellman. New directions in cryptography. *IEEE Trans. Info. Theory*, 22(6):644–654, 1976.

[19] C. Dwork, M. Naor, and A. Sahai. Concurrent zero-knowledge. In STOC'98 [48], pages 409–428.

[20] C. Dwork and A. Sahai. Concurrent zero-knowledge: Reducing the need for timing constraints. In Krawczyk [38], pages 442–457.

[21] T. ElGamal. A public key cryptosystem and a signature scheme based on discrete logarithm. *IEEE Trans. Info. Theory*, 31:465–472, 1985.

[22] U. Feige and A. Shamir. Zero knowledge proofs of knowledge in two rounds. In CRYPTO'89 [11], pages 526–545.

[23] Y. Frankel, P. Gemmell, P. D. MacKenzie, and M. Yung. Optimal-resilience proactive public-key cryptosystems. In FOCS'97 [35], pages 384–393.

[24] Y. Frankel, P. Gemmell, P. D. MacKenzie, and M. Yung. Proactive RSA. In CRYPTO'97 [14], pages 440–454.

[25] Y. Frankel, P. Gemmell, and M. Yung. Witness-based cryptographic program checking and robust function sharing. In STOC'96 [47], pages 499–508.

[26] Y. Frankel, P. D. MacKenzie, and M. Yung. Robust efficient distributed rsa-key generation. In STOC'98 [48], pages 663–672.

[27] Y. Frankel, P. D. MacKenzie, and M. Yung. Adaptively-Secure Distributed Public-Key Systems. Preliminary report of this work, Oct. 7, 1998 (STOC '99 submission).

[28] Y. Frankel and M. Yung. Distributed public-key cryptosystems. In H. Imai and Y. Zheng, editors, *Advances in Public Key Cryptography—PKC '98*, volume 1431 of *Lecture Notes in Computer Science*, pages 1–13. Springer-Verlag, Feb. 1998. invited talk.

[29] R. Gennaro, S. Jarecki, H. Krawczyk, and T. Rabin. Robust Threshold DSS Signatures. In *Advances in Cryptology—EUROYPTO '96*, volume 1070 of *Lecture Notes in Computer Science*, pages 354–371. Springer-Verlag, May. 1996.

[30] R. Gennaro, S. Jarecki, H. Krawczyk, and T. Rabin. Robust and efficient sharing of RSA functions. In *Advances in Cryptology—CRYPTO '96*, volume 1109 of *Lecture Notes in Computer Science*, pages 157–172. Springer-Verlag, 18–22 Aug. 1996.

[31] S. Goldwasser. Multi-party computations: Past and present. In PODC'97 [42], pages 1–6. invited talk.

[32] L. Harn. Group oriented (t,n) digital signature scheme. *IEEE Proc.-Comput. Digit. Tech.*, 141(5):307–313, Sept. 1994.

[33] A. Herzberg, M. Jakobsson, S. Jarecki, H. Krawczyk, and M. Yung. Proactive public-key and signature schemes. In *Proceedings of the Fourth Annual Conference on Computer and Communications Security*, pages 100–110, 1996.

[34] A. Herzberg, S. Jarecki, H. Krawczyk, and M. Yung. Proactive secret sharing, or: How to cope with perpetual leakage. In CRYPTO'95 [13], pages 339–352.

[35] IEEE. *38th Annual Symposium on Foundations of Computer Science*, Miami Beach, Florida, 20–22 Oct. 1997.

[36] S. Jarecki. *Proactive Secret Sharing and Public Key Cryptosystems*. PhD thesis, MIT, 1995.

[37] J. Kilian, E. Petrank, and C. Rackoff. Lower bounds for zero knowledge on the internet. In *39th Annual Symposium on Foundations of Computer Science*, pages 484–492. IEEE, Nov. 1998.
[38] H. Krawczyk, editor. *Advances in Cryptology—CRYPTO '98*, volume 1462 of *Lecture Notes in Computer Science*. Springer-Verlag, 17–21 Aug. 1998.
[39] R. Ostrovsky and M. Yung. How to withstand mobile virus attacks. In *Proceedings of the Tenth Annual ACM Symposium on Principles of Distributed Computing*, pages 51–61, 1991.
[40] T. P. Pedersen. Distributed provers with applications to undeniable signatures. In *Advances in Cryptology—EUROCRYPT 91*, volume 547 of *Lecture Notes in Computer Science*, pages 221–242. Springer-Verlag, 8–11 Apr. 1991.
[41] T. P. Pedersen. Non-interactive and information-theoretic secure verifiable secret sharing. In CRYPTO'91 [12], pages 129–140.
[42] *Proceedings of the Sixteenth Annual ACM Symposium on Principles of Distributed Computing*, 1997.
[43] T. Rabin. A simplified approach to threshold and proactive rsa. In Krawczyk [38], pages 89–104.
[44] R. Rivest, A. Shamir, and L. Adleman. A method for obtaining digital signature and public key cryptosystems. *Commun. ACM*, 21:120–126, 1978.
[45] C. P. Schnorr. Efficient identification and signatures for smart cards. In CRYPTO'89 [11], pages 239–252.
[46] A. Shamir. How to share a secret. *Commun. ACM*, 22:612–613, 1979.
[47] *Proceedings of the Twenty-Eighth Annual ACM Symposium on the Theory of Computing*, Philadelphia, Pennsylvania, 22–24 May 1996.
[48] *Proceedings of the Thirtieth Annual ACM Symposium on Theory of Computing*, Dallas, Texas, 23–26 May 1998.

A Basics of DL-based and RSA-based systems

A.1 Basics for DL-based systems

In these systems, we assume that p and q are two primes such that $p = mq + 1$ for some small integer m, such as 2 or 4, and that g and h are elements of Z_p^* of order q, so $g^q \equiv h^q \equiv 1 \bmod p$. El-Gamal public-key system and various signatures have been developed based on the intractability of computing discrete logs, which formally is stated as follows:

DLP Assumption Let k be the security parameter. The DLP assumption is as follows. Given primes p and q as discussed above with $|p| = k$, and given an element $g \in Z_p^*$ of order q, and the group G_g generated by g in Z_p^*, for any polynomial-time algorithm A, $\Pr[g^x \equiv y \bmod p : y \in_R G_g, x \leftarrow A(1^k, p, q, g, y)]$ is negligible.

We use various sharing techniques. We assume the reader is familiar with Shamir (t,l)-threshold polynomial secret sharing [46]. We will use the polynomial interpolation formula explicitly, so we will describe it here. For a t-degree polynomial $v(x)$, and a set $\Lambda = \{i_1, \dots, i_{t+1}\}$ of size $t+1$, $v(0)$ can be computed using polynomial interpolation. Define $z_{i,\Lambda} = \prod_{j \in \Lambda \setminus \{i\}} (i-j)^{-1}(0-j)$. Then $v(0) = \sum_{i \in \Lambda} v(i) z_{i,\Lambda}$.

We now describe an Unconditionally-Secure (t,l)-VSS ((t,l)-US-VSS) due to Pedersen [41] where secrets are drawn from Z_q and verification shares are computed in Z_p.

We assume the servers do not know the discrete log of h relative to g (this can be assured via proper initialization). The protocol begins with two (t,l)-threshold polynomial secret sharings, sharing secrets $s, s' \in Z_q$. Let $a(x) = \sum_{j=0}^{t} a_j x^j$ be the random polynomial used in sharing s and let $a'(x) = \sum_{j=0}^{t} a'_j x^j$ be the random polynomial used in sharing s'. For all i, S_i receives shares $s_i = a(i)$ and $s'_i = a'(i)$. (We refer to the pair $(a(i), a'(i))$ as *dual-share i*.) Also, the *verification shares* $\{\alpha_j (= g^{a_j} h^{a'_j})\}_{0 \leq j \leq t}$, are published.[2] Say *check share* $A_i = \prod_{j=0}^{t} \alpha_j^{i^j}$. S_i can verify the correctness of his shares by checking that $A_i = g^{s_i} h^{s'_i}$. Say s and s' are the shares computed using Lagrange interpolation from a set of $t+1$ shares that passed the verification step. If the dealer can reveal different secrets $\hat{s}$ and $\hat{s}'$ that also correspond to the zero coefficient verification share, then the dealer can compute a discrete log of h relative to g.

A.2 Basics for RSA-based systems

RSA-based systems rely on the intractability of computing RSA inverses, and hence, the intractability of factoring products of two large primes. Let k be the security parameter. Let key generator GE define a family of RSA functions to be $(e,d,N) \leftarrow GE(1^k)$ such that N is a composite number $N = P * Q$ where P, Q are prime numbers of $k/2$ bits each. The exponent e and modulus N are made public while $d \equiv e^{-1} \bmod \lambda(N)$ is kept private.[3] The **RSA encryption function** is public, defined for each message $M \in Z_N$ as: $C = C(M) \equiv M^e \bmod N$. The **RSA decryption function** (also called signature function) is the inverse: $M = C^d \bmod N$. It can be performed by the owner of the private key d. Formally the RSA Assumption is stated as follows.

RSA Assumption Let k be the security parameter. Let key generator GE define a family of RSA functions (i.e., $(e,d,N) \leftarrow GE(1^k)$ is an RSA instance with security parameter k). For any probabilistic polynomial-time algorithm A, $\Pr[u^e \equiv w \bmod N : (e,d,N) \leftarrow GE(1^k); w \in_R \{0,1\}^k; u \leftarrow A(1^k, w, e, N)]$ is negligible.

Next we describe variants of Shamir secret sharing and Pedersen VSS that we use in RSA-based systems. They differ in that operations on the shares are performed over the integers, instead of in a modular subgroup of integers.

(t,l)-secret sharing over the integers (INT-(t,l)-SS) [23] This is a variant of Shamir secret sharing [46]. Let $L = l!$ and let m be a positive integer. For sharing a secret $s \in [0, mK]$ (and K the size of an interval over the integers), a random polynomial $a(x) = \sum_{j=0}^{t} a_j x^j$ is chosen such that $a_0 = L^2 s$, and each other $a_j \in_R \{0, L, 2L, \ldots, mL^3K^2\}$.[4]

---

[2] In DL-based systems, we implicitly assume all verification operations are performed in Z_p^*.

[3] $\lambda(N) = \mathrm{lcm}(P-1, Q-1)$ is the smallest integer such that any element in Z_N^* raised by $\lambda(N)$ is the identity element. RSA is typically defined using $\phi(N)$, the number of elements in Z_N^*, but it is easy to see that $\lambda(N)$ can be used instead. We use it because it gives an explicit way to describe an element of maximal order in Z_N^*. Note that $\phi(N)$ is a multiple of $\lambda(N)$, and that knowing any value which is a multiple of $\lambda(N)$ implies breaking the system.

[4] We note that in our RSA-based systems, $L^2 s$ is actually the secret component of the RSA secret key, which when added to a public leftover component (in $[0, L^2-1]$), forms the RSA secret key.

Each shareholder $i \in \{1,\dots,l\}$ receives a secret share $s_i = a(i)$, and verifies[5] that (1) $0 \le s_i \le mL^3K^2l^{t+1}$, and (2) L divides s_i. Any set Λ of cardinality $t+1$ can compute s using Lagrange interpolation.

Unconditionally-Secure (t,l)-VSS over the Integers (INT-(t,l)-US-VSS)

This is a variant of Pedersen Unconditionally-Secure (t,l)-VSS [41], and is slightly different than the version in [26]. Let N be an RSA modulus and let g and h be generators whose discrete log modulo N with respect to each other is unknown. The protocol begins with two (t,l)-secret sharings over the integers, the first sharing secret s with $m=1$, and the second sharing s' with $m = NK$. Note that $s \in [0,K]$ and $s' \in [0,NK^2]$. Let $a(x) = \sum_{j=0}^{t} a_j x^j$ be the random polynomial used in sharing s and let $a'(x) = \sum_{j=0}^{t} a'_j x^j$ be the random polynomial used in sharing s'. For all i, S_i receives shares $s_i = a(i)$ and $s'_i = a'(i)$. (We refer to the pair $(a(i), a'(i))$ as *dual-share i*.) Also, the *verification shares* $\{\alpha_j (= g^{a_j} h^{a'_j})\}_{0 \le j \le t}$, are published.[6] Say *check share* $A_i = \prod_{j=0}^{t} \alpha_j^{i^j}$. S_i can verify the correctness of his shares by checking that $A_i = g^{s_i} h^{s'_i}$. Say s and s' are the shares computed using Lagrange interpolation from a set of $t+1$ shares that passed the verification step. If the dealer can reveal different secrets $\hat{s}$ and $\hat{s}'$ that also correspond to the zero coefficient verification share, then the dealer can compute an α and β such that $g^\alpha \equiv h^\beta$, which implies factoring (and thus breaking the RSA assumption).

Looking ahead, we will need to simulate an INT-(t,l)-US-VSS. Using Lemma 8, we can do this by constructing a random polynomial over an appropriate simulated secret (e.g., a random secret, or a secret obtained as a result of a previously simulated protocol) in the zero coefficient, and a random companion polynomial with a totally random zero coefficient. Note that the β value in the lemma will correspond to K, and the γ value in the lemma will correspond to the discrete log of g with respect to h, which is less than N. The probability of distinguishing a real VSS from the simulated VSS will be $(4t+2)/K$, which is exponentially small if the range of secrets K is exponentially large.

B Distributed Public-Key Systems - Formal Definitions

Definition 5. (Robustness of a Threshold System) *A (t,l)-threshold public-key system S is robust if for any polynomial-time t-restricted adaptive stationary adversary $\mathcal{A}$, with all but negligible probability, for each input m which is submitted to the* DISTAPPLY *protocol the resulting output s passes* VERIFY*(y,m,s).*

Definition 6. (Security of a Threshold System) *A (t,l)-threshold public-key system S is secure if for any polynomial-time t-restricted adaptive stationary adversary $\mathcal{A}$, after polynomially-many* DISTAPPLY *protocols performed during operational periods on given values, given a new value m and the view of $\mathcal{A}$, the probability of being able to produce an output s that passes* VERIFY*(y,m,s) is negligible.*

---

[5] These tests only verify the shares are of the correct form, not that they are correct polynomial shares.

[6] In RSA-based systems, we implicitly assume all verification operations are performed in Z_N^*.

Remark: The choice of the inputs to DISTAPPLY prior to the challenge m defines the *tampering power* of the adversary (i.e., "known message," "chosen message", "random message" attacks). The choice depends on the implementation within which the distributed system is embedded. In this work, we assume that the (centralized) cryptographic function is secure with respect to the tampering power of the adversary. We note that the provably secure signature and encryption schemes typically activate the cryptographic function on random values (decoupled from the message choice of the adversary).

For the definitions of security and robustness properties of a **distributed key generation** and **proactive maintenance** of a cryptosystem, see [33].

C ZK proofs

We use efficient ZK proofs of knowledge (POKs) derived from [26] and [10]. These are composed of combinations of Σ-protocols [9] (i.e., Schnorr-type proofs [45]). For each ZK proof that we need, we will have a separate "proof" protocol, but there will be a single "setup" protocol used for all ZK proofs. Say A wishes to prove knowledge of "X" to B. Then the setup protocol will consist of B making a commitment and proving that he can open it in a witness indistinguishable way [22], and the proof protocol will consist of A proving to B either the knowledge of "X" or that A can open the commitment. (See [10] for details) This construction allows the proof protocols to be run concurrently without any timing constraints, as long as they are run after all the setup protocols have completed. (For more on the problems encountered with concurrent ZK proofs see [37, 19, 20].)

The DL-based and RSA-based ZK-proof-setup protocols are exactly the Σ-protocols for commitments over q-one-way-group-homomorphisms (q-OWGH), given in [10]. Recall the q-OWGH for a DL-based system with parameters (p,q,g) is $f(x) = g^x \bmod p$, and the q-OWGH for an RSA-based system with parameters (N,e) is $f(x) = x^e \bmod N$ (with $q = e$ in this case).

Let KE denote the "knowledge error" of a POK.

Formally, we define ZKSETUP-DL$_{A,B}(p,q,g,h)$ as a protocol in which A generates a commitment C and engages B in a witness-hiding (WH) POK ($KE = 1/q$) of $\sigma, \sigma' \in Z_q$ where $C \equiv g^{\sigma}h^{\sigma'} \bmod p$.

We define ZKSETUP-RSA$_{A,B}(N,e,g)$ as a protocol in which A generates a commitment C and engages B in a WH POK ($KE = 1/e$)[7] of (σ, σ') (with $\sigma \in Z_e$, $\sigma' \in Z_N^*$) where $C \equiv g^{\sigma}(\sigma')^e \bmod N$.

We define ZKPROOF-DL$_{A,B}(p,q,g,h,D)$ as a protocol in which A engages B in a WH POK ($KE = 1/q$) of either $d \in Z_q$ where $D \equiv g^d \bmod p$, or $\tau, \tau' \in Z_q$ where $C_{B,A} \equiv g^{\tau}h^{\tau'} \bmod p$ and $C_{B,A}$ is the commitment generated in ZKSETUP-DL$_{B,A}(p,q,g,h)$.

We define ZKPROOF-DL-REP$_{A,B}(p,q,g,h,E,D)$ as a protocol in which A engages B in a WH POK ($KE = 1/q$) of either $d, d' \in Z_q$ where $D \equiv g^d h^{d'} \bmod p$ and $E \equiv$

[7] This implies e must be exponentially large in the security parameter k in order to obtain a sound proof. However, if e is small (say $e = 3$) we can use different setup and proof protocols described in [10] to obtain provably secure and robust RSA-based protocols.

$g^d \bmod p$, or $\tau, \tau' \in Z_q$ where $C_{B,A} \equiv g^\tau h^{\tau'} \bmod p$ and $C_{B,A}$ is the commitment generated in ZKSETUP-DL$_{B,A}(p,q,g,h)$.

We define[8] ZKPROOF-IF-REP$_{A,B}(N,e,m,g,h,E,D)$ as a protocol in which A, who knows integers $d \in (-a,a]$ and $d' \in (-b,b]$ such that $E \equiv m^d \bmod N$ and $D \equiv g^d h^{d'} \bmod N$, engages B in a WH POK ($KE = 1/e$) of either $\Delta \in Z_e$, $\delta \in (-2ae(N+1), 2ae(N+1)]$, and $\delta' \in (-2be(N+1), 2be(N+1)]$ where $D^\Delta \equiv g^\delta h^{\delta'} \bmod N$ and $E^\Delta \equiv g^\delta \bmod N$, or (τ, τ') (with $\tau \in Z_e$, $\tau' \in Z_N^*$) where $C_{B,A} \equiv g^\tau (\tau')^e \bmod N$ and $C_{B,A}$ is the commitment generated in ZKSETUP-RSA$_{B,A}(N,e,g)$. This protocol is honest-verifier statistical zero-knowledge with a statistical difference between the distribution of views produced by the simulator and in the real protocol bounded by $2/N$.

C.1 Proof of representations

Here we give the main Σ-protocol used in ZKPROOF-IF-REP$_{A,B}(N,e,m,g,h,E,D)$.[9]

1. Initially, the parameters (N,e,m,g,h,E,D) are public, and A knows integers $d \in (-a,a]$ and $d' \in (-b,b]$ such that $E \equiv m^d \bmod N$ and $D \equiv g^d h^{d'} \bmod N$.
2. A generates $r \in_R (-aeN, aeN]$ and $r' \in (-beN, beN]$, computes $V = m^r \bmod N$ and $W = g^r h^{r'} \bmod N$, and sends V, W to B.
3. B generates $c \in_R Z_e$ and sends c to A.
4. A computes $z = cd + r$ and $z' = cd' + r'$, and sends z, z' to B.
5. B checks that $m^z \equiv E^c V \bmod N$ and $g^z h^{z'} = D^c W \bmod N$.

In all steps, A and B also check that the values received are in the appropriate ranges.

The above is a POK of $\Delta \in Z_e$, $\delta \in (-2ae(N+1), 2ae(N+1)]$, and $\delta' \in (-2be(N+1), 2be(N+1)]$ in which $m^\delta \equiv E^\Delta \bmod N$ and $g^\delta h^{\delta'} = D^\Delta \bmod N$. The knowledge error is $1/e$, and the protocol is honest-verifier statistical zero-knowledge, with a statistical difference between views produced by the simulator and those in the real protocol bounded by $2/N$.

D Proofs of Techniques

The following lemma from [23] is used to prove the subsequent lemma, which in turn is used to prove the simulatability of INT-(t,l)-US-VSS.

Lemma 7. *Let $r(x) = r_0 + r_1 x + \cdots + r_t x^t$ be a random polynomial of degree t such that $r(0) = r_0 = L^2 k$ ($k \in [0,K]$) and $r_j \in_R \{0, L, \ldots, \beta L^3 K\}$ for $1 \leq j \leq t$. Let Λ' be a set of t servers. Then with probability at least $1 - 2t/\beta$, for any $\hat{k} \in [0,K]$, there exists a polynomial $r'(x) = r'_0 + r'_1 x + \cdots + r'_t x^t$ with $r'(0) = r'_0 = L^2 \hat{k}$ and $r'_j \in \{0, L, \ldots, \beta L^3 K\}$ for $1 \leq j \leq t$ such that $r(i) = r'(i)$ for $i \in \Lambda'$.*

[8] IF stands for "integer factorization."

[9] Recall that this main protocol is combined with a Σ-protocol proving knowledge of a commitment generated in a setup protocol, using an "OR" construction.

Lemma 8. *Let $\gamma \in [1,m]$. Let $r(x) = r_0 + r_1 x + \cdots + r_t x^t$ be a random polynomial of degree t such that $r(0) = r_0 = L^2 k$ ($k \in [0,K]$) and $r_j \in_R \{0, L, \ldots, \beta L^3 K\}$ for $1 \le j \le t$. Let $r'(x) = r'_0 + r'_1 x + \cdots + r'_t x^t$ be a random polynomial of degree t such that $r'(0) = r'_0 = L^2 k'$ ($k' \in [0, mK\beta]$) and $r_j \in_R \{0, L, \ldots, \beta L^3 mK\}$ for $1 \le j \le t$. Let Λ' be a set of t servers. Then with probability at least $1 - (4t+2)/\beta$, for any $\hat{k} \in [0,K]$, there exists polynomials $\hat{r}(x) = \hat{r}_0 + \hat{r}_1 x + \cdots + \hat{r}_t x^t$ with $\hat{r}(0) = \hat{r}_0 = L^2 \hat{k}$ and $\hat{r}_j \in \{0, L, \ldots, \beta L^3 K\}$ for $1 \le j \le t$, and $\hat{r}'(x) = \hat{r}'_0 + \hat{r}'_1 x + \cdots + \hat{r}'_t x^t$ with $\hat{r}'(0) = L^2(\gamma k + k' - \gamma \hat{k})$, $0 \le \hat{r}'(0) \le \beta L^2 mK$ and $\hat{r}'_j \in \{0, L, \ldots, \beta L^3 mK\}$ for $1 \le j \le t$, such that $r(i) = \hat{r}(i)$ and $r'(i) = \hat{r}'(i)$ for $i \in \Lambda'$, and $\gamma r(x) + r'(x) = \gamma \hat{r}(x) + \hat{r}'(x)$.*

Proof. Except for the last equation, we get from Lemma 7 that the probability that the polynomials $\hat{r}(x)$ and $\hat{r}'(x)$ (with coefficients in the correct ranges) do not exist is at most $[2/\beta] + [2t/\beta] + [2t/\beta]$, where the first $2/\beta$ arises from the probability that $\hat{r}'(0)$ is in the correct range, given an additive offset of $L^2(\gamma k - \gamma \hat{k}) \in [-L^2 mK, L^2 mK]$ from $L^2 k'$. If those polynomials do exist, then the last equation follows since (1) the degree of the polynomial on each side of the equivalence is t, (2) the polynomials obviously agree at the t locations in Λ', and (3) the polynomials agree at 0, since $\gamma r(0) + r'(0) = L^2 \gamma k + L^2 k' = L^2 \gamma \hat{k} + L^2(\gamma k + k' - \gamma \hat{k}) = \gamma \hat{r}(0) + \hat{r}'(0)$.

D.1 Useful RSA Lemmas

Lemma 9 ([26]). *Let h be the security parameter. Let modulus generator GE define a family of modulus generating functions (i.e., $N \leftarrow GE(1^h)$ be an RSA modulus with security parameter h). For any probabilistic polynomial-time adversary $\mathcal{A}$, the following is negligible:* $\Pr[u^e \equiv w^d \bmod N; (e \neq 0) \vee (d \neq 0) : N \leftarrow GE(1^h); u, w \in_R \{0,1\}^h; e, d \leftarrow A(1^h, w, u)]$

Proof. Similar to [2].

The following corollary follows from Lemma 9 and the RSA assumption (and hence from the RSA assumption).

Corollary 10. *Let k be the security parameter. Let GE be an RSA generator (that produces large public exponents), i.e., $(N,e) \leftarrow GE(1^k)$. For any probabilistic polynomial-time algorithm A,*

$$\begin{aligned} &\Pr[(g^\alpha \equiv h^\beta \bmod N; (\alpha \neq 0) \vee (\beta \neq 0)) \vee (g = u^e) : \\ &(N,e) \leftarrow GE(1^h); g, h \in_R \{0,1\}^h; (\alpha, \beta, u) \leftarrow A(1^h, g, h)] \end{aligned}$$

is negligible.

E Proofs of Protocols

Proof. **of Theorem 1**
We prove the robustness and security of the Basic DL-based threshold protocol. Both are based on the DLP Assumption. Recall that we assume the adversary is stationary and adaptive.

Robustness Say $P(k)$ is a polynomial bound on the number of messages m that the protocol signs. Say an adversary prevents the signing of a message with non-negligible probability ρ. We will show how to solve the DLP with probability

$$\rho - \frac{tP(k)+1}{q}.$$

Say we are given an instance of the DLP, namely for a given set of parameters (p,q,g), we are given a uniformly chosen value $h \in Z_p^*$. We create a simulator that runs the dealer as normal, except that h is taken from the DLP instance (when h is distributedly generated, we were able to inject our DLP instance as well). Then the servers are run as normal, except that the extractor for the ZKPROOF-DL-REP protocol is run whenever an incorrupted server is playing the verifier and a corrupted server is playing the prover. Note that since the simulator knows the secrets x, x', the normal operation of the servers can be simulated easily. We will show that if an adversary is able to prevent a message from being signed, we can (except with negligible probability) determine $\mathrm{DL}(h,g)$.

If a server is not corrupted, the probability of a failed extraction using that server is $1/q$. There will obviously be at least one incorrupted server that runs the extractor with every corrupted server, and thus the probability of any corrupted server not allowing a successful extraction during the protocol is at most $tP(k)/q$. Say S_i runs the extractor successfully on S_j. If S_i extracts a way to open the commitment $C_{i,j} = g^{\sigma_{i,j}} h^{\sigma'_{i,j}}$ (from the setup protocol), say with $(\tau_{i,j}, \tau'_{i,j})$, then except with probability $1/q$, this will give

$$\mathrm{DL}(h,g) = \frac{\sigma_{i,j} - \tau_{i,j}}{\sigma'_{i,j} - \tau'_{i,j}}.$$

Therefore, with probability at most

$$\frac{tP(k)+1}{q},$$

there was either an extraction that failed or an extraction that succeeded, but produced a way to open $C_{i,j}$ with the same pair of values used to create it.

Now say the adversary prevents a message m from being signed. It should be clear that after at most $t+1$ attempts, there will be a set Λ of $t+1$ servers that participate in signing m without any verification failures. This implies that the signature obtained must be incorrect. Let $\{E_i\}_{i\in\Lambda}$ be the check shares for the (single) additive shares in this signing attempt. If $\prod_{i\in\Lambda} E_i = y$, then the signature must be correct, so we may assume $\prod_{i\in\Lambda} E_i \neq y$. Let $\{(\delta_i, \delta'_i)\}_{i\in\Lambda}$ be the extracted (or simulator generated, for incorrupted servers) dual-shares. It is easy to see that

$$g^x h^{x'} = \prod_{i\in\Lambda} B_i = \prod_{i\in\Lambda} D_i = \prod_{i\in\Lambda} g^{\delta_i} h^{\delta'_i}.$$

We also have $g^x = y \neq \prod_{i\in\Lambda} E_i = \prod_{i\in\Lambda} g^{\delta_i}$. But then

$$g^x h^{x'} = g^{\sum_{i\in\Lambda}\delta_i} h^{\sum_{i\in\Lambda}\delta'_i},$$

with $x \neq \sum_{i\in\Lambda} \delta_i$ (and hence $x' \neq \sum_{i\in\Lambda} \delta'_i$), and thus

$$\mathrm{DL}(h,g) = \frac{(\sum_{i\in\Lambda} \delta_i) - x}{x' - \sum_{i\in\Lambda} \delta'_i} \bmod q.$$

Therefore, with probability $\rho - (tP(k)+1)/q$, $\mathrm{DL}(h,g)$ can be found.

Security Here we show that if the adversary can sign a new message, then it can break the security of the (non-distributed) signature scheme w.r.t. the same attack [33].

Say an adversary can sign a new message in the Basic DL-threshold Protocol with non-negligible probability ρ. We will show that we can sign a new message in the underlying signature scheme with probability ρ. Say the signature scheme has parameters (p,q,g,y). Then we create the following simulator:

1. **Initialization:** The signature scheme gives parameters (p,q,g,y)
2. Simulate the dealer by generating $h \in_R Z_p^*$, $x' \in_R Z_q$, and producing a (t,l)-US-VSS with polynomials $(\hat{a}(), \hat{a}'())$ on secrets $0, x'$ (i.e., $\hat{a}(0) = 0$ and $\hat{a}'(0) = x'$).
3. Each (ordered) pair of servers performs the ZKSETUP-DL protocol, using g and h as the generators, except that an incorrupted verifier interacting with a corrupted prover uses the extractor to determine how to open the commitment for that prover.
4. Each server maintains a list $\mathcal{G}$ of server indices for servers that have not misbehaved (i.e., they are considered good).
5. When a message m needs to be signed, the following DISTAPPLY protocol is run:
   (a) A set $\Lambda \subseteq G$ with $|\Lambda| = t+1$ is chosen in some public way.
   (b) 2poly-to-2sum is performed using the simulator-generated (t,l)-US-VSS with polynomials $(\hat{a}(), \hat{a}'())$, producing values $(\hat{b}_j, \hat{b}'_j)$ for $j \in \Lambda$, along with their associated check shares. If there are misbehaving servers, their indices are removed from $\mathcal{G}$ and the protocol loops to Step 5a.
   (c) A faking server, say S_i is picked at random from Λ. 2sum-to-2sum is performed using the simulator-generated values $\{(\hat{b}_j, \hat{b}'_j)\}_{j\in\Lambda}$ and their associated check shares from the previous step, producing values $(\hat{d}_j, \hat{d}'_j)$ for $j \in \Lambda$, along with their associated check shares. If there are misbehaving servers, their indices are removed from $\mathcal{G}$ and the protocol loops to Step 5a. If S_i is compromised, then the simulation rewinds to Step 5c, and is attempted again.
   (d) For each $j \in \Lambda \setminus \{i\}$, S_j performs 2sum-to-1sum using the simulator-generated values $(\hat{d}_j, \hat{d}'_j)$ and their associated check shares from the previous step. S_i, however, produces $\hat{E}_i = yg^{\hat{d}_j}$ and in each ZKPROOF-DL-REP actually proves knowledge of how to open the commitment, instead of knowledge of the discrete log of $\hat{E}_i$. If there are misbehaving servers, their indices are removed from $\mathcal{G}$ and the protocol loops to Step 5a. If S_i is compromised, then the simulation rewinds to Step 5c, and is attempted again.
   (e) The additive application step of the signature protocol is run, but with S_i simulated by using the signature on m obtained from the signature oracle. If there are misbehaving servers, their indices are removed from $\mathcal{G}$ and the protocol loops to Step 5a. If S_i is compromised, then the simulation rewinds to Step 5c, and is attempted again.

Note that the probability of rewinding is at most $t/(t+1)$, and thus the simulator requires on average a factor of at most $t+1$ more time than the real protocol. Thus the simulation is polynomial time. The simulation is perfect if all the extractors succeed. Since the extractors have to be run at most lt times, and the probability of a single extractor failing is $1/q$, the probability of distinguishing a simulated view from a real view is lt/q. Thus the simulation is statistically indistinguishable from the real protocol.

If the adversary is able to then generate a new signature, then it is clear that we would have an algorithm to break the signature scheme.

Proof. **of Theorem 2**
We prove the robustness and security of the Basic RSA-based threshold protocol. Both are based on Corollary 10. (We will call this the "Corollary 10 assumption.") Recall that we assume the adversary is stationary and adaptive. We will assume that the public key e is large ($\Theta(k)$ bits). (For small e we can use a technique from [10] to obtain ZK proofs that allow us to prove similar results.)

Robustness Say $P(k)$ is a polynomial bound on the number of messages m that the protocol signs. Say an adversary prevents the signing of a message with non-negligible probability ρ. We will show how to break the Corollary 10 assumption with probability

$$\rho - \frac{2tP(k)+1}{e}.$$

Say an RSA public key was generated $(N,e) \leftarrow GE(1^k)$ and we are given a uniformly chosen $g,h \in Z_N^*$, as in Corollary 10. We use the simulator that is used to prove security, except that g,h are taken from the RSA instance, and that the extractor for the ZKPROOF-IF-REP protocol is run whenever an incorrupted server is playing the verifier and a corrupted server is playing the prover. We will show that if an adversary is able to prevent a message from being signed, we can (except with negligible probability) either find α,β such that $g^\alpha \equiv h^\beta \bmod N$ or find u such that $u^e \equiv g \bmod N$.

If a server is not corrupted, the probability of a failed extraction using that server is $1/e$. There will obviously be at least one incorrupted server that runs the extractor with every corrupted server, and thus the probability of any corrupted server not allowing a successful extraction during the protocol is at most $tP(k)/e$. Say S_i runs the extractor successfully on S_j. If S_i extracts a way to open the commitment $C_{i,j} = g^{\sigma_{i,j}}(\sigma'_{i,j})^e$, say with $(\tau_{i,j},\tau'_{i,j})$, then this will give

$$g^{\sigma_{i,j}}(\sigma'_{i,j})^e = g^{\tau_{i,j}}(\tau'_{i,j})^e,$$

and thus

$$g^{\sigma_{i,j}-\tau_{i,j}} = (\tau'_{i,j}/\sigma'_{i,j})^e.$$

Except with probability $1/e$, $\gcd(\sigma_{i,j}-\tau_{i,j},e) = 1$, so using the Extended Euclidean algorithm one can compute α,β such that $\alpha e + 1 = \beta(\sigma_{i,j}-\tau_{i,j})$, and thus

$$g^{1/e} = \frac{\tau_{i,j}/(\sigma'_{i,j})^\beta}{g^\alpha}.$$

Therefore, assuming the adversary has not distinguished the simulation from the real protocol, with probability at most

$$\frac{tP(k)+1}{e},$$

there was either an extraction that failed or an extraction that succeeded, but produced a way to open $C_{i,j}$ with an RSA-REP pair that did not allow one to compute the RSA inverse of g. Recall that the probability the adversary can distinguish the simulation from the real protocol is at most

$$\frac{4t+2}{N}+\frac{tl}{e}.$$

Now say the adversary prevents a message m from being signed. It should be clear that after at most $t+1$ attempts, there will be a set Λ of $t+1$ servers that participate in signing m without any verification failures. This implies that the signature obtained must be incorrect. Let $\{E_j\}_{j\in\Lambda}$ be the check shares for the (single) additive shares in this signing attempt for message m. (Recall that for the faking server S_i, $E_i = m^{(1/e)-x^*} m^{d_i} \bmod N$.) Then $m^{x^*}\prod_{j\in\Lambda} E_j \neq m^{1/e}$. Let $\{(\delta_j, \delta'_j, \Delta_j)\}_{j\in\Lambda}$ be the extracted (or previously known to the simulator, for incorrupted servers) "dual shares," along with the exponent on D_j. (For faking server S_i, $\delta_i = d_i$, $\delta'_i = d'_i$ and $\Delta_i = 1$.) Let $\Delta = \prod_{j\in\Lambda}\Delta_j$. It is easy to see that

$$(h^{x'})^{\Delta} = \left(\prod_{j\in\Lambda} B_j\right)^{\Delta} = \left(\prod_{j\in\Lambda} D_j\right)^{\Delta} = \prod_{j\in\Lambda}(g^{\delta_j} h^{\delta'_j})^{\Delta/\Delta_j}.$$

We also have

$$(m^{1/e})^{\Delta} \neq \left(m^{x^*}\prod_{j\in\Lambda} E_j\right)^{\Delta} = (m^{x^*})^{\Delta}(m^{(1/e)-x^*}m^{\delta_i})^{\Delta} \prod_{j\in\Lambda\setminus\{i\}} (m^{\delta_j})^{\Delta/\Delta_j}.$$

But then

$$h^{x'\Delta} = g^{\sum_{j\in\Lambda}(\delta_j\Delta/\Delta_j)} h^{\sum_{j\in\Lambda}(\delta'_j\Delta/\Delta_j)},$$

with $0 \neq \sum_{j\in\Lambda}(\delta_j\Delta/\Delta_j)$ (and hence $x' \neq \sum_{j\in\Lambda}(\delta'_j\Delta/\Delta_j)$), and thus

$$g^{\sum_{j\in\Lambda}\delta_j\Delta/\Delta_j} = h^{x'\Delta - \sum_{j\in\Lambda}\delta'_j\Delta/\Delta_j}.$$

Thus, assuming the adversary can prevent a message from being signed, the probability of finding either α, β such that $g^{\alpha} \equiv h^{\beta} \bmod N$ or u such that $u^e \equiv g \bmod N$ is at least the non-negligible probability

$$\rho - \frac{tP(k)+1+tl}{e} + \frac{4t+2}{N},$$

contradicting Corollary 10.

Security Here we reduce the security of our Basic RSA-threshold protocol to the RSA assumption. Say an adversary, after watching polynomially many messages be signed in the Basic RSA-threshold protocol, can sign a new challenge message. with non-negligible probability ρ. Then we will give a polynomial-time algorithm to break RSA with probability close to ρ.

Say we are given an RSA key (N,e) and a challenge message m^* to be signed. We will run the adversary against a simulation of the protocol, and then present m^* to be signed. We will show that the probability that an adversary can distinguish the simulation from the real protocol is negligible, and thus the probability that it signs m^* is negligibly less than ρ.

The simulator is as follows:

1. **Initialization:** The RSA parameters (N,e) are given. We may also assume that we have a list of random message signature pairs $\{(m, m^{1/e})\}$.
2. Simulate the dealer by computing the public value x^* (using public values N,e,L, as in the real protocol) generating $g,h \in_R Z_N^*$, $x' \in_R Z_{N^3}$, and producing a INT-(t,l)-US-VSS with polynomials $(\hat{a}(), \hat{a}'())$ on secrets $0, x'$ (i.e., $\hat{a}(0) = 0$ and $\hat{a}'(0) = x'$).
3. Each (ordered) pair of servers performs the ZKSETUP-RSA protocol, using g and h as the generators, except that an incorrupted verifier interacting with a corrupted prover uses the extractor to determine how to open the commitment for that prover.
4. Each server maintains a list $\mathcal{G}$ of server indices for servers that have not misbehaved (i.e., they are considered good).
5. When a message m needs to be signed, the following DISTAPPLY protocol is run:
   (a) A set $\Lambda \subseteq G$ with $|\Lambda| = t+1$ is chosen in some public way.
   (b) 2poly-to-2sum is performed using the simulator-generated INT-(t,l)-US-VSS with polynomials $(\hat{a}(), \hat{a}'())$, producing values $(\hat{b}_j, \hat{b}'_j)$ for $j \in \Lambda$, along with their associated check shares. If there are misbehaving servers, their indices are removed from $\mathcal{G}$ and the protocol loops to Step 5a.
   (c) A faking server, say S_i is picked at random from Λ. 2sum-to-2sum is performed using the simulator-generated values $\{(\hat{b}_j, \hat{b}'_j)\}_{j \in \Lambda}$ and their associated check shares from the previous step, producing values $(\hat{d}_j, \hat{d}'_j)$ for $j \in \Lambda$, along with their associated check shares. If there are misbehaving servers, their indices are removed from $\mathcal{G}$ and the protocol loops to Step 5a. If S_i is compromised, then the simulation rewinds to Step 5c, and is attempted again.
   (d) For each $j \in \Lambda \setminus \{i\}$, S_j performs 2sum-to-1sum using the simulator-generated values $(\hat{d}_j, \hat{d}'_j)$ and their associated check shares from the previous step. S_i, however, produces $\hat{E}_i = m^{(1/e)-x^*} m^{\hat{d}_j}$ and in each ZKPROOF-IF-REP actually proves knowledge of how to open the commitment, instead of knowledge of the discrete log of $\hat{E}_i$. If there are misbehaving servers, their indices are removed from $\mathcal{G}$ and the protocol loops to Step 5a. If S_i is compromised, then the simulation rewinds to Step 5c, and is attempted again.

Note that the probability of rewinding is at most $t/(t+1)$, and thus the simulator requires on average a factor of at most $t+1$ more time than the protocol. Thus the simulation is polynomial time. The probability of distinguishing the simulation from the real protocol is at most the probability of distinguishing the simulated INT-(t,l)-US-VSS from the real one, plus the probability of an extractor failing. All of this can be bounded by $(4t+2)/N + tl/e$. Thus with probability negligibly less than ρ, we can generate a signature on m^*.

How Long Does a Bit Live in a Computer?

Bernhard Korte

Research Institute for Discrete Mathematics Lennestrasse 2, 53113 Bonn Germany
dm@or.uni-bonn.de

Computer industry announces that a gigahertz processor will be available around the year 2001. This is a chip which runs with a cycle time of one nanosecond which is not only a great challenge in technology, but even a greater challenge for application of mathematics in VLSI design. The timing graph of a micropro- cessor is a directed graph with several million edges, where each edge models the signal processing through combinatorial logic between two latches (registers). Latches itself are governed by clock signals. If the total travel time of a signal along one edge is at most one nanosecond, deviation of even a few pico- seconds due to design errors, technology, production etc. matter substantially.

This talk gives an overview of most recent approaches of discrete optimization to VLSI-design. More specifically, we describe methods to minimize the cycle time, i.e. the life span of a bit in a computer.

We have modeled the minimization problem of the cycle time of a microprocessor as an extended maximum mean weight cycle problem in a graph. By this, we are able to minimize the cycle time (or to maximize the frequency). Moreover, we can extend the mean weight cycle model in such a way that process variations, clock jitters, balancing problems, early mode problems and other additional constraints can be handled, too.

By this approach the global cycle time can be minimized under all constraints which are logically and technically given. It turned out that without this application of combinatorial optimization industry would be unable to produce microprocessors with a cycle time of one nanosecond.

Approximation Algorithms for the Traveling Purchaser Problem and Its Variants in Network Design

R. Ravi[1] and F. S. Salman[2]

[1] GSIA, Carnegie Mellon University,
5000 Forbes Avenue, Pittsburgh PA 15213, USA.
Supported in part by an NSF CAREER grant CCR-9625297.
ravi@cmu.edu
[2] GSIA, Carnegie Mellon University,
Supported by an IBM Corporate Fellowship.
fs2c+@andrew.cmu.edu

Abstract. The traveling purchaser problem is a generalization of the traveling salesman problem with applications in a wide range of areas including network design and scheduling. The input consists of a set of markets and a set of products. Each market offers a price for each product and there is a cost associated with traveling from one market to another. The problem is to purchase all products by visiting a subset of the markets in a tour such that the total travel and purchase costs are minimized. This problem includes many well-known NP-hard problems such as uncapacitated facility location, set cover and group Steiner tree problems as its special cases.
We give an approximation algorithm with a poly-logarithmic worst-case ratio for the traveling purchaser problem with metric travel costs. For a special case of the problem that models the ring-star network design problem, we give a constant-factor approximation algorithm. Our algorithms are based on rounding LP relaxation solutions.

1 Introduction

Problem. The traveling purchaser problem (TPP), originally proposed by Ramesh [Ram 81], is a generalization of the traveling salesman problem (TSP). The problem can be stated as follows. We are given a set $M = \{1,\dots,m\}$ of markets and a set $N = \{1,\dots,n\}$ of products. Also, we are given c_{ij}, cost of travel from market city i to city j, and nonnegative d_{ij}, the cost of product i at market j. A purchaser starts from his home city, say city 1, and travels to a subset of the m cities and purchases each of the n products in one of the cities he visits, and returns back to his home city. The problem is to find a tour for the purchaser such that the sum of the travel and purchase costs is minimized. It is assumed that each product is available in at least one market city. If a product i is not available at market j, then d_{ij} is set to a high value.

Applications. The traveling purchaser problem has applications in many areas including parts procurement in manufacturing facilities, warehousing, transportation, telecommunication network design and scheduling. An interesting scheduling application involves sequencing n jobs on a machine that has m states [Ong 82]. There is a set-up cost of c_{ij} to change the state of the machine from i to j. A cost d_{ij} is specified to process job i at state j. The objective is to minimize the sum of machine set-up and job processing costs.

The traveling purchaser problem contains the TSP, the prize collecting TSP, uncapacitated facility location problem, group Steiner tree problem and the set cover problem as its immediate special cases. The TSP is the case when each market city has a product available only at that city. In the uncapacitated facility location problem, let the fixed cost for opening facility j be f_j and the cost of servicing client i by facility j be d_{ij}. Then the problem is equivalent to a TPP with a market for each facility and a product for each client, where the travel cost between markets i and j is $c_{ij} = (f_i + f_j)/2$ and the purchase cost of product i at market j is d_{ij}. In the set cover problem, we are given a set S and subsets $S_1, \ldots, S_n \subset S$. The problem is to find a minimum size collection of subsets whose union gives S. This corresponds to a TPP where S is the set of products and there is a market j for each subset S_j. The cost of purchasing product i at market j (of S_j) is zero if $i \in S_j$ and is a large number otherwise. There is a unit cost of travel between each market. Then, there is a set cover of size k if and only if there is a TPP solution of cost k.

Hardness. Note that since there is no polynomial time approximation algorithm for the general TSP, TPP with no assumptions on the costs cannot be approximated in polynomial time unless $P = NP$ [GJ 79]. The TPP instance into which we reduce the set cover problem has metric travel costs. Therefore, from the above approximation-preserving reduction and current hardness results for set cover [F 96, RS 97, AS 97] it follows that there is no polynomial time approximation algorithm for the traveling purchaser problem even with metric travel costs whose performance ratio is better than $(1 - o(1)) \ln n$ unless $P = NP$.

Related Work. Due to the hardness of the problem, many researchers have focused on developing heuristics. Most of these algorithms are local search heuristics (Golden, Levy and Dahl [GLD 81], Ong [Ong 82], Pearn and Chien [PC 98]). Voss [V 96] generated solutions by tabu search. The exact solution methods are limited to the branch-and-bound algorithm of Singh and van Oudheusden [SvO 97], which solves relaxations in the form of the uncapacitated facility location problem.

Our Results. We give the first approximation results for the traveling purchaser problem. We give an approximation algorithm with a poly-logarithmic worst-case ratio for the TPP problem with metric travel costs (Corollary 6). In fact, this algorithm approximates a more general bicriteria version of the problem (Theorem 7). For a special case of the TPP problem that models the ring-star network design problem with proportional costs, we give a constant-factor approximation algorithm (Theorem 13 and Corollary 14).

2 Bicriteria Traveling Purchaser Problem

We consider a bicriteria version of the traveling purchaser problem, where minimizing the purchase costs and the travel costs are two separate objectives. The bicriteria problem is a generalization of the TPP, whose solutions provide the decision-maker insight into the tradeoffs between the two objectives.

We use the framework due to Marathe et al. [MRS+ 95] for approximating a bicriteria problem. We choose one of the criteria as the objective and bound the value of the other by a budget constraint. Suppose we want to minimize objectives A and B. We consider the problem

$$P: \min B \; s.t. \; A \leq a$$

Definition 1. *An (α, β)-approximation algorithm for the problem P outputs a solution with A-cost at most α times the budget a and B-cost at most β times the optimum value of P, where $\alpha, \beta \geq 1$.*

Our approximation algorithm rounds an LP relaxation solution. It uses the "filtering" technique of Lin and Vitter [LV 92] to obtain a solution feasible to the LP relaxation of a closely related Group Steiner Tree (GST) problem. Then, the LP rounding algorithm of Garg, Konjevod and Ravi [GKR 97] is utilized to obtain a feasible solution.

2.1 Formulation:

We represent the bicriteria TPP as the problem of minimizing the travel costs subject to a budget D on the purchasing costs. The following IP formulation is a relaxation of the TPP problem, where the market cities that the purchaser visits are connected by a 2-edge-connected subgraph instead of a tour. In the formulation, the variable x_{ij} indicates whether product i is purchased at market j, and variable z_{jk} indicates whether markets j and k are connected by an edge of the 2-connected subgraph.

$$\min \sum_{j,k \in M} c_{jk} z_{jk}$$

st

$$\sum_{i=1}^{n} \sum_{j=1}^{m} d_{ij} x_{ij} \leq D \qquad (1)$$

$$\sum_{j=1}^{m} x_{ij} = 1 \qquad i \in N \qquad (2)$$

$$\sum_{j \notin S} x_{ij} + \frac{1}{2} \sum_{j \in S, k \notin S} z_{jk} \geq 1 \qquad i \in N, S \subset M, 1 \notin S \qquad (3)$$

$$x_{ij} \in \{0,1\} \qquad i \in N, \; j \in M \qquad (4)$$

$$z_{jk} \in \{0,1\} \qquad j,k \in M \qquad (5)$$

Constraint (1) is the budget constraint on purchase cost. Constraints (2) enforce that each product is purchased. Constraint set (3) is intended to capture the requirement of crossing certain cuts in the graph by edges in the subgraph that connect the visited markets. Consider a set of markets S not including the traveler's start node 1, and a particular product i: Either i is purchased at a market not in S or the 2-edge-connected subgraph containing 1 must contain at least one market in S from where i is purchased, thus crossing at least two of the edges in the cut around S. This disjunction is expressed by constraints (3).

The LP relaxation relaxes the integrality of x_{ij} and z_{jk} variables. Although the LP has an exponential number of constraints, it can be solved in polynomial time using a separation oracle [GLS 88] based on a minimum cut procedure. To separate a given solution (z,x) over constraints (3) for a particular product i, we set up a capacitated undirected graph as follows: For every edge (i,j) of the complete graph on the market nodes, we assign an edge-capacity $z_{ij}/2$. We add a new node p_i and assign the capacity of the undirected edge between p_i and market node j to be x_{ij}. A polynomial-time procedure to determine the minimum cut separating 1 and p_i [AMO 93] can now be used to test violation of all constraints of type (3) for product i. Repeating this for every product i provides a polynomial-time separation oracle for constraints (3).

2.2 Filtering

Let $\hat{x}, \hat{z}$ be an optimal solution to the LP relaxation defined above. By filtering, we limit the set of markets a product can be purchased at. For each product, we filter out markets that offer a price substantially over the average purchase cost of the product in the LP solution.

Let D_i denote the purchase cost of product i in the solution $\hat{x}, \hat{z}$, i.e. $D_i = \sum_{j=1}^{m} d_{ij}\hat{x}_{ij}$. For a given $\varepsilon > 0$, define a group of markets for product i: $G_i = \{j \in M : d_{ij} \leq (1+\varepsilon)D_i\}$. Every group G_i gets at least a certain amount of fractional assignment of product i to its markets in the LP solution as shown by the next lemma.

Lemma 2. *For every product $i \in N$ and $\varepsilon > 0$, $\sum_{j \in G_i} \hat{x}_{ij} \geq \frac{\varepsilon}{1+\varepsilon}$.*

Proof. Suppose for a contradiction that $\sum_{j \in G_i} \hat{x}_{ij} < \frac{\varepsilon}{1+\varepsilon}$. Then, $\sum_{j \notin G_i} \hat{x}_{ij} \geq \frac{1}{1+\varepsilon}$. Note that $D_i = \sum_{j \in M} d_{ij}\hat{x}_{ij} \geq \sum_{j \notin G_i} d_{ij}\hat{x}_{ij} > (1+\varepsilon)D_i \sum_{j \notin G_i} \hat{x}_{ij}$ by the definition of G_i. Since $\sum_{j \notin G_i} \hat{x}_{ij} \geq \frac{1}{1+\varepsilon}$, we get the contradiction $D_i > D_i$.

2.3 Transformation to Group Steiner Tree Problem

For each product we identified a group of markets to purchase the product. We now need to select at least one market from each group and connect them by a tour. For this, we take advantage of the *Group Steiner Tree* (GST) problem which can be stated as follows. Given an edge-weighted graph with some subsets of vertices specified as groups, the problem is to find a minimum weight subtree which contains at least one vertex from each group. We assume without loss of generality that node 1 is required to be included in the tree. We define the following GST instance. Let G be a complete graph on vertex set equal to the market set M. The weight of edge (i,j) is set to c_{ij} (note that we assume c_{ij} is metric). Let the G_i defined as above for each product i be the groups.

Consider the LP relaxation of this GST problem, which we denote by LP-GST. The variables z_{jk} denote whether the edge between j and k is included in the tree.

$$\min \sum_{j,k \in M} c_{jk} z_{jk}$$

st

$$\sum_{j \in S, k \notin S} z_{jk} \geq 1 \qquad S \subset M,\ 1 \notin S \text{ and } G_i \subseteq S \text{ for some } i \quad (6)$$

$$0 \leq z_{jk} \leq 1 \qquad j,k \in M \quad (7)$$

The nontrivial constraints (6) enforce that there is a path from node 1 to some node in group G_i, for every i, in the solution.

Lemma 3. *Let* $\bar{z}_{jk} = (\frac{1+\varepsilon}{2\varepsilon})\hat{z}_{jk}$. *Then,* $\bar{z}$ *is feasible to LP-GST.*

Proof. Consider $S \subset M$ containing G_i but not city 1. By constraint (2), $\sum_{j\notin S}\hat{x}_{ij} + \frac{1}{2}\sum_{j\in S,k\notin S}\hat{z}_{jk} \geq 1$. Also, $\sum_{j\notin S}\hat{x}_{ij} \leq \sum_{j\notin G_i}\hat{x}_{ij} \leq \frac{1}{1+\varepsilon}$ by Lemma 2. Then, $\frac{1}{2}\sum_{j\in S,k\notin S}\hat{z}_{jk} \geq 1 - \frac{1}{1+\varepsilon} = \frac{\varepsilon}{1+\varepsilon}$. So, we have $\sum_{j\in S,k\notin S}\bar{z}_{jk} \geq 1$.

Garg, Konjevod and Ravi [GKR 97] gave a randomized approximation algorithm that rounds a solution to LP-GST. A de-randomized version can be found in [CCGG 98]. Using any of these algorithms to round the solution $\bar{z}$ provides a tree that includes at least one vertex from each group and has cost $O(\log^3 m \log\log m)$ times $\sum_{j,k\in M} c_{jk}\bar{z}_{jk}$.

We obtain a solution to the TPP as follows. Let T be the tree output by the GST rounding algorithm. Let v_i be a market in G_i included in T. We purchase product i at market v_i. We duplicate each edge in T and find an Eulerian tour. We obtain a Hamiltonian tour on the markets in T by short-cutting the Eulerian tour. That is, while traversing the Eulerian tour, when a node that has already been visited is next, we skip to the next unvisited node, say u, and include an edge that connects the current node to u.

The following lemmas are now immediate.

Lemma 4. *The TPP rounding algorithm outputs a solution with total purchase cost at most* $(1+\varepsilon)\sum_{i=1}^{n}\sum_{j=1}^{m} d_{ij}\hat{x}_{ij}$, *which is at most* $(1+\varepsilon)$ *times the budget D, for any chosen* $\varepsilon > 0$.

Lemma 5. *The TPP rounding algorithm outputs a solution with total travel cost at most* $O((1+\frac{1}{\varepsilon})(\log^3 m \log\log m))\sum_{j,k\in M} c_{jk}\hat{z}_{jk}$, *which is at most* $O((1+\frac{1}{\varepsilon})(\log^3 m \log\log m))$ *times the optimal TPP cost, for any chosen* $\varepsilon > 0$.

From Lemmas 4 and 5 we get the following theorem.

Theorem 6. *The TPP rounding algorithm outputs a* $((1+\varepsilon),(1+\frac{1}{\varepsilon})O(\log^3 m \log\log m))$-*approximate solution for the bicriteria TPP problem with metric travel costs in polynomial time, for any* $\varepsilon > 0$.

The same analysis gives a poly-logarithmic approximation for the TPP as well, where we relax the budget constraint on total purchase cost and add the cost to the objective function.

Corollary 7. *For any* $\varepsilon > 0$, *the TPP rounding algorithm finds a solution for the TPP with metric travel costs, whose cost is* $\max\{(1+\varepsilon),(1+\frac{1}{\varepsilon})\,O(\log^3 m \log\log m)\}$ *times the optimal TPP cost in polynomial time.*

We note that the TPP with metric costs can be directly transformed to a group Steiner *tour* problem[1] on a metric with $m+nm$ nodes, i.e., one of finding a tour that visits at least one node from each group. To construct this metric, we begin with the original metric c on the market nodes. To each market node, we attach n new nodes via "leaf" edges, one for each product - such an edge from market node j to its product node i is assigned cost $d_{ij}/2$. All other edges incident on the new nodes are given

[1] This is also called the generalized TSP in the literature; see [FGT 97].

costs implied by the triangle inequality. All the nodes corresponding to a product i specify a group - Thus, there are n groups, each with m nodes. It is now straightforward to verify that any group Steiner tour can be transformed to a solution to the original traveling purchaser instance with the same cost. Applying the rounding algorithms for group Steiner trees and short-cutting the tree obtained to a tour gives a direct $O(\log^3(m+nm)\log\log(m+nm))$ approximation to the metric TPP.

3 Network Design with Proportional Cost Metrics

In this section we consider a special case of the traveling purchaser problem, which models a telecommunication network design problem. A communication network consists of several local access network (LANs) that collect traffic of user nodes at the switching centers, and a backbone network that routes high-volume traffic among switching centers. We model this problem by requiring a ring architecture for the backbone network and a star architecture for the LANs. The ring structure is preferred for its reliability. Because of the "self-healing" properties associated with SONET rings, ring structures promise to be of increasing importance in future telecommunication networks ([Kli 98]). The formal model follows.

We are given a graph $G=(V,E)$, with length l_e on edge e. Without loss of generality, we use the metric completion of the given graph. That is, length of an edge e is replaced by the shortest-path length d_e between its endpoints. The problem is to pick a tour (ring backbone) on a subset of the nodes and connect the remaining nodes to the tour such that sum of the *tour cost* and the *access cost* is minimized. The access cost of connecting a non-tour node i to a tour node j is d_{ij}, i.e. the shortest-path length between i and j. The access cost includes the cost of connecting all non-tour nodes to the tour. On the other hand, the cost of including an edge e in the tour is $\rho\, d_e$, where the constant $\rho \geq 1$ reflects the more expensive cost of higher bandwidth connections in the backbone network.

This problem is a special case of TPP where the vertices of the graph correspond to both the set of markets and the set of products [V 90]. With the TPP terminology, the purchase cost of a product of node i at the market of node j is the shortest path length between nodes i and j. Thus, if node i is included in the tour, its product is purchased at its own market at zero cost. We consider a bicriteria version of this problem with the two objectives of minimizing the tour cost and minimizing the access cost. We use the following notation to denote the problems considered.

(A,T,ρ): Minimize tour cost T subject to a budget on the access cost A, where a tour edge costs ρ times the edge length.

$(A+T,\rho)$: Minimize sum of the tour and access costs, where a tour edge costs ρ times the edge length.

3.1 Hardness

The bicriteria problem (A,T,ρ) is NP-hard even when $\rho=1$. When the budget on the access cost A is set to zero, the problem reduces to the TSP since every node must be included in the tour. We show that it is NP-hard to approximate this problem with a sub-logarithmic performance ratio without violating the budget constraint. This result does not follow from the inapproximability of TSP since we assume that the distances d_{ij} are metric.

Theorem 8. *There exists no* $(1,\alpha)$*-approximation algorithm, for any* $\alpha = o(\log n)$*, for the* $(A,T,1)$ *problem unless* $P = NP$*. Here n is the number of nodes in the* $(A,T,1)$ *instance.*

The proof (omitted) is by an approximation preserving reduction from the connected dominating set problem. Note that since $(A,T,1)$ is a special case of (A,T,ρ), the same hardness result holds for (A,T,ρ).

Theorem 9. *The single criteria problem* $(A+T,1)$ *is NP-hard.*

The proof (omitted) is by a reduction from the Hamiltonian tour problem in an unweighted graph which is known to be NP-hard [GJ 79]. Again, since $(A,T,1)$ is a special case of (A,T,ρ), NP-hardness of the latter follows as well.

3.2 Approximation

There exists a simple 2-approximation algorithm for the $(A+T,1)$ problem. Find a minimum spanning tree of G, say MST, duplicate the edges of MST and shortcut this to a tour. Note that every node is included in the tour so that the access cost is zero. The cost of the tour is at most 2 times the cost of MST, which is a lower bound on the optimal cost.

Note that this heuristic is a 2ρ-approximation algorithm for $(A+T,\rho)$. However, we obtain a stronger constant factor approximation for both the bicriteria and single objective problems for arbitrary ρ by LP rounding. The LP rounding algorithm uses filtering to limit the set of tour nodes a node can be connected to, as in the TPP rounding algorithm. However, the construction of the tour differs from the TPP rounding algorithm. Tour nodes are chosen based on the access costs and the tour is built by shortcutting an MST on a graph obtained by contracting balls around the tour nodes.

We assume that a root node r is required to be included to the tour (this is similar to including the home city in the TPP). If no such node is specified, we can run the algorithm n times, each time with a different root node, and pick the best solution. We use the following relaxation of (A,T,ρ), which is very similar to the relaxation that we used in the TPP rounding algorithm.

$$\min \rho \sum_{e\in E} d_e z_e$$

st

$$\sum_{i\in V}\sum_{j\in V} d_{ij}x_{ij} \le D \qquad (1)$$

$$\sum_{j\in V} x_{ij} = 1 \qquad i\in V \qquad (2)$$

$$\sum_{j\notin S} x_{ij} + \frac{1}{2}\sum_{e\in\delta(S)} z_e \ge 1 \qquad i\in V,\ S\subset V,\ r\notin S \qquad (3)$$

$$x_{ij}\in\{0,1\} \qquad i\in N,\ j\in M \qquad (4)$$

$$z_{jk}\in\{0,1\} \qquad j,k\in M \qquad (5)$$

Variable x_{ij} indicates whether node i is connected to the tour at node j, and variable z_e indicates whether edge e is included in the tour. Constraint (1) is the budget constraint on access cost. Here, d_{ij} denotes the shortest path length between nodes i and j. Constraint (2) ensures that every node has access to the tour. For a node set S excluding r, constraint (3) ensures that at least two edges of the cut around S, denoted by $\delta(S)$, is included in the tour, if some node has been assigned to access the tour at a node in S. We obtain the LP relaxation (LPR) by relaxing the integrality in constraints (4) and (5).

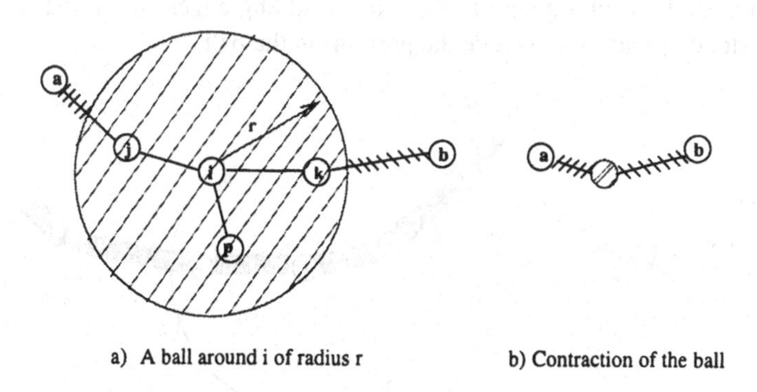

Fig. 1. The definition and contraction of a ball

We need a few definitions before we describe the algorithm. A *ball* of radius r around a node i is the set of all points in G that are within distance r from i under the length function d_e on the edges. The ball may include nodes, edges and partial edges as illustrated in Figure 1a. When we *contract* a ball around a node into a single node, (i) we delete edges with both ends in the ball; (ii) we connect the edges with exactly one endpoint in the ball to the new node and shorten their length by the length remaining in the ball (Figure 1b). Let $\varepsilon > 0$ and $\alpha > 1$ be input parameters. The algorithm is as follows:

(1) Solve LPR, let $\hat{x}, \hat{z}$ be an optimal solution.
(2) Let D_i denote the access cost of node i in this solution, i.e. $D_i = \sum_{j \in V} d_{ij}\hat{x}_{ij}$.
(3) Let $\widehat{D}_i = (1+\varepsilon)D_i$ and define a ball B_i around every node i of radius $\alpha\widehat{D}_i$.
(4) Preprocessing step: remove all balls containing r and connect their centers to r in the access network.
(5) While unprocessed balls remain:
    (5.1) Pick a ball with minimum radius, say B_k, and mark it as a "tour ball".
    (5.2) Remove all balls intersecting B_k and mark them as "connected via B_k".
(6) Contract each *tour ball* to a node. Let G' be a complete graph on the contracted nodes and r, with edge weights equal to shortest path lengths in G (after contractions).
(7) Find an MST of G' and construct H by replacing edges of the MST by shortest paths in G.
(8) Duplicate edges of H and shortcut them to a tour PT.

(9) Uncontract the balls. Construct tour T by connecting the center node i of each ball B_i to PT.
(10) Connect the center of every ball marked "connected via B_k" directly to k in the access network.

Before we analyze the worst-case performance of the algorithm, let us clarify how we process ball B_i in Step (9). Let b_i be the contracted node corresponding to B_i. Let e_1 and e_2 be the edges incident on b_i in PT. Let v_1 and v_2 be the endpoints of e_1 and e_2 in B_i. Connect the center node i to the tour by adding edges (i, v_1) and (i, v_2) (see Figure 2). Extend e_1 and e_2 to include the portions in the ball.

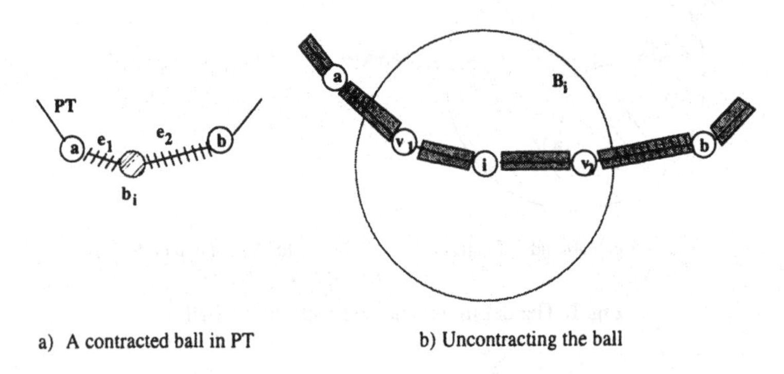

Fig. 2. Uncontracting a ball B_i to include in PT

Lemma 10. *The rounding algorithm outputs a solution with access cost at most* $2\alpha(1+\varepsilon)$ *times the budget D.*

Proof. Each nontour node i is connected to a tour node k such that $B_i \cap B_k$ is nonempty and $\widehat{D}_k \leq \widehat{D}_i$ by the choice of the tour balls in the algorithm. Then, the access cost of i is at most $\alpha\widehat{D}_i + \alpha\widehat{D}_k \leq 2\alpha\widehat{D}_i = 2\alpha(1+\varepsilon)\sum_{j\in V} d_{ij}\hat{x}_{ij}$. Since $\hat{x}$ is a solution to the relaxation LPR, it satisfies the budget constraint $\sum_{i\in V}\sum_{j\in m} d_{ij}\hat{x}_{ij} \leq D$. Thus, the access cost is at most $2\alpha(1+\varepsilon)D$.

Remark 11. The argument in the above proof is also valid for a problem where an access cost budget is specified separately for each node instead of a single budget constraint on the total access cost.

Lemma 12. *The rounding algorithm outputs a solution with tour cost at most* $\max\{2, \frac{\alpha}{\alpha-1}\}(1+\frac{1}{\varepsilon})$ *times the optimal cost.*

Proof. We use the following definitions. For an edge set M, let $c(M) = \sum_{e\in M} \rho d_e$. Let P be the set of nodes included in the tour T output by the algorithm. Let G_i be a ball around i of radius $\widehat{D}_i$. Let E_C denote the edge set of the contracted graph. That is, E_C excludes from E all edges with both ends in a tour ball as well as portions of the edges with one end point strictly inside a tour ball.

The proof follows from the following claims.
Claim 1: $c(T) \leq c(PT) + 2\alpha(1+\varepsilon)\rho \sum_{i \in P} D_i$.
Claim 2: $2(\alpha-1)\varepsilon\rho \sum_{i \in P} D_i \leq \rho \sum_{i \in P} \sum_{e \in (B_i - G_i)} d_e \hat{z}_e$.
Claim 3: $c(PT) \leq 2c(MST) \leq 2(1+\frac{1}{\varepsilon})\rho \sum_{e \in E_C} d_e \hat{z}_e$.

Proof of Claim 1: The cost of the tour T equals the cost of PT, the tour on the contracted nodes, plus the cost of the edges in the tour balls that connect the tour nodes to PT. For a tour ball B_i, suppose PT touches B_i at points k_1 and k_2. The path in B_i connecting k_1 to the center node i and i to k_2 has cost at most $2\alpha(1+\varepsilon)\rho D_i$ since B_i has radius $\alpha(1+\varepsilon)D_i$.

Proof of Claim 2: By an argument similar to the proof Lemma 2 it can be shown that for any $i \in V$, $\sum_{j \in G_i} \hat{x}_{ij} \geq \frac{\varepsilon}{1+\varepsilon}$. Then, by constraint (3) of LPR, it follows that $\sum_{e \in \delta(G_i)} \hat{z}_e \geq \frac{2\varepsilon}{(1+\varepsilon)}$ for any $i \in V$, and G_i excluding r. Note that a fractional $\hat{z}$ value of at least $\frac{2\varepsilon}{1+\varepsilon}$ must go a distance of at least $(\alpha-1)\widehat{D}_i$ to get out of the ball B_i. We can consider this distance as a moat around G_i of width $(\alpha-1)\widehat{D}_i$. So, we get $\rho \sum_{e \in (B_i - G_i)} d_e \hat{z}_e \geq \rho \frac{2\varepsilon}{(1+\varepsilon)}(\alpha-1)\widehat{D}_i = 2\rho\varepsilon(\alpha-1)D_i$ for any $i \in P$, since $\widehat{D}_i = (1+\varepsilon)D_i$.

Proof of Claim 3: The first inequality easily follows since we obtain PT by shortcutting MST. To show the second inequality, we show that $\bar{z} = \frac{(1+\varepsilon)}{2\varepsilon}\hat{z}$ is a feasible solution to an LP relaxation of a Steiner tree problem on the contracted graph $G_C = (V_C, E_C)$, with terminal nodes being the contracted balls and r.

Consider S containing B_i but not r. By constraint (3) of LPR, $\sum_{j \notin S} \hat{x}_{ij} + \frac{1}{2}\sum_{e \in \delta(S)} \hat{z}_e \geq 1$. By the definition of B_i, we also have $\sum_{j \notin S} \hat{x}_{ij} \leq \sum_{j \notin B_i} \hat{x}_{ij} \leq \frac{1}{1+\varepsilon}$. Then, $\frac{1}{2}\sum_{e \in \delta(S)} \hat{z}_e \geq 1 - \frac{1}{1+\varepsilon} = \frac{\varepsilon}{1+\varepsilon}$. So, $\sum_{e \in \delta(S)} \bar{z}_e \geq 1$. Thus, $\bar{z}$ is a feasible solution to the LP relaxation of the Steiner tree problem on G_C. Let $c(ST)$ be the cost of the LP relaxation of the Steiner tree problem on G_C with terminal set the contracted nodes plus r and edge costs ρd_e. Then, $c(MST) \leq 2c(ST)$ (see, e.g. [AKR 95]). Since $\bar{z}$ is a feasible solution, $c(ST) \leq \sum_{e \in E_C} \rho d_e \bar{z}_e = \frac{(1+\varepsilon)}{2\varepsilon}\sum_{e \in E_C} \rho d_e \hat{z}_e$. Thus, the claim follows.

From Claims 1, 2 and 3 we get,

$$C(T) \leq 2(1+\frac{1}{\varepsilon})(\rho \sum_{e \in E_C} d_e \hat{z}_e) + \frac{\alpha}{\alpha-1}(1+\frac{1}{\varepsilon})\rho \sum_{i \in P} \sum_{e \in (B_i - G_i)} d_e \hat{z}_e.$$

Since E_C excludes edges in B_i for any $i \in P$, $C(T) \leq \max\{2, \frac{\alpha}{\alpha-1}\}(1+\frac{1}{\varepsilon})\ \rho \sum_{e \notin G_i} d_e \hat{z}_e \leq \max\{2, \frac{\alpha}{\alpha-1}\}(1+\frac{1}{\varepsilon})OPT$, where OPT is the optimal cost to (A, T, ρ) problem.

From Lemmas 10 and 12, the next result follows immediately.

Theorem 13. *For any $\varepsilon > 0$, $\alpha > 1$ and any ρ, the rounding algorithm outputs a $(2\alpha(1+\varepsilon), \max\{2, \frac{\alpha}{\alpha-1}\}\ (1+\frac{1}{\varepsilon}))$ approximate solution for the bicriteria problem (A, T, ρ) in polynomial time.*

For minimizing the sum of the two objectives, the performance ratio of the algorithm is the maximum of the two ratios for the separate objectives. The best ratio is obtained by setting $\varepsilon = 1/\sqrt{2}$ and $\alpha = 1 + 1/\sqrt{2}$, yielding a performance ratio of $3 + 2\sqrt{2}$.

Corollary 14. *The rounding algorithm is a* $(3+2\sqrt{2})$*-approximation algorithm for* $(A+T,\rho)$ *problem.*

4 Acknowledgments

We are thankful to R. Hassin for proving the NP-hardness of the $(A+T,1)$ problem and other helpful discussions. We also thank G. Konjevod for pointing out the direct reduction of TPP to a group Steiner tour problem.

References

[AMO 93] R. K. Ahuja, T. L. Magnanti and J. B. Orlin, *Network flows: Theory, Algorithms and Applications*, Prentice Hall, Englewood Cliffs, NJ, 1993.

[AS 97] S. Arora and M. Sudan, "Improved low degree testing and its applications," Proc. 29th ACM Annual Symp. on Theory of Computing, 485-495, 1997.

[AKR 95] A. Agrawal, P. Klein and R. Ravi, "When trees collide: An approximation algorithm for the generalized Steiner problem on networks," SIAM J. Computing **24**, 440-456, 1995.

[CCGG 98] M. Charikar, C. Chekuri, A. Goel and S. Guha, "Rounding via tree: Deterministic approximation algorithms for group Steiner trees and k-median," Proc. 30th ACM Annual Symp. on Theory of Computing, 114-123, 1998.

[CS 94] J. R. Current and D. A. Schilling, "The median tour and maximal covering tour problems: Formulations and heuristics," European Journal of Operational Research, **73**, 114-126, 1994.

[F 96] U. Feige, "A threshold of ln n for approximating set cover," Proc. 28th ACM Annual Symp. on Theory of Computing, 314-318, 1996.

[FGT 97] M. Fischetti, J. S. Gonzalez and P. Toth, "A branch-and-cut algorithm for the symmetric generalized traveling salesman problem," Operations Research, **45**, 378-394, 1997.

[GJ 79] M. R. Garey and D. S. Johnson, *Computers and Intractability: A Guide to the Theory of NP-Completeness*, W. H. Freeman, San Francisco, 1979.

[GKR 97] N. Garg, G. Konjevod and R. Ravi, "A poly-logarithmic approximation algorithm for the group Steiner tree problem," Proc. of the 9th Ann. ACM-SIAM Symposium on Discrete Algorithms, 253-259, 1998.

[GLD 81] B. Golden, L. Levy and R. Dahl, "Two generalizations of the traveling salesman problem," OMEGA, **9**, 439-455, 1981.

[GLS 88] Martin Grötschel and Laszlo Lovász and Alexander Schrijver, *Geometric Algorithms and Combinatorial Optimization*, Springer-Verlag, 1988.

[Kli 98] J. G. Klincewicz, "Hub location in backbone/tributary network design: a review," To appear, Location Science, 1998.

[LV 92] J. H. Lin and J. S. Vitter, "ε-approximations with minimum packing constraint violation," In Proc. of the 24th Ann. ACM Symp. on Theory of Computing, 771-782, May 1992.

[MRS+ 95] M. V. Marathe, R. Ravi, R. Sundaram, S. S. Ravi, D. J. Rosenkratz and H. Hunt, "Bicriteria network design problems," J. Algorithms, **28**, 142-171, 1998.

[Ong 82] H. L. Ong, "Approximate algorithms for the traveling purchaser problem," Operations Research Letters, **1**, 201-205, 1982.

[PC 98] W. L. Pearn and R. C. Chien, "Improved solutions for the traveling purchaser problem," Computers and Operations Research, **25**, 879-885, 1998.

[Ram 81] T. Ramesh, "Traveling purchaser problem," OPSEARCH, **18**, 87–91, 1981.

[RS 97] R. Raz and S. Safra, "A sub-constant error-probability low-degree test, and a sub-constant error-probability PCP characterization of NP," Proc. 29th Annual ACM Symp. on Theory of Computing, 314-318, 1997.

[SvO 97] K. N. Singh and D. L. van Oudheusden, "A branch and bound algorithm for the traveling purchaser problem," European Journal of Operational Research, **97**, 571-579, 1997.

[V 90] S. Voss, "Designing special communication networks with the traveling purchaser problem," Proceedings of the First ORSA Telecommunications Conference, 106-110, 1990.

[V 96] S. Voss, "Dynamic tabu search strategies for the traveling purchaser problem," Annals of Operations Research, **63**, 253-275, 1996.

The Impact of Knowledge on Broadcasting Time in Radio Networks (Extended Abstract)

Krzysztof Diks[1] Evangelos Kranakis[24] Danny Krizanc[24] Andrzej Pelc[34]

[1] Instytut Informatyki, Uniwersytet Warszawski, Banacha 2, 02-097 Warszawa, Poland. diks@mimuw.edu.pl
Partially sponsored by KBN grant 8T11 C 03614. This work was done during the author's stay at the Université du Québec à Hull.
[2] School of Computer Science, Carleton University, Ottawa, Ontario, K1S 5B6, Canada. {kranakis,krizanc}@scs.carleton.ca

[3] Département d'Informatique, Université du Québec à Hull, Hull, Québec J8X 3X7, Canada. E-mail: Andrzej_Pelc@uqah.uquebec.ca
[4] Research supported in part by NSERC (Natural Sciences and Engineering Research Council of Canada)

Abstract. We consider the problem of distributed deterministic broadcasting in radio networks. Nodes send messages in synchronous time-slots. Each node v has a given transmission range. All nodes located within this range can receive messages from v. However, a node situated in the range of two or more nodes that send messages simultaneously, cannot receive these messages and hears only noise. Each node knows only its own position and range, as well as the maximum of all ranges. Broadcasting is adaptive: Nodes can decide on the action to take on the basis of previously received messages, silence or noise. We prove a lower bound on broadcasting time in this model and construct a broadcasting protocol whose performance matches this bound for the simplest case when nodes are situated on a line and the network has constant depth. We also show that if nodes do not even know their own range, every broadcasting protocol must be hopelessly slow.

While distributed randomized broadcasting algorithms, and, on the other hand, deterministic off-line broadcasting algorithms assuming full knowledge of the radio network, have been extensively studied in the literature, ours are the first results concerning broadcasting algorithms that are distributed and deterministic at the same time. We show that in this case the amount of knowledge available to nodes influences the efficiency of broadcasting in a significant way.

1 Introduction

Radio communication networks have recently received growing attention. This is due to the expanding applications of radio communication, such as cellular phones and wireless local area networks. The relatively low cost of infrastrucure and the flexibility of radio networks make them an attractive alternative to other types of communication media.

A radio network is a collection of transmitter-receiver devices (referred to as *nodes*). Nodes send messages in synchronous time-slots. Each node v has a given transmission range. All nodes located within this range can receive messages from v. However, a node situated in the range of two or more nodes that send messages simultaneously, cannot receive these messages and hears only noise.

One of the fundamental tasks in network communication is *broadcasting*. One node of the network, called the *source*, has a piece of information which has to be transmitted to all other nodes. Remote nodes get the source message via intermediate nodes, in several hops. One of the most important performance parameters of a broadcasting scheme is the total time it uses to inform all nodes of the network.

1.1 Previous work

In most of the research on broadcasting in radio networks [1, 4, 5, 7] the network is modeled as an undirected graph in which nodes are adjacent if they are in the range of each other. A lot of effort has been devoted to finding good upper and lower bounds on the broadcast time in radio networks represented as arbitrary graphs, under the assumption that nodes have full knowledge of the network. In [1] the authors proved the existence of a family of n-node networks of radius 2, for which any broadcast requires time $\Omega(\log^2 n)$, while in [5] it was proved that broadcasting can be done in time $O(D+\log^5 n)$ for any n-node network of diameter D. In [11] the authors restricted attention to communication graphs that can arise from actual geometric locations of nodes in the plane. They proved that scheduling optimal broadcasting is NP-hard even when restricted to such graphs and gave an $O(n\log n)$ algorithm to schedule an optimal broadcast when nodes are situated on a line. In [6] the authors discussed fault-tolerant broadcasting in radio networks arising from geometric locations of nodes on the line and in the plane. On the other hand, in [2] a randomized protocol was given for arbitrary radio networks where nodes have no topological knowledge of the network, not even about neighbors. This randomized protocol runs in expected time $O(D\log n+\log^2 n)$.

1.2 Our results

The novelty of our approach consists in considering broadcasting protocols that are distributed and deterministic at the same time. We assume that nodes have only local knowledge concerning their own position and range and additionally they know the maximum R of all ranges. This is a realistic assumption, as the transmitter-receiver devices can have varying power but usually belong to a set of *a priori* known standard types. Our aim is to show to what extent this restriction of knowledge concerning the network affects efficiency of broadcasting. We consider the simplest scenario when nodes are situated at integer points on the line. We prove the lower bound $\Omega(\frac{\log^2 R}{\log\log R})$ on broadcasting time of any deterministic protocol. (This lower bound is of course also valid for nodes situated in the plane.) Moreover, we show a broadcasting protocol running in time $O(D\frac{\log^2 R}{\log\log R})$, where D is the *depth* of the communication graph, i.e., the maximum length of a shortest path from the source to any node. Thus our protocol is asymptotically optimal for constant D. (In the full version of the paper we will also show another protocol running in time $O(D+\log^2 R)$, and thus optimal for $D=\Omega(\log^2 R)$.) We also consider the extreme scenario when nodes do not even know their own range. Under this assumption we show that every broadcasting protocol must use time $\Omega(R)$ for some networks of depth 2.

Our results lead to the problem of finding a protocol which is asymptotically optimal for any network on the line. It would be even more challenging to find good protocols for arbitrary networks in the plane. The lower bound $\Omega(\frac{\log^2 R}{\log\log R})$ seems weak in case of the plane and we would not expect protocols as efficient as the one presented in this paper.

2 Preliminaries and model description

Nodes are situated at integer points of the line and are identified with the respective integers. Every node v has a non-negative integer *range* $r(v)$. The set of pairs $(v, r(v))$, for all nodes v, with a distinguished node s called the *source*, is referred to as a *configuration*. If v sends a message, the signal reaches exactly those nodes that are in the segment $[v-r(v), v+r(v)]$. These nodes are said to be in the range of v. However, a node situated in the range of two or more nodes that send messages simultaneously, cannot receive these messages and hears only noise. In particular, a node u which sends a message in the same time as one or more nodes in whose range u is situated, hears noise. It should be stressed that noise is assumed to be different from silence, i.e., *collision detection* is available (cf. [2]).

Actions of nodes are performed in synchronous time-slots. We consider two models. In the main model (considered in sections 4 and 5) the *a priori* knowledge of every node v consists of its position v, its range $r(v)$ and the maximum R over all ranges. It is important to stress that nodes do not know positions or ranges of any other nodes. In the second model (considered in section 3) this knowledge is further reduced: a node v does not even know $r(v)$. We use the latter scenario to prove a strong lower bound on broadcasting time.

In each time-slot a node v receives one of the following inputs: either silence, (when neither itself nor any other node in whose range v is situated transmits), or a message (if a unique node in whose range v is situated transmits), or noise (if v is situated in the ranges of at least two simultaneously transmitting nodes). All nodes run a common broadcasting protocol. Broadcasting is *adaptive*: Every node can compute its action to be performed in a given time slot on the basis of previously received inputs. This action is either sending a particular message or keeping silent.

The *reachability graph* associated with a given configuration is the directed graph G whose vertices are nodes of the configuration and there is a directed edge from v to w, if w is in the range of v. We assume that there exists a directed path from the source s to any node of G. Let $d(v)$ denote the length of the shortest directed path in G, from s to v. The *depth* D of the graph G (or of the underlying configuration) is defined as the maximum of $d(v)$ over all nodes v of the configuration. The set of all nodes of a configuration can be partitioned into $D+1$ *layers* $L_0,\ldots,L_D$, where $L_i=\{v : d(v)=i\}$. Clearly, D is a lower bound on broadcasting time of any protocol. On the other hand, if all nodes know the positions and ranges of all other nodes, i.e., the entire configuration, it is easy to construct a distributed deterministic broadcasting protocol working in time $O(D)$.

3 Lower bound under range ignorance

We begin by considering the extreme scenario when the *a priori* knowledge of each node is limited to its position and the maximum R over all ranges. (A node does not even know its own range.) Under this assumption we show that the worst case broadcasting time is $\Omega(R)$ for some configurations of constant depth.

We consider the following situation: $n+1$ nodes are located at the points $0,1,\ldots,n$ on the line. We assume that n is divisible by 3. Node 0 is the source. It has range equal to $2n/3$. $k \geq 1$ out of the nodes $\{1,2,\ldots,n/3\}$ have range equal to n. For the remaining nodes in $\{1,2,\ldots,n/3\}$, node i has range equal to $2n/3-i$. The nodes $\{n/3+1,\ldots,n\}$ all have range equal to 1. Thus $R=n$ and $D=2$.

We will say a node in $\{1,\ldots,n/3\}$ is *strong* if its range is n, *weak* otherwise. There are $2^{n/3}-1$ possible configurations, corresponding to the possible settings of weak and strong nodes.

The protocol which has the nodes $0,1,\ldots,n/3$ broadcast in succession gives an upper bound of $n/3+1$ on broadcasting time. The remainder of this section is devoted to showing that any protocol must use at least $n/3+1$ steps, i.e., the above protocol is optimal.

In order to show this lower bound we must be more precise in our model of a protocol. We assume that the $n+1$ nodes are universal Turing machines running synchronously using a global clock initially set to 0. The input to a node i in $1,\ldots,n$ is a program P_i. Node 0 has as input P_0 and a string M on a special input tape that is to be broadcast. At the completion of the protocol, all nodes will have entered a terminal state and will have output M onto a special output tape. All steps of a protocol (except the first) consist of three phases: Receive, Compute, Broadcast. During the Receive phase, every node v reads from a special reception input tape the results of the Broadcast phase of the previous step which is determined by the rules for packet radio networks and in whose range v is situated in the given configuration (see the discussion below concerning the Broadcast phase). During the Compute phase, the nodes perform an arbitrary computation based upon the input they received, their current state (including the contents of all tapes) and the program they are running. As a result of this computation each node decides on one of two actions, either to broadcast or be silent during the Broadcast phase. If a processor decides to broadcast then it writes its state including the contents of all of its tapes, the positions of its heads, etc., to a special broadcast output tape. If it decides not to broadcast it writes a special symbol indicating "silence" to the broadcast output tape. After a Broadcast phase, a node's v reception input tape contains one of the following:
1. a special symbol representing "silence", if none of the nodes in whose range v is situated decided to broadcast;
2. a special symbol representing "noise", if two or more of the nodes in whose range v is situated decided to broadcast;

3. the contents of the broadcast tape of a node w, if w is the unique node that decided to broadcast among nodes in whose range v is situated.

Recall that every node is in its own range. The first step has no Receive phase. If a node enters a terminal state it no longer participates in the protocol. It is assumed that the same programs $P_0, \dots, P_n$ are used for all input configurations and for all messages M. We are now ready to state the main theorem of this section:

Theorem 1. *For any deterministic broadcast protocol there exists a configuration on which it requires $n/3+1$ steps.*

Proof. Assume to the contrary there exists a protocol $\mathcal{P}$ which on all configurations finishes in $t \le n/3$ steps. Let $|P_i|$ be the number of bits required to describe the input program to node i for $\mathcal{P}$. Let M be a string with Kolmogorov complexity [9] greater than $n + \sum_{i=0}^{n} |P_i|$ bits. The intuitive reason for choosing a message of this complexity is that it precludes the possibility of encoding it in at most $n/3$ time-slots by silence-noise bits. The proof of the theorem is based on the following three lemmas.

Lemma 2. *For input broadcast message M, for all configurations, there exists at least one step of protocol $\mathcal{P}$ during which precisely one strong node broadcasts.*

The second lemma shows that the source 0 must take one step to broadcast alone.

Lemma 3. *For input broadcast message M, for all configurations, there exists at least one step of protocol $\mathcal{P}$ during which which node 0 broadcasts and precisely zero nodes among $1, \dots, n/3$ broadcast.*

The third lemma shows that the actions taken by nodes $1, \dots, n/3$ for the first $t \le n/3$ steps of any protocol are the same for all configurations.

Lemma 4. *For input broadcast message M, at the end of step $i \le n/3$ of protocol $\mathcal{P}$, the state of each of the nodes $0, 1, \dots, 2n/3 - i + 1$ (including the contents of their broadcast tape) is the same for all configurations.*

As a consequence of Lemma 4, for protocol $\mathcal{P}$ running in $t \le n/3$ steps on broadcast input message M, for each of the nodes $0, 1, \dots, n/3$, its actions, i.e., whether it broadcasts or is silent, is the same during each of the steps of the protocol, for all configurations. Thus, we can consider the actions of these nodes as consisting of t sets, $S_1, S_2, \dots, S_t$ where set S_i is the subset of these nodes that broadcast during step i. We are now ready to complete the proof of the theorem.

Consider the protocol $\mathcal{P}$ on broadcast input message M. By lemma 3 one of the t steps of $\mathcal{P}$ must have node 0 broadcast while all nodes in $1, \dots, n/3$ are silent. Consider the remaining $t-1 < n/3$ steps, the only ones during which the nodes $1, \dots, n/3$ can broadcast. Assume that none of the sets associated with these steps are singletons. Then for the configuration consisting of all strong nodes, in all steps, either all strong nodes are silent or two or more strong nodes broadcast. This contradicts lemma 2. Therefore there must be at least one singleton set. For all singleton sets, assign the weak range to the node in the set. Now remove all nodes assigned weak range from the sets. If after this process, no singletons are created, then the configuration with all remaining nodes assigned the strong range again contradicts Lemma 2. Assign the resulting singletons weak range, and continue with this process. It must stop with all sets having been reduced to singletons or to the empty set. Since $t-1 < n/3$, there exists at least one node which does not appear in any singleton. Consider the configuration where that node is strong and all others are weak. In this configuration, the given node never broadcasts and therefore Lemma 2 is again contradicted. Therefore, no such protocol $\mathcal{P}$ exists and at least $t+1$ steps are required to solve the problem. □

It easily follows from considerations in the next section that in case when each node knows its position and range, as well as the maximum R over all ranges, broadcasting for configurations considered above can be done in time $O(\log R)$. (See Algorithm 1 – leader election in a cluster.)

4 Broadcast protocols with known range

In this section we show that if every node knows its own range and position, as well as the maximum R over all ranges, the lower bound from section 3 can be dramatically invalidated. We show a broadcast protocol running in worst-case time $O(\frac{\log^2 R}{\log\log R})$, for all configurations of depth 2. (Recall that without knowledge of nodes' own range we showed such a configuration requiring time $\Omega(R)$.) This protocol can be generalized to give time $O(D\frac{\log^2 R}{\log\log R})$, for all configurations of depth D. The lower bound to be proved in section 5 shows that the above protocol is asymptotically optimal for constant D.

For any nonempty set S of nodes, a set $S' \subseteq S$ is *right-equivalent* to S, if $\max\{v+r(v); v \in S'\} = \max\{v+r(v); v \in S\}$. *Left-equivalent* subsets are defined similarly. We will first restrict our considerations to informing nodes larger than the source, and we will use the term *equivalent* instead of right-equivalent.

For simplicity we assume that R is a power of 2. Modifications in the general case are obvious. For every integer j and every $l = 0,1,\ldots,\log R$, we define $I(j,l) = \{j2^l, j2^l+1,\ldots,(j+1)2^l-1\}$. Fix a layer L. Assume, without loss of generality, that all ranges of nodes in L are strictly positive. The *cluster* $C(j,l)$ is defined as the set of nodes $\{v \in L \cap I(j,l) : 2^l \leq r(v) < 2^{l+1}\}$, if this set is nonempty. The integer l is called the level of the cluster $C(j,l)$. A node $v \in L$ belongs to the cluster $C(j(v),l(v))$, where $l(v) = \lfloor \log r(v) \rfloor$ and $j(v) = \max\{j : j2^{l(v)} \leq v\}$.

Lemma 5. *Clusters form a partition of L. Every pair of nodes in a cluster are in each others range.*

The *leader* of a cluster is its node v with maximum value $v+r(v)$. If there are many such nodes then the leader is the one with maximum range among them. Notice that the singleton set of a leader is equivalent to the cluster. A node $u \in X \subseteq L$ is called X-*nonessential* if there exists $v \in X$ such that u is in the range of v and either (1) $v+r(v) > u+r(u)$ or (2) $v+r(v) = u+r(u)$ and $v < u$.
The set of all nodes X-nonessential is denoted by X^- and the set of all other nodes in X (called X-essential) is denoted by X^+.

Lemma 6. *The set X^+ is equivalent to X.*

We first construct a broadcasting protocol working in time $O(\frac{\log^2 R}{\log\log R})$, for configurations of depth 2. To this end we will solve the following problem. Consider a layer $L \subseteq \{0,...,R-1\}$, for which we have $v + r(v) \geq R$, for any node $v \in L$. We want to construct a small subset of L equivalent to L. We will show how to construct such a set of size $O(\log R)$ in time $O(\frac{\log^2 R}{\log\log R})$. This will yield the desired protocol for configurations of depth 2, with all nodes in this small equivalent subset broadcasting sequentially.

Lemma 7. *For every level* $l = 0,1,...,\log R$, *the set L contains at most two (consecutive) clusters* $C(2^{\log R - l} - 2, l)$ *and* $C(2^{\log R - l} - 1, l)$.

It follows that the number of clusters in L is at most $2\log R + 2$. The set of leaders of all clusters is equivalent to L. In fact, the small equivalent set that we seek, is the subset of the set of all leaders.

Let L_e (L_o) be the set of those nodes $u \in L$ for which $j(u)$ is even (odd).

Lemma 8. *The set* $L_e^+ \cup L_o^+$ *is equivalent to L and has size* $O(\log R)$.

We now show how to construct L_o^+. The construction of L_e^+ is similar. Consider clusters $C(j,l)$ with odd j.

Lemma 9. *Let* $l_1 \geq l_2 + 3$. *If there exists a node* $v \in C(j_1, l_1)$ *such that* $v \geq R - 3 \cdot 2^{l_1 - 3} + 1$ *then all nodes in* $C(j_2, l_2)$ *are* L_o*-nonessential.*

Clusters $C(j_2, l_2)$ satisfying Lemma 9 are called *useless*. All other clusters are called *useful*.

Lemma 10. *Let* $l_1 \geq l_2 + 3$. *If* $v \in C(j_1, l_1)$ *and* $v < R - 3 \cdot 2^{l_1 - 3} + 1$ *then v is not in the range of any node from* $C(j_2, l_2)$.

A sequence of clusters $\gamma = (C(j_1, l_1), ..., C(j_s, l_s))$ is called a *chain* if it satisfies the following conditions:

1. $l_1 > ... > l_s$,
2. for all i, all nodes from clusters of levels $l_i, l_{i+1}, ..., l_s$ are in the range of all nodes in cluster $C(j_i, l_i)$ but no node from clusters of levels $l_1, ..., l_{i-1}$ is in the range of a node from cluster $C(j_i, l_i)$.

It follows from Lemmas 9 and 10 that useful clusters in layer L_o can be partitioned into three chains, γ_0, γ_1, γ_2, where γ_i is the sequence of consecutive useful clusters on levels $l \bmod 3 = i$.

We can now formulate a high-level description of an algorithm constructing L_o^+. It consists of three phases.

1. Find and eliminate useless clusters.

2. In every chain γ_i, for $i = 0,1,2$, find leaders of clusters from this chain.

3. Eliminate nonessential nodes from the set of leaders found in phase 2.

We give the details only of phase 2 which is the most difficult. We show how to find leaders of all clusters in a chain in time $O(\frac{\log^2 R}{\log\log R})$. Define $b(j,l) = (j+1)2^{l+1} - 1$. We assign to every node $v \in C(j,l)$ its *label* defined as follows: $lab(v) = (v + r(v) - b(j,l) - 1)R + (b(j,l) - v)$. Notice that $0 \leq lab(v) \leq R^2 - R$.

Lemma 11. *Different nodes in a cluster have different labels. A node v is a leader in $C(j,l)$, if and only if, $lab(v) > lab(u)$, for all nodes u in $C(j,l)$, different from v.*

Every node v can compute parameters $j(v)$ and $l(v)$, as well as its label $lab(v)$ knowing its position and range. Labels will be used to elect a leader by binary search. Upon completion of the algorithm the value of a boolean variable leader(v) informs the node v if it is a leader.

Algorithm 1: Election of a leader in a cluster – explicit binary search. Algorithm for node v.

$l := 0$; $r := R^2 - R$;
while $r - l > 0$ **do**
    $m := \lfloor (l+r)/2 \rfloor$;
    if $m+1 \leq lab(v) \leq r$ **then** broadcast /*/ **else** keep silent;
    if silence **then** $r := m$ **else** $l := m+1$;
leader(v):=($lab(v) = l$);
if leader(v) **then** broadcast $(v, r(v))$. /**/

In step /*/ the node broadcasts any message (it is sufficient just to send a signal). In step /**/ messages may be different, depending on the purpose the leader is used for. In our case we want to identify nonessential leaders in clusters of the chain. Hence the leader broadcasts $(v, r(v))$.

For all nodes $v \in C(j,l)$, values of l and r are the same after each turn of the **while** loop. If v is a leader, we have $l \leq lab(v) \leq r$. Hence a leader in a cluster can be elected in time $\Theta(\log R)$. Algorithm 1 could be used to elect leaders in each cluster of the chain separately. However, this would result in time $\Theta(\log^2 R)$. In order to speed up the process, we need to elect leaders in many clusters simultaneously. However, in doing so, we need to avoid interference among nodes from different clusters broadcasting at the same time.

We start with a generalization of Algorithm 1. $P = R^2 - R + 1$ is the number of possible labels of nodes. Let $S \geq \lceil \log P \rceil$ be an integer and let A be an arbitrary set of size P of binary sequences of length S. Denote by $\alpha_0, \alpha_1, \ldots, \alpha_{P-1}$ the lexicographic ordering of sequences from A. Assign to every node u its *binary label*: $binlab(u) = \alpha_{lab(u)}$. Clearly $binlab(u) > binlab(v)$, if and only if, $lab(u) > lab(v)$. For a sequence α, $pref(\alpha, i)$ denotes its prefix of length i, and $\alpha[i]$ denotes the ith term of α. In particular, $pref(\alpha, 0)$ is the empty sequence ε.

Suppose that every node knows its binary label with respect to a given set A. The following algorithm elects a leader in a cluster, using binary labels.

Algorithm 2: Election of a leader in a cluster – implicit binary search. Algorithm for node v.

$\beta := \varepsilon$; $\alpha := binlab(v)$;
for $i := 1$ **to** S **do**
    if $pref(\alpha, i-1) = \beta$ and $\alpha[i] = 1$ **then** broadcast **else** keep silent;
    if silence **then** $\beta := \beta \bullet 0$ **else** $\beta := \beta \bullet 1$;
leader(v):=($\beta = \alpha$);
if leader(v) **then** broadcast $(v, r(v))$.

It can be easily shown by induction on the number of iterations of the **for** loop that the sequence β is the same for all $v \in C(j,l)$, and that $\beta = pref(binlab(v), i)$, if

v is a leader. The correctness of Algorithm 2 follows from this observation. Moreover, if $S = \lceil \log P \rceil$ and A is the set of all binary sequences of length S, Algorithm 2 is a restatement of Algorithm 1.

We will now use Algorithm 2 to find leaders in all clusters of a chain γ simultaneously. In each cluster a separate "copy" of the algorithm will perform election. Notice that if no node broadcasts in a given step, we can extend the sequence β by one bit: 0, in every cluster. Clearly, in nontrivial cases, exclusively silent steps cannot accomplish leader election. Nevertheless, we will keep the number of "noisy" steps small, and at the same time elect leaders fast. Noisy steps are a problem because nodes from a given cluster can be heard in all clusters of lower levels. In order to prevent this interference from disturbing computations in clusters of lower levels, we add, for each step of Algorithm 2 performed in any cluster, two steps verifying if noise heard by nodes in this cluster is not caused by nodes from clusters of higher levels. If it is, nodes from this cluster repeat the same step of Algorithm 2 and we say that the cluster is *delayed*. If we guarantee that each cluster is delayed only during $O(\frac{\log^2 R}{\log\log R})$ steps and that Algorithm 2 works in time $O(\frac{\log^2 R}{\log\log R})$ (i.e., $S = O(\frac{\log^2 R}{\log\log R})$) then leaders in all clusters will be elected in time $O(\frac{\log^2 R}{\log\log R})$.

To this end we show that the set A of sequences can be chosen to make the number of "noisy" steps $O(\frac{\log R}{\log\log R})$. Then no cluster will be delayed more than $O(\frac{\log^2 R}{\log\log R})$ times. Let $S = \lceil \frac{2\log^2 P}{\log\log P} \rceil$ and $H = \lceil \frac{\log P}{\log\log P} \rceil$.

Lemma 12. *There exist at least P binary sequences of length S containing at most H terms 1.*

Thus we can take as A any set of P binary sequences of length S containing at most H terms 1. It remains to show, how nodes that hear noise can determine that it is caused by broadcasting nodes from clusters of higher levels. Suppose that nodes know the consecutive number of their cluster in the chain. (This can be learned in $O(\log R)$ steps.) Fix a time unit i in which various steps of Algorithm 2 are performed in various clusters. In time unit $i+1$ (the first verifying step), all nodes that heard noise in time unit i and are in clusters with even number, broadcast. In time unit $i+2$ (the second verifying step), all nodes that heard noise in time unit i and are in clusters with odd number, broadcast. Notice that if some cluster heard noise in time unit i, caused by a higher level cluster, it must hear noise in *both* verifying steps. Such clusters repeat the step of Algorithm 2 performed in time unit i. On the other hand, clusters that hear noise in only one verifying step, can perform the next step of Algorithm 2 because all nodes of this cluster know that noise heard in time unit i was caused by nodes from this cluster. In order to keep synchrony, nodes that hear nothing in time unit i, perform the corresponding step of Algorithm 2 and wait two time units.

Hence it is possible to elect leaders in all clusters of a chain in time $O(\frac{\log^2 R}{\log\log R})$ and consequently phase 2 can be performed in time $O(\frac{\log^2 R}{\log\log R})$. Thus we have shown how to construct a set of size $O(\log R)$ equivalent to the layer L, in time $O(\frac{\log^2 R}{\log\log R})$. This implies that broadcasting in configurations of depth 2 can be done in time $O(\frac{\log^2 R}{\log\log R})$. This result can be easily generalized as follows.

Theorem 13. *Broadcasting in a configuration of depth D can be done in time* $O(D\frac{\log^2 R}{\log\log R})$.

5 Lower bound with known range

This section is devoted to establishing the lower bound $\Omega(\frac{\log^2 R}{\log\log R})$ on broadcasting time of any deterministic protocol in our main model, i.e., even when each node knows its own range. This lower bound will show that the protocol from section 4 is asymptotically optimal for constant D. For simplicity we use a more informal style than that from section 3. It is not difficult, however, to reformulate the argument using Turing machines, states of nodes, etc. As before, and for the same reasons, we need to take a broadcast message of sufficiently high Kolmogorov complexity.

Theorem 14. *For any deterministic broadcast protocol there exists a configuration of depth 2 on which it requires* $\Omega(\frac{\log^2 R}{\log\log R})$ *steps.*

Proof. Nodes are situated at nonnegative integers. 0 is the source and its range is R, which is a power of 2. Let $h = \lfloor\frac{1}{3}\log R\rfloor$ and $p = 2^h$. Let $x_0, x_1, ..., x_{h-1}$ be a decreasing sequence of integers defined by: $x_i = R + 1 - (3\cdot 2^i - 2)\cdot 2^h$. Notice that $x_i \geq 1$, for all $i = 0, 1, ..., h-1$.

For all $j = 0, 1, ..., h-1$, we define I_j as the segment $\{x_j, x_j+1, ..., x_j+p-1\}$. These segments are pairwise disjoint and a segment with higher index is to the left of a segment with lower index. Layer 1 is a subset of the union of these intervals. Denote $y_j = R + p - j$ and let $r(v) = y_j - v$, for all $v \in I_j$. Thus the range of every node in the first layer is at least p. Every pair of nodes in the same segment are in each other's range. Moreover, all nodes from segment I_j are in the range of nodes from segments I_k, for $k > j$ but not in the range of nodes from segments I_k, for $k < j$. Integers y_j form a descending sequence and y_j is in the range of nodes from $I_0, ..., I_j$ but not in the range of other nodes from layer 1.

The adversary will choose sets $C_j \subseteq I_j$ of nodes. Whenever C_j is nonempty, the adversary places a node in y_j and assigns $r(y_j) = 0$. Such a node must be informed but cannot inform any other node. The entire configuration consists of the source, of the union of sets C_j, for $j = 0, 1, ..., h-1$ and of the above nodes y_j. More precisely, layer 1 is equal to $C_0 \cup ... \cup C_{h-1}$ and layer 2 consists of corresponding nodes y_j. Hence $D = 2$.

Assume that all nodes in the first layer already know the source message. (This requires one step.) We will show that subsets C_j can be chosen in such a way that $\Omega(\frac{h^2}{\log h})$ steps are needed to inform all nodes of layer 2. Notice that the source is not in the range of any other node, hence it cannot modify its actions according to adversary decisions.

Let $t = \lfloor\frac{h^2}{\log h}\rfloor$ and let A be a broadcast protocol informing any configuration of the above type in at most t steps. A node $v \in C_j$ is called *solitaire* in C_j if, in some step of A, v is the only broadcasting node from C_j. The rest of the proof is based on the following lemmas.

Lemma 15. *Every nonempty set C_j must contain a solitaire.*

Lemma 16. *Fix j. For some nonempty set C_j, the number x of steps in which at least two nodes from C_j broadcast, according to protocol A, prior to the first step in which a solitaire in C_j broadcasts, is at least $m = \lfloor dh/\log h \rfloor$, where d is a constant independent of h.*

Now suppose that during the selection process of the solitaire in C_j some nodes from C_k, $k > j$, broadcast in steps $t_1, ..., t_s$. Is it possible to take advantage of these steps in order to reduce the number of remaining steps in which nodes in C_j broadcast? We will show that this is not the case, by constructing a set C_j with a stronger property than above:

Lemma 17. *The number of steps other than $t_1, ..., t_s$ in which at least two nodes from C_j broadcast, according to protocol A, prior to the first step in which a solitaire in C_j broadcasts, is at least m.*

In order to finish the proof of the theorem, we show that at least $\lfloor \frac{dh}{\log h} \rfloor h$ steps are required to inform all nodes in layer 2. Consider the segment I_{h-1}. We have shown that there exists a subset $C_{h-1} \subseteq I_{h-1}$ for which at least $\lfloor \frac{dh}{\log h} \rfloor$ "noisy" steps are required before a solitaire is chosen. If the latest of these steps exceeds $\lfloor \frac{dh}{\log h} \rfloor h$ then the proof is finished. Otherwise, we consider the segment I_{h-2}. There exists a subset $C_{h-2} \subseteq I_{h-2}$ for which at least $\lfloor \frac{dh}{\log h} \rfloor$ *additional* "noisy" steps are required before a solitaire is chosen. If the latest of these steps exceeds $\lfloor \frac{dh}{\log h} \rfloor h$ then the proof is finished. Otherwise, we proceed to the construction of C_{h-3}, and so on. After h stages the number of steps required will become at least $\lfloor \frac{dh}{\log h} \rfloor h = \Omega(\frac{\log^2 R}{\log\log R})$. □

References

[1] N. Alon, A. Bar-Noy, N. Linial and D. Peleg, A Lower Bound for Radio Broadcast, Journal of Computer and System Sciences 43 (1991), 290-298.

[2] R. Bar-Yehuda, O. Goldreich, and A. Itai, On the time complexity of broadcast in radio networks: An exponential gap between determinism and randomization, Proc. 6th ACM Symposium on Principles of Distributed Computing (1987), 98 - 108.

[3] R. Bar-Yehuda, A. Israeli, and A. Itai, Multiple Communication in Multi-Hop Radio Networks, Proc. 8th ACM Symposium on Principles of Distributed Computing (1989), 329 - 338.

[4] I. Chlamtac and S. Kutten, On broadcasting in radio networks - problem analysis and protocol design, IEEE Transactions on Communications 33 (1985).

[5] I. Gaber and Y. Mansour, Broadcast in Radio Networks, Proc. 6th Ann. ACM-SIAM Symp. on Discrete Algorithms, SODA'95, 577-585.

[6] E. Kranakis, D. Krizanc and A. Pelc, Fault-Tolerant Broadcasting in Radio Networks, Proc. 6th European Symp. on Algorithms, ESA'98, Venice, Italy, Aug. 1998, LNCS 1461, 283-294.

[7] E. Kushilevitz and Y. Mansour, An $\Omega(D\log n)$ Lower Bound for Broadcast in Radio Networks, Proc. 12th Ann. ACM Symp. on Principles of Distributed Computing (1993), 65-73.

[8] E. Kushilevitz and Y. Mansour, Computation in Noisy Radio Networks, Proc. 9th Ann. ACM-SIAM Symp. on Discrete Algorithms, SODA'98, 236-243.

[9] M. Li and P. Vitányi, Introduction to Kolmogorov Complexity and its Applications, Springer Verlag, 1993.
[10] K. Pahlavan and A. Levesque, Wireless Information Networks, Wiley-Interscience, New York, 1995.
[11] A. Sen and M. L. Huson, A New Model for Scheduling Packet Radio Networks, Proc. 15th Ann. Joint Conf. of the IEEE Computer and Communication Societies (1996), 1116 - 1124.

Multipacket Routing on 2-D Meshes and Its Application to Fault-Tolerant Routing

Kazuo Iwama[1] * and Eiji Miyano[2] **

[1] Department of Communications and Computer Engineering,
Kyoto University, Kyoto 606-8501, JAPAN
iwama@kuis.kyoto-u.ac.jp
[2] Common Technical Courses,
Kyushu Institute of Design, Fukuoka 815-8540, JAPAN
miyano@kyushu-id.ac.jp

Abstract. Our model in this paper is the standard, two-dimensional $n \times n$ mesh. The first result is a randomized algorithm for h-h routing which runs in $O(hn)$ steps with high probability using queues of constant size. The previous bound is $0.5hn + o(hn)$ but needs the queue-size of $\Omega(h)$. An important merit of this algorithm is to give us improved bounds by applying several schemes of faulty-mesh routing. For example, the scheme by [Rag95], originally $O(n \log n)$ time and $O(\log^2 n)$ queue-size, gives us an improved routing algorithm on p-faulty meshes ($p \leq 0.4$) which runs in $O(n\frac{\log^2 n}{k})$ time using $O(k)$ queue-size for any $k \leq \log n$. Thus, when $k = \log n$ it improves the queue-size by the factor of $\log n$ without changing the time bound and when k is constant, it needs only constant queue-size although the running time slows down by the factor of $\log n$.

1 Introduction

One of the most studied parallel models with a fixed interconnection network is a two-dimensional mesh-connected processor array (a *2-D mesh* for short). In this model, $n \times n$ processors are placed at intersections of horizontal and vertical grids, where each processor is connected to its four neighbors via point-to-point communication links and can communicate with them in a single step (more generally, each processor can transfer one message per link in a single time-unit).

Packet routing is clearly a fundamental problem in the area of parallel and/or distributed computing and a great deal of effort has been devoted to the design of efficient algorithms on meshes. A special case of the routing problem is *permutation routing*, in which every processor is a source and destination of precisely one packet. Among several variations of the routing problem, permutation routing has been the most popular one since it has been considered to be a standard benchmark to evaluate the overall efficiency of communication schemes. The efficiency of a routing algorithm is generally measured by its *running time*. However, efficiency in the *queue-size* of each processor,

* Supported in part by Scientific Research Grant, Ministry of Japan, 10558044, 09480055 and 10205215.

** Supported in part by Scientific Research Grant, Ministry of Japan, 10780198.

i.e., reducing the maximum number of temporal packets that can be held by a single processor at the same time, has been considered to be equally important. This is of course due to the practical reason, but another obvious reason is that the restriction of the queue-size makes the problem far more interesting from theoretical view-points.

Consequently, many researchers have been interested in routing algorithms with constant queue-size: Leighton, Makedon, and Tollis [LMT95] and Sibeyn, Chlebus, and Kaufmann [SCK97] gave deterministic algorithms with running time $2n-2$, matching the network diameter, and with constant queue-size. However, their algorithms involve a flavor of mesh-sorting algorithms and may be too complicated to implement on existing computers. Hence an *oblivious* path selection is another well-received approach [BH85, BRSU93, KKT91]. In the oblivious path selection, the entire path of each packet has to be completely determined by its source and destination before routing starts. A typical oblivious strategy for mesh network is called a *dimension-order* algorithm: A packet first moves horizontally to its destination column and then moves vertically to its destination row. It is well known that in spite of very regular paths, the algorithm can route any permutation on the meshes in $2n-2$ steps. However, unfortunately, some processor requires $\Omega(n)$–size queue in the worst case. In order to reduce the queue-size, the randomized techniques based on the Valiant-Brebner algorithm [VB81] are quite often used: The packets are first sent to random destinations and then they are routed to their final destinations. Valiant and Brebner gave a simple, randomized, oblivious routing algorithm which runs in $3n+o(n)$ steps with high probability. Rajasekaran and Tsantilas [RT92] reduced the time bound to $2n+O(\log n)$. However, those queue-sizes still grow up to $\Omega(\log n/\log\log n)$ large. Until very recently, little had been known whether the queue-sizes can be decreased to some constant without increasing the time bound or sacrificing the obliviousness. Last year, Iwama, Kambayashi and Miyano made a significant progress on this problem. Following their intermediate result in [IKM98], Iwama and Miyano finally gave an $O(n)$ deterministic oblivious algorithm on the 2-D mesh with constant queue-size [IM99].

The present paper deals with more general cases of packet routing in the following two senses: One is the *h-h* routing problem on 2-D meshes; i.e., at most h packets originate from any processor and at most h packets, whose original positions may be different, are destined for any processor. (The other is routing on faulty meshes mentioned later.) The *h-h* routing clearly reflects practical implementations better than permutation routing, or 1-1 routing, since each processor usually generates many packets during a specific computation. In the case of *h-h* routing on the 2-D meshes, an easy $hn/2$ lower bound comes from a fundamental nature of the model, i.e., this bound is known as the *bisection* bound. The first nontrivial upper bound for *h-h* routing was proposed by Kunde and Tensi [KT89]. Their deterministic algorithm runs in $5hn/4+o(hn)$ steps. Then, several progresses have been made: Kaufmann and Sibeyn [KS97] and Rajasekaran [Raj95] showed (originally in SPAA92) that randomized algorithms based on the Valiant-Brebner algorithm [VB81] can solve *h-h* routing in time $hn/2+o(hn)$, which almost matches the bisection bound. Then [KSS94] and [Kun93] developed deterministic algorithms of the similar performance. However, all these algorithms require $\Omega(h)$–size queues in the worst case.

In this paper, we present a randomized algorithm for h-h routing which needs only *constant queue-size*. It runs in $O(hn)$ steps with high probability. This algorithm follows the same line as the algorithm of Iwama and Miyano for permutation routing [IM99], but our randomization approach is completely different from the Valiant-Brebner algorithm. Although the running time is worse than above, our algorithm has the following two important merits: (i) The maximum queue-size is bounded by some constant that does not depend on h. (ii) The algorithm is *oblivious*, i.e., the path of any packet is determined by its source and destination, solely and deterministically.

The other generalization is to add the fault-tolerance capability to mesh-routing. In the *p-faulty* mesh, each processor may fail independently with some probability bounded above by a value p. Routing on p-faulty meshes has been also popular, for which many algorithms were developed [KKL+90, Mat92, Rag95]. Generally speaking, they are divided into two categories; the first type can only cope with static faults, i.e., faulty processors are fixed throughout the computation and their locations are known in advance. The second type of algorithms can give on-line adaptations to dynamically occurring faults.

Another important merit of our h-h routing algorithm is to give us improved bounds by applying several schemes of faulty-mesh routing. For example, the scheme by [Rag95] for dynamic faults, originally $O(n\log n)$ time and $O(\log^2 n)$ queue-size, gives us a routing algorithm on p-faulty meshes ($p \leq 0.4$) which runs in $O(n\frac{\log^2 n}{k})$ time using $O(k)$ queue-size for any $k \leq \log n$ (Raghavan first assumed that p is bounded above by 0.29 and then Mathies [Mat92] improved the probability bound to $p \leq 0.4$). Thus, when $k = \log n$, it improves the queue-size by the factor of $\log n$ and when k is constant, it needs only constant queue-size although the running time slows down by the factor of $\log n$. Another example is the scheme by [KKL+90] for static faults. In this case, no significant improvement is possible since their algorithm already runs in linear time and uses queues of constant size. However, it is possible to make the algorithm oblivious and much simpler; the original one is based on the sorting-like routing.

In what follows, we first describe our models and problems more formally in the next section. Our $h-h$ routing is presented in Section 3 including basic ideas, algorithms and analysis of their time complexities. In Section 4, we give how we can apply this $h-h$ routing in order to obtain improved routing algorithms on faulty meshes.

2 Our Models and Problems

Two-dimensional meshes are illustrated in Figure 1. A *position* is denoted by (i,j), $1 \leq i,j \leq n$ and a processor whose position is (i,j) is denoted by $P_{i,j}$. Each processor is connected to four neighbors and a connection between a neighboring processor is called a (*communication*) *link*. The processors operate in a synchronous fashion. In a single time-unit, say t, each processor can perform an arbitrary amount of internal computation, and communicate with all its neighbors. As usual, we assume that $t = 1$ and then the running time of algorithms is determined by the number of communication time-units. By the queue-size, we mean the working queue-size, i.e., the maximum number of packets the processor can *temporally* hold at the same time in the course of routing.

The problem of h-h routing on the meshes is defined as follows: Each processor initially holds at most (unordered) h packets. Each packet, (s,d), consists of two portions; s is a *source* address that shows the initial position of the packet and d is a *destination* address that specifies the processor to which the packet should be moved. (A real packet includes more information besides its source and destination such as its body data, but it is not important within this paper and is omitted.) Routing requires that each packet must be routed independently to its destination and every processor is the destination of at most h packets. The h-h routing is called *permutation routing* if $h = 1$.

If we fix an algorithm and an instance, then the path R of each packet is determined, which is a sequence of processors, $P_1(=\text{source}), P_2, \cdots, P_j(=\text{destination})$. A routing algorithm, A, is said to be *oblivious* if the path of each packet is completely determined by its source and destination.

As for the faulty-mesh routing, a mesh is called *p-faulty mesh* if each processor fails with a certain fixed probability p, independently of each other. In this paper, we assume that only processors fail, and that these failures are dynamic and their location are not known before routing starts. If a processor fails, it can neither compute nor communicate with its neighbors, i.e., the so-called crash type failure only happens. See [GHKS98] for more details on the faulty meshes.

3 h-h Randomized Routing

3.1 Basic Ideas

Kaufmann and Sibeyn [KS97], and Rajasekaran [Raj95] gave $O(hn)$ randomized algorithms for the h-h routing problem. Their randomization techniques are based on [VB81], i.e., all the packets are first distributed temporally to random destinations and then routed from there to their final destinations. This random selection of intermediate positions allows us to avoid serious path-congestion by distributing packets evenly, however, increases the maximum queue-sizes: Those algorithms require $\Omega(h)$ queue-size in the worst case.

Our algorithm is also randomized but its idea is different from the above. To see this, let us take a look at the following simple observation: As mentioned before, for permutation routing, there are several $O(n)$ algorithms with constant queue-size. Then if every processor picks one packet up among h initial ones and can route the packet (or n^2 packets in total) to its final destination within $O(n)$ steps, then, by repeating this process simply h times, we can achieve an $h \times O(n) = O(hn)$ step algorithm for h-h routing. Unfortunately, this observation is actually too optimistic: In each round, n^2 packets in total are routed on the whole plane. If hn ($h \leq n$) of those n^2 packets are destined for some single row/column and, to make matters worse, if $O(hn)$ packets should pass through a single communication link, then each round requires at least $O(hn)$ steps and hence the whole algorithm takes $O(h^2n)$ steps. So our basic strategy is quite simple: Each processor chooses one packet at random, independently of each other, and then moves it to the final destination. On average , the randomized selection chooses approximately n packets per row/column and it turns out that the above kind of bad path-congestion will be avoided with high probability.

Our routing method of each round is very similar to the algorithm of Iwama and Miyano shown in [IM99]: (i) As mentioned above, each processor chooses one packet at random and moves the packet in each round, i.e., n^2 packets move in total. (ii) Before routing those packets toward their destination, we change the order of packets in their flow to control the injecting ratio of packets into the *critical positions* where heavy path-congestion can occur, by using the idea based on the *bit-reversal permutation* ([Lei92], see below). (iii) Every packet is routed to its final destination.

Definition 1. Let $i_1 i_2 \cdots i_\ell$ denote the binary representation of an integer i. Then i^R denotes the integer whose binary representation is $i_\ell i_{\ell-1} \cdots i_1$. The *bit-reversal permutation (BRP)* π is a permutation from $[0, 2^\ell - 1]$ onto $[0, 2^\ell - 1]$ such that $\pi(i) = i^R$. Let $x = x_0 x_1 \cdots x_{2^\ell - 1}$ be a sequence of packets. Then $BRP(x)$ is defined to be $BRP(x) = x_{\pi(0)} x_{\pi(1)} \cdots x_{\pi(2^\ell - 1)}$.

When $\ell = 3$, i.e., when $x = x_0 x_1 \cdots x_7$, $BRP(x) = x_0 x_4 x_2 x_6 x_1 x_5 x_3 x_7$. Namely, x_j is placed at the $\pi(j)$th position in $BRP(x)$ (the leftmost position is the 0th position). The following lemma shown in [IM99] is important and is often used in the rest of the paper:

Lemma 1 [IM99]. Let $x = x_0 x_1 \cdots x_{n-1}$ be a sequence where $n = 2^\ell$ for some integer ℓ, and $z = x_i x_{i+1} \cdots x_{i+k-1}$ be its any subsequence of length k. Let x_{j_1}, x_{j_2} and x_{j_3} be any three symbols in z that appear in $BRP(x)$ in this order. Then the distance between x_{j_1} and x_{j_3} is at least $\lceil \frac{n}{2k} \rceil$.

3.2 Algorithms

Throughout this section, we assume that the side-length n of 2-D meshes is divided by six and the entire plane is divided into 36 subplanes, $SP_{1,1}$ through $SP_{6,6}$, as shown in Figure 2-(a). However, the following argument can be easily extended to the case that the length is not divided by six. For simplicity, the total number of processors in 2-D meshes is hereafter denoted not by n^2 but $36n^2$, i.e., each subplane consists of $n \times n$ processors.

It is important to define the following two notations on sequences of packets on linear arrays, which will play key roles in our algorithms:

Definition 2. For a sequence x of n packets, $SORT(x) = x_{s_0} x_{s_1} \cdots x_{s_{n-1}}$ denotes a sorted sequence according to the destination column. Namely, $SORT(x)$ is the sequence such that the destination column of x_{s_i} is farther than or the same as the destination column of x_{s_j} if $i > j$.

Definition 3. For technical reason partly due to Lemma 1, it it desirable if the length of a packet sequence is a power of two. Suppose that, for example, a sequence x includes only ten packets $x_0 x_1 \cdots x_9$. In this case we change the length of x into 16 ($= 2^4$) by padding x out with six spaces or "null" packets, ϕ_0 through ϕ_5. If a null packet exists at a processor P_i, then P_i's queue is just empty. Namely, $PAD_4(x)$ is defined to be $PAD_4(x) = \phi_0 \phi_1 \phi_2 \phi_3 \phi_4 \phi_5 x_0 x_1 x_2 x_3 x_4 x_5 x_6 x_7 x_8 x_9$ ("4" of PAD_4 is the 4 of $= 2^4$). Generalization is easy. Hence, $BRP(PAD_4(x)) = \phi_0 x_2 \phi_4 x_6 \phi_2 x_4 x_0 x_8 \phi_1 x_3 \phi_5 x_7 \phi_3 x_5 x_1 x_9$. Namely, x_j is placed at the $\pi((2^4 - 10) + j)$th position in $BRP(PAD_4(x))$ (or in general $\pi((2^\ell - n) + j)$th position in $BRP(PAD_\ell(x))$).

Now we present our *h-h* routing algorithm on the mesh. Before routing starts, each processor calculates integer value ℓ such that $2^{\ell-1}+1 \leq n \leq 2^{\ell}$. Then it repeats the similar process h rounds associated with the number of initial packets. Here is a detailed description of each round i:

(1) **Randomized Choice:** Each processor chooses exactly one packet among the remaining initial packets at random, independently of each other, and moves it in this round.

(2) **Routing [IM99]:** The routing process consists of the following 36×36 sequential *phases*. In the first phase only packets whose sources and destinations are both in $SP_{1,1}$ move (i.e., they may be only a small portion of the n^2 packets in $SP_{1,1}$). In the second phase only packets from $SP_{1,1}$ to $SP_{1,2}$ move, and so on. Here we only give an outline of each phase: Suppose that it is now the phase where packets from $SP_{4,3}$ to $SP_{2,4}$, called *active packets*, move. The paths of those packets are shown by arrows in Figure 2-(a). The entire phase is further divided into the following four stages:

Stage 1: Those active packets first move horizontally to $SP_{4,1}$ and then vertically to $SP_{5,1}$, i.e., they temporally move into the subplane which is located three subplanes away from the destination subplane both horizontally and vertically, without changing their relative positions within those subplanes. See Figure 2-(b): For example, a packet a initially placed on the upper-left corner in the source subplane always moves through the upper-left corner position. All the active packets can arrive at their temporal destinations in $SP_{5,1}$ exactly at the $3n$th step from the beginning of this phase.

Stage 2: They next go through three consecutive subplanes, from $SP_{5,1}$ to $SP_{5,3}$, called the *permutation zone*, where the packets change their order. Namely, if n processors on some row in $SP_{5,1}$ originally held a sequence x of n packets, then the rightmost 2^{ℓ} processors on the row in $SP_{5,2}$ and $SP_{5,3}$ eventually hold the sequence $BRP\ (PAD_{\ell}\ (SORT(x)))$ by using the method introduced in [IM99]. (The integer value ℓ has been computed in the precomputation stage.) The reason for the rather long paths of packets from $SP_{4,3}$ to $SP_{2,4}$ is to prepare this permutation zone.

Stage 3. The packets then move to $SP_{5,4}$, called the *critical zone*, where each packet enters its correct column position. Here we apply the *spacing* operation, i.e., $c_0 - 1$ steps are inserted between the first actions of any neighboring two processors (c_0 is some constant and will be fixed later). In the first step, only the rightmost processor in the permutation zone starts forwarding its active packet to the right, and then the packet keeps moving one position to the right at each step. However, the other active packets do not move at all during the first c_0 steps. In the (c_0+1)th step, the second rightmost processor starts forwarding its packet to the right, in the $(2c_0+1)$th step the third rightmost processor sends its packet, and so on. Once each packet starts, it keeps being shifted one position to the right if it still needs to move rightward, and then changes the direction from rightward to upward at the crossing of its correct destination column where turning packets are always given a higher priority.

Stage 4: At each step each processor moves upward its active packet if it still needs to move.

3.3 Time Complexities

Since each processor moves exactly one packet in each round, Stages 1 and 2 are linear in n. Then we shall investigate the time complexities of Stages 3 and 4 in the following.

Roughly speaking, if destinations of n^2 packets which move in each round are evenly distributed and hence the average number of packets which head for each single column is bounded above by cn for some constant c, then the expected time complexities of Stages 3 and 4 could become $O(n)$. Now we shall count the number of those packets and evaluate the probability of bad behaviors using the Chernoff bound (e.g., see [MR95]).

Lemma 2. Let $X_1, X_2, \cdots, X_{n^2}$ be independent Poisson trials such that $\Pr[X_i = 1] = p_i$, $\Pr[X_i = 0] = 1 - p_i$, $0 < p_i < 1$. Let $X = \Sigma_{i=1}^{n^2} X_i$ and $\mu = \Sigma_{i=1}^{n^2} p_i$. Then, for $0 < \delta < 2e - 1$,

$$\Pr[X > (1+\delta)\mu] < \exp(\frac{-\mu\delta^2}{4}).$$

The probabilistic behavior of the randomized choice can be modeled as follows: Suppose that it is now the phase of some round where packets move from SP_s to SP_d and that SP_s has n^2 processors, P_1 through P_{n^2}. Also suppose that processor P_i initially held m_i packets (out of h ones) whose destinations all are a single column, say COL, of the subplane SP_d. When P_i chooses one packet at random, let random variable $X_i = 1$ if the destination column of the packet is COL and $X_i = 0$ otherwise. Then $\Pr[X_i = 1] = \frac{m_i}{h}$, and $\mu = \Sigma_{i=1}^{n^2} \frac{m_i}{h} \le n$ since the number of packets whose destination column is COL is at most hn, or $\Sigma_{i=1}^{n^2} m_i \le hn$. Apply the Chernoff bound with $\delta = c_1\sqrt{\frac{\ln n}{\mu}}$ for some constant $c_1 > 0$. Then, for some constant $c_2 > 0$,

$$\Pr[X > n + c_1\sqrt{n \ln n} \ge \mu + c_1\sqrt{\mu \ln n}] < \exp(-c_2 \ln n) = n^{-c_2}.$$

Namely, the number of packets which are heading for a single column is at most $n + c_1\sqrt{n \ln n}$ with probability $1 - n^{-c_2}$.

Now consider the permutation zone, SP_1, SP_2 and SP_3, and the critical zone, SP_4. After Stage 2, the permuted sequence $BRP(PAD_\ell(SORT(x)))$ of 2^ℓ packets on every row is now placed on the rightmost 2^ℓ processors in SP_2 and SP_3. Suppose that the uppermost row of the permutation zone includes k_1 α_1's, the second uppermost row includes k_2 α_2's and so on. Here α_i's are packets whose column destinations are the same, say, the jth column in the critical zone. In general, the ith row includes k_i α_i's.

Consider a processor $P_{i,j}$ at the cross-point of the ith row and jth column. It is important to note that, from Lemma 1, $P_{i,j}$ receives at most two α_i's during some particular window Δ_i of $\frac{c_0 n}{2k_i}$ ($\le c_0 \times \left\lceil \frac{n}{2k_i} \right\rceil$) steps since $c_0 - 1$ spaces are inserted between any neighboring two packets in the third stage of the algorithm. If neither of those two packets α_i's can move up on at some step, then there must be some packet which "blocks" these α_i's that should be the packet which is now ready to enter the jth column by making a turn at some upper position than $P_{i,j}$. (Recall that turning packets have priority at crosspoints.) Let us call such a packet *blocking packet against* α_i*'s*. Note that the blocking packet against α_i's never block them again.

We shall count the total number of α_m's $(1 \leq m \leq i-1)$ which $P_{1,j}$ through $P_{i-1,j}$ can receive during the window Δ_i. Since $P_{m,j}$ can receive at most two α_m's in $c_0\lceil \frac{n}{2k_m} \rceil$ steps, the number of α_m's which $P_{m,j}$ receives during Δ_i of $\frac{c_0 n}{2k_i}$ steps is at most

$$(\frac{c_0 n}{2k_i}/c_0\left\lceil\frac{n}{2k_m}\right\rceil)\times 2 \leq \frac{c_0 n}{k_i}/\frac{c_0 n}{2k_m} = \frac{2k_m}{k_i}.$$

Hence, the total number of α's which $P_{1,j}$ through $P_{i-1,j}$ can receive is at most

$$\frac{2(k_1+k_2+\cdots+k_{i-1})}{k_i} \leq \frac{2(n+c_1\sqrt{n\ln n})}{k_i} - 2 \qquad (1)$$

since $k_1+k_2+\cdots+k_{i-1} \leq n+c_1\sqrt{n\ln n}-k_i$ with probability $1-n^{-c_2}$. Now we fix the value of c_0 such that

$$c_0 \geq 4(1+c_1\sqrt{\frac{\ln n}{n}}).$$

Hence the right side of the equation (1) is at most $\frac{c_0 n}{2k_i}-2$, i.e., there must be at least two time-slots such that no packets flow on the jth column during the window Δ_i. This means that there are no blocking packets at those time-slots, and therefore the two α_i's currently held in $P_{i,j}$ can move up during the window Δ_i. The same argument can apply for any j $(1 \leq j \leq n)$ and for any window Δ_i $(1 \leq i \leq n)$. As a result, if the queue-size of each processor is at least two, then any delay does not happen in the critical zone and hence Stages 3 and 4 can be also performed in linear time:

Theorem 1. There is an oblivious, randomized, *h*-*h* routing algorithm on 2-D meshes of queue-size two which runs in $O(hn)$ steps with high probability.

4 Application to Fault-Tolerant Routing

In this section we consider the problem of permutation routing on p-faulty meshes under the dynamic fault assumption, i.e., these failures are dynamic and their location are not known before routing starts. Under the same setting, Raghavan [Rag95] gave a randomized algorithm for solving permutation routing on p-faulty meshes $(p \leq 0.29)$ which runs in $O(n\log n)$ steps with high probability, using queues of $O(\log^2 n)$ size. This algorithm is based on the Valiant-Brebner randomized algorithm [VB81], in which each packet is first sent to the intermediate random destination and then it is routed to the final position along the dimension-order path. However, instead of sending each packet along its master path P as defined by Valiant-Brebner scheme, Raghavan's scheme broadcasts copies within a routing region $R(P)$ defined to contain all processors within distance $c\log n$ of P for some constant c: Namely, if there is a path between the source and destination of a packet through live processors, then its routing region contains the path and hence every packet is routed successfully to its destination with high probability as shown in [Rag95]. We can prove that by applying this broadcast scheme to our $h-h$ routing algorithm introduced in the previous section, we can achieve a randomized algorithm for permutation on p-faulty meshes whose running time slows at most by the factor of $\log n$ compared to the fault-free case.

Lemma 3. Suppose that 2-D meshes includes $\frac{n}{c \log n} \times \frac{n}{c \log n}$ processors, and each processor requires $c' \log n$ steps to communicate (to send/receive a single packet) with its neighboring processors for some constants c_0 and c'. Then, if the randomized algorithm of the previous section can solve the $c^2 \log^2 n$-$c^2 \log^2 n$ routing problem on such a mesh within T steps using queues of constant size, then there must exist a randomized algorithm for the permutation routing problem on the p-faulty mesh including $n \times n$ processors which runs within $O(T \log n)$ steps using queues of constant size.

Proof. Only an outline is given. The entire plane of the p-faulty mesh is divided into $\frac{n^2}{c^2 \log^2 n}$ submeshes, i.e., each submesh has $c \log n \times c \log n$ processors. Recall that in the algorithm of the previous section, each processor chooses one packet at random among $c^2 \log^2 n$ initial packets, and then route the packet to its destination. Then, we regard one submesh including $c \log n \times c \log n$ processors on the p-faulty mesh as one processor on the $\frac{n}{c \log n} \times \frac{n}{c \log n}$ fault-free mesh (referred to as the *smaller mesh* hereafter) and simulate both the random-choice and the routing process of a single processor on the smaller mesh by the whole submesh of $c \log n \times c \log n$ processors on the faulty mesh.

(1) We pick up one packet from $c \log n \times c \log n$ packets on the submesh as follows: Suppose that some submesh consists of P_1 through $P_{c^2 \log^2 n}$. However, several processors may fail. (i) First of all, we count how many processors are alive. P_1 (if alive) first broadcasts copies of its initial packet to those $c^2 \log^2 n$ processors by the method of Raghavan using the first $c^2 \log^2 n$ steps. After those steps, every processor surely receives the packet. Next P_2 broadcasts its packet also to all the $c^2 \log^2 n$ processors during the second $c^2 \log^2 n$ steps and so on. However, if some processor fails, then all the processors do nothing at all during the $c^2 \log^2 n$ steps. Thus every processor knows the number of live processors in the submesh. (ii) Suppose that now r processors are still alive. Then every processor chooses an integer from $\{1, 2, \cdots, r\}$ at random and keeps the value. (iii) Again all the processors broadcast their initial packets using the same as (i), where if some processor has chosen a value i in (ii), then the processor stores only the ith packet but does not store other packets, i.e., remove them from its queue (to reduce the queue-size). (iv) P_1 broadcasts copies of the packet which is stored in (iii) to all the processors of its submesh. However, if P_1 is dead then nothing happens. Namely, if no packet is received during the $c^2 \log^2 n$ steps, then P_2 takes over and broadcasts copies of the packet which P_2 chooses at random in (ii) and (iii). If P_2 is dead also, then P_3 takes over and so on. Hence only one packet is chosen at random from (live) packets within obviously polylog steps. Note that this packet, σ, is now shared by all the processors in the submesh and there must be a processor which originally held σ. That processor knows its packet was selected (and will be routed in the first round). In the second (and following) round, this processor behaves as it is dead. This completes the random selection process of the first round. Note that this process is completely repeated in each round.

(2) Now we go to the routing process of the first round. Each packet moves along the same path as the previous randomized algorithm, but this time the path consists of a sequence of submeshes. When a packet moves from one submesh to the next, it is broadcasted to all the $c \log n \times c \log n$ processors of each submesh. This can be done exactly the same as [Rag95] and needs $O(\log n)$ time.

(3) Repeat (1) and (2) $c^2\log^2 n$ rounds. For each round, we need polylog steps for (1) and $(T/c^2\log^2 n)\cdot c'\log n$ steps for (2). (Since $c^2\log^2 n$-$c^2\log^2 n$ routing needs T steps, its single round needs $(T/c^2\log^2 n)$ steps.) Therefore the total computation time is $O(T\log n)$. □

Theorem 2. There is a randomized permutation routing algorithm on p-faulty meshes of queue-size two which runs in $O(n\log^2 n)$ steps with high probability.

Proof. By Theorem 1, there is a randomized, $c\log^2 n$-$c\log^2 n$ routing algorithm on 2-D meshes, including $\frac{n}{c\log n}\times\frac{n}{c\log n}$ processors and with queue-size two which runs in $O(n\log n)$ steps with high probability. Now use Lemma 3 by substituting $T = O(n\log n)$. □

Theorem 3. There is a randomized permutation routing algorithm on p-faulty meshes of queue-size k which runs in $O(n\frac{\log^2 n}{k})$ steps with high probability.

Proof. Almost all processes are the same as above. However, k packets are chosen at random from $c\log n\times c\log n$ packets on every submesh by using the same idea as (1)-(i) in the proof of Lemma 3, and those k packets are moved in each round. It can be shown that each round can be performed again within $O(n\log n)$ steps since each queue-size is now k. Since there are $c\log n/k$ rounds, the total running time is $O(n\frac{\log^2 n}{k})$. Details are omitted. □

5 Concluding Remarks

Apparently there are several problems for future research. Among others the question of whether the time bound of Theorem 2 can be improved will be interesting. Note that our algorithm works only if we can broadcast each packet to all processors in a submesh. We set the size of the submesh some $\log n\times\log n$, but it then satisfies extra properties other than what we need. If we can reduce this size, then it would immediately improves the time bound.

References

[BH85] A. Borodin and J.E. Hopcroft, "Routing, merging, and sorting on parallel models of computation," *J. Computer and System Sciences* 30 (1985) 130-145.

[BRSU93] A. Borodin, P. Raghavan, B. Schieber and E. Upfal, "How much can hardware help routing?," In *Proc. ACM Symposium on Theory of Computing* (1993) 573-582.

[GHKS98] M. Grammatikakis, D. Hsu, M. Kraetzl and J. Sibeyn, "Packet routing in fixed-connection networks: A survey," *manuscript* (1998).

[IKM98] K. Iwama, Y. Kambayashi and E. Miyano, "New bounds for oblivious mesh routing," In *Proc. 6th European Symposium on Algorithms* (1998) 295-306.

[IM99] K. Iwama and E. Miyano, "An $O(\sqrt{N})$ oblivious routing algorithms for 2-D meshes of constant queue-size," In *Proc. 10th ACM-SIAM Symposium on Discrete Algorithms* (1999) 466-475.

[KKT91] C. Kaklamanis, D. Krizanc and A. Tsantilas, "Tight bounds for oblivious routing in the hypercube," *Mathematical Systems Theory 24* (1991) 223-232.

[KKL+90] C. Kaklamanis, A.R. Karlin, F.T. Leighton, V. Milenkovic, P. Raghavan, S. Rao, C. Thomborson, A. Tsantilas, "Asymptotically tight bounds for computing with faulty arrays of processors," In *Proc. 31st Symposium on Foundations of Computer Science* (1990) 285-296.

[KSS94] M. Kaufmann, J.F. Sibeyn, and T. Suel, "Derandomizing routing and sorting algorithms for meshes," In *Proc. 5th ACM-SIAM Symposium on Discrete Algorithms* (1994) 669-679.

[KS97] M. Kaufmann and J.F. Sibeyn, "Randomized multipacket routing and sorting on meshes," *Algorithmica*, 17 (1997) 224-244.

[KT89] M. Kunde and T. Tensi, "Multi-packet routing on mesh connected processor arrays," In *Proc. Symposium on Parallel Algorithms and Architectures* (1989) 336-343.

[Kun93] M. Kunde, "Block gossiping on grids and tori: deterministic sorting and routing match the bisection bound," In *Proc. European Symposium on Algorithms* (1993) 272-283.

[Lei92] F.T. Leighton, *Introduction to Parallel Algorithms and Architectures: Arrays, Trees, Hypercubes,* Morgan Kaufmann (1992).

[LMT95] F.T. Leighton, F. Makedon and I. Tollis, "A $2n-2$ step algorithm for routing in an $n \times n$ array with constant queue sizes," *Algorithmica* 14 (1995) 291-304.

[Mat92] T.R. Mathies, "Percolation theory and computing with faulty arrays of processors," In *Proc. 3rd ACM-SIAM Symposium on Discrete Algorithms* (1992) 100-103.

[MR95] R. Motwani and P. Raghavan, *Randomized Algorithms*, Cambridge University Press (1995).

[Rag95] P. Raghavan, "Robust algorithms for packet routing in a mesh," *Math. Systems Theory*, 28 (1995) 1-11.

[RT92] S. Rajasekaran and T. Tsantilas, "Optimal routing algorithms for mesh-connected processor arrays," *Algorithmica* 8 (1992) 21-38.

[Raj95] S. Rajasekaran, "*k-k* routing, *k-k* sorting, and cut-through routing on the mesh," *J. Algorithms*, 19, 3 (1995) 361-382.

[SCK97] J.F. Sibeyn, B.S. Chlebus and M. Kaufmann, "Deterministic permutation routing on meshes," *J. Algorithms* 22 (1997) 111-141.

[VB81] L.G. Valiant and G.J. Brebner, "Universal schemes for parallel communication," In *Proc. 13th Symposium on Theory of Computing* (1981) 263-270.

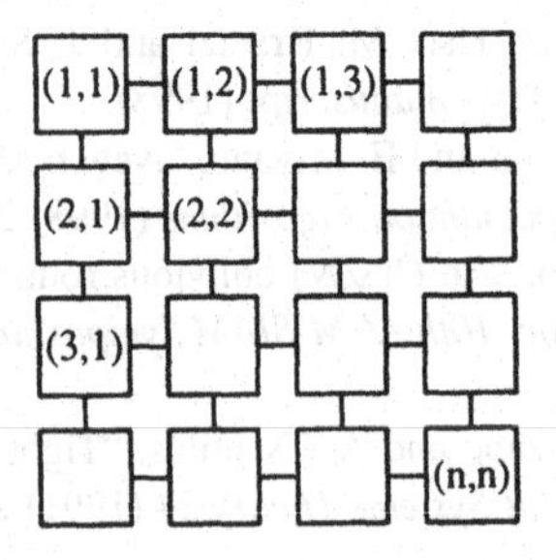

Figure 1: 2-D mesh

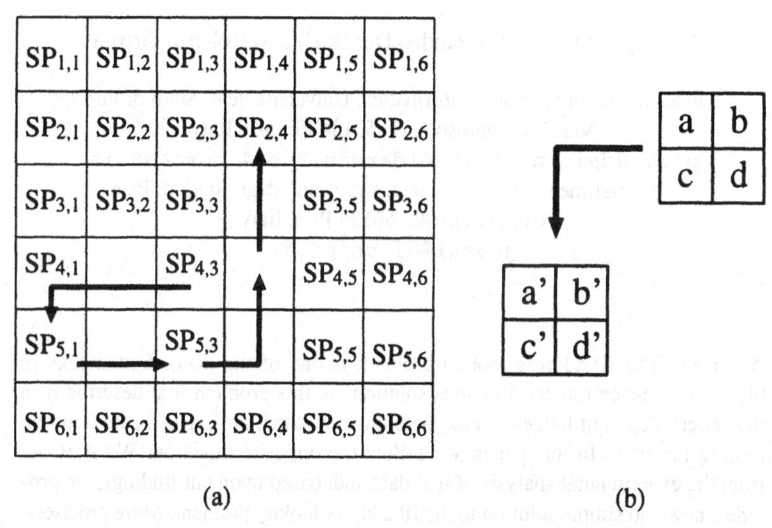

Figure 2: 36 subplanes

IP Address Lookup Made Fast and Simple

Pierluigi Crescenzi[1], Leandro Dardini[1], and Roberto Grossi[2]

[1] Dipartimento di Sistemi e Informatica, Università degli Studi di Firenze
Via C. Lombroso 6/17, 50134 Firenze, Italy
piluc@dsi.unifi.it, ldardini@usl4.toscana.it
[2] Dipartimento di Informatica, Università degli Studi di Pisa
Corso Italia 40, 56125 Pisa, Italy
grossi@di.unipi.it

Abstract. The IP address lookup problem is one of the major bottlenecks in high performance routers. Previous solutions to this problem first describe it in the general terms of longest prefix matching and, then, are experimented on real routing tables T. In this paper, we follow the opposite direction. We start out from the experimental analysis of real data and, based upon our findings, we provide a new and simple solution to the IP address lookup problem. More precisely, our solution for m-bit IP addresses is a reasonable trade-off between performing a binary search on T with $O(\log|T|)$ accesses, where $|T|$ is the number of entries in T, and executing a single access on a table of 2^m entries obtained by fully expanding T. While the previous results start out from space-efficient data structures and aim at lowering the $O(\log|T|)$ access cost, we start out from the expanded table with 2^m entries and aim at compressing it without an excessive increase in the number of accesses. Our algorithm takes *exactly three* memory accesses and occupies $O(2^{m/2} + |T|^2)$ space in the worst case. Experiments on real routing tables for $m = 32$ show that the space bound is overly pessimistic. Our solution occupies approximately one megabyte for the MaeEast routing table (which has $|T| \approx 44,000$ and requires approximately 250 KB) and, thus, takes three *cache* accesses on any processor with 1 MB of L2 cache. According to the measurement obtained by the VTune tool on a Pentium II processor, each lookup requires 3 additional clock cycles besides the ones needed for the memory accesses. Assuming a clock cycle of 3.33 nanoseconds and an L2 cache latency of 15 nanoseconds, search of MaeEast can be estimated in 55 nanoseconds or, equivalently, our method performs 18 millions of lookups per second.

1 Introduction

Computer networks are expected to exhibit very high performance in delivering data because of the explosive growth of Internet nodes (from 100,000 computers in 1989 to over 30 millions as of today's). The network bandwidth, which measures the number of bits that can be transmitted in a certain period of time, is thus continuously improved by adding new links and/or by improving the performance of the existing ones. Routers are at the heart of the networks in that they forward packets from input interfaces to output interfaces on the ground of the packets' destination Internet address, which we simply call IP address. They choose which output interface corresponds to a given packet by performing an *IP address lookup* at their routing table. As routers have to deal with an ever increasing number of links whose performance constantly improves, the address lookup is now becoming one of the major bottlenecks in high performance forwarding engines.

The IP address lookup problem was just considered a simple table lookup problem at the beginning of Internet. Now, it is unconceivable to store all existing IP addresses explicitly because, in this case, routing tables would contain millions of entries. In the early 1990s people realized that the amount of routing information would grow enormously, and introduced a simple use of prefixes to reduce space [3]. Specifically, IP protocols use hierarchical addressing, so that a network contains several subnets which in turn contain several host computers. Suppose that all subnets of the network with IP address 128.96.*.* have the same routing information apart from the subnet whose IP address is 128.96.34.*. We can succinctly describe this situation by just two entries (128.96 and 128.96.34) instead of many entries for all possible IP addresses of the network. However, the use of prefixes introduces a new dimension in the IP address lookup problem: For each packet, more than one table entry can match the packet's IP address. In this case, the applied rule consists of choosing the *longest prefix match*, where each prefix is a binary string that has a variable length from 8 to 32 in IPv4 [14].

For example, let us consider the routing table shown to the right, where ε denotes the empty sequence corresponding to the default output interface, and assume that the IP address of the packet to be forwarded is 159.213.37.2, that is, 10011111 11010101 00100101 00000010 in binary.

Prefix	Interface
ε	*A*
10011111	*B*
10011111 11010101	*C*
10011111 11010101 00	*D*
10011111 11110	*E*

Then, the longest prefix match is obtained with the fourth entry of the table and the packet is forwarded to output interface *D*. Instead, a packet whose IP address is 159.213.65.15, that is, 10011111 11010101 01000001 00001111, is forwarded to output interface *C*.

Looking for the longest matching prefix in IP routing tables represents a challenging algorithmic problem since lookups must be answered very quickly. In order to get a bandwidth of, say, 10 gigabits per second with an average packet length equal to 2,000, a router should forward 5 millions of packets per second. It means that each forwarding has to be performed in approximately 200 nanoseconds and, consequently, each lookup must be realized much faster.

1.1 Previous Results

Several approaches have been proposed in the last few years in order to solve the IP address lookup problem. Hardware solutions, though very efficient, are expensive and some of them may become outdated quite quickly [6, 10]. In this section, we will briefly review some of the most recent software solutions.

A traditional implementation of routing tables [16] use a version of *Patricia tries*, a very well-known data structure [8]. In this case, it is possible to show that, for tables with n random elements, the average number of examined bits is approximately $\lceil \log n \rceil$. Thus, for $n = 40,000$, this value is 16. In real routing tables entries are not random and have many initial bits in common, and so the average number of table accesses is higher than $\lceil \log n \rceil$. When compared to millions of address lookup requests served in one second, these accesses are too many. Another variation of Patricia tries has been proposed in order to examine k bit each time [13]. However, this variation deals with either the exact matching problem or with the longest prefix matching problem restricted to prefixes whose lengths are multiples of k. A more recent approach [11] inspired by the three-level data structure of [2], which uses a clever scheme to compress multibit trie nodes using a bitmap, is based on the compression of the routing tables by means of *level compressed tries*, which are a powerful and space efficient representation of binary tries. This approach seems to be very efficient from a memory size point of view but it requires many bit operations which, on the current technology, are time consuming. Another approach is based on *binary search on hash tables* organized by prefix lengths [17]. This technique is more memory consuming than the previous one but, according to the experimental evaluation presented by the authors, it seems to be very fast. A completely different way of using binary search is described in [5]. It is based on multi-way search on the number of possible prefixes rather than the number of possible prefix lengths and exploits the locality inherent in processor caches. The most recent (and fastest) software solution to the IP address lookup problem is the one based on *controlled prefix expansion* [15]. This approach, together with optimization techniques such as dynamic programming, can be used to improve the speed of most IP lookup algorithms. When applied to trie search, it results into a range of algorithms whose performance can be tuned and that, according to the authors, provide faster search and faster insert/delete times than earlier lookup algorithms.

While the code of the solution based on level compressed tries is public available along with the data used for the experimental evaluation, we could not find an analogous situation for the other approaches (this is probably due to the fact that the works related to some of these techniques have been patented). Thus, the only available experimental data for these approaches are those given by the authors.

1.2 Our Results

Traditionally, the proposed solutions to the IP address lookup problem aim at solving it in an efficient way, whichever is the table to be analyzed. In other words, these solutions do not solve the specific IP address lookup problem but the general longest prefix match problem. Subsequently, their practical behavior is experimented on real routing tables. In this paper we go the other way around. We start out from the real data and, as a consequence of the experimental analysis of these data, we provide a simple method whose performance depends on the statistical properties of routing tables T in a way different from the previous approaches. Our solution can also be used to solve the longest prefix match problem but its performance when applied to this more general problem is not guaranteed to be as good as in the case of the original problem.

Any solution for m-bit IP addresses can be seen as a trade-off between performing a binary search on T with $O(\log|T|)$ accesses (where $|T|$ is the number of prefixes in T) and executing a single access on a table of 2^m entries obtained by fully expanding T. The results described in the previous section propose space-efficient data structures and aim at lowering the $O(\log|T|)$ bound on the number of accesses. Our method, instead, starts out from the fully expanded table with 2^m entries and aim at compressing it without an excessive increase in the (constant) number of accesses. For this reason, we call it *an expansion/compression approach*.

We exploit the fact that the relation between the 2^m IP addresses and the very few output interfaces of a router is highly redundant for $m = 32$. By expressing this relation by means of strings (thus *expanding* the original routing table), these strings can be *compressed* (using the run-length encoding scheme) in order to provide an implicit representation of the expanded routing table which is memory efficient. More important, this representation allows us to perform an address lookup in exactly *three memory accesses* independently of the IP address. Intuitively, the first two accesses depend on the first and second half of the IP address, respectively, and provide an indirect access to a table whose elements specify the output interfaces corresponding to groups of IP addresses.

In our opinion, the approach proposed in this paper is valuable for several reasons. (i) *It should work for all routers belonging to Internet.* The analysis of the data has been performed on five databases which are made available by the IPMA project [7] and contain daily snapshots of the routing tables used at some major network access points. The largest database, called MaeEast, is useful to model a large backbone router while the smallest one, called Paix, can be considered a model for an enterprise router (see the first table of Sect. 3). (ii) *It should be useful for a long period of time.* The data have been collected over a period longer than six months and no significant change in the statistical properties we analyze has ever been encountered (see tables of Sect. 3). (iii) *Its dominant cost is really given by the number of memory accesses.* As already stated, the method requires *always* three memory accesses per lookup. It does not require anything else but addressing three tables (see Theorem 6). For example, on a Pentium II processor, each lookup requires only 3 additional clock cycles besides the ones needed for the memory accesses. (iv) *It can be easily implemented in hardware.* Our IP address lookup algorithm is very simple and can be implemented by few simple low-level instructions.

The counterpart of all the above advantages is that the theoretical upper bound on the worst-case memory size is $O(2^{m/2} + |T|^2)$ which may become infeasible. However, we experimentally show that in this case theory is quite far from practice and we believe that the characteristics of the routing tables will vary much slower than the rate at which cache memory size will increase.

In order to compare our method against the ones described in the previous section, we refer to data appeared in [15]. In particular, the first five rows of Table 1 are taken from [15, Table 2]. The last two rows reports the experimental data of our method that turns to be the fastest.

As for the methods proposed in [15] based on prefix controlled expansion, we performed the measurements by using the VTune tool [4] to compute the dynamic clock cycle counts. According to these measurements, on a Pentium II processor each lookup requires $3 \cdot clk + 3 \cdot M_D$ nanoseconds where *clk* denotes the clock cycle time and M_D is the memory access delay. If our data structure is small enough to fit into the L2 cache, then $M_D = 15$ nanoseconds, otherwise $M_D = 75$ nanoseconds. The comparison between our method and the best method of [15] is summarized in Table 2 where the first row is taken from [15, Tables 9 and 10].

Method	Average ns	Worst-case ns	Memory in KB
Patricia trie [16]	1500	2500	3262
6-way search [5]	490	490	950
Search on levels [17]	250	650	1600
Lulea [2]	349	409	160
LC trie [11]	1000	—	700
Ours (experimented)	172	—	960
Ours (estimated)	—	235	960

Table 1. Lookup times for various methods on a 300 MHz Pentium II with 512 KB of L2 cache (values refer to the MaeEast prefix database as on Sept. 12, 1997)

Method	512 KB L2 Cache	1 MB L2 Cache
Controlled prefix expansion [15]	196	181
Ours	235	55

Table 2. MaeEast lookup estimated times depending on cache size ($clk = 3.33$ nsec)

Due to lack of space, we focus on the lookup problem and we do not discuss the insert/delete operations. We only mention here that their realization may assume that the cache is divided into two banks. At any time, one bank is used for the lookup and the other is being updated by the network processor via a personal computer interface bus [12]. A preliminary version of our data structure construction on a 233 MHz Pentium II with 512 KB of L2 cache requires approximately 960 microseconds: We are confident that the code can be substantially improved thus decreasing this estimate.

We now give the details of our solution. The reader must keep in mind that the order of presentation of our ideas follows the traditional one (algorithms + experiments) but the methodology adopted for our study followed the opposite one (experiments + algorithms). Moreover, due to lack of space, we will not give the details of the implementation of our procedures. However, the C code of the implementation and a technical report are available via anonymous `ftp` (see Sect. 3).

2 The Expansion/Compression Approach

In this section, we describe our approach to solve the IP address lookup problem in terms of m-bit addresses. It runs in two phases, *expansion* and *compression*. In the expansion phase, we implicitly derive the output interfaces for all the 2^m possible addresses. In the compression phase, we fix a value $1 \leq k \leq m$ and find two statistical parameters α_k and β_k related to some combinatorial properties of the items in the routing table at hand. We then show that these two parameters characterize the space occupancy of our solution.

2.1 Preliminaries and Notations

Given the binary alphabet $\Sigma = \{0, 1\}$, we denote the set of all binary strings of length k by Σ_k, and the set of all binary strings of length at most m by $\Sigma_{\leq m} = \cup_{k=0}^{m} \Sigma_k$. Given two strings $\alpha, \beta \in \Sigma_{\leq m}$ of length $k_\alpha = |\alpha|$ and $k_\beta = |\beta|$, respectively, we say that α is a *prefix* of β (of length k_α) if the first $k_\alpha \leq k_\beta$ bits of β are equal to α (e.g., 101110 is a prefix of 1011101101110). Moreover, we denote by $\alpha \cdot \beta$ the *concatenation* of α and β, that is, the string whose first k_α bits are equal to α and whose last k_β bits are equal to β (e.g., the concatenation of 101110 and 1101110 is 1011101101110). Finally, given a string α and a subset S of $\Sigma_{\leq m}$, we define $\alpha \cdot S = \{x \mid x = \alpha \cdot \beta \text{ with } \beta \in S\}$.

A *routing table* T relative to m-bit addresses is a sequence of pairs (p,h) where the *route* p is a string in $\Sigma_{\leq m}$ and the *next-hop interface* h is an integer in $[1 \dots H]$, with H denoting the number of next-hop interfaces[1]. In the following we will denote by $|T|$ the size of T, that is, the number of pairs in the sequence. Moreover, we will assume that T always contains the pair $(\varepsilon, h_\varepsilon)$ where ε denotes the empty string and h_ε corresponds to the *default* next-hop interface.

The *IP address lookup problem* can be stated as follows. Given a routing table T and $x \in \Sigma_m$, compute the next-hop interface h_x corresponding to address x. That interface is uniquely identified by pair $(p_x, h_x) \in T$ for which (a) p_x is a prefix of x and (b) $|p_x| > |p|$ for any other pair $(p,h) \in T$, such that p is a prefix of x. In other words, p_x is the longest prefix of x appearing in T. The IP address lookup problem is well defined since T contains the default pair $(\varepsilon, h_\varepsilon)$ and so h_x always exists.

2.2 The Expansion Phase

We describe formally the intuitive process of extending the routes of T that are shorter than m in all possible ways by preserving the information regarding their corresponding next-hop interfaces. We say that T is in decreasing (respectively, increasing) order if the routes in its pairs are lexicographically sorted in that order. We take T in decreasing order and number its pairs according to their ranks, so that the resulting pairs are numbered $T_1, T_2, \dots, T_{|T|}$, where T_i precedes T_j if and only if $i < j$. As a result, if p_j is a prefix of p_i then T_i precedes T_j. We use this property to suitably expand the pairs.

With each pair $T_i = (p_i, h_i)$ we associate its expansion set, denoted $EXP(T_i)$, to collect all m-bit strings that have p_i as a prefix. Formally, $EXP(T_i) = (p_i \cdot \Sigma_{m-|p_i|}) \times \{h_i\}$, for $1 \leq i \leq |T|$. We then define the expansion of T on m bits, denoted T', as the union $T' = \cup_{i=1}^{|T|} T'_i$ where sets T'_i are inductively defined as follows: $T'_1 = EXP(T_1)$, and $T'_i = EXP(T_i) \ominus \cup_{1 \leq j < i} T'_j$, where the operator $\ominus$ removes from $EXP(T_i)$ all pairs whose routes already appear in the pairs of $\cup_{1 \leq j < i} T'_j$. In this way, we fill the entries of the expanded table T' consistently with the pairs in the routing table T, as stated by the following result.

Fact 1. *If* $(p_x, h_x) \in T$ *is the result of the IP address lookup for any* m*-bit string* x*, then* $(x, h_x) \in T'$.

It goes without saying that T' is made up of 2^m pairs and that, if we had enough space, we could solve the IP address lookup problem with a single access to T'. We therefore need to compress T' somehow.

[1] We are actually using the term "routing table" to denote what is more properly called "forwarding table." Indeed, a routing table contains some additional information.

2.3 The Compression Phase

This phase heavily relies on a parameter k to be fixed later on, where $1 \leq k \leq m$. We are given the expanded table T' and wish to build three tables `row_index`, `col_index` and `interface` to represent the same information as T' in less space by a simple *run length encoding* (RLE) scheme [9].

We begin by clustering the pairs in T' according to the first k bits of their strings. The cluster corresponding to a string $x \in \Sigma_k$ is $T'_{(x)} = \{(y, h_{xy}) \mid y \in \Sigma_{m-k} \text{ and } (x \cdot y, h_{xy}) \in T'\}$. Note that its size is $|T'_{(x)}| = 2^{m-k}$. We can define our first statistical parameter to denote the number of *distinct* clusters.

Definition 2. *Given a routing table T with m-bit addresses, the* row k-size *of T for $1 \leq k \leq m$ is*

$$\alpha_k = \left|\left\{T'_{(x)} \mid x \in \Sigma_k\right\}\right|.$$

Parameter α_k is the first measure of a simple form of compression. Although we expand the prefixes shorter than k, we do not increase the number of *distinct* clusters as the following fact shows.

Fact 3. *For a routing table T, let r_k be the number of distinct next-hop interfaces in all the pairs with routes of length at most k, and let n_k be the number of pairs whose routes are longer than k, where $1 \leq k \leq m$. We have $\alpha_k \leq r_k + n_k \leq |T|$.*

We now describe the compression based upon the RLE scheme. It takes a cluster $T'_{(x)}$ and returns an *RLE sequence* $s_{(x)}$ in two logical steps:

1. Sort $T'_{(x)}$ in ascending order with respect to the strings y and number its pairs according to their ranks, obtaining $T'_{(x)} = \{(y_i, h_i)\}_{1 \leq i \leq 2^{m-k}}$.
2. Transform $T'_{(x)}$ into $s_{(x)}$ by replacing each maximal run (y_i, h_i), (y_{i+1}, h_{i+1}), ..., (y_j, h_{i+l}), such that $h_i = h_{i+1} = \cdots = h_{i+l}$, by a pair $\langle h_i, l+1\rangle$, where $l+1$ is called the *run length* of h_i.

The previous steps encode the 2^{m-k} pairs of strings and interfaces of each cluster $T'_{(x)}$ into a single and (usually) shorter sequence $s_{(x)}$. Note that, by Definition 2, α_k is the number of distinct RLE sequences $s_{(x)}$ so produced. We further process them to obtain an equal number of equivalent RLE sequences $s'_{(x)}$. The main goal of this step is to obtain sequences such that, for any i, the i-th pair of any two such sequences have the same run length value. We show how to do it by means of an auxiliary function $\varphi(s,t)$ defined on two nonempty RLE sequences $s = \langle a, f\rangle \cdot s_1$ and $t = \langle b, g\rangle \cdot t_1$, whose total sum of run lengths is equal, as follows:

$$\varphi(s,t) = \begin{cases} t & \text{if } s_1 = t_1 = \varepsilon, \\ \langle b, f\rangle \cdot \varphi(s_1, t_1) & \text{if } f = g \text{ and } s_1, t_1 \neq \varepsilon, \\ \langle b, f\rangle \cdot \varphi(s_1, \langle b, g-f\rangle \cdot t_1) & \text{if } f < g \text{ and } s_1, t_1 \neq \varepsilon, \\ \langle b, g\rangle \cdot \varphi(\langle a, f-g\rangle \cdot s_1, t_1) & \text{if } f > g \text{ and } s_1, t_1 \neq \varepsilon. \end{cases}$$

The purpose of φ is to "unify" the run lengths of two RLE sequences by splitting some pairs $\langle b,f\rangle$ into $\langle b,f_1\rangle,\ldots,\langle b,f_r\rangle$, such that $f=f_1+\cdots+f_r$. The unification defined by φ is a variant of standard merge, except that it is not commutative as it only returns the (split) pairs in the second RLE sequence.

In order to apply unification to a set of RLE sequences $s_1,s_2,\ldots,s_q$, we define function $\Phi(s_1,\ldots,s_q)$ that returns RLE sequences $s'_1,s'_2,\ldots,s'_q$ as follows. First,

$$s'_q=\varphi(\varphi(\ldots\varphi(\varphi(s_1,s_2),s_3)\ldots,s_{q-1}),s_q).$$

As a result of this step, we obtain that the run lengths in s'_q are those common to all the input sequences. Then, $s'_i=\varphi(s'_q,s_i)$ for $i<q$: in this way, the pairs of the set of RLE sequences $s_1,s_2,\ldots,s_q$ are *equally* split.

We are now ready to define the second statistical parameter. Given an RLE sequence s, let its *length* $|s|$ be the number of pairs in it. Regarding the routing table T, let us take the α_k RLE sequences $s_1,s_2,\ldots,s_{\alpha_k}$ obtained by the distinct clusters $T'_{(x)}$ with $x\in\Sigma_k$, and apply the unification $\Phi(s_1,s_2,\ldots,s_{\alpha_k})$ defined above.

Definition 4. *Given a routing table T with m-bit addresses, the* column k-size β_k *of T, for $1\le k\le m$, is the (equal) length of the RLE sequences resulting from $\Phi(s_1,s_2,\ldots,s_{\alpha_k})$. That is,*

$$\beta_k=|s'_{\alpha_k}|.$$

Although we increase the length of the original RLE sequences, we have that β_k is still linear in $|T|$.

Fact 5. *For $1\le k\le m$, $\beta_k\le 3|T|$.*

Proof. Let us number the distinct clusters from 1 to α_k, and let n_j be the number of routes in cluster j, where $\sum_{j=1}^{\alpha_k}n_j\le|T|$. The initial RLE sequence for cluster j has length $2n_j+1$. Indeed, the first route gives a length of 1, and each subsequent route increases that length by 2. After Φ has been applied, each $\phi(t,s)$ produced an RLE sequence of length at most $|s|+|t|$. As a result, β_k is the length of the last sequence, and so β_k is bounded by $\sum_{j=1}^{\alpha_k}(2n_j+1)\le 2|T|+\alpha_k\le 3|T|$.

2.4 Putting All Together

We can, finally, prove our result on storing a routing table T in a sufficiently compact way to guarantee always a constant number of accesses. We let *#bytes*(n) denote the number of bytes necessary to store a positive integer n, and a *word* be sufficiently large to hold $\max\{\log|T|+2,\log H\}$ bits.

Theorem 6. *Given a routing table T with m-bit addresses and H next-hop interfaces, we can store T into three tables* `row_index`, `col_index` *and* `interface` *of total size*

$$2^k\cdot\#bytes(\alpha_k)+2^{m-k}\cdot\#bytes(\beta_k)+\alpha_k\cdot\beta_k\cdot\#bytes(H)$$

in bytes or $O(2^k + 2^{m-k} + |T|^2)$ *words, for* $1 \le k \le m$*, so that an IP address lookup for an m-string x takes* exactly three *accesses given by*

$$h_x = \texttt{interface}\big[\texttt{row\_index}[x[1\dots k]], \texttt{col\_index}[x[k+1\dots m]]\big]$$

where $x[i\dots j]$ *denotes the substring of x starting from the i-th bit and ending at the j-th bit.*

Proof. We first find the α_k distinct clusters (without their explicit construction) and produce the corresponding RLE sequences, which we unify by applying Φ. The resulting RLE sequences, of length β_k, are numbered from 1 to α_k. At this point, we store the β_k next-hop values of the j-th sequence in row j of table `interface`, which hence has α_k rows and β_k columns. We then set $\texttt{row\_index}[x[1\dots k]] = j$ for each string x such that cluster $T_{(x[1\dots k])}$ has been encoded by the j-th RLE sequence. Finally, let $f_1, \dots, f_{\beta_k}$ be the run lengths in any such sequence. We set $\texttt{col\_index}[x[k+1\dots m]] = \ell$ for each string x, such that $x[k+1\dots m]$ has rank q in Σ_{m-k} and $\sum_{t=1}^{\ell-1} f_t < q \le \sum_{t=1}^{\ell} f_t$. The space and memory access bounds immediately follow. □

We wish to point out that the Theorem 6 represents a reasonable trade-off between performing a binary search on T with $O(\log|T|)$ accesses and executing a single access on a table of 2^m entries obtained by T.

3 Data Analysis and Experimental Results

In this section, we show that the space bound stated by Theorem 6 is practically feasible for real routing tables for Internet. In particular, we have analyzed the data of five prefix databases which are made available by the IPMA project [7]: these data are daily snapshots of the routing tables used at some major network access points. The largest database, called MaeEast, is useful to model a large backbone router while the smallest one, called Paix, can be considered a model for an enterprise router. The following table shows the sizes of these routing tables on 7/7/98 and on 1/11/99 (the third column indicates the minimum/maximum size registered between the previous two dates while the fourth column indicates the number H of next-hop interfaces).

Router	7/7/98	1/11/99	Min/Max	H
MaeEast	41231	43524	37134/44024	62
MaeWest	18995	23411	17906/23489	62
Aads	23755	24050	18354/24952	34
PacBell	22416	22849	21074/23273	2
Paix	3106	5935	1519/5935	21

Besides suggesting the approach described in the previous section, the data of these databases have also been used to choose the appropriate value of k (that is, the way of splitting an IP address). The following table shows the percentages of routes of length 16 and 24 on 7/7/98 (the third column denotes the most frequent among the remaining prefix lengths).

Router	Length 16	Length 24	Next
MaeEast	13%	56%	8% (23 bits)
MaeWest	14%	53%	7% (23 bits)
Aads	25%	54%	8% (23 bits)
PacBell	13%	56%	7% (23 bits)
Paix	12%	53%	8% (23 bits)

As it can be seen from the table, these two lengths are the two most frequent in all observed routing tables. (In other words, even though it is now allowed to use prefixes of any length, it seems that network managers are still using the original approach of using multiples of 8 according to the class categorization of IP addresses.) If we also consider that choosing k multiple of 8 is more adequate to the current technology in which bit operations are still time consuming, the table's data suggests to use either $k = 16$ or $k = 24$. *We choose* $k = 16$ to balance the size of the tables `row_index` and `col_index` (see Theorem 6). It now remains to estimate the value of the two statistical parameters α_k and β_k. The next table shows these values measured on 7/7/98 and on 1/11/99 (the third column shows the maximum values registered between the previous two dates).

Router	7/7/98 (α_k/β_k)	1/11/99 (α_k/β_k)	Maximum (α_k/β_k)
MaeEast	2577/277	2745/299	2821/299
MaeWest	2017/263	2335/268	2335/285
Aads	1903/259	2100/269	2140/273
PacBell	1399/256	1500/256	1500/260
Paix	722/256	984/261	989/261

From the previous table's data and from the registered values of H, it is then possible to argue that the $\#bytes(\alpha_k) = \#bytes(\beta_k) = 2$ and $\#bytes(H) = 1$, so that, according to the bound of Theorem 6, the memory occupancy of our algorithm is equal to $M_1 = 2^{16} \cdot 2 + 2^{16} \cdot 2 + \alpha_k \cdot \beta_k$ bytes. Actually, the above memory size can be slightly increased in order to make faster the access to the table `interface`: to this aim, the elements of table `row_index` (respectively, `column_index`) can be seen as memory pointers (respectively, offsets) instead of as indexes. In this way, the actual value of $\#bytes(\alpha_k)$ becomes 4 and the corresponding memory occupancy increases to $M_2 = 2^{16} \cdot 4 + 2^{16} \cdot 2 + \alpha_k \cdot \beta_k$ bytes. According to these two formulae, we have then the real memory occupancy of our algorithm. These values are shown in the following table.

Router	7/7/98 (M_1/M_2)	1/11/99 (M_1/M_2)	Maximum (M_1/M_2)
MaeEast	975973/1107045	1082899/1213971	1082899/1213971
MaeWest	792615/923687	887924/1018996	887924/1018996
Aads	755021/886093	827044/958116	837804/968876
PacBell	620288/751360	646144/777216	646424/777496
Paix	446976/578048	518968/650040	518968/650040

The data relative to MaeEast seem to show that we need a little bit more than 1 Megabyte of cache memory. However, this is not true in practice: indeed, only a small fraction (approximately, 1/5) of table `row_index` is filled with used values from actual IP addresses.

Since only these values are loaded in the cache memory, this implies that the real cache memory occupancy is much smaller than the one shown in the previous table.

We have implemented our algorithm in ANSI C and we have compiled it by using the GCC compiler under the Linux operating system and by using the Microsoft Visual C compiler under the Windows 95 operating system. The C source and the complete data-tables of our analysis are available at the URL
`http://www.dsi.unifi.it/~piluc/IPLookup/`, where we show that the generated code is highly tuned and very few can be done to improve the performance of our lookup process.

4 Conclusions

In this paper, we have proposed a new method for solving the IP address lookup problem which originated from the analysis of the real data. This analysis may lead to different solutions. For example, we have observed that a well-known binary search approach (based on secondary memory access techniques) requires on the average a little bit more than one memory access for address lookup. However, this approach does not always perform better than the expansion/compression technique presented in this paper: the reason for this is that the implementation of the binary search requires too many arithmetic and branch operations which become the actual bottleneck of the algorithm. Nevertheless, there exist computer architectures for which this new solution performs better: actually, it is our intention to realize a serious comparison of the two approaches (comparison that fits into the recently emerging area of algorithm engineering).

Theorem 6 can be generalized in order to obtain, for any integer $r > 1$, a space bound $O(r2^{m/r} + |T|^r)$ and $r+1$ memory accesses. This could be the right way to extend our approach to the coming IPv6 protocol family [1]. It would be interesting to find a method for unifying the RLE sequences, so that the final space is provably less than the pessimistic $O(|T|^r)$ one.

Finally, we believe that our approach arises several interesting theoretical questions. For example, is it possible to derive combinatorial properties of the routing tables in order to give better bounds on the values of α_k and β_k? What is the lower bound on space occupancy in order to achieve a (small) constant number of memory accesses? What about compression schemes other than RLE?

References

[1] S. Deering and R. Hinden. Internet protocol, version 6 (IPv6). RFC 1883 (`http://www.merit.edu/internet/documents/rfc/rfc1883.txt`), 1995.

[2] M. Degermark, A. Brodnik, S. Carlsson, and S. Pink. Small forwarding tables for fast routing lookups. *ACM Computer Comm. Review*, 27(4):3–14, 1997.

[3] V. Fuller, T. Li, J. Yu, and K. Varadhan. Classless inter-domain routing (CIDR): an address assignment and aggregation strategy. RFC 1519 (`http://www.merit.edu/internet/documents/rfc/rfc1519.txt`), 1993.

[4] Intel. Vtune. `http://developer.intel.com/design/perftool/ vtune/`, 1997.

[5] B. Lampson, V. Srinivasan, and G. Varghese. IP lookups using multi-way and multicolumn search. In *ACM INFOCOM Conference*, April 1998.

[6] A. McAuley, P. Tsuchiya, and D. Wilson. Fast multilevel hierarchical routing table using content-addressable memory. U.S. Patent Serial Number 034444, 1995.

[7] Merit. IPMA statistics. `http://nic.merit.edu/ipma`, 1998.

[8] D.R. Morrison. PATRICIA – Practical Algorithm To Retrieve Information Coded In Alfanumeric. *Journal of ACM*, 15(4):514–534, October 1968.

[9] M. Nelson. *The Data Compression Book.* M&T Books, San Mateo, CA, 1992.

[10] P. Newman, G. Minshall, T. Lyon, and L. Huston. IP switching and gigabit routers. *IEEE Communications Magazine*, January 1997.

[11] S. Nilsson and G. Karlsson. Fast address look-up for internet routers. In *ALEX*, pages 9–18. Università di Trento, February 1998.

[12] C. Partridge. A 50-Gb/s IP router. *IEEE/ACM Transactions on Networking*, 6(3):237–247, June 1998.

[13] T.-B. Pei and C. Zukowski. Putting routing tables into silicon. *IEEE Network*, pages 42–50, January 1992.

[14] J. Postel. Internet protocol. RFC 791 (`http://www.merit.edu/internet/documents/rfc/rfc0791.txt`), 1981.

[15] V. Srinivasan and G. Varghese. Fast address lookups using controlled prefix expansion. In *ACM SIGMETRICS Conference*, September 1998. Full version to appear in *ACM TOCS*.

[16] W.R. Stevens and G.R. Wright. *TCP/IP Illustrated, Volume 2 The Implementation.* Addison-Wesley Publishing Company, Reading, Massachusetts, 1995.

[17] M. Waldvogel, G. Varghese, J. Turner, and B. Plattner. Scalable high speed IP routing lookups. *ACM Computer Communication Review*, 27(4):25–36, 1997.

On-Line Load Balancing in a Hierarchical Server Topology

Amotz Bar-Noy[1], Ari Freund[2], and Joseph (Seffi) Naor[3,*]

[1] Department of Electrical Engineering, Tel Aviv University, Tel Aviv 69978, Israel
amotz@eng.tau.ac.il
[2] Department of Computer Science, Technion, Haifa 32000, Israel
arief@cs.technion.ac.il
[3] Bell Laboratories, Lucent Technologies, 600 Mountain Ave. Murray Hill, NJ 07974
naor@research.bell-labs.com

Abstract. In a hierarchical server environment, jobs must be assigned in an on-line fashion to a collection of servers forming a hierarchy of capability. Each job requests a specific server meeting its needs but the system is free to assign it either to that server or to any other server higher in the hierarchy. Each job carries a certain load, which it imparts to the server to which it is assigned. The goal is to minimize the maximum total load on a server.
We consider the *linear* hierarchy, where the servers are totally ordered in terms of capability, and the *tree* hierarchy, where ancestors are more powerful than their descendants. We investigate several variants of the problem, differing from one another by whether jobs are *weighted* or *unweighted*; whether they are *permanent* or *temporary*; and whether assignments are *fractional* or *integral*. We derive upper and lower bounds on the competitive ratio.

1 Introduction

One of the most basic on-line load-balancing problems is the following. Jobs arrive one at a time and each must be scheduled on one of n servers. Each job has a certain load associated with it and a subset of the servers on which it may be scheduled. These servers are said to be eligible for the job. The goal is to assign jobs to servers so as to minimize the cost of the assignment, which is defined as the maximum load on a server.

The nature of the load balancing problem considered here is on-line: decisions must be made without any knowledge of future jobs, and previous decisions may not be revoked. We compare the performance of an on-line algorithm with the performance of an optimal off-line scheduler—one that knows the entire sequence of jobs in advance. The efficacy parameter of an on-line scheduler is its *competitive ratio*, roughly defined as the maximum ratio, taken over all possible sequences of jobs, between the cost incurred by the algorithm and the cost of an optimal assignment.

1.1 The Hierarchical Servers Problem

In the general setting, studied in [6, 4, 5], the sets of eligible servers are completely arbitrary. In practical situations, however, this is almost never the case; very often we

* On leave from the Department of Computer Science, Technion, Haifa 32000, Israel.

find a mixed system in which certain servers are more powerful than others. In the *hierarchical servers* problem the servers form a hierarchy of capability such that any job which may run on a given server may also run on any server higher in the hierarchy. We consider the *linear* hierarchy, in which the servers are numbered 1 through n and we imagine them to be physically ordered along a straight line running from left to right, with server 1 leftmost and server n rightmost. Leftward servers are more capable than rightward ones. We say that servers $1,\ldots,s$ are *to the left* of s (note that s is to the left of itself), and that servers $s+1,\ldots,n$ are *to the right* of s.

The input is a sequence of jobs, each carrying a positive *weight* and *requesting* one of the servers. A job requesting server s can be assigned to any of the servers to the left of s. Thus, the job's *eligible* servers are $1,\ldots,s$. The assignment of a job with weight w to server s increases the *load* on s by w (initially, all loads are 0). We use the terms 'job' and 'request' interchangeably. The *cost* of a given assignment is $COST = \max_s \{l_s\}$, where l_s is the load on server s. We use OPT to denote the cost of an optimal offline assignment. An algorithm is *c-competitive* if there exists a constant b such that $COST \leq c \cdot OPT + b$ for all input sequences.

We consider variants, or *models*, of the problem according to three orthogonal dichotomies. In the *integral* model each job must be assigned in its entirety to a single server, whereas in the *fractional* model a job's weight may be split among several eligible servers. In the *weighted* model jobs may have arbitrary positive weights, whereas in the *unweighted* model all jobs have weight equal to unity. Our results for the fractional model hold for both the unweighted and weighted cases, so we do not distinguish between the unweighted fractional model and the weighted fractional model. Finally, *permanent* jobs continue to load the servers to which they are assigned indefinitely, whereas *temporary* jobs are only *active* for a finite duration, at the end of which they depart. The duration for which a temporary job is active is *not* known upon its arrival. When temporary jobs are allowed, the cost of an assignment is defined as $COST = \max_t \max_s \{l_s(t)\}$, where $l_s(t)$ is the load on server s at time t. The version of the problem which we view as basic is the weighted integral model with permanent jobs only.

A natural (and more realistic) generalized setting is one in which the servers form a (rooted) *tree* hierarchy: a job requesting a certain server may be assigned to any of its ancestors in the tree. The various models pertain to this generalization as well.

In practical systems there are often several groups of identical servers (e.g. a computer network consisting of fifty identical PC's and three identical file servers). To model such systems we can extend our formulation of the problem to allow each level in the hierarchy (be it linear or tree) to be populated by several *equivalent* servers. In this model, requests refer to *equivalence classes* of servers rather than to individual ones. For lack of space we omit the details, but it can be seen quite easily that both models are equivalent.

The *hierarchical servers* problem is an important paradigm in the sense that it captures many interesting applications from diverse areas. Among these are assigning classes of service quality to calls in communication networks, routing queries to hierarchical databases, signing documents by ranking executives, and upgrading classes of cars by car rental companies.

The *hierarchical servers* problem is an instance of the general assignment problem considered in [6], obtained by restricting the class of allowable eligible sets. (The class we consider is all sets of the form $\{1,\ldots,s\}$ with $1 \le s \le n$.) Since an $\Omega(\log n)$ lower bound is known for the general setting [6], it is natural to ask for which (interesting) classes of eligible sets can we get competitive factors better than $\Omega(\log n)$. The *hierarchical servers* problem admits constant competitive factors and seems to be unique in this respect; we know of no other non-trivial "natural" classes for which sub-logarithmic competitiveness is attainable. In fact, in the full paper we analyze three such classes, showing a logarithmic lower bound for each.

A further motivation for studying the problem of *hierarchical servers* is its relation to the problem of *related machines* introduced in [2]. In this problem all servers are eligible for all jobs, but the servers may have different *speeds*: assigning a job of weight w to a server with speed v increases its load by w/v. Without loss of generality, assume $v_1 \ge v_2 \ge \cdots \ge v_n$, where v_i is the speed of server i. Consider a set of jobs to be assigned at a cost bounded by C and let us focus on a particular job whose weight is w. To achieve $COST \le C$ we must refrain from assigning this job to any server i for which $w/v_i > C$. In other words, there exists a rightmost server to which we may assign the job. Thus, restricting the cost yields eligibility constraints similar to those in the *hierarchical servers* problem.

1.2 Background

Graham [10] explored the problem of assignment to *identical machines*, where each job may be assigned to any of the servers. He showed that the greedy algorithm has a competitive ratio of $2 - \frac{1}{n}$. Later work [7, 8, 11, 1] investigated the exact competitive ratio achievable for this problem for general n and for various special cases. The best results to date, for general n, are a lower bound of 1.852 and an upper bound of 1.923 [1].

Over the years many other load balancing problems were studied; see [3, 12] for surveys. The assignment problem in which arbitrary sets of eligible servers are allowed was considered in [6]. They showed upper and lower bounds of $\Theta(\log n)$ for several variants of this problem. Permanent jobs were assumed. Subsequent papers generalized the problem to allow temporary jobs: in [4] a lower bound of $\Omega(\sqrt{n})$ and an upper bound of $O(n^{2/3})$ were shown. The upper bound was later tightened to $O(\sqrt{n})$ [5].

The *related machines* problem was investigated in [2]. They showed an 8-competitive algorithm based on the doubling technique. This result was improved in [9], where a more refined doubling algorithm was shown to be $3 + 2\sqrt{5} = 5.828$-competitive. By randomizing this algorithm, they were able to improve the bound to 4.311. They also showed lower bounds of 2.438 (deterministic) and 1.837 (randomized). In [5] the problem was generalized to allow temporary jobs; they showed an upper bound of 20, achieved by a doubling algorithm, and a lower bound of 3. Both the upper and lower bounds are deterministic.

1.3 Our Results

A significant portion of our work is devoted to developing a continuous framework in which we recast the problem. The continuous framework is a fully fledged model, in which a new variant of the problem is defined. The novelty of our approach lies in the realization that *weight* and *load* are qualitatively distinct notions. Although the term *load* is defined as a sum of weights, it is in fact more accurately interpreted as the *density* of weight ocurring at a server in the weight distribution defined by the assignment. This distinction between *load* and *weight*, made explicit in the continuous model, is unobserved in the problem definition due to the fact that the "volume" of each server is equal to unity, and thus the numerical value of *weight density* (i.e. weight/volume) coincides with that of *total weight*.

In Sect. 2 we define a *semi*-continuous model and construct an e-competitive algorithm. We then show how to transform any algorithm for the semi-continuous model into an algorithm for the fractional model, and how to transform any algorithm for the fractional model into an algorithm for the integral models (both weighted and unweighted). We thus obtain an e-competitive algorithm for the fractional model and an algorithm for the integral models which is e and $(e+1)$-competitive in the unweighted and weighted cases, respectively.

In Sect. 3 we develop a procedure for deriving lower bounds in the context of the continuous model. The lower bounds obtained with our procedure are also valid in the discrete models (fractional as well as integral), even in the unweighted case with permanent jobs only, and even with respect to randomized algorithms. Using our procedure we find that e is a tight lower bound.

In Sect. 4 we consider temporary jobs in the integral model. We show an algorithm which is 4 and 5-competitive in the unweighted and weighted cases, respectively. In the full paper we also show a deterministic lower bound of 3.

In the full paper we extend the problem to the tree hierarchy. We show an algorithm which is respectively 4, 4 and 5-competitive for the fractional, unweighted integral, and weighted integral models. Randomizing this algorithm improves its competitiveness to e, e and $e+1$ respectively. We show deterministic and randomized lower bounds of $\Omega(\sqrt{n})$ for all models when temporary jobs are allowed. Our lower bound constructions, which also apply in the general setting of arbitrary eligible sets, are considerably simpler than those presented in [4]. They are, however, less tight by small constant factors.

In the full paper we also investigate the effect of restricting the sets of eligible servers in ways other than the linear and tree hierarchies. Namely, we consider the following three models: (1) the servers are on a line and eligible servers must be contiguous, (2) the servers form a tree such that if a server is eligible then so are its descendants, and (3) The eligible sets have some fixed cardinality. We show a logarithmic lower bound in each case.

2 Upper Bounds

We show an algorithm whose respective versions for the fractional, unweighted integral and weighted integral models are e, e and $(e+1)$-competitive. The fractional version admits temporary jobs; the integral versions do not. We build up to the algorithm by introducing and studying the *semi-continuous* model and the class of *memoryless* algorithms. We begin with the Optimum Lemma, which characterizes *OPT* in terms of the input sequence.

Lemma 1 (Optimum Lemma). *For a given input sequence, denote by W_s the total weight of jobs requesting servers to the left of s and let $H = \max_s\{W_s/s\}$. Let $w_{\max}$ be the maximum weight of a job in the input sequence. Then,*

1. *In the fractional model, $OPT = H$.*
2. *In the unweighted integral model, $OPT = \lceil H \rceil$.*
3. *In the weighted integral model, $\max\{H, w_{\max}\} \leq OPT < H + w_{\max}$.*

2.1 Memoryless Algorithms

Generally speaking, a *memoryless* algorithm is an algorithm which assigns each job independently of previous jobs. Clearly, memoryless algorithms are only of interest in the fractional model, which shall therefore be the model on which we focus in this section. Note that the competitiveness of memoryless algorithms is immune to the presence of temporary jobs. If a memoryless algorithm is c-competitive with respect to permanent jobs, then it must remain c-competitive in the presence of temporary jobs, since at all times the momentary cost of its assignment cannot exceed c times the optimal cost of the active jobs.

We focus on a restricted type of memoryless algorithms, which we name *uniform* algorithms. Uniform memoryless algorithms are instances of the generic algorithm shown below. Each instance is characterized by a function $u : \mathbb{N} \to (0,1]$ satisfying $u(1) = 1$.

Algorithm GenericUniform

When a job of weight w, requesting server s, arrives:

1. Let $r \leftarrow w$ and $i \leftarrow s$.
2. While $r > 0$:
   3. Assign $a = \min\{w \cdot u(i), r\}$ units of weight to server i.
   4. $r \leftarrow r - a$.
   5. $i \leftarrow i - 1$.

The algorithm starts with the server requested by the job and proceeds leftward as long as the job is not fully assigned. The fraction of the job's weight assigned to server i is $u(i)$, unless $w \cdot u(i)$ is more than what is left of the job when i is reached. The condition $u(1) = 1$ ensures that the job's weight will always be assigned in full by the algorithm.

Note that the assignment generated by a uniform memoryless algorithm is independent of both the number of servers and the order of the jobs in the input. We therefore assume that exactly one job requests each server (we allow jobs of zero weight) and that the number of servers is infinite. We allow infinite request sequences for which the cost is finite. Such sequences represent the limit behavior of the algorithm over a sequence of finite input sequences. We denote the weight of the job requesting some server s by w_s.

Consider a job of weight w requesting a server to the right of some server s. If the requested server is close to s the job will leave $w \cdot u(s)$ units of weight on s regardless of the exact server requested. At some point, however, if the request is made far enough from s the weight assigned to s will begin to diminish as the distance of the request from s grows. Finally, if the request is made "very" far away, it will have no effect on s. We denote by p_s the point beyond which the effect on s begins to diminish and by p'_s the point at which it dies out completely, namely, $p_s = \max\left\{s' \middle| \sum_{i=s}^{s'} u(i) \leq 1\right\}$ and $p'_s = \max\left\{s' \middle| \sum_{i=s+1}^{s'} u(i) < 1\right\}$.

Note that p_s and p'_s may be undefined, in which case we take them to be infinity. We are only interested in functions u for which p_s is finite. The importance of p_s stems from the fact that the load on s due to jobs requesting servers in the range $s, \ldots, p_s$ is simply $u(s)$ times the total weight of these jobs.

Lemma 2 (Worst Case Lemma). *Let $\mathcal{A}$ be a uniform memoryless algorithm. The problem:*

> *Given $K > 0$ and some server s, find an input sequence that maximizes the load on s in $\mathcal{A}$'s assignment, subject to $OPT = K$*

is solved by the following sequence of jobs:

$$w_i = \begin{cases} 0 & 1 \leq i < p_s \\ p_s K & i = p_s \\ K & p_s < i \leq p'_s \\ 0 & i > p'_s \text{ (if } p'_s < \infty) \end{cases}, \tag{1}$$

and l_s—the resultant load on s—satisfies $p_s K u(s) \leq l_s \leq p'_s K u(s)$.

Corollary 3. *Let $\mathcal{A}$ be a uniform memoryless algorithm defined by u whose competitive ratio is $C_{\mathcal{A}}$. Then,* $\sup_s \{p_s u(s)\} \leq C_A \leq \sup_s \{p'_s u(s)\}$.

2.2 The Semi-continuous Model

In both the fractional and integral versions of the problem the servers and the jobs are discrete objects. We therefore refer to these models as the *discrete* models. In this section we introduce the *semi-continuous* model, in which the servers are made continuous. In section 3 we make the jobs continuous as well, resulting in the *continuous* model.

The semi-continuous model is best understood through a physical metaphor. Consider the bottom of a vessel filled with some non-uniform fluid applying varying degrees of pressure at different points. The force acting at any single point is zero, but any region of non-zero area suffers a net force equal to the integral of the pressure over the region. Similarly, in the semi-continuous model we do not speak of individual servers; rather, we have a continuum of servers, analogous to the bottom of the vessel. An arriving job is analogous to a quantity of fluid which must be added to the vessel. The notions of *load* and *weight* become divorced; *load* is analogous to pressure and *weight* is analogous to force.

Formally, the *server interval* is $(0, \infty)$, to which jobs must be assigned. Job j has weight w_j and it requests a point s_j in the server interval. The assignment of job j is specified by an integrable function $g_j : (0,\infty) \to [0,\infty)$ satisfying $\int_0^{s_j} g_j(x)\,dx = w_j$ and $x > s_j \Rightarrow g_j(x) = 0$. The assignment of a sequence of jobs is $g = \sum_j g_j$. The load l_I on a non-empty interval $I = (x_0, x_1)$ is defined as $l_I = \frac{1}{x_1 - x_0} \int_{x_0}^{x_1} g(x)\,dx$—the mean weight density over I. The load at a point x is defined as $\sup_{x \in I} \{l_I\}$. The cost of an assignment is $COST = \sup_I \{l_I\}$.

Lemma 4 (Optimum Lemma—Semi-continuous Model). *Let $W(x)$ be the total weight of requests made to the left of x (including x itself). Then $OPT = \sup_x \{W(x)/x\}$.*

Let us adapt the definition of uniform memoryless algorithms to the semi-continuous model. In this model a uniform algorithm is characterized by a function $u : (0,\infty) \to (0,\infty)$ as follows. For a given point x, let $q(x)$ be the point such that $\int_{q(x)}^{x} u(z)\,dz = 1$. Then the assignment of job j is $g_j(x) = w_j u(x)$ for $q(s_j) \le x \le s_j$ and $g_j(x) = 0$ elsewhere. For this to work we must require that $\int_0^{\varepsilon} u(x)\,dx = \infty$ for all $\varepsilon > 0$. We define $p(x)$ as the point such that $\int_x^{p(x)} u(z)\,dz = 1$. If such a point does not exist then the algorithm's competitive ratio cannot be bounded, as demonstrated by the infinite request sequence in which all jobs have weight equal to unity and the j'th job requests the point $s_j = j$. Clearly, $OPT = 1$, whereas the algorithm's assignment places an infinite load at x. We shall therefore allow only algorithms such that $\int_M^{\infty} u(x)\,dx = \infty$ for all $M \ge 0$.

Lemma 5 (Worst Case Lemma—Semi-continuous Model). *Let $\mathcal{A}$ be uniform algorithm defined by $u(x)$. The problem:*

> *Given $K > 0$ and some point s, find an input sequence that maximizes the load on s in $\mathcal{A}$'s assignment, subject to $OPT = K$*

is solved by a single job of weight $p(s)K$ requesting the point $p(s)$. The resultant load at s is $p(s)Ku(s)$.

Corollary 6. *The competitive ratio of $\mathcal{A}$ is $\sup_x \{p(x)u(x)\}$.*

2.3 An e-competitive Algorithm

Consider Algorithm Harmonic—the uniform memoryless algorithm defined by $u(x) = 1/x$. Let us calculate $p(x)$:

$$1 = \int_x^{p(x)} \frac{dz}{z} = \ln \frac{p(x)}{x} \ . \tag{2}$$

$$p(x) = ex \ . \tag{3}$$

Thus, the competitive ratio of Algorithm Harmonic is $\sup_x \{ex\frac{1}{x}\} = e$. In Sect. 3 we show that e is a lower bound, hence Algorithm Harmonic is optimal.

2.4 Application to the Discrete Models

We show how to transform any algorithm for the semi-continuous model into an algorithm for the (discrete) fractional model, and how to transform any algorithm for the fractional model into an algorithm for the integral models.

Semi-continuous to Fractional. Let $\mathcal{A}$ be a c-competitive online algorithm for the semi-continuous model. Define algorithm $\mathcal{B}$ for the fractional model as follows. When job j arrives, B assigns $\int_{i-1}^{i} g_j(x)\,dx$ units of weight to server i, for all i, where g_j is the assignment function generated by $\mathcal{A}$ for the job. Clearly, the cost incurred by $\mathcal{B}$ is bounded by the cost incurred by $\mathcal{A}$, and thus $\mathcal{B}$ is c-competitive.

An important observation is that if $\mathcal{A}$ is memoryless, then so is $\mathcal{B}$. Thus, even if temporary jobs are allowed, the assignment generated by $\mathcal{B}$ will be c-competitive at all times, compared to an optimal (off-line) assignment of the active jobs.

Fractional to Integral. Let $\mathcal{A}$ be an algorithm for the fractional model. Define algorithm $\mathcal{B}$ for the integral model (both weighted and unweighted) as follows. As jobs arrive, $\mathcal{B}$ keeps track of the assignments $\mathcal{A}$ would make. A server is said to be *overloaded* if its load in $\mathcal{B}$'s assignment exceeds its load in $\mathcal{A}$'s assignment. When a job arrives, $\mathcal{B}$ assigns it to the rightmost eligible server which is not overloaded (after $\mathcal{A}$ is allowed to assign the job).

Proposition 7. *Algorithm $\mathcal{B}$ is well defined, i.e. whenever a job arrives, at least one of its eligible servers is not overloaded. Moreover, if $\mathcal{A}$ is c-competitive then $\mathcal{B}$ is c and $(c+1)$-competitive in the unweighted and weighted models, respectively.*

3 Lower Bounds

In this section we outline a technique for proving lower bounds. The bounds obtained are valid in both the fractional and integral models—even in the unweighted case. In fact, they remain valid even at the presence of randomization with respect to oblivious adversaries. Using this technique, we obtain a tight constant lower bound of e. The success of our approach is facilitated by transporting the problem from the discrete setting into the *continuous* model, in which both jobs and servers are continuous.

3.1 A Simple Lower Bound

We consider the fractional model, restricting our attention to *right-to-left* input sequences: sequences in which for all $i < j$, all requests for server j are made before any request for server i. We further restrict our attention to sequences in which each server is requested exactly once (although we now allow jobs of zero weight).

Let $\mathcal{A}$ be a k-competitive algorithm, and consider some input sequence. Denote by r_s the weight of the job requesting server s, and by l_s the load on server s at a given moment. Recall the definition of H in the Optimum Lemma. Suppose the first $n-i+1$ jobs (culminating with the job requesting server i) have been assigned by $\mathcal{A}$. Define h_i as the value of H for this prefix of the input sequence. For $j \geq i$, define $h_{i,j} = \frac{1}{j}\sum_{s=i}^{j} r_s$. We have $h_i = \max_{i \leq j \leq n}\{h_{i,j}\}$. Since $\mathcal{A}$ is k-competitive, the loads must obey $l_s \leq kh_i$, for all s.

Now consider the specific input sequence defined by $r_1 = \cdots = r_n = w$, for some $w > 0$. For this sequence we have $h_i = (n-i+1)w/n$ for all i. Thus, after the first job is assigned we have $l_n \leq kw/n$. After the second job is assigned we have $l_{n-1} \leq 2kw/n$, but $l_n \leq kw/n$ still holds, because the new job could not be assigned to server n. In general, after the request for server s is processed, we have $l_i \leq (n-i+1)kw/n$ for all $i \geq s$. Noting that the total weight of jobs in the input equals the total load on servers in the cumulative assignment, we get,

$$nw = \sum_{i=1}^{n} r_i = \sum_{i=1}^{n} l_i \leq \sum_{i=1}^{n} (n-i+1)\frac{kw}{n} = \sum_{j=1}^{n} j\frac{kw}{n} = \frac{kwn(n+1)}{2n} \; . \quad (4)$$

Hence, $k \geq 2 \cdot \frac{n}{n+1}$. Thus, 2 is a constant lower bound.

3.2 Discussion

Figure 1 depicts the request sequence and the resultant kh_i's in histogram-like fashion, with heights of bars indicating the respective values. The bars are of equal width, so we can equivalently consider their area rather than height. To be precise, let us redraw the histograms with bars of width 1 and height equal to the numerical values they represent. Then, the total weight to be assigned is the total area of the job bars, and the total weight actually assigned is bounded from above by the total area of the kh_i bars. Now, instead of drawing a histogram of kh_i, let us draw a histogram of h_i. Then, the lower bound is found by solving

$$\text{total area of job bars} \leq k \cdot \text{total area of } h_i \text{ bars} \; .$$

$$k \geq \frac{\text{total area of job bars}}{\text{total area of } h_i \text{ bars}} \; .$$

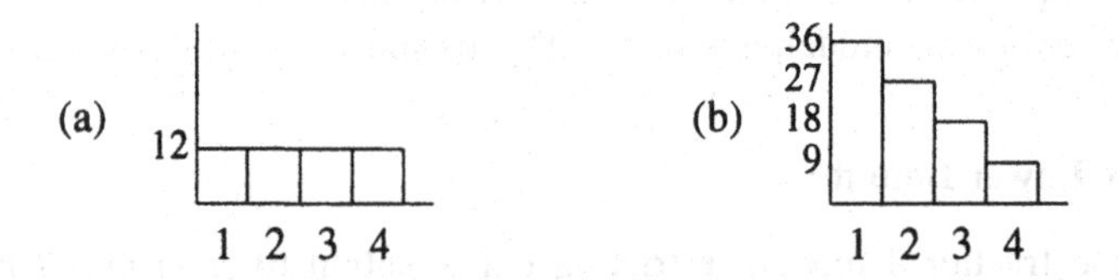

Fig. 1. (a) Histogram of job weights; $w = 12$. (b) Histogram of kh_i; k=3.

These considerations are independent of the specific input sequence at hand, so we have actually developed a general procedure for obtaining lower bounds. Select an input sequence and plot its histogram and the histogram of the resultant h_i's. Divide the area of the former by the area of the latter to obtain a lower bound.

Scaling both histograms by the same factor does not affect the ratio of areas, so we can go one step further and cast the procedure in purely geometric terms. Take as input a histogram where the width of each bar is $\frac{1}{n}$ and the height of the tallest bar is 1. Let $h_{i,j}$ be the area of bars i through j divided by j/n (the width of j bars), and let $h_i = \max_{i \leq j \leq n}\{h_{i,j}\}$. Divide the area of the input histogram by the area of the h_i histogram to obtain the corresponding lower bound.

3.3 The Continuous Model

The continuous model is motivated by the geometric interpretation outlined in the previous section. It differs from the semi-continuous model introduced in Section 2 in two ways. First, in contrast with the semi-continuous model, we now use a finite server interval $[0,S]$. The second, more important, difference is that in the continuous model requests are not discrete; rather, they arrive over time in a continuous flow.

Formally, we have a server interval $[0,S]$, and a time interval $[0,T]$ during which the request flow arrives. Instead of a request sequence we have a *request function* $f(x,t)$. For $t \in [0,T]$, $f(x,t)$ is an integrable non-negative real function of x defined over $[0,S]$. The interpretation of $f(x,t)$ is by means of integration, i.e. $\int_{x_0}^{x_1} f(x,t)\,dx$ represents the total amount of weight requesting points in the interval $[x_0,x_1]$ up to time t. We express the fact that requests accumulate over time by requiring that $t' > t \Rightarrow f(x,t') \geq f(x,t)$ for all x. The *assignment function* $g(x,t)$ is defined similarly: for $t \in [0,T]$, $g(x,t)$ is an integrable non-negative real function of x defined over $[0,S]$, and $\int_{x_0}^{x_1} g(x,t)\,dx$ is the total weight assigned to interval $[x_0,x_1]$ up to time t. Assigned weight accumulates too, so g must obey $t' > t \Rightarrow g(x,t') \geq g(x,t)$. We express the fact that weight may be assigned to the left of the point it requested but not to its right, by demanding that for all x' and t, $\int_{x'}^{S} g(x,t)\,dx \leq \int_{x'}^{S} f(x,t)\,dx$, with equality for $x' = 0$ (which expresses our desire not to assign more weight than was requested). To make assignments irrevocable, we require that for all x' and $t < t'$, $\int_{x'}^{S} g(x,t')\,dx - \int_{x'}^{S} g(x,t)\,dx \leq \int_{x'}^{S} f(x,t')\,dx - \int_{x'}^{S} f(x,t)\,dx$.

An on-line algorithm in the continuous model is an algorithm which, given $f(x,t)$, outputs $g(x,t)$ such that for all $\tau \in [0,T]$, $g(x,t)$ in the region $[0,S] \times [0,\tau]$ is independent of $f(x,t)$ outside that region. The cumulative assignment is $g(x) = g(x,T)$. An offline assignment is a function $g(x)$ that satisfies the conditions for representing an assignment at time T (those conditions relating assignments at different times notwithstanding). The definitions of *load* and *cost* are identical to those in the semi-continuous model.

Lemma 8 (Optimum Lemma—Continuous Model).

$$OPT = \sup_{x' \in (0,S]} \left\{ \frac{1}{x'} \int_0^{x'} f(x,T)\,dx \right\} . \tag{5}$$

While the continuous model is an interesting construction in its own right, we focus here on the aspects relevant to the discrete problem. Assume $S = T = 1$. A *right-to-left* request function is one that satisfies $f(x,t) = 0$ for $t < 1-x$, and $f(x,t) = f(x,1-x)$ for $t > 1-x$. Thus, $f(x,t)$ is completely defined by specifying $f(x,1-x)$ for all $x \in [0,1]$. We abbreviate and use $f(x)$. Any online assignment generated in response to a right-to-left request function clearly satisfies $g(x,t) = g(x,1-x)$ for $t > 1-x$. We extend the definition of right-to-left request functions to cases other than $S = T = 1$ in the obvious way.

Consider a right-to-left request function $f(x)$ and the corresponding assignment $g(x)$ generated by some k-competitive on-line algorithm. We wish to bound the value of $g(x)$ at some point a. Since the interpretation of g is only by means of its integral, we may assume without loss of generality that g is continuous from the left. Define a new request function f_a by: $f_a(x) = f(x)$ for $a \leq x \leq S$, and $f_a(x) = 0$ elsewhere. Define, for $b > a$, $h_a(b) = \frac{1}{b}\int_a^b f(x)\,dx$, and $h(a) = \sup_{a<b\leq S}\{h_a(b)\}$. Then $h(a) = OPT$ with

respect to f_a. (Note the analogy with $h_{i,j}$ and h_i in the discrete model.) We denote $W = \int_0^S f(x)\,dx$ and $W' = \int_0^S h(a)\,da$. The value of g in $[a,1]$ must be the same for f and f_a, as g is produced by an on-line algorithm, thus $g(a) \le kh(a)$. Hence,

$$W = \int_0^S f(x)\,dx = \int_0^S g(x)\,dx \le k \int_0^S h(a)\,da = kW' \ , \tag{6}$$

from which the lower bound W/W' is readily obtained.

Claim. A lower bound of e can be obtained with our method by considering the request function $e^{-kx^{1/k}}$ in the limit $k \to \infty$ and $S \to \infty$.

Theorem 9. *The lower bounds obtained by our method in the continuous model are valid in all discrete models as well, even in the presence of randomization with respect to oblivious adversaries.*

4 Temporary Jobs

In this section we allow temporary jobs in the input. In Sect. 2 we saw an e-competitive algorithm for the fractional model; here we present an algorithm for the integral model which is 4-competitive in the unweighted case and 5-competitive in the weighted case. In the full paper we also show a lower bound of 3 for the unweighted integral model.

Recall the definition of H in the Optimum Lemma. Consider the jobs which are active upon job j's arrival (including job j). Let $H(j)$ be the value of H defined with respect to these jobs. A server is *saturated* on the arrival of job j if its load is at least $t \cdot H(j)$, where t is a constant to be determined below.

Algorithm PushRight

Assign each job it to its rightmost unsaturated eligible server.

Proposition 10. *If $t \ge 4$, then whenever a job arrives, at least one of its eligible servers is unsaturated. Thus, by taking $t = 4$ we get an algorithm which is 4 and 5-competitive for the unweighted and weighted models, respectively.*

References

[1] S. Albers. Better bounds for online scheduling. *Proc. 29th ACM Symp. on Theory of Computing*, pp. 130–139, 1997.

[2] J. Aspnes, Y. Azar, A. Fiat, S. Plotkin, and O. Waartz. On-line machine scheduling with applications to load balancing and virtual circuit routing. *Journal of the ACM*, 44:486–504, 1997.

[3] Y. Azar. On-line load balancing. *On-line Algorithms: The State of the Art*, Eds. A. Fiat and G. Woeginger, LNCS 1442, pp. 178–195, Springer, 1998.

[4] Y. Azar, A. Broder, and A. Karlin. On-line load balancing. *Proc. 33rd IEEE Symp. on Foundations of Computer Science*, pp. 218–225, 1992.

[5] Y. Azar, B. Kalyanasundaram, S. Plotkin, K. Pruhs, and O. Waarts. On-line load balancing of temporary tasks. *Journal of Algorithms,* Vol. 22, pp. 93–110, 1997.

[6] Y. Azar, J. Naor, and R. Rom. The competitiveness of on-line assignments. *Journal of Algorithms,* Vol. 18, pp. 221–237, 1995.

[7] Y. Bartal, A. Fiat, H. Karloff, and R. Vohra. New algorithms for an ancient scheduling problem. *Proc. 24th ACM Symp. on Theory of Computing,* pp. 51–58, 1992.

[8] Y. Bartal, H.J. Karloff, and Y. Rabani. A Better Lower Bound for On-line Scheduling. *Information Processing Letters,* 50:113–116, 1994.

[9] P. Berman, M. Charikar, and M. Karpinski. On-line load balancing for related machines. *5th International Workshop on Algorithms and Data Structures '97,* LNCS 1272, pp. 116–125, Springer, 1997.

[10] R. Graham. Bounds for certain multiprocessor anomalies. *Bell System Technical Journal,* Vol. 45, pp. 1563–1581, 1966.

[11] D. R. Karger, S. J. Phillips, and E. Torng. A better algorithm for an ancient scheduling problem. *Proc. 5th ACM-SIAM Symp. on Discrete Algorithms,* pp. 132–140, 1994.

[12] J. Sgall. On-line scheduling. *On-line Algorithms: The State of the Art,* Eds. A. Fiat and G. Woeginger, LNCS 1442, pp. 196–231, Springer, 1998.

Provably Good and Practical Strategies for Non-uniform Data Management in Networks*

Friedhelm Meyer auf der Heide[1], Berthold Vöcking[2], and Matthias Westermann[1]

[1] Department of Mathematics and Computer Science, and Heinz Nixdorf Institute,
University of Paderborn,
D-33102 Paderborn, Germany
{fmadh,marsu}@uni-paderborn.de

[2] International Computer Science Institute,
Berkeley, CA 94709, USA
voecking@icsi.berkeley.edu

Abstract. This paper deals with the on-line allocation of shared data objects to the local memory modules of the nodes in a network. We assume that the data is organized in indivisible objects such as files, pages, or global variables. The data objects can be replicated and discarded over time in order to minimize the communication load for read and write accesses done by the nodes in the network. Non-uniform data management is characterized by a different communication load for accesses to small pieces of the data objects and migrations of whole data objects.

We introduce on-line algorithms that minimize the congestion, i.e., the maximum communication load over all links. Our algorithms are evaluated in a competitive analysis comparing the congestion produced by an on-line algorithm with the congestion produced by an optimal off-line algorithm. We present the first deterministic and distributed algorithm that achieves a constant competitive ratio on trees. Our algorithm minimizes not only the congestion but minimizes simultaneously the load on each individual edge up to a optimal factor of 3.

Algorithms for trees are of special interest as they can be used as a subroutine in algorithms for other networks. For example, using our tree algorithm as a subroutine in the recently introduced "access tree strategy" yields an algorithm that is $O(d \cdot \log n)$-competitive for d-dimensional meshes with n nodes. This competitive ratio is known to be optimal for meshes of constant dimension.

1 Introduction

Large parallel and distributed systems – such as massively parallel processor systems (MPPs), networks of workstations (NOWs), or the Internet – consist of a set of nodes each having its own local memory module. In this paper, we consider the problem of

* F. Meyer auf der Heide and M. Westermann are supported in part by DFG-Sonderforschungsbereich 376 and EU ESPRIT Long Term Research Project 20244 (ALCOM-IT). B. Vöcking is supported by a grant of the "Gemeinsames Hochschulsonderprogramm III von Bund und Ländern" through the DAAD.

managing shared data that is read and written from the nodes in the network. Usually, the data is organized in blocks, which we call data objects. The objects are, e.g., files on a distributed file server, pages in a virtual shared memory system, or global variables in a parallel program. The communication overhead for accessing the shared data can be reduced by storing copies of the data objects in the local storage of some processors. However, the data of an object should be kept together in order to reduce the bookkeeping overhead.

Clearly, creating many copies reduces the communication overhead for read accesses. However, it increases the overhead for maintaining the copies consistent when a write access is issued. A data management strategy has to answer the following questions.

- How many copies of an object should be made?
- On which nodes should these copies be placed?
- How should read and write requests be served?

Typically, words or other small blocks of data that can be accessed or updated over the communication links are much smaller than the data objects. For example, large files usually consist of several small records, and pages of virtual memory consist of cache lines that can be accessed and updated individually. Thus, moving a data object can be much more expensive than accessing only a small piece of data from the object.

The file allocation problem (FAP) is an abstract formulation of this non-uniform data management problem, which was introduced by Bartal et al. in [1]. The algorithms for FAP are evaluated in a competitive analysis that compares the communication cost of an on-line algorithm with the cost of an optimal off-line algorithm. For a given application A, let $C_{\text{opt}}(A)$ denote the minimum cost expended by an optimal off-line strategy. A deterministic strategy is said to be *c-competitive* if it expends cost of at most $c \cdot C_{\text{opt}}(A)$, for any application A.

If the on-line algorithm uses randomization one has to describe the power given to the adversary more precisely. This is studied intensively in [2]. We always assume that the adversary is *oblivious*, i.e., the adversary is assumed to specify the whole sequence (e.g., in advance) without knowing the random bits of the on-line algorithm. A randomized strategy is said to be *c-competitive* if it expends expected cost of at most $c \cdot C_{\text{opt}}(A)$, for any application A.

1.1 Formal definition of FAP

We are given a weighted undirected graph $G = (V, E)$, where each node represents a processor with local memory module, the edges represent communication links, and the edge weights represent the bandwidths of these links. For $e \in E$, let $b(e)$ denote the bandwidth of e. Let X denote the set of data objects. At any time, for every object $x \in X$, let $R(x) \subseteq V$, the residence set, represent the set of nodes that hold a copy of x. We always require $R(x) \neq \emptyset$. Initially, only a single node contains a copy of x, this node is known by all nodes in the system.

As time goes on, read and write requests occur at the processors. The requests are assumed to be generated by an adversary. The adversary initiates a sequence of requests

$\sigma = \sigma_1\sigma_2\cdots$, where σ_i corresponds to a read or write request issued by one of the nodes. Initially, we assume that the on-line algorithm has to serve these requests one after the other, that is, it is assumed that σ_{i+1} is issued not before the reallocation for σ_i is finished. At the end of this paper, in Section 5, we will show that all of the algorithms that we have developed in the sequential FAP model are able to handle parallel and overlapping requests, too.

A read request to an object x requires access to a copy of x, a write request requires updating all copies of x. After a request is served, the on-line algorithm can decide how to reallocate the multiple copies of x. Any communication that proceeds along an edge increases the *communication load* of that edge by some amount, which depends on whether a read, write, or migration operation is performed. The increase is defined as follows.

- Read operation: A read request for x issued by a node v can be served by any processor u holding a copy of x. A path has to be allocated through the network from v to u. The communication load on each edge e on this path increases by $1/b(e)$.
- Write operation: A write request for x issued by a node v requires to update all copies of x. A multicast tree connecting v with all nodes holding a copy of x, i.e., a Steiner tree, has to be allocated. The communication load on each edge e in the Steiner tree increases by $1/b(e)$.
- Object Migration: The algorithm can replicate a copy of x from one node to another along an arbitrary path. The communication load on each edge e on this path increases by $D(x)/b(e)$, where $D(x) \geq 1$ is assumed to be an arbitrary integer representing the ratio between the load induced by the migration of x and the load induced by accessing only a unit of data of x.

Efficient algorithms for distributed data management have to work in a distributed fashion. In particular, the processors do not have knowledge about the global state of the system, that is, each processor notices only the read and write accesses and the copy migrations that pass the node. In order to accumulate additional knowledge a processor has to communicate with other processors, which also increases the load on the involved edges.

- Exchange of Information: Information about the global state of the system, e.g. the actual residence set, can be exchanged by sending messages along a path from one node to another. It is assumed that the messages have small size, e.g., they include an ident-number of a data object and a tag for an action that should be performed on the receiving node. The communication load on each edge e on this path increases by $1/b(e)$.

The original formulation of FAP does not consider communication overhead for the exchange of information between processors. Apart from that, it considers only a very simple cost measure, the *total communication load*, i.e., the sum of the load over all edges in the network. Additionally, we investigate the *congestion*, i.e., the maximum communication load over all links. We believe that the congestion is of special interest for practical algorithms as it prevents that some of the links become bottlenecks.

1.2 Previous and related work

Data management algorithms for trees are of special interest since many networks have a tree-like topology, e.g., Ethernet-connected NOWs. Besides algorithms for trees can be used as a subroutine for data management strategies for other networks. Therefore, much research deals with data management on trees. For example, Bartal et al. [1] describe distributed strategies for FAP on trees that aim to minimize the total communication load. They introduce a randomized 3-competitive algorithm and a deterministic 9-competitive algorithm. Both competitive ratios are with respect to the total communication load.

Lund et al. [6] describe a 3-competitive deterministic but centralized strategy for the same problem. The algorithm makes use of global knowledge about a work function which is influenced by any request issued in the network. Ignoring cost for information exchange, the algorithm minimizes the load on any edge up to a factor of 3, and, hence, it achieves competitive ratio 3 with respect to the total communication load and the congestion. Unfortunately, this algorithm is inherently centralized.

Data management strategies for trees and meshes that aim to minimize the congestion in a uniform cost model (i.e., $D(x) = 1$) are given by Maggs et al. [7]. The uniform costs simplify the problem significantly. They present a 3-competitive strategy for trees, and an $O(d \cdot \log n)$-competitive strategy for d-dimensional meshes with n nodes. Both competitive ratios are with respect to the congestion in the uniform model. Further, they present strategies for Internet-like clustered networks, and a lower bound of $\Omega(\log n/d)$ on the best possible competitive ratio for data management on meshes.

A lower bound that holds for any network including at least one edge is shown by Black and Sleator [3]. Properly said, they give a lower bound of 3 on the competitive ratio of data migration algorithms on two processors connected by a single edge, which induces that the best possible competitive ratio for the total load and the congestion in any network is 3. Bartal et al. [1] show that the bound holds also for randomized algorithms.

A difficult problem that has to be solved by any distributed data management strategy is the *data tracking*, i.e., the problem of how to locate the copies of a particular object. To our knowledge, data tracking mechanisms that aim to minimize the congestion have not been investigated previously. Some results (see, e.g., [1]) are known for the total communication load.

1.3 Our results

At first, we describe a strategy for FAP on two nodes connected by a single edge. This strategy is 3-competitive, and it can be computed in a distributed fashion by the two connected nodes. The key feature of the edge strategy is that it can be extended to work on trees just by simulating it on any edge in the tree. The result of this approach is a simple, deterministic, and distributed strategy for FAP on trees which minimizes the load on any edge up to a factor of 3. This result is optimal because of the lower bound in [3]. Obviously, the bound on the load of any edge induces that our tree strategy is 3-competitive with respect to the total communication load and the congestion simultaneously.

Further, we present a distributed strategy for FAP on meshes. We use a variant of the locality preserving embedding of so-called "access trees" introduced in [7], and simulate our tree strategies on the access trees. We show that this simulation approach yields a data management strategy having competitive ratio $O(d \cdot \log n)$ with respect to the congestion, where d denotes the dimension and n the number of nodes of the mesh. The strategy is randomized, and the bound on the congestion does not only hold for the expectation but holds *with high probability (w.h.p.)*, i.e., with probability $1 - n^{-\alpha}$, where α is an arbitrary constant. The lower bound presented in [7] shows that this competitive ratio is optimal for meshes of constant dimension.

The 3-competitive tree strategy can not only be used as a subroutine for data management on meshes but also on other networks, e.g., Internet-like clustered network as defined in [7]. Plugging in our tree strategy for FAP as a subroutine in the data management strategy on clustered networks instead of the tree strategy used in [7] yields close-to-optimal congestion for FAP on clustered networks. In fact, all bounds shown there for data management with uniform migration cost can be extended immediately to FAP with non-uniform cost.

At the end of this paper, we show that all presented bounds hold also for "data-race free programs" including overlapping and parallel requests, which illustrates the practical usage on real world computer networks. The major restriction of data-race free programs is that parallel write accesses to the same data object have to be protected by synchronization mechanisms like barriers or locks. In fact, all of our strategies can be used even in a fully asynchronous setting allowing arbitrary overlappings of read and write accesses, only the theoretical model breaks down if applications with data-race conditions are used. Practical experiments for uniform variants (i.e., D(x) = 1) of the presented strategies for meshes have been implemented in the DIVA (Distributed Variables) library [5]. Results of an experimental evaluation showing that these strategies are very competitive also in practice can be found in [4, 9].

2 File allocation on a single edge

We describe a deterministic and distributed file allocation strategy for a single edge e connecting two nodes a and b. The strategy uses a simple counting mechanism which records read and write accesses that are issued on the two nodes. Later on we will use this strategy for building our tree strategy. Then the nodes a and b correspond to the two connected components in which the tree is divided if e is removed.

Initially, we do not care about how the processors a and b exchange information about the residence set, and how the counters are distributed among them. We assume that node a always knows whether or not node b holds a copy and vice versa. Afterwards we show how our strategy can be adapted to the distributed setting.

2.1 The centralized edge strategy

Each object x is handled independently from the other objects. Let us fix an object x. Define $D = D(x)$. Concerning this object there are two counters c_a and c_b. Informally, these counters represent saving accounts for cost referring to x.

- Node a issues a read request for object x:
  If $c_a < D$ then $c_a := c_a + 1$;
  If $c_a = D$ and a holds no copy of x then
  - Move a new copy of x onto a;
  - If $c_b = 0$ then delete the copy of x on b;
- Node a issues a write request for object x:
  If $c_b > 0$ then
  - $c_b := c_b - 1$;

  else
  - If $c_a < D$ then $c_a := c_a + 1$;

  If $c_a = D$ and a holds no copy of x then move a new copy of x onto a;
  If $c_b = 0$ and a holds a copy of x then delete the copy of x on b;

Fig. 1. The edge strategy for requests issued by node a.

Initially, one copy of x is placed on one of the two nodes. Assume that it is placed on a. Then c_a is set to D, and c_b is set to 0. In Fig. 1 the *edge strategy* is described for requests issued by node a. The strategy works analogously for requests issued by node b. Note that the edge strategy always keeps one copy of x, since it only deletes a copy on a node if the other node also holds a copy.

Lemma 1. *The centralized edge strategy minimizes the load up to a factor of 3.*

Proof. We use a potential function argument (cf. [8]). First, let us fix an optimal off-line strategy, which is denoted the *optimal strategy* in the following. We assume that the optimal strategy, in contrast to the on-line strategy, reallocates its residence set before serving a request. W.l.o.g., the optimal strategy fulfills the following properties.

- If a node v issues a read request to x, then the optimal strategy does not delete a copy, that is, the only possible change of the residence set is that a new copy of x is moved to v.
- If a node v issues a write request to x, then the only possible changes of the residence set are that a new copy of x is moved to v and/or a copy of x is deleted on the neighbor of v.

Fix a sequence of read and write requests $\sigma = \sigma_1\sigma_2\cdots$. Let $L_{\text{edge}}(t)$ and $L_{\text{opt}}(t)$ denote the load of the edge strategy and the optimal strategy, respectively, after serving σ_t, and let $\Phi(t)$ denote the value of a potential function after serving σ_t, which is defined in detail later. In order to prove the lemma, we have to show that

(a) $L_{\text{edge}}(t) + \Phi(t) \leq 3 \cdot L_{\text{opt}}(t)$ and
(b) $\Phi(t) \geq 0$.

Let $c_a(t)$ and $c_b(t)$ denote the value of the counter c_a and the value of the counter c_b, respectively, after serving σ_t, and let $R_{\text{edge}}(t)$ and $R_{\text{opt}}(t)$ denote the residence set of the edge strategy and the optimal strategy, respectively, after serving σ_t. We define

$$\Phi(t) = \Phi_a(t) + \Phi_b(t) - D \ ,$$

$a \in R_{\text{edge}}(t)$	$a \in R_{\text{opt}}(t)$	$a \in R_{\text{edge}}(t+1)$	$a \in R_{\text{opt}}(t+1)$	ΔL_{edge}	$\Delta\Phi = \Delta\Phi_a \leq$	ΔL_{opt}
no	no	no	no	1	2	1
no	no	no	yes	1	$3D-1$	D
no	yes	no	yes	1	-1	0
no	no	yes	no	$1+D$	$2-D$	1
no	no	yes	yes	$1+D$	$2-D$	D
no	yes	yes	yes	$1+D$	$-1-D$	0
yes	no	yes	no	0	1	1
yes	no	yes	yes	0	$3D-2$	D
yes	yes	yes	yes	0	0	0

Table 1. Possible changes of configuration if node a issues a read request.

where, for $v \in \{a,b\}$,

$$\Phi_v(t) = \begin{cases} 2c_v(t), & \text{if } v \notin R_{\text{edge}}(t) \text{ and } v \notin R_{\text{opt}}(t), \\ 3D - c_v(t), & \text{if } v \notin R_{\text{edge}}(t) \text{ and } v \in R_{\text{opt}}(t), \\ 3D - 2c_v(t), & \text{if } v \in R_{\text{edge}}(t) \text{ and } v \in R_{\text{opt}}(t), \\ c_v(t), & \text{if } v \in R_{\text{edge}}(t) \text{ and } v \notin R_{\text{opt}}(t). \end{cases}$$

First, we prove (b). The optimal strategy always holds a copy of x on a node. Hence, $v \in R_{\text{opt}}(t)$, for some node $v \in \{a,b\}$, and, consequently, $\Phi_v(t) \geq D$, which implies $\Phi_a(t) + \Phi_b(t) \geq D$. Hence, (b) is shown.

Now we prove (a) by induction on the length of σ. Obviously, (a) holds for the initial setting. For the induction step suppose that $L_{\text{edge}}(t) + \Phi(t) \leq 3 \cdot L_{\text{opt}}(t)$. Let $\Delta L_{\text{edge}} = L_{\text{edge}}(t+1) - L_{\text{edge}}(t)$, $\Delta L_{\text{opt}} = L_{\text{opt}}(t+1) - L_{\text{opt}}(t)$, $\Delta\Phi_a = \Phi_a(t+1) - \Phi_a(t)$, and $\Delta\Phi_b = \Phi_b(t+1) - \Phi_b(t)$. In order to prove the induction step, we show that

$$\Delta L_{\text{edge}} + \Delta\Phi_a + \Delta\Phi_b \leq 3 \cdot \Delta L_{\text{opt}}. \tag{1}$$

We distinguish between read and write requests.

- σ_{t+1} is a read request issued by node a. In this case equation (1) can be checked with Table 1 containing all possible changes of configuration. Note that, if a issues a read request, the only possible changes of the residence sets are that one of the strategies moves a copy to a, or $c_b = 0$ and the edge strategy deletes the copy on b. In both cases, $\Delta\Phi_b = 0$, and, hence, $\Delta\Phi = \Delta\Phi_a$.
- σ_{t+1} is a write request from node a. We distinguish between the cases $c_b(t) > 0$ and $c_b(t) = 0$. Note that, if a issues a write request, the only possible changes of the residence set of the optimal strategy is that a new copy of x is moved to a and/or a copy of x is deleted on b.
  - Suppose $c_b(t) > 0$. In this case equation (1) can be checked with Table 2 containing all possible changes of configuration. Note that, if a issues a write request and $c_b(t) > 0$, the only possible transitions of the edge strategy are from $\{a\}$ to $\{a\}$, from $\{b\}$ to $\{b\}$, or from $\{a,b\}$ to $\{a,b\}$ or $\{a\}$.

$R_{edge}(t)$	$R_{opt}(t)$	$R_{edge}(t+1)$	$R_{opt}(t+1)$	ΔL_{edge}	$\Delta\Phi_a \leq$	$\Delta\Phi_b \leq$	ΔL_{opt}
a	a	a	a	0	0	−2	0
a	a,b	a	a,b	0	0	1	1
a	a,b	a	a	0	0	−2	0
a	b	a	b	0	0	1	1
a	b	a	a,b	0	$3D$	1	$1+D$
a	b	a	a	0	$3D$	−2	D
b	a	b	a	1	0	−1	0
b	a,b	b	a,b	1	0	2	1
b	a,b	b	a	1	0	−1	0
b	b	b	b	1	0	2	1
b	b	b	a,b	1	$3D$	2	$1+D$
b	b	b	a	1	$3D$	−1	D
a,b	a	a,b	a	1	0	−1	0
a,b	a,b	a,b	a,b	1	0	2	1
a,b	a,b	a,b	a	1	0	−1	0
a,b	b	a,b	b	1	0	2	1
a,b	b	a,b	a,b	1	$3D$	2	$1+D$
a,b	b	a,b	a	1	$3D$	−1	D
a,b	a	a	a	1	0	−1	0
a,b	a,b	a	a,b	1	0	2	1
a,b	a,b	a	a	1	0	$2-3D$	0
a,b	b	a	b	1	0	2	1
a,b	b	a	a,b	1	$3D$	2	$1+D$
a,b	b	a	a	1	$3D$	$2-3D$	D

Table 2. Possible changes of configuration if node a issues a write request and $c_b(t) > 0$.

- Suppose $c_b(t) = 0$. In this case equation (1) can be checked with Table 3 containing all possible changes of configuration. Note that, if a issues a write request and $c_b(t) = 0$, the only possible transitions of the edge strategy are from $\{a\}$ to $\{a\}$ or from $\{b\}$ to $\{b\}$ or $\{a\}$.

2.2 The distributed edge strategy

Next, we describe how the edge strategy can be adapted from the centralized to the distributed setting. Each node keeps always the current value of both counters c_a and c_b. It is obvious that this assignment makes it possible for the nodes to make the right decisions according to the edge strategy. Now we specify how each node keeps track of both counters c_a and c_b. W.l.o.g., consider node a.

- A read request for x is issued by b. If b holds no copy of x then a is able to update its counters. In the other case, b sends an information message along e if and only if b has increased its counter c_b.

$R_{\text{edge}}(t)$	$R_{\text{opt}}(t)$	$R_{\text{edge}}(t+1)$	$R_{\text{opt}}(t+1)$	ΔL_{edge}	$\Delta\Phi_a \leq$	$\Delta\Phi_b \leq$	ΔL_{opt}
a	a	a	a	0	0	0	0
a	a,b	a	a,b	0	0	0	1
a	a,b	a	a	0	0	$-3D$	0
a	b	a	b	0	1	0	1
a	b	a	a,b	0	1	0	$1+D$
a	b	a	a	0	1	$-3D$	D
b	a	b	a	1	-1	0	0
b	a,b	b	a,b	1	-1	0	1
b	a,b	b	a	1	-1	$-3D$	0
b	b	b	b	1	2	0	1
b	b	b	a,b	1	$3D-1$	0	$1+D$
b	b	b	a	1	$3D-1$	$-3D$	D
b	a	a	a	$1+D$	$-1-D$	0	0
b	a,b	a	a,b	$1+D$	$-1-D$	0	1
b	a,b	a	a	$1+D$	$-1-D$	$-3D$	0
b	b	a	b	$1+D$	$2-D$	0	1
b	b	a	a,b	$1+D$	$2-D$	0	$1+D$
b	b	a	a	$1+D$	$2-D$	$-3D$	D

Table 3. Possible changes of configuration if node a issues a write request and $c_b(t) = 0$.

- A write request for x is issued by b. If a holds a copy of x then a is able to update its counters. In the other case, b sends an information message along e if and only if b has decreased its counter c_a or increased its counter c_b.

In this way, each node is able to keep both of its counters up-to-date. The following lemma shows that the additional information messages are sent very rarely.

Lemma 2. *The distributed edge strategy minimizes the load up to a factor of 3.*

Proof. We adopt the notations and definitions of Lemma 1. An information message is only sent if a request changes the value of a counter. Thus, the tables used for proving Lemma 1 change only slightly.

- σ_{t+1} is a read request issued by node a. Then an additional message is sent only if $a \in R_{\text{edge}}(T)$. In this case, ΔL_{edge} equals 1 rather than 0. Besides, $\Delta\Phi$ equals -2 rather than 0 if $a \in R_{\text{opt}}(t)$ and $c_a(t) < D$. It is easy to check that equation (1) is still satisfied by applying these changes to Table 1.
- σ_{t+1} is a write request issued by node a. Then an additional message is sent only if $b \notin R_{\text{edge}}(T)$. In this case, $\Delta L_{\text{edge}} = 1$ rather than 0. Further, $\Delta\Phi_a$ equals -2 rather than 0 if $a \in R_{\text{opt}}(t)$ and $c_a(t) < D$. It is easy to check that equation (1) is still satisfied by applying these changes to the Tables 2 and 3.

3 File allocation on trees

The distributed edge strategy can be extended to tree-connected networks. The network is modeled by a graph $T = (V,E)$ without cycles, i.e., a tree. The edges in the tree are allowed to have arbitrary bandwidths.

The *tree strategy* is composed out of $|E|$ individual edge strategies. The idea is to simulate the distributed edge strategy on each edge. Consider an arbitrary edge $e = (a,b)$. The removal of an edge e divides T into two subtrees T_a and T_b, containing a and b, respectively. The two nodes a and b execute the algorithm described in Fig. 1. The phrases "if a holds a (no) copy of x" and "node a issues a read (write) request for object x" are just replaced by "if a node in T_a holds a (no) copy of x" and "a node in T_a issues a read (write) request for object x", respectively.

The simulation works properly as long as the nodes in the residence set $R(x)$ build a connected component in the tree. A key feature of our edge strategy is that it fulfills this condition. This is shown in the following lemma.

Lemma 3. *The graph induced by the residence set $R(x)$ is always a connected component.*

Proof. (Sketch) Via induction on the length of the sequence of requests it can be shown that those counters on any simple path in the tree that are responsible for moving a copy along an edge towards the first node of the path are non-decreasing from the first to the last node on the path. Hence, all these counters "agree" about the distribution of copies. This ensures that all copies stay in a connected component.

Note that the tree algorithm also does not need any additional information exchange apart from the one done by the distributed edge strategy. Therefore, the following theorem follows immediately from Lemma 2.

Theorem 4. *The tree strategy minimizes the load on any edge up to a factor of 3.*

4 File allocation on meshes

In this section, we consider strategies for the mesh $M = M(m_1,\ldots,m_d)$, i.e., the d-dimensional mesh-connected network with side length $m_i \geq 2$ in dimension i. The number of processors is denoted by n, i.e., $n = m_1 \cdots m_d$. Each edge is assumed to have bandwidth 1.

The strategy uses a locality preserving embedding of "access trees" introduced in [7]. It is based on a hierarchical decomposition of M, which we describe recursively. Let i be the smallest index such that $m_i = \max\{m_1,\ldots,m_d\}$. If $m_i = 1$ then we have reached the end of the recursion. Otherwise, we partition M into two non-overlapping submeshes $M_1 = M(m_1,\ldots,\lceil m_i/2\rceil,\ldots,m_d)$ and $M_2 = M(m_1,\ldots,\lfloor m_i/2\rfloor,\ldots,m_d)$. M_1 and M_2 are then decomposed recursively according to the same rules.

The hierarchical decomposition has associated with it a *decomposition tree* $T(M)$, in which each node corresponds to one of the submeshes, i.e., the root of $T(M)$ corresponds to M itself, and the children of a node v in the tree correspond to the two submeshes into which the submesh corresponding to v is divided. Thus, $T(M)$ is a binary tree of height $O(\log n)$ in which the leaves correspond to submeshes of size one, i.e., to the processors of M. For each node v in $T(M)$, let $M(v)$ denote the corresponding submesh.

For each object $x \in X$, define an *access tree* $T_x(M)$ to be a copy of the decomposition tree $T(M)$. We embed the access trees randomly into M, i.e., for each $x \in X$, each interior node v of $T_x(M)$ is mapped uniformly at random to one of the processors in $M(v)$, and each leaf v of $T_x(M)$ is mapped onto the only processor in $M(v)$.

The remaining description of our data management strategy is very simple: For object $x \in X$, we simulate the tree strategy on the access tree $T_x(M)$. All messages that should be sent between neighboring nodes in the access trees are sent along the *dimension-by-dimension order paths* between the associated nodes in the mesh, i.e., the unique shortest path between the two nodes using first edges of dimension 1, then edges of dimension 2, and so on.

The access tree nodes have to be remapped dynamically when too many *access messages*, i.e., messages that simulate messages of the tree strategy, traverse a node. The remapping is done as follows. For every object x, and every node v of the access tree $T_x(M)$ we add a counter $\tau(x,v)$. Initially, this counter is set to 0, and the counter $\tau(x,v)$ is increased by 1 whenever an access message for object x traverses node v, starts at node v, or arrives at node v. When the counter $\tau(x,v)$ reaches K the node v is remapped randomly to another node in $M(v)$, where K is some integer of suitable size, i.e., $K = O(D(x))$. Remapping v to a new host means that we have to send a *migration message* that informs the new host about the migration and, if the old host holds a copy of x, moves the copy to the new host. Migration messages reset the counter $\tau(x,v)$ to 0. Furthermore, we have to send *notification messages* including information about the new host to the mesh nodes that hold the access tree neighbors of v. These notification messages also increase the counters at their destination nodes. The counter mechanism ensures that the load due to messages that are directed to an access tree node embedded on a randomly selected host is $O(K) = O(D(x))$.

The following theorem gives the competitive ratio of the access tree strategy for d-dimensional meshes with n nodes. It can be obtained from the analysis in [7].

Theorem 5. *The access tree strategy is $O(d \cdot \log n)$-competitive with respect to the congestion, w.h.p., for meshes of dimension d with n nodes.*

5 Extending the results to data-race free applications

An important class that allows concurrent read accesses is the class of *data–race free applications*, which is defined as follows. We assume that an adversary specifies a parallel application running on the nodes of the network, i.e., the adversary initiates read and write requests on the nodes of the network. A write access to an object is not allowed to overlap with other accesses to the same object, and there is some order among the accesses to the same object such that, for each read and write access, there is a unique least recent write. Note that this still allows arbitrary concurrent accesses to different objects and concurrent read accesses to the same object. An execution using a dynamic data management strategy is called *consistent* if it ensures that a read request directed to an object always returns the value of the most recent write access to the same object.

A data management strategy is allowed to migrate, create, and invalidate copies of an object during execution time. We use the same cost metric as defined in the Introduction, that is, any message except for migration messages increases the load of an edge e by $1/b(e)$. Migration messages of an object x increase the load by $D(x)/b(e)$.

We have to describe how parallel accesses are handled by the presented strategy such that the execution is consistent. On trees this works as follows. Since we consider only data-race free applications, write accesses do not overlap with other accesses. Overlapping read accesses are handled in the following way. Consider a request message M arriving on a node u that does not hold a copy of x. Let e denote the next edge on the path to the nearest copy. Suppose another request message M' directed to x has been sent already along e but a data message has not yet been sent back. Then the request message M is blocked on node u until the data message corresponding to M' passes e. When this message arrives, either a new copy is created on node u, and u serves the request message M, or M continues its path to the connected component of copies. As meshes and clustered networks simulate the tree strategy on access trees that are embedded in the network, they can follow the same approach.

We can conclude that all competitive ratios given in this paper hold also for data-race free applications, which indicates that the introduced strategies are well suited for practical usage.

References

[1] Y. Bartal, A. Fiat, and Y. Rabani. Competitive algorithms for distributed data management. In *Proc. of the 24th ACM Symp. on Theory of Computing (STOC)*, pages 39–50, 1992.

[2] S. Ben-David, A. Borodin, R. M. Karp, G. Tardos, and A. Widgerson. On the power of randomization in online algorithms. In *Proc. of the 22th ACM Symp. on Theory of Computing (STOC)*, pages 386–379, 1992.

[3] D. L. Black and D. D. Sleator. Competitive algorithms for replication and migration problems. Technical Report CMU–CS–89–201, Department of Computer Science, Carnegie–Mellon University, 1989.

[4] C. Krick, F. Meyer auf der Heide, H. Räcke, B. Vöcking, and M. Westermann. Data management in networks: Experimental evaluation of a provably good strategy. In *Proc. of the 11th ACM Symp. on Parallel Algorithms and Architectures (SPAA)*, 1999.

[5] C. Krick, H. Räcke, B. Vöcking, and M. Westermann. *The DIVA (Distributed Variables) Library*. University of Paderborn, www.uni-paderborn.de/sfb376/a2/diva.html, 1998.

[6] C. Lund, N. Reingold, J. Westbrook, and D. Yan. On-line distributed data management. In *Proc. of the 2nd European Symposium on Algorithms (ESA)*, 1996.

[7] B. M. Maggs, F. Meyer auf der Heide, B. Vöcking, and M. Westermann. Exploiting locality for networks of limited bandwidth. In *Proc. of the 38th IEEE Symp. on Foundations of Computer Science (FOCS)*, pages 284–293, 1997.

[8] D. D. Sleator and R. E. Tarjan. Amortized efficiency of list update and paging rules. *Communication of the ACM*, 28(2):202–208, 1985.

[9] B. Vöcking. *Static and Dynamic Data Management in Networks*. PhD thesis, University of Paderborn, Germany, 1998.

Approximation Algorithms for Restoration Capacity Planning

Steven J. Phillips and Jeffery R. Westbrook

AT&T Labs–Research
AT&T Shannon Laboratory, 180 Park Ave, Florham Park, NJ 07932, USA.
{phillips,jeffw}@research.att.com

Abstract. A major task of telecommunication network planners is deciding where spare capacity is needed, and how much, so that interrupted traffic may be rerouted in the event of a failure. Planning the spare capacity so as to minimize cost is an NP-hard problem, and for large networks, even the linear relaxation is too large to be solved with existing methods. The main contribution of this paper is a fast algorithm for restoration capacity planning with a proven performance ratio of at most $2+\varepsilon$, and which generates solutions that are at most 1% away from optimal in empirical studies on a range of networks, with up to a few hundred nodes.
As a preliminary step, we present the first $(1+\varepsilon)$-approximation algorithm for restoration capacity planning. The algorithm could be practical for moderate-size networks. It requires the solution of a multicommodity-flow type linear program with $O(m|G|)$ commodities, however, where G is the set of distinct traffic routes, and therefore $O(m^2|G|)$ variables. For many networks of practical interest, this results in programs too large to be handled with current linear programming technology. Our second result, therefore, has greater practical relevance: a $(2+\varepsilon)$-approximation algorithm that requires only the solution of a linear program with $O(m)$ commodities, and hence $O(m^2)$ variables. The linear program has been of manageable size for all practical telecommunications network instances that have arisen in the authors' applications, and we present an implementation of the algorithm and an experimental evaluation showing that it is within 1% of optimal on a range of networks arising practice.
We also consider a more general problem in which both service and restoration routes are computed together. Both approximation algorithms extend to this case, with approximation ratios of $1+\varepsilon$ and $4+\varepsilon$, respectively.

1 Introduction

Modern telecommunications networks are designed to be highly fault tolerant. Customers expect to see uninterrupted service, even in the event of faults such as power outages, equipment failures, natural disasters and cable cuts. Typically networks are engineered to guarantee complete restoration of disrupted services in the event of any single catastrophic failure (such as a fiber cut). For this to be possible, spare capacity must be added to the network so that traffic that has been interrupted by a fault can be rerouted.

Restoration capacity is a sizable fraction of total network capacity, and hence accounts for a large part of the infrastructure cost of telecommunications networks. It is therefore essential for network planners to have efficient and effective algorithms for deciding where restoration capacity is needed, and how much. This is known variously as restoration capacity planning, capacitated survivable network design, resilient capacity reservation, or spare capacity assignment. Typical applications require that capacity be allocated in discrete units, and that traffic flows are indivisible. With these requirements, the problem becomes NP hard [4].

Modern telecommunications networks may involve hundreds of offices and fiber routes. For such large networks, even the linear relaxation of the natural integer program formulations cannot be solved, as it is too large for current linear program solvers. However, advances in linear programming methods and in processor power have recently allowed near-optimal solution of the restoration capacity planning problem for some reasonably large networks. Cwilich *et al.* [3] describe a system based on column generation that exactly solves a linear relaxation of the restoration capacity planning problem, and that has been used in the field. While column generation works very well in practice, and is typically faster than solution of equivalent multicommodity-flow based formulations, it has not been proven to run in polynomial time. In addition, there are existing telecommunications networks that are too large for the column generation method to solve effectively.

The first contribution of this paper is an application of randomized rounding [13] to the multicommodity flow formulation to derive the first polynomial-time $(1+\varepsilon)$-approximation algorithm for the restoration capacity planning problem. This result is of theoretical interest only, since in practice the column generation approach, with suitable rounding, is faster and yields comparable results.

The major contribution of this paper is a new approximation algorithm, LBALG, that is fast and effective on large networks. Cwilich *et al.* [4] describe a linear program that gives a lower bound for the restoration capacity planning problem, and find empirically that the lower bound is extremely close to upper bound found by column generation. We show here that the lower bound is tight within a factor of two, and we use the lower bound as the basis of LBALG. We prove that LBALG is a $(2+\varepsilon)$-approximation algorithm. It much faster both theoretically and in practice than both the flow-based $(1+\varepsilon)$ approximation and the column-generation algorithm, so it is useful for large networks that are beyond the reach of the existing methods. We present an empirical study in which LBALG consistently produces solutions that are within 1% of the lower bound, and hence within 1% of optimal, on a range of networks that have arisen in practice. (Column generation acheives the same quality solutions on the smaller networks in the study.) The running time of LBALG ranged from a few minutes on a network with 58 nodes, to a few hours on a network with 452 nodes.

The third contribution of this paper concerns situations when service routing and restoration capacity may be optimized together. [1] In the standard restoration capacity assignment problem, the service routes are taken as given, but it may be possible to reduce the total network cost if the service and restoration routes are optimized at the same time. We give a polynomial-time $1+\varepsilon$ approximation algorithm for this more gen-

[1] This seems to happen infrequently in practice; service routes are generally optimized according to other measures, such as minimal delay or customer requirements, and then restoration capacity is added separately.

eral problem and show that the $2+\varepsilon$ approximation algorithm for fixed service routes can be combined with shortest-path routing to give a $4+\varepsilon$-approximation algorithm.

1.1 Related Work

Alevras, Grötschel and Wessäly [1] give a good survey of models for restoration capacity planning (which they term capacitated survivable network design). Herzberg, Bye and Utano [8] and Iraschko, MacGregor and Grover [9] give integer programming formulations for various restoration capacity planning problems, based on enumerating all possible restoration paths between demand endpoints. These formulations are turned into practical methods by restricting the set of restoration paths by various heuristics. Cwilich *et al.* [3] substantially improve these methods by the use of LP column generation techniques, which allow intelligent search of the space of restoration paths. No approximation guarantees or estimates of the quality of the solutions are given in any of these papers, though Cwilich *et al.* provably find the optimal solution to the linear relaxation of their formulation. Brightwell, Oriolo and Shepherd [2] study the problem of designing restoration capacity for a single demand, and give a $1+\frac{1}{14}$ approximation algorithm for this problem. Cwilich *et al.* [4] present an empirical comparison of column generation with a heuristic approach that is fast enough for use on large networks, but without performance guarantees. We know of no prior approximation guarantees for the restoration capacity planning problem with multiple demand pairs.

Alevras, Grötschel and Wessäly [1] mention some computational results on solving mixed integer linear programs that model restoration capacity planning problems related to ours. Their largest instance has 17 nodes, 64 edges, and 106 demand pairs. They report "reasonable" solutions in times ranging from a few seconds to several hours. Iraschko, MacGregor and Grover [9] report computational results for their integer programming heuristic, but give no estimate of the absolute quality of the solutions. Their running times vary from minutes to 2.7 days, and their largest test networks are smaller than our smallest test networks.

The restoration capacity planning problem is rather different from purely graph-theoretic network-reliability problems such as finding disjoint paths (see for example Kleinberg's thesis [11]); connectivity augmentation (for example adding a min-cost set of edges to a graph to increase its connectivity [5, 6]); or the generalized Steiner problem (see eg. [10]), in which a subgraph must be found that contains r_{uv} edge-disjoint paths between a collection of required $\{u,v\}$ pairs. (Grötschel, Monma and Stoer [7] survey methods to solve exactly these NP-Hard topological connectivity problems.) The restoration capacity planning problem differs in being capacitated and in having a fixed underlying network and explicit representation of the set of failures and the affected demands.

2 The Restoration Capacity Planning Problem

The input to the restoration capacity planning problem is an undirected network $N=(V,E)$, where V is a set of nodes and E is a set of edges, and a set of routed demands, R. We use $n=|V|$ and $m=|E|$. We assume that N is two-edge connected; otherwise, we work within each 2-edge-connected component. Each edge has an associated cost per unit capacity. Each demand $d \in R$ is described by a pair of terminations in V, a size or required capacity, and a service route, which is a simple path between the terminations. A *failure* is the deletion of a single edge $f \in E$. A failure *affects* a demand if the failing edge occurs in the service route of that demand. The output of a restoration capacity planning algorithm is a set of *restoration routes* for each demand. For each failure that affects the demand, there must be a restoration route that connects the endpoints and bypasses the failure. The same restoration route may be used for different failures, as long as it avoids all the failing edges.

The objective is to minimize the cost of the solution, which is determined by the edge capacities. These are determined in turn from the restoration routes. First, for each failure f, determine the collection of restoration paths that will be used. For each edge e in the network, determine how much capacity, $cap(e, f)$, is needed to carry the protection paths that are in use for the failure f. The required restoration capacity of edge e is the maximum over failures f of $cap(e, f)$. In this way restoration capacity is shared between different failures.

There are two variants to the cost model: in the *total cost* model we seek to minimize the cost of service and restoration capacity combined, while in the *restoration cost* model, we consider only the cost of restoration capacity. This distinction is useful in describing the performance of approximation algorithms, and it represents two extremes in practical applications. When planning restoration capacity for an existing network, it is important to minimize just the restoration capacity. If a network design is to be generated as part of an architectural study, for example to investigate the total impact of a particular architectural decision, then the total cost may be a more suitable metric.

For the applications that inspired this paper, the edges are long-haul optical fiber routes, and the nodes are cities or fiber junctions. Due to advances in optical multiplexing the number of wavelengths that can be transmitted down a fiber optic cable is effectively unlimited, but each transmitted wavelength requires a pair of lasers at either endpoint of the cable and a mileage-dependent number of repeaters along the line. This determines the cost per unit capacity.

Our formulation has several aspects that model actual network restoration techniques. First, the route that a demand follows cannot change unless it is actually interrupted by a failure (this is termed "strong resilience" by Brightwell *et al.* [2]). This precludes arbitrary rearrangements of the traffic after a failure. The latter approach might yield cheaper overall cost, but it is both undesirable to interrupt customer service and hard to do so reliably and safely. Second, a restoration route for a given demand can use as much of its own service capacity as is useful in the restoration route, but it cannot use any other service capacity that may have been freed up by moving other demands from their service routes, for similar reasons of reliability and safety. Finally, the models and algorithms in this paper can be generalized to multiple failures, but in practice such events occur with sufficiently low probability that the cost of building enough capacity to handle them is not justified.

3 A $(1+\varepsilon)$-Approximation Algorithm

This section describes an algorithm, based on linear programming and randomized rounding [13], that produces restoration plans whose cost is provably within a $(1+\varepsilon)$ factor of optimal.

The linear program essentially conjointly solves a collection of multicommodity flows, one per failure. To reduce the total number of commodities, all demands that have the same service route are aggregated together into a *demand group*. Let G denote the resulting set of demand groups. Let F denote the set of failures, which in our case is isomorphic to the set of edges E.

The linear program is shown in Figure 1. For each pair (g, f) where g is a demand group whose service route is affected by the failure f, there is a commodity. Variables $\text{flow}(g, f, u, v)$ and $\text{flow}(g, f, v, u)$, give the flow between u and v in either direction, subject to flow conservation (constraint 4) and the additional constraint that there is no flow on a failed edge (constraint 1). The non-negative variable $\text{cap}(e)$ represents the restoration capacity of edge e, which is the maximum required for any failure (constraint 3). A demand group can use its own service capacity for free. The (arbitrary) choice of source and sink for the endpoints of the demand group determines a direction of service flow.

Constants:

$\text{supply}(g, v)$ Supply for demand group g at node v
$\text{service}(g, u, v)$ Service flow of demand group g from u to v on $e = \{u, v\}$
$\text{cost}(e)$ Unit cost of provisioning capacity on edge e

Variables, all non-negative:

$\text{flow}(g, f, u, v)$ Flow of group g on edge $e = \{u, v\}$ from u to v, under failure f
$\text{restcap}(g, f, u, v)$ Restoration capacity on $e = \{u, v\}$ required for group g under failure f
$\text{cap}(e)$ Total restoration capacity of edge e

Minimize $\sum_e \text{cost}(e) \cdot \text{cap}(e)$ subject to:

$$\text{flow}(g, f, u, v) = \text{flow}(g, f, v, u) = 0 \quad \forall \{u, v\} = f \in F, g \in G \tag{1}$$

$$\text{restcap}(g, f, u, v) \geq \text{flow}(g, f, u, v) - \text{service}(g, u, v) \quad \forall g \in G, f \in F, u, v \mid \{u, v\} \in E \tag{2}$$

$$\text{cap}(e = \{u, v\}) \geq \sum_g (\text{restcap}(g, f, u, v) + \text{restcap}(g, f, v, u)) \quad \forall e \in E, f \in F \tag{3}$$

$$\sum_{x:\{x,v\} \in E} \text{flow}(g, f, x, v) = \sum_{x:\{v,x\} \in E} \text{flow}(g, f, v, x) + \text{supply}(g, v) \quad \forall f \in F, g \in G \tag{4}$$

Fig. 1. The linear program `large-LP`

The non-negativity of variables $\text{restcap}(g, f, u, v)$ enforces the requirement that a given demand cannot use (for restoration) service capacity that has been freed up by other demands that have been rerouted. If this requirement were removed, the linear program could be simplified by eliminating the restcap variables.

After solving the linear program, we perform the following randomized rounding step. The flow is first partitioned between the demands in each demand group in the natural way. Then for each demand we perform a random walk guided by the flow values. Specifically, for a demand d between s and t, let A be the set of edges leaving s. For $e \in A$ let $f(e)$ be the flow of d on e. We choose an edge e with probability $f(e)/\sum_A f(e)$. The random walk is continued from the other end of the chosen edge, until t is reached.

3.1 Analysis

The number of variables in the linear program is $\Theta(m^2|G|)$, since there are m edges, m failures and $|G|$ demand groups. Clearly this is polynomial in the size of the input.

We now analyze the performance of the algorithm. For simplicity we assume that all demands have size 1; the analysis generalizes to the general case. By induction one can show that the probability that the random walk for a demand d crosses an edge e is equal to the flow of d on e. Now consider the indicator variable $\chi(d,e,f)$ that is 1 iff demand d crosses edge e on failure f: for fixed e and f the variables $\chi(d,e,f)$ are independent Bernoulli variables. We use the following Chernoff bound [12].

Lemma 1. *Let $X_1,\ldots,X_n$ be a set of independent Bernoulli variables such that $P[X_i = 1] = p_i$ and $P[X_i = 0] = 1 - p_i$. Let $Y = \sum X_i$, so $E[Y] = \sum p_i$. Then for $\varepsilon \in [0,1]$,*

$$P[|Y - E[Y]| > \varepsilon E[Y]] \leq 2e^{-0.38\varepsilon^2 E[Y]}.$$

Applying this bound gives the following result:

Theorem 2. *Suppose that the service capacity of each edge is at least $\frac{5.3}{\varepsilon^2}\log 2m$. Then with probability at least $1/2$ the cost of the restoration plan produced by the above algorithm is at most $1+\varepsilon$ times that of the optimal restoration plan.*

Proof. For an edge e let cap(e) be the capacity given in the solution of the linear program. Consider the event that after the randomized rounding a particular edge e requires capacity more than $(1-\varepsilon)$cap(e) under a particular failure f. Applying Lemma 1, we have that the probability of this event is at most $1/2m^2$. There are m^2 such bad events, so with probability at least $1/2$ none of them occurs, and the capacity required on each edge after randomized rounding is at most $1-\varepsilon$ times that of the linear program solution.

It is not too unreasonable to have a restriction on the service capacity, since telecommunications network use discrete capacities. For example, in a network of OC48s designed to carry T3s, the minimum capacity of an edge is 48. If the network has 30 edges then we get $\varepsilon = 0.66$, and if the linear program solution has at least 2 OC48s per edge we can use $\varepsilon = 0.47$.

The bound proven in Theorem 2 is a worst-case bound, and it is worth noting that the algorithm does not not depend on the choice of ε, and would perform better than the theoretical bound in practice. Note also that the solution of the linear program provides a lower bound on the cost of the optimal restoration plan. This lower bound is at least as tight as the linear program of Section 4, at the price of requiring the solution of a linear program with a much larger number of variables.

4 A Linear Program Lower Bound

This section describes a linear program that gives a lower bound on the cost of the optimal routing of both service and protection. The linear program was first presented in [4]. The number of variables in the linear program is $O(m^2)$, independent of G, so it is substantially smaller than the LP of Section 3. In the following section, we use this lower-bound LP to construct a solution whose *total* cost is within $(2+\varepsilon)$ of the optimal total cost for a given demand set R.

The intuition behind the lower bound LP is as follows. Choose a particular demand d and its service route p. Suppose we restrict restoration routes so that if p is cut by the failure of edge f, then the restoration route p' must consist of the prefix of p up to the tail of f, some path from the tail to the head of f, and then remaining suffix of p from the head of f onwards. In this restricted setting, a lower bound on the restoration cost can be computed by aggregating all traffic through each possible failure f, and conjointly computing flows from the tail to the head of each f so as to minimize the maximum over flows of the cost. To extend this to the general case, we do not charge for flow that is traveling along service paths in the reverse direction.

The solution to the linear program is only a lower bound on required restoration capacity, because the aggregation of demands for each failure allows feasible solutions that do not correspond to any set of restoration routes, and because the solution may be fractional.

The lower bound linear program `lowerbound-LP` is shown in Figure 2. There is a commodity for each edge f in the network, which represents the service traffic that must be rerouted when edge f fails. This is set equal to the flow of service traffic across the edge. The source of the commodity is arbitrarily chosen to be one endpoint of f, and the sink the other endpoint. The choice of source and sink implies a direction of flow for all service routes crossing the edge (the source of the service route is connected to the source of commodity f. The constant $\text{shared}(f,e)$ denotes the service traffic that uses both the failing edge f and the edge e. Let $\text{service}(e)$ refer to all service traffic on e.

Standard network flow constraints (constraint 8) generate a flow between source and sink. The variable $\text{cap}(e)$ represent the restoration capacity of edge e, and constraint 5 ensures the restoration capacity is enough to handle any edge failure. The objective of the linear program is to minimize the total cost of restoration capacity.

Theorem 3. *The optimum solution of the lower bound LP is a lower bound on the optimum solution of the large LP.*

Proof. [4] Let R_f denote the set of service paths crossing edge f. ¿From the restoration routes for R_f we construct a feasible solution to the linear program whose cost is no more than the cost of the restoration routes. Consider $d \in R_f$ with service path p and restoration path p_f avoiding f. For the commodity for edge f in the linear program, we route the contribution from d as follows. ¿From either the endpoints of f the commodity follows p in the reverse directions until hitting a node on p_f. Between those two nodes the commodity flows forward along p_f. Only the latter part of the route contributes to the $\text{cap}(\cdot)$ variables, because along the rest of the route d contributes equally to both $\text{flow}(f,\cdot)$ and $\text{shared}(f,\cdot)$.

Constants:

demand(f,v)	Supply for edge f commodity at node v
shared$(f,\{u,v\})$	Service flow of commodity f on edge $e=\{u,v\}$
cost(e)	Unit cost of provisioning capacity on edge e

Restoration flow variables, all non-negative:

flow(f,u,v), flow(f,v,u)	Forward and backward flow on edge $e=\{u,v\}$ of commodity f
cap(e)	Restoration capacity of edge e

Minimize $\sum_e \text{cost}(e)\cdot\text{cap}(e)$ subject to

$$\text{cap}(e=\{u,v\}) \geq \text{flow}(f,u,v)+\text{flow}(f,v,u)-\text{shared}(f,\{u,v\}) \quad (5)$$

$$\forall e\in E, f\in F \quad (6)$$

$$\text{flow}(f,u,v) = \text{flow}(f,v,u)=0 \quad \forall f=\{u,v\}\in F \quad (7)$$

$$\sum_{x:\{x,v\}\in E} \text{flow}(f,x,v) = \sum_{x:\{v,x\}\in E} \text{flow}(f,v,x)+\text{demand}(f,v) \quad \forall f\in F, v\in V \quad (8)$$

$$(9)$$

Fig. 2. The linear program `Lower-LP`

The linear program has $O(m)$ commodities, each of which has $O(m)$ flow variables. Hence the LP has $O(m^2)$ variables and $O(mn)$ constraints. Note that we could alternatively use a path-generation formulation of the the lower bound. Such a formulation may be faster in practice than the LP presented here, but it has exponential behavior in the worst case.

5 From Lower Bound to Approximation Algorithm

The algorithm LBALG starts with an optimum (fractional) solution of `lowerbound-LP`, and produces a restoration plan. The bounds below work for demands in the range $(0,1]$, but for simplicity of exposition we simply assume unit demands. We first present a simple version that allows a simple proof of the approximation ratio, followed by the efficient version, optimized for performance in practice, that is used in the empirical evaluation of the next section.

The simple version consists of the following two steps:

1. Path extraction. For each edge failure f, the flow variables flow$(f,\cdot,\cdot)$ from `lowerbound-LP` are used to generate a set of paths between the endpoints of e as follows.
   (a) First for each edge $e=(u,v)$ set the capacity of e to be cap$(e) = \lceil \text{flow}(f,u,v)+\text{flow}(f,v,u)\rceil$.
   (b) Since the edge capacities are integral, we can then find an integral flow of size demand(f,u) from u to v, satisfying the edge capacities.
   (c) From the integer flow it is simple to produce a set of demand(f,u) paths from u to v, such that the number of paths on an edge e is at most cap(e).

2. Path-assignment. The restoration paths are arbitrarily matched to the service paths that use f
3. Splicing. Each restoration path is spliced into its matched service path. See Figure 3.

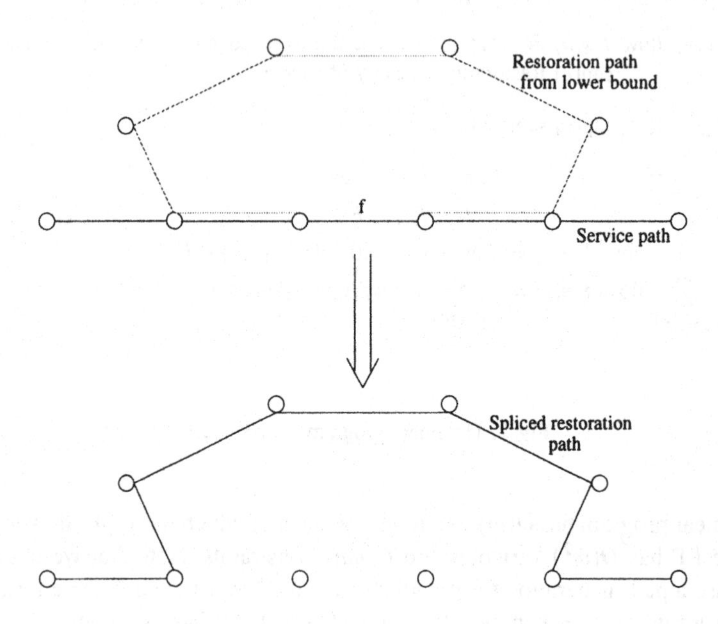

Fig. 3. Splicing a restoration path between endpoints of failed edge f into service path

Theorem 4. *Assume the minimum service capacity on an edge is at least* $1/\varepsilon$. *Then algorithm* LBALG *is a* $(2+\varepsilon)$*-approximation algorithm under the total-cost measure.*

Proof. During path extraction at most $\lceil flow(f,u,v) + flow(f,v,u) \rceil$ restoration paths are generated that cross edge $e = \{u,v\}$. Applying constraint (5), the restoration capacity required by algorithm LBALG on failure f is at most

$$\begin{aligned} &\lceil \mathrm{cap}(e) \rceil + \mathrm{shared}(f,e) \\ &\leq \mathrm{cap}(e) + 1 + \mathrm{service}(e) \\ &\leq \mathrm{cap}(e) + \mathrm{service}(e)(1+\varepsilon) \qquad \text{(by hypothesis)} \end{aligned}$$

and therefore the cost due to restoration capacity of algorithm LBALG is at most the cost of `lowerbound-LP` plus $(1+\varepsilon)$ times the cost of the service capacity.

The efficient version of LBALG uses only the per-edge restoration capacities $\mathrm{cap}(\cdot)$ from the `lowerbound-LP`, rather than the paths, and is structured as follows.

1. Capacity Rounding. Obtain a solution to `lowerbound-LP` with integer edge capacities:

   (a) Sort the edges in increasing order of the fractional fraction part of cap(·)
   (b) For each edge e in this order, do the following:
      i. Round down cap(e)
      ii. For each edge failure f, test whether the demand for f can still be routed within the edge capacities.
      iii. If the test fails for some f, increment cap(e)
2. Path Generation.
   (a) Start with the edge capacities from Step 1.
   (b) Then for each edge failure f in turn, generate restoration routes for all demands affected by f so as to minimize the cost of the implied increase in edge capacities.

For a single edge failure f, the path problem of generating restoration routes (Step 2) can be represented by an integer min-cost multicommodity flow problem, with a commodity for each demand group affected by f. Define the restoration flow for commodity d on an edge e to be the amount that the flow of d on e exceeds that of the service routing. The cost of a unit of restoration flow crossing an edge e is 0 for the first c units of flow, where c is the current capacity of e, and the edge cost of e for further units of flow.

There are a number of exact and approximate algorithms for min-cost multicommodity flow. We chose a three-stage implementation, and found it to work very well in practice (see Section 6). However, there is room for improvement in the running time of our implementation, and it is worth pursuing other approaches to solving the flow problem, for example combinatorial approximation algorithms or linear program with a column generation formulation.

Our implementation uses three steps to solve the min-cost multicommodity flow problem, in increasing order of running time and power:

1. Greedy. Attempt to route each demand in turn, using an integer capacitated max-flow subroutine and decrementing the available edge capacities after each demand. If all demands are routed within the current edge capacities, we have found a zero-cost solution to the multicommodity flow problem. Repeat this process for a number of random permutations of the demands.
2. Short Multicommodity Flow. If the greedy routing fails, construct the following aggregated and constrained version of the multicommodity flow problem.
   (a) Solve a capacitated flow problem to generate a flow of the appropriate size between the endpoints of the failed edge (similar to the flows in `lowerbound-LP`).
   (b) Break each demand d into three segments d_1, d_2, d_3, with d_1 and d_3 maximal such that none of their internal nodes are reached by the flow from 2a.
   (c) Create a graph h containing the union of the edges in all d_2 and those containing flow from 2a.
   (d) Aggregate demands between the same endpoints of h, and solve the reduced multicommodity flow problem.
   (e) If a zero-cost solution is found, then combine the resulting restoration paths with the appropriate d_1 and d_3 to produce a zero-cost solution to the large problem.
3. Long Multicommodity Flow. If the short multicommodity flow does not produce a zero-cost solution, solve the full multicommodity flow.

To solve the short and long multicommodity flows, our implementation generates flow-based formulations using AMPL and solves them using an interior algorithm in CPLEX. We find empirically that the greedy step is successful for about 90% of the edges, and the short multicommodity flow has a zero-cost solution in the majority of the remaining cases. Thus the number of long multicommodity flow problems that must be solved is very small, under 5% of the number of edges.

6 Experimental Results

This section describes an experimental evaluation of algorithm LBALG. The data sets (also used in [4]) are for four two-connected networks, sized as follows:

data set	number of nodes	number of links
A	58	92
B	78	121
C	102	165
D	452	577

The unit capacity cost for an edge is 300 plus the edge length in miles. The constant 300 was chosen to represent the relative cost of equipment placed only at the end of a link, such as switch ports, compared to equipment which is placed at regular intervals along a link, such as optical amplifiers. Edge lengths ranged from 0.1 to 992 (measured in miles), giving edge costs in the range 300.1 to 1292.

For each network, an actual matrix of traffic forecasts between approximately 600 US cities and towns was first mapped to the network nodes, then concentrated into large units of bandwidth, resulting in problems of moderate size, with unit demand sizes. Two different forecasts were used for each of A, B and C, and one for D, resulting in the following demand sets:

demand set	number of demands
A 1	178
A 2	258
B 1	260
B 2	465
C 2	679
D 1	2120

The algorithm LBALG was run on these demands sets, giving the following results. The table shows the cost of the service routing, the cost of the extra capacity required for restoration in the lower bound and for LBALG, the ratio of the cost of restoration capacity required by LBALG and the lower bound, and lastly the ratio of the total cost (service plus restoration) of LBALG compared to the lower bound.

demand set	LB	LBALG	LBALG / LB	LB Runtime	Extraction Runtime
A 1	971636	975339	1.003	30	3:01
A 2	1043934	1046977	1.003	29	3:55
B 1	1139824	1147916	1.007	1:02	3:28
B 2	1367741	1371754	1.003	1:04	7:13
C 1	1245997	1257945	1.006	28:15	9:32
C 2	1656629	1667173	1.006	27:33	35:04
D 1	2271209	2291555	1.009	1:15:32	4:04:46

As can be seen, the restoration plans produced by LBALG are within 1% of the lower bound, and therefore within 1% of optimal. This is far better than Theorem 4 would suggest. There is some opportunity to reduce the running time of the extraction process by using a more efficient combinatorial algorithm for exact or approximate multicommodity flow, or by using column generation instead of a flow-based representation of the multicommodity flow linear program.

For comparison, the next table (extracted from [4]) presents the performance of the column generation algorithm of [3] on the three smaller networks. This algorithm exactly solves a linear relaxation of the restoration capacity planning problem, then applies some heuristics to generate an integer solution. Network D is too large for the column generation approach to handle.

demand set	Col Gen / LB	Col Gen runtime (minutes)
A 1	1.010	38:00
A 2	1.010	53:00
B 1	1.007	2:06:00
B 2	1.003	4:08:00
C 1	1.004	23:43:00
C 2	1.005	59:33:00

The restoration plans produced by LBALG are as good as those produced by column generation, while the large reduction in running time enables the solution of much larger problems than was previously possible.

7 Combined Optimization of Service and Restoration

In this section we consider the problem of building capacity to handle both service and restoration at minimum total cost. That is, we compute service and restoration paths together.

The linear program of Section 3 can be extended without much difficulty to handle the more general case. The idea is to add a service multicommodity flow to the collection of restoration multicommodity flows. The service flow constants in `large-LP` are replaced by service flow variables.

The linear program is shown in Figure 4. A demand group now contains all demands between a given pair of nodes. Applying an analysis similar to that of section 3 gives the following theorem.

Constants:

supply(g,v) Supply for demand group g at node v
cost(e) Unit cost of provisioning capacity on edge e

Variables, all non-negative:

sflow(g,u,v) Service flow of demand group g from u to v on $e=\{u,v\}$
rflow(g,f,u,v) Flow of group g on edge $e=\{u,v\}$ from u to v, under failure f
restcap(g,f,u,v) Restoration capacity on $e=\{u,v\}$ required for group g under failure f
cap(e) Total service and restoration capacity of edge e

Minimize $\sum_e \text{cost}(e)\cdot\text{cap}(e)$ subject to:

$$\text{rflow}(g,f,u,v)=\text{rflow}(g,f,v,u)=0 \quad \forall\{u,v\}=f\in F, g\in G \tag{10}$$

$$\text{restcap}(g,f,u,v)\geq \text{rflow}(g,f,u,v)-\text{sflow}(g,u,v) \quad \forall g\in G, f\in F, u,v \mid \{u,v\}\in E \tag{11}$$

$$\text{cap}(e=\{u,v\})\geq \sum_g(\text{sflow}(g,u,v)+\text{sflow}(g,v,u)) \tag{12}$$

$$+\sum_g(\text{restcap}(g,f,u,v)+\text{restcap}(g,f,v,u))\,\forall e\in E, f\in F \tag{13}$$

$$\sum_{x:\{x,v\}\in E}\text{sflow}(g,x,v)=\sum_{x:\{x,v\}\in E}\text{sflow}(g,v,x)+\text{supply}(g,v) \quad \forall f\in F, g\in G \tag{14}$$

$$\sum_{x:\{x,v\}\in E}\text{rflow}(g,f,x,v)=\sum_{x:\{x,v\}\in E}\text{rflow}(g,f,v,x)+\text{supply}(g,v) \quad \forall f\in F, g\in G \tag{15}$$

(16)

Fig. 4. The linear program gen-LP

Theorem 5. *Suppose that the service capacity of each edge is at least $\frac{5.3}{\varepsilon^2}\log 2m$. Then with probability at least $1/2$ the cost of the service and restoration routes produced by the above algorithm is at most $1+\varepsilon$ times that of the optimal cost.*

The above LP uses $\Theta(m^2|G|)$ flow variables. We can use the lower-bound algorithm to get a $(4+\varepsilon)$-approximate solution with many fewer variables. The idea is simple: for the service routes, simply use shortest paths in the network according to the cost per unit capacity values. Then input these service routes into the lower-bound algorithm. The bound on the approximation ratio follows from the following lemma.

Lemma 6. *Let OPT denote the optimum cost of a solution for the service and restoration planning problem. There is a solution in which the service routing follows shortest paths (according to cost per unit capacity) with total cost at most $2\cdot OPT$.*

Proof. Let O be an optimum solution for the service and restoration planning problem. We will construct a solution to the restricted problem, at most doubling the cost.

Consider a demand d and a fault f on the shortest path between the endpoints of d. The demand d is affected by f in the restricted problem, so we must specify a restoration route for it.

If d is affected by f in O, then we use the restoration path for d on failure f in O as the restoration path for the restricted problem. Otherwise, the service route for d in O does not contain f, so we use it as the restoration path for the restricted problem.

The cost of the restoration capacity for the restricted problem is at most the sum of the service and restoration costs of O. Furthermore, the service cost of the restricted problem is at most the service cost of O, so the total cost of the restricted problem is at most double that of O.

The bound of Lemma 6 is tight, as can be seen from the following example. Let the network consist of $k+1$ parallel edges from s to t and the demands consist of k unit demands from s to t. Let one edge have cost $1-\varepsilon$ and the others have cost 1. Then in the shortest paths solution, all traffic routes over the edge of weight $1-\varepsilon$, necessitating a total of k units of restoration on other edges. An optimal solution routes 1 unit of demand on each of the k edges, necessitating only a single unit of restoration capacity on the $k+1$st edge. As k grows the cost ratio tends to 2.

Theorem 7. *Assume the minimum service capacity under shortest-paths routing is at least $1/\varepsilon$. Then algorithm* LBALG *is a $(4+\varepsilon)$-approximation algorithm under the total-cost measure.*

Proof. The proof follows by combining theorem 4 with lemma 6.

8 Remarks

Theorem 7 can be strengthened to a $2+\varepsilon$-approximation guarantee by a lower bound LP that more carefully combines a service flow calculation with restoration capacity planning.

As stated in the introduction, our model limits restoration plans in certain ways that correspond to standard practice in network planning. Our methods can be adapted to handle variations in the model. For example, we can allow service routes to change upon an edge failure even if the old route is unaffected by the failure, or we can allow a restoration path for one demand to use service capacity belong to another demand, if the other demand has also been rerouted onto a service route.

The specifics of the model sometimes allow us to reduce the number of variables in the $(1+\varepsilon)$-approximate flow formulation. We are also generally able to achieve a constant-approximate solution for these variations, for a small constant, with $O(m^2)$ variables, using the lower-bound approach. Details are omitted from this extended abstract.

9 Acknowledgements

The authors thank the anonymous referee for a number of helpful comments.

References

[1] D. Alevras, M. Grötschel, and R. Wessäly. Capacity and survivability models for telecommunication networks. Technical Report ZIB Preprint SC 97-24, Konrad-Zuse-Zentrum für Informationstechnik Berlin, 1997. http://www.zib.de/pub/zib-publications/reports/SC-97-24.ps.

[2] G. Brightwell, G. Oriolo, and F. B. Shepherd. Some strategies for reserving resilient capacity. Technical Report LSE-CDAM-98-04, London School of Economics Centre for Discrete and Applicable Mathematics, 1998. http://www.cdam.lse.ac.uk/Reports/Files/cdam-98-04.ps.gz.

[3] S. Cwilich, M. Deng, D. Houck, and D. Lynch. An LP-based approach to restoration network design. In *ITC16, International Teletraffic Congress*, 1999.

[4] S. Cwilich, M. Deng, D. Lynch, S. Phillips, and J. Westbrook. Algorithms for restoration planning in a telecommunications network. In *Proc. ACM Workshop on Experimental Analysis of Algorithms (ALENEX 99)*. Springer-Verlag, 1999.

[5] K. Eswaran and R. Tarjan. Augmentation problems. *SIAM Journal on Computing*, 5(4):653–665, 1976.

[6] A. Frank. Augmenting graphs to meet edge-connectivity requirements. *SIAM J. DISC. MATH.*, 5(1):25–53, 1992.

[7] M. Grötschel, C. L. Monma, and M. Stoer. Design of survivable networks. In *Handbooks in Operations Research and Management Science, Network Models*, chapter 10, pages 617–672. North-Holland, 1995.

[8] M. Herzberg, S. J. Bye, and A. Utano. The hop-limit approach for spare-capacity assignment in survivable networks. *IEEE/ACM Trans. on Networking*, 3(6):775–783, 1995.

[9] R. R. Iraschko, M. H. MacGregor, and W. D. Grover. Optimal capacity placement for path restoration in STM or ATM mesh-survivable networks. *IEEE/ACM Trans. on Networking*, 6(3):325–336, 1998.

[10] K. Jain. A factor 2 approximation algorithm for the generalized network steiner problem. In *Proc. 39th IEEE Symp. on Foundation of Computer Science*, pages 448–457, 1998.

[11] J. Kleinberg. *Approximation algorithms for disjoint paths problems*. PhD Thesis, Department of EECS, MIT, 1996.

[12] R. Motwani and P. Raghavan. *Randomized Algorithms*. Cambridge University Press, New York, 1995.

[13] P. Raghavan and C. Thompson. Randomized rounding. *Combinatorica*, 7:365–374, 1987.

Efficient Algorithms for Integer Programs with Two Variables per Constraint (Extended Abstract)

Reuven Bar-Yehuda[1] and Dror Rawitz[2]

[1] Computer Science Department,
Technion - IIT, Haifa 32000, Israel
e-mail: reuven@cs.technion.ac.il
http://www.cs.technion.ac.il/~reuven
[2] Computer Science Department,
Technion - IIT, Haifa 32000, Israel
e-mail: rawitz@cs.technion.ac.il
http://www.cs.technion.ac.il/~rawitz

Abstract. Given a bounded integer program with n variables and m constraints each with 2 variables we present an $O(mU)$ time and $O(m)$ space feasibility algorithm for such integer programs (where U is the maximal variable range size). We show that with the same complexity we can find an optimal solution for the positively weighted minimization problem for *monotone* systems. Using the local-ratio technique we develop an $O(nmU)$ time and $O(m)$ space 2-approximation algorithm for the positively weighted minimization problem for the general case. We further generalize all results to non linear constraints (called *axis-convex constraints*) and to non linear (but monotone) weight functions.
Our algorithms are not only better in complexity than other known algorithms, but they are also considerably simpler, and contribute to the understanding of these very fundamental problems.

Keywords: Combinatorial Optimization, Integer Programming, Approximation Algorithm, Local Ratio Technique, 2SAT, Vertex Cover.

1 Introduction

This paper is motivated by a recent paper of Hochbaum, Megiddo, Naor and Tamir [10], which discusses integer programs with two variables per constraint. The problem is defined as follows:

$$\begin{aligned} \text{(2VIP)}\ \min \textstyle\sum_{i=1}^{n} w_i x_i & \\ \text{s.t. } a_k x_{i_k} + b_k x_{j_k} \geq c_k & \quad \forall k \in \{1,\dots,m\} \\ \ell_i \leq x_i \leq u_i & \quad \forall i \in \{1,\dots,n\} \end{aligned}$$

where $1 \leq i_k, j_k \leq n$, $w_i \geq 0$, $a,b,c \in \mathbb{Z}^m$ and $\ell, u \in \mathbb{N}^n$.

Obviously this problem is a generalization of the well known *minimum weight vertex cover problem* (VC) and the *minimum weight 2 satisfiability problem* (2SAT). Both problems are known to be NP-hard [6] and the best known approximation ratio for VC [3, 9, 11] and 2SAT [8] is 2. Both results are best viewed via the local ratio technique (see [2, 4]).

Hochbaum et al. [10] presented a 2-approximation algorithm for the 2VIP problem. Their algorithm uses a maximum flow algorithm, therefore the time complexity of their algorithm is relatively high, i.e., when using Goldberg and Tarjan's maximum flow algorithm [7] it is $O(nmU^2 \log(\frac{n^2U}{m}))$, where $U = \max_i \{u_i - \ell_i\}$. By using the local-ratio technique we present a more natural and simpler $O(nmU)$ time and $O(m)$ space 2-approximation algorithm.

In order to develop an approximation algorithm it seems natural to first study the feasibility problem. Indeed this is done by Hochbaum et al. [10] for the 2VIP problem and by Gusfield and Pitt [8] for the 2SAT problem. In Sect. 2 we present our $O(mU)$ time and $O(m)$ space feasibility algorithm for 2VIP systems. Section 3 includes the 2-approximation algorithm for linear integer systems. In Sect. 4[1] we show that the feasibility algorithm and the approximation algorithm presented in this paper can be generalized to some non-linear systems with the same time and space complexity. We define a generalization of linear inequalities, called *axis-convex* constraints, and show that the algorithms can be generalized to work with such constraints. We also generalize the 2-approximation algorithm to objective functions of the form $\sum_{i=1}^{n} w_i(x_i)$, where all the w_i's are *monotone weight functions*. The optimality algorithm for *monotone* linear systems appears in Sect. 5[1]. We show that this algorithm can work with some non-linear constraints, and we generalize the algorithm to monotone weight functions, as well.

Table 1 summarizes our results for 2VIP systems.

Table 1. Summary of Results

Problem	Previous results (time,space)	Our results (time,space)
2SAT Feasibility	$O(m), O(m)$ Even, Itai and Shamir [5]	
2VIP Feasibility	$O(mU), O(mU)$ by using reduction to 2SAT [10]	$O(mU), O(m)$
2SAT 2-approximation	$O(nm), O(n^2+m)$ Gusfield and Pitt [8]	$O(nm), O(m)$
2VIP 2-approximation	$O(nmU^2 \log(n^2U/m)), O(mU)$ Hochbaum, Megiddo, Naor and Tamir [10]	$O(mnU), O(m)$
Monotone 2VIP Optimization[1]	$O(mU), O(mU)$ by using reduction to 2SAT	$O(mU), O(m)$

[1] Omitted from this extended abstract.

2 Feasibility Algorithm

Given a 2VIP system we are interested in developing an algorithm which finds a feasible solution if one exists. Since the special case when $\ell = 0^n$ and $u = 1^n$ is the known 2SAT feasibility problem, it is natural to try to extend the well known $O(m)$ time and $O(m)$ space algorithm of Even et al. [5]. It is possible to transform the given 2VIP system to an equivalent 2SAT instance with nU variables and $(m+n)U$ constraints (this transformation by Feder appears in [10]). By combining this transformation with the linear time and space algorithm of Even et al. we get an $O(mU)$ time and $O(mU)$ space feasibility algorithm. In this section we present an $O(mU)$ time and $O(m)$ space feasibility algorithm which generalizes the algorithm by Even et al.

The main idea of the algorithm of Even et al. is as follows: we choose a variable x_i and discover the force values for other variables by assigning $x_i = 0$ and by assigning $x_i = 1$. If one of these assignments does not lead to a contradiction, we can assign x_i this value and make the corresponding forced assignments. The correctness of their approach is achieved by proving that a non contradictory assignment preserve the feasibility property. The efficiency of their algorithm is achieved by discovering the forced assignments of $x_i = 0$ and those of $x_i = 1$ in parallel.

The purpose of this section is not only to show the factor $\Omega(U)$ improvement in space complexity but also to put the foundations for the 2-approximation algorithm presented in the next section.

Definition 1. *For a given 2VIP instance*

$$sat(\ell,u) = \{x : \ell \leq x \leq u \text{ and } x \text{ satisfies all 2VIP constraints}\} \ .$$

Definition 2. *For a given constraint k on the variables* x_i, x_j

$$constraint(k) = \{(\alpha,\beta) : x_i = \alpha, x_j = \beta \text{ satisfy constraint } k\} \ .$$

Definition 3. *Given* $\alpha,\beta \in \mathbb{Z}$ *we define* $[\alpha,\beta] = \{z \in \mathbb{Z} : \alpha \leq z \leq \beta\}$.

For a constraint k on the variables x_i and x_j:

Observation 4. *If* $(\alpha,\beta),(\alpha,\gamma) \in constraint(k)$ *then* $(\alpha,\delta) \in constraint(k)$ *for all* $\delta \in [\beta,\gamma]$.

Observation 5. *If* $(\alpha_1,\alpha_2),(\beta_1,\beta_2),(\gamma_1,\gamma_2) \in constraint(k)$ *then all points inside the triangle induced by* $(\alpha_1,\alpha_2),(\beta_1,\beta_2)$ *and* (γ_1,γ_2) *satisfy constraint k.*

We present a routine in Fig. 1 which will be repeatedly used for constraint propagation[2]. It receives as input two arrays ℓ and u of size n (passed by reference), two variables indices i, j and a constraint index k on these two variables. The objective of this routine is to find the impact of constraint k and the bounds ℓ_i, u_i on the bounds ℓ_j, u_j.

We denote by ℓ^{after} and u^{after} the values of ℓ and u after calling *OneOnOneImpact* (ℓ, u, i, j, k). Hence we get:

[2] Constraint propagation was used for the LP version of the problem, e.g., see [1].

```
Routine OneOnOneImpact(ℓ,u,i,j,k)

  Let constraint k be ax_i + bx_j ≥ c.
  If b > 0 then
      if a > 0 then ℓ'_j ← ⌈(c−au_i)/b⌉
      else ℓ'_j ← ⌈(c−aℓ_i)/b⌉
  else
      if a > 0 then u'_j ← ⌊(c−aℓ_i)/b⌋
      else u'_j ← ⌊(c−au_i)/b⌋
  ℓ_j ← max{ℓ_j, ℓ'_j}
  u_j ← min{u_j, u'_j}
```

Fig. 1. Routine *OneOnOneImpact*.

Observation 6. $sat(\ell^{after}, u^{after}) = sat(\ell, u)$.

Observation 7. *If* $\beta \in [\ell_j^{after}, u_j^{after}]$ *there exists* $\alpha \in [\ell_i^{after}, u_i^{after}]$ *such that* $(\alpha, \beta) \in$ *constraint*(k).

The routine in Fig. 2, which is called *OneOnOneImpact*, receives as input two arrays ℓ and u of size n (passed by reference) and a variable index t, and change ℓ and u according to the impact of ℓ_t and u_t on all the intervals.

```
Routine OneOnAllImpact(ℓ,u,t)

  Stack ← {t}
  While Stack ≠ ∅ do
     i ← POP(Stack)
     For each constraint k involving x_i and another variable x_j
         OneOnOneImpact(ℓ,u,i,j,k)
         If u_j < ℓ_j then return "fail"
         If ℓ_j or u_j changed then PUSH j into Stack
```

Fig. 2. Routine *OneOnAllImpact*.

We now prove that we do not lose feasible solutions after activating *OneOnAllImpact*.

Lemma 8. *If* ℓ^{after} *and* u^{after} *are the values of* ℓ *and* u *after calling* OneOnAllImpact *then* $sat(\ell^{after}, u^{after}) = sat(\ell, u)$.

Proof. All changes made to ℓ and u are done by routine *OneOnOneImpact*. It is easy to prove the lemma by induction using Observation 6. □

Lemma 9. *If* OneOnAllImpact(ℓ,u,t) *terminates without failure with the bounds* ℓ^{after}, u^{after} *and* $sat((\ell_1,\ldots,\ell_{t-1},-\infty,\ell_{t+1},\ldots,\ell_n),(u_1,\ldots,u_{t-1},\infty,u_{t+1},\ldots,u_n)) \neq \emptyset$ *then* $sat(\ell^{after},u^{after}) \neq \emptyset$.

Proof. Let $y \in \text{sat}((\ell_1,\ldots,\ell_{t-1},-\infty,\ell_{t+1},\ldots,\ell_n),(u_1,\ldots,u_{t-1},\infty,u_{t+1},\ldots,u_n))$. We define a vector y' as:

$$y'_t = \begin{cases} y_t, & y_t \in [\ell_t^{\text{after}},u_t^{\text{after}}] \\ \ell_t^{\text{after}}, & y_t < \ell_t^{\text{after}} \\ u_t^{\text{after}}, & y_t > u_t^{\text{after}} \end{cases}$$

Consider constraint k on x_i and x_j. We need to show that $y'_i, y'_j \in \text{constraint}(k)$.

Case 1: $y_i \in [\ell_i^{\text{after}},u_i^{\text{after}}]$ and $y_j \in [\ell_j^{\text{after}},u_j^{\text{after}}]$.
$(y'_i,y'_j) = (y_i,y_j) \in \text{constraint}(k)$.

Case 2: $y_i < \ell_i^{\text{after}}$ and $y_j \in [\ell_j^{\text{after}},u_j^{\text{after}}]$.
y is a feasible solution, thus $(y_i,y_j) \in \text{constraint}(k)$. When we changed the lower bound of x_i to ℓ_i^{after} we called *OneOnOneImpact* for all constraints involving x_i including constraint k. By Observation 7 there exists $\alpha \in [\ell_i^{\text{after}},u_i^{\text{after}}]$ for which $(\alpha,y_j) \in \text{constraint}(k)$. Thus by Observation 4 we get that $(\ell_i^{\text{after}},y_j) \in \text{constraint}(k)$.

Case 3: $y_i < \ell_i^{\text{after}}$ and $y_j < \ell_j^{\text{after}}$.
y is a feasible solution, thus $(y_i,y_j) \in \text{constraint}(k)$. When we changed the lower bound of x_i to ℓ_i^{after} we called *OneOnOneImpact* for all constraint involving x_i including constraint k. By Observation 7 there exists $\alpha \in [\ell_i^{\text{after}},u_i^{\text{after}}]$ for which $(\alpha,\ell_j^{\text{after}}) \in \text{constraint}(k)$. From the same arguments we get that there exists $\beta \in [\ell_j^{\text{after}},u_j^{\text{after}}]$ for which $(\ell_i^{\text{after}},\beta) \in \text{constraint}(k)$ as well. Thus by Observation 5 we get that $(\ell_i^{\text{after}},\ell_j^{\text{after}}) \in \text{constraint}(k)$.

Other cases are similar to Cases 2 and 3. □

The algorithm in Fig. 3 returns a feasible solution if such a solution exists.

Theorem 10. *Algorithm* Feasibility *returns a feasible solution if such a solution exists.*

Proof. Each recursive call reduces at least one of the ranges (the t'th), thus the algorithm must terminate. By Lemma 8 if $\text{sat}(\ell,u) = \emptyset$ the algorithm returns "fail". On the other hand, if $\text{sat}(\ell,u) \neq \emptyset$ we prove by induction on $\sum_{i=1}^{n}(u_i - \ell_i)$ that the algorithm finds a feasible solution.

Base: $\sum_{i=1}^{n}(u_i - \ell_i) = 0$ implies $\ell = u$, thus $x = \ell$ is a feasible solution.

Step: By Lemma 8 at least one of the calls to *OneOnAllImpact* terminates without failure. If call left was chosen then by Lemma 9 we know that $\text{sat}(\ell^{\text{left}},u^{\text{left}}) \neq \emptyset$, therefore by the induction hypothesis we can find a feasible solution for $\ell^{\text{left}},u^{\text{left}}$. Obviously a feasible solution $x \in \text{sat}(\ell^{\text{left}},u^{\text{left}})$ satisfies $x \in \text{sat}(\ell,u)$. The same goes for call right.

Algorithm ***Feasibility***$(\ell, u \in \mathbb{Z}^n)$

```
If ℓ = u then
    If x = ℓ is a feasible solution then return ℓ
    else return "fail"
Choose a variable x_t, for which ℓ_t < u_t
α ← ⌊½(ℓ_t + u_t)⌋      /* An arbitrary value α ∈ [ℓ_i, u_i − 1] suffices as well */
(ℓ^left, u^left) ← (ℓ, (u_1, ..., u_{t−1}, α, u_{t+1}, ..., u_n))
(ℓ^right, u^right) ← ((ℓ_1, ..., ℓ_{t−1}, α + 1, ℓ_{t+1}, ..., ℓ_n), u)
Call OneOnAllImpact(ℓ^left, u^left, t) and OneOnAllImpact(ℓ^right, u^right, t)
If both calls fail then return "fail"
Choose a successful run of OneOnAllImpact
If call left was chosen
    then return Feasibility(ℓ^left, u^left)
    else return Feasibility(ℓ^right, u^right)
```

Fig. 3. Feasibility Algorithm

This concludes the proof. □

Theorem 11. *Algorithm* Feasibility *can be implemented in time* $O(mU)$ *and space* $O(m)$.

Proof. To achieve time complexity of $O(mU)$, we run both calls to *OneOnAllImpact* in parallel (this approach was used for 2SAT by Even et al. [5]), and prefer the faster option of the two, if a choice exists. After every change in the range of a variable x_i, we need to check the m_i constraints involving this variable, in order to discover the impact of the change. To perform this task efficiently we can store the input in an incidence list, where every variable has its constraints list. As x_i can be changed up to $(u_i - \ell_i)$ times, we conclude that the total time complexity of the changes is $O(\sum_{i=1}^{n} m_i(u_i - \ell_i)) = O(mU)$ (the time wasted on unfinished trials is bounded by the time complexity of the chosen trials). The algorithm uses $O(m)$ space for the input and a constant number of arrays of size n, thus uses linear space. □

3 From Feasibility to Approximation

Before presenting our approximation algorithm, let us first discuss the special case where $|U| = 2$ which is the minimum 2SAT problem. The approach of Gusfield and Pitt [8] can be viewed as follows. The 2CNF formula can be presented as a digraph where each vertex represent a Boolean variable or its negation, and an edge represent an *OneOnOneImpact* propagation (logical "→"). A propagation of an assignment can be viewed as a traversal (e.g., BFS, DFS) in the digraph. In order to get the *OneOnAllImpact* mechanism, Gusfield and Pitt's algorithm starts with a preprocess of constructing a transitive closure. This preprocess uses $\Omega(n^2)$ extra memory, which is expensive. It is much more critical when we try to generalize Gusfield and Pitt's algorithm to 2VIP, in this case the preprocess uses $\Omega(n^2U^2)$ extra memory. As far as we know, every algorithm which relies upon direct 2SAT transformation suffers from this drawback.

We present an $O(nmU)$ time and $O(m)$ space 2-approximation algorithm, which is a specific implementation of our feasibility algorithm. Not only does this seem natural, but also its complexity of $O(nm)$ time and $O(m)$ space in the case of 2SAT dominates that of Gusfield and Pitt's 2SAT algorithm.

In order to use the local-ratio technique [2] we extend the problem definition. Given $\ell, u \in \mathbb{N}^n$ and $\hat{\ell}, \hat{u} \in \mathbb{R}^n$ for which $\ell \le \hat{\ell} \le \hat{u} \le u$ we define the following Extended 2VIP problem:

$$\begin{array}{lll} \text{(E2VIP)} & \min \sum_{i=1}^{n} \Delta(x_i, \hat{\ell}_i, \hat{u}_i) w_i & \\ & \text{s.t. } a_k x_{i_k} + b_k x_{j_k} \ge c_k & \forall k \in \{1, \ldots, m\} \\ & x_i \in [\ell_i, u_i] & \forall i \in \{1, \ldots, n\} \end{array}$$

where

$$\Delta(x, \hat{\ell}_i, \hat{u}_i) = \begin{cases} (x_i - \hat{\ell}_i), & x_i \in [\hat{\ell}_i, \hat{u}_i] \\ (\hat{u}_i - \hat{\ell}_i), & x_i > \hat{u}_i \\ 0, & x_i < \hat{\ell}_i \end{cases}$$

and $1 \le i_k, j_k \le n$, $w_i \ge 0$, $a, b, c \in \mathbb{Z}^m$, and $\ell, u \in \mathbb{N}^n$.

We define $W(x, \hat{\ell}, \hat{u}) = \sum_{i=1}^{n} \Delta(x_i, \hat{\ell}_i, \hat{u}_i) w_i$. A feasible solution x^* is called an *optimal solution* if for every feasible solution x: $W(x^*, \hat{\ell}, \hat{u}) \le W(x, \hat{\ell}, \hat{u})$. We define $W^*(\hat{\ell}, \hat{u}) = W(x^*, \hat{\ell}, \hat{u})$. A feasible solution x is called an *r-approximation* if $W(x, \hat{\ell}, \hat{u}) \le r \cdot W^*(\hat{\ell}, \hat{u})$.

Observation 12. *Given $\hat{\ell}, \hat{u}, \hat{m} \in \mathbb{R}^n$ for which $\hat{\ell} \le \hat{m} \le \hat{u}$ we get*

$$W(x, \hat{\ell}, \hat{u}) = W(x, \hat{\ell}, \hat{m}) + W(x, \hat{m}, \hat{u})$$

Similarly to the Decomposition Observation from [2] we have:

Observation 13. **(Decomposition Observation)**
Given $\hat{\ell}, \hat{u}, \hat{m} \in \mathbb{R}^n$ such that $\hat{\ell} \le \hat{m} \le \hat{u}$ then

$$W^*(\hat{\ell}, \hat{m}) + W^*(\hat{m}, \hat{u}) \le W^*(\hat{\ell}, \hat{u}) \ .$$

Proof. We denote by x^* an optimal solution for the system with regard to $\hat{\ell}, \hat{m}$, by y^* an optimal solution with regard to $\hat{m}, \hat{u}$, and by z^* an optimal solution with regard to $\hat{\ell}, \hat{u}$.

$$\begin{array}{rll} W^*(\hat{\ell}, \hat{m}) + W^*(\hat{m}, \hat{u}) & = W(x^*, \hat{\ell}, \hat{m}) + W(y^*, \hat{m}, \hat{u})) & \text{[By definition]} \\ & \le W(z^*, \hat{\ell}, \hat{m}) + W(z^*, \hat{m}, \hat{u})) & \text{[Optimality of } x^*, y^*\text{]} \\ & \le W(z^*, \hat{\ell}, \hat{u}) & \text{[Observation 12]} \\ & = W^*(\hat{\ell}, \hat{u}) & \text{[By definition]} \end{array}$$

□

The following is this paper's version of the Local-Ratio Theorem (see [2, 4]):

Theorem 14. *If x is an r-approximation with regard to $\hat{\ell}, \hat{m}$ and an r-approximation with regard to $\hat{m}, \hat{u}$ then x is an r-approximation with regard to $\hat{\ell}, \hat{u}$.*

Proof.

$$\begin{aligned} W(x,\hat{\ell},\hat{u}) &= W(x,\hat{\ell},\hat{m}) + W(x,\hat{m},\hat{u}) && \text{[Observation 12]} \\ &\le r\cdot W^*(\hat{\ell},\hat{m}) + r\cdot W^*(\hat{\ell},\hat{u}) && \text{[Given]} \\ &\le r\cdot W^*(\hat{\ell},\hat{u}) && \text{[Decomposition Observation]} \end{aligned}$$

□

Definition 15. *Given $a,b \in \mathbb{R}^n$ we define*

$$\max\{a,b\} = \{\max\{a_1,b_1\},\ldots,\max\{a_n,b_n\}\}$$
$$\min\{a,b\} = \{\min\{a_1,b_1\},\ldots,\min\{a_n,b_n\}\} \ .$$

We are ready to present the 2-approximation algorithm – see Fig. 4.

Algorithm *Approximate*$(\ell, u \in \mathbb{N}^n; \hat{\ell}, \hat{u} \in \mathbb{R}^n)$

If $\hat{\ell} \not\geq \ell$ then return *Approximate*$(\ell, u, \max\{\hat{\ell},\ell\}, \hat{u})$
If $\hat{u} \not\leq u$ then return *Approximate*$(\ell, u, \hat{\ell}, \min\{\hat{u},u\})$
If $\ell = u$ then
  If $x = \ell$ is a feasible solution
    then return ℓ
    else return "fail"
Choose a variable x_t, for which $\ell_t < u_t$
$\alpha \leftarrow \lfloor \frac{1}{2}(\ell_t + u_t) \rfloor$
$(\ell^{\text{left}}, u^{\text{left}}) \leftarrow (\ell, (u_1,\ldots,u_{t-1},\alpha,u_{t+1},\ldots,u_n))$
$(\ell^{\text{right}}, u^{\text{right}}) \leftarrow ((\ell_1,\ldots,\ell_{t-1},\alpha+1,\ell_{t+1},\ldots,\ell_n), u)$
Call *OneOnAllImpact*$(\ell^{\text{left}}, u^{\text{left}}, t)$ and *OneOnAllImpact*$(\ell^{\text{right}}, u^{\text{right}}, t)$
If both calls failed then return "fail"
If call right failed then return *Approximate*$(\ell^{\text{left}}, u^{\text{left}}, \hat{\ell}, \hat{u})$
If call left failed then return *Approximate*$(\ell^{\text{right}}, u^{\text{right}}, \hat{\ell}, \hat{u})$
If $W(\ell^{\text{left}}, \hat{\ell}, \hat{u}) \le W(\ell^{\text{right}}, \hat{\ell}, \hat{u})$
  then
    Find $\hat{m} \in \mathbb{R}^n$ such that:
      $\hat{\ell} \le \hat{m} \le \max\left\{\ell^{\text{right}}, \hat{\ell}\right\}$ and $W(\ell^{\text{right}}, \hat{\ell}, \hat{m}) = W(\ell^{\text{left}}, \hat{\ell}, \hat{u})$
      $\hat{m} \leftarrow \max\left\{\hat{m}, \ell^{\text{left}}\right\}$
    Return *Approximate*$(\ell^{\text{left}}, u^{\text{left}}, \hat{m}, \hat{u})$
  else
    Find $\hat{m} \in \mathbb{R}^n$ such that:
      $\hat{\ell} \le \hat{m} \le \max\left\{\ell^{\text{left}}, \hat{\ell}\right\}$ and $W(\ell^{\text{left}}, \hat{\ell}, \hat{m}) = W(\ell^{\text{right}}, \hat{\ell}, \hat{u})$
      $\hat{m} \leftarrow \max\left\{\hat{m}, \ell^{\text{right}}\right\}$
    Return *Approximate*$(\ell^{\text{right}}, u^{\text{right}}, \hat{m}, \hat{u})$

Fig. 4. 2-Approximation Algorithm

Observation 16. *Algorithm* Approximate *is a specific implementation of Algorithm* Feasibility.

Theorem 17. *Algorithm* Approximate *is a 2-approximation algorithm for E2VIP systems.*

Proof. By Observation 16 when Algorithm *Approximate* returns a solution it is a feasible solution. If $\text{sat}(\ell,u) \neq \emptyset$ we prove by induction that the algorithm finds a 2-approximation.

Base: $u = \ell$ implies $W(\ell,\hat{\ell},\hat{u}) = 0$
Step: There are several cases:

Case 1: $\hat{\ell} \not\geq \ell$:
A 2-approximation solution with respect to $\hat{\ell}$ is obviously a 2-approximation solution with respect to $\max\{\hat{\ell},\ell\}$.

case 2: $\hat{u} \not\leq u$:
Trivial.

Case 3: Call right failed:
By Lemma 8 there is no feasible solution which satisfies $x_t \geq \alpha+1$, therefore we do not change the problem by adding the constraint $x_y \leq \alpha$. By Lemma 8 calling *OneOnAllImpact*$(\ell^{\text{left}}, u^{\text{left}}, t)$ does not change the problem, as well.

Case 4: Call left failed:
Similar to Case 3.

Case 5: Both calls succeeded and $W(\ell^{\text{left}},\hat{\ell},\hat{u}) \leq W(\ell^{\text{right}},\hat{\ell},\hat{u})$:
We first show that every feasible solution is a 2-approximation with regard to $\hat{\ell}$ and $\hat{m}$. We examine an optimal solution x^*. If $x^* \geq \ell^{\text{left}}$ then $\hat{m} \geq \ell^{\text{left}}$ implies $W(x^*,\hat{\ell},\hat{m}) \geq W(\ell^{\text{left}},\hat{\ell},\hat{m})$. If $x^* \geq \ell^{\text{right}}$ then by the definition of $\hat{m}$ we get that $W(x^*,\hat{\ell},\hat{m}) \geq W(\ell^{\text{right}},\hat{\ell},\hat{m}) \geq W(\ell^{\text{left}},\hat{\ell},\hat{u}) \geq W(\ell^{\text{left}},\hat{\ell},\hat{m})$. On the other hand, $W(\hat{m},\hat{\ell},\hat{u}) \leq 2\cdot W(\ell^{\text{left}},\hat{\ell},\hat{m})$, therefore $W(x,\hat{\ell},\hat{m}) \leq 2\cdot W(\ell^{\text{left}},\hat{\ell},\hat{m})$ for every feasible solution x. Therefore by Theorem 14 a 2-approximation with regard to $\hat{m}$ and $\hat{u}$ is a 2-approximation with regard to $\hat{\ell}$ and $\hat{u}$.
We need to show that there exists an optimal solution x^* for which $x^*_i \leq \alpha$. For every feasible solution y such that $y_i \geq \alpha+1$ we define y' as:

$$y'_i = \begin{cases} y_i, & y_i \in [\ell_i^{\text{left}}, u_i^{\text{left}}] \\ \ell_i^{\text{left}}, & y_i < \ell_i^{\text{left}} \\ u_i^{\text{left}}, & y_i > u_i^{\text{left}} \end{cases}$$

By Lemma 9 y' is a feasible solution. $\ell^{\text{left}} \leq \hat{m}$ implies $W(y',\hat{m},\hat{u}) \leq W(y,\hat{m},\hat{u})$, thus there is an optimal solution with regard to $\hat{m}$ and $\hat{u}$ within the bounds $\ell^{\text{left}}, u^{\text{left}}$. Therefore a 2-approximation within the bounds $\ell^{\text{left}}, u^{\text{left}}$ is a 2-approximation with regard to $\hat{m}$ and $\hat{u}$.

Case 6: Both calls succeeded and $W(\ell^{\text{left}},\hat{\ell},\hat{u}) > W(\ell^{\text{right}},\hat{\ell},\hat{u})$:
Similar to Case 5.

□

Corollary 18. *Algorithm* Approximate *is a 2-approximation algorithm for 2VIP systems.*

Theorem 19. *Algorithm* Approximate *can be implemented in time $O(nmU)$ and space $O(m)$.*

Proof. In order to get the required time complexity, we must choose the x_i's wisely. One possibility is to choose the variables in an increasing order, i.e. $x_1, x_2, \dots, x_n$, and to restart again from the beginning after reaching x_n. We call such n iterations on all n variables a *pass*. As stated before, changing the range of x_i might cause changes in the ranges of other variables. The existence of a constraint on x_i and another variable x_j makes x_j a candidate for a range update. This means that we have to check the m_i constraints involving x_i, to discover the consequences of changing its range each time this range changes. x_i can be changed up to $u_i - \ell_i$ times, therefore we get that the time complexity of a single iteration is $O(mU+n) = O(mU)$. One pass may involve all n variables, so the time complexity of one pass is $O(nmU)$. By choosing $\alpha = \lfloor \frac{1}{2}(\ell_t + u_t) \rfloor$, we reduce the possible range for x_t at least by half. Therefore in a single pass we reduce the possible ranges for all variables at least by half. Thus we get that the total time complexity is: $\sum_{k=1}^{\log U} O(mn\frac{U}{2^k}) = O(mnU)$. As before, the algorithm uses an incidence list data structure, thus uses linear space. □

Acknowledgments

We would like to thank Avigail Orni and Hadas Heier for their careful reading and suggestions.

References

[1] B. Aspvall and Y. Shiloach. A polynomial time algorithm for solving systems of linear inequalities with two variables per inequality. *SIAM Journal on Computing*, 9:827–845, 1980.

[2] R. Bar-Yehuda. One for the price of two: A unified approach for approximating covering problems. In *APPROX'98 1st International Workshop on Approximation Algorithms for Combinatorial Optimization Problems*, pages 49–62, July 1998. To appear in Algorithmica.

[3] R. Bar-Yehuda and S. Even. A linear time approximation algorithm for the weighted vertex cover problem. *Journal of Algorithms*, 2:198–203, 1981.

[4] R. Bar-Yehuda and S. Even. A local-ratio theorem for approximating the weighted vertex cover problem. *Annals of Discrete Mathematics*, 25:27–46, 1985.

[5] S. Even, A. Itai, and A. Shamir. On the complexity of timetable and multi-commodity flow problems. *SIAM Journal on Computing*, 5(4):691–703, 1976.

[6] M. R. Garey and D. S. Johnson. *Computers and Intractability; A Guide to the Theory of NP-Completeness*. W.H. Freeman and Company, 1979.

[7] A. V. Goldberg and R. E. Tarjan. A new approach to the maximum flow problem. *Journal of the ACM*, 35:921–940, 1988.

[8] D. Gusfield and L. Pitt. A bounded approximation for the minimum cost 2-SAT problem. *Algorithmica*, 8:103–117, 1992.

[9] D. S. Hochbaum. Approximation algorithms for the set covering and vertex cover problems. *SIAM Journal on Computing*, 11(3):555–556, 1982.
[10] D. S. Hochbaum, N. Megiddo, J. Naor, and A. Tamir. Tight bounds and 2-approximation algorithms for integer programs with two variables per inequality. *Mathematical Programming*, 62:69–83, 1993.
[11] G. L. Nemhauser and L. E. Trotter. Vertex packings: structural properties and algorithms. *Mathematical Programming*, 8:232–248, 1975.

Convex Quadratic Programming Relaxations for Network Scheduling Problems

Martin Skutella*

Technische Universität Berlin
skutella@math.tu-berlin.de
http://www.math.tu-berlin.de/~skutella/

Abstract. In network scheduling a set of jobs must be scheduled on unrelated parallel processors or machines which are connected by a network. Initially, each job is located on some machine in the network and cannot be started on another machine until sufficient time elapses to allow the job to be transmitted there. This setting has applications, e. g., in distributed multi-processor computing environments and also in operations research; it can be modeled by a standard parallel machine environment with machine-dependent release dates. We consider the objective of minimizing the total weighted completion time.
The main contribution of this paper is a provably good convex quadratic programming relaxation of strongly polynomial size for this problem. Until now, only linear programming relaxations in time- or interval-indexed variables have been studied. Those LP relaxations, however, suffer from a huge number of variables. In particular, the best previously known relaxation is of exponential size and can therefore not be solved exactly in polynomial time. As a result of the convex quadratic programming approach we can give a very simple and easy to analyze randomized 2–approximation algorithm which slightly improves upon the best previously known approximation result. Furthermore, we consider preemptive variants of network scheduling and derive approximation results and results on the power of preemption which improve upon the best previously known results for these settings.

1 Introduction

We study the following parallel machine scheduling problem. A set J of n jobs has to be scheduled on m unrelated parallel machines which are connected by a network. The jobs continually arrive over time and each job originates at some node of the network. Therefore, before a job can be processed on another machine, it must take the time to travel there through the network. This is modeled by machine-dependent release dates $r_{ij} \geq 0$ which denote the earliest point in time when job j may be processed on machine i. Together with each job j we are given its positive processing requirement which also depends on the machine i job j will be processed on and is therefore denoted by p_{ij}. Each job j must be processed for the respective amount of time without interruption on

* This research was partially supported by DONET within the frame of the TMR Programme (contract number ERB FMRX-CT98-0202) while the author was staying at C.O.R.E., Louvain-la-Neuve, Belgium, for the academic year 1998/99.

one of the m machines, and may be assigned to any of them. However, for a given job j it may happen that $p_{ij} = \infty$ for some (but not all) machines i such that job j cannot be scheduled on those machines. Every machine can process at most one job at a time. This network scheduling model has been introduced in [4, 1].

We denote the completion time of job j by C_j. The goal is to minimize the total weighted completion time: a weight $w_j \geq 0$ is associated with each job j and we seek to minimize $\sum_{j \in J} w_j C_j$. In scheduling, it is quite convenient to refer to the respective problems using the standard classification scheme of Graham, Lawler, Lenstra, and Rinnooy Kan [7]. The problem $R \mid r_{ij} \mid \sum w_j C_j$, just described, is strongly NP-hard, even for the special case of two identical parallel machines without nontrivial release dates, see [2, 12].

Since we cannot hope to be able to compute optimal schedules in polynomial time, we are interested in how close one can approach the optimum in polynomial time. A (randomized) α–approximation algorithm computes in polynomial time a feasible solution to the problem under consideration whose (expected) value is bounded by α times the value of an optimal solution; α is called the performance guarantee or performance ratio of the algorithm. All randomized approximation algorithms that we discuss or present can be derandomized by standard methods; therefore we will not go into the details of derandomization.

The first approximation result for the scheduling problem $R \mid r_{ij} \mid \sum w_j C_j$ was obtained by Phillips, Stein, and Wein [15] who gave an algorithm with performance guarantee $O(\log^2 n)$. The first constant factor approximation was developed by Hall, Shmoys, and Wein [9] (see also [8]) whose algorithm achieves performance ratio $\frac{16}{3}$. Generalizing a single machine approximation algorithm of Goemans [6], this result was then improved by Schulz and Skutella [18] to a $(2+\varepsilon)$–approximation algorithm. All those approximation results rely somehow on (integer) linear programming formulations or relaxations in time-indexed variables. In the following discussion we assume that all processing times and release dates are integral; furthermore, we define $p_{\max} := \max_{i,j} p_{ij}$.

Phillips, Stein, and Wein modeled the network scheduling problem as a hypergraph matching problem by matching each job j to p_{ij} consecutive time intervals of length 1 on a machine i. The underlying graph contains a node for each job and each pair formed by a machine and a time interval $[t, t+1)$ where t is integral and can achieve values in a range of size np_{max}. Therefore, since $p_{\max}$ may be exponential in the input size, the corresponding integer linear program contains exponentially many variables as well as exponentially many constraints. Phillips et al. eluded this problem by partitioning the set of jobs into groups such that the jobs in each group can be scaled down to polynomial size. However, this complicates both the design and the analysis of their approximation algorithm.

The result of Hall, Shmoys, and Wein is based on a polynomial variant of time-indexed formulations which they called *interval-indexed*. The basic idea is to replace the intervals of length 1 by time intervals $[2^k, 2^{k+1})$ of geometrically increasing size. The decision variables in the resulting linear programming relaxation then indicate on which machine and in which time interval a given job completes. Notice, however, that one looses already at least a factor of 2 in this formulation since the interval-indexed variables do not allow a higher precision for the completion times of jobs. The approximation algorithm of Hall et al. relies on Shmoys and Tardos' rounding technique for the generalized assignment problem [20].

Schulz and Skutella generalized an LP relaxation in time-indexed variables that was introduced by Dyer and Wolsey [5] for the corresponding single machine scheduling problem. It contains a decision variable for each triple formed by a job, a machine, and a time interval $[t, t+1)$ which indicates whether the job is being processed in this time interval on the respective machine. The resulting LP relaxation is a 2–relaxation of the scheduling problem under consideration, i. e., the optimum LP value is within a factor 2 of the value of an optimal schedule. However, as the formulation of Phillips et al., this relaxation suffers from an exponential number of variables and constraints. One can overcome this drawback by turning again to interval-indexed variables. However, in order to ensure a higher precision, Schulz and Skutella used time intervals of the form $[(1+\varepsilon)^k, (1+\varepsilon)^{k+1})$ where $\varepsilon > 0$ can be chosen arbitrarily small; this leads to a $(2+\varepsilon)$–relaxation of polynomial size. Notice, however, that the size of the relaxation still depends substantially on $p_{\max}$ and may be huge for small values of ε. The approximation algorithm based on this LP relaxation uses a randomized rounding technique.

For the problem of scheduling unrelated parallel machines in the absence of nontrivial release dates $R||\sum w_j C_j$, the author has introduced a convex quadratic programming relaxation that leads to a simple $\frac{3}{2}$–approximation algorithm [22]. One of the basic observations for this result is that in the absence of nontrivial release dates the parallel machine problem can be reduced to an assignment problem of jobs to machines; for a given assignment of jobs to machines the sequencing of the assigned jobs can be done optimally on each machine i by applying Smith's Ratio Rule [24]: schedule the jobs in order of nonincreasing ratios w_j/p_{ij}. Therefore, the problem can be formulated as an integer quadratic program in assignment variables. An appropriate relaxation of this program together with randomized rounding leads to the approximation result mentioned above. Independently, the same result has later also been derived by Jay Sethuraman and Mark S. Squillante [19].

Unfortunately, for the general network scheduling problem including release dates the situation is more complicated; for a given assignment of jobs to machines, the sequencing problem on each machine is still strongly NP-hard, see [12]. However, we know that in an optimal schedule a 'violation' of Smith's Ratio Rule can only occur after a new job has been released; in other words, whenever two successive jobs on machine i can be exchanged without violating release dates, the job with the higher ratio w_j/p_{ij} will be processed first in an optimal schedule. Therefore, the sequencing of jobs that are being processed between two successive release dates can be done optimally by Smith's Ratio Rule. We make use of this insight by partitioning the processing on each machine i into n *time slots* which are essentially defined by the n release dates r_{ij}, $j \in J$; since the sequencing of jobs in each time slot is easy, we have to solve an assignment problem of jobs to time slots and can apply similar ideas as in [22]. In particular, we derive a convex quadratic programming relaxation in n^2m assignment variables and $O(nm)$ constraints. Randomized rounding based on an optimal solution to this relaxation finally leads to a very simple and easy to analyze 2–approximation algorithm for network scheduling.

Our technique can be extended to network scheduling problems with preemptions. In preemptive scheduling, a job may repeatedly be interrupted and continued later on another (or the same) machine. In the context of network scheduling it is reasonable to assume that after a job has been interrupted on one machine, it cannot immediately be continued on another machine; it must again take the time to travel there through the network. We call the delay caused by such a transfer *communication delay*. In a similar context, communication delays between precedence constrained jobs have been studied, see, e. g., [14].

We give a 3–approximation algorithm for the problem $R \mid r_{ij}, pmtn \mid \sum w_j C_j$ that, in fact, does not make use of preemptions but computes nonpreemptive schedules. Therefore, this approximation result also holds for preemptive network scheduling with arbitrary communication delays. Moreover, it also implies a bound on the power of preemption, i. e., one cannot gain more than a factor 3 by allowing preemptions. For the problem without nontrivial release dates $R \mid pmtn \mid \sum w_j C_j$, the same technique yields a 2–approximation algorithm. For the preemptive scheduling problems without communication delays, Phillips, Stein, and Wein [16] gave an $(8+\varepsilon)$–approximation. In [21] the author has achieved slightly worse results than those presented here, based on a time-indexed LP relaxation in the spirit of [18].

The paper is organized as follows. In the next section we introduce the concept of scheduling in time slots. We give an integer quadratic programming formulation of the network scheduling problem in Section 3 and show how it can be relaxed to a convex quadratic program. In Section 4 we present a simple 2–approximation algorithm and prove a bound on the quality of the convex quadratic programming relaxation. Finally, in Section 5, we briefly sketch the results and techniques for preemptive network scheduling.

Due to space limitations, we do not provide proofs in this extended abstract; we refer to the full paper [23] which combines [22] and the paper at hand and can be found on the authors homepage.

2 Scheduling in time slots

The main idea of our approach for the scheduling problem $R \mid r_{ij} \mid \sum w_j C_j$ is to somehow get rid of the release dates of jobs. We do this by partitioning time on each machine i into several time slots. Each job is being processed on one machine in one of its time slots and we make sure that job j can only be processed in a slot that starts after its release date.

Let $\rho_{i_1} \le \rho_{i_2} \le \cdots \le \rho_{i_n}$ be an ordering of the release dates r_{ij}, $j \in J$; moreover, we set $\rho_{i_{n+1}} := \infty$. For a given feasible schedule we say that i_k, the kth time slot on machine i, contains all jobs j that are started within the interval $[\rho_{i_k}, \rho_{i_{k+1}})$ on machine i; we denote this by $j \in i_k$. We may assume that there is no idle time between the processing of jobs in one time slot, i. e., all jobs in a slot are processed one after another without interruption.

Moreover, as a consequence of Smith's Ratio Rule we can restrict to schedules where the jobs in time slot i_k are sequenced in order of nonincreasing ratios w_j/p_{ij}. Throughout the paper we will use the following convention: whenever we apply Smith's

Ratio Rule in a time slot on machine i and $w_k/p_{ik} = w_j/p_{ij}$ for a pair of jobs j,k, the job with smaller index is scheduled first. For each machine $i = 1,\dots,m$ we define a corresponding total order $(J, \prec_i)$ on the set of jobs by setting $j \prec_i k$ if either $w_j/p_{ij} > w_k/p_{ik}$ or $w_j/p_{ij} = w_k/p_{ik}$ and $j < k$.

Lemma 1. *In an optimal solution to the scheduling problem under consideration, the jobs in each time slot i_k are scheduled without interruption in order of nondecreasing ratios w_j/p_{ij}. Furthermore, there exists an optimal solution where the jobs are sequenced according to $\prec_i$ in each time slot i_k.*

Notice that there may be several empty time slots i_k. This happens in particular if $\rho_{i_k} = \rho_{i_{k+1}}$. Therefore it would be sufficient to introduce only q_i time slots for machine i where q_i is the number of different values r_{ij}, $j \in J$. For example, if there are no nontrivial release dates (i. e., $r_{ij} = 0$ for all i and j), we only need to introduce one time slot $[0,\infty)$ on each machine. The problem $R||\sum w_j C_j$ has been considered in [22]; for this special case our approach coincides with the one given there.

Up to now we have described how a feasible schedule can be interpreted as a feasible assignment of jobs to time slots. We call an assignment *feasible* if each job j is being assigned to a time slot i_k with $\rho_{i_k} \geq r_{ij}$. On the other hand, for a given feasible assignment of the jobs in J to time slots we can easily construct a corresponding feasible schedule: Sequence the jobs in time slot i_k according to $\prec_i$ and start it as early as possible after the jobs in the previous slot on machine i are finished but not before ρ_{i_k}; in other words, the starting time s_{i_k} of time slot i_k is given by $s_{i_1} := \rho_{i_1}$ and $s_{i_{k+1}} := \max\{\rho_{i_{k+1}}, s_{i_k} + \sum_{j \in i_k} p_{ij}\}$, for $k = 1,\dots,n-1$.

Lemma 2. *Given its assignment of jobs to time slots, we can reconstruct an optimal schedule meeting the properties described in Lemma 1.*

We close this section with one final remark. Notice that several feasible assignments of jobs to time slots may lead to the same feasible schedule. Consider, e. g., an instance consisting of three jobs of unit length and unit weight that have to be scheduled on a single machine. Jobs 1 and 2 are released at time 0, while job 3 becomes available at time 1. We get an optimal schedule by processing the jobs without interruption in order of increasing numbers. This schedule corresponds to five different feasible assignments of jobs to time slots. We can assign job 1 to one of the first two slots, job 2 to the same or a later slot, and finally job 3 to slot 3.

3 A convex quadratic programming relaxation

As a consequence of Lemma 2 we have reduced the scheduling problem under consideration to finding an optimal assignment of jobs to time slots. Therefore we can give a formulation of $R|r_{ij}|\sum w_j C_j$ in assignment variables $a_{i_k j} \in \{0,1\}$ where $a_{i_k j} = 1$ if job j is being assigned to time slot i_k, and $a_{i_k j} = 0$ otherwise. This leads to the following integer quadratic program:

$$
\begin{aligned}
\text{minimize} \quad & \sum_j w_j C_j & & \\
\text{subject to} \quad & \sum_{i,k} a_{i_k j} = 1 & \text{for all } j & \quad (1)\\
& s_{i_1} = \rho_{i_1} & \text{for all } i & \quad (2)\\
& s_{i_{k+1}} = \max\{\rho_{i_{k+1}}, s_{i_k} + \sum_j a_{i_k j} p_{ij}\} & \text{for all } i,\, k & \quad (3)\\
& C_j = \sum_{i,k} a_{i_k j}\Big(s_{i_k} + p_{ij} + \sum_{j' \prec_i j} a_{i_k j'} p_{ij'}\Big) & \text{for all } j & \quad (4)\\
& a_{i_k j} = 0 & \text{if } \rho_{i_k} < r_{ij} & \quad (5)\\
& a_{i_k j} \in \{0,1\} & \text{for all } i,\, k,\, j &
\end{aligned}
$$

Constraints (1) ensure that each job is being assigned to exactly one time slot. In constraints (2) and (3) we set the starting times of the time slots as described in Section 2. If job j is being assigned to time slot i_k, its completion time is the sum of the starting time s_{i_k} of this slot, its own processing time p_{ij}, and the processing times of other jobs $j' \prec_i j$ that are also scheduled in this time slot. The right hand side of (4) is the sum of these expressions over all time slots i_k weighted by $a_{i_k j}$; it is thus equal to the completion time of j. Finally, constraints (5) ensure that no job is being processed before its release date.

It follows from our considerations in Section 2 that we could replace (5) by the stronger constraint

$$a_{i_k j} = 0 \qquad \text{if } \rho_{i_k} < r_{ij} \text{ or } \rho_{i_k} = \rho_{i_{k+1}}$$

which reduces the number of available time slots on each machine. For the special case $R| |\sum w_j C_j$ this leads to the integer quadratic program that has been introduced in [22]. It is also shown there that it is still NP-hard to solve the continuous relaxation of this integer quadratic program; however, it can be solved in polynomial time if the term p_{ij} on the right hand side of (4) is replaced by $p_{ij}(1 + a_{i_k j})/2$.

Observe that this replacement does not affect the value of the integer quadratic program since the new term is equal to p_{ij} whenever $a_{i_k j} = 1$. This motivates the study of the following quadratic programming relaxation (QP) for the general problem including release dates:

$$
(QP)\quad
\begin{array}{lll}
\text{minimize} & \sum_j w_j C_j & \\
\text{subject to} & \sum_{i,k} a_{i_k j} = 1 & \text{for all } j \\
& s_{i_1} = \rho_{i_1} & \text{for all } i \quad (6) \\
& s_{i_{k+1}} = \max\{\rho_{i_{k+1}}, s_{i_k} + \sum_j a_{i_k j} p_{ij}\} & \text{for all } i, k \quad (7) \\
& C_j = \sum_{i,k} a_{i_k j} \left(s_{i_k} + \frac{1 + a_{i_k j}}{2} p_{ij} + \sum_{j' \prec_i j} a_{i_k j'} p_{ij'} \right) & \text{for all } j \quad (8) \\
& a_{i_k j} = 0 & \text{if } \rho_{i_k} < r_{ij} \\
& a_{i_k j} \geq 0 & \text{for all } i, k, j
\end{array}
$$

Notice that a solution to this program is uniquely determined by giving the values of the assignment variables $a_{i_k j}$. In contrast to the case without nontrivial release dates, we cannot directly prove that this quadratic program is convex. Nevertheless, in the remaining part of this section we will show that it can be solved in polynomial time. The main idea is to show that one can restrict to solutions satisfying $s_{i_k} = \rho_{i_k}$ for all i and k. Adding these constraints to (QP) then leads to a convex quadratic program.

Lemma 3. *For all instances of* $R|r_{ij}|\sum w_j C_j$ *there exists an optimal solution to* (QP) *satisfying* $s_{i_k} = \rho_{i_k}$ *for all* i *and* k.

As a consequence of Lemma 3 we can replace the variables s_{i_k} in (QP) by the constants ρ_{i_k} by changing constraints (7) to

$$\sum_j a_{i_k j} p_{ij} \leq \rho_{i_{k+1}} - \rho_{i_k} \qquad \text{for all } i, k.$$

Furthermore, if we remove constraints (8) and replace C_j in the objective function by the right hand side of (8), we can reformulate the quadratic programming relaxation as follows:

$$
(CQP)\quad
\begin{array}{lll}
\text{minimize} & b^T a + \frac{1}{2} a^T D a & \quad (9) \\
\text{subject to} & \sum_{i,k} a_{i_k j} = 1 & \text{for all } j \quad (10) \\
& \sum_j a_{i_k j} p_{ij} \leq \rho_{i_{k+1}} - \rho_{i_k} & \text{for all } i, k \quad (11) \\
& a_{i_k j} = 0 & \text{if } \rho_{i_k} < r_{ij} \\
& a \geq 0 &
\end{array}
$$

Here, $a \in \mathbb{R}^{mn^2}$ denotes the vector consisting of all variables $a_{i_k j}$ lexicographically ordered with respect to the natural order $1_1, 1_2, \ldots, m_n$ of the time slots and then, for each slot i_k, the jobs ordered according to $\prec_i$. The vector $b \in \mathbb{R}^{mn^2}$ is given by $b_{i_kj} = \frac{1}{2}w_jp_{ij} + w_jp_{i_k}$, and $D = \left(d_{(i_kj)(i'_{k'}j')}\right)$ is a symmetric $mn^2 \times mn^2$–matrix given through

$$d_{(i_kj)(i'_{k'}j')} = \begin{cases} 0 & \text{if } i_k \neq i'_{k'}\,, \\ w_{j'}p_{ij} & \text{if } i_k = i'_{k'} \text{ and } j \prec_i j'\,, \\ w_jp_{ij'} & \text{if } i_k = i'_{k'} \text{ and } j' \prec_i j\,, \\ w_jp_{ij} & \text{if } i_k = i'_{k'} \text{ and } j = j'\,. \end{cases}$$

Because of the lexicographic order of the indices the matrix D is decomposed into mn diagonal blocks corresponding to the mn time slots. If we assume that the jobs are indexed according to $\prec_i$ and if we denote p_{ij} simply by p_j, each block corresponding to a time slot on machine i has the following form:

$$\begin{pmatrix} w_1p_1 & w_2p_1 & w_3p_1 & \cdots & w_np_1 \\ w_2p_1 & w_2p_2 & w_3p_2 & \cdots & w_np_2 \\ w_3p_1 & w_3p_2 & w_3p_3 & \cdots & w_np_3 \\ \vdots & \vdots & \vdots & \ddots & \vdots \\ w_np_1 & w_np_2 & w_np_3 & \cdots & w_np_n \end{pmatrix}$$

It has been observed in [22] that those matrices are positive semidefinite and therefore the whole matrix D is positive semidefinite. In particular, the objective function (9) is convex and the quadratic programming relaxation can be solved in polynomial time, see, e. g., [11, 3].

The convex quadratic programming relaxation (CQP) is in some sense similar to the linear programming relaxation in time-indexed variables that has been introduced in [18]. Without going into the details, we give a rough idea of the common underlying intuition of both relaxations: a job may be split into several parts (corresponding to fractional values a_{i_kj} in (CQP)) who can be scattered over the machines and over time. The completion time of a job in such a 'fractional schedule' is somehow related to its mean busy time; the mean busy time of a job is the average point in time at which its fractions are being processed (see (8) where C_j is set to the average over the terms in brackets on the right hand side weighted by a_{i_kj}). However, in contrast to the time-indexed LP relaxation, the construction of the convex quadratic program (CQP) contains more insights into the structure of an optimal schedule. As a result, (CQP) is of strongly polynomial size while the LP relaxation contains an exponential number of time-indexed variables and constraints.

4 A simple 2–approximation algorithm

The value of an optimal solution to the convex quadratic programming relaxation (CQP) of the last section is a lower bound on the value of an optimal schedule. Moreover, from the structure of an optimal solution to the relaxation we can gain important insights that turn out to be useful in the construction of a provably good solution to the scheduling problem under consideration. In this context, randomized rounding has proved to be a powerful algorithmic tool. On the one hand, it yields very simple and easy to analyze algorithms; on the other hand, it is able to minutely capture the structure of the solution to the relaxation and to carry it over to a feasible schedule. The idea of using randomized rounding in the study of approximation algorithms was introduced by Raghavan and Thompson [17], an overview can be found in [13].

For a given optimal solution a to (CQP), we compute an integral solution $\bar{a}$ by setting for each job j exactly one of the variables $\bar{a}_{i_k j}$ to 1 with probabilities given through $a_{i_k j}$. Notice that $0 \leq a_{i_k j} \leq 1$ and the sum of the $a_{i_k j}$ for job j is equal to one by constraints (10). Although the integral solution $\bar{a}$ does not necessarily fulfill constraints (11), it represents a feasible assignment of jobs to time slots, i. e., a feasible solution to (QP), and thus a feasible schedule. For our analysis we require that the random choices are performed pairwise independently for the jobs.

Theorem 4. *Computing an optimal solution to (CQP) and using randomized rounding to turn it into a feasible schedule is a 2–approximation algorithm for the problem* $R|r_{ij}|\sum w_j C_j$.

Theorem 4 follows from the next lemma which gives a slightly stronger result including job-by-job bounds.

Lemma 5. *Using randomized rounding in order to turn an arbitrary feasible solution to (QP) into a feasible assignment of jobs to time slots yields a schedule such that the expected completion time of each job is bounded by twice the corresponding value* (8) *in the given solution to (QP).*

Since the value of an optimal solution to (CQP) is a lower bound on the value of an optimal schedule, Theorem 4 follows from Lemma 5 and linearity of expectations.

Our result on the quality of the computed schedule described in Theorem 4 also implies a bound on the quality of the quadratic programming relaxation that served as a lower bound in our estimations.

Corollary 6. *For instances of $R|r_{ij}|\sum w_j C_j$, the value of an optimal solution to the relaxation (CQP) is within a factor 2 of the value of an optimal schedule. This bound is tight even for the case of identical parallel machines without release dates* $P||\sum w_j C_j$.

5 Extensions to scheduling with preemptions

In this section we discuss the preemptive problem $R|r_{ij}, pmtn|\sum w_j C_j$ and generalizations to network scheduling. In contrast to the nonpreemptive setting, a job may now repeatedly be interrupted and continued later on another (or the same) machine. In the context of network scheduling, it is reasonable to assume that after a job has been interrupted on one machine it cannot be continued on another machine until a certain communication delay is elapsed that allows the job to travel through the network to its new machine.

The ideas and techniques presented in the last section can be generalized to this setting. However, since we have to use a somewhat weaker relaxation in order to capture the possibility of preemptions, we only get a 3–approximation algorithm. This result can be improved to performance guarantee 2 in the absence of nontrivial release dates $R|pmtn|\sum w_jC_j$ but with arbitrary communication delays. For reasons of brevity we only give a brief sketch of the main differences to the nonpreemptive setting.

Although the quadratic program (QP) allows to break a job into fractions and thus to preempt it by choosing fractional values a_{i_kj}, it is not a relaxation of $R|r_{ij}, pmtn|\sum w_jC_j$. However, we can turn it into a relaxation by replacing (8) with the weaker constraint

$$C_j = \sum_{i,k} a_{i_kj}\Big(s_{i_k} + \frac{a_{i_kj}}{2}p_{ij} + \sum_{j' \prec_i j} a_{i_kj'}p_{ij'}\Big) \qquad \text{for all } j.$$

Moreover, we restrengthen the relaxation by adding the following constraint

$$\sum_j w_jC_j \geq \sum_j w_j \sum_{i,k} a_{i_kj}p_{ij}$$

which bounds the objective value from below by the weighted sum of processing times (a similar constraint has already been used in [22]). Since Lemma 3 can be carried over to the new setting, we again get a convex quadratic programming relaxation.

In order to turn an optimal solution to this relaxation into a feasible schedule, we apply exactly the same randomized rounding heuristic as in the nonpreemptive case. In particular, we do not make use of the possibility to preempt jobs but compute a nonpreemptive schedule. Therefore, our results hold for the case of arbitrary communication delays.

Theorem 7. *Randomized rounding based on an optimal solution to the convex quadratic programming relaxation yields a 3–approximation algorithm for* $R|r_{ij}, pmtn|\sum w_jC_j$ *and a 2–approximation algorithm for* $R|pmtn|\sum w_jC_j$*, even for the case of arbitrary communication delays. The same bounds hold for the quality of the relaxation.*

Theorem 7 also implies bounds on the power of preemption. Since we can compute a nonpreemptive schedule whose value is bounded by 3 respectively 2 times the value of an optimal preemptive schedule, we have derived upper bounds on the ratios of optimal nonpreemptive to optimal preemptive schedules.

Corollary 8. *For instances of* $R|r_{ij}|\sum w_jC_j$*, the value of an optimal nonpreemptive schedule is at most a factor* 3 *above the value of an optimal preemptive schedule. In the absence of nontrivial release dates, this bound can be improved to* 2.

6 Conclusion

We have presented convex quadratic programming relaxations of strongly polynomial size which lead to simple and easy to analyze approximation algorithms for preemptive and nonpreemptive network scheduling. Although our approach and the presented results might be at first sight of mainly theoretical interest, we hope that nonlinear relaxations like the one we discuss in this paper will also prove useful in solving real world scheduling problems in the near future. With the development of better algorithms that solve convex quadratic programs more efficiently in practice, the results obtained by using such relaxations might become comparable or even better than those based on linear programming relaxations with a huge number of time-indexed variables and constraints.

Precedence constraints between jobs play a particularly important role in most real world scheduling problems. Therefore it would be both of theoretical and of practical interest to incorporate those constraints into our convex quadratic programming relaxation.

Hoogeveen, Schuurman, and Woeginger [10] have shown that the problems $R \mid r_j \mid \sum C_j$ and $R \mid \mid \sum w_j C_j$ cannot be approximated in polynomial time within arbitrarily good precision, unless P=NP. It is an interesting open problem to close the gap between this negative result and the 2–approximation algorithm presented in this paper.

References

[1] B. Awerbuch, S. Kutten, and D. Peleg. Competitive distributed job scheduling. In *Proceedings of the 24th Annual ACM Symposium on the Theory of Computing*, pages 571 – 581, 1992.

[2] J. L. Bruno, E. G. Coffman Jr., and R. Sethi. Scheduling independent tasks to reduce mean finishing time. *Communications of the Association for Computing Machinery*, 17:382 – 387, 1974.

[3] S. J. Chung and K. G. Murty. Polynomially bounded ellipsoid algorithms for convex quadratic programming. In O. L. Mangasarian, R. R. Meyer, and S. M. Robinson, editors, *Nonlinear Programming 4*, pages 439 – 485. Academic Press, 1981.

[4] X. Deng, H. Liu, J. Long, and B. Xiao. Deterministic load balancing in computer networks. In *Proceedings of the 2nd Annual IEEE Symposium on Parallel and Distributed Processing*, pages 50 – 57, 1990.

[5] M. E. Dyer and L. A. Wolsey. Formulating the single machine sequencing problem with release dates as a mixed integer program. *Discrete Applied Mathematics*, 26:255 – 270, 1990.

[6] M. X. Goemans. Improved approximation algorithms for scheduling with release dates. In *Proceedings of the 8th Annual ACM–SIAM Symposium on Discrete Algorithms*, pages 591 – 598, 1997.

[7] R. L. Graham, E. L. Lawler, J. K. Lenstra, and A. H. G. Rinnooy Kan. Optimization and approximation in deterministic sequencing and scheduling: A survey. *Annals of Discrete Mathematics*, 5:287 – 326, 1979.

[8] L. A. Hall, A. S. Schulz, D. B. Shmoys, and J. Wein. Scheduling to minimize average completion time: Off–line and on–line approximation algorithms. *Mathematics of Operations Research*, 22:513 – 544, 1997.

[9] L. A. Hall, D. B. Shmoys, and J. Wein. Scheduling to minimize average completion time: Off–line and on–line algorithms. In *Proceedings of the 7th Annual ACM–SIAM Symposium on Discrete Algorithms*, pages 142 – 151, 1996.

[10] H. Hoogeveen, P. Schuurman, and G. J. Woeginger. Non-approximability results for scheduling problems with minsum criteria. In R. E. Bixby, E. A. Boyd, and R. Z. Ríos-Mercado, editors, *Integer Programming and Combinatorial Optimization*, volume 1412 of *Lecture Notes in Computer Science*, pages 353 – 366. Springer, Berlin, 1998.

[11] M. K. Kozlov, S. P. Tarasov, and L. G. Hačijan. Polynomial solvability of convex quadratic programming. *Soviet Mathematics Doklady*, 20:1108 – 1111, 1979.

[12] J. K. Lenstra, A. H. G. Rinnooy Kan, and P. Brucker. Complexity of machine scheduling problems. *Annals of Discrete Mathematics*, 1:343 – 362, 1977.

[13] R. Motwani, J. Naor, and P. Raghavan. Randomized approximation algorithms in combinatorial optimization. In D. S. Hochbaum, editor, *Approximation algorithms for NP-hard problems*, chapter 11, pages 447 – 481. Thomson, 1996.

[14] C. H. Papadimitriou and M. Yannakakis. Towards an architecture-independent analysis of parallel algorithms. *SIAM Journal on Computing*, 19:322 – 328, 1990.

[15] C. Phillips, C. Stein, and J. Wein. Task scheduling in networks. *SIAM Journal on Discrete Mathematics*, 10:573 – 598, 1997.

[16] C. Phillips, C. Stein, and J. Wein. Minimizing average completion time in the presence of release dates. *Mathematical Programming*, 82:199 – 223, 1998.

[17] P. Raghavan and C. D. Thompson. Randomized rounding: A technique for provably good algorithms and algorithmic proofs. *Combinatorica*, 7:365 – 374, 1987.

[18] A. S. Schulz and M. Skutella. Scheduling–LPs bear probabilities: Randomized approximations for min–sum criteria. In R. Burkard and G. J. Woeginger, editors, *Algorithms – ESA '97*, volume 1284 of *Lecture Notes in Computer Science*, pages 416 – 429. Springer, Berlin, 1997.

[19] J. Sethuraman and M. S. Squillante. Optimal scheduling of multiclass prallel machines. In *Proceedings of the 10th Annual ACM–SIAM Symposium on Discrete Algorithms*, pages 963 – 964, 1999.

[20] D. B. Shmoys and É. Tardos. An approximation algorithm for the generalized assignment problem. *Mathematical Programming*, 62:461 – 474, 1993.

[21] M. Skutella. *Approximation and Randomization in Scheduling*. PhD thesis, Technical University of Berlin, Germany, 1998.

[22] M. Skutella. Semidefinite relaxations for parallel machine scheduling. In *Proceedings of the 39th Annual IEEE Symposium on Foundations of Computer Science*, pages 472 – 481, 1998.

[23] M. Skutella. Convex Quadratic and Semidefinite Programming Relaxations in Scheduling. Manuscript, 1999.

[24] W. E. Smith. Various optimizers for single–stage production. *Naval Research and Logistics Quarterly*, 3:59 – 66, 1956.

Resource-Constrained Project Scheduling: Computing Lower Bounds by Solving Minimum Cut Problems*

Rolf H. Möhring[1], Andreas S. Schulz[2], Frederik Stork[1], and Marc Uetz[1]

[1] Technische Universität Berlin, Fachbereich Mathematik
Sekr. MA 6–1, Straße des 17. Juni 136, 10623 Berlin, Germany
{moehring,stork,uetz}@math.tu-berlin.de
[2] MIT, Sloan School of Management and Operations Research Center
E53-361, 30 Wadsworth St, Cambridge, MA 02139
schulz@mit.edu

Abstract. We present a novel approach to compute Lagrangian lower bounds on the objective function value of a wide class of resource-constrained project scheduling problems. The basis is a polynomial-time algorithm to solve the following scheduling problem: Given a set of activities with start-time dependent costs and temporal constraints in the form of time windows, find a feasible schedule of minimum total cost. In fact, we show that any instance of this problem can be solved by a minimum cut computation in a certain directed graph.
We then discuss the performance of the proposed Lagrangian approach when applied to various types of resource-constrained project scheduling problems. An extensive computational study based on different established test beds in project scheduling shows that it can significantly improve upon the quality of other comparably fast computable lower bounds.

1 Introduction and Problem Formulation

Resource-constrained project scheduling problems usually comprise several activities or jobs which have to be scheduled subject to both temporal and resource constraints in order to minimize a certain objective. Temporal constraints often consist of precedence constraints, that is, certain activities must be completed before others can be processed, but sometimes even arbitrary minimal and maximal time lags, so-called time windows between pairs of activities have to be respected. Moreover, activities require resources while being processed, and the resource availability is limited. Also time-varying resource requirements and resource availabilities may occur. Most frequently, the project makespan is to be minimized, but also other, even non-regular objective functions are considered in the literature. For a detailed account of the various problem settings, most relevant references as well as a classification scheme for resource-constrained project scheduling problems we refer to [2].

* Research partially supported by the Bundesministerium für Bildung, Wissenschaft, Forschung und Technologie (bmb+f), grant No. 03-MO7TU1-3.

In general, resource-constrained project scheduling problems are among the most intractable problems, and in the case of time windows even the problem of finding a feasible solution is NP-hard, cf., e.g., [1]. The intractability of these problems motivates the search for good and fast computable lower bounds on the objective value, which may be used to improve both heuristics and exact procedures. However, it is very unlikely that provably good lower bounds can be computed within polynomial time, since project scheduling problems contain node coloring in graphs as a special case [20]. Thus, as for node coloring, there is no polynomial-time approximation algorithm with a performance guarantee less that n^{ε} for some $\varepsilon > 0$, unless $P = NP$. This negative results also implies limits on the computation of good lower bounds.

Problem formulation. Let $V = \{0, \ldots, n+1\}$ be a set of activities j with integral activity durations p_j. All activities must be scheduled non-preemptively, and by $S = (S_0, \ldots, S_{n+1})$ we denote a schedule, where S_j is the start time of activity j. Activities 0 and $n+1$ are assumed to be dummy activities indicating the project start and the project completion, respectively. Temporal constraints in the form of minimal and maximal time lags between pairs of activities are given. By d_{ij} we denote a time lag between two activities $i, j \in V$, and $L \subseteq V \times V$ is the set of all time lags. We assume that the temporal constraints always refer to the start times, thus every schedule S has to fulfill $S_j \geq S_i + d_{ij}$ for all $(i, j) \in L$. Note that $d_{ij} \geq 0$ ($d_{ji} < 0$) implies a minimal (maximal) positive time lag of S_j relative to S_i, thus so-called *time windows* of the form $S_i + d_{ij} \leq S_j \leq S_i - d_{ji}$ between any two activities can be modeled. Ordinary precedence constraints can be represented by letting $d_{ij} = p_i$ if activity i must precede activity j. Additionally, we suppose that a time horizon T as an upper bound on the project makespan is given. It can be checked in polynomial time by longest path calculations if such a system of temporal constraints has a feasible solution. Throughout the paper we will assume that a schedule exists that satisfies all temporal constraints. We then obtain for each activity a set of (integral) feasible start times $I_j := \{ES_j, ..., LS_j\}$, $j \in V$, where ES_j and LS_j denote the earliest and latest start time of activity j, respectively. Activities need resources for their processing. In the model with constant resource requirements, we are given a finite set $\mathcal{R}$ of different, renewable resources, and the availability of resource $k \in \mathcal{R}$ is denoted by R_k, that is, an amount of R_k units of resource k is available throughout the project. Every activity j requires an amount of r_{jk} units of resource k, $k \in \mathcal{R}$. The activities have to be scheduled such as to minimize a given measure of performance, usually the project makespan.

Project scheduling problems are often formulated as integer linear programs with time-indexed binary variables x_{jt}, $j \in V$, $t \in \{0, ..., T\}$, which are defined by $x_{jt} = 1$ if activity j starts at time t and $x_{jt} = 0$ otherwise. This leads to the following, well known integer linear programming formulation.

$$\text{minimize} \quad \sum_t t\, x_{n+1,t} \tag{1}$$

$$\text{subject to} \quad \sum_t x_{jt} = 1, \qquad j \in V, \tag{2}$$

$$\sum_{s=t}^{T} x_{is} + \sum_{s=0}^{t+d_{ij}-1} x_{js} \leq 1, \qquad (i,j) \in L,\ t \in \{0,\dots,T\}, \tag{3}$$

$$\sum_j r_{jk}\Big(\sum_{s=t-p_j+1}^{t} x_{js}\Big) \leq R_k, \qquad k \in \mathcal{R},\ t \in \{0,\dots,T\}, \tag{4}$$

$$x_{jt} \in \{0,1\}, \qquad j \in V,\ t \in \{0,\dots,T\}. \tag{5}$$

Constraints (2) indicate that each activity is started exactly once, and inequalities (3) represent the temporal constraints given by the time lags L. Inequalities (4) assure that the activities processed simultaneously at time t do not consume more resources than available. Note that this formulation can easily be generalized to time-dependent resource profiles, i.e., $R_k = R_k(t)$ and $r_{jk} = r_{jk}(t)$. In fact, although we discuss in the following the case of time-independent resource profiles only, the presented results carry over to the general case. Computational results for both models are discussed in Sect. 3.

Related work. The above time-indexed formulation for project scheduling problems has been used before by various authors (e.g. [18, 7, 5, 4]), sometimes with a weaker formulation of temporal constraints given by $\sum_t t(x_{jt} - x_{it}) \geq d_{ij}$, $(i,j) \in L$. Most relevant to our work is the paper by Christofides, Alvarez-Valdes, and Tamarit [7]. They have investigated a Lagrangian relaxation of the above integer program in order to obtain lower bounds on the makespan. They solve the Lagrangian relaxation with the help of a branch and bound algorithm, apparently unaware that it can be solved in polynomial time by purely combinatorial methods (see Sect. 2). As a matter of fact, the LP relaxation of (2), (3), and (5) is known to be integral. This important structural result is due to Chaudhuri, Walker, and Mitchell [5]. For problems with precedence constraints and time-varying resources, this has also been shown by Cavalcante, De Souza, Savelsbergh, Wang, and Wolsey [4]. The latter authors solve the linear programming relaxation of (1) – (5) in order to exploit its solution for ordering heuristics to construct good feasible schedules. Another technique to compute lower bounds on the project makespan for resource-constrained project scheduling problems has been proposed by Mingozzi, Maniezzo, Ricciardelli, and Bianco [14]. Their approach relies on a different mathematical formulation that is based on variables $y_{\ell t}$ which indicate if a (resource feasible) subset of activities $V_\ell \subseteq V$ is in process at a certain time t. Clearly, this formulation is of exponential size, since there are exponentially many such feasible subsets V_ℓ. They derive different lower bounds by considering several relaxations, including a very fast computable lower bound, usually referred to as LB_3, which is based on the idea to sum up the processing times of activities which pairwise cannot be scheduled simultaneously. Their bounds have then been evaluated and modified by various authors. In particular, Brucker and Knust [3] solve the following relaxation: Feasible subsets of activities must be scheduled (preemptively) such that every activity receives at least its total processing time. Brucker and Knust apply column generation where the pricing is done by branch and bound. They obtain the best known bounds on the majority of instances of a well known test bed [13]. However, their approach often requires extremely large computation times.

Results. In the spirit of Christofides, Alvarez-Valdes, and Tamarit [7], we propose a Lagrangian relaxation of (1) – (5) to compute lower bounds for resource-constrained project scheduling problems. Within a subgradient optimization algorithm we solve a series of project scheduling problems given by (2), (3), and (5), subject to start time dependent costs for each activity. The core of our approach is a direct transformation of this problem to a minimum cut problem in an appropriately defined directed graph which can then be solved by a standard maximum flow algorithm. The potential of this approach is demonstrated by computational results. We have used widely accepted test beds in project scheduling, namely problems with ordinary precedence constraints [13] as well as arbitrary minimal and maximal time lags [21], and labor-constrained scheduling problems with a time varying resource profile modeled after chemical production processes within BASF AG, Germany [12]. The experiments reveal that our approach is capable of computing very good lower bounds at very short computation times. We thus improve previous, fast computable lower bounds, and in the setting with time windows we even obtain best known lower bounds for quite a few instances. Compared to other approaches which partially require prohibitive running times, our algorithm offers a good tradeoff between quality and computation time. It also turns out that it is especially suited for problems with extremely scarce resources, which are the problems that tend to be intractable. For the instances stemming from BASF, Cavalcante et al. [4] report on tremendous computation times for solving the corresponding linear programming relaxations. Our experiments show that one can obtain essentially the same value as with the LP relaxation much more efficiently.

Organization of the paper. In Sect. 2 we present the Lagrangian relaxation of the integer program (1) – (5), and introduce a direct transformation of the resulting subproblems to minimum cut problems in an appropriate directed graph. Section 3 is then concerned with an extensive computational study of this approach. We analyze our algorithm in comparison to both the solution of the corresponding LP relaxations, and other lower bounding algorithms. We conclude with some remarks on future research in Sect. 4. Due to space limitations, quite some details have been omitted from this extended abstract. They will be presented in the full version [17].

2 The Lagrangian Relaxation

Christofides, Alvarez-Valdes, and Tamarit [7] have proposed the following Lagrangian relaxation of the time indexed integer programming formulation of resource-constrained project scheduling given by (1) – (5). They dualize the resource constraints (4), and introduce Lagrangian multipliers $\lambda_{tk} \geq 0$, $t \in \{0, \ldots, T\}$, $k \in \mathcal{R}$.

$$\text{minimize} \sum_t t\, x_{n+1,t} + \sum_j \sum_t \Big(\sum_{k \in \mathcal{R}} r_{jk} \sum_{s=t}^{t+p_j-1} \lambda_{sk} \Big) x_{jt} - \sum_t \sum_{k \in \mathcal{R}} \lambda_{tk} \cdot R_k \tag{6}$$

subject to (2), (3), and (5).

If one omits the constant term $\sum_t \sum_{k \in \mathcal{R}} \lambda_{tk} \cdot R_k$ and introduces weights $w_{jt} = \sum_{k \in \mathcal{R}} r_{jk} \sum_{s=t}^{t+p_j-1} \lambda_{sk}$ for all $j \neq n+1$ and $w_{n+1,t} = t$, (6) can be reformulated as

$$\text{minimize} \qquad c(x) := \sum_j \sum_t w_{jt} \cdot x_{jt} \qquad \text{subject to (2), (3), and (5).} \tag{7}$$

This formulation specifies a project scheduling problem where the activities have start-time dependent costs, and where the aim is to minimize the overall cost subject to minimal and maximal time lags between activities. We refer to this problem as *project scheduling problem with start-time dependent costs*. Note that all weights can without loss of generality be assumed to be positive, since due to (2), any additive transformation of the weights only affects the solution value, but not the solution itself. (In (7) the weights are non-negative by definition.) The problem can trivially be solved by longest path calculations if the w_{jt} are non-decreasing in t. However, this is not the case for general weights.

Observe also that the above Lagrangian relaxation is not restricted to makespan minimization, but can as well be applied to any other regular, and even non-regular objective function. Thus the procedure proposed below is applicable to a variety of project scheduling problems like, for instance, the minimization of the weighted sum of completion times, problems that aim at minimizing lateness, or resource investment problems [15, 8].

The transformation. We now present a reduction of the project scheduling problem with start time dependent costs given in (7) to a minimum cut problem in a directed graph $D = (N, A)$ which is defined as follows.

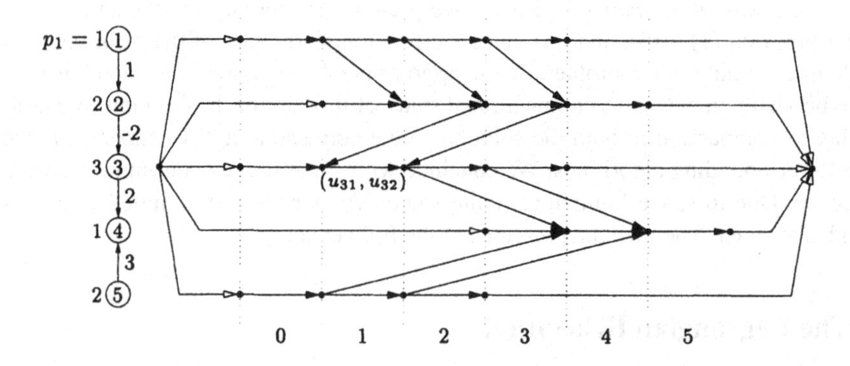

Fig. 1. The left digraph represents the relevant data of the underlying example: Each node represents an activity, each arc represents a temporal constraint. The right digraph D is the corresponding graph obtained by the transformation. Each assignment arc of D corresponds to a binary variable x_{jt}. Arcs marked by a white arrow head are dummy arcs that connect the source a and the sink b with the remaining network.

• *Nodes.* The set N of nodes contains for each activity $j \in V$ the nodes u_{jt}, $t \in \{ES_j, \ldots, LS_j + 1\}$. Furthermore, it contains two auxiliary nodes, a dummy source a and a dummy sink b. Hence, $N := \{a,b\} \cup \{u_{jt} | j \in V, t \in \{ES_j,...,LS_j+1\}\}$.

• *Arcs.* The arc set A can be divided into three disjoint subsets. The set of *assignment arcs* is induced by the binary variables x_{jt} of the integer program (7) and contains all arcs $(u_{jt}, u_{j,t+1})$ for all $j \in V$ and $t \in I_j$, where x_{jt} corresponds to $(u_{jt}, u_{j,t+1})$. The set of *temporal arcs* is defined as the set $\{(u_{is}, u_{j,s+d_{ij}}) | (i,j) \in L, s \in I_i\}$. It will later be needed to guarantee that no temporal constraint is violated. Finally, a set of *dummy arcs* connects the source and the sink nodes a and b with the remaining network. The dummy arcs are given by $\{(a, u_{i,ES_i} | i \in V\} \cup \{(u_{i,LS_i+1}, b) | i \in V\}$.

• *Capacities.* The capacity of an assignment arc $(u_{jt}, u_{j,t+1})$, $j \in V$, $t \in I_j$, equals the weight w_{jt} of its associated binary variable x_{jt}, the capacity of temporal arcs and dummy arcs is infinite, and all *lower* capacities are 0.

Figure 1 shows an example of the graph D based on an instance with 5 activities $V = \{1,...,5\}$. The set of time lags is $\{d_{12} = 1, d_{23} = -2, d_{34} = 2, d_{54} = 3\}$. The activity durations are $p_1 = p_4 = 1$, $p_2 = p_5 = 2$, and $p_3 = 3$, and $T = 6$ is a given upper bound on the project makespan. Thus the earliest start vector is $ES = (0,1,0,3,0)$ and the latest start vector is $LS = (3,4,3,5,2)$.

We use the following notation. Given a directed graph, an *a,b-cut* is a pair $(X, \bar{X})$ of disjoint sets $X, \bar{X} \subset N$ with $X \cup \bar{X} = N$, and $a \in X$, $b \in \bar{X}$. The *capacity* $c(X, \bar{X})$ of a cut $(X, \bar{X})$ is the sum of capacities of the forward arcs in the cut, $c(X, \bar{X}) := \sum_{(u,\bar{u}) \in (X,\bar{X})} c(u, \bar{u})$.

Theorem 1. *A minimum a,b–cut $(X^*, \bar{X}^*), a \in X^*, b \in \bar{X}^*$ of the digraph D corresponds to an optimal solution x^* of the integer program (7) of the project scheduling problem with start time dependent costs by setting*

$$x^*_{jt} = \begin{cases} 1 & \text{if } (u_{jt}, u_{j,t+1}) \text{ is a forward arc of the cut } (X^*, \bar{X}^*), \\ 0 & \text{otherwise.} \end{cases} \tag{8}$$

Moreover, the value $c(x^*)$ of that solution equals the capacity $c(X^*, \bar{X}^*)$ of the minimum cut $(X^*, \bar{X}^*)$.

The proof crucially uses the fact that each minimum a,b–cut of the digraph D consists of exactly one forward arc $(u_{jt}, u_{j,t+1})$ for every activity j. Note that this only holds since the weights w_{jt} are strictly positive and thus also the capacities of the arcs are strictly positive. Furthermore, it is essential that the given instance has a feasible solution and thus a minimum cut has finite capacity.

Since D has $O(n \cdot T)$ nodes and $O((n+m) \cdot T)$ arcs, a minimum cut in D can be computed in $O(nmT^2 log(T))$ time with the classical push-relabel-algorithm for maximum flows [9]. Here, m is the number of given time lags L.

A related transformation has been investigated by Chaudhuri, Walker, and Mitchell [5]. They transform the integer program (7) into a cardinality-constrained stable set problem in comparability graphs, with the objective to identify a stable set of minimum weight among all stable sets of maximum cardinality. The weighted stable set problem in comparability graphs can be transformed in polynomial time to a maximum cut problem on a digraph, in which the maximum cut corresponds to the maximum weighted stable set, cf. [10, 16]. However, the resulting digraph is dense while the digraph resulting from our transformation has a very sparse structure, since the set L of temporal constraints is usually sparse. Moreover, the directed graph D as defined above need not be acyclic, and thus cannot be derived from a transitive orientation of the comparability graph defined in [5].

3 Experimental Study

We first compare the performance of the Lagrangian relaxation approach with the LP-relaxation of (1) – (5). We then empirically analyze how the running time depends on the time horizon and the number of activities. We also analyze the dependency of both running time and solution quality on the scarceness of resources. Next, we compare our bounds with those computed by other lower bounding algorithms. We finally briefly investigate the computation of feasible schedules from the solution of the Lagrangian relaxation.

We use a standard subgradient method to attack the Lagrangian dual. It is aborted if the objective value was not improved significantly over five consecutive iterations. If this happens within the first 10 iterations we restart the procedure with another choice of step sizes.

For the computations we have used strengthened resource inequalities which have been proposed in [7]. They guarantee that no activity is scheduled parallel to the dummy sink $n+1$.

$$\sum_j r_{jk}\Big(\sum_{s=t-p_j+1}^{t} x_{js}\Big) + R_k \sum_{s=ES_{n+1}}^{t} x_{n+1,s} \leq R_k, \quad k \in \mathcal{R}, t = 0, \ldots, T \tag{9}$$

3.1 Benchmark Instances

We have applied our algorithm to the test beds of the ProGen and the ProGen/max library [13, 21], and to a small test bed of problems modeled after chemical production processes with labor constraints [11].

The ProGen library [13] provides instances for precedence-constrained scheduling with multiple resource constraints and with 30, 60, 90, and 120 activities, respectively. They are generated by modifying three parameters, the *network complexity*, which reflects the average number of direct successors of an activity, the *resource factor*, which describes the average number of resources required in order to process an activity, and the *resource strength*, which is a measure of the scarcity of the resources. The resource strength varies between 0.1 and 0.7 where a small value indicates very scarce resources. This variation results into 480 instances of each of the first three instance sizes (30, 60, 90), and 600 instances of 120 activities. The activity durations were chosen randomly between 1 and 10 and the maximum number of different resources is 4 per activity. The library also contains best known upper bounds on the makespan of these instances, which we have used as time horizon T. ¿From the whole set of instances we only took those for which the given upper bound is larger than the trivial lower bound LB_0 which is the earliest start time ES_{n+1} of the dummy activity $n+1$. The number of instances then reduces to 264 (30 activities), 185 (60 activities), 148 (90 activities), and 432 (120 activities). The time horizon of these instances varies between 35 and 306. For further details we refer to [13].

The ProGen/max library [21] provides 1080 instances for scheduling problems with time windows and multiple resource requirements, each of which consists of 100 activities. The parameters are similar to those of the ProGen library, but an additional parameter controls the number of cycles in the digraph of temporal constraints. 21 of the 1080 instances are infeasible and for another 693 instances there exists a feasible solution with a project makespan which equals the trivial lower bound LB_0. Thus, the number of instances of interest within this test bed reduces to 366. For these instances, the time horizon varies between 253 and 905. For further details we refer to [21].

Finally, we consider instances which have their origin in a labor-constrained scheduling problem (LCSP instances) from BASF AG, Germany, which can briefly be summarized as follows: The production process for a set of orders has to be scheduled. Every order represents the output of a constant amount of a chemical product, and the aim is to minimize the project makespan. The production process for an order consists of a sequence of identical activities, each of which must be scheduled non-preemptively. Due dates for individual orders are given, and due to technical reasons there may be precedence constraints between activities of different orders. Additionally, resource constraints have to be respected, which are imposed by a limited number of available workers: An activity usually consists of several consecutive tasks which require a certain amount of personnel. Thus, the personnel requirement of an activity is a piecewise constant function. More details can be found in [12]. The instances considered here are taken from [11].

3.2 Computational Results

Computing Environment. Our experiments were conducted on a Sun Ultra 2 with 200 MHz clock pulse operating under Solaris 2.6 with 512 MB of memory. The code is written in C++ and has been compiled with the GNU g++ compiler version 2.7.2. We use Cherkassky and Goldberg's maximum flow code [6]. It is written in C and has been compiled with the GNU gcc compiler version 2.7.2. Both compilers used the -O3 optimization option. All reported CPU times have been averaged over three runs.

LP relaxation versus Lagrangian relaxation. Since the LP relaxation of (7) is integral, the lower bounds obtained by the Lagrangian relaxation are bounded from above by the optimal solution of the LP relaxation. Since the LPs are usually very large we have compared the running times to solve these LPs with the Lagrangian approach. For the ProGen test bed with 30 activities [13], the LPs are solved within 18 seconds on average (max. 516 sec.) with CPLEX version 4.0.8, while the Lagrangian relaxation plus the subgradient optimization requires only one second (max. 6.2 sec.) at an average number of 104 iterations. The average deviation of the solution values turned out to be less than 1%.

For the LCSP instances Cavalcante et al. [4] have solved LP-relaxations of different integer programming formulations in order to obtain both lower bounds and ideas for generating feasible schedules. They report on excessive computation times to solve the LP-relaxation of (1) – (5); particularly for large instances they require more than 5 hours on average on a RS6000 model 590 (see [4], Table 1). To solve the corresponding Lagrangian relaxation (including the subgradient optimization) it requires only one minute on average for the same instances, while the lower bounds obtained are only slightly inferior (less than 2%). It turns out that for these instances the same bounds can also be obtained by solving a weaker LP-relaxation based on a weaker formulation of the temporal constraints (see [4], Table 2); here the computation times correspond to those of the Lagrangian approach. However, note that the Lagrangian relaxation is based on the strong formulation of the temporal constraints (3), and thus has the potential of providing better bounds.

Problem characteristics. We have empirically analyzed the running time and the performance of the Lagrangian method with respect to varying problem parameters. Figures 2 (a) and (b) display plots that show how the running time depends on both the time horizon and the number of activities. Each plot additionally contains a corresponding regression curve. Since other algorithms often require large running times when dealing with instances consisting of very hard resource constraints, we investigate the dependency of the running time of our algorithm on the resource strength. Recall that a low resource strength parameter is an indicator for such instances. As depicted in Fig. 2 (c), the running time of our algorithm seems only slightly affected by the resource strength parameter.

Other lower bounding algorithms. In comparison to other lower bounding procedures, our approach behaves quite reasonable with respect to the tradeoff between quality and computation time. For the scenario with precedence constraints, we compare our algorithm with two other approaches which are both based on [14]. First, we consider the lower bounds reported by Brucker and Knust [3] which are the strongest known bounds for the ProGen instances. Second, we have implemented the $O(|V|^2)$ lower bound LB_3 (cf. Sect. 1). The average results on the running time and the quality of the lower bounds are provided in Table 1, while Fig. 2 (d) displays the quality of the bounds depending on the resource strength parameter. Compared to the bound LB_3, our algorithm produces far better bounds in most of the cases. While the computation time for LB_3 is negligible (< 0.5 sec.), the algorithm of Brucker and Knust provides better bounds, but in exchange for much larger running times. To obtain the lower bounds for the instances which consist of 120 activities, their algorithm occasionally requires a couple of days per instance (on a Sun Ultra 2 with 167 MHz clock pulse), as reported to us in private communication. We could solve all of these instances within an average of less than a minute and a maximum of 362 seconds using 12 MB memory.

For the instances with time windows, the algorithm proposed by Brucker and Knust [3] cannot be applied, since it is developed for the model with precedence constraints only. The best known lower bounds collected in the library are computed by different algorithms, mostly by a combination of preprocessing steps and a generalization of the lower bound LB_3. As indicated in Table 1, the results of our algorithm on this test bed are less satisfactory with respect to quality of the bounds as well as running times. The reason might be a weaker average resource strength which leads to bounds of low quality, and also the larger time horizons which result in large running times. However, we were able to improve 38 of the best known lower bounds among the 366 instances.

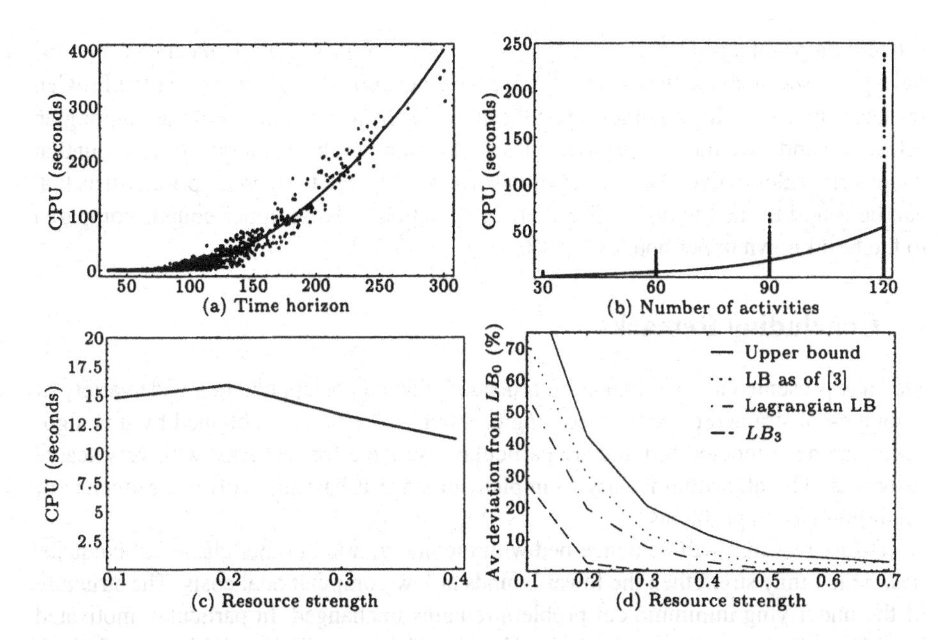

Fig. 2. Plots (a) and (b) show the running time depending on the time horizon and the number of activities (based on all instances of the ProGen library). Graphic (c) displays the effect of the resource strength on the running time for fixed $n = 120$ and $T \in [100, 120]$, and Graphic (d) visualizes the quality of the different bounds depending on the resource strength (with respect to the critical path lower bound LB_0).

For the LCSP instances, our computational experiences coincide with the above observation that the quality of our lower bounds increases when the availability of personnel (resources) is very low and vice versa.

Table 1. Lower bounds obtained by the Lagrangian relaxation for the different test beds as described in Sect. 3.1. *prec* and *temp* indicates whether the instances consist of precedence constraints [13] or of arbitrary time lags [21], respectively.

								best known	
Type	#act.	#inst.	LB_0	LB	UB	CPU	LB_3	LB	CPU
prec	60	185	71.3	78.8	90.6	6.1	74.2	85.6	13.5
prec	90	148	86.3	99.6	115.8	20.0	86.8	106.1	170.8
prec	120	432	94.6	116.7	137.6	56.9	102.0	124.9	n.a.
temp	100	366	431.4	435.9	499.0	72.1	434.2	452.2	n.a.

Computing feasible schedules. Besides the computation of lower bounds, both LP and Lagrangian relaxations allow the construction of good upper bounds by exploiting the structure of the corresponding solution within heuristics. For labor-constrained scheduling problems, Cavalcante et al. [4] as well as Savelsbergh, Uma, and Wein [19] have proposed such techniques. So far, we have performed experiments for the ProGen instances by extracting an ordering on the activities from the solution of the Lagrangian relaxation and used them as priority rules to generate feasible solutions. It turns out that the priority rules deliver the best schedules when compared to standard priority lists that can be found in the literature. The average deviation of these upper bounds compared to the best known upper bounds is 18%.

4 Concluding Remarks

We have presented a lower bounding procedure that can be applied to a wide variety of resource-constrained project scheduling problems. The bounds obtained by this algorithm can be computed fast and are particularly suitable for scenarios with very scarce resources. The algorithm is easy to implement since it basically solves a sequence of minimum s-t-cut problems.

Future research will be concerned with the integration of other classes of inequalities, which may strengthen the lower bounds at low computational costs. The structure of the underlying minimum cut problem remains unchanged. In particular, motivated by [14], such inequalities can be derived by identifying sets W of activities out of which not more than $\ell < |W|$ activities can be scheduled simultaneously. Furthermore, it could be valuable to adapt the maximum flow algorithm for our specific application, and also to recycle the flow (cut) data of the previous iteration.

Acknowledgements. The authors are grateful to Matthias Müller-Hannemann for many helpful discussions on maximum flow algorithms, and to Olaf Jahn for his technical support.

References

[1] M. Bartusch, R. H. Möhring, and F. J. Radermacher. Scheduling project networks with resource constraints and time windows. *Annals of Operations Research*, 16:201–240, 1988.

[2] P. Brucker, A. Drexl, R. H. Möhring, K. Neumann, and E. Pesch. Resource-constrained project scheduling: Notation, classification, models, and methods. *European Journal of Operational Research*, 112:3–41, 1999.

[3] P. Brucker and S. Knust. A linear programming and constraint propagation-based lower bound for the RCPSP. Technical Report 204, Osnabrücker Schriften zur Mathematik, 1998.

[4] C. C. B. Cavalcante, C. C. De Souza, M. W. P. Savelsbergh, Y. Wang, and L. A. Wolsey. Scheduling projects with labor constraints. CORE Discussion Paper 9859, Université Catholique de Louvain, Louvain-la-Neuve, Belgium, 1998.

[5] S. Chaudhuri, R. A. Walker, and J. E. Mitchell. Analyzing and exploiting the structure of the constraints in the ILP approach to the scheduling problem. *IEEE Transactions on Very Large Scale Integration (VLSI) Systems*, 2:456 – 471, 1994.

[6] B. Cherkassky and A. V. Goldberg. On implementing push-relabel method for the maximum flow problem. In *Proceedings of the 4th Conference on Integer Programming and Combinatorial Optimization*, pages 157–171, 1995.

[7] N. Christofides, R. Alvarez-Valdes, and J. Tamarit. Project scheduling with resource constraints: A branch-and-bound approach. *European Journal of Operational Research*, 29:262–273, 1987.

[8] A. Drexl and A. Kimms. Optimization guided lower and upper bounds for the Resource Investment Problem. Technical Report 481, Manuskripte aus den Instituten für Betriebswirtschaftslehre der Universität Kiel, 1998.

[9] A. V. Goldberg and R. E. Tarjan. A new approach to the maximum-flow problem. *Journal of the ACM*, 35:921–940, 1988.

[10] M. C. Golumbic. *Algorithmic Graph Theory and Perfect Graphs*. Academic Press, New York, 1980.

[11] `http://www.dcc.unicamp.br/~cris/SPLC.html`.

[12] J. Kallrath and J. M. Wilson. *Business Optimisation using Mathematical Programming*. Macmillan Business, London, U.K., 1997.

[13] R. Kolisch and A. Sprecher. PSPLIB - a project scheduling problem library. *European Journal of Operational Research*, 96:205–216, 1996.

[14] A. Mingozzi, V. Maniezzo, S. Ricciardelli, and L. Bianco. An exact algorithm for the multiple resource-constraint project scheduling problem based on a new mathematical formulation. *Management Science*, 44:714–729, 1998.

[15] R. H. Möhring. Minimizing costs of resource requirements subject to a fixed completion time in project networks. *Operations Research*, 32:89–120, 1984.

[16] R. H. Möhring. Algorithmic aspects of comparability graphs and interval graphs. In I. Rival, editor, *Graphs and Order*, pages 41–101. D. Reidel Publishing Company, Dordrecht, 1985.

[17] R. H. Möhring, A. S. Schulz, F. Sork, and M. Uetz. In preparation, 1999.

[18] A. A. B. Pritsker, L. J. Watters, and P. M. Wolfe. Multi project scheduling with limited resources: A zero-one programming approach. *Management Science*, 16:93–108, 1969.

[19] M. W. P. Savelsbergh, R. N. Uma, and J. Wein. An experimental study of LP-based approximation algorithms for scheduling problems. In *Proceedings of the Ninth Annual ACM-SIAM Symposium on Discrete Algorithms*, pages 453–462, 1998.

[20] M. Schäffter. Scheduling with respect to forbidden sets. *Discrete Applied Mathematics*, 72:141–154, 1997.

[21] C. Schwindt. Generation of resource constrained project scheduling problems with minimal and maximal time lags. Technical Report 489, WIOR, University of Karlsruhe, Germany, 1996.

Approximation Schemes for Scheduling on Uniformly Related and Identical Parallel Machines

Leah Epstein[1] and Jiří Sgall[2]

[1] Dept. of Computer Science, Tel-Aviv University, Israel,
lea@math.tau.ac.il
[2] Mathematical Inst., AS CR, Žitná 25, CZ-11567 Praha 1, Czech Republic and
Department of Applied Mathematics, Faculty of Mathematics and Physics, Charles University, Praha,
sgall@math.cas.cz, http://www.math.cas.cz/~sgall

Abstract. We give a polynomial approximation scheme for the problem of scheduling on uniformly related parallel machines for a large class of objective functions that depend only on the machine completion times, including minimizing the l_p norm of the vector of completion times. This generalizes and simplifies many previous results in this area.

1 Introduction

We are given n jobs with processing times p_j, $j \in J = \{1, \dots, n\}$ and m machines M_1, $M_2, \dots, M_m$, with speeds $s_1, s_2, \dots, s_m$. A schedule is an assignment of the n jobs to the m machines. Given a schedule, for $1 \le i \le m$, T_i denotes the weight of machine M_i which is the total processing time of all jobs assigned to it, and C_i denotes the completion time of M_i, which is T_i/s_i. (Each job is assigned to exactly one machine, i.e., we do not allow preemption.)

Our objective is, for some fixed function $f : [0, +\infty) \to [0, +\infty)$, one of the following:

(I) minimize $\sum_{i=1}^{m} f(C_i)$,	**(III)** maximize $\sum_{i=1}^{m} f(C_i)$, or
(II) minimize $\max_{i=1}^{m} f(C_i)$,	**(IV)** maximize $\min_{i=1}^{m} f(C_i)$.

Most of such problems are NP-hard, see [9, 14]. Thus we are interested in approximation algorithms. Recall that a *polynomial time approximation scheme (PTAS)* is a family of polynomial time algorithms over $\varepsilon > 0$ such that for every ε and every instance of the problem, the corresponding algorithm outputs a solution whose value is within a factor of $(1+\varepsilon)$ of the optimum value [9].

We give a PTAS for scheduling on uniformly related machines for a rather general set of functions f, covering many natural functions studied before. Let us give some examples covered by our results and references to previous work.

Example 1. Problem (I) with $f(x) = x$ is the basic problem of minimizing the maximal completion time (makespan). It was studied for identical machines in [10, 11, 15, 12]; the last paper gives a PTAS. Finally, Hochbaum and Shmoys [13] gave a PTAS for uniformly related parallel machines. Thus our result can be seen as a generalization of that result. In fact our paper is based on their techniques, perhaps somewhat simplified.

Example 2. Problem (I) with $f(x) = x^p$ for any $p > 1$. This is equivalent to minimizing the l_p norm of the vector $(C_1, \ldots, C_m)$, i.e., $(\sum C_i^p)^{1/p}$. For $p = 2$ and identical machines this problem was studied in [6, 5], motivated by storage allocation problems. For a general p a PTAS for identical machines was given in [1]. For related machines this problem was not studied before. Note that for $p \leq 1$ the minimization problem is trivial: it is optimal to schedule all jobs on the fastest machine (choose one arbitrarily if there are more of them). In this case a more interesting variant is the maximization version, i.e., (III).

Example 3. Problem (IV) with $f(x) = x$. In this problem the goal is to maximize the time when all the machines are running; this corresponds to keeping all parts of some system alive as long as possible. This problem on identical machines was studied in [8, 7], and [16] gave a PTAS. For uniformly related machines a PTAS was given in [3].

Example 4. Scheduling with rejection. In this problem each job has associated certain penalty. The schedule is allowed to schedule only some subset of jobs, and the goal is to minimize the maximal completion time plus the total penalty of all rejected jobs. This problem does not conform exactly to any of the categories above, nevertheless our scheme can be extended to work for it as well. This problem was studied in [4] where also a PTAS for the case of identical machines is given. For related machines this problem was not studied.

Our paper is directly motivated by Alon et al. [2], who proved similar general results for scheduling on identical machines. We generalize it to uniformly related machines and a similar set of functions f. Even for identical machines, our result is stronger than that of [2] since in that case we allow a more general class of functions f.

The basic idea is to round the size of all jobs to a constant number of different sizes, to solve the rounded instance exactly, and then re-construct an almost optimal schedule for the original instance. This rounding technique traces back to Hochbaum and Shmoys [12]. We also use an important improvement from [1, 2]: the small jobs are clustered into blocks of jobs of small but non-negligible size. The final ingredient is that of [13, 3]: the rounding factor is different for each machine, increasing with the weight of the machine (i.e., the total processing time of jobs assigned to it). Such rounding is possible if we assign the jobs to the machines in the order of non-decreasing weight. This is easy to do for identical machines. For uniformly related machines we prove, under a reasonable condition on the function f, that in a good solution the weight of machines increases with their speed, which fixes the order of machines.

2 Our assumptions and results

Now we state our assumptions on the function f. Let us note at the beginning that the typical functions used in scheduling problems satisfy all of them.

The first condition says that to approximate the contribution of a machine up to a small multiplicative error, it is sufficient to approximate the completion time of the machine up to a small multiplicative error. This condition is from Alon et al. [2], and it is essential for all our results.

$$(F*): \quad (\forall\varepsilon > 0)(\exists\delta > 0)(\forall x, y \geq 0)$$
$$(|y - x| \leq \delta x \rightarrow |f(y) - f(x)| \leq \varepsilon f(x)).$$

If we transform f so that both axes are given a logarithmic scale, the condition naturally translates into uniform continuity, as $|y - x| \leq \delta x$ iff $(1-\delta)x \leq y \leq (1+\delta)x$ iff $\ln(1-\delta) + \ln x \leq \ln y \leq \ln(1+\delta) + \ln x$, and similarly for $f(x)$. More formally, (F$*$) is equivalent to the following statement:

(F$**$): The function $h_f : (-\infty, +\infty) \rightarrow (-\infty, +\infty)$
defined by $h_f(z) = \ln f(e^z)$
is defined everywhere and uniformly continuous.

We also need a condition that would guarantee that in the optimal schedule, the weights of the non-empty machines are monotone. These conditions are different for the cases when the objective is maximize or minimize $\sum f(C_i)$, and for the cases of min-max or max-min objectives.

Recall that a function g is convex iff for every $x \leq y$ and $0 \leq \Delta \leq y - x$, $f(x + \Delta) + f(x - \Delta) \leq f(x) + f(y)$. For the cases of min-sum and max-sum we require this condition:

(G$*$): The function $g_f : (-\infty, +\infty) \rightarrow [0, +\infty)$
defined by $g_f(z) = f(e^z)$ is convex.

Note that $f(x) = g_f(\ln x)$. Thus, the condition says that the function f is convex, if plotted in a graph with a logarithmic scale on the x-axis and a linear scale on the y-axis. This is true for example for any non-decreasing convex function f. However, g_f is convex, e.g., even for $f(x) = \ln(1 + x)$. On the other hand, the condition (G$*$) implies that f is either non-decreasing or it is unbounded for x approaching 0.

For the case of min-max or max-min we require that the function f is bimodal on $(0, +\infty)$, i.e., there exists an x_0 such that f is monotone (non-decreasing or non-increasing) both on $(0, x_0]$ and $[x_0, +\infty)$. (E.g., $(x-1)^2 + 1$ is bimodal, decreasing on $(0, 1]$ and increasing on $[1, +\infty)$.) This includes all convex functions as well as all non-decreasing functions.

Note that none of the conditions above puts any constraints on $f(0)$.

Last, we need the function f to be computable in the following sense: for any $\varepsilon > 0$ there exists an algorithm that on any rational x outputs a value between $(1-\varepsilon)f(x)$ and $(1+\varepsilon)f(x)$, in time polynomial in the size of x. To simplify the presentation, we will assume in the proofs that f is computable exactly; choosing smaller ε and computing the approximations instead of the exact values always works. Typical functions f that we want to use are computable exactly, but for example if $f(x) = x^p$ for non-integral p then we can only approximate it.

Our main results are:

Theorem 1. *Let f be a non-negative computable function satisfying the conditions* (F$*$) *and* (G$*$). *Then the scheduling problems of minimizing and maximizing $\sum f(C_i)$ on uniformly related machines both possess a PTAS.*

Theorem 2. *Let f be a non-negative computable bimodal function satisfying the condition* (F$*$). *Then the scheduling problems of minimizing* $\max f(C_i)$ *and of maximizing* $\min f(C_i)$ *on uniformly related machines both possess a PTAS.*

Theorem 3. *Let f be a non-negative computable function satisfying the condition* (F$*$). *Then the scheduling problems of minimizing or maximizing $\sum f(C_i)$, of minimizing* $\max f(C_i)$, *and of maximizing* $\min f(C_i)$ *on identical machines all possess a PTAS.*

All our PTAS are running in time $O(n^c p(|I|))$, where c is a constant depending on the desired precision and f, p is the polynomial bounding the time of the computation of f, and $|I|$ is the size of the input instance. Thus the time is polynomial, but the exponent in the polynomial depends on f and ε. This should be contrasted with Alon et al [2], where for the case of identical machines they are able to achieve linear time (i.e., the exponential dependence on ε is only hidden in the constant) using integer programming in fixed dimension. It is an open problem if such an improvement is also possible for related machines; this is open even for the case of minimizing the makespan (Example 1 above).

3 Ordering of the machines

The following lemma implies that, depending on the type of the problem and f, we can order the machines in either non-decreasing or non-increasing order of speeds and then consider only the schedules in which the weights of the machines are non-decreasing (possibly with the exception of the empty machines). It shows why the conditions (G$*$) and bimodality are important for the respective problems; it is the only place where the conditions are used. Note that for identical machines the corresponding statements hold trivially without any condition.

Lemma 4. *Let the machines be ordered so that the speeds s_i are non-decreasing.*

(i) *Under the same assumptions as in Theorem 1, there exists a schedule with minimal (maximal, resp.) $\sum f(C_i)$ in which the non-zero weights of the machines are monotone non-decreasing (monotone non-increasing, resp.). I.e., for any $1 \le i < j \le m$ such that $T_i, T_j > 0$, we have $T_i \le T_j$ ($T_i \ge T_j$, resp.).*

(ii) *Under the same assumptions as in Theorem 2, there exist schedules both with minimal* $\max f(C_i)$ *and with maximal* $\min f(C_i)$ *in which the non-zero weights of the machines are either monotone non-decreasing or monotone non-increasing.*

Proof. (i) We prove the lemma for minimization of $\sum f(C_i)$; the case of maximization is similar. We prove that if $T_i > T_j$ then switching the assigned jobs between M_i and M_j leads to at least as good schedule. This is sufficient, since given any optimal schedule we can obtain an optimal schedule with ordered weights by at most $m-1$ such transpositions.

Denote $s = s_j/s_i$, $\Delta = \ln s$, $X = \ln C_j = \ln(T_j/s_j)$, and $Y = \ln C_i = \ln(T_i/s_i)$. From the assumptions we have $X < Y$ and $0 \le \Delta < X - Y$. The difference of the original cost and the cost after the transposition is $f(C_i) + f(C_j) - f(C_i \cdot s) - f(C_j/s) = g_f(X) + g_f(Y) - g_f(X+\Delta) - g_f(Y-\Delta) \ge 0$, using convexity of g_f. Thus the transposed schedule is at least as good as the original one.

(ii) Suppose that we are maximizing $\min f(C_i)$ and f is bimodal, first non-increasing then non-decreasing. We prove that if $T_i > T_j$ for $i < j$, then switching the assigned jobs between M_i and M_j can only improve the schedule. We have $C_j/s \le C_i, C_j \le C_i \cdot s$, for $s = s_j/s_i$. By bimodality we have $\min\{C_i, C_j\} \le \min\{C_j/s, C_i \cdot s\}$, hence the transposed schedule can only be better. The cases of minimization of $\max f(C_i)$ and f first non-decreasing then non-increasing are similar, with the order reversed as needed.

4 Preliminaries and definitions

Let $\delta > 0$ and λ be such that $\lambda = 1/\delta$ is an even integer; we will choose it later. The meaning of δ is the (relative) rounding precision.

Given w, either 0 or an integral power of two, intuitively the order of magnitude, we will represent a set of jobs with processing times not larger than w as follows. For each job of size more than δw we round its processing time to the next higher multiple of $\delta^2 w$; for the remaining small jobs we add their processing times, round up to the next higher multiple of δw, and treat these jobs as some number of jobs of processing time δw. Now it is sufficient to remember the number n_i of modified jobs of processing time $i\delta^2 w$, for each i, $\lambda \le i \le \lambda^2$. Such a vector together with w is called a configuration.

In our approximation scheme, we will proceed machine by machine, and use this representation for two types of sets of jobs. First one is the set of all jobs scheduled so far; we represent them always with the least possible w (principal configurations below). The second type are the sets of jobs assigned to individual machines; we represent them with w small compared to the total processing time of their jobs (heavy configurations below), and this is sufficient to guarantee that the value of f can be approximated well.

Definition 5. *Let $A \subseteq J$ be a set of jobs.*

- *The* weight of a set of jobs *is $W(A) = \sum_{j \in A} p_j$.*
- *A* configuration *is a pair $\alpha = (w, (n_\lambda, n_{\lambda+1}, \dots, n_{\lambda^2}))$, where $w = 0$ or $w = 2^i$ for some integer i (possibly negative) and n is a vector of nonnegative integers.*
- *A configuration (w, n)* represents *A if*
  (i) *no job $j \in A$ has processing time $p_j > w$,*
  (ii) *for any i, $\lambda < i \le \lambda^2$, n_i equals the number of jobs $j \in A$ with $p_j \in ((i-1)\delta^2 w, i\delta^2 w]$, and*
  (iii) *$n_\lambda = \lceil W(A')/(\delta w) \rceil$ where $A' = \{j \in A \mid p_j \le \delta w\}$.*
- *The* principal configuration *of A is the configuration $\alpha(A) = (w, n)$ with the smallest w that represents A.*
- *The* weight of a configuration *(w, n) is defined by $W(w, n) = \sum_{i=\lambda}^{\lambda^2} n_i i\delta^2 w$.*
- *A configuration (w, n) is called* heavy *if its weight $W(w, n)$ is at least $w/2$.*
- *The* successor *of a configuration (w, n) is $\mathrm{succ}(w, n) = (w, n + (1, 0, \dots, 0))$.*

Note that given an $A \subseteq J$ and w, the definition gives a linear-time procedure that either finds the unique configuration (w,n) representing it or decides that no such configuration exists.

Lemma 6.

(i) *Let $A \subseteq J$ be a set of jobs and let (w,n) be any configuration representing it. Then*

$$(1-\delta)W(w,n)-\delta w \leq W(A) \leq W(w,n).$$

(ii) *Any $A \subseteq J$ has a unique principal configuration, and it can be constructed in linear time. The number of principal configurations is bounded by $(n+1)^{\lambda^2}$ and they can be enumerated efficiently.*

(iii) *Let $A, A' \subseteq J$ be both represented by (w,n). Then for any $w' > w$ they are both represented by (w',n') for some n'.*

Proof. (i) Any job $j \in A$ with $p_j > \delta w$ contributes to $W(w,n)$ some r, $p_j \leq r < p_j + \delta^2 w$. Its contribution to $W(A)$ is $p_j > r - \delta^2 w \geq r - \delta p_j \geq (1-\delta)r$. The small jobs $j \in A$, $p_j \leq \delta w$, can cause a total additive error of at most δw. Summing over all jobs, the bound follows.

(ii) The principal configuration of $A = \emptyset$ has $w = 0$. For a nonempty A, find a job $j \in A$ with the largest processing time p_j, and round up p_j to a power of two to obtain w. It follows that there are at most $n+1$ possible values of w in the principal configurations. To enumerate all principal configurations with a given w, find a representation (w,n) of $J(w) = \{j \in J \mid p_j \leq w\}$ and enumerate all the vectors n' bounded by $0 \leq n' \leq n$ (coordinatewise), such that $n'_i > 0$ for some $i > \lambda^2/2$. (Here we use the fact that λ is even.)

(iii) If $w = 0$ then $A = A' = \emptyset$ and the statement is trivial. Otherwise define

$$n_i^+ = \begin{cases} \left\lceil \frac{W(0,(n_\lambda,n_{\lambda+1},\ldots,n_{2\lambda},0,\ldots,0))}{2\delta} \right\rceil & \text{for } i = \lambda, \\ n_{2i-1} + n_{2i} & \text{for } \lambda < i \leq \lambda^2/2, \\ 0 & \text{for } i > \lambda^2/2. \end{cases}$$

It is easy to verify that if A is represented by (w,n) then it is represented by $(2w,n^+)$. Iterating this operation sufficiently many times proves that the representation with any $w' > w$ is the same for both A and A'.

Next we define a difference of principal configurations and show how it relates to a difference of sets. This is essential for our scheme. (It is easy to define difference of any configurations, but we do not need it.)

Definition 7. *Let (w,n) and (w',n') be two principal configurations. Their difference is defined as follows. First, let (w',n'') be the configuration that represents the same sets of jobs as (w,n) (using Lemma 6 (iii)). Now define $(w',n') - (w,n) = (w',n'-n'')$. If $w' < w$, or the resulting vector has some negative coordinate(s), the difference is undefined.*

Lemma 8. *Let $A \subseteq J$ be a set of jobs and (w,n) its principal configuration.*

(i) *Let (w',n') be a principal configuration such that $(w',n')-(w,n)$ is defined. Then there exists a set of jobs B represented by (w',n') such that $A \subseteq B \subseteq J$, and it can be constructed in linear time.*

(ii) *Let B be any set of jobs such that $A \subseteq B \subseteq J$, and let (w',n') be its principal configuration. Then $\gamma = (w',n')-(w,n)$ is defined and $B-A$ is represented by γ or* succ(γ). *Furthermore, if $\gamma = (w',n')-(w,n)$ is heavy then the weight of $B-A$ is bounded by $|W(B-A)-W(\gamma)| \leq 3\delta W(\gamma)$.*

Proof. Let (w',n'') be the configuration representing A; it can be computed from A and w' in linear time.

(i) Since the difference is defined, $n'' \leq n'$. For each i, $\lambda < i \leq \lambda^2$, add $n'_i - n''_i$ jobs with processing time $p_j \in ((i-1)\delta^2 w, i\delta w]$; since (w',n') is a principal configuration we are guaranteed that a sufficient number of such jobs exists. Finally, add jobs with $p_j \leq \delta w$ one by one until n''_λ increases to n'_λ. Each added job increases the coordinate by at most 1, and we have sufficiently many of them since (w',n') is principal.

(ii) It is easy to see that $A \subseteq B$ implies that $w \leq w'$ and $n'' \leq n'$, thus the difference is defined. For i, $\lambda < i \leq \lambda^2$, $n'_i - n''_i$ is the number of jobs in $B-A$ with the appropriate processing times. Let W_A and W_B be the weight of jobs in A and B with $p_j \leq \delta w'$. We have $n''_\lambda - 1 < W_A/(\delta w') \leq n''_\lambda$ and $n'_\lambda - 1 < W_B/(\delta w') \leq n'_\lambda$, thus $n'_\lambda - n''_\lambda - 1 < (W_B - W_A)/(\delta w') < n'_\lambda - n''_\lambda + 1$. This is rounded to $n'_\lambda - n''_\lambda$ or $n'_\lambda - n''_\lambda + 1$, hence $B-A$ is represented by γ or γ'. By Lemma 6 (i) it follows that

$$(1-\delta)W(\gamma) - \delta w' \leq W(B-A) \leq W(\gamma') = W(\gamma) + \delta w'.$$

If γ is heavy then $\delta w' \leq 2\delta W(\gamma)$, and the lemma follows.

¿From the part (i) of the lemma it also follows that given a principal configuration, it is easy to find a set it represents: just set $A = \emptyset$. Thus also the difference can be computed in linear time, using the procedure in the definition.

5 The approximation scheme

Given an $\varepsilon \in (0,1]$, we choose δ using (F∗) so that $\lambda = 1/\delta$ is an even integer and

$$(\forall x,y \geq 0)(|y-x| \leq 3\delta x \rightarrow |f(y)-f(x)| \leq \frac{\varepsilon}{3} f(x)).$$

Definition 9. *We define the graph G of configurations as follows.*

The vertices of G are $(i,\alpha(A))$, for any $1 \leq i < m$ and any $A \subseteq J$, the source vertex $(0,\alpha(\emptyset))$, and the target vertex $(m,\alpha(J))$.

For any i, $1 \leq i \leq m$, and any configurations α and β, there is an edge from $(i-1,\alpha)$ to (i,β) iff either $\beta = \alpha$, or $\beta - \alpha$ is defined and succ($\beta-\alpha$) *is heavy. The cost of this edge is defined as $f(W(\beta-\alpha)/s_i)$. There are no other edges.*

Definition 10. *Let $J_1, \ldots, J_m$ be an assignment of jobs J to machines $M_1, \ldots, M_m$. Its* representation *is a sequence of vertices of G $\{(i,\alpha_i)\}_{i=0}^m$, where $\alpha_i = \alpha(\bigcup_{i'=1}^{i} J_{i'})$.*

Note that we really obtain vertices of G, as $\alpha_0 = \alpha(\emptyset)$ and $\alpha_m = \alpha(J)$.

The approximation scheme performs the following steps:

(1) Order the machines with speeds either non-decreasing or non-increasing, according to the type of the problem and f so that by Lemma 4 there exists an optimal schedule with non-decreasing non-zero weights of the machines.
(2) Construct the graph G.
(3) Find an optimal path in G from source $(0,\alpha(\emptyset))$ to $(m,\alpha(J))$. The cost of the path is defined as the sum, maximum or minimum of the costs of the edges used, and an optimal path is one with the cost minimized or maximized, as specified by the problem.
(4) Output an assignment represented by the optimal path constructed as follows: Whenever the path contains an edge of the form $((i-1,\alpha),(i,\alpha))$, put $J_i = \emptyset$. For every other edge, apply Lemma 8 (i), starting from the beginning of the path.

Lemma 6 (ii) shows that we can construct the vertices of G in time $O(n^{1/\lambda^2})$. Computing the edges of G and their costs is also efficient. Since the graph G is layered, finding an optimal path takes linear time in the size of G. Given a path in a graph, finding a corresponding assignment is also fast. Hence the complexity of our PTAS is as claimed.

Lemma 11.

(i) *If $\{J_i\}$ is an assignment with non-decreasing weights of the machines with non-zero weights (cf. Lemma 4), then its representation $\{(i,\alpha_i)\}_{i=0}^{m}$ is a path in G.*
(ii) *Let $\{J_i\}$ be an assignment whose representation $\{(i,\alpha_i)\}_{i=0}^{m}$ is a path in G and such that if $\alpha_{i-1} = \alpha_i$ then $J_i = \emptyset$. Let C be the cost of the schedule given by the assignment, and let $C^\#$ be the cost of the representation as a path in the graph. Then $|C - C^\#| \le \varepsilon C^\#/3$.*

Proof. (i) For any $i = 1,\dots,m$, if $\alpha_{i-1} = \alpha_i$ then $((i-1,\alpha_{i-1}),(i,\alpha_i))$ is an edge by definition. Otherwise by Lemma 8 (ii), the difference $\gamma = \alpha_i - \alpha_{i-1}$ is defined and J_i is represented by γ or $\mathrm{succ}(\gamma)$, and thus $W(J_i) \le W(\mathrm{succ}(\gamma))$. We need to show $\mathrm{succ}(\gamma)$ is heavy. Let w be the order of α_i, and thus also of γ. If $w = 0$, the statement is trivial. Otherwise some job with $p_j > w/2$ was scheduled on one of the machines $M_1, \dots, M_i$. Since the assignment has non-decreasing weights, it follows that $w/2 < W(J_i) \le W(\mathrm{succ}(\gamma))$, and γ is heavy.

(ii) Let $X_i = W(\alpha_i - \alpha_{i-1})$ and let $Y_i = f(X_i/s_i)$ be the cost of the ith edge of the path. If $\alpha_{i-1} = \alpha_i$, then $J_i = \emptyset$ and $f(C_i) = Y_i = f(0)$. Otherwise by Lemma 8 (ii), $|W(J_i) - X_i| \le 3\delta X_i$. Thus $|C_i - X_i/s_i| = |W(J_i)/s_i - X_i/s_i| \le 3\delta X_i/s_i$ and by the condition (F*) and our choice of δ we get $|f(C_i) - Y_i| \le \varepsilon Y_i/3$. Summing over all edges of the path we get the required bound.

We now finish the proof of Theorems 1 and 2 for the minimization versions; the case of maximization is similar. Let C^* be the optimal cost, let $C^\#$ be the cost of an optimal path in G, and let C be the cost of the output solution of the PTAS. By Lemma 4 there exists an optimal schedule with non-decreasing weights. Thus by Lemma 11 (i) it is represented by a path, which cannot be cheaper than the optimal path, and by Lemma 11 (ii) $C^* \leq (1-\frac{\varepsilon}{3})C^{\#}$. Using Lemma 11 (ii) for the output assignment we get

$$C \leq (1+\frac{\varepsilon}{3})C^{\#} \leq \frac{1+\frac{\varepsilon}{3}}{1-\frac{\varepsilon}{3}}C^* \leq (1+\varepsilon)C^*.$$

Thus we have found a required approximate solution. (Note that the output solution need not have non-decreasing weights, due to rounding.)

6 Discussion

To get more insight in the meaning of the condition (F$*$), we prove the following characterization for convex functions. We omit the proof.

Observation 12. *Suppose $f : [0,+\infty) \to [0,+\infty)$ is on $(0,+\infty)$ convex and not identically 0. Then f satisfies* (F$*$) *if and only if the following conditions hold:*

- *$f(x) > 0$ for any $x > 0$,*
- *for $x \to \infty$, $f(x)$ is polynomially bounded both from above and below (i.e., for some constant c, $f(x) \leq O(x^c)$ and $f(x) \geq \Omega(1/x^c)$).*
- *for $x \to 0$, $f(x)$ is polynomially bounded both from above and below (i.e., for some constant c, $f(x) \leq O(1/x^c)$ and $f(x) \geq \Omega(x^c)$).*

This characterization is related to Conjecture 4.1 of Alon et al. [2] which we now disprove. The conjecture says that for a convex function f, and for the problem of minimizing $\sum f(C_i)$ on identical machines the following three conditions are equivalent: (i) it has a PTAS, (ii) it has a polynomial approximation algorithm with a finite performance guarantee, (iii) the heuristic LPT, which orders the jobs according to non-increasing processing times and schedules them greedily on the least loaded machine, has a finite performance guarantee.

We know that if (F$*$) holds, there is a PTAS (for a computable f) [2]. Observation 12 implies that if (F$*$) does not hold, then LPT does not have a finite performance guarantee; the proof is similar to Observation 4.1 of [2] (which says that no such algorithm exists for an exponentially growing function, unless $P = NP$, by a reduction to KNAPSACK).

Now consider $f(x) = x^{t(x)}$ where t is some slowly growing unbounded function; $t(x) = \log\log\log\log x$ will work. It is easy to verify that any such f is convex and does not satisfy (F$*$). However, it is possible to find a PTAS on identical machines using the integer programming approach of [2]. The function f does not satisfy (F$*$) on $[0,+\infty)$, but it satisfies (F$*$) for any interval $[0,T]$, moreover for a fixed ε the value of δ can be bounded by $\varepsilon/O(t(T))$. The PTAS algorithm now proceeds in the following way. It computes the bound on the completion time T as the sum of all processing times and chooses δ and $\lambda = 1/\delta$ accordingly. Since M is at most singly exponential in the size of the instance, λ is proportional to a triple logarithm of the instance size. Now we use the integer programming approach from [2]. Resulting algorithm has time complexity doubly exponential in λ, which is bounded by the size of the instance. Thus the algorithm is polynomial.

Let us conclude by a few remarks about the problem of minimizing $\max f(C_i)$. It is easy to approximate it for any increasing f satisfying (F*): just approximate the minimum makespan, and then apply f to that. Thus our extension to bimodal functions is not very strong. However, our techniques apply to a wider range of functions f. Suppose for example that the function f is increasing between 0 and 1, and then again from between c and $+\infty$, with an arbitrary behavior between 1 and c. Then it is possible to prove a weaker version of Lemma 4, saying that for some (almost) optimal schedule, for any $i < j$, $T_i < \mu T_j$ (if $T_i, T_j > 0$). The constant μ will depend only on the function f. This is sufficient for the approximation scheme, if we redefine the heavy edges to be the ones with weight $\mu w/2$ rather than $w/2$, and choose the other constants appropriately smaller. We omit the details and precise statement, since this extension of our results does not seem to be particularly interesting.

7 Scheduling with rejection

In this section we study the problem from Example 4 in the introduction.

Theorem 13. *Let f be a non-negative computable function satisfying the conditions* (F*) *and* (G*). *Then the problem of scheduling with rejection on uniformly related machine with the objection to minimize the sum of weights of rejected jobs plus $\sum f(C_i)$ possesses a polynomial approximation scheme.*

Let f be a non-negative computable bimodal function satisfying the condition (F*). *Then the problem of scheduling with rejection on uniformly related machines with the objection to minimize the sum of weights of rejected jobs plus* $\max f(C_i)$ *possesses a polynomial approximation scheme.*

If the machines are identical, then the same is true (in both cases above) even if f is computable and satisfies only the condition (F*).

The proof is a modification of our general PTAS. We give only a brief sketch. We start with the first case, i.e., the objective is penalty plus $\sum f(C_i)$.

We modify the graph G used in our PTAS in the following way. We add n auxiliary levels between any two levels of the original graph, as well as after the last level. Each level will again have nodes corresponding to all principal configurations, the target node will now be the node $\alpha(J)$ on the last auxiliary level. The edges entering the original nodes and their values will be as before. The edges entering the auxiliary levels will be as follows. There will be an edge from a configuration (w,n) to (w',n') iff the following holds: $w < w'$, $(w',n'') = (w',n') - (w,n)$ is defined, and $n''_i = 0$ for all $i \leq \lambda^2/2$. (The last condition says that (w',n'') represents only sets of jobs with all processing times greater than $w'/2$.) The value of the edge will be the smallest total penalty of a set of jobs represented by (w',n''). Additionally, there will be edges between identical configurations, with weight 0.

The proof that the cost of the shortest path is a good approximation of the optimum is along the same lines as for our general PTAS. The rest is omitted.

The case of minimizing the penalty plus $\max f(C_i)$ is somewhat different. The obstacle is that the cost of a path in the graph should be sum of the costs of edges on certain levels plus the maximum of the costs of edges on the remaining levels; for such a problem we are not able to use the usual shortest path algorithm.

Let M be some bound on $\max f(C_i)$. We use a similar graph as above, with the following modification. We include an edge entering an original node only if its value would be at most M; we set its value to 0. Now the cost of the shortest path is an approximation of the minimal penalty among all schedules with $\max f(C_i) \leq M$. More precisely, similarly as in Lemma 11, if there is a schedule with $\max f(C_i) \leq (1-\varepsilon/3)M$ and total penalty P, the shortest path has cost at most P; on the other hand, from a path with cost P we may construct a schedule with $\max f(C_i) \leq (1+\varepsilon/3)M$ and penalty P.

Now we solve the optimization problem by using the procedure above polynomially many times. Let s_{min} and s_{max} be the smallest and the largest machine speeds, respectively. Let p_{min} be the minimal processing time of a job, and let T be the total processing time of all jobs. In any schedule, any non-zero completion time is between $b = p_{min}/s_{max}$ and $B = T/s_{min}$. Now we cycle through all values $x = (1+\delta)^i b$, $i = 0, 1, \ldots$, such that $x \leq B$; the constant δ is chosen by the condition (F$*$), as in Section 5. In addition, we consider $x = 0$. The number of such x is polynomial in the size of the number B/b, which is polynomial in the size of the instance. For each x, we compute $M = f(x)$, and find a corresponding schedule with the smallest penalty P by the procedure above. (As a technical detail, we have to round each x to a sufficient precision so that the length of x is polynomial in the size of the instance; this is possible to do so that the ratio between successive values of x never exceeds $1+2\delta$, and that is sufficient.) We chose the best of these schedules, and possibly the schedule rejecting all jobs. Since the relative change between any two successive non-zero value of x is at most 3δ, the relative change between the successive values of M is at most $\varepsilon/3$ (by our choice of δ using (F$*$)), and we cover all relevant values of M with sufficient density.

Acknowledgements. We thank Yossi Azar and Gerhard Woeginger for useful discussions and introduction to this problem. Work of the second author was partially supported by grants A1019602 and A1019901 of GA AV ČR, postdoctoral grant 201/97/-P038 of GA ČR, and cooperative research grant INT-9600919/ME-103 from the NSF and MŠMT ČR.

References

[1] N. Alon, Y. Azar, G. J. Woeginger, and T. Yadid. Approximation schemes for scheduling. In *Proc. of the 8th Ann. ACM-SIAM Symp. on Discrete Algorithms*, pages 493–500. ACM-SIAM, 1997.

[2] N. Alon, Y. Azar, G. J. Woeginger, and T. Yadid. Approximation schemes for scheduling on parallel machines. *J. of Scheduling*, 1:55–66, 1998.

[3] Y. Azar and L. Epstein. Approximation schemes for covering and scheduling on related machines. In *Proc. of the 1st Workshop on Approximation Algorithms for Combinatorial Optimization Problems (APPROX '98), Lecture Notes in Comput. Sci. 1444*, pages 39–47. Springer-Verlag, 1998.

[4] Y. Bartal, S. Leonardi, A. Marchetti-Spaccamela, J. Sgall, and L. Stougie. Multiprocessor scheduling with rejection. In *Proc. of the 7th Ann. ACM-SIAM Symp. on Discrete Algorithms*, pages 95–103. ACM-SIAM, 1996.

[5] A. K. Chandra and C. K. Wong. Worst-case analysis of a placement algorithm related to storage allocation. *SIAM J. Comput.*, 4:249–263, 1975.

[6] R. A. Cody and E. G. Coffman. Record allocation for minimizing expected retrieval costs on drum-like storage devices. *J. Assoc. Comput. Mach.*, 23:103–115, 1976.

[7] J. Csirik, H. Kellerer, and G. J. Woeginger. The exact LPT-bound for maximizing the minimum completion time. *Oper. Res. Lett.*, 11:281–287, 1992.

[8] D. K. Friesen and B. L. Deuermeyer. Analysis of greedy solutions for a replacement part sequencing problem. *Mathematics of Operations Research*, 6:74–87, 1981.

[9] M. R. Garey and D. S. Johnson. *Computers and Intractability: A Guide to the Theory of NP–completeness*. Freeman, 1979.

[10] R. L. Graham. Bounds for certain multiprocessor anomalies. *Bell System Technical J.*, 45:1563–1581, Nov. 1966.

[11] R. L. Graham. Bounds on multiprocessor timing anomalies. *SIAM J. Appl. Math.*, 17(2):416–429, 1969.

[12] D. S. Hochbaum and D. B. Shmoys. Using dual approximation algorithms for scheduling problems: Theoretical and practical results. *J. Assoc. Comput. Mach.*, 34:144–162, 1987.

[13] D. S. Hochbaum and D. B. Shmoys. A polynomial approximation scheme for scheduling on uniform processors: Using the dual approximation approach. *SIAM J. Comput.*, 17:539–551, 1988.

[14] E. L. Lawler, J. K. Lenstra, A. H. G. Rinnooy Kan, and D. B. Shmoys. Sequencing and scheduling: Algorithms and complexity. In S. C. Graves, A. H. G. Rinnooy Kan, and P. Zipkin, editors, *Handbooks in Operations Research and Management Science, Vol. 4: Logistics of Production and Inventory*, pages 445–552. North-Holland, 1993.

[15] S. Sahni. Algorithms for scheduling independent tasks. *J. Assoc. Comput. Mach.*, 23:116–127, 1976.

[16] G. J. Woeginger. A polynomial time approximation scheme for maximizing the minimum machine completion time. *Oper. Res. Lett.*, 20:149–154, 1997.

Off-Line Temporary Tasks Assignment

Yossi Azar[1] and Oded Regev[2]

[1] Dept. of Computer Science, Tel-Aviv University, Tel-Aviv, 69978, Israel.
azar@math.tau.ac.il***

[2] Dept. of Computer Science, Tel-Aviv University, Tel-Aviv, 69978, Israel.
odedr@math.tau.ac.il

Abstract. In this paper we consider the temporary tasks assignment problem. In this problem, there are m parallel machines and n independent jobs. Each job has an arrival time, a departure time and some weight. Each job should be assigned to one machine. The load on a machine at a certain time is the sum of the weights of jobs assigned to it at that time. The objective is to find an assignment that minimizes the maximum load over machines and time.
We present a polynomial time approximation scheme for the case in which the number of machines is fixed. We also show that for the case in which the number of machines is given as part of the input (i.e., not fixed), no algorithm can achieve a better approximation ratio than $\frac{4}{3}$ unless $P = NP$.

1 Introduction

We consider the off-line problem of non-preemptive load balancing of temporary tasks on m identical machines. Each job has an arrival time, departure time and some weight. Each job should be assigned to one machine. The load on a machine at a certain time is the sum of the weights of jobs assigned to it at that time. The goal is to minimize the maximum load over machines and time. Note that the weight and the time are two separate axes of the problem.

The load balancing problem naturally arises in many applications involving allocation of resources. As a simple concrete example, consider the case where each machine represents a communication channel with bounded bandwidth. The problem is to assign a set of requests for bandwidth, each with a specific time interval, to the channels. The utilization of a channel at a specific time t is the total bandwidth of the requests, whose time interval contains t, which are assigned to this channel.

Load balancing of permanent tasks is the special case in which jobs have neither an arrival time nor a departure time. This special case is also known as the classical scheduling problem which was first introduced by Graham [5, 6]. He described a greedy algorithm called "List Scheduling" which has a $2 - \frac{1}{m}$ approximation ratio where m is the number of machines. Interestingly, the same analysis holds also for load balancing of temporary tasks. However, until now, it was not known whether better approximation algorithms for temporary tasks exist.

*** Research supported in part by the Israel Science Foundation and by the US-Israel Binational Science Foundation (BSF).

For the special case of permanent tasks, there is a polynomial time approximation scheme (PTAS) for any fixed number of machines [6, 10] and also for arbitrary number of machines by Hochbaum and Shmoys [7]. That is, it is possible to obtain a polynomial time $(1+\varepsilon)$-approximation algorithm for any fixed $\varepsilon > 0$.

In contrast we show in this paper that the model of load balancing of temporary tasks behaves differently. Specifically, for the case in which the number of machines is fixed we present a PTAS. However, for the case in which the number of machines is given as part of the input, we show that no algorithm can achieve a better approximation ratio than $\frac{4}{3}$ unless $P = NP$.

Note that similar phenomena occur at other scheduling problems. For example, for scheduling (or equivalently, load balancing of permanent jobs) on unrelated machines, Lenstra et al. [9] showed on one hand a PTAS for a fixed number of machines. On the other hand they showed that no algorithm with an approximation ratio better than $\frac{3}{2}$ for any number of machines can exist unless $P = NP$.

In contrast to our result, in the on-line setting it is impossible to improve the performance of Graham's algorithm for temporary tasks even for a fixed number of machines. Specifically, it is shown in [2] that for any m there is a lower bound of $2 - \frac{1}{m}$ on the performance ratio of any on-line algorithm (see also [1, 3]).

Our algorithm works in four phases: the rounding phase, the combining phase, the solving phase and the converting phase. The rounding phase actually consists of two subphases. In the first subphase the jobs' active time is extended: some jobs will arrive earlier, others will depart later. In the second subphase, the active time is again extended but each job is extended in the opposite direction to which it was extended in the first subphase. In the combining phase, we combine several jobs with the same arrival and departure time and unite them into jobs with higher weights. Solving the resulting assignment problem in the solving phase is easier and its solution can be converted into a solution for the original problem in the converting phase.

The novelty of our algorithm is in the rounding phase. Standard rounding techniques are usually performed on the weights. If one applies similar techniques to the time the resulting algorithm's running time is not polynomial. Thus, we had to design a new rounding technique in order to overcome this problem.

Our lower bound is proved directly by a reduction from exact cover by 3-sets. It remains as an open problem whether one can improve the lower bound using more sophisticated techniques such as PCP reductions.

2 Notation

We are given a set of n jobs that should be assigned to one of m identical machines. We denote the sequence of events by $\sigma = \sigma_1, ..., \sigma_{2n}$, where each event is an arrival or a departure of a job. We view σ as a sequence of times, the time σ_i is the moment after the ith event happened. We denote the weight of job j by w_j, its arrival time by a_j and its departure time by d_j. We say that a job is active at time τ if $a_j \leq \tau < d_j$. An assignment algorithm for the temporary tasks problem has to assign each job to a machine.

Let $Q_i = \{j | a_j \leq \sigma_i < d_j\}$ be the active jobs at time σ_i. For a given algorithm A let A_j be the machine on which job j is assigned. Let

$$l_k^A(i) = \sum_{\{j|A_j=k, j\in Q_i\}} w_j$$

be the load on machine k at time σ_i, which is the sum of weights of all jobs assigned to k and active at this time. The cost of an algorithm A is the maximum load ever achieved by any of the machines, i.e., $C_A = max_{i,k} l_k^A(i)$. We compare the performance of an algorithm to that of an optimal algorithm and define the approximation ratio of A as r if for any sequence $C_A \leq r \cdot C_{opt}$ where C_{opt} is the cost of the optimal solution.

3 The Polynomial Time Approximation Scheme

Assume without loss of generality that the optimal makespan is in the range $(1,2]$. That is possible since Graham's algorithm can approximate the optimal solution up to a factor of 2, and thus, we can scale all the jobs' weights by $\frac{2}{l}$ where l denotes the value of Graham's solution.

We perform a binary search for the value λ in the range $(1,2]$. For each value we solve the $(1+\varepsilon)$ relaxed decision problem, that is, either to find a solution of size $(1+\varepsilon)\lambda$ or to prove that there is no solution of size λ. From now on we fix the value of λ.

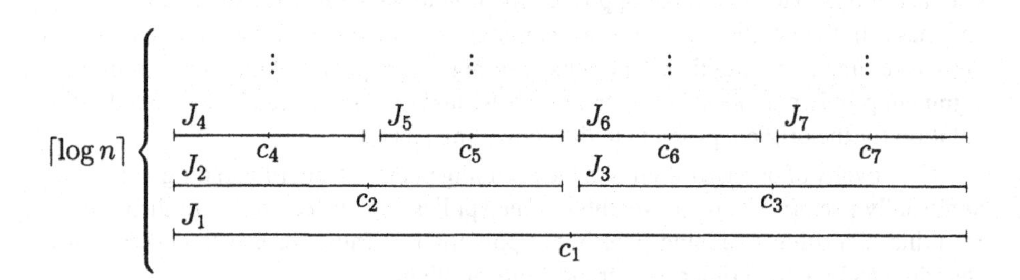

Fig. 1. Partitioning J into $\{J_i\}$

In order to describe the rounding phase with its two subphases we begin with defining the partitions based on which the rounding will be performed. We begin by defining a partition $\{J_i\}$ of the set of jobs J. Let M_i be a set of jobs and consider the sequence of times σ in which jobs of M_i arrive and depart. Since the number of such times is $2r$ for some r, let c_i be any time between the r-th and the $r+1$-st elements in that set. The set J_i contains the jobs in M_i that are active at time c_i. The set M_{2i} contains the jobs in M_i that depart before or at c_i and the set M_{2i+1} contains the jobs in M_i that arrive after c_i. We set $M_1 = J$ and define the M's iteratively until we reach empty sets. The important property of that partition is that the set of jobs that exist at a certain time is partitioned into at most $\lceil \log n \rceil$ different sets J_i.

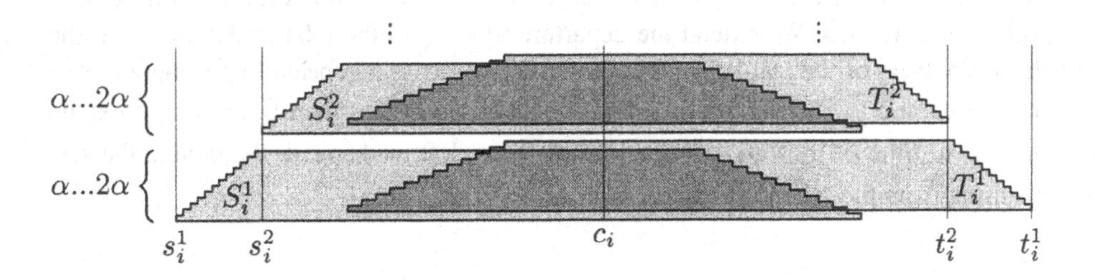

Fig. 2. Partitioning J_i into $\{S_i^j, T_i^j\}$ (R_i is not shown)

We continue by further partitioning the set J_i. We separate the jobs whose weight is greater than a certain constant α and denote them by R_i. We order the remaining jobs according to their arrival time. We denote the smallest prefix of the jobs whose total weight is at least α by S_i^1. Note that its total weight is less than 2α. We order the same jobs as before according to their departure time. We take the smallest suffix whose weight is at least α and denote that set by T_i^1. Note that there might be jobs that are both in S_i^1 and T_i^1. We remove the jobs and repeat the process with the jobs left in J_i and define $S_i^2, T_i^2, ..., S_i^{k_i}, T_i^{k_i}$. The last pair of sets $S_i^{k_i}$ and $T_i^{k_i}$ may have a weight of less than α. We denote by s_i^j the arrival time of the first job in S_i^j and by t_i^j the departure time of the last job in T_i^j. Note that $s_i^1 \leq s_i^2 \leq ... \leq s_i^{k_i} \leq c_i \leq t_i^{k_i} \leq ... \leq t_i^2 \leq t_i^1$.

The first subphase of the rounding phase creates a new set of jobs J' which contains the same jobs as in J with slightly longer active times. We change the arrival time of all the jobs in S_i^j for $j = 1, ..., k_i$ to s_i^j. Also, we change the departure time of all the jobs in T_i^j to t_i^j. The jobs in R_i are left unchanged. We denote the sets resulting from the first subphase by J', J_i', S'^j_i, T'^j_i.

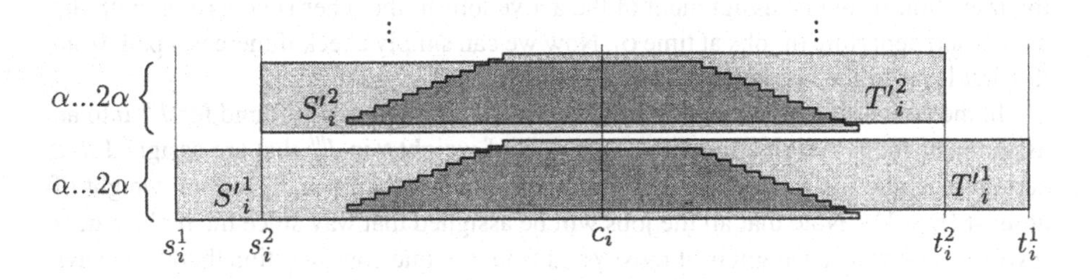

Fig. 3. The set J_i' (after the first subphase)

The second subphase of the rounding phase further extends the active time of the jobs of the first subphase. We take one of the sets J_i' and the partition we defined earlier to $R_i, S'^1_i \cup T'^1_i, S'^2_i \cup T'^2_i, ..., S'^{k_i}_i \cup T'^{k_i}_i$. We order the jobs in S'^j_i according to an increasing order of departure time. We take the smallest prefix of this ordering whose total weight is at least β. We extend the departure time of all the jobs in that prefix to the departure time of the last job in that prefix. The process is repeated until there are no more jobs in S'^j_i. The last prefix may have a weight of less than β. Similarly, extend the arrival time of jobs in T'^j_i. The jobs in R_i are left unchanged. We denote the sets resulting from the second subphase by J'', J''_i, S''^j_i, T''^j_i.

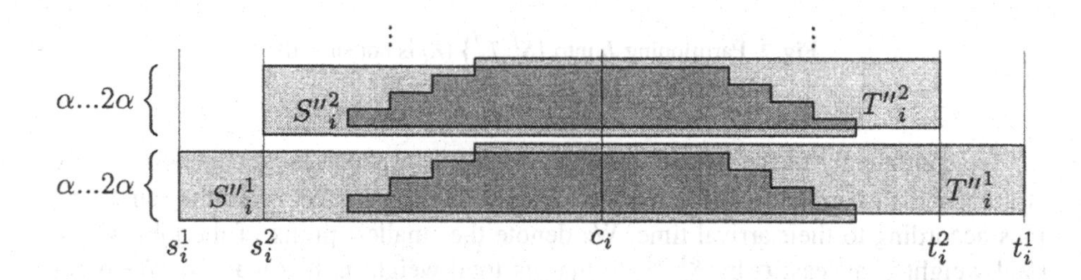

Fig. 4. The set J''_i (after the rounding phase)

The combining phase of the algorithm involves the weight of the jobs. Let J''_{st} be the set of jobs in J'' that arrive at s and depart at t. Assume the total weight of jobs whose weight is at most γ in J''_{st} is $a\gamma$. The combining phase replaces these jobs by $\lceil a \rceil$ jobs of weight γ. We denote the resulting sets by J'''_{st}. The set J''' is created by replacing every J''_{st} with its corresponding J'''_{st}, that is, $J''' = \cup_{s<t} J'''_{st}$.

The solving phase of the algorithm solves the modified decision problem of J''' by building a layered graph. Every time $\sigma_i \in \sigma$ in which jobs arrive or depart has its own set of vertices called a layer. In each layer we hold a vertex for every possible assignment of the current active jobs to machines; that is, an assignment whose makespan is at most λ for a certain λ. Two vertices of adjacent layers are connected by an edge if the transition from one assignment of the active jobs to the other is consistent with the arrival and departure of jobs at time σ_i. Now we can simply check if there is a path from the first layer to the last layer.

In the converting phase the algorithm converts the assignment found for J''' into an assignment for J. Assume the number of jobs of weight γ in J'''_{st} that are assigned to a certain machine i is r_i. Replace these with jobs smaller than γ in J''_{st} of total weight of at most $(r_i + 1)\gamma$. Note that all the jobs will be assigned that way since the replacement involves jobs whose weight is at most γ and from volume consideration there is always at least one machine with a load of at most r_i of these jobs. The assignment for J'' is also an assignment for J' and J.

4 Analysis

Lemma 1. *For $\alpha = \frac{\varepsilon}{2\lceil \log n \rceil}$, given a solution whose makespan is λ to the original problem J, the same solution applied to J' has a makespan of at most $\lambda + \varepsilon$. Also, given a* *solution whose makespan is λ to J', the same solution applied to J has a makespan of at most λ.*

Proof. The second claim is obvious since the jobs in J are shorter than their perspective jobs in J'. As for the first claim, every time τ is contained in at most $\lceil \log n \rceil$ sets J_i. Consider the added load at τ from jobs in a certain set J_i. If $\tau < s_i^1$ or $\tau \geq t_i^1$ then the same load is caused by J_i' and J_i. Assume $\tau < c_i$ and define $s_i^{k_i+1} = c_i$, the other case is symmetrical. Then for some j, $s_i^j \leq \tau < s_i^{j+1}$ and the added load at τ is at most the total load of S_i^j which is at most 2α. Summing on all sets J_i, we conclude that the maximal load has increased by at most $2\alpha\lceil \log n \rceil = \varepsilon$.

Lemma 2. *For $\beta = \frac{\varepsilon^2}{4m\lceil \log n \rceil}$, given a solution whose makespan is λ to the problem J', the same solution applied to J'' has a makespan of at most $\lambda(1+\varepsilon)$. Also, given a solution whose makespan is λ to J'', the same solution applied to J' has a makespan of at most λ.*

Proof. The second claim is obvious since the jobs in J' are shorter than their perspective jobs in J''. As for the first claim, given a time τ and a pair of sets S'^j_i, T'^j_i from J_i' we examine the increase in load at τ. If $\tau < s_i^j$ or $\tau \geq t_i^j$ it is not affected by the transformation because no job in $T'^j_i \cup S'^j_i$ arrives before s_i^j or departs after t_i^j. Assume that $\tau < c_i$, the other case is symmetrical. So τ is affected by the decrease in arrival time of jobs in T'^j_i. It is clear that the way we extend the jobs in T'^j_i increases the load at τ by at most β. Also, since $\tau \geq s_i^j$, we know that the load caused by S'^j_i is at least α if $j < k_i$. Thus, an extra load of at most β is created by every pair S'^j_i, T'^j_i for $1 \leq j < k_i$ only if the pair contributes at least α to the load. Also, $S_i^{k_i}$ for all i contributes an extra load of at most $\beta\lceil \log n \rceil$. Since the total load on all machines at any time is at most λm, the increase in load and therefore in makespan is at most $\frac{\beta}{\alpha}\lambda m + \beta\lceil \log n \rceil = \frac{\varepsilon\lambda}{2} + \frac{\varepsilon^2}{4m} \leq \varepsilon\lambda$.

Lemma 3. *For $\gamma = \frac{\varepsilon\beta}{m} = \frac{\varepsilon^3}{4m^2\lceil \log n \rceil}$, given a solution whose makespan is λ to the problem J'', the modified problem J''' has a solution with a makespan of $\lambda(1+\varepsilon)$. Also, given a solution whose makespan is λ to the modified problem J''', the solution given by the converting phase for the problem J'' has a makespan of at most $\lambda(1+\varepsilon)$.*

Proof. Consider a solution whose makespan is λ to J''. If the load of jobs smaller than γ in a certain J''_{st} on a certain machine i is $r_i\gamma$, we replace it by at most $\lceil r_i \rceil$ jobs of weight γ. Note that this is an assignment to J''' and that the increase in load on every machine is at most γ times the number of sets J''_{st} that contain jobs which are scheduled on that machine. As for the other direction, consider a solution whose makespan is λ to J'''. The increase in load on every machine by the replacement described in the algorithm is also at most γ times the number of sets J''_{st} that contain jobs which are scheduled on that machine.

The number of sets J''_{st} that can coexist at a certain time is at most $\frac{\lambda m}{\beta}$ since the weight of each set is at least β and the total load at any time is at most λm. Therefore, the increase in makespan is at most $\gamma\frac{\lambda m}{\beta} = \varepsilon\lambda$.

Lemma 4. *The running time of the algorithm for solving the relaxed decision problem for* λ *is bounded by* $O(n^{16\lambda m^3 \log m \varepsilon^{-3}+1})$. *The running time of the PTAS is the above bound times* $O(\log 1/\varepsilon)$.

Proof. Every layer in the graph stores all the possible assignments of jobs to machines. Since the smallest job is of weight γ, the maximum number of active jobs at a certain time is $\frac{\lambda m}{\gamma}$. So, the maximum number of edges in the graph is $nm^{2\lambda\frac{m}{\gamma}}$ and the running time of the relaxed decision problem algorithms is $O(nm^{2\lambda\frac{m}{\gamma}}) = O(nm^{8\lambda m^3 \lceil \log n \rceil \frac{1}{\varepsilon^3}}) = O(n^{16\lambda m^3 \log m \varepsilon^{-3}+1})$. The running time of the PTAS is the above bound times $O(\log 1/\varepsilon)$ since there are $O(\log 1/\varepsilon)$ phases in the binary search for the appropriate λ.

5 The Unrestricted Number of Machines Case

In this section we show that in case the number of machines is given as part of the input, the problem cannot be approximated up to a factor of $4/3$ in polynomial time unless $P = NP$. We show a reduction from the exact cover by 3-sets ($X3C$) which is known to be NP-complete [4, 8]. In that problem, we are given a set of $3n$ elements, $A = \{a_1, a_2, ..., a_{3n}\}$, and a family $F = \{T_1, ..., T_m\}$ of m triples, $F \subseteq A * A * A$. Our goal is to find a covering in F, i.e. a subfamily F' for which $|F'| = n$ and $\cup_{T_i \in F'} T_i = A$.

Given an instance for the $X3C$ problem we construct an instance for our problem. The number of machines is m, the number of triples in the original problem. There are three phases in time. First, there are times $1, ..., m$, each corresponding to one triple. Then, times $m+1, ..., m+3n$ each corresponding to an element of A. And finally, the two times $m+3n+1, m+3n+2$.

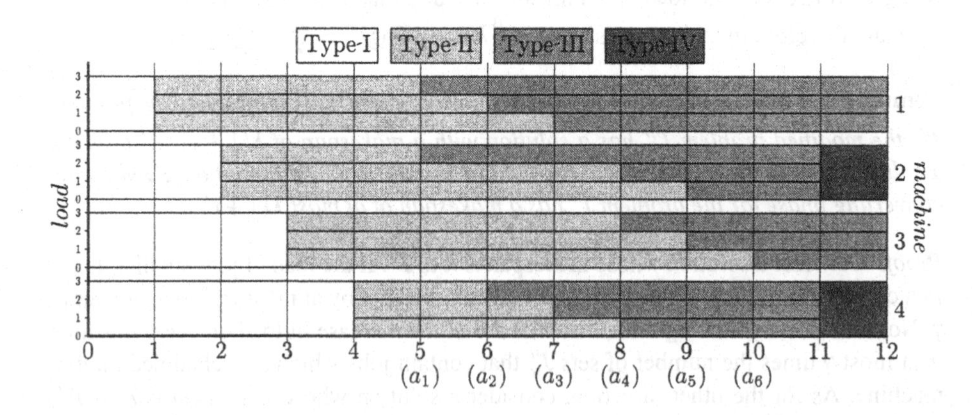

Fig. 5. An assignment for the scheduling problem corresponding to $m = 4$, $n = 2$, $F = \{(1,2,3), (1,4,5), (4,5,6), (2,3,4)\}$

There are four types of jobs. The first type are m jobs of weight 3 starting at time 0. Job r, $1 \le r \le m$ ends at time r. For any appearance of a_j in a triple T_i corresponds a job of the second type of weight 1 that starts at i and ends at $m+j$ and another job of the third type of weight 1 that starts at time $m+j$. Among all the jobs that start at time $m+j$, one ends at $m+3n+2$ while the rest end at $m+3n+1$. The fourth type of jobs are $m-n$ jobs of weight 3 that start at $m+3n+1$ and end at $m+3n+2$.

We show that there is a schedule with makespan at most 3 if and only if there is an exact cover by 3-sets. Suppose there is a cover. We schedule a job of the first type that ends at time i to machine i. We schedule the three jobs of the second type corresponding to T_i to machine i. At time $m+j$, some jobs of type two depart and the same number of jobs of type three arrives. One of these jobs is longer than the others since it ends at time $m+3n+2$. We schedule that longer job to machine i where T_i is the triple in the covering that contains j. At time $m+3n+1$ many jobs depart. We are left with $3n$ jobs, three jobs on each of the n machines corresponding to the 3-sets chosen in the cover. Therefore, we can schedule the $m-n$ jobs of the fourth type on the remaining machines.

Now, assume that there is a schedule whose makespan is at most 3. One important property of our scheduling problem is that at any time τ, $0 \leq \tau < m+3n+2$ the total load remains at $3m$ so the load on each machine has to be 3. We look at the schedule at time $m+3n+1$. Many jobs of type three depart and only the long ones stay. The number of these jobs is $3n$ and their weight is 1. Since $m-n$ jobs of weight 3 arrive at time $m+3n+1$, the $3n$ jobs must be scheduled on n machines. We take the triples corresponding to the n machines to be our covering. Assume by contradiction that this is not a covering. Therefore, there are two 3-sets that contain the same element, say a_j. At time $m+j$ only one long job arrives. The machine in which a shorter job was scheduled remains with a load of 3 until time $m+3n+1$ and then the short job departs and its load decreases to at most 2. This is a contradiction since at time $m+3n+1$ there are n machines each with 3 long jobs.

Corollary 5. *For every $\rho < \frac{4}{3}$, there does not exist a polynomial ρ-approximation algorithm for the temporary tasks assignment problem unless $P = NP$.*

6 Acknowledgments

We are grateful to Jiří Sgall and Gerhard J. Woeginger for their helpful discussions.

References

[1] Y. Azar. On-line load balancing. In A. Fiat and G. Woeginger, editors, *Online Algorithms - The State of the Art*, chapter 8, pages 178–195. Springer, 1998.

[2] Y. Azar and L. Epstein. On-line load balancing of temporary tasks on identical machines. In *5th Israeli Symp. on Theory of Computing and Systems*, pages 119–125, 1997.

[3] A. Borodin and R. El-Yaniv. *Online Computation and Competitive Analysis*. Cambridge University Press, 1998.

[4] M.R. Garey and D.S. Johnson. *Computers and Intractability*. W.H. Freeman and Company, San Francisco, 1979.

[5] R.L. Graham. Bounds for certain multiprocessor anomalies. *Bell System Technical Journal*, 45:1563–1581, 1966.

[6] R.L. Graham. Bounds on multiprocessing timing anomalies. *SIAM J. Appl. Math*, 17:263–269, 1969.

[7] D. S. Hochbaum and D. B. Shmoys. Using dual approximation algorithms for scheduling problems: Theoretical and practical results. *J. of the ACM*, 34(1):144–162, January 1987.

[8] R.M. Karp. *Reducibility among Combinatorial Problems, R.E. Miller and J.W. Thatcher (eds.), Complexity of Computer Computations*. Plenum Press, 1972.

[9] J.K. Lenstra, D.B. Shmoys, and E. Tardos. Approximation algorithms for scheduling unrelated parallel machines. *mathprog*, 46:259–271, 1990.

[10] S. Sahni. Algorithms for scheduling independent tasks. *Journal of the Association for Computing Machinery*, 23:116–127, 1976.

Load Balancing Using Bisectors – A Tight Average-Case Analysis

Stefan Bischof, Thomas Schickinger, and Angelika Steger

Institut für Informatik, Technische Universität München
D-80290 München, Germany
{bischof|schickin|steger}@in.tum.de
http://www14.in.tum.de/

Abstract. In parallel computation we often need an algorithm for dividing one computationally expensive job into a fixed number, say N, of subjobs, which can be processed in parallel (with reasonable overhead due to additional communication). In practice it is often easier to repeatedly bisect jobs, i.e., split one job into exactly two subjobs, than to generate N subjobs at once. In order to balance the load among the N machines, we want to minimize the size of the largest subjob (according to some measure, like cpu-time or memory usage).
In this paper we study a recently presented load balancing algorithm, called Heaviest First Algorithm (Algorithm HF), that is applicable to all classes of problems for which bisections can be computed efficiently. This algorithm implements a very natural strategy: During $N-1$ iterations we always bisect the largest subproblem generated so far.
The maximum load produced by this algorithm has previously been shown to differ from the optimum only by a constant factor even in the worst-case. In this paper we consider the average-case, assuming a natural and rather pessimistic random distribution for the quality of the bisections. Under this model the heaviest load generated by Algorithm HF is proved to be only twice as large as the optimum with high probability. Furthermore, our analysis suggests a simpler version of Algorithm HF which can easily be parallelized.

1 Introduction

Dynamic load balancing for irregular problems is a major research issue in the context of parallel and distributed computing. Often it is essential to achieve a balanced distribution of load in order to reduce the execution time of an application or to maximize system throughput.

We consider a general scenario, where an irregular problem is generated at runtime and must be split into subproblems that can be processed on different processors. If N processors are available, the problem is to be split into N subproblems and the goal is to balance the weight of the subproblems assigned to the individual processors. The weight of a problem (or subproblem) represents the load (CPU load, for example) caused by the problem on the processor to which it is assigned. It is assumed that the weight of a problem can be calculated (or approximated) easily once it is generated.

Instead of splitting the original problem into N subproblems in a single step, we use repeated *bisections* for generating the N subproblems, because in practice it is often easier to find efficient bisection methods than to find a method for splitting a problem into an arbitrary number of subproblems in one step. A bisection subdivides a problem into exactly two subproblems. Optimization problems in planar graphs (and their generalizations) are a classical field where a similar approach is commonly used (see for example [LT79, AST90]), and which may be adapted to fit into this model.

Load balancing using bisectors can and has been applied to a variety of practical problems. Our research is motivated by an application from distributed numerical simulations, namely a parallel solver for partial differential equations using a variant of the finite element method (FEM) with adaptive, recursive substructuring [HS94, BEE98]. Other possible application domains include chip layout [RR93] or multi-dimensional adaptive numerical quadrature [Bon93].

In this paper, we will not be concerned with details regarding any particular application. Instead, we study parallel load balancing from a more abstract point of view. We only assume that bisectors can be computed efficiently for the problems under consideration.

The remainder of the paper is organized as follows. In Sect. 2 we introduce our load balancing model and briefly review previous results regarding the worst-case behavior of Algorithm HF. In Sect. 3 we present our main result and the theoretical analysis. Finally, Sect. 4 contains some concluding remarks.

2 Load Balancing Model and Previous Work

As in [BEE98], we study the following simplified model for dynamic load balancing. The parallel system, on which we want to solve a problem p, consists of N processors or machines. The goal of the load balancing is to split p into N subproblems $p_1, \dots, p_N$, which can be solved individually. In our analysis we neglect the overhead which results from the combination of the partial solutions to the solution of the initial problem. This simplification is justified if the bisections produce only loosely coupled subproblems. Assuming a weight function w that measures the resource demand (for example, CPU load or memory requirement, depending on the application at hand) of subproblems, the goal is to minimize the maximum weight among the resulting subproblems, i.e., to minimize $\max_{1 \leq i \leq N} \mathsf{w}(p_i)$.

As mentioned before, we assume that the load balancing algorithm can only split a subproblem into *two* subproblems in a single step, using a bisection method. Repeated bisections must then be used to split a problem into N subproblems. If problem q is bisected into subproblems q_1, q_2 we assume that $\mathsf{w}(q) = \mathsf{w}(q_1) + \mathsf{w}(q_2)$, i.e., the sum of the weights of the subjobs equals the weight of the initial job.

In general, it can not be expected that bisections always split a problem p of weight $\mathsf{w}(p)$ into two subproblems with weight $\mathsf{w}(p)/2$ each. For many classes of problems, however, there are bisection methods that guarantee that the weights of the two obtained subproblems do not differ too much. The following definition captures this concept more precisely.

Definition 1. *Let* $0 < \alpha \leq \frac{1}{2}$. *A class* $\mathcal{P}$ *of problems with weight function* $\mathsf{w} : \mathcal{P} \to \mathbb{R}^+$ has α-bisectors *if every problem* $p \in \mathcal{P}$ *can be efficiently divided into two problems* $p_1 \in \mathcal{P}$ *and* $p_2 \in \mathcal{P}$ *with* $\mathsf{w}(p_1) + \mathsf{w}(p_2) = \mathsf{w}(p)$ *and* $\mathsf{w}(p_1), \mathsf{w}(p_2) \in [\alpha \mathsf{w}(p); (1-\alpha)\mathsf{w}(p)]$.

```
algorithm HF(p,N)
begin
    P := {p};
    while |P| < N do
        begin
        q := a problem in P with maximum weight;
        bisect q into q1 and q2;
        P := (P ∪ {q1,q2}) \ {q};
        end;
    return P;
end.
```

Fig. 1. Algorithm HF (Heaviest Problem First)

Fig. 1 shows Algorithm HF, which receives a problem p and a number N of processors as input and divides p into N subproblems by repeated bisection of the heaviest remaining subproblem. There is also an efficient parallel version of this algorithm as shown in [BEE99].

A perfectly balanced load distribution on N processors would be achieved if a problem p of weight $\mathrm{w}(p)$ was divided into N subproblems of weight exactly $\mathrm{w}(p)/N$ each.

The following theorem shows that the worst-case behavior of Algorithm HF differs from the optimum only by a factor depending on α but not on N.

Theorem 2. [BEE98] *Let $\mathcal{P}$ be a class of problems with weight function $\mathrm{w} : \mathcal{P} \to \mathbb{R}^+$ that has α-bisectors. Given a problem $p \in \mathcal{P}$ and a positive integer N, Algorithm HF uses $N-1$ bisections to partition p into N subproblems denoted by $p_1, \ldots, p_N$ such that*

$$\max_{1 \le i \le N} \mathrm{w}(p_i) \le \frac{\mathrm{w}(p)}{N} \cdot r_\alpha, \quad \text{where} \quad r_\alpha = \left\lfloor \frac{1}{\alpha} \right\rfloor \cdot (1-\alpha)^{\lfloor \frac{1}{\alpha} \rfloor - 2}.$$

Note that r_α is equal to 2 for $\alpha \ge 1/3$, below 3 for $\alpha \ge 1 - 1/\sqrt[4]{2} \approx 0.159$, and below 10 for $\alpha \ge 0.04$. Hence, Algorithm HF achieves provably good load balancing for classes of problems with α-bisectors for a surprisingly large range of α.

3 Average-Case Analysis of Algorithm HF

3.1 Main Result

The following stochastic model for an average-case scenario that may arise from practical applications seems reasonable: If a problem q is bisected into q_1 and q_2 we can find a *bisection parameter* $\alpha \in [0, \frac{1}{2}]$ such that $\mathrm{w}(q_1) = \alpha\mathrm{w}(q)$ and $\mathrm{w}(q_2) = (1-\alpha)\mathrm{w}(q)$. Assume that the actual bisection parameter α is drawn uniformly at random from the interval $[\hat{\alpha}, \hat{\beta}]$, $0 \le \hat{\alpha} \le \hat{\beta} \le 1/2$, and that all $N-1$ bisection steps are independent and identically distributed. We will write $\alpha \sim U[\hat{\alpha}, \hat{\beta}]$ if α is uniformly distributed on $[\hat{\alpha}, \hat{\beta}]$. In this paper, we study the case $\hat{\alpha} = 0, \hat{\beta} = \frac{1}{2}$, which seems rather pessimistic for practical applications, where in many cases α-bisectors (for $\alpha > 0$) are available. Note that the assumption $\alpha \sim U[0,\frac{1}{2}]$ could be changed to $\alpha \sim U(0,\frac{1}{2}]$ throughout our analysis without any modifications in the proofs.

Using this model we obtain the following result:

Theorem 3. *If Algorithm HF starts with a problem of weight W and produces a set P of N subproblems after N − 1 bisections then for* $\varepsilon = 9\sqrt{\ln(N)/N}$

$$\Pr\left[(2-\varepsilon)\frac{W}{N} \leq W^N_{\max} < (2+\varepsilon)\frac{W}{N}\right] = 1 - o(1/N) \ ,$$

where $W^N_{\max} := \max\{\mathrm{w}(p) \ : \ p \in P\}$ *denotes the weight of the heaviest subproblem in P.*

Thus, with high probability the maximum load differs from the best attainable value $\frac{W}{N}$ only by a factor of roughly two.

3.2 Outline of the Analysis

For our analysis we assume that Algorithm HF executes infinitely many iterations of the loop and consider the *infinite bisection tree* (IBT) generated by this process. This tree grows larger with each iteration of Algorithm HF. There is a one-to-one mapping from the subproblems produced by Algorithm HF to the nodes in the IBT. At a every point in time the subproblems in P correspond to the leaves in the part of the IBT which has been generated so far. The root is the initial problem p. If a node/subproblem q is split two new nodes/subproblems q_1 and q_2 are appended to q and, thus, the IBT is an infinite binary tree. One can also imagine that the IBT exists a priori and Algorithm HF visits all nodes one by one. When we say that Algorithm HF visits or expands a node in the IBT we adopt this view on the model. The IBT is the infinite version of the *bisection tree* (BT) employed in [BEE98].

Now let's define the probability space more formally: We set $\Omega := [0,1]^{\mathbb{N}}$ with the uniform distribution. As usual in the continuous case, the set of events is restricted to the Borel σ-field on $[0,1]^{\mathbb{N}}$, which we denote by $\mathbb{F}$.

An element $s := (\alpha_1, \alpha_2, \ldots) \in \Omega$ has the following interpretation: $\alpha_i \sim U[0,1]$ corresponds to the relative weight of the, say, left successor of the node which is split by the i-th bisection. We do not consider the actual bisection parameter (which is equal to α_i or $(1-\alpha_i)$), since studying $U[0,1]$ instead of $U[0,\frac{1}{2}]$ simplifies some calculations and we do not care which successor node is heavier. We call α_i the *generalized bisection parameter* of bisection i. Note that for every $s \in \Omega$ the corresponding IBT can be computed by simulating Algorithm HF.

We call a node v in the IBT *d-heavy* iff $\mathrm{w}(v) \geq d$, and accordingly we use the term *d-light*. If the value of d is obvious from the context we just say *heavy* and *light* for short.

For every node v in the complete binary tree T we define the random variables H^d_v (for $d > 0$)

$$H^d_v := \begin{cases} 1 & \text{if } \mathrm{w}(v) \geq d \\ 0 & \text{otherwise} \end{cases} \ .$$

Since T is a superset of the nodes in any IBT, the random variable $H^d := \sum_{v \in T} H_v^d$ counts the number of d-heavy nodes in the IBT. If a node v is not part of the IBT, H_v^d is zero.

Additionally, we define for each $v \in T$ a random variable X_v. If v is part of the IBT X_v denotes the weight of v relative to its ancestor p in the tree, i.e., $X_v := \mathsf{w}(v)/\mathsf{w}(p)$. Thus, X_v corresponds to the generalized bisection parameter α and, as mentioned above, we will assume that $X_v \sim U[0,1]$. We won't be interested in the value of X_v when v is not part of the IBT, so we don't specify how X_v behaves in this case.

The random variables H^d directly correspond to the performance of Algorithm HF, since H^d equals the number of iterations after which the set P of processes generated by Algorithm HF contains only light nodes. This is due to the fact that Algorithm HF visits all d-heavy nodes before it expands any d-light node.

Our analysis proceeds as follows: First we show that the expected number of heavy nodes $\mathbb{E}[H^d]$ is comparatively small. Then we prove that H^d is sharply concentrated around $\mathbb{E}[H^d]$ and thus H^d is small with high probability. Finally, the desired results for Algorithm HF follow easily.

3.3 Expected Number of Heavy Nodes

If we look at a node v_l on level l in the IBT (level 0 contains only the root, which has weight W, level i contains all nodes at BFS-distance i) and denote its ancestors on the levels $0,\dots,l-1$ by $v_0,\dots,v_{l-1}$ we get

$$\mathsf{w}(v_l) = W \cdot \prod_{i=1}^{l} X_{v_i} .$$

The following lemma enables us to analyze the distribution of products of $U[0,1]$-distributed random variables exactly.

Lemma 4. *Let $X_1,\dots,X_n$ be independent random variables with the exponential distribution and $\mathbb{E}[X_1] = \dots = \mathbb{E}[X_n] = 1/\lambda$. Then for $X := \sum_{i=1}^n X_i$*

$$\Pr[X \le t] = \begin{cases} 1 - \sum_{i=0}^{n-1} \frac{(\lambda t)^i}{i!} \cdot e^{-\lambda t} = e^{-\lambda t} \cdot \sum_{i=n}^{\infty} \frac{(\lambda t)^i}{i!} & \text{if } t \ge 0 \\ 0 & \text{otherwise} \end{cases} .$$

Proof. See [Fel71, p. 11].

It is easy to show (see [Fel71, pp. 25]) that a random variable $X_i \sim U[0,1]$ can be transformed to a random variable $Y_i := -\ln X_i$ with exponential distribution and $\mathbb{E}[Y_i] = 1$. If we want to analyze $X := \prod_{i=1}^n X_i$ we can equivalently analyze $Y := -\ln X := \sum_{i=1}^n Y_i$. Combining this fact with Lemma 4 yields the following lemma.

Lemma 5. *Let $X := \prod_{i=1}^n X_i$ be the product of independent random variables with $X_i \sim U[0,1]$. If we define Y as $Y := -\ln X$ we obtain for $t \in [0,1]$*

$$\Pr[X \ge t] = \Pr[Y \le -\ln t] = t \cdot \sum_{i=n}^{\infty} \frac{(-\ln t)^i}{i!} .$$

Proof. See [Fel71, p. 25].

In the following discussion we will often be concerned with the probability p_l^d that an arbitrary node on level l in the IBT is d-heavy.

Lemma 6. *Let v_l be an arbitrary node on level l of the IBT and denote the ancestors of v_l in the tree by $v_0, \ldots, v_{l-1}$. If the root of the IBT has weight W ($W \geq d$ for a fixed $d > 0$) then for the probability*

$$p_l^d := \Pr[\mathsf{w}(v_l) \geq d]$$

the following holds (with $l \geq \ln(W/d)$ and $l \geq 1$):

$$p_l^d = \frac{d}{W} \cdot \sum_{i=l}^{\infty} \frac{(\ln(W/d))^i}{i!} \leq \frac{d}{W} \cdot \left(\frac{\ln(W/d) \cdot e}{l}\right)^l .$$

Proof. Since $\mathsf{w}(v_l)$ is the product of W and the generalized bisection parameters of the ancestors of v_l in the IBT we obtain using Lemma 5

$$p_l^d = \Pr\left[W \cdot \prod_{i=1}^{l} X_{v_i} \geq d\right] = \Pr\left[\sum_{i=1}^{l} Y_{v_i} \leq -\ln\left(\frac{d}{W}\right)\right] = \frac{d}{W} \cdot \sum_{i=l}^{\infty} \frac{(\ln(W/d))^i}{i!} .$$

This proves the first equation. The second inequality can be shown as follows:

$$\begin{aligned} \sum_{i=l}^{\infty} \frac{(\ln(W/d))^i}{i!} &= (\ln(W/d))^l \cdot \sum_{i=0}^{\infty} \frac{(\ln(W/d))^i}{(i+l)!} = \frac{(\ln(W/d))^l}{l^l} \cdot \sum_{i=0}^{\infty} \frac{(\ln(W/d))^i \, l^l}{(i+l)!} \\ &\leq (\ln(W/d))^l \cdot \frac{1}{l^l} \cdot \sum_{i=0}^{\infty} \frac{l^{i+l}}{(i+l)!} \leq (\ln(W/d))^l \cdot \frac{e^l}{l^l} . \end{aligned}$$

Now we are in a position to state our first main result, namely the expected value for the number of heavy nodes in the IBT.

Theorem 7. *For an IBT where the root has weight W with $W > d$, it holds that*

$$\mathbb{E}[H^d] = \frac{2}{d} W - 1 .$$

Proof. Let $T(l)$ denote the set of nodes on level l in the complete binary tree. For a single node $v \in T(l)$ we obtain

$$\mathbb{E}[H_v^d] = p_l^d \cdot \Pr[v \in \text{IBT}] = p_l^d ,$$

because $\Pr[v \in \text{IBT}] = 1$. We just sketch how this can be shown: If $v \notin \text{IBT}$ we can find a node $v' \in \text{IBT}$ which is never bisected. This implies that there are infinitely many nodes in the tree with weight at least $d' := \mathsf{w}(v')$. Furthermore, we observe that sequences $s \in [0,1]^{\mathbb{N}}$ where only finitely many components belong to $[\varepsilon, 1-\varepsilon]$ (for $0 < \varepsilon < 0.5$) occur with probability zero. If, on the other hand, infinitely many components belong to $[\varepsilon, 1-\varepsilon]$ there are only finitely many d'-heavy nodes in the tree.

The theorem follows by linearity of expectation using the value of p_l^d which we derived in Lemma 6.

$$\begin{aligned}\mathbb{E}[H^d] &= \sum_{l=0}^{\infty}\sum_{v\in T(l)} \mathbb{E}[H_v^d] = \sum_{l=0}^{\infty} 2^l \cdot p_l^d \\ &= \frac{d}{W}\cdot\sum_{l=0}^{\infty}\sum_{i=l}^{\infty} 2^l \cdot \frac{(\ln(W/d))^i}{i!} = \frac{d}{W}\cdot\sum_{i=0}^{\infty}\sum_{l=0}^{i} 2^l \cdot \frac{(\ln(W/d))^i}{i!} \\ &= \frac{d}{W}\cdot\sum_{i=0}^{\infty}\frac{(\ln(W/d))^i}{i!}(2^{i+1}-1) = \frac{d}{W}\cdot\left[2\left(\frac{W}{d}\right)^2 - \frac{W}{d}\right] = \frac{2}{d}W - 1\ .\end{aligned}$$

3.4 Sharp Concentration

Theorem 7 gives us an idea how good Algorithm HF performs. If we set $d = 2W/N$ we get $\mathbb{E}[H^d] = N-1$. This is exactly the number of bisections that are necessary to produce N subproblems. Thus, on the average, we have bisected every heavy subproblem after $N-1$ iterations of Algorithm HF, and all the N generated subproblems have weight smaller than $2W/N$, which exceeds the optimal value W/N only by a factor of two. In the following part we will show that Algorithm HF really behaves the way this intuitive argument suggests. This is due to the fact that H^d is sharply concentrated around its expectation.

The following lemma shows that with high probability all heavy nodes reside rather close to the root of the IBT.

Lemma 8. *With probability* $1-o(1/N)$ *all nodes of the IBT on level* l *with* $l \geq k\cdot\ln((WN)/c)$ *and* $k > 2e$ *are* (c/N)*-light (for* $W \geq c/N$ *and* $c > 0$*).*

Proof. Setting $l := k\cdot\ln((WN)/c)$ for $k>1$ such that $l\in\mathbb{N}$ and applying Lemma 6 for $d = c/N$, we obtain

$$\begin{aligned}&\Pr[\exists\ d\text{-heavy node on level } l] \\ &\quad\leq 2^l\cdot p_l^d \leq 2^{k\ln((WN)/c)}\cdot\frac{c}{WN}\cdot\left(\frac{e}{k}\right)^{k\ln((WN)/c)} = \left(\frac{c}{WN}\right)^{1+k\ln k-k-k\ln 2}\ .\end{aligned}$$

A simple analysis of the exponent shows that $\Pr[\exists\ d\text{-heavy node on level } l] = o(1/N)$ for $k > 2e$.

Remark 9. The error term $o(1/N)$ in Lemma 8 is chosen rather arbitrarily and could be changed to $o(1/\text{poly}(N))$ without essential changes in the proofs.

Lemma 8 immediately yields a rather weak upper bound on the total number of heavy nodes, which we will improve later.

Corollary 10. *With probability* $1-o(1/N)$ *it holds that* $H^{c/N} = O(N\log N)$ *(for* $c>0$*).*

Proof. Lemma 8 shows that with high probability only the first, say, $6\cdot\ln(WN/c)$ levels contain heavy nodes. On every level the number of (c/N)-heavy nodes is at most WN/c since the weights of all nodes on the same level must have sum W.

Since we want to show that with high probability $H^{2W/N} \approx N$, Corollary 10 is still far off from our desired result, but we already know that the possibly infinite runtime of Algorithm HF until all subproblems are $(2W/N)$-light is $O(N \log N)$ with probability $1 - o(1/N)$. In order to improve this result, we define a martingale and apply the method of bounded differences to show a sharp concentration around the mean.

For the definition of the martingale we denote by $\mathbb{F}_i := \sigma(\alpha_1, \ldots, \alpha_i)$ the σ-algebra generated by the random variables α_i (for $i \geq 1$) (see [Fel71, p. 116f]). Furthermore we set $\mathbb{F}_0 := \{\emptyset, \Omega\}$. The sequence $\mathbb{F}_0 \subset \mathbb{F}_1 \subset \cdots \subset \mathbb{F}$ forms a *filter*. As mentioned in [Fel71, p. 212f]

$$Z_i^d := \mathbb{E}[H^d \mid \mathbb{F}_i] = \mathbb{E}[H^d \mid \alpha_1, \ldots, \alpha_i]$$

defines a martingale, which is also called a *Doob-martingale*. It holds that $Z_0^d = \mathbb{E}[H^d]$.

Z_i^d is a function from Ω to $\mathbb{R}$. The intuitive interpretation of Z_i^d is as follows: Given a sequence $s := (\alpha_1, \ldots, \alpha_i, \alpha_{i+1}, \ldots) \in \Omega$ of general bisection parameters, $Z_i^d(s)$ tells us how many heavy nodes we expect in the IBT if the generalized bisections parameters of the first i bisections correspond to $s_i := (\alpha_1, \ldots, \alpha_i)$.

Let T_i denote the part of the IBT visited by Algorithm HF up to the i-th iteration (for example T_0 contains only the root of the tree). When we know s_i we can simulate Algorithm HF and compute T_i. Therefore, evaluating Z_i^d corresponds to calculating the expected value of H^d, given the tree T_i which is generated by the first i bisections of Algorithm HF. In order to capture this intuition, we use the following notation:

$$Z_i^d = \mathbb{E}[H^d \mid \mathbb{F}_i] =: \mathbb{E}[H^d \mid T_i] \ .$$

In the following lemma we show that $|Z_{i+1}^d - Z_i^d|$ is small. This a prerequisite for the application of the method of bounded differences.

Lemma 11. *For all $i \geq 0$ it holds that $|Z_{i+1}^d - Z_i^d| \leq 2$.*

Proof. Throughout this proof we assume that $s \in \Omega$ is fixed, and show that $|Z_{i+1}^d(s) - Z_i^d(s)| \leq 2$.

Since we know the prefixes s_i and s_{i+1} of s, we can compute the trees T_i and T_{i+1}. Let v denote the node that is bisected at the $(i+1)$-st iteration.

If v is light ($\mathsf{w}(v) < d$) the claim follows easily: All nodes not yet visited by Algorithm HF must also be light, because v is a heaviest leaf in the expanded part of the tree. Thus, we have already seen all heavy nodes and obtain $Z_{i+1}^d = Z_i^d = H^d$.

Now we assume that $\mathsf{w}(v) \geq d$ (in the sequel we write $w := \mathsf{w}(v)$ for short). Let v_1 and v_2 be the two nodes generated by the bisection of v. In addition, let $\mathsf{w}(v_1) = \alpha w$ and $\mathsf{w}(v_2) = (1-\alpha)w$ (without loss of generality $0.5 \leq \alpha \leq 1$).

We set $T_i = I \cup L \cup \{v\}$ and $T_{i+1} = I \cup L \cup \{v\} \cup \{v_1\} \cup \{v_2\}$, where I denotes the set of interior nodes of T_i and L denotes the set of leaves of T_i except v. T is the complete binary tree (as before) and $T(u)$ is the subtree of T beginning at node u. Moreover, we introduce the abbreviation $\overline{H}^d(u \mid T_i) := \sum_{x \in T(u)} \mathbb{E}[H_x^d \mid T_i]$ for the expected number of heavy nodes in $T(u)$ under the condition that we already know the prefix T_i of the IBT.

We observe that all nodes in I must be d-heavy. This yields

$$Z_i^d = \mathbb{E}[H^d \mid T_i] = \sum_{u\in T} \mathbb{E}[H_u^d \mid T_i] = |I| + \sum_{u\in L} \overline{H}^d(u \mid T_i) + \overline{H}^d(v \mid T_i)$$

$$Z_{i+1}^d = |I| + \sum_{u\in L} \overline{H}^d(u \mid T_i) + 1 + \overline{H}^d(v_1 \mid T_{i+1}) + \overline{H}^d(v_2 \mid T_{i+1}) ,$$

since $\mathbb{E}[H_v^d \mid T_{i+1}] = 1$, because v is heavy.

We bound $|Z_i^d - Z_{i+1}^d|$ by considering three cases and using estimates for $\overline{H}^d(\tau \mid T_i)$ (for $\tau \in \{v, v_1, v_2\}$). Note that, if u is a d-heavy leaf in T_i, we can use Theorem 7 and obtain $\overline{H}^d(u \mid T_i) = \frac{2}{d}\mathrm{w}(u) - 1$.

Case 1: $\alpha w < d$ In this case v_1 and v_2 are light. Therefore, it holds that $w < 2d$ and $\overline{H}^d(v_1 \mid T_{i+1}) = \overline{H}^d(v_2 \mid T_{i+1}) = 0$. It follows that

$$Z_i^d - Z_{i+1}^d = \overline{H}^d(v \mid T_i) - (1 + \overline{H}^d(v_1 \mid T_{i+1}) + \overline{H}^d(v_2 \mid T_{i+1})) = \frac{2}{d}w - 1 - 1 < 2 .$$

Furthermore, we obtain due to $w \geq d$ that $Z_i^d - Z_{i+1}^d \geq 0$.

The cases **Case 2:** $(1-\alpha)w < d$ **and** $\alpha w \geq d$ and **Case 3:** $(1-\alpha)w \geq d$ are similar.

The next lemma is basic probability theory, but we prefer to state it separately in order to render the proof of the main theorem more readable.

Lemma 12. *Let A and B be two events over a probability space Ω. If* $\Pr[B] = 1 - o(1/N)$ *then*

$$\Pr[A] \leq \Pr[A \cap B] + o(1/N) .$$

Proof. Bound the terms in $\Pr[A] = \Pr[A|B]\Pr[B] + \Pr[A|\bar{B}]\Pr[\bar{B}]$.

We will use the following theorem from the method of bounded differences (see [McD89]):

Theorem 13. *Let $Z_0, Z_1, \ldots$ be a martingale sequence such that $|Z_k - Z_{k-1}| \leq c_k$ for each k, where c_k may depend on k. Then, for all $t \geq 0$ and any $\lambda > 0$*

$$\Pr[|Z_t - Z_0| \geq \lambda] \leq 2\exp\left(-\frac{\lambda^2}{2\sum_{k=1}^t c_k^2}\right) .$$

Using this inequality we prove the next theorem:

Theorem 14. *With probability $1 - 2e^{-k^2/9} - o(1/N)$ it holds for $d = c/N$ with $c = O(1)$ and $k > 0$ that*

$$|H^d - \mathbb{E}[H^d]| < k\sqrt{\frac{2W}{c}} \cdot \sqrt{N} .$$

Proof. First we note that if $H^d \leq t$ then $Z_t^d = H^d$, because after t steps Algorithm HF has certainly bisected all H^d heavy nodes. Hence, only light nodes remain and we know the exact value of H^d, since we have already seen all heavy nodes. Consequently, $H^d = \mathbb{E}[H^d \mid T_t] = Z_t^d$ in this case.

We have shown in Lemma 11 that $|Z_i^d - Z_{i+1}^d| \leq 2 =: c_i$. Now we apply Theorem 13 for $\lambda = N^\gamma$ with $\gamma = 0.5 + \varepsilon$ ($\varepsilon > 0$). For $t' = O(N \log N)$ we obtain from Corollary 10 and Lemma 12:

$$\begin{aligned}
\Pr[|H^d - \mathbb{E}[H^d]| \geq N^\gamma] &\leq \Pr[|H^d - \mathbb{E}[H^d]| \geq N^\gamma \wedge H^d \leq t'] + o(1/N) \\
&= \Pr[|Z_{t'}^d - Z_0^d| \geq N^\gamma \wedge H^d \leq t'] + o(1/N) \\
&\leq \Pr[|Z_{t'}^d - Z_0^d| \geq N^\gamma] + o(1/N) \\
&\leq 2 \cdot \exp\left(-\frac{N^{2\gamma}}{2\sum_{i=1}^{t'} c_i^2}\right) + o(1/N) = o(1/N) \ .
\end{aligned}$$

Now we know that with high probability $H^d < t'' := 2W/d - 1 + N^\gamma = (2W/c)N - 1 + N^\gamma$. If we apply Theorem 13 one more time using this estimate for H^d and performing similar calculations we get (for N large enough)

$$\begin{aligned}
\Pr[|H^d - \mathbb{E}[H^d]| \geq k\sqrt{(2W/c)N}] &\leq \Pr[|Z_{t''}^d - Z_0^d| \geq k\sqrt{(2W/c)N}] + o(1/N) \\
&< 2e^{-k^2/9} + o(1/N) \ .
\end{aligned}$$

The results for the random variable $H^{c/N}$ immediately yield the desired consequences for the performance of Algorithm HF.

Theorem 3. *If Algorithm HF starts with a problem of weight W and produces a set P of N subproblems after $N-1$ bisections then for $\varepsilon = 9\sqrt{\ln(N)/N}$*

$$\Pr\left[(2-\varepsilon)\frac{W}{N} \leq W_{\max}^N < (2+\varepsilon)\frac{W}{N}\right] = 1 - o(1/N) \ ,$$

where $W_{\max}^N := \max\{w(p) \ : \ p \in P\}$ denotes the weight of the heaviest subproblem in P.

Proof. First we show that $p^+ := \Pr\left[W_{\max}^N \geq (2+\varepsilon)\frac{W}{N}\right] = o(1/N)$.

For $c' := (2+\varepsilon)W$ and $\beta := \frac{c'}{N}$ it holds that $p^+ = \Pr\left[W_{\max}^N \geq \beta\right] = \Pr[H^\beta > N-1]$, because Algorithm HF expands heavy subproblems before light subproblems. Due to Theorem 7 we have $\mathbb{E}[H^\beta] = \frac{2}{\beta}W - 1 = \frac{2}{2+\varepsilon}N - 1$ and, thus, we can rewrite p^+ as $p^+ = \Pr\left[H^\beta - \mathbb{E}[H^\beta] > \frac{\varepsilon}{2+\varepsilon}N\right]$.

Since (for N sufficiently large) $\frac{\varepsilon}{2+\varepsilon}N \geq \frac{\varepsilon}{2.5}N \geq 3.6 \cdot \sqrt{\ln(N) \cdot N}$ and $1 \geq \sqrt{2/(2+\varepsilon)} = \sqrt{2W/c'}$, we obtain using Theorem 14 for $k = 3.6\sqrt{\ln(N)}$ that

$$p^+ \leq \Pr\left[|H^\beta - \mathbb{E}[H^\beta]| \geq k\sqrt{2W/c'} \cdot \sqrt{N}\right] = o(1/N) \ .$$

The proof for $p^- := \Pr\left[W_{\max}^N < (2-\varepsilon)\frac{W}{N}\right] = o(1/N)$ is similar.

Simulation results complement Theorem 3 and show that also for small N Algorithm HF exhibits a good average-case behavior. Already for $N \approx 100$ the predicted value for $W_{\max}^N$ from Theorem 3 and the 'real' value seen in simulations differ only by a relative error of less than one percent. The standard deviation is tiny, as one could expect due to Theorem 14.

3.5 Simplified Parallel Version of Algorithm HF

The proofs did not depend on the exact order in which Algorithm HF processes the nodes in the tree. We only made use of the fact that heavy nodes are processed before light nodes. Hence, it would suffice to run Algorithm HF until all subproblems are d-light with $d = (2+\varepsilon)\frac{W}{N}$ and $\varepsilon = 9\sqrt{\ln(N)/N}$. In this case we know that the number of generated subproblems is at most N with high probability. This variant of Algorithm HF does not need priority queues and requires only constant time per iteration to find the node q which shall be bisected next. Furthermore, there is a natural parallel version of the modified Algorithm HF, because for each subproblem it can be decided independently if further bisections should be applied to it. The sharp concentration result shows that this modification makes no substantial difference for the quality of the load distribution.

4 Conclusion

In this paper we have analyzed the Heaviest First Algorithm for dynamic load balancing. From a practitioners point of view this algorithm is applicable to a wide range of problems, because it only depends on the fact that some bisection algorithm is available. Additionally, it is very easy to implement and can be efficiently parallelized.

Our analysis focused on the average-case performance of Algorithm HF. Under rather pessimistic assumptions concerning the distribution of the bisection parameters we showed that the size of the maximum subproblem produced by Algorithm HF differs from the best attainable value only by a factor of roughly two. We believe that this provides a reasonable explanation for the good performance of Algorithm HF in simulations and practical applications. In order to improve the understanding of Algorithm HF further it would be interesting to transfer our results to more general distributions.

References

[AST90] N. Alon, P. Seymour, and R. Thomas. A separator theorem for nonplanar graphs. *Journal of the AMS*, 3(4):801–808, 1990.

[BEE98] Stefan Bischof, Ralf Ebner, and Thomas Erlebach. Load Balancing for Problems with Good Bisectors, and Applications in Finite Element Simulations. In *Proceedings of the Fourth International EURO-PAR Conference on Parallel Processing EURO-PAR'98*, volume 1470 of *LNCS*, pages 383–389, Berlin, 1998. Springer-Verlag. Also available as technical report (http://www14.in.tum.de/berichte/1998/TUM-I9811.ps.gz).

[BEE99] Stefan Bischof, Ralf Ebner, and Thomas Erlebach. Parallel Load Balancing for Problems with Good Bisectors. In *Proceedings of the 2nd Merged International Parallel Processing Symposium and Symposium on Parallel and Distributed Processing IPPS/SPDP'99*, Los Alamitos, 1999. IEEE Computer Society Press. To appear.

[Bon93] T. Bonk. A New Algorithm for Multi-Dimensional Adaptive Numerical Quadrature. In W. Hackbusch, editor, *Adaptive Methods – Algorithms, Theory and Applications: Proceedings of the 9th GAMM Seminar, Kiel, January 22–24, 1993*, pages 54–68, 1993.

[Fel71] William Feller. *An Introduction to Probability Theory and its Applications. Volume II.* Wiley Series in Probability and Mathematical Statistics. John Wiley & Sons, Chichester, second edition, 1971.

[HS94] Reiner Hüttl and Michael Schneider. Parallel Adaptive Numerical Simulation. SFB-Bericht 342/01/94 A, Technische Universität München, 1994.

[LT79] R.J. Lipton and R.E. Tarjan. A separator theorem for planar graphs. *SIAM J. Appl. Math.*, 36(2):177–189, 1979.

[McD89] C. McDiarmid. On the method of bounded differences. In J. Siemons, editor, *Surveys in Combinatorics*, volume 141 of *London Math. Soc. Lecture Note Series*, pages 148–188. Cambridge University Press, Cambridge, 1989.

[RR93] H. Regler and U. Rüde. Layout optimization with Algebraic Multigrid Methods (AMG). In *Proceedings of the Sixth Copper Mountain Conference on Multigrid Methods, Copper Mountain, April 4–9, 1993*, Conference Publication, pages 497–512. NASA, 1993.

On the Analysis of Evolutionary Algorithms — A Proof That Crossover Really Can Help*

Thomas Jansen and Ingo Wegener

FB Informatik, LS 2, Univ. Dortmund, 44221 Dortmund, Germany
{jansen, wegener}@ls2.cs.uni-dortmund.de

Abstract. There is a lot of experimental evidence that crossover is, for some functions, an essential operator of evolutionary algorithms. Nevertheless, it was an open problem to prove for some function that an evolutionary algorithm using crossover is essentially more efficient than evolutionary algorithms without crossover. In this paper, such an example is presented and its properties are proved.

1 Introduction

Stochastic search strategies have turned out to be efficient heuristic optimization techniques, in particular, if not much is known about the structure of the function and intense algorithmic investigations are not possible. The most popular among these algorithms are simulated annealing (van Laarhoven and Aarts [12]) and evolutionary algorithms, which come in great variety (evolutionary programming (Fogel, Owens, and Walsh [4]), genetic algorithms (Holland [7], Goldberg [6]), evolution strategies (Schwefel [22])). There is a lot of "experimental evidence" that these algorithms perform well in certain situations but there is a lack of theoretical analysis. It is still a central open problem whether, for some function, a simulated annealing algorithm with an appropriate cooling schedule is more efficient than the best Metropolis algorithm, i. e., a simulated annealing algorithm with fixed temperature (Jerrum and Sinclair [9]). Jerrum and Sorkin [10] perform an important step towards an answer of this question. Here, we solve a central open problem of similar flavor for genetic algorithms based on mutation, crossover, and fitness based selection. All these three modules are assumed to be essential but this has not been proved for crossover. Evolutionary algorithms without crossover are surprisingly efficient. Juels and Wattenberg [11] report that even hill climbing (where the population size equals 1) outperforms genetic algorithms on nontrivial test functions. For a function called "long path", which was introduced by Horn, Goldberg, and Deb [8], Rudolph [20] has proved that a hill climber performs at least comparable to genetic algorithms. Indeed, the following central problem considered by Mitchell, Holland, and Forrest [15] is open:

- Define a family of functions and prove that genetic algorithms are essentially better than evolutionary algorithms without crossover.

* This work was supported by the Deutsche Forschungsgemeinschaft (DFG) as part of the Collaborative Research Center "Computational Intelligence" (531).

One cannot really doubt that such examples exist. Several possible examples have been proposed. Forrest and Mitchell [5] report for the well-known candidate called Royal Road function (Mitchell, Forrest, and Holland [14]) that some random mutation hill climber outperforms genetic algorithms. So the problem is still open (Mitchell and Forrest [13]). The problem is the difficulty to analyze the consequences of crossover, since crossover creates dependencies between the objects. Hence, the solution of the problem is a necessary step to understand the power of the different genetic operators and to build up a theory on evolutionary algorithms.

There are some papers dealing with the effect of crossover. Baum, Boneh, and Garrett [1] use a very unusual crossover operator and a population of varying size which not really can be called genetic algorithm. Another approach is to try to understand crossover without fitness based selection. Rabinovich, Sinclair, and Wigderson [18] model such genetic algorithms as quadratical dynamic systems, and Rabani, Rabinovich, and Sinclair [17] investigate the isolated effects of crossover for populations. These are valuable fundamental studies. Here, we use a less general approach but we investigate a typical genetic algorithm based on mutation, uniform crossover, and fitness based selection. The algorithm is formally presented in Section 2. In Section 3, we prove, for the chosen functions, that algorithms without crossover necessarily are slow and, in Section 4, we prove that our genetic algorithm is much faster.

Definition 1. *The function* $\text{JUMP}_{m,n}\colon \{0,1\}^n \to \mathbb{R}$ *is defined by*

$$\text{JUMP}_{m,n}(x_1,\ldots,x_n) = \begin{cases} m + \|x\|_1 & \text{if } \|x\|_1 \le n-m \text{ or } \|x\|_1 = n \\ n - \|x\|_1 & \text{otherwise} \end{cases}$$

where $\|x\|_1 = x_1 + \cdots + x_n$ *denotes the number of ones in* x.

The value of JUMP_m (the index n is usually omitted) grows linearly with the number of ones in the input but there is a gap between the levels $n-m$ and n. We try to maximize JUMP_m. Then, inputs in the gap are the worst ones. We expect that we have to create the optimal input $(1,1,\ldots,1)$ from inputs with $n-m$ ones. This "jump" is difficult for mutations but crossover can help. More precisely, we prove that time $\Omega(n^m)$ is necessary without crossover while a genetic algorithm can optimize JUMP_m with large probability in time $O(n^2 \log n + 2^{2m} n \log n)$ and the same bound holds for the expected time. The gap is polynomial for constant m and even superpolynomial for $m = \Theta(\log n)$.

2 Evolutionary Algorithms

We discuss the main operators of evolutionary algorithms working on the state space $S = \{0,1\}^n$ where we maximize a fitness function $f\colon S \to \mathbb{R}$. We use the operators initialization, mutation, crossover, and selection.

$X :=$ **initialize**(S,s). Choose randomly and independently s objects from S to form the population X.

$(y,b) :=$ **mutate**(X,p). Choose randomly an object $x \in X$ and, independently for all positions $i \in \{1,\ldots,n\}$, set $y_i := 1 - x_i$ with probability p and $y_i := x_i$ otherwise. Set $b := 0$, if $x = y$, and $b := 1$ otherwise.

The most common choice is $p = 1/n$ ensuring that, on average, one bit of x is flipped. The optimality of this choice has been proved for linear functions by Droste, Jansen, and Wegener [3]. For some evolutionary algorithms it is not unusual to abstain from crossover. Evolutionary programming (Fogel, Owen, and Walsh [4]) and evolution strategies (Schwefel [22]) are examples. The so-called $(\mu+\lambda)$-evolution strategy works with a population of size μ. Then λ children are created independently by mutation and the best μ objects among the parents and children are chosen as next population. Ties are broken arbitrarily.

Genetic algorithms typically use crossover. For the function at hand, uniform crossover is appropriate.

(y,b) := **uniform-crossover-and-mutate**(X,p). Choose randomly and independently $x', x'' \in X$ and, independently for all positions $i \in \{1,\dots,n\}$, set $z_i := x'_i$ with probability $1/2$ and $z_i := x''_i$ otherwise. Then $y :=$ mutate$(\{z\},p)$. Set $b := 0$, if $y \in \{x',x''\}$, and $b := 1$ otherwise.

Ronald [19] suggests to avoid duplicates in the population in order to prevent populations with many indistinguishable objects. We adopt this idea and only prevent replications (see below). Moreover, we use a variant of genetic algorithms known as steady state (Sarma and De Jong [21]). This simplifies the analysis, since, in one step, only one new object is created. Now we are able to describe our algorithm.

Algorithm 1.

1. $X :=$ **initialize**$(\{0,1\}^n, n)$.
2. Let r be a random number from $[0,1]$ (uniform distribution).
3. If $r \le 1/(n \log n)$,
   $(y,b) :=$ **uniform-crossover-and-mutate**$(X, 1/n)$.
4. If $r > 1/(n \log n)$,
   $(y,b) :=$ **mutate**$(X, 1/n)$.
5. Choose randomly one of the objects $x \in X$ with smallest f-value.
6. If $b = 1$ and $f(y) \ge f(x)$,
   $X := (X - \{x\}) \cup \{y\}$.
7. Return to Step 2.

Steps 5 and 6 are called steady state selection preventing replications. We do not care about the choice of an appropriate stopping rule by using as most of the authors the following complexity measure. We count the number of evaluations of f on created objects until an optimal object is created. In the following we discuss in detail the evolution strategies described above and the genetic algorithm described in Algorithm 1. We are able to obtain similar results for the following variants of the algorithm (details can be found in the full version).

1. Evolutionary algorithms without crossover may use subpopulations which work independently for some time and may exchange information sometimes.
2. Evolutionary algorithms without crossover as well as the genetic algorithm may choose objects based on their fitness (for mutation, crossover, and/or selection) as long as objects with higher fitness get a better chance to be chosen and objects with the same fitness get the same chance to be chosen.
3. The genetic algorithm may refuse to include any duplicate into the population.

4. The genetic algorithm may accept replications as well as duplicates. In this case, all our results qualitatively still hold. The actual size of the upper bounds for the genetic algorithm changes in this case, though. At the end of Section 4 we discuss this in more detail.
5. The genetic algorithm may replace the chosen object by y even if $f(y) < f(x)$.

3 Evolutionary Algorithms without Crossover on JUMP$_m$

Evolutionary algorithms without crossover create new objects by mutations only. If x contains i zeros, the probability of creating the optimal object equals $p^i(1-p)^{n-i}$. Let $m \leq (\frac{1}{2}-\varepsilon)n$. It follows by Chernoff's inequality that, for populations of polynomial size, the probability to have an object x, where $||x||_1 > n-m$, in the first population is exponentially small. Then, the expected time to reach the optimum is bounded below by $t_{m,n} = \min\{p^{-i}(1-p)^{i-n} \mid m \leq i \leq n\}$. This holds since we do not select objects x where $n-m < ||x||_1 < n$. It is obvious that $t_{m,n} = \Theta(n^m)$, if $p = 1/n$. The following result follows by easy calculations.

Proposition 2. *Let $m \leq (\frac{1}{2}-\varepsilon)n$ for some constant $\varepsilon > 0$. Evolutionary algorithms without crossover need expected time $\Omega(n^m)$ to optimize JUMP$_m$ if mutations flip bits with probability $1/n$. For each mutation probability p, the expected time is $\Omega(n^{m-c})$ for each constant $c > 0$.*

Droste, Jansen, and Wegener [2] have proved that, for population size 1 and $p = 1/n$, the expected time of an evolutionary algorithm on JUMP$_m$, $m > 1$, equals $\Theta(n^m)$.

4 The Genetic Algorithm as Optimizer of JUMP$_m$

The main result of this paper is the following theorem.

Theorem 3. *Let m be a constant. For each constant $k \in \mathbb{N}$ the genetic algorithm creates an optimal object for* JUMP$_m$ *within $O(n^2 \log n)$ steps with probability $1-\Omega(n^{-k})$. With probability $1-e^{-\Omega(n)}$ the genetic algorithm creates an optimal object for* JUMP$_m$ *within $O(n^3)$ steps.*

Proof. Our proof strategy is the following. We consider different phases of the algorithm and "expect" in each phase a certain behavior. If a phase does not fulfill our expectation, we estimate the probability of such a "failure" and may start the next phase under the assumption that no failure has occurred. We also assume to have not found an optimal object, since otherwise we are done. Finally, the failure probability can be estimated by the sum of the individual failure probabilities. The constants $c_1, c_2, c_3 > 0$ will be chosen appropriately.

Phase 0: Initialization. We expect to obtain only objects x where $||x||_1 \leq n-m$ or $||x||_1 = n$.

Phase 1: This phase has length $c_1 n^2 \log n$, if we want $O(n^{-k})$ as upper bound for the probability of a failure. Setting its length to $c_1 n^3$, we can guarantee an error bound of even $e^{-\Omega(n)}$. We expect to create an optimal object or to finish with n objects with $n-m$ ones.

Remark: If Phase 1 is successful, the definition of the genetic algorithm ensures that the property is maintained forever.

Phase 2: This phase has length $c_2 n^2 \log n$. We expect to create an optimal object or to finish with objects with $n-m$ ones where the zeros are not too concentrated. More precisely, for each bit position i, there are at most $\frac{1}{4m}n$ of the n objects with a zero at position i.

Phase 3: This phase has length $c_3 n^2 \log n$. We expect that, as long as no optimal object is created, the n objects contain at each position $i \in \{1,\ldots,n\}$ altogether at most $\frac{1}{2m}n$ zeros. Moreover, we expect to create an optimal object.

Analysis of Phase 1. We apply results on the coupon collector's problem (see Motwani and Raghavan [16]). There are N empty buckets and balls are thrown randomly and independently into the buckets. Then the probability that, after $(\beta+1)N \ln N$ throws, there is still an empty bucket is $N^{-\beta}$. This result remains true if, between the throws, we may rename the buckets.

We consider the n^2 bit positions of the n objects as buckets, so we have $N = n^2$. Buckets corresponding to zeros are called empty. The genetic algorithm never increases the number of zeros. Hence, we slow down the process by ignoring the effect of new objects created by crossover or by a mutation flipping more than one bit. We further slow down the process by changing the fitness to $||x||_1$ and waiting for ones at all positions. If a mutation flips a single bit from 0 to 1, we obtain a better object which is chosen. The number of empty buckets decreases at least by 1. If a single bit flips from 1 to 0, we ignore possible positive effects (perhaps we replace a much worse object). Hence, by the result on the coupon collector's problem, the failure probability is bounded by $N^{-k} = n^{-2k}$ after $(k+1)N\ln N = 2(k+1)n^2 \ln n$ good steps and bounded by $e^{-\sqrt{N}} = e^{-n}$ after $((\sqrt{N}/\ln N)+1)N \ln N = n^3 + 2n^2 \ln n$ good steps.

A step is not good if we choose crossover (probability $1/(n \log n)$) or we flip not exactly one bit. The probability of the last event equals $1 - n \cdot \frac{1}{n} \cdot (1-\frac{1}{n})^{n-1}$ and is bounded by a constant $a < 1$. Hence, by Chernoff's bound, we can bound the probability of having enough good steps among $c_1 n^2 \log n$ (or $c_1 n^3$) steps by $1 - e^{-\Omega(n^2)}$, if c_1 is large enough.

Analysis of Phase 2. We have n objects with m zeros each. We cannot prove that the mn zeros are somehow nicely distributed among the positions. Good objects tend to create similar good objects, at least in the first phase. The population may be "quite concentrated" at the end of the first phase. Then, crossover cannot help. We prove that mutations ensure in the second phase that the zeros become "somehow distributed".

We only investigate the first position and later multiply the failure probability by n to obtain a common result for all positions. Let z be the number of zeros in the first position of the objects. Then $z \leq n$ in the beginning and we claim that, with high probability, $z \leq \frac{1}{4m}n$ at the end. We look for an upper bound $p^+(z)$ on the probability to increase the number of zeros in one round and for a lower bound $p^-(z)$ on the probability to decrease the number of zeros in one round. The number of zeros at a fixed position can change at most by 1 in one round. As long as we do not create an optimal object, all objects contain $n-m$ ones.

Let A^+ resp. A^- be the event that (given z) the number of zeros (at position 1) increases resp. decreases in one round. Then

$$A^+ \subseteq B \cup \left(\overline{B} \cap C \cap \left[\left(D \cap E \cap \bigcup_{1 \leq i \leq m-1} F_i^+\right) \cup \left(\overline{D} \cap \overline{E} \cap \bigcup_{1 \leq i \leq m} G_i^+\right)\right]\right)$$

with the following meaning of the events:

- B: crossover is chosen as operator, $\text{Prob}(B) = 1/(n \log n)$.
- C: an object with a one at position 1 is chosen for replacement, $\text{Prob}(C) = (n-z)/n$.
- D: an object with a zero at position 1 is chosen for mutation, $\text{Prob}(D) = z/n$.
- E: the bit at position 1 does not flip, $\text{Prob}(E) = 1 - 1/n$.
- F_i^+: there are exactly i positions among the $(m-1)$ 0-positions $j \neq 1$ which flip and exactly i positions among the $(n-m)$ 1-positions which flip, $\text{Prob}(F_i^+) = \binom{m-1}{i}\binom{n-m}{i}(1/n)^{2i}(1-1/n)^{n-2i-1}$. We can exclude the case $i = 0$ which leads to a replication.
- G_i^+: there are exactly i positions among the m 0-positions which flip and exactly $i-1$ positions among the $(n-m-1)$ 1-positions $j \neq 1$ which flip, $\text{Prob}(G_i^+) = \binom{m}{i}\binom{n-m-1}{i-1}(1/n)^{2i-1}(1-1/n)^{n-2i}$.

Hence,

$$\begin{aligned}
p^+(z) \leq{}& \frac{1}{n \log n} + \left(1 - \frac{1}{n \log n}\right)\frac{n-z}{n} \\
& \left[\frac{z}{n}\sum_{i=1}^{m-1}\binom{m-1}{i}\binom{n-m}{i}\left(\frac{1}{n}\right)^{2i}\left(1-\frac{1}{n}\right)^{n-2i}\right. \\
& \left. + \frac{n-z}{n}\sum_{i=1}^{m}\binom{m}{i}\binom{n-m-1}{i-1}\left(\frac{1}{n}\right)^{2i}\left(1-\frac{1}{n}\right)^{n-2i}\right] \\
\leq{}& \frac{1}{n \log n} + \left(1 - \frac{1}{n \log n}\right)\frac{n-z}{n} \\
& \left[\frac{z}{n}(m-1)(n-m)\frac{1}{n^2}\left(1-\frac{1}{n}\right)^{n-2} + O\left(\frac{m^2}{n^2}\right)\right. \\
& \left. + \frac{n-z}{n}\left(m\frac{1}{n^2}\left(1-\frac{1}{n}\right)^{n-2} + O\left(\frac{m^2}{n^3}\right)\right)\right].
\end{aligned}$$

Similarly, we get

$$A^- \supseteq \overline{B} \cap \overline{C} \cap \left[\left(D \cap \overline{E} \cap \bigcup_{1 \leq i \leq m} F_i^-\right) \cup \left(\overline{D} \cap E \cap \bigcup_{1 \leq i \leq m} G_i^-\right)\right]$$

where

- F_i^-: there are exactly $i-1$ positions among the $(m-1)$ 0-positions $j \neq 1$ which flip and exactly i positions among the $(n-m)$ 1-positions which flip, $\text{Prob}(F_i^-) = \binom{m-1}{i-1}\binom{n-m}{i}(1/n)^{2i-1}(1-1/n)^{n-2i}$.

- G_i^-: there are exactly i positions among the m 0-positions which flip and exactly i positions among the $(n-m-1)$ 1-positions $j \neq 1$ which flip (if $i=0$, we get the case of a replication), $\text{Prob}(G_i^-) = \binom{m}{i}\binom{n-m-1}{i}(1/n)^{2i}(1-1/n)^{n-2i-1}$.

Hence,

$$p^-(z) \geq \left(1 - \frac{1}{n\log n}\right)\frac{z}{n}\left[\frac{z}{n}\sum_{i=1}^{m}\binom{m-1}{i-1}\binom{n-m}{i}\left(\frac{1}{n}\right)^{2i}\left(1-\frac{1}{n}\right)^{n-2i} + \frac{n-z}{n}\sum_{i=1}^{m}\binom{m}{i}\binom{n-m-1}{i}\left(\frac{1}{n}\right)^{2i}\left(1-\frac{1}{n}\right)^{n-2i}\right]$$
$$\geq \left(1 - \frac{1}{n\log n}\right)\frac{z}{n}\left[\frac{z}{n}(n-m)\frac{1}{n^2}\left(1-\frac{1}{n}\right)^{n-2} + \frac{n-z}{n}m(n-m-1)\frac{1}{n^2}\left(1-\frac{1}{n}\right)^{n-2}\right].$$

Since m is a constant, we obtain, if $z \geq \frac{1}{8m}n$, that $p^-(z) \geq p^-(z) - p^+(z) = \Omega(\frac{1}{n})$ and it is also easy to see that $p^-(z) = O(\frac{1}{n})$.

We call a step essential, if the number of zeros in the first position changes. The length of Phase 2 is $c_2 n^2 \log n$. The following considerations work under the assumption $z \geq \frac{1}{8m}n$. The probability that a step is essential is $\Omega(\frac{1}{n})$. Hence, for come $c_2' > 0$, the probability of having less than $c_2' n \log n$ essential steps, is bounded by Chernoff's bound by $e^{-\Omega(n)}$. We assume that this failure does not occur. Let $q^+(z)$ resp. $q^-(z)$ be the conditional probability of increasing resp. decreasing the number of zeros in essential steps. Then $q^+(z) = p^+(z)/(p^+(z)+p^-(z))$, $q^-(z) = p^-(z)/(p^+(z)+p^-(z))$, and $q^-(z) - q^+(z) = (p^-(z) - p^+(z))/(p^+(z)+p^-(z)) = \Omega(1)$. Hence, for some $c_2'' > 0$, the probability of decreasing the number of zeros by less than $c_2'' n$ is bounded by Chernoff's bound by $e^{-\Omega(n)}$. We obtain $c_2'' = 1$ by choosing c_2 large enough. But this implies that we have at some point of time less than $z^* = \frac{1}{8m}n$ zeros at position 1 and our estimations on $p^+(z)$ and $p^-(z)$ do not hold. We investigate the last point of time with z^* zeros at position 1. Then there are t essential steps left. If $t \leq \frac{1}{8m}n$, it is sure that we stop with at most $\frac{1}{4m}n$ zeros at position 1. If $t > \frac{1}{8m}n$, we can apply Chernoff's bound and obtain a failure probability of $e^{-\Omega(n)}$. Altogether the failure probability is $ne^{-\Omega(n)} = e^{-\Omega(n)}$.

Analysis of Phase 3. First, we investigate the probability that the number of zeros at position 1 reaches $\frac{1}{2m}n$. For this purpose, we consider subphases starting at points of time where the number of zeros equals $\frac{1}{4m}n$. A subphase where the number of zeros is less than $\frac{1}{4m}n$ cannot cause a failure. The same holds for subphases whose length is bounded by $\frac{1}{4m}n$. In all other cases we can apply Chernoff's bound and the assumption $z \geq \frac{1}{4m}n$. Hence, the failure probability for each subphase is bounded by $e^{-\Omega(n)}$ and the same holds for all subphases and positions altogether. We create an optimal object if we perform crossover (probability $\frac{1}{n\log n}$) if the following mutation does not flip any bit (probability $(1-\frac{1}{n})^n$), if the chosen bit strings do not share a zero at some position (probability at least $\frac{1}{2}$, see below) and the crossover chooses at each of the $2m$ positions, where the objects differ, the object with the one at this position (probability $(\frac{1}{2})^{2m}$). We prove the open claim. We fix one object of the population. The m zeros are w. l. o. g. at the positions $1,\ldots,m$. There are at most $\frac{1}{2m}n$ objects with a zero at some fixed position $j \in \{1,\ldots,n\}$, altogether at most $\frac{1}{2}n$ colliding objects. Hence, the probability of choosing a second object without collision with the first one is at least $\frac{1}{2}$. Hence, the success probability is at least $\frac{1}{n\log n}(1-\frac{1}{n})^n 2^{-2m-1}$ and the failure probability for $c_3 n^2 \log n$ steps is bounded by $e^{-\Omega(n)}$.

Combining all estimations we have proved the theorem. □

Corollary 4. *Let m be a constant. The expected time of the genetic algorithm on $JUMP_m$ is bounded by $O(n^2 \log n)$.*

Proof. We remark that Theorem 1 can be proved for arbitrary starting populations instead of random ones. The analysis of Phase 1 can be easily adjusted to populations that may contain bit strings with at least one but less than m zeros. We note, that the number of such bit strings can never be increased. It follows, that the probability of a failure in Phase 1 is still bounded by $O(n^{-k})$, if we choose the length of this phase as $c_1 n^2 \log n$. Therefore, the sum of all failure probabilities can be bounded by $O(1/n)$. In case that after one superround, i. e. the four phases, the optimum is not found, the process is repeated starting with the current population at the end of Phase 4. The expected number of repetitions is $(1-O(1/n))^{-1} = O(1)$, implying that the expected running time is bounded by $O(n^2 \log n)$. □

In the following, we generalize our results to the case $m = O(\log n)$ where we reduce the crossover probability to $\frac{1}{n\log^3 n}$. Nothing has to be changed for Phase 1 (and Phase 0). In Phase 2, we obtain $p^-(z) - p^+(z) = \Omega(\frac{1}{n\log^2 n})$, $p^-(z) = O(\frac{\log n}{n})$, and $q^-(z) - q^+(z) = \Omega(\frac{1}{\log^3 n})$, if $z \geq \frac{1}{8m}n$. Then $O(n^2 \log^5 n)$ steps are enough to obtain the desired properties with a probability bounded by $e^{-\Omega(n^\delta)}$ for each $\delta < 1$. The same arguments work for the first property of Phase 3 as long as the length is polynomially bounded. In order to have an exponentially small failure probability for the event to create an optimal object, we increase the number of steps to $\Theta(n^2 2^{2m})$. If we are satisfied with a constant success probability, $\Theta(n(\log^3 n)2^{2m})$ steps are sufficient. We summarize these considerations.

Theorem 5. *Let $m = O(\log n)$. For each constant $\delta < 1$, with probability $1 - e^{-\Omega(n^\delta)}$, the genetic algorithm creates an optimal object for $JUMP_m$ within $O(n^3 + n^2(\log^5 n + 2^{2m}))$ steps. The expected run time is bounded by*

$$O(n\log^3 n(n\log^2 n + 2^{2m})).$$

If we allow replications as well as duplicates things change a little. We concentrate on the case where m is a constant. Nothing changes for phase 0. Our analysis of Phase 1 remains valid, too. We remark that replications occur with probability at least $(1 - \frac{1}{n\log n})(1-\frac{1}{n})^n$, so the tendency of good objects to create similar or equal objects in the first phase is enlarged. We change the length of Phase 2 to $a(n) \cdot c_2 n^3 \log n$ and discuss the role of $a(n)$ later. We have to adapt our considerations to the circumstance that replications are allowed. For A^+ we have to include the event F_0^+, for A^- we include

G_0^-. We still have $p^-(z) \geq p^-(z) - p^+(z) = \Omega(\frac{1}{nm^2}) = \Omega(\frac{1}{n})$, but $p^-(z) = O(\frac{1}{n})$ does not hold, now. We consider only essential steps that still occur with probability $\Omega(\frac{1}{n})$. Let again $q^-(z) = p^-(z)/(p^-(z)+p^+(z))$ resp. $q^+(z) = p^+(z)/(p^-(z)+p^+(z))$ be the conditional probabilities for decreasing resp. increasing the number of zeros in one essential step. Now, we have $q^-(z) - q^+(z) = \Omega(\frac{1}{n})$. If in exactly d of $a(n) \cdot c_2' n^2 \log n$ essential steps the number of zeros is decreased, we end up with $z + a(n) \cdot c_2' n^2 \log n - 2d$ zeros. Applying Chernoff's inequality yields that the probability not to decrease the number of zeros to at most $\frac{1}{4m} n$ in $a(n) \cdot c_2' n^2 \log n$ essential steps is $e^{-\Omega(a(n) \log n)}$.

Choosing $a(n) = \frac{n}{\log n}$ yields that with probability $1 - e^{-\Omega(n)}$ the number of zeros is at most $\frac{1}{4m} n$ at all positions after Phase 2, which now has a length of $c_2 n^3$ steps. With $a(n) = \log n$, Phase 2 needs only $c_2 n^3 \log^2 n$ steps, but the probability of a failure is increased to $e^{-\Omega(\log^2 n)}$, which is still subpolynomial.

In Phase 3 replications change the probabilities for changing the number of zeros in the same way as in Phase 2. Therefore, we can adapt our proof to the modified algorithm the same way as we did for Phase 2. Since we are satisfied, if the number of zeros is not increasing too much, compared to Phase 2, where we need a decreasement, it is not necessary to adjust the length of Phase 3. In order to have the bounds for the probability of a failure in Phase 1 and Phase 2 in the same order of magnitude, we choose $c_1 a(n) n^2 \log n$ as length of Phase 1. We conclude that a genetic algorithm that allows replications finds a optimum of JUMP$_m$, for constant m, in $O(a(n) n^3 \log n)$ steps with probability $1 - e^{-\Omega(a(n) \log n)}$. We remark that the analysis of this variant of a genetic algorithm can be adapted to $m = O(\log n)$, too.

5 Conclusion

Evolutionary and genetic algorithms are often used in applications but the theory on these algorithms is in its infancy. In order to obtain a theory on evolutionary and genetic algorithms, one has to understand the main operators. This paper contains the first proof that, for some function, genetic algorithms with crossover can be much more efficient (polynomial versus superpolynomial) than all types of evolutionary algorithms without crossover. The specific bounds are less important than the fact that we have analytical tools to prove such a result. The difference in the behavior of the algorithms can be recognized in experiments already for small parameters, e. g. $n = 50$ and $m = 3$.

References

[1] Baum, E. B., Boneh, D., and Garret, C.: On genetic algorithms. In Proceedings of the 8th Conference on Computational Learning Theory (COLT '95), (1995) 230–239.

[2] Droste, S., Jansen, Th., and Wegener, I.: On the Analysis of the $(1+1)$ Evolutionary Algorithm. Tech. Report CI-21/98. Collaborative Research Center 531, Reihe Computational Intelligence, Univ. of Dortmund, Germany, (1998).

[3] Droste, S., Jansen, Th., Wegener, I.: A rigorous complexity analysis of the $(1+1)$ evolutionary algorithm for separable functions with Boolean inputs. Evolutionary Computation **6(2)** (1998) 185–196.

[4] Fogel, L. J., Owens, A. J., and Walsh, M. J.: Artificial Intelligence Through Simulated Evolutions. (1966) Wiley, New York.

[5] Forrest, S. and Mitchell, M.: Relative building block fitness and the building block hypothesis. In D. Whitley (Ed.): Foundations of Genetic Algorithms 2, (1993) 198–226, Morgan Kaufmann, San Mateo, CA.

[6] Goldberg, D. E.: Genetic Algorithms in Search, Optimization, and Machine Learning. (1989) Addison Wesley, Reading, Mass.

[7] Holland, J. H.: Adaption in Natural and Artificial Systems. (1975) Univ. of Michigan.

[8] Horn, J., Goldberg, D. E., and Deb, K.: Long Path problems. In Y. Davidor, H.-P. Schwefel, and R. Männer (Eds.): Parallel Problem Solving from Nature (PPSN III), (1994) 149–158, Springer, Berlin, Germany.

[9] Jerrum, M. and Sinclair, A.: The Markov Chain Monte Carlo method: An approach to approximate counting and integration. In D. S. Hochbaum (Ed.): Approximation Algorithms for NP-hard Problems. (1997) 482–520, PWS Publishers, Boston, MA.

[10] Jerrum, M. and Sorkin, G. B.: The Metropolis algorithm for graph bisection. Discrete Applied Mathematics **82** (1998) 155–175.

[11] Juels, A. and Wattenberg, M.: Stochastic Hillclimbing as a Baseline Method for Evaluating Genetic Algorithms, Tech. Report CSD-94-834, (1994), Univ. of California.

[12] van Laarhoven, P. J. M. and Aarts, E. H. L.: Simulated Annealing. Theory and Applications, (1987), Reidel, Dordrecht, The Netherlands.

[13] Mitchell, M. and Forrest, S.: Royal Road functions. In Th. Bäck, D. B. Fogel and Z. Michalewicz (Eds.): Handbook of Evolutionary Computation, (1997) B2.7:20–B2.7:25, Oxford University Press, Oxford UK.

[14] Mitchell, M., Forrest, S., and Holland, J. H.: The Royal Road function for genetic algorithms: Fitness landscapes and GA performance. In F. J. Varela and P. Bourgine (Eds.): Proceedings of the First European Conference on Artificial Life, (1992) 245–254, MIT Press, Cambridge, MA.

[15] Mitchell, M., Holland, J. H., and Forrest, S.: When will a genetic algorithm outperform hill climbing? In J. Cowan, G. Tesauro, and J. Alspector (Eds.): Advances in Neural Information Processing Systems, (1994), Morgan Kaufmann, San Francisco, CA.

[16] Motwani, R. and Raghavan, P.: Randomized Algorithms. (1995) Cambridge University Press, Cambridge.

[17] Rabani, Y., Rabinovich, Y., and Sinclair, A.: A computational view of population genetics. Random Structures and Algorithms **12(4)** (1998) 314–334.

[18] Rabinovich, Y., Sinclair, A., and Wigderson, A.: Quadratical dynamical systems (preliminary version). In Proceedings of the 33rd IEEE Symposium on Foundations of Computer Science (FOCS '92), (1992) 304–313, IEEE Press Piscataway, NJ.

[19] Ronald, S.: Duplicate genotypes in a genetic algorithm. In Proceedings of the IEEE International Conference on Evolutionary Computation (ICEC '98), (1998) 793–798, IEEE Press Piscataway, NJ.

[20] Rudolph, G.: How mutation and selection solve long path problems in polynomial expected time. Evolutionary Computation **4(2)** (1997) 195–205.

[21] Sarma J. and De Jong, K.: Generation gap methods. In Th. Bäck, D. B. Fogel and Z. Michalewicz (Eds.): Handbook of Evolutionary Computation, (1997) C2.7, Oxford University Press, UK.

[22] Schwefel, H.-P.: Evolution and Optimum Seeking. (1995) Wiley, New-York, NY.

Motif Statistics*
Abstract**

Pierre Nicodème[1], Bruno Salvy[2], and Philippe Flajolet[2]

[1] DKFZ Theoretische Bioinformatik. Germany.
[2] Algorithms Project. Inria Rocquencourt. France.

Abstract. We present a complete analysis of the statistics of number of occurrences of a regular expression pattern in a random text. This covers "motifs" widely used in computational biology. Our approach is based on: *(i)* classical constructive results in theoretical computer science (automata and formal language theory); *(ii)* analytic combinatorics to compute asymptotic properties from generating functions; *(iii)* computer algebra to determine generating functions explicitly, analyse generating functions and extract coefficients efficiently. We provide constructions for overlapping or non-overlapping matches of a regular expression. A companion implementation produces: multivariate generating functions for the statistics under study; a fast computation of their Taylor coefficients which yields exact values of the moments with typical application to random texts of size 30,000; precise asymptotic formulæ that allow predictions in texts of arbitrarily large sizes. Our implementation was tested by comparing predictions of the number of occurrences of motifs against the 7 megabytes aminoacid database PRODOM. We handled more than 88% of the standard collection of PROSITE motifs with our programs. Such comparisons help detect which motifs are observed in real biological data more or less frequently than theoretically predicted.

1 Introduction

The purpose of molecular biology is to establish relations between chemical form and function in living organisms. From an abstract mathematical or computational standpoint, this gives rise to two different types of problems: processing problems that, broadly speaking, belong to the realm of pattern-matching algorithmics, and probabilistic problems aimed at distinguishing between what is statistically significant and what is not, at discerning "signal" from "noise". The present work belongs to the category of probabilistic studies originally motivated by molecular biology. As we shall see, however, the results are of a somewhat wider scope.

Fix a finite alphabet, and take a large random *text* (a sequence of letters from the alphabet), where randomness is defined by either a Bernoulli model (letters are drawn independently) or a Markov model. Here, a *pattern* is specified by an *unrestricted regular expression R* and occurrences anywhere in a text file are considered. The problem is to quantify precisely what to expect about the *number of occurrences* of pattern R in a random text of size n. We are interested first of all in moments of the distributions—*what is*

* This work has been partially supported by the Long Term Research Project Alcom-IT (#20244) of the European Union.

** An extended version of this abstract is available as INRIA Research Report 3606.

the mean and the variance?—but also in asymptotic properties of the distribution—*does the distribution have a simple asymptotic form?*—as well as in computational aspects—*are the characteristics of the distribution effectively accessible?*

We provide positive answers to these three questions. Namely, for all "non-degenerate" pattern specifications[1] R, we establish the following results: *(i).* The number of occurrences has a mean of the form $\mu \cdot n + O(1)$, with a standard deviation that is of order $\sqrt{n}$; in particular, concentration of distribution holds; *(ii).* The number of occurrences, once normalized by the mean and standard deviation, obeys in the asymptotic limit a Gaussian law; *(iii).* The characteristics of the distribution are effectively computable, both exactly and asymptotically, given basic computer algebra routines. The resulting procedures are capable of treating fairly large "real-life" patterns in a reasonable amount of time.

Though initially motivated by computational biology considerations, these results are recognizably of a general nature. They should thus prove to be of use in other areas, most notably, the analysis of complex string matching algorithms, large finite state models of computer science and combinatorics, or natural language studies. (We do not however pursue these threads here and stay with the original motivation provided by computational biology.)

The basic mathematical objects around which the paper is built are counting *generating functions*. In its bivariate version, such a generating function encodes exactly all the information relative to the frequency of occurrence of a pattern in random texts of all sizes. We appeal to a combination of classical results from the theory of *regular expressions and languages* and from basic combinatorial analysis (marking by auxiliary variables) in order to determine such generating functions systematically. Specifically, we use a chain from regular expression patterns to bivariate generating functions that goes through nondeterministic and deterministic finite automata. Not too unexpectedly, the generating functions turn out to be rational (Th. 1), but also computable at a reasonable cost for most patterns of interest (§6). Since coefficients of univariate rational GF's are computable in $O(\log n)$ arithmetic operations, this provides the exact statistics of matches in texts of several thousands positions in a few seconds, typically. Also, asymptotic analysis of the coefficients of rational functions can be performed efficiently [13]. Regarding multivariate asymptotics, a perturbation method from analytic combinatorics then yields the Gaussian law (Th. 2).

In the combinatorial world, the literature on pattern statistics is vast. It originates largely with the introduction of correlation polynomials by Guibas and Odlyzko [14] in the case of patterns defined by one word. The case of several words in Bernoulli or Markov texts was studied by many authors, including [14, 10, 4, 24, 25]; see also the review in [31, Chap. 12]. As a result of these works, the number of occurrences of any *finite set of patterns* in a random Bernoulli or Markov text is known to be asymptotically normal. Several other works are motivated by computational biology considerations [20, 28, 22, 26, 29, 1]. Our distributional results that deal with arbitrary regular expression patterns, including *infinite word sets*, extend the works of these authors.

[1] Technically, non-degeneracy is expressed by the "primitivity" condition of Th. 2. All cases of interest can be reduced to this case; see the discussion at the end of §4.

The effective character of our results is confirmed by a *complete implementation* based on symbolic computation, the Maple system in our case. Our implementation has been tested against real-life data provided by a collection of patterns, the frequently used PROSITE collection[2] [2]. We apply our results to compute the statistics of matches and compare with what is observed in the PRODOM database[3].

In its most basic version, string-matching considers one or a few strings that are searched for in the text. *Motifs* appear in molecular biology as signatures for families of similar sequences and they characterize structural functionalities of sequences derived from a common ancestor. For instance, a typical motif of PROSITE is [LIVM](2)-x-D-D-x(2,4)-D-x(4)-R-R-[GH], where the capital letters represent aminoacids, 'x' stands for any letter, brackets denote a choice and parentheses a repetition. Thus x(2,4) means two to four consecutive arbitrary aminoacids, while [LIVM](2) means two consecutive elements of the set {L,I,V,M}. Put otherwise, a motif is a regular expression of a restricted form that may be expanded, in principle at least, into a *finite* set of words. Our analysis that addresses general regular expression patterns, including a wide class of *infinite* sets of words, encompasses the class of all motives.

On the practical side, it is worthwhile to remark that the automaton description for a motif tends to be much more compact than what would result from the expansion of the language described by the motif, allowing for an exponential reduction of size in many cases. For instance, for motif PS00844 from PROSITE our program builds an automaton which has 946 states while the number of words of the finite language generated by the motif is about 2×10^{26}. In addition, regular expressions are able to capture long range dependencies, so that their domain of application goes far beyond that of standard motifs.

Contributions of the paper. This work started when we realized that computational biology was commonly restricting attention to what seemed to be an unnecessarily constrained class of patterns. Furthermore, even on this restricted class, the existing literature often had to rely on approximate probabilistic models. This led to the present work that demonstrates, both theoretically and practically, that a more general framework is fully workable. On the theory side, we view Th. 2 as our main result, since it appears to generalize virtually everything that is known regarding probabilities of pattern occurrences. On the practical side, the fact that we can handle *in an exact way* close to 90% of the motifs of a standard collection that is of common use in biological applications probably constitutes the most striking contribution of the paper.

2 Main results

We consider the number of occurrences of a pattern (represented by a fixed given regular expression R) in a text under two different situations: in the *overlapping* case, all the positions in the text where a match with the regular expression can occur are counted (once); in the *non-overlapping* case, the text is scanned from left to right, and every

[2] At the moment, Prosite comprises some 1,200 different patterns, called "motifs", that are regular expressions of a restricted form and varying structural complexity.

[3] Prodom is a compilation of "homologous" domains of proteins in SWISS-PROT, and we use it as a sequence of length 6,700,000 over the alphabet of aminoacids that has cardinality 20.

time a match is found, the count is incremented and the search starts afresh at this position. These cases give rise to two different statistics for the number X_n of matches in a random text of size n, and we handle both of them. Without loss of generality, we assume throughout that R does not contain the empty word ε.

In each context, the method we describe gives an algorithm to compute the bivariate probability generating function $P(z,u) = \sum_{n,k\geq 0} p_{n,k} u^k z^n$, with $p_{n,k} = \Pr\{X_n = k\}$. This generating function specializes in various ways. First, $P(z,0)$ is the probability generating function of texts that do not match against the motif, while $R(z) = 1/(1-z) - P(z,0)$ is the probability generating function of texts with at least one occurrence. More generally, the coefficient $[u^k]P(z,u)$ is the generating function of texts with k occurrences. Partial derivatives $M_1(z) = \frac{\partial F}{\partial u}(z,1)$ and $M_2(z) = \frac{\partial}{\partial u} u \frac{\partial F}{\partial u}(z,u)\Big|_{u=1}$, are generating functions of the first and second moments of the number of occurrences in a random text of length n.

Our first result characterizes these generating functions as effectively computable rational functions.

Theorem 1. *Let R be a regular expression, X_n the number of occurrences of R in a random text of size n, and $p_{n,k} = \Pr\{X_n = k\}$ the corresponding probability distribution. Then, in the overlapping or in the non-overlapping case, and under either the Bernoulli model or the Markov model, the generating functions $P(z,u)$, $R(z)$, $M_1(z)$, $M_2(z)$, corresponding to probabilities of number of occurrences, existence of a match, and first and second moment of number of occurrences, are rational and can be computed explicitly given R.*

Our second result provides the corresponding asymptotics. Its statement relies on the fundamental matrix $T(u)$ defined in §4, as well as the notion of primitivity, a technical but nonrestrictive condition, that is defined there.

Theorem 2. *Under the conditions of Th. 1, assume that the "fundamental matrix" $T(1)$ defined by (3) is primitive. Then, the mean and variance of X_n grow linearly, $\mathrm{E}(X_n) = \mu n + c_1 + O(A^n)$, $\mathrm{Var}(X_n) = \sigma^2 n + c_2 + O(A^n)$, where $\mu \neq 0$, $\sigma \neq 0$, c_1, c_2 are computable constants.*

The normalized variable, $(X_n - \mu n)/(\sigma\sqrt{n})$, converges with speed $O(1/\sqrt{n})$ to a Gaussian law. A local limit and large deviation bounds also hold.

3 Algorithmic Chain

In order to compute the probability generating function of the number of occurrences of a regular expression, we use classical constructions on non-deterministic and deterministic finite automata. For completeness, we state all the algorithms, old and new, leading to the probability generating functions of Th. 1. References for this section are [19, 17, 15, 23] among numerous textbooks describing regular languages and automata.

Regular Languages. We consider a *finite alphabet* $\Sigma = \{\ell_1, \ldots, \ell_r\}$. A *word* over Σ is a finite sequence of *letters*, that is, elements of Σ. A *language* over Σ is a set of words. The *product* $A = A_1 \cdot A_2$ of two languages A_1 and A_2 is $A = \{w_1 w_2, w_1 \in A_1, w_2 \in A_2\}$, where w_1w_2 is the concatenation of words w_1 and w_2. Let A^n be the set of products of n words belonging to A, then the *star closure* $A^\star$ of a language A is the infinite union $A^\star = \bigcup_{n\geq 0} A^n$. The language $\Sigma^\star$ is thus the collection of all possible words over Σ.

Regular languages over Σ are defined inductively. Such a language is either the empty word, or it reduces to a single letter, or it is obtained by union, product or star closure of simpler regular languages. The formula expressing a regular language in terms of these operations and letters is called a *regular expression*. As notational convenience, ℓ denotes the singleton language $\{\ell\}$, $+$ represents a union, and $\cdot$ is freely omitted. The order of precedence for the operators is $\star, \cdot, +$.

A Nondeterministic Finite Automata (or NFA) is formally specified by five elements. (1) An input alphabet Σ; (2) A finite collection of states Q; (3) A start state $s \in \Sigma$; (4) A collection of final states $F \subset Q$; (5) A (possibly partial) transition function δ from $Q \times \Sigma$ to S_Q the set of subsets of Q. There exists a *transition* from state q_i to state q_j if there is a letter $\ell \in \Sigma$ such that $q_j \in \delta(q_i, \ell)$. A word $w = w_1w_2 \cdots w_n \in \Sigma^\star$ is *accepted* or *recognized* by a NFA $A = (\Sigma, Q, s, F, \delta)$ if there exists a sequence of states $q_0, q_1, q_2, \ldots, q_n$ such that $q_0 = s$, $q_j \in \delta(q_{j-1}, w_j)$ and $q_n \in F$.

Kleene's theorem states that a language is regular if and only if it is recognized by a NFA. Several algorithms are known to construct such a NFA. We present below an algorithm due to [6] as improved by [8] that constructs a NFA called the Glushkov automaton.

ALGORITHM

1. [Berry & Sethi] Input: a regular expression R over an alphabet Σ. Output: a NFA recognizing the corresponding language.

1 Give increasing indices to the occurrences of each letter of Σ occurring in R. Let Σ' be the alphabet consisting of these indexed letters.
2 For each letter $\ell \in \Sigma'$, construct the subset $\mathsf{follow}(\ell)$ of Σ' of letters that can follow ℓ in a word recognized by R.
3 Compute the sets $\mathsf{first}(R)$ and $\mathsf{last}(R)$ of letters of Σ' that can occur at the beginning and at the end of a word recognized by R.
4 The automaton has as states the elements of Σ' plus a start state. The transitions are obtained using follow and erasing the indices. The final states are the elements of $\mathsf{last}(R)$.

Steps 2 and 3 are performed by computing inductively four functions "first", "last", "follow" and "nullable". Given a regular expression r over Σ', first returns the set of letters that can occur at the beginning of a match; last returns those that can occur at the end of a match; nullable returns true if r recognizes the empty word and false otherwise; for each $\ell \in \Sigma'$ that occurs in r, follow returns the set of letters that can follow ℓ in a word recognized by r. The computation of these functions is a simple induction [8]; the whole algorithm has a quadratic complexity.

Deterministic Finite Automata (or DFAs) are special cases of NFAs where the images of the transition function are singletons. By a classical theorem of Rabin & Scott, NFAs are equivalent to DFAs in the sense that they recognize the same class of languages. This is made effective by the powerset construction.

ALGORITHM

2. [Rabin & Scott] Input: a NFA $A = (\Sigma, Q, s, F, \delta)$. Output: a DFA recognizing the same language.

1 Define a transition function $\Delta : S_Q \times \Sigma \to S_Q$ by: $\forall V \in S_Q$, $\forall \ell \in \Sigma$, $\Delta(V, \ell) = \bigcup_{q\in V} \delta(q, \ell)$, where S_Q is the set of subsets of Q.

2 Define Q_F as the set of subsets of Q that contain at least one element of F.
3 Return the automaton $(\Sigma, S_Q, \{s\}, Q_F, \Delta)$.

The number of states of the DFA constructed in this way is not necessarily minimal. In the worst case, the construction is of exponential complexity in the number of states of the NFA. For applications to motifs however, this construction is done in reasonable time in most cases (see §6).

Generating Functions. Let $\mathcal{A}$ be a language over Σ. The generating function of the language is obtained by summing formally all the words of $\mathcal{A}$ and collecting the resulting monomials with the letters being allowed to commute. The *generating function* of the language $\mathcal{A}$ is then defined as the formal sum $A(\ell_1, \ldots, \ell_r) = \sum_{w \in \mathcal{A}} \mathrm{com}(w)$, with $\mathrm{com}(w) = w_1 w_2 \cdots w_n$ the monomial associated to $w = w_1 w_2 \cdots w_n \in \mathcal{A}$, and $\mathrm{com}(\varepsilon) = 1$. We use the classical notation $[\ell_1^{i_1} \cdots \ell_r^{i_r}]A$ to denote the coefficient of $\ell_1^{i_1} \cdots \ell_r^{i_r}$ in the generating function A. (There is a slight abuse of notation in using the same symbols for the alphabet and the variables, which makes notation simpler.)

ALGORITHM

3. [Chomsky & Schützenberger] Input: A regular expression. Output: Its generating function.
1 Construct the DFA recognizing the language. For each state q, let $\mathcal{L}_q$ be the language of words recognized by the automaton with q as start state. These languages are connected by linear relations, $\mathcal{L}_q = (\varepsilon +) \bigcup_{\ell \in \Sigma} \ell \mathcal{L}_{\delta(q,\ell)}$, where ε is present when q is a final state. The automaton being deterministic, the unions in this system are disjoint.
2 Translate this system into a system of equations for the associated generating functions: $L_q = (1+) \sum_{\ell \in \Sigma} \ell L_{\delta(q,\ell)}$.
3 Solve the system and get the generating function $F = L_s$, where s is the start state.

The resulting generating is rational, as it is the solution of a linear system [9]. Naturally, the algorithm specializes in various ways when numerical weights (probabilities) are assigned to letters of the alphabet.

Regular Expression Matches. We first consider the Bernoulli model. The letters of the text are drawn independently at random, each letter ℓ_i of the alphabet having a fixed probability p_i, and $\sum p_i = 1$. The basis of the proof of Th. 1 is the following construction.

ALGORITHM

4. [Marked automaton] Input: A regular expression R over the alphabet Σ. Output: A DFA recognizing the (regular) language of words over $\Sigma \cup \{m\}$ where each match of the regular expression R is followed by the letter $m \notin \Sigma$, which occurs only there.
1 Construct a DFA $A = (Q, s, F, \Sigma, \delta)$ recognizing $\Sigma^\star R$.
2 Initialize the resulting automaton: set $A' = (Q', s, Q, \Sigma + m, \delta')$ with initial values $\delta' = \delta$ and $Q' = Q$.
3 Mark the matches of R: for all $q \in Q$ and all $\ell \in \Sigma$ such that $\delta(q, \ell) = f \in F$, create a new state q_ℓ in Q', set $\delta'(q, \ell) := q_\ell$ and $\delta'(q_\ell, m) := f$.
4 Restart after match (non-overlap case only): for all $f \in F$, and all $\ell \in \Sigma$ set $\delta'(f, \ell) := \delta(s, \ell)$.
5 Return A'.

We note that the automaton constructed in this way is deterministic since all the transitions that have been added are either copies of transitions in A, or start from a new state, or were missing.

This automaton recognizes the desired language. Indeed, the words of $\Sigma^\star R$ are all the words of $\Sigma^\star$ ending with a match of R. Thus the final states of A are reached only at the end of a match of R. Conversely, since no letter is read in advance, every time a match of R has just been read by A, the state which has been reached is a final state. Thus inserting a non-final state and a marked transition "before" each final state corresponds to reading words with the mark m at each position where a match of R ends. Then by making all the states final except those intermediate ones, we allow the words to end without it being the end of a match of R. In the non-overlapping case, the automaton is modified in step 4 to start afresh after each match. (This construction can produce states that are not reachable. While this does not affect the correctness of the rest of the computation, suppressing these states saves time.)

The proof of Th. 1 is concluded by the following algorithm in the Bernoulli model.

ALGORITHM

5. [Number of matches—Bernoulli] Input: A regular expression R over an alphabet Σ and the probabilities p_i of occurrence of each letter $\ell_i \in \Sigma$. Output: The bivariate generating function for the number of occurrences of R in a random text according to the Bernoulli model.

1 Construct the marked automaton for R.

2 Return the generating function $F(p_1 z, \ldots, p_r z, u)$ of the corresponding language, as given by the Chomsky-Schützenberger Algorithm.

The proof of Th. 1 in the Markov model follows along similar lines. It is based on an automaton that keeps track of the letter most recently read.

ALGORITHM

6. [Markov automaton] Input: A DFA A over an alphabet Σ. Output: A DFA over the alphabet $(\ell_0 + \Sigma)^2$, where $\ell_0 \notin \Sigma$. For each word $w_1 \cdots w_n$ recognized by A, this DFA recognizes the word $(\ell_0, w_1)(w_1, w_2) \cdots (w_{n-1}, w_n)$.

1 Duplicate the states of A until there are only input transitions with the same letter for each state. Let $(Q, s, F, \Sigma, \delta)$ be the resulting automaton.

2 Define a transition function $\Delta : Q \times (\ell_0 + \Sigma)^2 \to Q$ by $\Delta(\delta(q, \ell), (\ell, \ell')) = \delta(\delta(q, \ell), \ell')$ for all $q \in Q \setminus \{s\}$, and $\ell, \ell' \in \Sigma$; and $\Delta(\delta(s, \ell), (\ell_0, \ell)) = \delta(s, \ell)$ for all $\ell \in \Sigma$.

3 Return $(Q, s, F, (\ell_0 + \Sigma)^2, \Delta)$.

This construction then gives access to the bivariate generating function.

ALGORITHM

7. [Number of matches—Markov] Input: A regular expression R over an alphabet Σ, the probabilities q_{ij} of transition from letter ℓ_i to ℓ_j and the probabilities q_{0j} of starting with letter ℓ_j for all $\ell_i, \ell_j \in \Sigma$. Output: The bivariate generating function for the number of occurrences of R in a random text according to the Markov model.

1 Apply the algorithm "Marked automaton" with "Markov automaton" as an extra step between steps 1 and 2.

2 Return the generating function $F(q_{01} z, \ldots, q_{rr} z, u)$ of the corresponding language.

This concludes the description of the algorithmic chain, hence the proof of Th. 1, as regards the bivariate generating function $P(z, u)$ at least. The other generating functions then derive from P in a simple manner. □

4 Limiting Distribution

In this section, we establish the limiting behaviour of the probability distribution of the number of occurrences of a regular expression R in a random text of length n and prove that it is asymptotically Gaussian, thereby establishing Th. 2. Although this fact could be alternatively deduced from limit theorems for Markov chains, the approach we adopt has the advantage of fitting nicely with the computational approach of the present paper. In this abstract, only a sketch of the proof is provided.

Streamlined proof. The strategy of proof is based on a general technique of singularity perturbation, as explained in [11], to which we refer for details. This technique relies on an analysis of the bivariate generating function $P(z,u)$. The analysis reduces to establishing that in a fixed neighbourhood of $u = 1$, $P(z,u)$ behaves as

$$\frac{c(u)}{1 - z\lambda(u)} + g(z,u), \tag{1}$$

with $c(1) \neq 0$, $c(u)$ and $\lambda(u)$ analytic in the neighbourhood of $u = 1$ and $g(z,u)$ analytic in $|z| < \delta$ for some $\delta > 1/\lambda(1)$ independent of u. Indeed, if this is granted, there follows

$$[z^n]P(z,u) = c(u)\lambda(u)^n(1 + O(A^n)), \tag{2}$$

for some $A < 1$. The last equation says that X_n has a generating function that closely resembles a large power of a fixed function, that is, the probability generating function of a sum of independent random variables. Thus, we are close to a case of application of the central limit theorem and of Levy's continuity theorem for characteristic functions [7]. This part of our treatment is in line with the pioneering works [3, 5] concerning limit distributions in combinatorics. Technically, under the "variability condition", namely $\lambda''(1) + \lambda'(1) - \lambda'(1)^2 \neq 0$, we may conveniently appeal to the *quasi-powers theorem* [16] that condenses the consequences drawn from analyticity and the Berry-Esseen inequalities. This implies convergence to the Gaussian law with speed $O(1/\sqrt{n})$, the expectation and the variance being

$$\mathrm{E}(X_n) = n\lambda'(1) + c_1 + O(A^n), \quad \mathrm{Var}(X_n) = n(\lambda''(1) + \lambda'(1) - \lambda'(1)^2) + c_2 + O(A^n),$$
$$c_1 = c'(1), c_2 = c''(1) + c'(1) - c'(1)^2.$$

Linear structure. We now turn to the analysis leading to (1). Let A be the automaton recognizing $\Sigma^\star R$ and let m be its number of states. In accordance with the developments of §3, the matrix equation computed by Algorithm 3 for the generating functions can be written $L = zT_0L + \varepsilon$, where ε is a vector whose ith entry is 1 if state i is final and zero otherwise. The matrix T_0 is a stochastic matrix (i.e., the entries in each of its lines add up to 1). The entry $t_{i,j}$ in T_0 for $i, j \in \{1,\ldots,m\}$, is the probability of reaching state j from state i of the automaton in one step. In the overlapping case, the construction of Algorithm 5 produces a system equivalent to $L = zT_0\,\mathrm{diag}(\phi_i)L + \mathbf{1}$, $\phi_i \in \{1,u\}$, where $\mathbf{1}$ is a vector of ones since all the states of the new automaton are final, and $\phi_i = u$ when state i of A is final, and 1 otherwise. In the non-overlapping case, the system has the same shape; the transitions from the final states are the same as the transitions from the start state, which is obtained by replacing the rows corresponding to the final state by that corresponding to the start state.

Thus, up to a renumbering of states, the generating function $P(z,u)$ is obtained as the first component of the vector L in the vector equation

$$L = zT(u)L + \mathbf{1}, \tag{3}$$

with $T(u) = T_0\,\mathrm{diag}(1,\ldots,1,u,\ldots,u)$, the number of u's being the number of final states of A. Eq. (3) implies $P(z,u) = (1,0,\ldots,0)L = B(z,u)/\det(I - zT(u))$, for some polynomial $B(z,u)$, where I denotes the $m \times m$ identity matrix. The matrix $T(u)$ is called the *fundamental matrix* of the pattern R.

Perron-Frobenius properties. One can resort to results on matrices with nonnegative entries [12, 21] to obtain precise information on the location of the eigenvalue of $T(u)$ of largest modulus. Such eigenvalues determine dominant asymptotic behaviours and in particular they condition (1).

The Perron-Frobenius theorem states that if the matrix $T(u)$ $(u > 0)$ is *irreducible* and additionally *primitive*, then it has a unique eigenvalue $\lambda(u)$ of largest modulus, which is real positive.

(For an $m \times m$–matrix A, irreducibility mean that $(I+A)^m \gg \mathbf{0}$ and primitivity means $A^e \gg \mathbf{0}$, for some e, where $X \gg \mathbf{0}$ iff all the entries of X are positive.) In the context of automata, irreducibility means that from any state, any other state can be reached (possibly in several steps); primitivity means that there is a large enough e such that for any pair (i,j) of states, the probability of reaching j from i in exactly e steps is positive. (Clearly, primitivity implies irreducibility.) In the irreducible case, if the matrix is not primitive, then there is a periodicity phenomenon and an integer $k \le m$ such that $T(u)^k$ is "primitive by blocks". Irreducibility and primitivity are easily tested algorithmically.

Gaussian distribution. Consider the characteristic polynomial of the fundamental matrix, $Q(\lambda) \equiv Q(\lambda,u) = \det(\lambda I - T(u))$, where $T(u)$ is assumed to be primitive. By the Perron-Frobenius theorem, for each $u > 0$, there exists a unique root $\lambda(u)$ of $Q(\lambda)$ of maximal modulus that is a positive real number. The polynomial Q has roots that are algebraic in u and therefore continuous. Uniqueness of the largest eigenvalue of $T(u)$ then implies that $\lambda(u)$ is continuous and is actually an algebraic function of u for $u > 0$. Thus there exists a $\varepsilon > 0$ and $\eta_1 > \eta_2$ two real numbers such that for u in a neighbourhood $(1-\varepsilon, 1+\varepsilon)$ of 1, $\lambda(u) > \eta_1 > \eta_2 > |\mu(u)|$, for any other eigenvalue $\mu(u)$.

The preceding discussion shows that in the neighbourhood $u \in (1-\varepsilon, 1+\varepsilon)$,

$$P(z,u) = \frac{B(\lambda^{-1}(u),u)}{\lambda^{1-m}(u)Q'(\lambda(u))(1-z\lambda(u))} + g(z,u),$$

where g is analytic in z with radius of convergence at least $1/\eta_2$. This proves (1). Then, the residue theorem applied to the integral $I_n(u) = \frac{1}{2i\pi}\oint_\gamma P(z,u)dz/z^{n+1}$, where γ is a circle around the origin of radius $\delta = 2/(\eta_1+\eta_2)$, yields (2).

The variability condition is now derived by adapting an argument in [30] relative to analytic dynamic sources in information theory, which reduces in our case to using the Cauchy-Schwartz inequality. For the L_1 matrix norm, $||T(u)^n||$ is a polynomial in u with nonnegative coefficients. It follows that $||T^n(uv)|| \le ||T^n(u^2)||^{1/2}||T^n(v^2)||^{1/2}$. Since for any matrix T, the modulus of the largest eigenvalue of T is $\lim_{n\to\infty}||T^n||^{1/n}$, we get $\lambda(uv) \le \lambda(u^2)^{1/2}\lambda(v^2)^{1/2}$, $\forall u,v > 0$. This inequality reads as a concavity property for $\phi(t) := \log\lambda(e^t)$: $\phi((x+y)/2) \le (\phi(x)+\phi(y))/2$, for any real x and y. If this inequality is strict in a neighbourhood of 0, then $\phi'' < 0$. (The case where $\phi''(0) = 0$ is discarded since $\lambda(u)$ is nondecreasing.) Otherwise, if there exist $x < 0$ and $y > 0$ such that the equality holds in the concavity relation for ϕ, then necessarily equality also holds in the interval (x,y) and ϕ is actually affine in this interval. This in turn implies $\lambda(u) = au^b$ for some real a and b and u in an interval containing 1, and therefore equality holds for all $u > 0$ from the Perron-Frobenius theorem as already discussed. Since $\lambda(1) = 1$, necessarily $a = 1$. From the asymptotic behaviour (2) follows that $b \le 1$. Now λ being a root of $Q(\lambda)$, if $\lambda(u) = u^b$ with $b < 1$, then b is a rational number p/q and the conjugates $e^{2ik\pi/q}\lambda$, $k = 1,\ldots,q-1$ are also solutions of $Q(\lambda)$, which contradicts the Perron-Frobenius theorem. Thus the only possibility for b is 1. Now, u is an eigenvalue of $uT(1)$ and another property of nonnegative matrices [21, Th. 37.2.2] shows that the only way u can be an eigenvalue of $T(u)$ is when $T(u) = uT(1)$, which can happen only when all the states of the automaton are final, i.e., $\Sigma^\star R = \Sigma^\star$, or, equivalently $\varepsilon \in R$. This concludes the proof of Th. 2 in the Bernoulli case.

Markov model. The Markov case requires a tensor product construction induced by Algorithms 6 and 7. This gives rise again to a linear system that is amenable to singularity perturbation. The condition of primitivity is again essential but it is for instance satisfied as soon as both the Markov model and the pattern automaton are primitive. (Details omitted in this abstract.) This discussion concludes the proof of Th. 2. □

We observe that the quantities given in the statement are easily computable. Indeed, from the characteristic polynomial Q of $T(u)$, the quantities involved in the expectation and variance of the statement of Th. 2 are

$$\lambda'(1) = -\left.\frac{\frac{\partial Q}{\partial u}}{\frac{\partial Q}{\partial \lambda}}\right|_{z=\lambda=1} \qquad \lambda''(1) = -\left.\frac{\frac{\partial^2 Q}{\partial u^2} + 2\lambda'(1)\frac{\partial^2 Q}{\partial u \partial \lambda} + \lambda'(1)^2 \frac{\partial^2 Q}{\partial \lambda^2}}{\frac{\partial Q}{\partial \lambda}}\right|_{z=\lambda=1}$$

We end this section with a brief discussion showing how the "degenerate" cases in which $T(1)$ is not primitive are still reducible to the case when Th. 2 applies.

Irreducibility. The first property we have used is the irreducibility of $T(1)$. It means that from any state of the automaton, any other state can be reached. In the non-overlapping case, this property is true except possibly for the start state, since after a final state each of the states following the start state can be reached. In the overlapping case, the property is not true in general, but since the generating function $P(z,u)$ does not depend on the choice of automaton recognizing $\Sigma^* R$, we can assume that the automaton is minimal (has the minimum number of states), and then the property becomes true after a finite number of steps by an argument we omit in this abstract. Thus in both cases, $T(u)$ is either irreducible or decomposes as $\begin{pmatrix} P & L \\ 0 & A(u) \end{pmatrix}$ where $A(u)$ is irreducible and it can be checked that the largest eigenvalue arises from the A-block for u near 1. It is thus sufficient to consider the irreducible case.

Primitivity. When $T(u)$ is not primitive, there is an integer $k \leq m$ such that $T^k(u)$ is primitive. Thus our theorem applies to each of the variables $X_n^{(i)}$ counting the number of matches of the regular expression R in a text of length $kn+i$ for $i = 0, \ldots, k-1$. Then, the theorem still holds once n is restricted to any congruence class modulo k.

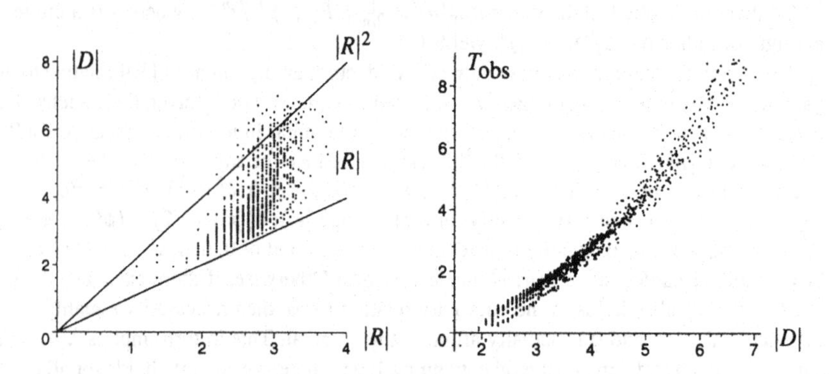

Fig. 1. The correlations between $|R|, |D|$ and $|D|, T_{\mathrm{obs}}$ in logarithmic scales.

5 Processing Generating Functions

Once a bivariate generating function of probabilities has been obtained explicitly, several operations can be performed efficiently to retrieve information.

First, differentiating with respect to u and setting $u = 1$ yields univariate generating functions for the moments of the distribution as explained in §2. By construction, these generating functions are also rational.

Fast coefficient extraction. The following algorithm is classical and can be found in [18]. It is implemented in the Maple package `gfun` [27].

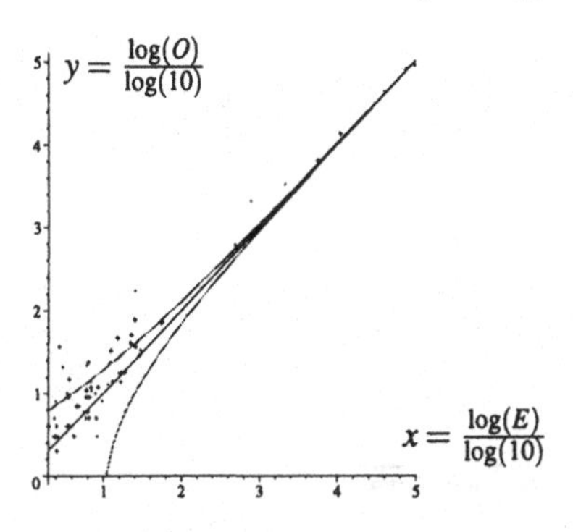

Fig. 2. Motifs with theoretical expectation $E \geq 2$. Each point corresponds to a motif with coordinates (E, O) plotted on a log-log scale. The two curves represent an approximation of ± 3 standard deviations.

Fig. 3. Histograms of motifs with 1 (dark gray), 2 (medium gray) and 3 (white) observed matches. Coordinates: $x = \log_{10} E, y =$ number of motifs.

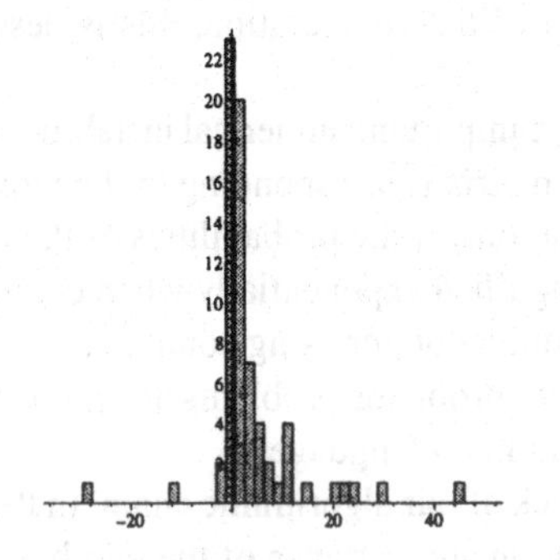

Fig. 4. Motifs with theoretical expectation $E \geq 2$: Histogram of the Z-scores $Z = \frac{O-E}{\sigma}$.

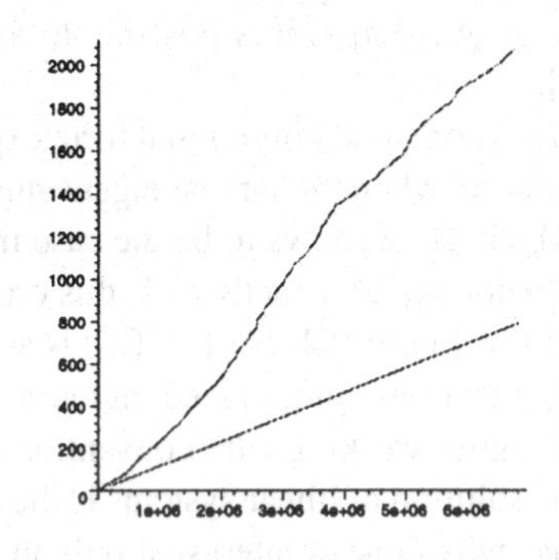

Fig. 5. Scanning PRODOM with motif PS00013. Observed matches versus expectation.

ALGORITHM

8. [Coefficient extraction] Input: a rational function $f(z) = P(z)/Q(z)$ and an integer n. Output: $u_n = [z^n]f(z)$ computed in $O(\log n)$ arithmetic operations.

1 Extract the coefficient of z^n in $Q(z)f(z) = P(z)$, which yields a linear recurrence with constant coefficients for the sequence u_n. The order m of this recurrence is $\deg(Q)$.

2 Rewrite this recurrence as a linear recurrence of order 1 relating the vector $U_n = (u_n, \ldots, u_{n-m+1})$ to U_{n-1} by $U_n = AU_{n-1}$ where A is a constant $m \times m$ matrix.

3 Use binary powering to compute the power of A in $U_n = A^{n-m}U_m$.

As an example, Fig. 6 displays the probability that the pattern ACAGAC occurs exactly twice in a text over the alphabet {A,C,G,T} against the length n of the text. The probabilities assigned to each of the letters are taken from a viral DNA ($\phi X174$). The shape of the curve is typical of that expected in the non-asymptotic regime.

Asymptotics of the coefficients of a rational function can be obtained directly. Since the recurrence satisfied by the coefficients is linear with constant coefficients, a solution can be found in the form of an exponential polynomial: $u_n = p_1(n)\lambda_1^n + \cdots + p_k(n)\lambda_k^n$,

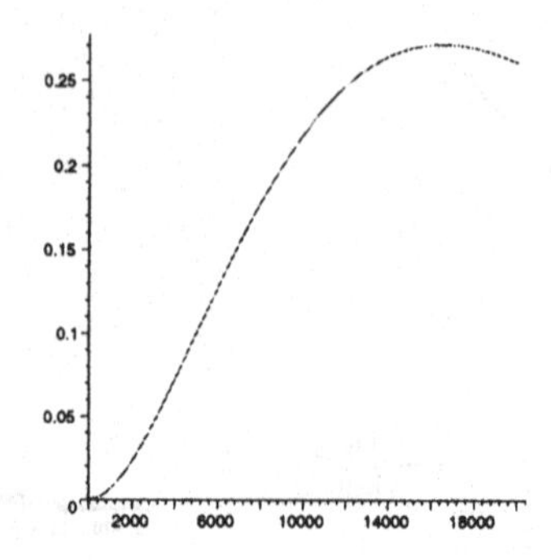

Fig. 6. Probability of two occurrences of ACAGAC in a text of length up to 20,000

where the λ_i's are roots of the polynomial $z^m Q(1/z)$ and the p_i's are polynomials. An asymptotic expression follows from sorting the λ_i's by decreasing modulus. When the degree of Q is large, it is possible to avoid part of the computation, this is described in [13].

The exponential polynomial form explains the important numerical instability of the computation when the largest eigenvalue of the matrix (corresponding to the largest λ) is 1, which Th. 2 shows to be the case in applications: if the probabilities of the transitions do not add up exactly to 1, this error is magnified exponentially when computing moments for large values of n. This is another motivation for using computer algebra in such applications, and, indeed, numerical stability problems problems are encountered by colleagues working with conventional programming languages.

The solution of linear systems is the bottleneck of our algorithmic chain. In the special case when one is interested only in expectation and variance of the number of occurrences of a pattern, it is possible to save time by computing only the local behaviour of the generating function. This leads to the following algorithm for the expectation, the variance is similar.

ALGORITHM

9. [Asymptotic Expectation] Input: the bivariate system $(I - zT(u))L - \mathbf{1} = 0$ from (3). Output: first two terms of the asymptotic behaviour of the expectation of the number of occurrences of the corresponding regular expression.

1 Let $A_1 = T(1), A_0 = I - T(1), C_0 = -\frac{\partial T}{\partial u}(1)$.
2 Solve the system $A_0 X_1 + \alpha\mathbf{1} = -C_0$, whence a value for α and a line $\tilde{X}_1 + \beta\mathbf{1}$ for X_1.
3 Solve the system $A_0 X_2 + \beta\mathbf{1} = C_0 - A_1 \tilde{X}_1$ for β. The expectation is asymptotically $E = \alpha n + \alpha - x + O(A^n)$ for some $A < 1$ and x the coordinate of X_1 corresponding to the start state of the automaton.

Algorithm 9 reduces the computation of asymptotic expectation to the solution of a few linear systems with constant entries instead of one linear system with polynomial entries. This leads to a significant speed-up of the computation. Moreover, with due care, the systems could be solved using floating-point arithmetic. (This last improvement will be tested in the future; the current implementation relies on safe rational arithmetics.)

As can be seen from the exponential polynomial form a nice feature of the expansion of the expectation to two terms is that the remainder is exponentially small.

6 Implementation

The theory underlying the present paper has been implemented principally as a collection of routines in the Maple computer algebra system. Currently, only the Bernoulli model and the non-overlapping case have been implemented. The implementation is based mainly on the package `combstruct` (developed at Inria and a component of the Maple V.5 standard distribution) devoted to general manipulations of combinatorial specifications and generating functions. Use is also made of the companion Maple library `gfun` which provides various procedures to deal with generating functions and recurrences. About 1100 lines of dedicated Maple routines have been developed by one of us (P. N.) on top of `combstruct` and `gfun` [4].

This raw analysis chain does not include optimizations and it has been assembled with the sole purpose of testing the methodology we propose. It has been tested on a collection of 1118 patterns described below and whose processing took about 10 hours when distributed over 10 workstations. The computation necessitates an average of 6 minutes per pattern, but this average is driven up by a few very complex patterns. In fact, *the median of the execution times is only 8 seconds.*

There are two main steps in the computation: construction of the automaton and asymptotic computation of expectation and variance. Let R be the pattern, D the finite automaton, and T the arithmetic complexity of the underlying linear algebra algorithms. Then, the general bounds available are: $|R| \leq |D| \leq 2^{|R|}$, $T = O(|D|^3)$, as results from the previous sections. (Sizes of R and D are defined as number of states of the corresponding NFA or DFA.) Thus, the driving parameter is $|D|$ and, eventually, the computationally intensive phase is due to linear algebra. In practice, the exponential upper bound on $|D|$ appear to be *extremely* pessimistic. Statistical analysis of the 1118 experiments indicates that the automaton is constructed in time slightly worse than linear in $|D|$ and that $|D|$ is almost always between $|R|$ and $|R|^2$. The time taken by the second step behaves roughly quadratically (in $O(|D|^2)$), which demonstrates that the sparseness of the system is properly handled by our program. For most of the patterns, the overall "pragmatic" complexity T_{obs} thus lies somewhere around $|R|^3$ or $|R|^4$ (Fig. 1).

7 Experimentation

We now discuss a small campaign of experiments conducted on PROSITE motifs intended to test the soundness of the methodological approach of this paper. No immediate biological relevance is implied. Rather, our aim is to check whether the various quantities computed do appear to have statistical relevance.

The biological target database, the "text", is built from the consensus sequences of the multi-alignments of PRODOM34.2. This database has 6.75 million positions, each occupied by one of 20 aminoacids, so that it is long enough to provide matches for rare motifs. Discarding a few motifs constrained to occur at the beginning or at the end of a sequence (a question that we do not address here) leaves 1260 unconstrained

[4] `Combstruct` and `gfun` are available at `http://algo.inria.fr/libraries`. The motif-specific procedures are to be found at `http://www.dkfz.de/tbi/people/nicodeme`.

motifs. For 1118 of these motifs (about 88% of the total) our implementation produces complete results. With the current time-out parameter, the largest automaton treated has 946 states. It is on this set of 1118 motifs that our experiments have been conducted.

For each motif, the computer algebra tools of the previous section have been used to compute exactly the (theoretical) *expectation* E and *standard deviation* σ of the statistics of number of matches. The letter frequencies that we use in the mathematical and the computational model are the empirical frequencies in the database. Each theoretical expectation E is then compared to the corresponding number of observed matches (also called observables), denoted by O, that is obtained by a straight scan of the 6.75 million position PRODOM data base[5].

Expectations. First, we discuss expectations E versus observables O. For our reference list of 1118 motifs, the theoretical expectations E range from 10^{-23} to 10^5. The observed occurrences O range from 0 to 100,934, with a median at 1, while 0 is observed in about 12% of cases. Globally, we thus have a collection of motifs with fairly low expected occurrence numbers, though a few do have high expected occurrences. Consider a motif to be "frequent" if $E \geq 2$. Fig. 2 is our main figure: it displays in log-log scale points that represent the 71 pairs (E, O) for the frequent motifs, $E \geq 2$. The figure shows a good agreement between the *orders of growths* of predicted E and observed O values: (*i*) the average value of $\log_{10} O / \log_{10} E$ is 1.23 for these 71 motifs; (*ii*) the two curves representing 3 standard deviations enclose most of the data.

Fig. 3 focusses on the classes of motifs observed $O = 1, 2, 3$ times in PRODOM. For each such class, a histogram of the frequency of observation versus $\log_{10} E$ is displayed. These histograms illustrate the fact that some motifs with very small expectation are still observed in the database. However, there is a clear tendency for motifs with smaller (computed) expectations E to occur less often: for instance, no motif whose expectation is less than 10^{-6} occurs 3 times.

***Z*-scores.** Another way to quantify the discrepancy between the expected and the observed is by means of the Z-score that is defined as $Z = (O - E)/\sigma$. Histograms of the Z-scores for the frequent motifs ($E \geq 2$) should converge to a Gaussian curve if the Bernoulli model would apply strictly and if there would be a sufficient number of data corresponding to large values of E. None of these conditions is satisfied here, but nonetheless, the histogram displays a sharply peaked profile tempered by a small number of exceptional points.

Standard deviations. We now turn to a curious property of the Bernoulli model regarding standard deviations. At this stage this appears to be a property of the model alone. It would be of interest to know whether it says something meaningful about the way occurrences tend to fluctuate in a large number of observations.

Theoretical calculations show that when the expectation of the length between two matches for a pattern is large, then $\sigma \approx \sqrt{E}$ is an excellent approximation of the standard deviation. Strikingly enough, computation shows that for the 71 "frequent" patterns, we have $0.4944 \leq \log(\sigma)/\log(E) \leq 0.4999$. (Use has been made of this approximation when plotting (rough) confidence intervals of 3 standard deviations in Fig. 2.)

[5] The observed quantities were determined by the PROSITE tools contained in the IRSEC motif toolbox `http://www.isrec.isb-sib.ch/ftp-server/`.

Table 1. Motifs with large Z-scores

Index	Pattern	E	O	Z	$\frac{O-E}{E}$
2	S-G-x-G	2149	3302	25	0.54
4	[RK](2)-x-[ST]	11209	13575	22	0.21
13	DERK(6)-[LIVMFWSTAG](2)-[LIVMFYSTAGCQ]-[AGS]-C	788	2073	46	1.63
36	[KR]-x(1,3)-[RKSAQ]-N-x(2)-[SAQ](2)-x-[RKTAENQ]-x-R-x-[RK]	2.75	37	20	12.45
190	C-CPWHF-CPWR-C-H-CFYW	25	173	29	5.86
5	[ST]-x-[RK]	99171	90192	-30	-0.09

Discussion. The first blatant conclusion is that predictions (the expectation E) tend to underestimate systematically what is observed (O). This was to be expected since the PROSITE patterns do have an *a priori* biological significance. A clearer discussion of this point can be illustrated by an analogy with words in a large *corpus* of natural language, such as observed with Altavista on the Web. The number of occurrences of a word such as 'deoxyribonucleic' is very large (about 7000) compared to the probability (perhaps 10^{-15}) assigned to it in the Bernoulli model. Thus, predictions on the category of patterns that contain long (hence unlikely) words that can occur in the *corpus* are expected to be gross underestimations. However, statistics for a pattern like "A $\langle$ any_word $\rangle$ IS IN" (590,000 matches) are more likely to be realistic. This naive observation is consistent with the fact that Fig. 2 is more accurate for frequent patterns than for others, and it explains why we have restricted most of our discussion to patterns such that $E \geq 2$. In addition, we see that the scores computed are meaningful as regards orders of growth, at least. This is supported by the fact that $\log O/\log E$ is about 1.23 (for the data of Fig. 2), and by the strongly peaked shape of Fig. 4.

Finally we discuss the patterns that are "exceptional" according to some measure. The largest automaton computed has 946 states and represents the expression $\Sigma^* R$ for the motif PS00844 ([LIV]-x(3)-[GA]-x-[GSAIV]-R-[LIVCA]-D-[LIVMF](2)-x(7,9)-[LI]-x-E-[LIVA]-N-[STP]-x-P-[GA]). Expectation for this motif is 1.87×10^{-6}, standard-deviation 0.00136, while $O = 0$. This automaton corresponds to a finite set of patterns whose cardinality is about 1.9×10^{26}. The pattern with largest expectation is PS0006 ([ST]-x(2)-[DE]) for which $E = 104633$ (and $O = 100934$) and the renewal time between two occurrences is as low as 64 positions. The motifs with very exceptional behaviours $|Z| > 19$ are listed in Table 1. The motif PS00005 ([ST]-x-[RK]) is the only motif that is clearly observed significantly less than expected.

We plot in Fig. 5 the number of observed and expected matches of PS00013 against the number of characters of PRODOM that have been scanned. The systematic deviation from what is expected is the type of indication on the possible biological significance of this motif that our approach can give.

8 Directions for Future Research

There are several directions for further study: advancing the study of the Markov model; enlarging the class of problems in this range that are guaranteed to lead to Gaussian laws; conducting sensitivity analysis of Bernoulli or Markov models. We briefly address each question in turn.

The Markov model. Although the Markov model on letters is in principle analytically and computationally tractable, the brute-force method given by algorithm "Markov automaton" probably leaves room for improvements. We wish to avoid having to deal with finite-state models of size the product $|\Sigma| \times |Q|$, with $|\Sigma|$ the alphabet cardinality and $|Q|$ the number of states of the automaton. This issue appears to be closely related to the areas of Markov chain decomposability and of Markov modulated models.

Gaussian Laws. Our main theoretical result, Th. 2, is of wide applicability in all situations where the regular expression under consideration is "nondegenerate". Roughly, as explained in §4, the overwhelming majority of regular expression patterns of interest in biological applications are expected to be nondegenerate. (Such is for instance the case for *all* the motifs that we have processed.) Additional work is called for regarding sufficient structural conditions for nondegeneracy in the case of Markov models. It is at any rate the case that the conditions of Th. 2 can be tested easily in any specific instance.

Model sensitivity and robustness. An inspection of Table 1 suggests that the exceptional motifs in the classification of Z-scores cover very different situations. While a ratio O/E of about 3 and an observable O that is > 2000 is certainly significant, some doubt may arise for other situations. For instance, is a discrepancy of 5% only on a motif that is observed about 10^5 times equally meaningful? To answer this question it would be useful to investigate the way in which small changes in probabilities may affect predictions regarding pattern occurrences. Our algebraic approach supported by symbolic computation algorithms constitutes an ideal framework for investigating model sensitivity, that is, the way predictions are affected by small changes in letter or transition probabilities.

References

[1] ATTESON, K. Calculating the exact probability of language-like patterns in biomolecular sequences. In *Sixth International Conference on Intelligent Systems for Molecular Biology* (1998), AAAI Press, pp. 17–24.

[2] BAIROCH, A., BUCHER, P., AND HOFMAN, K. The PROSITE database, its status in 1997. *Nucleic Acids Res. 25* (1997), 217–221. MEDLINE: 97169396, `http://expasy.hcuge.ch/sprot/prosite.html`.

[3] BENDER, E. A. Central and local limit theorems applied to asymptotic enumeration. *Journal of Combinatorial Theory 15* (1973), 91–111.

[4] BENDER, E. A., AND KOCHMAN, F. The distribution of subword counts is usually normal. *European Journal of Combinatorics 14* (1993), 265–275.

[5] BENDER, E. A., RICHMOND, L. B., AND WILLIAMSON, S. G. Central and local limit theorems applied to asymptotic enumeration. III. Matrix recursions. *Journal of Combinatorial Theory 35*, 3 (1983), 264–278.

[6] BERRY, G., AND SETHI, R. From regular expressions to deterministic automata. *Theoretical Computer Science 48*, 1 (1986), 117–126.

[7] BILLINGSLEY, P. *Probability and Measure*, 2nd ed. John Wiley & Sons, 1986.

[8] BRÜGGEMANN-KLEIN, A. Regular expressions into finite automata. *Theoretical Computer Science 120*, 2 (1993), 197–213.

[9] CHOMSKY, N., AND SCHÜTZENBERGER, M. P. The algebraic theory of context-free languages. In *Computer programming and formal systems*. North-Holland, Amsterdam, 1963, pp. 118–161.

[10] FLAJOLET, P., KIRSCHENHOFER, P., AND TICHY, R. F. Deviations from uniformity in random strings. *Probability Theory and Related Fields 80* (1988), 139–150.

[11] FLAJOLET, P., AND SEDGEWICK, R. The average case analysis of algorithms: Multivariate asymptotics and limit distributions. Research Report 3162, Institut National de Recherche en Informatique et en Automatique, 1997. 123 pages.

[12] GANTMACHER, F. R. *The theory of matrices. Vols. 1, 2.* Chelsea Publishing Co., New York, 1959. Translated by K. A. Hirsch.

[13] GOURDON, X., AND SALVY, B. Effective asymptotics of linear recurrences with rational coefficients. *Discrete Mathematics 153*, 1–3 (1996), 145–163.

[14] GUIBAS, L. J., AND ODLYZKO, A. M. String overlaps, pattern matching, and nontransitive games. *Journal of Combinatorial Theory. Series A 30*, 2 (1981), 183–208.

[15] HOPCROFT, J. E., AND ULLMAN, J. D. *Introduction to automata theory, languages, and computation.* Addison-Wesley Publishing Co., Reading, Mass., 1979. Addison-Wesley Series in Computer Science.

[16] HWANG, H. K. *Théorèmes limites pour les structures combinatoires et les fonctions arithmétiques.* PhD thesis, École polytechnique, Palaiseau, France, Dec. 1994.

[17] KELLEY, D. *Automata and formal languages.* Prentice Hall Inc., Englewood Cliffs, NJ, 1995. An introduction.

[18] KNUTH, D. E. *The art of computer programming. Vol. 2. Seminumerical algorithms*, second ed. Computer Science and Information Processing. Addison-Wesley Publishing Co., Reading, Mass., 1981.

[19] KOZEN, D. C. *Automata and computability.* Springer-Verlag, New York, 1997.

[20] PEVZNER, P. A., BORODOVSKI, M. Y., AND MIRONOV, A. A. Linguistic of nucleotide sequences: The significance of deviation from mean statistical characteristics and prediction of the frequencies of occurrence of words. *Journal of Biomolecular Structure Dyn. 6* (1989), 1013–1026.

[21] PRASOLOV, V. V. *Problems and theorems in linear algebra.* American Mathematical Society, Providence, RI, 1994. Translated from the Russian manuscript by D. A. Leĭtes.

[22] PRUM, B., RODOLPHE, F., AND DE TURCKHEIM, É. Finding words with unexpected frequencies in deoxyribonucleic acid sequences. *Journal of the Royal Statistical Society. Series B 57*, 1 (1995), 205–220.

[23] RAYWARD-SMITH, V. J. *A first course in formal language theory.* Blackwell Scientific Publications Ltd., Oxford, 1983.

[24] RÉGNIER, M. A unified approach to words statistics. In *Second Annual International Conference on Computational Molecular Biology* (New-York, 1998), ACM Press, pp. 207–213.

[25] RÉGNIER, M., AND SZPANKOWSKI, W. On pattern frequency occurrences in a Markovian sequence. *Algoritmica* (1998). To appear.

[26] REINERT, G., AND SCHBATH, S. Compound Poisson approximations for occurrences of multiple words in Markov chains. *Journal of Computational Biology 5*, 2 (1998), 223–253.

[27] SALVY, B., AND ZIMMERMANN, P. Gfun: a Maple package for the manipulation of generating and holonomic functions in one variable. *ACM Transactions on Mathematical Software 20*, 2 (1994), 163–177.

[28] SCHBATH, S., PRUM, B., AND DE TURCKHEIM, É. Exceptional motifs in different Markov chain models for a statistical analysis of DNA sequences. *Journal of Computational Biology 2*, 3 (1995), 417–437.

[29] SEWELL, R. F., AND DURBIN, R. Method for calculation of probability of matching a bounded regular expression in a random data string. *Journal of Computational Biology 2*, 1 (1995), 25–31.

[30] VALLÉE, B. Dynamical sources in information theory: Fundamental intervals and word prefixes. Les cahiers du GREYC, Université de Caen, 1998. 32p.

[31] WATERMAN, M. S. *Introduction to Computational Biology: Maps, sequences and genomes*. Chapman & Hall, 1995.

Approximate Protein Folding in the HP Side Chain Model on Extended Cubic Lattices (Extended Abstract)

Volker Heun

Fakultät für Informatik der TU München, D-80290 München, Germany
International Computer Science Institute, 1947 Center St., Berkeley, CA 94704, U.S.A.
heun@in.tum.de heun@icsi.berkeley.edu
http://www.in.tum.de/~heun

Abstract. One of the most important open problems in computational molecular biology is the prediction of the conformation of a protein based on its amino acid sequence. In this paper, we design approximation algorithms for structure prediction in the so-called HP side chain model. The major drawback of the standard HP side chain model is the bipartiteness of the cubic lattice. To eliminate this drawback, we introduce the extended cubic lattice which extends the cubic lattice by diagonals in the plane. For this lattice, we present two linear algorithms with approximation ratios of $59/70$ and $37/42$, respectively. The second algorithm is designed for a 'natural' subclass of proteins, which covers more than 99.5% of all sequenced proteins. This is the first time that a protein structure prediction algorithm is designed for a 'natural' subclass of all combinatorially possible sequences.

1 Introduction

One of the most important open problems in molecular biology is the prediction of the spatial conformation of a protein from its sequence of amino acids. The classical methods for structure analysis of proteins are X-ray crystallography and NMR-spectroscopy. Unfortunately, these techniques are too slow and complex for a structure analysis of a large number of proteins. On the other hand, due to the technological progress, the sequencing of proteins is relatively fast, simple, and inexpensive. Therefore, it becomes more and more important to develop efficient algorithms for determining the 3-dimensional structure of a protein based on its sequence of amino acids.

1.1 Protein Folding and the HP Model

A protein is a linear chain of amino acids linked together by peptide bonds. An amino acid consists of a common main chain part and one of twenty residues, which determines its characteristic. The sequence of amino acids for a given protein is called its primary structure. Each natural protein folds into a unique spatial conformation called its tertiary structure. From the thermodynamic hypothesis it is assumed that the unique tertiary structure of a protein is the conformation with the minimal free energy. Experiments have shown that the folding process *in vitro* is independent of external influence (by folding *in vivo* sometimes helper-molecules called chaperones are involved). It seems that the tertiary structure of a protein is encoded in its primary structure. Under this hypothesis, the spatial conformation of a protein can be computationally determined from its sequence of amino acids.

It is assumed that the hydrophobicity of amino acids is the main force for the development of a unique conformation. All natural proteins form one or more hydrophobic cores, i.e., the more hydrophobic amino acids are concentrated in compact cores whereas the more hydrophilic amino acids are located at the surface of the protein. This leads to a more simplified model, the so-called *HP model* (see, e.g., Dill [4] and Dill et al. [5]). Here, we distinguish only between two types of amino acids: hydrophobic (or non-polar) and hydrophilic (or polar). Therefore, a protein is modeled as a string over $\{H,P\}$, where each hydrophobic amino acid is represented by an H and each polar is represented by a P. In the following, a string in $\{H,P\}^*$ will also be called an *HP-sequence*.

The 3-dimensional space will be discretized by a cubic lattice. More formally, let $\mathcal{L}_k$, for $k \in \mathbb{N}$, be the following graph

$$\mathcal{L}_k = \left(\mathbb{Z}^3, \left\{ \{x,x'\} \in \mathbb{Z}^3 \times \mathbb{Z}^3 \,\middle|\, |x-x'|_2 \leq \sqrt{k} \right\}\right) ,$$

where $|\cdot|_2$ is the usual Euclidean norm. Then $\mathcal{L}_1$ is the cubic lattice. A folding of a protein can be viewed as a self-avoiding path in the cubic lattice. More formally, a *folding* of an HP-sequence $\sigma = \sigma_1 \cdots \sigma_n$ is a one-to-one mapping $\phi:[1:n] \to \mathcal{L}_k$ such that $|\phi(i-1) - \phi(i)|_2 \leq \sqrt{k}$ for all $i \in [2:n]$. The *score* of a folding is the number of adjacent pairs of hydrophobic amino acids in the cubic lattice which are not adjacent in the given primary structure. Thus, the expected spatial conformation of a given protein is a folding with the largest score, since the negative score models the free energy. Therefore, a folding of a protein with a maximal score is called a *conformation*.

The major disadvantage of the the HP model is the representation of the 3-dimensional space by a cubic lattice because it is a bipartite graph. Thus, two hydrophobic amino acids with an even distance in the protein cannot contribute to the score, since they cannot be adjacent in the cubic lattice. In particular, all foldings of the sequence $(HP)^n$ are optimal, although each folding on the cubic lattice has score 0. Hence, we are interested in a more natural discretization of the 3-dimensional space. In this paper, we consider the *extended cubic lattice*. In the extended cubic lattice we add to each lattice point 12 neighbors using diagonals in the plane, i.e., each lattice point has 18 neighbors. More formally, $\mathcal{L}_2$ is the mathematical description of the extended cubic lattice. Note that in $\mathcal{L}_2$ lattice points along a space diagonal are not connected.

A natural extension of the HP model is the *HP side chain model*. This is a more realistic model where the residues will be explicitly represented. In terms of graph theory, a protein is modeled as a caterpillar graph instead of a linear chain. A *caterpillar* of length n is the following graph $C = (B \cup L, E)$, where

$$B = \{b_1, \ldots, b_n\} , \qquad L = \{\ell_1, \ldots, \ell_n\} ,$$
$$E = \{(b_i, \ell_i) \mid i \in [1:n]\} \cup \{(b_{i-1}, b_i) \mid i \in [2:n]\} .$$

Here, the set B represents the nodes in the *backbone* and L the so-called *legs*. A backbone node represents the α carbon atom together with the main chain part of the amino acid whereas the leg represents its characteristic residue. This is still a simplification, since the residue can be as simple as a hydrogen atom in Alanine and as complex as two aromatic rings in Tryptophan. Note that we only mark the legs as hydrophobic or polar. Hence, a backbone node cannot increase the score of a folding.

1.2 Related Results

It is widely believed that the computational task of predicting the spatial structure of a given polymer (or, in particular, a protein) requires exponential time. First evidence for this assumption has been established by proving that the prediction of the conformation of a polymer for some more or less realistic combinatorial models is $\mathcal{NP}$-hard (see, e.g., Ngo and Marks [11], Unger and Moult [15], and Fraenkel [6]). For a comprehensive discussion of these lower bounds, we refer the reader to the survey of Ngo, Marks, and Karplus [12].

In [13], Paterson and Przytycka show that for an extended HP model with an infinite number of different hydrophobic amino acids it is $\mathcal{NP}$-hard to determine the conformation. In the extended HP model a protein will be modeled as a string over the (arbitrarily large) alphabet $\{P, H_1, H_2, H_3, \ldots\}$. Here only pairs of adjacent hydrophobic amino acids of the same type (i.e., contacts of the form H_i–H_i) contribute to the score. Recently, Nayak, Sinclair, and Zwick [10] improved this result. Even for a constant (but quite large) number of different types of amino acids the problem remains $\mathcal{NP}$-hard. Moreover, they proved that this problem is hard to approximate by showing its MAXSNP-hardness. More recently, Crescenzi et al. [3] as well as Berger and Leighton [2] have shown independently that it is $\mathcal{NP}$-hard to determine the conformation the HP Model.

On the other hand, there is also progress on positive results on protein structure prediction. As a first milestone, Hart and Istrail exhibit in [7, 8] an approximation algorithm for protein folding reaching at least $3/8$ of the optimal score in the HP model on the usual cubic lattice $\mathcal{L}_1$. In [9], the same authors present an approximation with a ratio of at least $2/5$ in the HP side chain model on the cubic lattice. In [1], Agarwala et al. presented an algorithm with an approximation ratio of $3/5$ in the HP model on the so-called triangular lattice (also known as face centered cubic lattice). This was the first approach to investigate non-bipartite lattices. Although the triangular lattice is differently defined, it can be topologically viewed as a superset of $\mathcal{L}_1$ and a subset of $\mathcal{L}_2$. An extension of the cubic lattice by just one plane diagonal direction in all three 2-dimensional subspaces is topologically isomorphic to the triangular lattice. Thus, in the triangular lattice each lattice point has 12 neighbors. Later, Hart and Istrail constructed in [9] a $31/36$ approximation for the HP side chain model on triangular lattices. Note that the quality for all these approximation algorithms are measured with asymptotic approximation ratios.

1.3 Our Results

In this paper, we investigate protein folding on extended cubic lattices. The extended cubic lattice is a natural extension of the cubic lattice which bypasses its major drawback, its bipartiteness. First we present a general folding algorithm A which achieves for all protein sequences an approximation ratio of $59/70$ ($\approx$84.3%). Then we describe a special folding algorithm B which can be applied to a restricted subset of HP-sequences. With the second algorithm we obtain an approximation ratio of $37/42$ ($\approx$88.1%). Although it is difficult to compare the approximation ratios for protein structure prediction algorithms on different lattice models, it should be mentioned that this is the best known approximation ratio for such algorithms.

Former protein structure prediction algorithms construct 'layered' foldings. This means that the algorithms constructs in reality a folding in the 2-dimensional sublattice from which the final folding in the 3-dimensional lattice will be generated. Therefore, only a few bonds use the third dimension. To obtain the high quality of the presented folding algorithm B, it is substantial to construct non-layered foldings in most parts of the conformation. Moreover, this construction does not only depend on the distribution of the hydrophobic amino acids in the protein as former algorithms. It also strongly depends on the length of contiguous subsequences of polar residues. This is strong evidence that the predicted folding is not too artificial.

On the other hand, this is the first time that folding algorithms for a 'natural' subclass of HP-sequences have been investigated. A strong indication that the considered subclass of HP-sequences is a 'natural choice' is the fact that more than 99.5% of all known sequences of proteins in the protein data base SWISS-PROT [16] belong to the considered subclass. Finally, the running time of both approximation algorithms are linear.

2 The General Folding Algorithm

In this section, we present a general folding algorithm in the HP side chain model on extended cubic lattices. Let $s{=}s_1 \cdots s_n$ be an HP-sequence. A sequence of HP-sequences $(\sigma_1, \ldots, \sigma_m)$ is called a *k-decomposition* of s iff the following four conditions hold:

1. $s = \sigma_1 \cdots \sigma_m$,
2. $|\sigma_i|_H {=} k$ for all $i \in [2{:}m{-}1]$,
3. $0 {<} |\sigma_1|_H \leq k$ and $|\sigma_m|_H \leq k$, and
4. the last symbol in each σ_i is an H for all $i {\in} [1{:}m{-}1]$.

Here $|s|_H$ is the number of H's in the sequence s. The strings σ_i of a k-decomposition $(\sigma_1, \ldots, \sigma_m)$ are called *k-fragments*. If $|\sigma_1|_H {=} k$, we call σ the *canonical k-decomposition*.

Let s be an HP-sequence and let $\sigma{=}(\sigma_1, \ldots, \sigma_m)$ be the canonical 5-decomposition of s. First we fold each σ_i as shown in Fig. 1. Here, the nodes on the backbone of the protein are drawn as circles. More precisely, a backbone node is drawn black if it represents a hydrophobic amino acid and white otherwise. Hydrophobic residues are drawn as black squares, whereas the polar residues are not explicitly marked. The numbers in front of the squares represent the order of the hydrophobic residues in the sequence

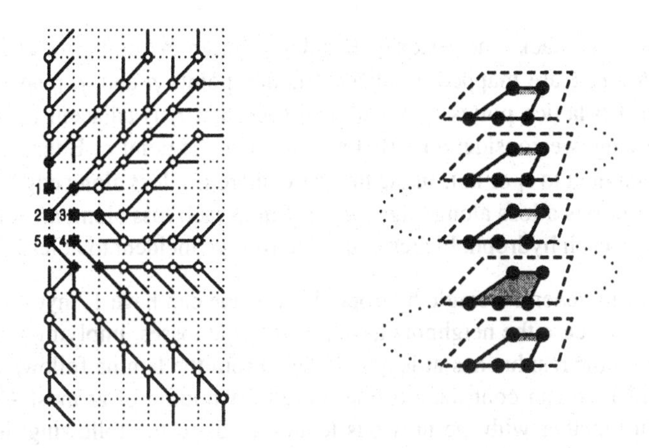

Fig. 1. Folding of a single 5-fragment and arrangement to a pole

of amino acids. The contiguous block of polar amino acids between two hydrophobic amino acids are not connected in Fig. 1. From the numbering of the hydrophobic residues, it should be clear which strands have to be connected and in which way.

We observe that for each 5-fragment consecutive backbone nodes with a hydrophobic residue are placed at neighbored lattice points, with the exception of the third and fourth backbone node. Therefore, the folding of the 5-fragment is still admissible even if the P-sequence between two hydrophobic residues is empty. If there is no polar residue between the third and fourth hydrophobic residue of a 5-fragment, we just remap the backbone node of the fourth hydrophobic residue one position up in the vertical direction.

In what follows, we show how to combine this folding of 5-fragments to obtain a folding in the 3-dimensional space. Using the third dimension, we combine the 5-fragments to a pole of height m such that the corresponding hydrophobic residues form a vertical column. This will be achieved by arranging the layers in a zig-zag-style in the third dimension. This is sketched in Fig. 1 where only the hydrophobic residues are drawn explicitly as black circles. Note that at the front half of this pole the three hydrophobic residues have no neighbors outside the pole. Using a turn after $m/2$ layers, we combine the two halves to a new pole such that each layer contains 10 hydrophobic residues. A simple computation shows that each layer of 10 hydrophobic residues contributes 59 to the score: 23 H–H contacts within a layer and 36 H–H contacts to the two neighboring layers.

Clearly, each lattice point has exactly 18 neighbors. Thus, each hydrophobic residue can have at most 17 contacts with other hydrophobic neighbors. This upper bound on the number of hydrophobic neighbors of a hydrophobic residue can be improved as follows. We denote by a *loss* an edge in the lattice with the property that a hydrophobic residue is mapped to exactly one of its endpoints.

Lemma 1. *For each folding in the extended cubic lattice, a single hydrophobic residue is on average incident to at least* 3 *losses.*

Proof. Consider a backbone vertex $b \in B$ and its adjacent hydrophobic residue $\ell \in L$. Assume that b and ℓ are mapped to adjacent lattice points p and q, respectively. There exists at least 6 lattice points r_i, for $i \in [1{:}6]$ such that r_i is adjacent to both p and q. In the following, we consider a fix (but arbitrary) lattice point r_i. Either a hydrophobic residue is assigned to r_i or not. In the first case, there is a loss along edge $\{p, r_i\}$; in the latter case, there is a loss along edge $\{q, r_i\}$. Since each loss along an edge is counted at most twice, each hydrophobic residue is on average incident to at least 3 losses. □

Note that in general a single hydrophobic residue can have 17 hydrophobic neighbors. But in this case the neighbors have 3 additional losses, implying that on average each hydrophobic residue has at least 3 losses. From the lemma follows that each hydrophobic residue can contribute to the score of a folding of at most $\frac{17-3}{2} = 7$. Our construction together with the previous lemma leads to the following theorem. Note that we consider asymptotic approximation ratios in this paper.

Theorem 2. *Algorithm A constructs a folding in the HP side chain model on extended cubic lattices for an arbitrary HP-sequence with an approximation ratio of at least* $59/70$ *(≈84.3%). Moreover, this folding can be computed in linear time.*

3 The Improved Folding Algorithm

In this section, we describe an improved folding algorithm B. This algorithm is designed for a special subset of HP-sequences. Let s be an HP-sequence and let $\sigma = (\sigma_1, \ldots, \sigma_m)$ be a 6-decomposition of s. Further, let $\sigma_v = P^{\ell_1} H \cdots P^{\ell_6} H$ be 6-fragment. We call σ_v *perfect* iff there exists $i \in [2{:}6]$ such that $\ell_i = 0$, or there exists $i \neq j \in [1{:}6]$ such that $\ell_i + \ell_j \leq 3$. An HP-sequence is called *perfect* if it has a 6-decomposition such that all its 6-fragments are perfect. If it has a 6-decomposition such that all but one of its 6-fragments are perfect, the HP-Sequence is called *nearly perfect.* The substrings P^{ℓ_i} for $i \in [1{:}6]$ are called an ℓ_i-block at position i. For example, the 6-fragment $\sigma = P^{27} H P^2 H P^{11} H P^{12} H P^1 H P^4 H$ is perfect and has a 12-block at position 4.

Again, we first describe how to fold a single 6-fragment. We will use two adjacent 2-dimensional planes to achieve the folding. In each plane, we will place 3 hydrophobic residues. We distinguish three cases depending on whether the 6-fragment is perfect because of a 0-block at position greater than 1, a combination of a 0-block at position 1 and a 3-block, or a combination of a 1- and a 2-block.

Case 1: First, we assume that the 6-fragment is perfect because of a 0-block at position $i > 1$. The folding is illustrated in Fig. 2. In Fig. 2a and 2b the foldings for a 6-fragment with a 0-block at position 2 and 3, respectively, are shown. The folding will be completed as illustrated in Fig. 2d. In Fig. 2c the first part of the folding of a 6-fragment with a 0-block at position 4 is shown. This folding will be completed by a reverse traversal of the same folding given in Fig. 2c in the next layer. The case where the 0-block is at position 5 or 6 is symmetric to the cases where the 0 block is at position 2 or 1, respectively. In contrast to the folding in the previous section, the folding of a 6-fragment consists of two layers with three hydrophobic residues each. In both layers the hydrophobic residues form a triangle. The narrow dotted horizontal lines in Fig. 2 indicate where the 6-fragment will be folded to obtain this construction.

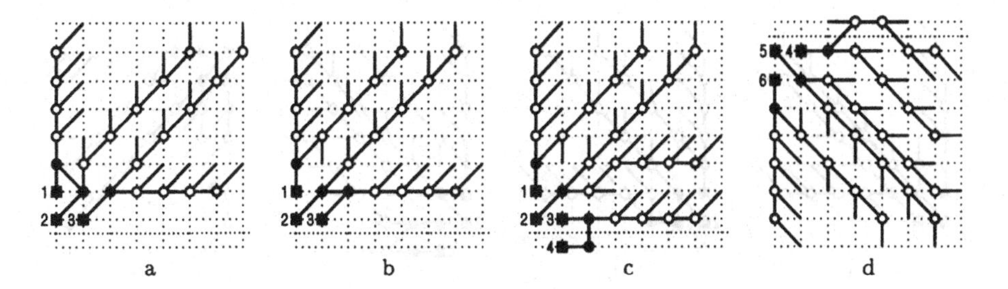

Fig. 2. Case 1: Folding of a 6-fragment with a 0-block

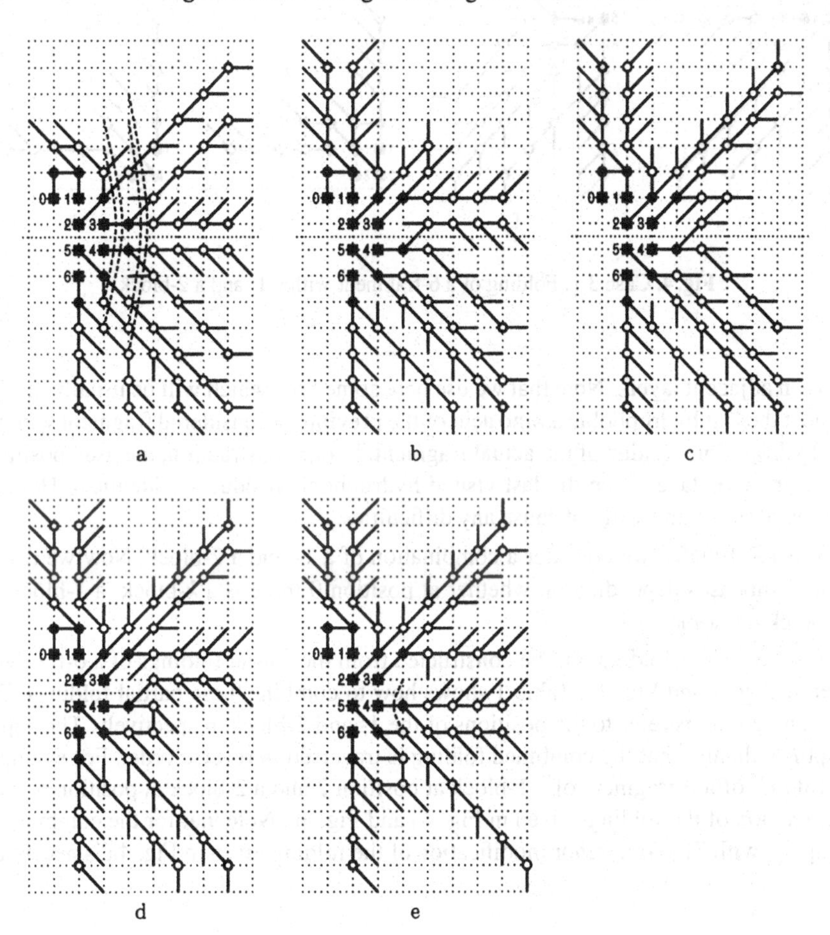

Fig. 3. Case 2: Folding of a 6-fragment with a 0-block at position 1 and a 3-block

Case 2: Now we consider the case of a 0-block at position 1. The Figs. 3a, 3b, 3c, 3d, and 3e illustrate the folding if the 3-block is at position 2, 3, 4, 5, and 6, respectively. Again, the narrow dotted horizontal line indicates where the folding will be folded to obtain two layers. The dashed lines indicates edges of the caterpillar which arise

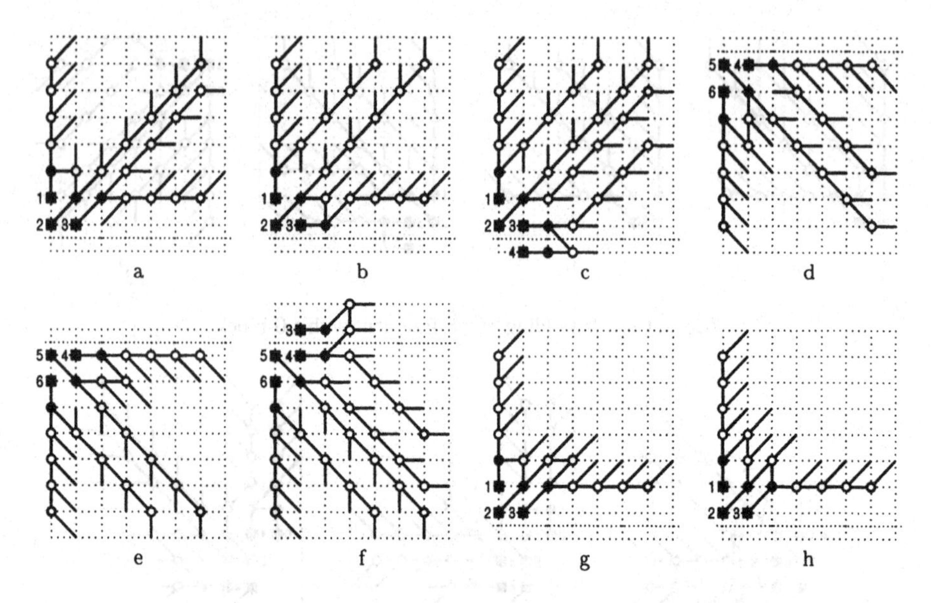

Fig. 4. Case 3.1: Folding of a 6-fragment with a 1- and a 2-block

between adjacent layers. Note that we use here some area which will be usually used to connect the last hydrophobic amino acid of the previously considered fragment with the first hydrophobic residue of the actual fragment. In our construction, the used positions in the previous layer from the last visited hydrophobic residue are identical. Hence a reuse is possible and will not cause any difficulties.

Case 3: Finally, we consider a combination of a 1- and a 2-block. Now we distinguish 3 subcases depending on whether at position 0 there is a k-block, a 1-block, or a 2-block for some $k>2$.

Case 3.1: The folding will be constructed from the partial foldings of a 6-fragment given in Fig. 4 and Fig. 2d. Table 1 shows how to combine these partial foldings. The rows and columns refer to the positions of the 1- and 2-block, respectively. The superscript R indicates that the combined folding is traversed in reverse order. For example, the folding of a 6-fragment of a 1-block at position 2 and a 2-block at position 5 is the combination of the foldings given in Fig. 4a and Fig. 4e. Note that for the combination of Fig. 4a with Fig. 4f a minor modification of the folding given in Fig. 4a is necessary.

Table 1. Combinations of subfoldings to a folding of a 6-fragment

1- \ 2-block	2	3	4	5	6
2	—	4g+2d	4a+4f	4a+4e	4a+4d
3	4h+2d	—	4b+4f	4b+4e	4b+4d
4	$(4a+4f)^R$	$(4b+4f)^R$	—	4c+4e	4c+4d
5	$(4a+4e)^R$	$(4b+4e)^R$	$(4c+4f)^R$	—	$(4h+2d)^R$
6	$(4a+4d)^R$	$(4b+4d)^R$	$(4c+4d)^R$	$(4g+2d)^R$	—

The backbone node of the hydrophobic residue labeled with 3 has to be remapped just to the right of the hydrophobic residue which is obviously possible.

Case 3.2: Now we consider the case that the 1-block is at position 1 in the 6-fragment. Fig. 5a illustrates the folding if the 2-block is at position 2. The folding for a 2-block at position 3 is obtained by a combination of the foldings given in Fig. 5b and Fig. 2d. If the 2-block is at position 5 or 6, the folding will be combined from the foldings given in Fig. 5c and Fig. 4e or Fig. 4d, respectively. If the 2-block is at position 4, the folding is more complex and is illustrated in Fig. 5d. Here, the dotted curves indicate connected subsequences of polar residues. Observe that the order of the traversed six hydrophobic residues is different from that in the other foldings. The last visited node is directly above the fourth visited node of this fragment instead of the first one.

Case 3.3: It remains the case where the 2-block is at position 1 in the 6-fragment. These are the most complex foldings and they are explicitly illustrated in Figs. 6a through 6e depending on the position of the 1-block.

Note that all foldings are drawn for the case that the subsequences of contiguous polar residues may be arbitrarily long. But nevertheless our construction is also valid for any length of subsequences of contiguous polar residues with some minor modifications.

It remains to construct a complete folding based on the presented foldings of the 6-fragments. First we combine the foldings of the 6-fragments to a long pole and break it into 4 parts $P_1, \ldots, P_4$ of equal height. Then the four parts will be arranged as shown in Fig. 7. In Fig. 7 only the hydrophobic residues are represented by gray quarters of a cylinder. For example, a folding of a single 6-fragment is illustrated in this figure by six black circles. The connections between these four quarters are drawn as dashed curves. In the final folding, each layer consists of 12 hydrophobic residues. Each layer of 12 hy-

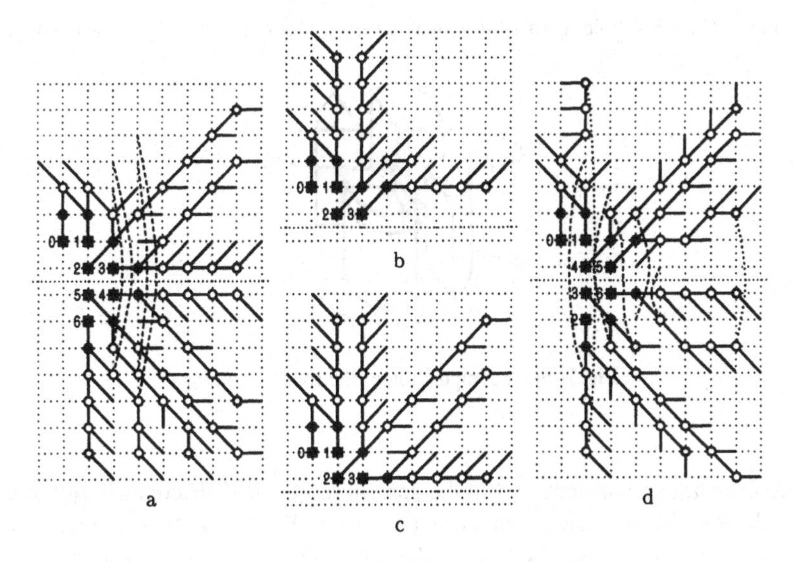

Fig. 5. Case 3.2: Folding of a 6-fragment with a 1-block at position 1 and a 2-block

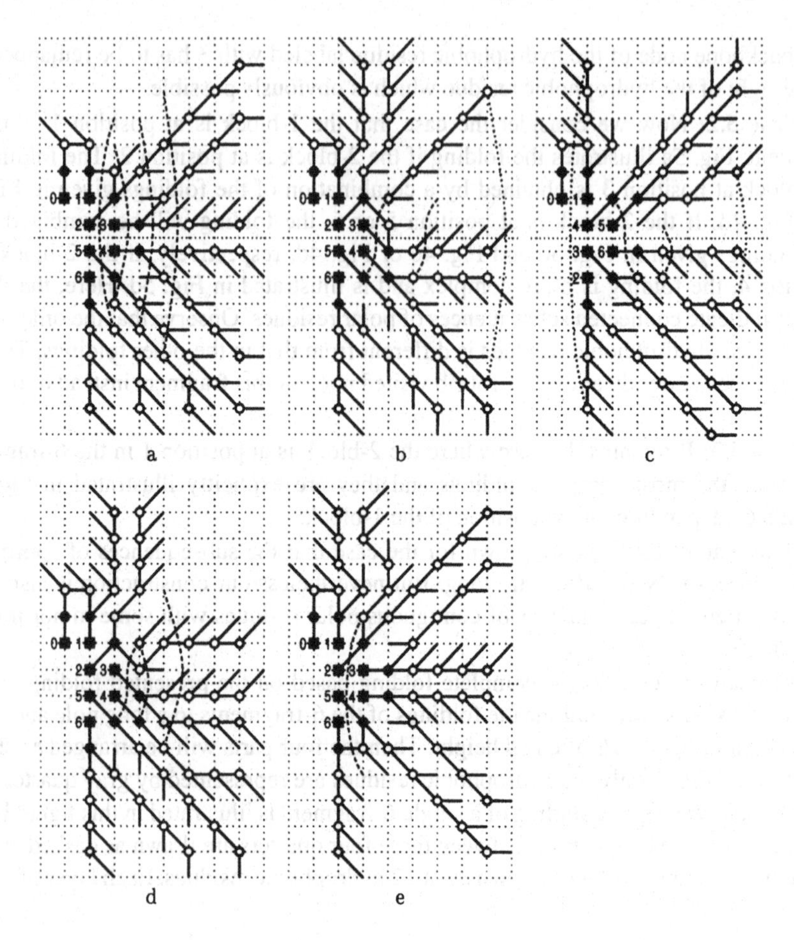

Fig. 6. Case 3.3: Folding of a 6-fragment with a 2-block at position 1 and a 1-block)

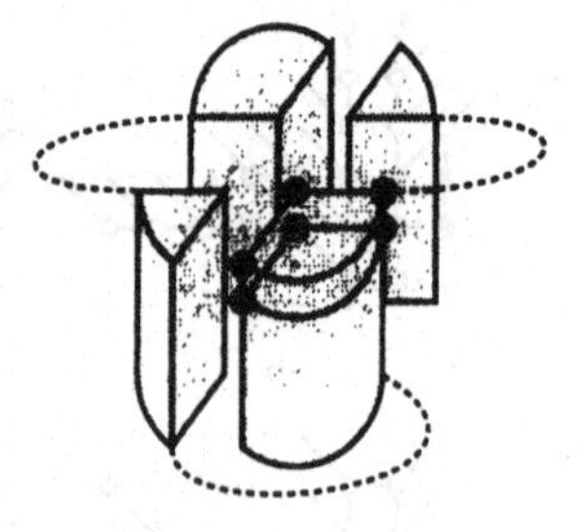

Fig. 7. Final composition of the four subfoldings

drophobic residues contributes 74 to the general score: 30 H–H contacts within a layer and 44 H-H contacts to the neighboring two layers. By Lemma 1, each layer can contribute on average at most 12∗7=84 to the score. Thus, we have proved the following theorem.

Theorem 3. *Algorithm B constructs a folding in the HP side chain model on extended cubic lattices for perfect HP-sequences with an approximation ratio of at least* $37/42$ *(*$\approx 88,1\%$*). Moreover, this folding can be computed in linear time.*

It is possible to extend this embedding for nearly perfect HP-sequences.

Theorem 4. *Algorithm B constructs a folding in the HP side chain model on extended cubic lattices for nearly perfect HP-sequences with an approximation ratio of at least* $37/42$ *(*$\approx 88,1\%$*). Moreover, this folding can be computed in linear time.*

An inspection of the protein data base SWISS-PROT [16] shows that more than 97.5% of all stored proteins have a perfect 6-decomposition and more than 99.5% have a nearly perfect 6-decomposition. Thus, algorithm B is applicable to nearly all natural proteins. In our analysis, we marked the amino acids Ala, Cys, Phe, Ile, Leu, Met,Val, Trp, and Tyr as hydrophobic and all other amino acids as polar. This classification follows Sun et al. [14] and is a conservative classification in the sense that other classifications mark more amino acids as hydrophobic. Obviously, the more amino acids are marked as hydrophobic the more proteins have a (nearly) perfect HP-sequence. The detailed analysis of amino acids in SWISS-PROT 36 as of July 1998 can be found in Table 2. Here, $N(i)$ is the number of amino acids which have a optimal 6-decomposition with i imperfect 6-fragments. An *optimal* k-decomposition is a k-decomposition with a minimal number of imperfect k-fragments.

References

[1] R. Agarwala, S. Batzoglou, V. Dančík, S. Decatur, M. Farach, S. Hannenhalli, S. Muthukrishnan, S. Skiena: Local Rules for Protein Folding on a Triangular Lattice and generalized Hydrophobicity in the HP Model, *Proceedings of the 8th Symposium on Discrete Algorithms*, 390–399, 1997, also in *Proceedings of the First Conference on Computational Molecular Biology*, 1–2, 1997.

[2] B. Berger, F.T. Leighton: Protein Folding in the Hydrophobic-Hydrophilic (HP) Model is $\mathcal{NP}$-Complete, *Proceedings of the 2nd Conference on Computational Molecular Biology*, 30–39, 1998.

[3] P. Crescenzi, D. Goldman, C. Papadimitriou, A. Piccolboni, M. Yannakakis: On the Complexity of Protein Folding, *Proceedings of the 30th Symposium on Theory of Computing*, 597-603, 1998, also in *Proceedings of the 2nd Conference on Computational Molecular Biology*, 61–62, 1998.

[4] K.A. Dill: Dominant Forces in Protein Folding, *Biochemistry*, **29**(31):7133–7155, 1990.

[5] K.A. Dill, S. Bromberg, K. Yue, K.M. Fiebig, D. Yee, P. Thomas, H. Chan: Principles of Protein Folding: A Perspective From Simple Exact Models, *Prot. Sci.*, **4**:561–602, 1995.

[6] A. Fraenkel: Complexity of Protein Folding, *Bull. Math. Biol.*, **55**(6):1199–1210, 1993.

Table 2. Statistics of proteins in SWISS-PROT 36 with optimal 6-decompositions

i	0	1	2	3	4	5	>5	≥0
$N(i)$	71265	1315	194	86	33	24	46	72963
%	97.7%	1.8%	0.3%	0.1%				

[7] W.E. Hart, S. Istrail: Fast Protein Folding in the Hydrophobic-Hydrophilic Model Within Three-Eights of Optimal, *Proceedings of the 27th Symposium on Theory of Computing*, 157–167, 1995.

[8] W.E. Hart, S. Istrail: Fast Protein Folding in the Hydrophobic-Hydrophilic Model Within Three-Eights of Optimal, *J. Comp. Biol.*, **3**(1):53–96, 1996.

[9] W.E. Hart, S. Istrail: Lattice and Off-Lattice Side Chain Models of Protein Folding: Linear Time Structure Prediction Better Than 86% of Optimal, *Proceedings of the 2nd Conference on Computational Molecular Biology*, 137–146, 1997.

[10] A. Nayak, A. Sinclair, U. Zwick: Spatial Codes and the Hardness of String Folding Problems, *Proceedings of the 9th Symposium on Discrete Algorithms*, 639–648, 1998.

[11] J.T. Ngo, J. Marks: Computational Complexity of a Problem in Molecular Structure Prediction, *Prot. Engng.*, **5**(4):313–321, 1992.

[12] J.T. Ngo, J. Marks, M. Karplus: Computational Complexity, Protein Structure Prediction, and the Levinthal Paradox, in *The Protein Folding Problem and Tertiary Structure Prediction*, K. Merz Jr., S. LeGrand (Eds.), Birkhäuser, 1994.

[13] M. Paterson, T. Przytycka: On the Complexity of String Folding, *Proceedings of the 23rd Int'l Colloquium on Automata, Languages, and Programming*, 658–669, 1996.

[14] S. Sun, R. Brem, H.D. Chan, K.A. Dill: Designing Amino Acid Sequences to Fold with Good Hydrophobic Cores, *Prot. Engng.* **8**(12):1205–1213, 1995.

[15] R. Unger, J. Moult: Finding the Lowest Free Energy Conformation of a Protein is an $\mathcal{NP}$-Hard Problem: Proof and Implications, *Bull. Math. Biol.*, **55**(6):1183–1198, 1993.

[16] SWISS-PROT Protein Sequence Data Bank: `http://www.expasy.ch/sprot/`, `ftp://www.expasy.ch/databases/swiss-prot/sprot36.dat` (as of July 21, 1998).

On Constructing Suffix Arrays in External Memory *

Andreas Crauser and Paolo Ferragina

[1] Max-Planck-Institut für Informatik, Saarbrücken, Germany.
[2] Dipartimento di Informatica, Università di Pisa, Italy
crauser@mpi-sb.mpg.de, ferragin@di.unipi.it

Abstract. The construction of full-text indexes on very large text collections is nowadays a hot problem. The suffix array [16] is one of the most attractive full-text indexing data structures due to its simplicity, space efficiency and powerful/fast search operations supported. In this paper we analyze theoretically and experimentally, the I/O-complexity and the working space of six algorithms for constructing large suffix arrays. Additionally, we design a new external-memory algorithm that follows the basic philosophy underlying the algorithm in [13] but in a significantly different manner, thus combining its good practical qualities with efficient worst-case performances. At the best of our knowledge, this is the first study which provides a wide spectrum of possible approaches to the construction of suffix arrays in external memory, and thus it should be helpful to anyone who is interested in building full-text indexes on very large text collections.

1 Introduction

Full-text indexes—like suffix trees [17], suffix arrays [16] (cfr. PAT-arrays [13]), PAT-trees [13] and String B-trees [12], just to cite a few—have been designed to deal with arbitrary (unstructured) texts and to support powerful string-search queries (cfr. word-indexes [8]). They have been successfully applied to fundamental string-matching problems as well text compression, analysis of genetic sequences and recently to the indexing of special linguistic texts [11]. The most important complexity measures for evaluating their efficiency are [24]: (i) the time and the extra space required to build the index, (ii) the time required to search for a string, and (iii) the space used to store the index. Points (ii) and (iii) have been largely studied in the scientific literature (see e.g. [5, 12, 13, 16, 17]). In this paper, we will investigate the efficient *construction* of these data structures on very large text collections. This is nowadays a hot topic [1] because the construction phase may be a bottleneck that can even prevent these indexing tools to be used in large-scale applications. In fact, known construction algorithms are very fast when employed on textual data that fit in the internal memory of computers [3, 16] but their performance immediately degenerates when the text size becomes

* Part of this work was done while the second author had a Post-Doctoral fellowship at the Max-Planck-Institut für Informatik, Saarbrücken, Germany. The work has been supported by EU ESPRIT LTR Project N. 20244 (ALCOM-IT)

[1] Zobel *et al.* [24] say that: "*We have seen many papers in which the index simply *is*, without discussion of how it was created. But for an indexing scheme to be useful it must be possible for the index to be constructed in a reasonable amount of time, ...*".

so large that the texts must be arranged on (slow) external storage devices [5, 12]. These algorithms suffer from the so called *I/O bottleneck*: They spend most of the time in moving data to/from the disk.

To study the efficiency of algorithms that operate on very large text collections, we refer to the classical *Parallel Disk Model* [21, 23]. Here a computer is abstracted to consist of a *two-level memory*: a fast and small internal memory, of size M, and a slow and arbitrarily large external memory, called *disk*. Data between the internal memory and the disk are transfered in blocks of size B (called *disk pages*). It is well known [20, 24] that accessing one page from the disk decreases the cost of accessing the page succeeding it, so that "bulk" I/Os are less expensive per page than "random" I/Os. This difference becomes much more prominent if we also consider the reading-ahead/buffering/caching optimizations which are common in current disks and operating systems. To deal with these disk specialties we therefore adopt the simple accounting scheme introduced in [10]: Let $c < 1$ be a constant, a *bulk I/O* is the reading/writing of a contiguous sequence of cM/B disk pages; a *random* I/O is any single disk-page access which is not part of a bulk I/O. The performance of the external-memory algorithms is therefore evaluated by measuring: (a) the number of I/Os (bulk and random), (b) the internal running time (CPU time), and (c) the number of disk pages used during the construction process (working space).

Previous Work. For simplicity of exposition, we use N to denote the size of the whole text collection and assume throughout the paper that it consists of only one long text. The most famous indexing data structure is the *suffix tree*. In internal memory, a suffix tree can be constructed in $O(N)$ time [17, 9]; in external memory, Farach *et al.* [10] showed that a suffix tree can be optimally constructed within the same I/O-bound as sorting N atomic items; nonetheless, known practical construction algorithms for external memory still operate in a brute-force manner requiring $\Theta(N^2)$ total I/Os in the worst-case [5]. Their working space is not predictable in advance, since it depends on the text structure, and requires between $15N$ and $25N$ bytes [15, 16].

Since space occupancy is a crucial issue, Manber and Myers [16] proposed the *suffix array* data structure, which consists of an array of pointers to text positions and thus occupies overall $4N$ bytes. Suffix arrays can be efficiently constructed in $O(N \log_2 N)$ time [16] and $O((N/B)(\log_2 N) \log_{M/B}(N/B))$ I/Os [1]. The motivation of the recent interest in suffix arrays has to be found in their simplicity, reduced space occupancy and in the small constants hidden in the big-Oh notation, which make them suitable to index very-large text collections in practice. Suffix arrays also present some natural advantages over the other data structures for what concerns the construction phase. Indeed, their simple topology (i.e., an array of pointers) avoids at construction time the problems related to the efficient management of tree-based data structures (like suffix trees and String B-trees) on external storage devices [14]. Furthermore, efficient practical procedures for building suffix arrays are definitively useful for efficiently constructing suffix trees, String B-trees and the other full-text indexing data structures.

Our Contribution. With the exception of some preliminary and partial experimental works [16, 13, 18], to the best of our knowledge, no full-range comparison exists among the known algorithms for building large suffix arrays. This will be the main goal of our paper, where we will theoretically study and experimentally analyze six suffix-array construction algorithms. Some of them are the state-of-the-art in the practical setting [13], others are the most efficient theoretical ones [16, 1], whereas three other algorithms are our new proposals obtained either as slight variations of the previous ones or as a careful combination of known techniques which were previously employed only in the theoretical setting. In the design of the new algorithms we will address mainly two issues: (i) simple algorithmic structure, and (ii) reduced working space. The first issue has clearly an impact on the predictability and practical efficiency of the proposed algorithms. The second issue is important because the real disk size is limited and "*space optimization is closely related to time optimization in a disk memory*" [14][Sect. 6.5].

We will discuss all the algorithms according to these two resources and we will pose particular attention to differentiate between random and bulk I/Os in our theoretical analysis. This adopted accounting scheme allows to reasonably explain some interesting I/O-phenomena which arise during the experiments and which would be otherwise meaningless in the light of other simpler external-memory models. As a result, we will give a precise hierarchy of suffix-array construction algorithms according to their working-space vs. construction-time tradeoff; thus providing a wide spectrum of possible approaches for anyone who is interested in building large full-text indexes.

The experimental results have finally driven us to deeply study the intriguing, and apparently counterintuitive, "contradiction" between the effective practical performance of one of the experimented algorithms, namely the algorithm in [13], and its unappealing (i.e., cubic) worst-case behavior. This study has lead us to devise a new construction algorithm that follows the basic philosophy of [13] but in a significantly different manner, thus resulting in a novel approach which combines good practical qualities with efficient worst-case performances.

2 The Suffix Array Data Structure

The suffix array SA built on a text $T[1,N]$ is an array containing the lexicographically ordered sequence of suffixes of T, represented via pointers to their starting positions (i.e., *integers*). For instance, if $T = ababc$ then $SA = [1,3,2,4,5]$. SA occupies $4N$ bytes if $N \leq 2^{32}$. In this paper we consider three well-known algorithms for constructing suffix arrays, called Manber-Myers [16] (MM), BaezaYates–Gonnet–Snider [13] (BGS) and Doubling [1], and we refer the reader to the corresponding literature for their algorithmic details, due to space limitations. We now concentrate on our three new proposals, describe their features (Section 2.1), evaluate their complexities (Table 1), and finally compare all of them via a set of experiments on two texts collections (Section 3). The last Section 4 will be dedicated to the description of an improvement to the algorithm in [13] (called *new BGS*).

2.1 Three new algorithms

The three proposed algorithms asymptotically improve the previously known ones [16, 13, 1] by offering better trade-offs between total number of I/Os and working space. Their algorithmic structure is simple because it is based only upon *sorting* and *scanning*

Algorithm	Working space	CPU–time	total number of I/Os
MM [16]	$8N$	$N\log_2 N$	$N\log_2 N$
BGS [13]	$8N$	$(N^3\log_2 M)/M$	$n^2(N\log_2 M)/m$
Doubl [1]	$24N$	$N(\log_2 N)^2$	$n(\log_m n)\log_2 N$
Doubl+Disc	$24N$	$N(\log_2 N)^2$	$n(\log_m n)\log_2 N$
Doubl+Disc+Radix	$12N$	$N(\log_2 N)^2$	$n(\log_{m/\log N} N)\log_2 N$
Constr. L pieces	$\max\{\frac{24N}{L}, \frac{2NL+8N}{L}\}$	$N(\log_2 N)^2$	$n(\log_m n)\log_2 N$
New BGS	$8N$	$N^2(\log_2 M)/M$	(n^2/m)

Table 1. *CPU-time and I/Os are expressed in big-Oh notation; the working space is evaluated exactly; L is an integer constant ≥ 1; $n = N/B$ and $m = M/B$. BGS and new-BGS operate via sequential disk scans, whereas all the other algorithms mainly execute random I/Os. Notice that with a tricky implementation, the working space of BGS can be reduced to $4N$. The last four lines of the table indicate our new proposals.*

routines. This feature has two immediate advantages: The algorithms are expected to be fast in practice because they can benefit from the prefetching/caching of the disk; and they can be easily adapted to work efficiently on D-disk arrays and clusters of P workstations by plugging in proper sorting/scanning routines [21, 23] (cfr. [18]).

Doubling combined with a discarding stage. The main idea of the doubling algorithm [1] is to assign names (i.e. small integers) to the power-of-two length substrings of $T[1,N]$ in order to satisfy the so called *lexicographic naming property*: Given two text substrings α and β of length 2^h, it is $\alpha \leq_L \beta$ if and only if the name of α is smaller than the name of β. These names are computed inductively in [1] by exploiting the following observation: the lexicographic order between any two substrings of length 2^h can be obtained by exploiting the (inductively known) lexicographic order between their two (disjoint) substrings of length 2^{h-1}. After $q = \Theta(\log_2 N)$ stages, the order between any two suffixes of T, say $T[i,N]$ and $T[j,N]$, can be derived in constant time by comparing the names of $T[i, i+2^q-1]$ and $T[j, j+2^q-1]$.

Our first new algorithm is based on the observation that: *In each stage h of the doubling approach, all the text substrings of length 2^h are considered although the final position in SA of some of their corresponding suffixes might be already known.* Our main idea is therefore to identify and "discard" all those substrings from the naming process thus reducing the overall number of items ordered at each stage. However, this discarding step is not easy to be implemented because some of the discarded substrings might be necessary in next stages for computing the names of other longer substrings. In what follows, we describe how to cope with this problem.

The algorithm inductively keeps two lists of tuples: FT (finished tuples) and UT (unfinished tuples). The former is a list of tuples $\langle pos, -1, i\rangle$ denoting the suffixes $T[i,N]$ whose final position in SA is known: $SA[pos] = i$. UT is a list of tuples $\langle x, y, i\rangle$ denoting the suffixes $T[i,N]$ whose final position is not yet known. Initially, UT contains the tuples $\langle 0, T[i], i\rangle$, for $1 \leq i \leq N$; FT is empty. The algorithm executes $\leq \log_2 N$ stages; each stage j consists of six steps:

1. Sort the tuples in UT according to their first two components. If UT is empty go to step 6.

2. Scan UT, identify the "finished" tuples and assign new names to all tuples in UT. Formally, a tuple is considered "finished" if it is preceded and followed by two tuples which are different in at least one of their first two components; in this case, the algorithm sets the second component of this tuple to -1. The new names for all tuples are computed differently from [1] by setting the first component of a tuple $t = \langle x, y, * \rangle$ equals to $(x + c)$, where c is the number of tuples that precede t in UT and have the form $\langle x, y', * \rangle$ with $y' \neq y$.
3. Sort UT according to the third component of its tuples.
4. Merge the lists UT and FT according to the third component of their tuples. UT contains the final merged sequence, whereas FT is emptied.
5. Scan UT and for each *not-finished* tuple $t = \langle x, y, i \rangle$ (i.e. $y \neq -1$), take the next tuple at distance 2^j (say $\langle x', *, i + 2^j \rangle$) and set t equal to $\langle x, x', i \rangle$. If a tuple is marked "finished" (i.e., $y = -1$), then it is discarded from UT and put into FT. Finally, set $j = j + 1$ and go to step 2.
6. Sort FT according to the first component of its tuples (UT is empty); and derive SA by reading rightwards the third component of the sorted tuples.

The correctness follows from the invariant (proof in the full version): *At a generic stage j and after step 2, we have that in any tuple $t = \langle x, y, i \rangle$ the parameter x denotes the number of text suffixes whose prefix of length 2^j is strictly smaller than $T[i, i + 2^j - 1]$.* The algorithm has the same I/O-complexity as the Doubling algorithm (see Table 1), but we expect that the discarding step helps in improving its practical performance by reducing the overall number of tuples on which the algorithm is called to operate at each stage. In our implementation, we stuff four characters $T[i, i + 4]$ into each tuple (instead of the single $T[i]$) when constructing UT, thus initially saving four sorting and four scanning steps.

Doubling+Discard and Radix Heaps. Although the doubling technique gives the two most I/O-efficient algorithms for constructing large suffix arrays, it has the major drawback that its working space is large (i.e. $24N$ bytes) compared to the other known approaches (see Table 1). This is due to the fact that it uses external mergesort [14] to sort the list of tuples, and this algorithm requires an auxiliary array to store the intermediate results (see Section 3). Our new idea is to reduce the overall working space using an external version of the radix heap data structure introduced in [2]. Radix heaps are space efficient but their I/O-performance degenerates when the maximum priority value is large. The new algorithm replaces the mergesort routine in steps 1 and 3 above with a sorting routine based on external radix heaps [6]. This reduces the overall required space to $12N$ bytes, but at the cost of increasing the I/O–complexity (see Table 1). We will experiment this algorithm on real data to check whether the reduction in the number of processed tuples, induced by the discarding strategy, compensates the time increase of the radix heap approach (see Section 3).

Construction in L pieces. This algorithm improves over all the previous ones in terms of both I/O-complexity, CPU-time and working space. It constructs the suffix array into *pieces of equal size* and thus turns out useful either when the underlying application *does not* need the suffix array as a unique data structure, but allows to keep it in a distributed fashion [4]; or when we operate in a distributed-memory environment [18].

Unlike the approaches in [18], our algorithm *does not* need the careful setting of system parameters. It is very simple and applies in a different way, useful for practical purposes, a basic idea known so far only in the theoretical setting (see e.g. [9]).

Let L be a positive integer parameter (to be fixed later), and assume that T is *logically* padded with L blank characters. The algorithm constructs L suffix arrays, say $SA_1, SA_2, \ldots, SA_L$ each of size N/L. Array SA_i stores the lexicographically ordered sequence of suffixes $\{T[i,N], T[i+L,N], T[i+2L,N], \ldots,\}$. The logic underlying our new algorithm is to first construct SA_L, and then derive all the others arrays $SA_{L-1}, SA_{L-2}, \ldots, SA_1$ (in that order) by means of a simple algorithm for sorting *triple of integers*.

SA_L is built in two main stages: First, the string set $S = \{T[L,2L-1], T[2L,3L-1], T[3L,4L-1], \ldots\}$ is formed and lexicographically sorted by means of any external string-sorting algorithm [1]; the compressed text $T'[1,N/L]$ is then derived from $T[L,N+L-1]$ by replacing each string $T[iL,(i+1)L-1]$ with its *rank* in the sorted set S. Subsequently, any known construction algorithm is used to build the suffix array SA' of T'; and then SA_L is derived by setting $SA_L[j] = SA'[j] \times L$. The other $L-1$ suffix arrays are constructed by exploiting the observation: *Any suffix $T[i+kL,N]$ in SA_i can be seen as the concatenation of the character $T[i+kL]$ and the suffix $T[i+1+kL,N]$, which actually occurs in SA_{i+1}*. It follows that given SA_{i+1}, the construction of SA_i can be reduced to the sorting of $\Theta(N/L)$ triples (details in the full paper).

Sorting the set S takes $O(Sort(N))$ random I/Os and $2N + 8N/L$ bytes, where $Sort(N) = (N/B)\log_{M/B}(N/B)$ [1]. Building the L suffix arrays takes $O(Sort(N)\log_2 N)$ random I/Os, $O(N/M \log_2(N/M))$ bulk I/Os and $24N/L$ bytes. Of course the larger is L, the bigger is the number of suffix arrays to be constructed, but the smaller is the working space required. By setting $L = 4$, we get an interesting trade-off: $6N$ working space, $O(Sort(N)\log_2 N)$ random I/Os and $O(N/M \log_2(N/M))$ bulk I/Os. The practical performance of this algorithm will be evaluated in Section 3.

3 Experimental Results

We implemented the algorithms above using a recently developed external-memory library of algorithms and data structures called LEDA-SM [7] (an acronym for "LEDA for Secondary Memory"). [2] This library is an extension of the internal-memory library *LEDA* [19] and follows LEDAs main ideas: portability, efficiency and high level specification of data structures. The specialty of LEDA-SM's data structures is that we can specify (and therefore control) the maximum amount of internal memory that they are allowed to use; furthermore we can count the number of I/Os performed. This way, library LEDA-SM allows the programmer to experimentally investigate how the model parameters M and B influence the performance of an external-memory algorithm. For what concerns our experiments, we used the external-array data structure and the external sorting/scanning algorithms provided by LEDA-SM; the other in-core algorithms and data structures are taken from LEDA. In particular, we used an implementation of external mergesort that needs $2Xb$ bytes for sorting X items of b bytes each.

The computer used in our experiments is a SUN ULTRA-SPARC 1/143 with 64 Mbytes of internal memory running the SUN Solaris 2.5.1 operating system. It is connected

[2] For another interesting external-memory library see [22].

to one single Seagate Elite-9 SCSI disk via a fast-wide differential SCSI controller ($B = 8$ Kbytes). According to the adopted accounting scheme (see Section 1), we have chosen `bulk_size` $= 64$ disk pages, for a total of 512 Kbytes. This way, the seek time is 15% of a bulk I/O and we achieve the 81% of the maximum transfer rate of our disk while keeping the service time of the requests still low. Of course, other values for the `bulk_size` might be chosen and experimented, thus achieving different trade-offs between random/bulk disk accesses. However, the *qualitative* considerations on the algorithmic performance drawn in the next section will remain mostly unchanged.

For our experiments we collected over various WEB sites two textual datasets: the Reuters corpus[3] of about 26 Mbytes; and a set of amino-acid sequences taken from a SWISSPROT database[4] of about 26 Mbytes. These datasets have some nice features: the former set is structured and presents long repeated substrings, whereas the latter set is unstructured and thus suitable for full-text indexing. Notice that for $N = 26$ Mbytes the suffix array *SA* occupies 104 Mbytes, and the working space of all tested algorithms is more than 200 Mbytes. Hence, the datasets are large enough to evaluate the I/O-performance of the studied algorithms, and investigate their *scalability* in an external-memory setting. The overall results are reported in Tables 2 and 3. (For further comments and results we refer the reader to the full paper.)

Results for the MM-algorithm. It is not astonishing to observe that the construction time of the MM-algorithm is outperformed by every other algorithm studied in this paper as soon as its working space exceeds the internal memory size. This worse behavior is due to the fact that the algorithm accesses the suffix array in an unstructured and unpredictable way. When $N > 8$ Mbytes, its time complexity is still *quasi-linear* but the constant hidden in the big-Oh notation is very large due to the paging activity, thus making the algorithmic behavior unacceptable.

Results for the BGS-algorithm. If we double the text size, the running time and I/Os increase by nearly a factor of four. The number of total and bulk I/Os is nearly identical for all datasets, so that the practical behavior is actually *quadratic*. It is not astonishing to verify experimentally that BGS is the fastest algorithm for building a (unique) suffix array when $N \leq 25$ Mbytes. This scenario probably remains unchanged for text collections which are slightly larger than the ones we experimented in this paper; however, for larger and larger sizes, the quadratic behavior of BGS will be probably no longer *"hidden"* by its nice algorithmic properties (i.e., sequential scans of disk, small hidden constants, etc. [13]). In Table 3 we notice that (i) only the 1% of all disk accesses are random I/Os; (ii) the algorithm performs the least number of random I/Os on both the datasets; (iii) BGS is the fastest algorithm to construct one unique suffix array, and it is the second fastest algorithm in general. Additionally we observe that its quadratic time complexity heavily influences its practical efficiency, so that disk-I/Os are not the only bottleneck for BGS.

Results for the Doubling algorithm. The doubling algorithm performs 11 stages on the Reuters corpus, hence 21 scans and 21 sorting steps. Consequently, we can conclude

[3] We used the text collection "Reuters-21578, Distribution 1.0" available from David D. Lewis' professional home page, currently: `http://www.research.att.com/~lewis`

[4] See the site: `http://www.bic.nus.edu.sg/swprot.html`

that there is a repeated substring in this text collection of length about 2^{12} (indeed, we detected a duplicated article). The Doubling algorithm scales well in the tested input range; however, due to the high number of random I/Os and to the large working space, we expect that this algorithm surpasses the performance of BGS only for *very large* values of N. Hence, although theoretically interesting and almost asymptotically optimal, the Doubling algorithm is not much appealing in practice; this motivated our development of Doubl+Disc and Doubl+Disc+Radix algorithms.

Results for Doubl+Disc algorithm. The number of discarded tuples is nearly the same as the size of the test set increases, and the gain induced is approximately the 32% of the running time of doubling. In our experimental datasets, we save approximately 19% of the I/Os compared to Doubling. The percentage of random I/Os is 28%, which is much less than Doubling (42%), and drives us to conclude that discarding helps in reducing mainly the random I/Os. The saving induced by the discarding strategy is expected to pay much more on larger text collections, because of the significant reduction in the number of manipulated tuples, which should facilitate caching and prefetching operations.

Results for Doubl+Disc+Radix algorithm. This algorithm is not as fast as we conjectured. The reason is that we cannot fully exploit the good qualities of radix heaps by keeping the maximum priority value small. Step 2 in Doubl+Disc-algorithm must be implemented via two sorting steps and this naturally *doubles* the overall work. It is therefore not surprising to observe in Table 3 that the Doubl+Disc+Radix algorithm performs *twice* the I/Os of the other Doubling variants, and it is the slowest among all the tested algorithms. Consequently, the *"compensation"* conjectured in Section 2.1 between the number of discarded tuples and the increase in the I/O-complexity of heap-sorting does not actually occur in practice. If we consider space vs. time trade-off, we can reasonably claim that Doubl+Disc+Radix is *worse* than BGS because the former requires larger working space and it is expected to surpass the BGS-performance only for *very large* text collections.

Results for L-pieces algorithm. We fixed $L = 4$, used multi-way mergesort for string-sorting and Doubling for constructing SA_L. Looking at Table 3 we notice that 40% of the total I/Os are random, and that the present algorithm executes slightly more I/Os than BGS. Nonetheless, it is the fastest algorithm (see Table 2): It is three to four times faster than BGS (due to its quadratic CPU-time) and four times faster than the Doubl+Disc algorithm (due to the larger number of I/Os). The running time distributes as follows: 63% is used to build SA_L; 4% to sort the set S; the rest is used to build the other three suffix arrays. It must be said that for our test sizes, the short strings fit in internal memory at once thus making their sorting stage very fast. However, it is also clear that sorting short strings takes no more time than the one needed by *one* stage of the Doubl-algorithm. Consequently, it is not hazardous to *conjecture* that this algorithm is still significantly faster than all the other approaches when working on S and the SA_i's entirely on disk. The only "limit" of this algorithm is that it constructs the suffix array in four distinct pieces. If the underlying text-retrieval applications does not impose to have *one unique* suffix array [4] then this approach turns out to be de-facto 'the' choice for constructing such a data structure.

4 The New BGS-Algorithm

We conclude our paper by addressing one further issue related to the intriguing, and apparently counterintuitive, "contradiction" between the effective practical performance of the BGS-algorithm and its unappealing (i.e., cubic) worst-case complexity. We briefly sketch (details in the full paper) a new approach that follows the *basic philosophy* underlying the BGS-design but in a significantly different manner, thus resulting in a novel algorithm which combines good practical qualities with efficient worst-case performance.

Let us set $m = \ell M$, where $\ell < 1$ is a positive constant to be fixed later. We divide the text T into k non-overlapping substrings of length m each, namely $T = T_k T_{k-1} \cdots T_2 T_1$. The algorithm executes $k = \Theta(N/M)$ stages (like BGS) and processes the text *from the right to the left* (unlike BGS). The following invariant is kept inductively before stage h starts: *String* $S = T_{h-1} T_{h-2} \cdots T_1$ *is the text part processed in the previous* $(h-1)$ *stages. The algorithm has computed and stored on disk two data structures: The suffix array* SA_{ext} *of the string* S *and its "inverse" array* Pos_{ext}, *which keeps at each entry* $Pos_{ext}[j]$ *the position in* SA_{ext} *of the suffix* $S[j, |S|]$. After all k stages are executed, we have $S = T$ and thus $SA = SA_{ext}$.

The main idea underlying the *leftward*-scanning of the text is that when the h-th stage processes the text suffixes starting in T_h, it has already accumulated into SA_{ext} and Pos_{ext} some informations about the text suffixes starting to the right of T_h. This way, the comparison of the former text suffixes can be done by exploiting these two arrays, and thus using only *localized* information which eliminates the need of *random I/Os* (cfr. construction of SA_{int} in BGS [13]). The next Lemma formalizes this intuition (proof in the full version):

Lemma 1. *A suffix* $T[i,N]$ *starting into the text piece* T_h *can be represented succinctly via the pair* $(T[i, i+m-1], Pos_{ext}[((i+m-1) \bmod m) + 1])$. *Consequently, all text suffixes starting into* T_h *can be represented using overall* $O(m)$ *space.*

Stage h preserves the invariant above and thus updates SA_{ext} and Pos_{ext} by *properly* inserting into them the "information" regarding the text suffixes of T which start in the text piece T_h (currently processed). After that, the new SA_{ext} and Pos_{ext} will correctly refer to the "extended" string $T_h \cdot S$, thus preserving the invariant for the next $(h+1)$th stage (where $S = T_h \cdot S = T_h T_{h-1} \cdots T_2 T_1$). The algorithmic details are not obvious but due to space limitations we refer the interested reader to the full paper, where we will prove that:

Theorem 2. *The suffix array of a text* $T[1,N]$ *can be constructed in* $O(N^2/M^2)$ *bulk-I/Os, no random-I/Os, and* $8N$ *disk space in the worst case. The overall CPU time is* $O(\frac{N^2}{M} \log_2 M)$.

The value of the parameter ℓ is properly chosen to fit the auxiliary data structures into internal memory. The practical behavior of the new-BGS algorithm is guaranteed on any indexed text independently of its structure, thus overcoming the (theoretical) limitations of the classical BGS [13] and still keeping its attractive practical properties.

5 Conclusions

It is often observed that practitioners use algorithms which tend to be different from what is claimed as optimal by theoreticians. This is doubtless because theoretical models tend to be simplifications of reality, and theoretical analysis need to use conservative assumptions. In the present paper we actually tried to "bridge" this difference by analyzing more deeply some suffix-array construction algorithms taking more into account the specialties of current disk systems, without going into much technological details but still referring to an (abstract) I/O-model. As it appears clear from the experiments, the final choice of the "best" algorithm depends on the available disk space, on the disk characteristics (which induce different trade-offs between random and bulk I/Os), on the structural features of the indexed text, and also on the patience of the user to wait for the completion of the suffix-array construction. However, it must be noticed that the running-time evaluations indicated in our tables and pictures are not clearly intended to be definitive. Algorithmic engineering and software tuning of the C++-code might definitively lead to improvements without anyway changing the features of the experimented algorithms, and therefore without affecting significantly the scenario that we have depicted in these pages. The qualitative analysis developed in the previous sections should, in our opinion, route and clarify to the software developers which algorithm best fits their needs.

The results in this paper suggest some other directions of research that deserve further investigation. The most notable one is, in our opinion, an adaptation and simplification of the I/O-optimal algorithm for building *suffix trees* [10] to the I/O-optimal (*direct*) construction of suffix arrays.

References

[1] L. Arge, P. Ferragina, R. Grossi and J. S. Vitter. On sorting Strings in External Memory. In *ACM Symp. on Theory of Computing*, pp. 540-548, 1997.

[2] R. Ahuja, K. Mehlhorn, J. B. Orlin and R. E. Tarjan. Faster Algorithms for the Shortest Path Problem. *Journal of the ACM*(2), pp. 213-223, 1990.

[3] A. Andersson and S. Nilsson. Efficient implementation of Suffix Trees. *Software Practice and Experience*, 2(25): 129-141, 1995.

[4] S. Burkhard, A. Crauser, P. Ferragina, H. Lenhof, E. Rivals and M. Vingron. *q*-gram based database searching using a suffix array (QUASAR). *International Conference on Computational Molecular Biology*, 1999.

[5] D. R. Clark and J. I. Munro. Efficient Suffix Trees on Secondary Storage. In *ACM-SIAM Symp. on Discrete Algorithms*, pp.383-391, 1996.

[6] A. Crauser, P. Ferragina and U. Meyer. Practical and Efficient Priority Queues for External Memory. Technical Report MPI, see WEB pages of the authors.

[7] A. Crauser and K. Mehlhorn. LEDA-SM: A Library Prototype for Computing in Secondary Memory. Technical Report MPI, see WEB pages of the authors.

[8] C. Faloutsos. Access Methods for text. *ACM Computing Surveys*, 17, pp.49-74, March 1985.

[9] M. Farach. Optimal suffix tree construction with large alphabets. In *IEEE Foundations of Computer Science*, pp. 137–143, 1997.

[10] M. Farach, P. Ferragina and S. Muthukrishnan. Overcoming the Memory Bottleneck in Suffix Tree Construction. In *IEEE Foundations of Computer Science*, 1998.

[11] C.L. Feng. PAT-Tree-Based Keyword Extraction for Chinese Information Retrieval. *ACM SIGIR* ,pp. 50-58, 1997.
[12] P. Ferragina and R. Grossi. A Fully-Dynamic Data Structure for External Substring Search. In *ACM Symp. Theory of Computing*, pp. 693-702, 1995. Also *Journal of the ACM* (to appear).
[13] G. H. Gonnet, R. A. Baeza-Yates and T. Snider. New indices for text:PAT trees and PAT arrays. In *Information Retrieval – Data Structures and Algorithms*, W.B. Frakes and R. BaezaYates Editors, pp. 66-82, Prentice-Hall, 1992.
[14] D. E. Knuth. *The Art of Computer Programming: Sorting and Searching.* Vol. 3, Addison-Wesley Publishing Co. 1973.
[15] S. Kurtz. Reducing the Space Requirement of Suffix Trees. *Technical Report 98-03*, University of Bielefeld, 1998.
[16] U. Manber and G. Myers. Suffix arrays: a new method for on-line string searches. *SIAM Journal of Computing* 22, 5,pp. 935-948, 1993.
[17] E. M. McCreight. A space-economical suffix tree construction algorithm. *Journal of the ACM 23*, 2,pp. 262-272, 1976.
[18] G. Navarro, J.P. Kitajima, B.A. Ribeiro-Neto and N. Ziviani. Distributed Generation of Suffix Arrays. In *Combinatorial Pattern Matching Conference*, pp. 103–115, 1997.
[19] S. Näher and K. Mehlhorn. LEDA: A Platform for Combinatorial and Geometric Computing. *Communications of the ACM (38)*, 1995.
[20] C. Ruemmler and J. Wilkes. An introduction to disk drive modeling. *IEEE Computer*, 27(3):17–29, 1994.
[21] E. A. Shriver and J. S. Vitter. Algorithms for parallel memory I: two-level memories. *Algorithmica*, 12(2-3), pp. 110-147, 1994.
[22] D. E. Vengroff and J. S. Vitter. I/O-efficient scientific computing using TPIE. In *IEEE Symposium on Parallel and Distributed Computing*, 1995.
[23] J. Vitter. External memory algorithms. Invited Tutorial in *17th Ann. ACM Symp. on Principles of Database Systems (PODS '98)*, 1998. Also Invited Paper in *European Symposium on Algorithms (ESA '98)*, 1998.
[24] J. Zobel, A. Moffat and K. Ramamohanarao. Guidelines for presentation and comparison of indexing techniques. *SIGMAD Record 25*, 3:10–15, 1996.

The Reuters corpus						
N	MM	BGS	Doubl	Doubl+Disc	Doubl+Disc+Radix	L-pieces
1324350	67	125	828	982	1965	331
2578790	141	346	1597	1894	3739	582
5199134	293	1058	3665	4070	7833	1119
10509432	223200	4808	8456	8812	16257	2701
20680547	–	16670	23171	20434	37412	5937
26419271	–	27178	42192	28937	50324	7729
The Amino-Acid Dataset						
26358530	–	20703	37963	24817	41595	6918

Table 2. *Construction time (in seconds) of all experimented algorithms on the two text collections, whose size N is expressed in bytes. The symbol '–' indicates that the test was stopped after* 63 *hours.*

N	BGS	Doubl	Doubl+Disc	Doubl+Disc+Radix	L-pieces
The Reuters corpus					
1324350	120/7865	2349/256389	2242/199535	4872/377549	837/57282
2578790	317/20708	4517/500151	4383/395018	10075/787916	1693/177003
5199134	929/60419	9095/1009359	8916/809603	22466/1761273	3386/360210
10509432	4347/282320	18284/2041285	18126/1655751	47571/3728159	6849/730651
20680547	14377/933064	35935/4017664	35904/3293234	96292/7550794	14243/1530995
26419271	24185/1568947	45911/5132822	45842/4202902	129071/10001152	18178/1956557
The Amino-Acid Dataset					
26358530	24181/1568773	41709/4656578	39499/3539148	105956/8222236	16118/1719883

Table 3. *Number of I/Os (bulk/total) of all experimented algorithms on the two text collections, whose size N is expressed in bytes.*

Strategies for Searching with Different Access Costs

Eduardo Sany Laber, Ruy Luiz Milidiú, and Artur Alves Pessoa

Departamento de Informática, PUC-Rio, Brazil.
{laber,milidiu,artur}@inf.puc-rio.br

Abstract. Let us consider an ordered set of elements $A = \{a_1 < \cdots < a_n\}$ and a list of positive costs $c_1, \ldots, c_n$, where c_i is the access cost of the element a_i. Any search strategy for the set A can be represented by a binary search tree (BST) with n nodes, where each node corresponds to an element of A. The search strategy with minimum expected access cost is given by the BST that minimizes $\sum_{i=1}^{n} c_i n(a_i)$ among all binary trees, where $n(a_i)$ denotes the number of nodes in the subtree rooted by the node that corresponds to the element a_i.
In this paper, we prove that the cost of an optimal search tree is bounded above by $4C\ln(1+n)$, where $C = \sum_{i=1}^{n} c_i$. Furthermore, we show the this upper bound is asymptotically optimal. The proof of this upper bound is constructive and generates a $4\ln(1+n)$-approximate algorithm for constructing search trees. This algorithm runs in $O(nH)$ time and requires $O(n)$ space, where H is the height of the tree produced at the end of the algorithm. We also prove some combinatorial properties of the optimal search trees, and based on them, we propose two heuristics to constructing search trees. We report some experimental results that indicates a good performance of these heuristics. The algorithms devised in this paper can be useful for practical cases, since the best known exact algorithm for this problem runs in $O(n^3)$ time, requiring $O(n^2)$ space.

1 Introduction

Let us consider an ordered set of elements $A = \{a_1 < \cdots < a_n\}$, a list of positive costs $c_1, \ldots, c_n$ and a list of probabilities $p_1, \ldots, p_n$, where c_i and p_i are, respectively, the access cost and the access probability of element a_i. Any search strategy for the set A can be represented by a binary search tree (BST) with n nodes, where each node corresponds to an element of set A. The search strategy with minimum expected access cost is given by the BST that minimizes

$$\sum_{i=1}^{n} c_i P_i \tag{1}$$

among all BST with n nodes, where P_i is the probability of the subtree rooted by the node corresponding to the element a_i in the tree. In fact, P_i is given by the sum of the probabilities of all the elements in this subtree.

The best exact algorithm for finding the optimal search strategy is based on dynamic programming and runs in $O(n^3)$, with $O(n^2)$ space requirement [1]. The case with uniform costs and different access probabilities has been extensively studied [2, 3, 4]. The best exact algorithm for this case is due to Knuth [2]. It runs in $O(n^2)$ time, with $O(n^2)$ space requirement.

In this paper we consider the case where the costs are different and all

elements are equiprobable [5, 6, 1]. This case has applications in filter design [6] and on searching in hierarchical memories [5]. In [6], Knight proposed a simple dynamic programming algorithm to build the optimal search strategy. This algorithm runs in $O(n^3)$ time, requiring $O(n^2)$ space. Currently, this is the best exact algorithm for this problem. Knight also analyzed the expected cost of the search for the cost structure $c_i = i^k$, where k is a fixed constant. This cost structure arises in filter design problems.

Nevertheless, the time and space requirements of the exact algorithm, makes it prohibitive for big values of n. Motivated by this fact, we consider alternatives for constructing good search trees. In this paper, we obtain the following results. We present a necessary condition for a BST be an optimal search tree. Based on this condition, we give a non-trivial upper bound for the height of an optimal search tree. In fact, we prove that the height of an optimal search tree is $O(\sqrt{n})$ for practical cases. We present the algorithm Ratio, a $4\ln(1+n)$-approximate algorithm that runs in $O(nH)$ time, with $O(n)$ space requirement, where H is the height of the tree obtained at the end of the algorithm. The analysis of this algorithm shows that $4C\ln(1+n)$ is an upper bound for the cost of an optimal search tree, where $C = \sum_{i=1}^{n} c_i$. We also prove that this upper bound is asymptotically optimal. Finally, we propose two practical heuristics for constructing search trees and we report some experiments comparing their results with the results obtained by other algorithms proposed in the literature. These experiments indicate a good performance of the heuristics devised in this paper.

Since all the elements are equiprobable, (1) can be rewritten as

$$\frac{1}{n}\sum_{i=1}^{n} c_i n(a_i)$$

where $n(a_i)$ denotes the number of nodes in the subtree rooted by the node that corresponds to the element a_i. For convenience, we omit the $1/n$ factor.

Throughout this paper, as an abuse of notation, we also use a_i to denote the node in a BST that corresponds to the element a_i, for $i = 1,\dots,n$. In addition, given an ordered set $a_1 < \cdots < a_n$ and a corresponding BST T, we observe that the ordered set $a_i < \cdots < a_j$ that corresponds to a given subtree T' of T is always a contiguous subset of $a_1 < \cdots < a_n$. Hence, this subset can be represented by the interval of indexes $[i,\dots,j]$. This representation is extensively used throughout this paper. Finally, we use C to denote $\sum_{i=1}^{n} c_i$.

This paper is divided as follows. In section 2, we show a necessary condition for a binary tree be an optimal search tree and we give an upper bound for the height of an optimal search tree. In section 3, we present the algorithm Ratio and an an upper bound for the cost of an optimal search tree. In section 4, we propose two heuristics based in the properties proved throughout the paper and we report some experiments involving these heuristics and other algorithms proposed in the literature.

2 Combinatorial Properties

In this section, we show some combinatorial properties of optimal search trees. We show a necessary condition for a given node be the root of an optimal search tree and we give a necessary condition for the height of an optimal tree be equal to H.

2.1 Optimality Condition

Theorem 1. *If the node a_k is the root of the optimal search tree T for the nodes in the interval $[1,\ldots,m]$ then:*

(i) For all $j < k$ we have $c_j \geq \frac{jc_k}{(m-n(a_j))}$,

(ii) For all $l > k$ we have $c_l \geq \frac{(m-l+1)c_k}{m-n(a_l)}$.

Proof. The idea of the proof is showing that if these conditions do not hold, then the cost of the tree T can be improved. We only give a proof that condition (i) holds, since the proof for condition (ii) is entirely analogous.

Let us assume that a_k is the root of the optimal search tree T for the interval $[1,\ldots,m]$. Now, let $j < k$ and let T^* be the tree obtained after the application of the procedure below.

While a_j is not the root of the T, obtain a new tree T through a rotation involving a_j and its parent

Figure 1.(a) shows a tree T and Figure 1.(b) shows the tree T^* obtained after applying the transformation proposed.

The transformation proposed implies in two facts that we mention without proving.

(a) If $a_{j'}$ is not an ancestor of node a_j in T, then its number of descendants does not modify from T to T^*.

(b) If $a_{j'}$ is an ancestor of node a_j in T and $j' \neq j$, then its number of descendants in T is greater than or equal its number of descendants in T^*.

Let $n^*(a_h)$ be the number of descendants of node a_h in the tree T^*. We have that $n^*(a_j) = m$ and $n(a_k) - n^*(a_k) = j$. Furthermore, it follows from (a) and (b) that $n^*(a_h) \leq n(a_h)$ for $h \neq j$. Hence, we have that

$$c(T^*) - c(T) \leq (m - n(a_j))c_j - j.c_k.$$

Hence, we must have $c_j \geq \frac{jc_k}{m-n(a_j)}$, otherwise we would have $c(T^*) < c(T)$, which contradicts the optimality of T. □

We point out that this result can be easily generalized for the case with different access probabilities. Now, we present a corollary of the previous theorem that will be used in the designs of two heuristics in section 4.

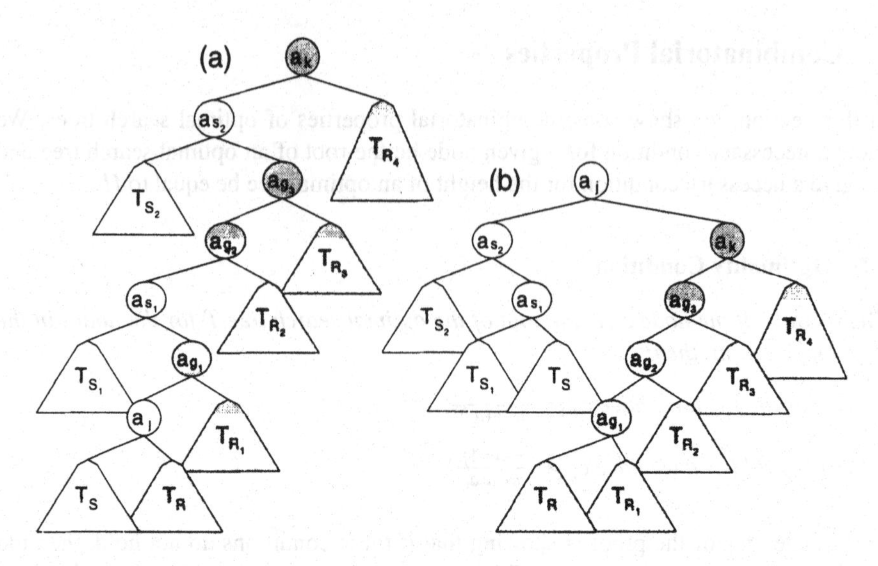

Fig. 1. Figure (a) shows a tree T with root a_k and figures (b) shows the tree T^* obtained applying the transformation.

Corollary 2. *If the node a_k is the root of the optimal search tree for the interval $[i,\ldots,m]$ then:*

(i) For all $j < k$ we have $c_j \geq \frac{(j-i+1)c_k}{m-i}$

(ii) For all $l > k$ we have $c_l \geq \frac{(m-l+1)c_k}{m-i}$.

Proof. This result follows immediately from the previous theorem since $n(a_j) \geq 1$ and $n(a_l) \geq 1$ □

2.2 Bounding the Height of an Optimal Search Tree

Now, we apply Theorem 1 to give a necessary condition for the height of an optimal search tree be equal to H.

Theorem 3. *If the height of the optimal search tree for the interval $[1,\ldots,n]$ is H, then there are two elements a_i and a_j such that*

$$\frac{c_i}{c_j} > (\lceil H/2 \rceil - 1)! \quad \left(\frac{\lceil H/2 \rceil - 1}{n}\right)^{\lceil H/2 \rceil - 1}$$

Proof. See appendix A □

By using Stirling approximation for $\lceil H/2 \rceil !$, we can obtain the following corollary.

Corollary 4. *If the height of the optimal search tree for the interval* $[1,\dots,n]$ *is* H, *then, for large* n, *there are two elements* a_i *and* a_j, *such that*

$$\frac{c_i}{c_j} \geq \sqrt{\pi H}\ \left(\frac{H^2}{4ne}\right)^{H/2}$$

A consequence of this last result is that the height of an optimal search tree for practical cases must be $O(\sqrt{n})$, otherwise there must be costs differing by enormous factor as $f^{\sqrt{n}}$, where f is $\omega(1)$.

3 An Approximate Algorithm

In [6], Knight proved upper and lower bounds for the cost of optimal trees for a special structure of costs. He considered the structure $c_i = i^k$ for $i = 1,\dots,n$, where k is a fixed constant. In this section, we prove that the cost of an optimal search tree is bounded above by $4C\ln(1+n)$. It must be observed that we do not assume anything concerning the costs.

The proof of this result is based on the analysis of the algorithm Ratio presented in figure 2. This algorithm uses a top-down approach combined with a simple rule to select the root of the search tree for the current interval of nodes. In the pseudo-code $a_k.left$ and $a_k.right$ are, respectively, the left and the right children of the node a_k. If a node a_k does not have a left(right) children, then *null* is assigned to $a_k.left(right)$.

Algorithm Ratio ;
  root ← Root(1,n)

Function Root (i,m): integer ;
  If $i \leq m$ then
    1. Find the node a_k that minimizes $c_k/(\min\{k-i+1, m-k+1\})$ on the interval $[i,..,m]$
    2. $a_k.left$ ← Root$(i,k-1)$; $a_k.right$ ← Root$(k+1,m)$.
    3. Return k ;
  Else
    Return *null*;

Fig. 2. The Ratio Algorithm.

3.1 Cost Analysis

In this section we analyze the cost of the tree produced by algorithm Ratio. In order to bound this cost, we need the following proposition.

Proposition 5. *If the node* a_k *is selected by the algorithm Ratio to be the root of the nodes interval* $[1,..,n]$, *then* $c_k \leq \frac{4kC}{n(n+2)}$.

Proof. We assume without loss of generality that $k \leq \lceil n/2 \rceil$, since the other case is symmetric. Since k is selected, then we have the following inequalities

$$c_i \geq \frac{c_k \times \min\{i, n-i+1\}}{k} \quad \text{for } i = 1, \ldots, n$$

By summing the inequalities above in i, we obtain that

$$\sum_{i=1}^{n} c_i \geq \sum_{i=1}^{\lceil n/2 \rceil} \frac{c_k \times i}{k} + \sum_{i=\lceil n/2 \rceil+1}^{n} \frac{c_k \times (n-i+1)}{k} \geq \frac{c_k(n+2)n}{4k}$$

Since $\sum_{i=1}^{n} c_j = C$, then we obtain that

$$c_k \leq \frac{4kC}{n(n+2)}$$

□

Now, we prove that the cost of the tree constructed by algorithm Ratio is bounded above by $4C\ln(1+n)$.

Theorem 6. *The cost of the tree T constructed by the algorithm Ratio for the interval of nodes $[1,..,n]$, with $\sum_{i=1}^{n} c_i = C$, is bounded above by $4C\ln(1+n)$.*

Proof. If $n = 1$, the result obviously holds. Now, we assume that the result holds for $k < n$ and we prove that the result holds for $k = n$.

Let us assume that k is the index selected by the algorithm Ratio to be the root of the interval of nodes $[1,..,n]$. Moreover, we assume without loss of generality that $k \leq \lceil n/2 \rceil$. Hence, the cost $c(T)$ of the tree T constructed by the algorithm Ratio is given by

$$c(T) = nc_k + c(T(1,k-1)) + c(T(k+1,n)),$$

where $c(T(1,k-1))$ and $c(T(k+1,n))$ are, respectively, the costs of the trees constructed by the algorithm Ratio for the intervals of nodes $[1,..,k-1]$ and $[k+1,..,n]$. If $C_1 = \sum_{i=1}^{k-1} c_i$ and $C_2 = \sum_{i=k+1}^{n} c_i$, then it follows from induction that

$$c(T) \leq nc_k + 4C_1\ln(k) + 4C_2\ln(n-k+1),$$

Since $\ln(k) \leq \ln(n-k+1)$ for $1 \leq k \leq \lceil n/2 \rceil$ and $C_1 + C_2 < C$, then we obtain that

$$c(T) \leq nc_k + 4(C_1+C_2)\ln(n-k+1) < nc_k + 4C\ln(n-k+1),$$

On the other hand, proposition 5 assures that $c_k \leq (4kC)/n(n+2)$. Hence,

$$c(T) < \frac{4kC}{n+2} + 4C\ln(n-k+1).$$

Nevertheless, we can show by using some calculus techniques that $f(k) = \frac{4kC}{n+2} + 4C\ln(n-k+1)$ reaches the maximum in the interval $[0, \lceil n/2 \rceil]$ when $k = 0$. Hence, we have that

$$c(T) < \frac{4kC}{n+2} + 4C\ln(n-k+1) \leq 4C\ln(1+n),$$

what establish the theorem. □

Now, we give an immediate corollary of the previous result.

Corollary 7. *The cost of an optimal search tree for the elements $a_1 < \cdots < a_n$, with corresponding costs $c_1, \ldots, c_n$ is bounded above by $4C\ln(1+n)$, where $C = \sum_{i=1}^{n} c_i$. Furthermore, this bound is asymptotically optimal up to constant factors.*

Proof. The upper bound for the cost of an optimal search tree follows immediately from the previous theorem. The asymptotical optimality follows from the fact that the cost of an optimal search tree for the list of costs $c_1 = \cdots = c_n$ is given by $C\log n$. □

The expected cost of the tree produced by Ratio algorithm is proved to be no greater than to $O((C/n)\ln(1+n))$. This is the best that we can have if all the costs are equal. It would be interesting to determine if there is a constant k such that $c(T_R)/c(T^*) \leq k$ for any list of costs, where $c(T_R)$ and $c(T^*)$ are, respectively, the cost of the tree produced by Ratio Algorithm and the cost of an optimal tree. Until now, we do not know how to respond this question.

The algorithm Ratio runs in $O(nH)$ time, where H is the height of the tree obtained at the end of the algorithm. In order to have a better idea of the running time, the value of H must be bounded in terms of the input costs as we did for the optimal search tree.

4 Search Heuristics and Experimental Results

In this section, we present some heuristics to construct search strategies and we report some experimental results obtained by comparing the performance of these heuristics with the optimal search strategy.

We introduce two heuristics that we call *Candidate Heuristics*. First, we define the concept of a candidate node.

Definition 8. *If the node a_k respects the conditions of Corollary 2 for a given interval $[i, \ldots, m]$, then we say that this node is a candidate node.*

For example, let us consider the nodes $a_3, \ldots, a_8$, with respective costs $c_3 = 3, c_4 = 6, c_5 = 20, c_6 = 30, c_7 = 5, c_8 = 8$. In this case, the candidates nodes are the nodes

a_3, a_4, a_7. As an example, the node a_5 cannot be a candidate node since $c_3 < 20/(6-1) = 4$.

Now, we are ready to explain the idea of candidate heuristics. We say that a heuristic A is a candidate heuristic if it satisfies the two items below.

(i) *A uses a top down approach, that is, at each step A selects the root of the current interval through some criterion and, after that, it recursively considers the left and the right intervals.*

(ii) *A always selects a candidate node to be the root of the current interval.*

We must observe that the difference between two candidate heuristics is the criterion used to select the root node among the candidate nodes, for the current interval. We point out that the algorithm Ratio presented in the previous section is a candidate heuristic, since we can prove that the node selected by the algorithm to be the root of the interval $[i, ..., m]$ is always a candidate node. Next, we list two possible criteria for the selection of the root, each of them defining a different candidate heuristic.

1. Select the candidate node that minimize the absolute difference between the sum of the costs of the nodes on its left side and the sum of the costs of the nodes on its right side.
2. Select the candidate node with index closest to the median of the current interval.

Let us consider the example presented in this section. If we choose criterion 1, then the heuristic selects the node a_7, since it minimizes the absolute difference between the sum of its right and left nodes. If we choose criterion 2, then the heuristic selects either the node a_4 or the node a_7. Figure 3 presents a pseudo-code for a generic candidate heuristic.

```
Heuristic Candidate ;

  root ← Root(1,n)

Function Root (i,m): integer ;
  If i ≤ m then
        1. Find the set of candidates for the interval [i,...,m]
        2. Among the set of candidates nodes, choose the node a_k that satisfies the desired
criterion.
        3. a_k.left ← Root(i,k−1); a_k.right ← Root(k+1,m); Return k
  Else
        Return null;
```

Fig. 3. The Heuristic Candidate.

Now, we show how to find the set of candidates for a given interval $[i, \ldots, m]$ in $O(m-i)$ The pseudo-code is presented in Figure 4. First, the procedure scans the vector from left to right checking for each new index k whether the node a_k satisfies the condition (i) of Corollary 2 or not. This check can be done in $O(1)$ by comparing a_k

```
CandidateSet(i,m) ;
    For j=i to m do candidate[j]=true
    j* ← i
    For j = i+1 to m do
        If c_{j*}/(j*-i+1) < c_j/(m-i) then candidate[j]=false
        If c_j/(j-i+1) < c_{j*}/(j*-i+1) then j* = j
    End For
    j* ← m
    For j = m-1 downto i do
        If c_{j*}/(m-j*+1) < c_j/(m-i) then candidate[j]=false
        If c_j/(m-j+1) < c_{j*}/(m-j*+1) then j* = j
    End For
```

Fig. 4. Finding the set of candidates.

with a_{j^*}, where a_{j^*} is the node that minimizes $a_j/(j-i+1)$ in the interval $[i,..,k-1]$. Next, the procedure scans the vector from right to left testing if a_k satisfies the condition (ii) of Corollary 2 or not. This test is analogous to that performed in step the first scan. At the end of the procedure the nodes that satisfies conditions (i) and (ii) of Corollary 2 are the candidates nodes.

4.1 Complexity of Candidate Heuristics

The time complexity of a candidate heuristic depends on the criterion uses to decide which candidate node will be the root of the current interval. The two criteria proposed in the previous section can be implemented to run in linear time. Since the set of candidate nodes can be found in $O(m-i)$, then the two candidate heuristics run in $O(nH)$, where H is the height of the tree produced by each heuristic. Since H is $O(n)$, then the heuristics are $O(n^2)$.

4.2 Experimental Results

In this section, we report some experiments performed to evaluate the quality of the heuristics proposed in this paper. We consider six different algorithms: the optimal algorithm (Opt) [6], the algorithm Ratio, the candidate heuristic with criterion 1 (Cand1), the candidate heuristic with criterion 2 (Cand2), a greedy algorithm that always choose the node with minimum cost (Small) [1], and an ordinary binary search (BinS).

We compare the behavior of these algorithms for two kind of costs structures. The first one, is the structure that arises in a filter design problem [6]. The cost c_i is given by i^k, where k is a fixed constant. The second one is a random structure of costs. The costs are generated by choosing c_i, randomly, in the interval $[1,..,c^*]$, where c^* is chosen as a function of n.

Table 1 presents the results obtained by the algorithms for the structure of costs $c_i = i^k$ for $k = 0,1,2,3$ and $n = 50,200,500$. For equal costs ($k = 0$), all the heuristics obtain an optimal search strategy. This fact is easy to explain analytically. For $k \geq 1$, we observe that the algorithms Ratio, Cand2 and BinS achieved very good results. The relative error of these algorithms, comparing to the optimal one, was smaller than 4% for all the cases. The algorithm Cand1 also obtained good results. Its relative error oscillated from 5% to 12%. The algorithm Small obtained very poor results, as we could predict analytically. In fact, the algorithm Small was designed for a random structure of costs and for the case where the access cost depends on the previous access [1].

Table 1. Results for the cost structure $c_i = i^k$ for $n = 50, 200, 500$ and $k = 1, 2, 3$.

k	0			1			2			3	
n	50	200	500	50	200	500	50	200	500	50	200
Opt	4.86	6.76	8.00	4.75	6.66	7.98	4.39	6.27	7.56	4.09	5.94
Ratio	4.86	6.76	8.00	4.85	6.76	8.00	4.47	6.35	7.60	4.20	5.99
Cand1	4.86	6.76	8.00	5.34	7.29	8.60	4.82	6.58	7.95	4.47	6.41
Cand2	4.86	6.76	8.00	4.83	6.76	7.99	4.52	6.39	7.65	4.19	6.05
Small	4.86	6.76	8.00	17.33	67.33	167.33	13.13	50.63	125.63	10.61	40.60
BinS	4.86	6.76	8.00	4.83	6.76	7.99	4.54	6.44	7.66	4.24	6.11

Table 2 presents the results obtained by the algorithms for the random cost structure. We choose the value of the maximum cost c^* as a function of n. For each n, we consider 3 values for c^*, $c^* = n/10, n, 10n$. In order to obtain more stable results, we run the experiments hundred times for each pair (n, c^*). After, we evaluated the average cost for each experiment. Looking at 2, we observe that the algorithm *Ratio* obtained excellent results in all experiments. Its relative error was smaller than 1% in all cases. The algorithms Cand1, Cand2 and Small obtained good results for all cases. The maximum relative error of these algorithms was about 15%. Finally, the ordinary Binary Search obtained poor results.

Table 2. Results for the random cost structure

c^*	n/10			n			10n		
n	50	200	500	50	200	500	50	200	500
Opt	3.06	2.88	2.80	2.53	2.65	2.68	2.50	2.64	2.67
Ratio	3.08	2.89	2.82	2.54	2.66	2.70	2.51	2.65	2.68
Cand1	3.46	3.19	3.08	2.72	2.87	2.92	2.68	2.85	2.90
Cand2	3.50	3.30	3.21	2.79	2.99	2.99	2.76	2.95	2.99
Small	3.17	3.08	3.05	2.76	2.92	2.97	2.75	2.92	2.96
BinS	4.86	6.73	8.09	4.95	6.86	7.87	4.81	6.75	7.98

Comparing all the experiments, we realize that the Ratio algorithm got a very impressive result, with relative error smaller than 3% in all cases. Nevertheless, we must observe that we just consider two cost structures and we do not have strong approximation results for any of the proposed algorithms.

Acknowledgement. We would like to thank Gonzalo Navarro that introduced us this problem. We also thank the referees for their helpful comments, in particular the one that realized a mistake in our implementation.

References

[1] G. Navarro, E. Barbosa, R. Baeza-Yates, W. Cunto, and N. Ziviani, "Binary searching with non-uniform costs and its application to text retrieval," *Algorithmica*, 1998. To appear. `ftp//ftp.dcc.uchile.cl/pub/users/gnavarro/algor98.2.ps.gz`.

[2] D. E. Knuth, "Optimum binary search trees," *Acta Informatica, Springer Verlag (Heidelberg, FRG and NewYork NY, USA) Verlag*, vol. 1, 1971.

[3] L. L. Larmore, "A subquadratic algorithm for constructing approximately optimal binary search trees," *Journal of Algorithms*, vol. 8, pp. 579–591, 1987.

[4] R. De Prisco and A. De Santis, "On binary search trees," *Information Processing Letters*, vol. 45, pp. 249–253, 1993.

[5] A. Aggarwal, B. Alpern, K. Chnadra, and M. Snir, "A model for hierarchical memory," in *Proceedings of the 19th Annual ACM Symposium on Theory of Computing*, pp. 305–314, ACM Press, 1987.

[6] W. J. Knight, "Search in an ordered array having variable probe cost," *SIAM Journal on Computing*, vol. 17, pp. 1203–1214, 1988.

A Proof of of theorem 3

Proof. Let us consider an optimal search tree T with height H and let P be a path from some leaf to the root of T that contains $H+1$ nodes. Then, there is a sequence of nodes of P $a_{i_1}, a_{i_2}, \ldots, a_{i_{\lceil H/2 \rceil}}$ satisfying the two conditions below.

(i) a_{i_j} is descendant of $a_{i_{j+1}}$ for $j = 1, \ldots, \lceil H/2 \rceil - 1$.

(ii) Either $i_1 < i_2 \ldots < i_{\lceil H/2 \rceil}$ or $i_1 > i_2 \ldots > i_{\lceil H/2 \rceil}$.

We assume without loss of generality that $i_1 < i_2 \ldots < i_{\lceil H/2 \rceil}$.

Now, let a_{ℓ_j} be the descendant of the subtree rooted at $a_{i_{j+1}}$ with the smallest index. Hence, $a_{i_{j+1}}$ is the root of the interval $[\ell_j, .., i_{j+2} - 1]$. Observe that $i_{\lceil H/2 \rceil} + 1 \leq n + 1$. Furthermore, a_{i_j} is the root of the interval $[\ell_{j-1}, .., i_{j+1} - 1]$. Since T is an optimal search tree and a_{i_j} is descendant of $a_{i_{j+1}}$, it follows from Theorem 1 that

$$c_{i_j} \geq \frac{c_{i_{j+1}}(i_j - \ell_j + 1)}{(i_{j+2} - i_{j+1}) + (\ell_{j-1} - \ell_j)} \quad \text{for } j = 1, \ldots, \lceil H/2 \rceil - 1$$

By multiplying the set of inequalities above, we obtain that

$$\frac{c_{i_1}}{c_{i_{\lceil H/2 \rceil}}} \geq \prod_{j=1}^{\lceil H/2 \rceil - 1} \frac{i_j - \ell_j + 1}{(i_{j+2} - i_{j+1}) + (\ell_{j-1} - \ell_j)}, \tag{2}$$

Hence, in order to give a lower bound for $c_{i_1}/c_{i_{\lceil H/2 \rceil}}$, we must minimize the right side of (2) constrained to $1 \leq \ell_{\lceil H/2 \rceil} \leq \cdots \leq l_0 \leq i_1 < \cdots < i_{\lceil H/2 \rceil + 1} \leq n + 1$.

First, we have that

$$\prod_{j=1}^{\lceil H/2 \rceil - 1} (i_j - \ell_j + 1) \geq (\lceil H/2 \rceil - 1)! \tag{3}$$

On the hand, since

$$\sum_{j=1}^{\lceil H/2 \rceil - 1} (i_{j+2} - i_{j+1}) + (\ell_{j-1} - \ell_j) \leq n,$$

it follows from means inequality that.

$$\prod_{j=1}^{\lceil H/2 \rceil - 1} (i_{j+2} - i_{j+1}) + (\ell_{j-1} - \ell_j) \leq \left(\frac{n}{\lceil H/2 \rceil - 1} \right)^{\lceil H/2 \rceil - 1} \tag{4}$$

By combining (3) and (4), we obtain that

$$\frac{c_{i_1}}{c_{i_{\lceil H/2 \rceil}}} \geq \prod_{j=1}^{\lceil H/2 \rceil - 1} \frac{i_j - \ell_j + 1}{(i_{j+2} - i_{j+1}) + (\ell_{j-1} - \ell_j)} \geq (\lceil H/2 \rceil - 1)! \left(\frac{\lceil H/2 \rceil - 1}{n} \right)^{\lceil H/2 \rceil - 1}$$

□

On the Informational Asymmetry between Upper and Lower Bounds for Ultrametric Evolutionary Trees

Ting Chen[1] * and Ming-Yang Kao[2] **

[1] Department of Genetics, Harvard Medical School
Boston, MA 02115 USA
tchen@salt2.med.harvard.edu

[2] Department of Computer Science, Yale University
New Haven, CT 06520-8285 USA
kao-ming-yang@cs.yale.edu

Abstract. This paper addresses the informational asymmetry for constructing an ultrametric evolutionary tree from upper and lower bounds on pairwise distances between n given species. We show that the tallest ultrametric tree exists and can be constructed in $O(n^2)$ time, while the existence of the shortest ultrametric tree depends on whether the lower bounds are ultrametric. The tallest tree construction algorithm gives a very simple solution to the construction of an ultrametric tree. We also provide an efficient $O(n^2)$-time algorithm for checking the uniqueness of an ultrametric tree, and study a query problem for testing whether an ultrametric tree satisfies both upper and lower bounds.

1 Introduction

Constructing the evolutionary tree of a species set is a fundamental problem in computational biology. Such trees describe how species are related to one another in terms of common ancestors. A useful computational problem for constructing evolutionary tree is that given an $n \times n$ distance matrix M where M_{ij} is the observed distance between two species i and j, find an edge-weighted evolutionary tree in which the *distance* d_{ij} in the tree between the leaves i and j, equals M_{ij}. Pairwise distance measures carry some uncertainty in practice. Thus, one seeks a tree that is close to the distance matrix, as measured by various choices of optimization objectives [2, 3, 6, 8].

This paper focuses on the class of *ultrametric trees* [5, 7, 8, 9]. An ultrametric tree is a rooted tree whose edges are weighted by a non-negative number such that the lengths of all the root-to-leaf paths, measured by summing the weights of the edges, are equal. A distance matrix is *ultrametric* if an ultrametric tree can be constructed from this matrix. Figure 1 shows an example of an ultrametric matrix and an ultrametric tree constructed from this matrix.

In practice, when distance measures are uncertain, a distance is expressed as an interval, defined by a lower bound and an upper bound for the true distance. From such data, we obtain two distance matrices M^{ℓ} and M^{h}, representing pairwise distance lower and upper bounds. The tree construction problem becomes that of constructing

* Supported in part by the Lipper Foundation.

** Supported in part by NSF Grant 9531028.

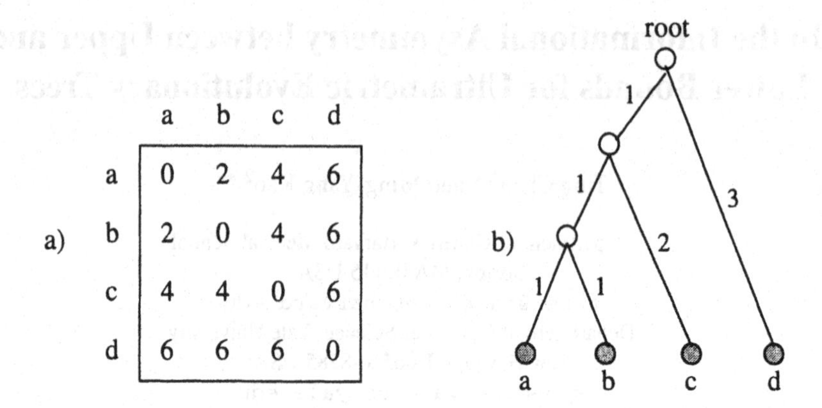

Fig. 1. a. An ultrametric matrix M. b. An ultrametric tree for M.

an ultrametric tree whose pairwise distances fit between two bounds, i.e., $M^{\ell}_{ij} \leq d_{ij} \leq M^{h}_{ij}$. Farah, Kannan and Warnow [4] gave the first known algorithm for constructing ultrametric trees from the sandwich distances.

This paper studies the informational asymmetry between lower and upper bounds in the construction of ultrametric trees. Our results are as follows:

- Given an upper bound matrix, the *tallest* ultrametric tree, where the distance of any two leaves reaches the maximum among all satisfied ultrametric trees, can be constructed in $O(n^2)$ time. This result immediately leads to a new and simpler tree construction algorithm than that of [4].
- Given a lower bound matrix, the *shortest* ultrametric tree, defined similarly to the tallest ultrametric tree, exists if and only if the matrix is ultrametric.
- We provide an $O(n^2)$-time algorithm to check the uniqueness of an ultrametric tree satisfying given upper and lower bounds.
- We study a query problem: if a lower bound matrix and an upper bound matrix are given, how fast can we determine whether an ultrametric tree satisfies both constraints? This problem is useful, for example, for developing an interactive software for finding the most suitable tree among many. We present an algorithm to test the satisfaction of the upper bound constraints in $O(n)$ time after preprocessing the upper bound matrix. A similarly fast algorithm for testing the lower bounds remains open.

2 Notation

Let M represent a distance matrix, and M^{ℓ} and M^{h} represent the lower bound and upper bound matrices for M. Let d_{ij} represent the distance between leaf i and leaf j in a tree, defined as the length of the path connecting two leaves. Given a tree U, d^{U}_{ij} represents that distance in U. If U is ultrametric, the height of U is denoted by function $\hbar(U)$, or $\hbar(r)$ if r is the root of U.

An $n \times n$ distance matrix M corresponds to an undirected edge-weighted complete graph G with n vertices, where a vertex represents a species and the weight $w(i, j)$ of edge (i,j) is M_{ij}. G is also said to be *the corresponding graph* to M. For the lower bound matrix M^{ℓ} and the upper bound matrix M^h, their corresponding graphs are denoted by G^{ℓ} and G^h, and the weight functions are $w^{\ell}()$ and $w^h()$.

3 Constructing the tallest ultrametric tree

This section discusses the problem of finding the tallest ultrametric tree for any upper bound matrix. We give a simple $O(n^2)$-time algorithm that takes advantages of the minimum weight spanning tree.

Fact 1 (see [1, 4, 8]). *A matrix is ultrametric if and only if in the corresponding complete weighted graph, the largest weight edge in any cycle of more than one node is not unique.*

Proof. Straightforward. □

Suppose we have an upper bound matrix M^h on the pairwise leaf-leaf distances of an ultrametric tree. There are many ultrametric trees satisfying M^h, but which one is the tallest? The following algorithm gives the answer. Let G^h be the corresponding graph of M^h.

Algorithm Compute_Tallest_Tree

1. Construct the minimum weight spanning tree T of G^h.
2. Sort the edges of T in decreasing order as $e_1, e_2, \ldots, e_{n-1}$.
3. Return the tree U constructed by the following procedure:
   (a) If T is empty, return a leaf with zero height.
   (b) Otherwise, remove the first edge e_1 from T, leaving two subtrees, T_1 and T_2, each of which maintains its edges in decreasing order.
   (c) Recursively construct a tree U_1 from T_1, and a tree U_2 from T_2.
   (d) Construct a root r for U with height $\hbar(U) = \frac{1}{2}w^h(e_1)$, and attach U_1 to r with an edge weighted $\frac{1}{2}w^h(e_1) - \hbar(U_1)$ and U_2 to r with an edge weighted $\frac{1}{2}w^h(e_1) - \hbar(U_2)$.
   (e) Return U.

This algorithm constructs an ultrametric tree in $O(n^2)$ time:

Lemma 2. *Algorithm Compute_Tallest_Tree runs $O(n^2)$ time and returns an ultrametric satisfying the given upper bound matrix.*

Proof. The algorithm takes $O(n^2)$ time to build the minimum weight spanning tree T, and $O(n \log n)$ time to sort the edges. Maintaining the edges in the subtrees of T in decreasing order (at Step 3b) takes $O(n\alpha(n,n))$ time by using the disjoint-set forest data structure, where $\alpha()$ is the inverse of Ackermann's function, and $\alpha(n,n) \leq 4$ in any conceivable application. Here we implicitly use the fact that any connected subtree of a minimum spanning tree is still a minimum spanning tree. Therefore, the total time complexity is $O(n^2)$. At Step 3d, if both U_1 and U_2 are ultrametric, U is ultrametric because the distance from root r to any leaf is $\frac{1}{2}w^h(e_1)$. Step 3a returns a tree that has only one leaf and is ultrametric, so by recursion U is ultrametric. U also satisfies the upper bound matrix: for any two leaves x of U_1 and y of U_2, the weight of the edge $w^h(x,y)$ in graph G^h is at least $w^h(e_1)$, which equals d^U_{xy}; otherwise, we could replace e_1 by (x,y) in T. □

Theorem 3. *The tree U constructed by Algorithm Compute_Tallest_Tree is the tallest ultrametric tree.*

Proof. Let x and y be any two leaves of U, and c be their lowest common ancestor. Let (a,b) be the edge in the minimum spanning tree T whose removal generates c at Step 3d of Algorithm Compute_Tallest_Tree. Then,

$$d^U_{xy} = d^U_{ab} = M^h_{ab}.$$

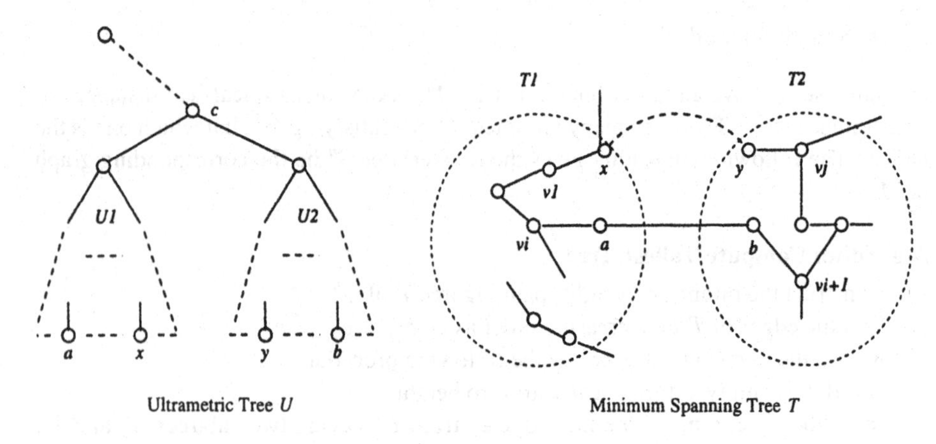

Fig. 2. Constructing an ultrametric tree using a minimum spanning tree

Let U_1 and U_2 be two immediate subtrees of c, which are constructed from two subtrees T_1 and T_2 of T, respectively. Assume that leaf a and leaf x are under U_1, and leaf b and leaf y are under U_2, as shown in Figure 2. Note that T_1 is a connected component and thus contains a path from x to a, and similarly T_2 contains a path from b to y. Combining them with the edge (a,b), we obtain a path P from x to y, *i.e.*,

$$P = x - v_1 - ... - v_i - a - b - v_{i+1} - ... - v_j - y.$$

Note that (a,b) is selected in the algorithm because it has a larger weight than any other edge in $T_1 \cup T_2$. Thus,

$$\max_{(w,z)\in P} M^h_{wz} = M^h_{ab}.$$

Let V be any other satisfied ultrametric tree. By Fact 1, d^V_{xy} should not be the sole largest distance in the cycle $P \cup (x,y)$. Thus,

$$d^V_{xy} \leq \max_{(w,z)\in P} d^V_{wz} \leq \max_{(w,z)\in P} M^h_{wz}.$$

Combining all the inequalities, $d^V_{xy} \le d^U_{xy}$ and U is the tallest ultrametric tree. □

Theorem 3 immediately leads to the following theorem:

Theorem 4. *Algorithm Compute_Tallest_Tree constructs an ultrametric tree satisfying both lower bound and upper bound constraints, if one exists.*

Proof. Algorithm Compute_Tallest_Tree constructs the tallest ultrametric tree U for upper bounds. If lower bounds are also given, either U satisfies the lower bounds, or no tree satisfies them. □

4 Asymmetry between upper and lower bounds

We have shown how to construct the tallest ultrametric tree in Section 3. However, the shortest ultrametric tree may not exist. The following lemma explains the asymmetry between upper and lower bounds.

Lemma 5. *There exists a lower bound matrix L such that for any ultrametric tree V that satisfies L, there is an upper bound matrix H which V cannot satisfy but some ultrametric tree U satisfies both L and H.*

Proof. Consider a lower bound matrix L having three elements a, b and c, whose distances are x, y and z, satisfying $x > y$ and $x > z$. Let d_m be the maximum distance in L. We construct two upper bound matrices H_1 and H_2, where every distance equals d_m except the three distances shown below.

L	b	c
a	x	y
b		z

H_1	b	c
a	x	x
b		z

H_2	b	c
a	x	y
b		x

Every element in H_1 is at least as large as the corresponding element in L, and H_1 is ultrametric by Fact 1. Thus, there are ultrametric trees satisfying L and H_1. Let V be such a tree. So d^V_{ab} and d^V_{bc} are fixed: $d^V_{ab} = x$ and $d^V_{bc} = z$. By Fact 1, $d^V_{ac} = x$, which is larger than the corresponding element y in H_2. V does not satisfy H_2. Similarly, H_2 is ultrametric and any ultrametric tree satisfying L and H_2 does not satisfy H_1. Therefore, no ultrametric tree satisfies both H_1 and H_2 at the same time. □

By Lemma 5, a lower bound matrix does not have a shortest ultrametric tree if there exists a three-element cycle whose largest value is unique. On the other hand, if the largest value of any of its three-element cycle is non-unique, then,

Lemma 6. *If any three element cycle in a matrix has a non-unique largest value, so does any cycle in the matrix.*

Proof. Let M be a matrix where the largest value in any three-element cycle is not unique. Assume there exists a cycle $C = (v_1, v_2, ..., v_k, v_1)$, in which $k > 3$ and $M_{v_1 v_2}$ is the unique largest value. We decompose C into three-element cycles: (v_1, v_2, v_3, v_1), (v_1, v_3, v_4, v_1), ..., and (v_1, v_{k-1}, v_k, v_1). Since the largest value in any of them is not unique, the first cycle has $M_{v_1v_2} = M_{v_1v_3}$ because $M_{v_1v_2} > M_{v_2v_3}$ by the assumption. Similarly, $M_{v_1v_i} = M_{v_1v_{i+1}}$ in the ith cycle, for $i = 2,...,k-1$. Combining these results, $M_{v_1v_2} = M_{v_1v_k}$, which contradicts our assumption. Thus the largest value in any cycle of M is not unique. □

Theorem 7. *A lower bound matrix has a shortest ultrametric tree if and only if it is ultrametric.*

Proof. If a lower bound matrix is ultrametric, by definition, an ultrametric tree can be constructed from this matrix. This tree is the shortest. If a lower bound matrix has a shortest ultrametric tree, by the proof in Lemma 5, the largest value in any three-element cycle is not unique. So does any cycle in the matrix, by Lemma 6. Finally, by Fact 1 this matrix is ultrametric. □

5 Uniqueness of ultrametric tree

Given M^ℓ and M^h, is the ultrametric tree built by Algorithm Compute_Tallest_Tree topologically unique? Knowing the uniqueness is of vital importance because it indicates the significance of the tree.

We first define what kind of trees have the same topological structure. An edge-weighted tree is *compact* if it has no zero-weight edge. A tree can be *compacted* by merging any two internal nodes into a single node if they are connected by a zero weight edge. A *compact* ultrametric tree is one that is compact. Any ultrametric tree can be converted into a compact ultrametric tree by merging, but the resulting tree may not be binary. Assume all the trees have the same set of labeled leaves. An internal node can be represented by a *leaf set*, consisting of all the descendent leaves of the node. A tree can be represented as the *superset* of leaf sets, where every element in the superset corresponds to an internal node, and vice versa. Two trees are *equivalent* or have the same *topological structure* if their representing supersets are the same, in other words, if there is a one-to-one mapping between all nodes that preserve the parent-child relation. Two compact ultrametric trees may have the same topological structure even though their edge weights are completely different. In discussing topological structures, the edge weights are ignored, because equivalent evolutionary trees give the same evolutionary process, and the difference in distances are usually caused by measuring errors.

Given M^ℓ and M^h, we construct an ultrametric tree from M^h by Algorithm Compute_Tallest_Tree, and then convert it into a compact tree U. We can check the uniqueness of U by the following lemma:

Lemma 8. *For given M^ℓ and M^h, the compact ultrametric tree U, constructed from Algorithm Compute_Tallest_Tree, is unique if and only if for every internal node u, any two children u_i and u_j of u satisfy that* $\max M^\ell_{xy} = \hbar(u)$, *for any leaf x under u_i and any leaf y under u_j.*

Proof. Suppose that M^ℓ and M^h define a unique compact ultrametric tree. Since U is the tallest compact ultrametric tree for M^h, it satisfies M^ℓ and is unique. Assume the lemma does not hold, then there exist two different children u_i and u_j of u, such that

$\max M^{\ell}_{xy} < \hbar(u)$, for any two leaves x under u_i and y under u_j. Let $d_{u_iu_j} = \max M^{\ell}_{xy}$ and $d_m = \max(d_{u_iu_j}, \hbar(u_i), \hbar(u_j))$. We can construct a different tree from U by deleting two children u_i and u_j from u, and replacing by a child u' who has two children u_i and u_j, and $\hbar(u') = \frac{1}{2}(\hbar(u) + d_m)$. This tree is a compact ultrametric tree because $\hbar(u) > \hbar(u') > d_m$, and it is topologically different from U, contradicting the uniqueness of U.

Conversely, if any two children u_i and u_j of u satisfy that $\max M^{\ell}_{xy} = \hbar(u)$ for any leaf x under u_i and any leaf y under u_j, we state that U is unique. Assume there exists another topologically different tree V. Let d^U_{xy} be the minimum value among those who satisfy $d^V_{xy} < d^U_{xy}$ (U is the tallest). d^U_{xy} must exist, otherwise V equals U because all pairwise distances of V are equal to those of U. Let u be the least common ancestor of x and y in U, u_x be the child of u that contains x as a descendant, and u_y be the child of u that contains y as a descendant. Let S_x be the set of leaves under u_x and S_y be the set of leaves under u_y.

For any $w \in S_x$, $d^V_{xw} \le d^U_{xw} = h^U(u_x) < d^U_{xy}$, and for any $z \in S_y$, $d^V_{yz} \le d^U_{yz} = h^U(u_y) < d^U_{xy}$. By Lemma 6 and Fact 1, $d^V_{xz} \le \max(d^V_{xy}, d^V_{yz}) < d^U_{xy}$. Thus, in V, the distance between x and any other leaf in $S_x \cup S_y$ is less than d^U_{xy}. So is the distance of any pair of leaves in $S_x \cup S_y$. However, it contradicts the condition in the lemma that there exist a pair of leaves $w \in S_x$ and $z \in S_y$ with $M^{\ell}_{wz} = h^U(u) = d^U_{xy}$. Hence, V can not exist, and U is unique. □

Theorem 9. *Given M^{ℓ} and M^h as input, we can determine the uniqueness of ultrametric trees in $O(n^2)$ time.*

Proof. We can construct a compact ultrametric tree U in $O(n^2)$ time by Algorithm Compute_Tallest_Tree, and then check the conditions of every internal node in a child-parent topological order in $O(n^2)$ time by Lemma 8, because every pair of leaves is visited exactly once. □

6 Query of ultrametric trees

Assuming that we have obtained new ways to estimate the evolutionary distances between species (by some new evidence), and there are many previously computed ultrametric trees (by other evidences), finding the suitable trees among them may suggest new relations between these evidences. This section studies this query problem: given M^{ℓ} and M^h, and an ultrametric tree V, does V satisfy M^{ℓ} and M^h? A naive algorithm runs in $O(n^3)$ time by checking $O(n^2)$ pairs of leaves in V and calculating the distance of each pair in $O(n)$ time. We can improve $O(n^3)$ to $O(n^2)$ by pre-calculating the *lowest common ancestor* in linear time, and thus finding the distance of each pair of leaves in constant time. If preprocessing is permitted, is there a faster algorithm than $O(n^2)$?

Lemma 10. *We can preprocess the upper bound matrix M^h in $O(n^2)$ time, so that for any given ultrametric tree V, whether V satisfies M^h can be determined in $O(n)$ time.*

Proof. Assume we have built the tallest ultrametric tree U by Algorithm Compute_Tallest_Tree, and have calculated the lowest common ancestor (lca) function for V in linear time. We define a recursive function f to map a node in U to a node in V:

$$f(u) = \begin{cases} v & \text{if } u \text{ is a leaf in } U, v \text{ is a leaf in } V, u = v \\ lca(f(u_l), f(u_r)) & \text{if } u \text{ has children } u_l \text{ and } u_r \end{cases}$$

For each internal node u, f returns an internal node v under which its leaves form a superset of the leaves under u. By lca function, v is the lowest node whose leaves form the minimum superset.

We sort all the leaves and internal nodes of U into a topological order $u_1, u_2, \ldots$, where a node appears before its parent, in $O(n)$ time by a depth-first search. For each u_i, if $h^V(f(u_i)) \leq h^U(u_i)$, V satisfies M^h. We next show this in two steps. First, because $u_1, u_2, \ldots$ follow the topological order (child to parent), we can calculate $v_1 = f(u_1), v_2 = f(u_2), \ldots$ in this order in $O(n)$ time. Second, if u has two children u_1 and u_2 and $v = f(u)$, $h^V(v) \leq h^U(u)$ indicates that for any pair of leaves w under u_1 and z under u_2, $d^V_{f(w)f(z)} \leq 2h^V(v) \leq 2h^U(u) = d^U_{wz} \leq M^h_{wz}$. Visiting every node in U and checking the corresponding inequality is equivalent to comparing the distance of every pair of leaves in U with that in V. If all of them hold, V satisfies M^h. Otherwise, if $h^V(v) > h^U(u)$, there must exist a leaf w under u_1 and z under u_2 such that $d^V_{f(w)f(z)} > d^U_{wz}$, violating the assumption that U is the tallest (by construction) in Theorem 3.

The preprocessing takes $O(n^2)$ time and $O(n)$ space to construct U and sort a topological order. To answer a query, it takes $O(n)$ time and $O(n)$ space to calculate function f, and the same for visiting $O(n)$ nodes of U and evaluating $O(n)$ inequalities. □

However, for lower bound matrices, whether there is an algorithm better than $O(n^2)$ time remains an open question. Unlike upper bound matrices having the tallest ultrametric tree, lower bound matrices may have multiple minimal trees as shown in Section 4. This asymmetry prevents the linear time checking algorithm in Lemma 10 from being applicable to the lower bounds.

Open Problem *By preprocessing, is there an algorithm that can test whether a tree satisfies a lower bound matrix faster than $O(n^2)$ time?*

7 Acknowledgments

We thank Raphael Ryger for useful comments.

References

[1] J.-P. Barthélémy and A. Guénoche. *Trees and proximity representations.* Wiley-Interscience Series in Discrete Mathematics and Optimization. Wiley, New York, NY, 1991.

[2] V. Berry and O. Gascuel. Inferring evolutionary trees with strong combinatorial evidence. In T. Jiang and D. T. Lee, editors, *Lecture Notes in Computer Science 1276: Proceedings of the 3rd Annual International Computing and Combinatorics Conference*, pages 111–123. Springer-Verlag, New York, NY, 1997.

[3] J. C. Culbertson and P. Rudnicki. A fast algorithm for constructing trees from distance matrices. *Information Processing Letters*, 30(4):215–220, 1989.

[4] M. Farach, S. Kannan, and T. Warnow. A robust model for finding optimal evolutionary trees. *Algorithmica*, 13(1/2):155–179, 1995.
[5] D. Gusfield. *Algorithms on Strings, Trees, and Sequences: Computer Science and Computational Biology*. Cambridge University Press, New York, NY, 1997.
[6] J. J. Hein. An optimal algorithm to reconstruct trees from additive distance data. *Bulletin of Mathematical Biology*, 51:597–603, 1989.
[7] D. M. Hillis, C. Moritz, and B. K. Mable, editors. *Molecular Systematics*. Sinauer Associates, Sunderland, Ma, 2nd edition, 1996.
[8] J. C. Setubal and J. Meidanis. *Introduction to Computational Molecular Biology*. PWS Publishing Company, Boston, MA, 1997.
[9] M. S. Waterman. *Introduction to Computational Biology: Maps, Sequences and Genomes*. Chapman & Hall, New York, NY, 1995.

Optimal Binary Search with Two Unreliable Tests and Minimum Adaptiveness

Ferdinando Cicalese*,[1] and Daniele Mundici**[2]

[1] Dipartimento di Informatica ed Applicazioni, University of Salerno.
Via S. Allende, Baronissi (Salerno), Italy
cicalese@dia.unisa.it
[2] Dipartimento di Informatica, University of Milano.
Via Comelico, 39–41, Milano, Italy
mundici@imiucca.csi.unimi.it

Abstract. What is the minimum number of yes-no questions needed to find an m bit number x in the set $S = \{0, 1, \ldots, 2^m - 1\}$ if up to ℓ answers may be erroneous/false ? In case when the $(t+1)$th question is *adaptively* asked after receiving the answer to the tth question, the problem, posed by Ulam and Rényi, is a chapter of Berlekamp's theory of error-correcting communication with feedback. It is known that, with finitely many exceptions, one can find x asking Berlekamp's minimum number $q_\ell(m)$ of questions, i.e., the smallest integer q such that $2^q \geq 2^m(\binom{q}{\ell} + \binom{q}{\ell-1} + \cdots + \binom{q}{2} + q + 1)$. At the opposite, *nonadaptive* extreme, when all questions are asked in a unique batch before receiving any answer, a search strategy with $q_\ell(m)$ questions is the same as an ℓ-error correcting code of length $q_\ell(m)$ having 2^m codewords. Such codes in general do not exist for $\ell > 1$. Focusing attention on the case $\ell = 2$, we shall show that, with the exception of $m = 2$ and $m = 4$, one can always find an unknown m bit number $x \in S$ by asking $q_2(m)$ questions in two nonadaptive batches. Thus the results of our paper provide shortest strategies with as little adaptiveness/interaction as possible.

1 Introduction

Consider the following game: Two players, Paul and Carole, first fix a *search space* $S = \{0, 1, \ldots, 2^m - 1\}$. Now Carole thinks of a number $x \in S$, and Paul must find out x by asking questions to which Carole can only answer yes or no. Assuming Carole is allowed to lie (or just to be inaccurate) in up to ℓ answers, what is the minimum number of questions needed by Paul to infallibly guess x?

When the ith question is adaptively asked knowing the answer to the $(i-1)$th question, the problem is generally referred to as the Ulam-Rényi problem, [13, p.47], [15, p.281], and naturally fits into Berlekamp's theory of error-correcting communication with feedback [3] (also see [7] for a survey). Optimal solutions can be found in [11], [6], [10], and [14], respectively for the case $\ell = 1$, $\ell = 2$, $\ell = 3$, and for the general case. If one allows queries having k possible answers, the corresponding Ulam-Rényi

* Partially supported by ENEA Phd Grant
** Partially supported by COST ACTION 15 on Many-valued logics for computer science applications, and by the Italian MURST Project on Logic.

problems on k-ary search with lies are solved in [1] for the case $\ell = 1$, and in [5], for the case $\ell = 2$.

At the other *nonadaptive* extreme, when all questions are asked in advance, the Ulam-Rényi problem amounts to finding an ℓ-error correcting code, with $|S|$ many codewords of shortest length, where $|S|$ denotes the number of elements of S. As is well known, for $\ell = 1$ Hamming codes yield optimal searching strategies—indeed, Pelc [12] shows that adaptiveness is irrelevant even under the stronger assumption that repetition of the same question is forbidden. By contrast, for $\ell > 1$ a multitude of results (see, e.g., [8]) shows that nonadaptive search over a search space with 2^m elements generally cannot be implemented by strategies using $q_2(m)$ questions.

In this paper we shall prove that when $\ell = 2$, searching strategies with $q_2(m)$ questions do exist with the least possible degree of adaptiveness/interaction. Specifically, Paul can infallibly guess Carole's secret number $x \in S$ by asking a first batch of m *nonadaptive* questions, and then, only depending on the m-tuple of Carole's answers, asking a second mini-batch of n *nonadaptive* questions. Our strategies are the *shortest* possible, in that $m+n$ coincides with Berlekamp's lower bound $q_2(m)$, the number of questions that are *necessary* to accommodate all possible answering strategies of Carole.

Since Paul is allowed to adapt his strategy only once, the results of our paper yield *shortest* 2-fault tolerant search strategies with minimum adaptiveness.

2 The Ulam-Rényi Game

Questions, answers, states, strategies

By a *question* T we understand an arbitrary subset T of S. The *opposite question* is the complementary set $S \setminus T$. In case Carole's answer is "yes", numbers in T are said to *satisfy* Carole's answer, while numbers in $S \setminus T$ *falsify* it. Carole's negative answer to T has the same effect as a positive answer to the opposite question $S \setminus T$. Suppose questions $T_1, \ldots, T_t$ have been asked, and answers $b_1, \ldots, b_t$ have been received from Carole ($b_i \in \{no, yes\}$). Then a number $y \in S$ must be rejected from consideration if, and only if, it falsifies 3 or more answers. The remaining numbers of S still are possible candidates for the unknown x. All that Paul knows (Paul's *state* of knowledge) is a triplet $\sigma = (A_0, A_1, A_2)$ of pairwise disjoint subsets of S, where A_i is the set of numbers falsifying i answers, $i = 0, 1, 2$. The *initial* state is naturally given by $(S, \emptyset, \emptyset)$. A state (A_0, A_1, A_2) is *final* iff $A_0 \cup A_1 \cup A_2$ is empty, or has exactly one element. For any state $\sigma = (A_0, A_1, A_2)$ and question $T \subseteq S$, the two states σ^{yes} and σ^{no} respectively resulting from Carole's positive or negative answer, are given by

$$\sigma^{yes} = (A_0 \cap T,\ (A_0 \setminus T) \cup (A_1 \cap T),\ (A_1 \setminus T) \cup (A_2 \cap T)) \tag{1}$$

$$\sigma^{no} = (A_0 \setminus T,\ (A_0 \cap T) \cup (A_1 \setminus T),\ (A_1 \cap T) \cup (A_2 \setminus T)). \tag{2}$$

Turning attention to questions $T_1, \ldots, T_t$ and answers $b = b_1, \ldots, b_t$, iterated application of the above formulas yields a sequence of states

$$\sigma_0 = \sigma,\ \ \sigma_1 = \sigma_0^{b_1},\ \ \sigma_2 = \sigma_1^{b_2},\ \ldots,\ \sigma_t = \sigma_{t-1}^{b_t}. \tag{3}$$

By a *strategy S with q questions* we mean the full binary tree of depth q, where each node ν is mapped into a question T_ν, and the two edges $\eta_{left}, \eta_{right}$ generated by ν are respectively labeled *yes* and *no*. Let $\eta = \eta_1, \ldots, \eta_q$ be a path in S, from the root to a leaf, with respective labels $b_1, \ldots, b_q$, generating nodes $\nu_1, \ldots, \nu_q$ and associated questions $T_{\nu_1}, \ldots, T_{\nu_q}$. Fixing an arbitrary state σ, iterated application of (1)-(2) naturally transforms σ into σ^η (where the dependence on the b_j and T_{ν_j} is understood). We say that strategy S is *winning* for σ iff for every path η the state σ^η is final. A strategy is said to be *nonadaptive* iff all nodes at the same depth of the tree are mapped into the same question.

Type, weight, character, Berlekamp's lower bound

Let $\sigma = (A_0, A_1, A_2)$ be a state. For each $i = 0,1,2$ let $a_i = |A_i|$ be the number of elements of A_i. Then the triplet (a_0, a_1, a_2) is called the *type* of σ. The *Berlekamp weight of σ before q questions*, $q = 0,1,2,\ldots$, is given by

$$w_q(\sigma) = a_0\left(\binom{q}{2} + q + 1\right) + a_1(q+1) + a_2. \tag{4}$$

The *character* $ch(\sigma)$ is the smallest integer $q \geq 0$ such that $w_q(\sigma) \leq 2^q$. By abuse of notation, the weight of *any* state σ of type (a_0, a_1, a_2) before q questions shall be denoted $w_q(a_0, a_1, a_2)$. Similarly, the character of a state $\sigma = (A_0, A_1, A_2)$ of type (a_0, a_1, a_2) will also be denoted $ch(a_0, a_1, a_2)$. As an immediate consequence of the definition we have the following monotonicity properties: For any two states $\sigma' = (A'_0, A'_1, A'_2)$ and $\sigma'' = (A''_0, A''_1, A''_2)$ respectively of type (a'_0, a'_1, a'_2) and (a''_0, a''_1, a''_2), if $a'_i \leq a''_i$ for all $i = 1,2,3$ then

$$ch(\sigma') \leq ch(\sigma'') \text{ and } w_q(\sigma') \leq w_q(\sigma'') \tag{5}$$

for each $q \geq 0$. Note that $ch(\sigma) = 0$ iff σ is a final state.

The proof of the following results goes back to [3].

Lemma 1. *Let σ be an arbitrary state and $T \subseteq S$ a question. Let σ^{yes} and σ^{no} be as in (1)-(2). We then have*

(i) (Conservation Law). *For any integer $q \geq 1$,*

$$w_q(\sigma) = w_{q-1}(\sigma^{yes}) + w_{q-1}(\sigma^{no}).$$

(ii) (Berlekamp's lower bound) *If σ has a winning strategy with q questions then $q \geq ch(\sigma)$.*

A winning strategy for σ with q questions is said to be *perfect* iff $q = ch(\sigma)$. We shall often write $q_2(m)$ instead of $ch(2^m, 0, 0)$. Let $\sigma = (A_0, A_1, A_2)$ be a state. Let $T \subseteq S$ be a question. We then say that T is *balanced for* σ iff for each $j = 0,1,2$, we have $|A_j \cap T| = |A_j \setminus T|$.

Lemma 2. *Let T be a balanced question for a state $\sigma = (A_0, A_1, A_2)$. Let $n = ch(\sigma)$. Let σ^{yes} and σ^{no} be as in (1)-(2) above. Then, for each integer $q \geq 0$,*

(i) $w_q(\sigma^{yes}) = w_q(\sigma^{no})$,

(ii) $ch(\sigma^{yes}) = ch(\sigma^{no}) = n - 1$.

3 Background from Coding Theory

Let $x,y \in \{0,1\}^n$. The *Hamming distance* $d_H(x,y)$ is defined by $d_H(x,y) = |\{i \in \{1,\ldots,n\} \mid x_i \neq y_i\}|$, where, as above, $|A|$ denotes the number of elements of A. The *Hamming sphere* $\mathcal{B}_r(x)$ *with radius* r *and center* x is the set of elements of $\{0,1\}^n$ whose Hamming distance from x is $\leq r$, in symbols, $\mathcal{B}_r(x) = \{y \in \{0,1\}^n \mid d_H(x,y) \leq r\}$. The *Hamming weight* $w_H(x)$ of x is the number of nonzero digits of x.

We refer to [8] for background in coding theory. Throughout this paper, by a code we shall mean a binary code, in the following sense:

Definition 3. *A* (binary) code C of length n *is a nonempty subset of* $\{0,1\}^n$. *Its elements are called codewords. The* minimum distance *of* C *is given by* $\delta(C) = \min\{d_H(x,y) \mid x,y \in C, x \neq y\}$. *We say that* C *is an* (n,m,d) *code iff* C *has length* n, $|C| = m$ *and* $\delta(C) = d$. *By definition, the* minimum weight *of* C *is the minimum of the Hamming weights of its codewords, in symbols,* $\mu(C) = \min\{w_H(x) \mid x \in C\}$. *The* minimum distance *between two codes* C_1 *and* C_2 *is defined by* $\Delta(C_1,C_2) = \min\{d_H(x,y) \mid x \in C_1, y \in C_2\}$.

Lemma 4. *Let* e,n,m_1,m_2 *be integers* > 0. *For each* $i = 1,2$ *suppose* C_i *to be an* (n,m_i,d_i) *code such that* $\mu(C_i) \geq e$, $d_i \geq 3$ *and* $\Delta(C_1,C_2) \geq 1$. *Then there exist two* $(n+2,m_1+m_2,d_i')$ *codes* $\mathcal{D}_i$ $(i=1,2)$ *such that* $d_i' \geq 3$, $\mu(\mathcal{D}_i) \geq e$ *and* $\Delta(\mathcal{D}_1,\mathcal{D}_2) \geq 1$.

Proof. Let us define [1]

$$\mathcal{D}_1 = C_1 \otimes (1,0) \cup C_2 \otimes (0,1) \qquad \mathcal{D}_2 = C_1 \otimes (0,0) \cup C_2 \otimes (1,1). \tag{6}$$

Obviously the length of $\mathcal{D}_1$ and $\mathcal{D}_2$ is $n+2$, and $\mu(\mathcal{D}_1), \mu(\mathcal{D}_2) \geq e$. Because of $C_1 \otimes (1,0) \cap C_2 \otimes (0,1) = \emptyset$ and $C_1 \otimes (0,0) \cap C_2 \otimes (1,1) = \emptyset$, we get

$$|\mathcal{D}_1| = |C_1 \otimes (1,0)| + |C_2 \otimes (0,1)| = m_1 + m_2 = |C_1 \otimes (0,0)| + |C_2 \otimes (1,1)| = |\mathcal{D}_2|.$$

In order to show that $\delta(\mathcal{D}_1) \geq 3$, note that any two codewords $x,y \in \mathcal{D}_1$ can be written as $x = (x_1,\ldots,x_n,a_1,a_2)$ and $y = (y_1,\ldots,y_n,b_1,b_2)$, for some codewords $(x_1,\ldots,x_n)$, $(y_1,\ldots,y_n) \in C_1 \cup C_2$ and (a_1,a_2), $(b_1,b_2) \in \{(1,0),(0,1)\}$. It follows that

$$d_H(x,y) = d_H((x_1,\ldots,x_n),(y_1,\ldots,y_n)) + d_H((a_1,a_2),(b_1,b_2)). \tag{7}$$

If either $(x_1,\ldots,x_n)$, $(y_1,\ldots,y_n) \in C_1$ or $(x_1,\ldots,x_n)$, $(y_1,\ldots,y_n) \in C_2$, then $(a_1,a_2) = (b_1,b_2)$, whence $d_H(x,y) = d_H((x_1,\ldots,x_n),(y_1,\ldots,y_n)) \geq 3$, because $\delta(C_1)$, $\delta(C_2) \geq 3$. If $(x_1,\ldots,x_n) \in C_1$ and $(y_1,\ldots,y_n) \in C_2$, then $(a_1,a_2) = (1,0)$ and $(b_1,b_2) = (0,1)$, whence $d_H((a_1,a_2),(b_1,b_2)) = 2$ and $d_H((x_1,\ldots,x_n),(y_1,\ldots,y_n)) \geq 1$, because $\Delta(C_1,C_2) \geq 1$. From (7) it follows that $d_H(x,y) \geq 3$. The case $(x_1,\ldots,x_n) \in C_2$, $(y_1,\ldots,y_n) \in C_1$ is symmetric. Thus $\delta(\mathcal{D}_1) \geq 3$, as required. A similar argument shows that $\delta(\mathcal{D}_2) \geq 3$.

To conclude we must show $\Delta(\mathcal{D}_1,\mathcal{D}_2) \geq 1$. For any $x \in D_1$ and $y \in D_2$ let us write $x = (x_1,\ldots,x_n,a_1,a_2)$ and $y = (y_1,\ldots,y_n,b_1,b_2)$, for some codewords $(x_1,\ldots,x_n)$, $(y_1,\ldots,y_n) \in C_1 \cup C_2$ and $(a_1,a_2) \in \{(1,0),(0,1)\}$, $(b_1,b_2) \in \{(0,0),(1,1)\}$. Then the desired result follows from $d_H(x,y) \geq d_H((a_1,a_2),(b_1,b_2)) = 1$.

---

[1] For any code C of length n, we denote with $C \otimes (a_1,\ldots,a_k)$ the code of length $n+k$ whose codewords are obtained by adding the suffix $(a_1,\ldots,a_k)$ to all codewords of C, i.e., $C \otimes (a_1,\ldots,a_k) = \{(x_1,\ldots,x_n,a_1,\ldots,a_k) \mid (x_1,\ldots,x_n) \in C\}$.

4 Perfect Strategies with Minimum Interaction

The first batch of questions

Recall that $q_2(m) = ch(2^m,0,0)$ is the smallest integer $q \geq 0$ such that $2^q \geq 2^m(\binom{q}{2} + q+1)$. By Lemma 1(ii), at least $q_2(m)$ questions are *necessary* for Paul to guess the unknown number $x_{Carole} \in S = \{0,1,\ldots,2^m-1\}$, if up to two answers may be erroneous. Aim of this paper is to prove that, conversely, (with the exception of $m=2$ and $m=4$) $q_2(m)$ questions are *sufficient* under the following constraint: Paul first sends to Carole a predetermined batch of m questions $D_1,\ldots,D_m$, and then, only depending on Carole's answers, he sends the remaining $q_2(m)-m$ questions in a second batch.

The *first batch of questions* is easily described, as follows: For each $i = 1,2,\ldots,m$, let $D_i \subseteq S$ denote the question "Is the ith binary digit of x_{Carole} equal to 1 ?" Thus a number $y \in S$ belongs to D_i iff the ith bit y_i of its binary expansion $y = y_1 \ldots y_m$ is equal to 1. Identifying $1 = yes$ and $0 = no$, let $b_i \in \{0,1\}$ be Carole's answer to question D_i. Let $b = b_1 \ldots b_m$. Repeated application of (1)-(2) beginning with the initial state $\sigma = (S,\emptyset,\emptyset)$, shows that Paul's state of knowledge as an effect of Carole's answers is a triplet $\sigma^b = (A_0,A_1,A_2)$, where

A_0 = the singleton containing the number whose binary expansion equals b

$A_1 = \{y \in S \mid d_H(y,b) = 1\}$

$A_2 = \{y \in S \mid d_H(y,b) = 2\}$.

By direct verification we have $|A_0| = 1$, $|A_1| = m$, $|A_2| = \binom{m}{2}$. Thus the state σ^b has type $(1,m,\binom{m}{2})$. As in (3), let σ_i be the state resulting after Carole's first i answers, beginning with $\sigma_0 = \sigma$. Since each question D_i is balanced for σ_{i-1}, an easy induction using Lemma 2 yields $ch(\sigma^b) = q_2(m) - m$.

The critical index m_n

For each m-tuple $b \in \{0,1\}^m$ of Carole's answers, we shall construct a nonadaptive strategy $\mathcal{S}_b$ with $ch(1,m,\binom{m}{2})$ questions, which turns out to be winning for the state σ^b. To this purpose, let us consider the values of $ch(1,m,\binom{m}{2})$ for $m \geq 1$. A direct computation yields $ch(1,1,0) = 4$, $ch(1,2,1) = 5$, $ch(1,3,3) = ch(1,4,6) = 6$, $ch(1,5,10) = \cdots = ch(1,8,28) = 7$, $ch(1,9,36) = \cdots = ch(1,14,91) = 8, \ldots$

Definition 5. *Let $n \geq 4$ be an arbitrary integer. The critical index m_n is the largest integer $m \geq 0$ such that $ch(1,m,\binom{m}{2}) = n$. Thus,*

$$ch\left(1,m_n,\binom{m_n}{2}\right) = n \text{ and } ch\left(1,m_n+1,\binom{m_n+1}{2}\right) > n. \tag{8}$$

Lemma 6. *Let $n \geq 4$ be an arbitrary integer.*

(i) If n is odd then $m_n = 2^{\frac{n+1}{2}} - n - 1$.

(ii) If n is even then, letting $m^* = \lfloor 2^{\frac{n+1}{2}} \rfloor - n - 1$, we either have $m_n = m^*$, or $m_n = m^* + 1$.

Proof. The case $n = 4$ is settled by direct verification, recalling that $m_4 = 1$. For $n \geq 5$ see [9, Lemma 4.2].

Strategies vs. Codes

Lemma 7. *Let $m = 1,2,3,\ldots$ and $n = ch(1,m,\binom{m}{2})$. Then the following conditions are equivalent:*

(a) For every *state $\sigma = (A_0,A_1,A_2)$ of type $(1,m,\binom{m}{2})$ there is a nonadaptive winning strategy for σ with n questions;*
(b) For some *state $\sigma = (A_0,A_1,A_2)$ of type $(1,m,\binom{m}{2})$ there is a nonadaptive winning strategy for σ with n questions;*
(c) For some integer $d \geq 3$ there exists an (n,m,d) code with minimum Hamming weight ≥ 4.

Proof. The implication (a) $\Rightarrow$ (b) is trivial. To prove (b) $\Rightarrow$ (c) assume $\sigma = (A_0,A_1,A_2)$ to be a state of type $(1,m,\binom{m}{2})$ having a nonadaptive winning strategy $\mathcal{S}$ with n questions $T_1,\ldots,T_n$. Let the map $z \in A_0 \cup A_1 \cup A_2 \mapsto z^{\mathcal{S}} \in \{0,1\}^n$ send each $z \in A_0 \cup A_1 \cup A_2$ into the n-tuple of bits $z^{\mathcal{S}} = z_1^{\mathcal{S}} \ldots z_n^{\mathcal{S}}$ arising from the sequence of "true" answers to the questions "does z belong to T_1 ?", "does z belong to T_2 ?", ..., "does z belong to T_n? " More precisely, for each $j = 1,\ldots,n$, $z_j^{\mathcal{S}} = 1$ iff $z \in T_j$. Let $C \subseteq \{0,1\}^n$ be the range of the map $z \mapsto z^{\mathcal{S}}$. By hypothesis, $A_0 = \{h\}$, for a unique element $h \in S$. Let $h^{\mathcal{S}}$ be the corresponding codeword in C. We shall first prove that, for some $d \geq 3$ the set $C_1 = \{y^{\mathcal{S}} \in C \mid y \in A_1\}$ is an (n,m,d) code, and for every $z \in A_1$ we have the inequality $d_H(z^{\mathcal{S}},h^{\mathcal{S}}) \geq 4$. The set C_1 will be finally used to build a code satisfying all conditions in (c).

Since $\mathcal{S}$ is winning, the map $z \mapsto z^{\mathcal{S}}$ is one-one, whence in particular $|C_1| = m$. By definition, C_1 is a subset of $\{0,1\}^n$.

Claim 1. $\delta(C_1) \geq 3$.

For otherwise (absurdum hypothesis) assuming c and d to be two distinct elements of A_1 such that $d_H(c^{\mathcal{S}},d^{\mathcal{S}}) \leq 2$, we will prove that $\mathcal{S}$ is not a winning strategy. We can safely assume $c_j^{\mathcal{S}} = d_j^{\mathcal{S}}$ for each $j = 1,\ldots,n-2$. Suppose Carole's secret number is equal to c. Suppose Carole's answer to question T_j is "yes" or "no" according as $c_j^{\mathcal{S}} = 1$ or $c_j^{\mathcal{S}} = 0$, respectively. Then after Carole's $n-2$ answers, Paul's state of knowledge has the form $\sigma' = (A'_0,A'_1,A'_2)$, with $\{c,d\} \subseteq A'_1$, whence the type of σ' is (a'_0,a'_1,a'_2) with $a'_1 \geq 2$. Since $ch(\sigma') \geq ch(0,2,0) = 3$, Lemma 1(ii) shows that the remaining two questions/answers will not suffice to reach a final state, thus contradicting the assumption that $\mathcal{S}$ is winning.

Claim 2. For each $y \in A_1$ we have the inequality $d_H(y^{\mathcal{S}},h^{\mathcal{S}}) \geq 4$.

For otherwise (absurdum hypothesis) let $y \in A_1$ be a counterexample, and $d_H(y^{\mathcal{S}},h^{\mathcal{S}}) \leq 3$. Writing $y^{\mathcal{S}} = y_1^{\mathcal{S}} \ldots y_n^{\mathcal{S}}$ and $h^{\mathcal{S}} = h_1^{\mathcal{S}} \ldots h_n^{\mathcal{S}}$, it is no loss of generality to assume $h_j^{\mathcal{S}} = y_j^{\mathcal{S}}$, for all $j = 1,\ldots,n-3$. Suppose Carole's secret number coincides with h. Suppose further that Carole's answer to question T_j is "yes" or "no" according as $h_j^{\mathcal{S}} = 1$ or $h_j^{\mathcal{S}} = 0$, respectively. Then Paul's state of knowledge after these answers will have the form $\sigma'' = (A''_0,A''_1,A''_2)$, where $A''_0 = \{h\}$, and $y \in A''_1$. Therefore, since $ch(\sigma'') \geq ch(1,1,0) = 4$, Lemma 1(ii) again shows that three more questions will not suffice to find the unknown number. This contradicts the assumption that $\mathcal{S}$ is a winning strategy.

Having thus settled our two claims, let $C_2 = \{y \oplus h \mid y \in C_1\}$, where $\oplus$ stands for bitwise sum modulo 2. For any two distinct codewords $a, b \in C_2$ we have $a = c \oplus h$ and $b = d \oplus h$, for uniquely determined elements $c, d \in C_1$. From Claim 1 we get $d_H(a,b) = d_H(c \oplus h, d \oplus h) = d_H(c,d) \geq 3$, whence $\delta(C_2) \geq 3$. Using the abbreviation $0 = \underbrace{0 \cdots 0}_{n \text{ times}}$, by Claim 2 we have

$$w_H(a) = d_H(a,0) = d_H(c \oplus h, h \oplus h) = d_H(c,h) \geq 4,$$

whence $\mu(C_2) \geq 4$. In conclusion, C_2 is an (n,m,d) code with $d \geq 3$ and $\mu(C_2) \geq 4$, as required.

(c) $\Rightarrow$ (a) Let C be an (n,m,d) code, with $d \geq 3$ and $\mu(C) \geq 4$. Let $\mathcal{H} \subseteq \{0,1\}^n$ be the union of the Hamming spheres of radius 1 centered at the codewords of C, together with the Hamming sphere of radius 2 centered at 0, in symbols, $\mathcal{H} = \mathcal{B}_2(0) \cup \bigcup_{x \in C} \mathcal{B}_1(x)$. Notice that $\mathcal{B}_2(0), \mathcal{B}_1(x), \mathcal{B}_1(y)$ are pairwise disjoint. Therefore $|\mathcal{H}| = \binom{n}{2} + n + 1 + m(n+1) = \binom{n}{2} + (m+1)(n+1)$.
Let $C_1 = \{0,1\}^n \setminus \mathcal{H}$. Then $|C_1| = 2^n - \binom{n}{2} - (m+1)(n+1) \geq \binom{m}{2}$. The second inequality follows for $2^n \geq \binom{n}{2} + n + 1 + m(n+1) + \binom{m}{2}$, since $n = ch(1, m, \binom{m}{2})$.
Let $\sigma = (A_0, A_1, A_2)$ be an arbitrary state of type $(1, m, \binom{m}{2})$, and write $A_0 = \{h\}$ for a unique element $h \in S$. Let us now fix, once and for all, two one-one maps $f_1 : A_1 \to C$ and $f_2 : A_2 \to C_1$. The existence of f_1 and f_2 is ensured by our assumptions about C, together with $|C_1| \geq \binom{m}{2}$.

Let the map $f : A_0 \cup A_1 \cup A_2 \to \{0,1\}^n$ be defined by cases as follows:

$$f(y) = \begin{cases} 0, & y \in A_0 \\ f_1(y), & y \in A_1 \\ f_2(y), & y \in A_2 \end{cases} \tag{9}$$

Note that f is one-one. For each $y \in A_0 \cup A_1 \cup A_2$ and $j = 1, \ldots, n$ let $f(y)_j$ be the jth bit of the binary vector corresponding to y via f.

Let the set $T_j \subseteq S$ be defined by $T_j = \{z \in S \mid f(z)_j = 1\}$, where $j = 1, \ldots, n$. This is Paul's *second batch of questions*. Intuitively, letting x_{Carole} denote Carole's secret number, T_j asks "is the jth bit of $f(x_{Carole})$ equal to one ?" Again writing *yes* $= 1$ and *no* $= 0$, Carole's answers to questions $T_1, \ldots, T_n$ determine an n-tuple of bits $a = a_1 \ldots a_n$.

We shall show that the sequence $T_1, \ldots, T_n$ yields a perfect nonadaptive winning strategy for σ. Let $\sigma_1 = \sigma^{a_1}$, $\sigma_2 = \sigma_1^{a_2}, \ldots, \sigma_n = \sigma_{n-1}^{a_n}$. Arguing by cases we shall show that $\sigma_n = (A_0^*, A_1^*, A_2^*)$ is a final state. By (1)-(2), any $z \in A_2$ that falsifies > 0 answers does not survive in σ_n—in the sense that $z \notin A_0^* \cup A_1^* \cup A_2^*$. Further, any $y \in A_1$ that falsifies > 1 answers does not survive in σ_n. And, of course, the same holds for h if it turns out to falsify > 2 answers.

Case 1. $a \notin \mathcal{B}_2(0) \cup f(A_2) \cup \bigcup_{y \in A_1} \mathcal{B}_1(f(y))$.

Then $h \notin A_0^* \cup A_1^* \cup A_2^*$. As a matter of fact, from $a \notin \mathcal{B}_2(0)$, it follows that $d_H(f(h), a) = d_H(0, a) > 2$, whence h falsifies > 2 of the answers to $T_1, \ldots, T_n$, and h does not survive in σ_n. Similarly, for each $y \in A_1$ we must have $y \notin A_0^* \cup A_1^* \cup A_2^*$. Indeed, the assumption $a \notin \mathcal{B}_1(f(y))$ implies $d_H(f(y), a) > 1$, whence y falsifies > 1 of the answers to

$T_1,\ldots,T_n$, and y does not survive in σ_n. Finally, for all $z \in A_2$ we have $z \notin A_0^* \cup A_1^* \cup A_2^*$, because the assumption $a \neq f(z)$ implies $d_H(f(z),a) > 0$, whence z falsifies at least one of the answers to $T_1,\ldots,T_n$, and z then does not survive in σ_n. We have proved that $A_0^* \cup A_1^* \cup A_2^*$ is empty, and σ_n is a final state.

Case 2. $a \in \mathcal{B}_2(0)$.

Then $h \in A_0^* \cup A_1^* \cup A_2^*$, because $d_H(f(h),a) = d_H(0,a) \leq 2$, whence h falsifies ≤ 2 answers. Our assumptions about C ensure that, for all $y \in A_1$, $a \notin \mathcal{B}_1(f(y))$. Thus, $d_H(f(y),a) > 1$ and y falsifies > 1 of the answers to $T_1,\ldots,T_n$, whence y does not survive in σ_n. This shows that $y \notin A_0^* \cup A_1^* \cup A_2^*$. A similar argument shows that, for all $z \in A_2$, $z \notin A_0^* \cup A_1^* \cup A_2^*$. Therefore, $A_0^* \cup A_1^* \cup A_2^*$ only contains the element h, and σ_n is a final state.

Case 3. $a \in \mathcal{B}_1(f(y))$ for some $y \in A_1$.

Arguing as in the previous cases, we see that y is an element of $A_0^* \cup A_1^* \cup A_2^*$, but neither h, nor any $z \in A_2$, nor any $y' \in A_1$ with $y' \neq y$ can be an element of $A_0^* \cup A_1^* \cup A_2^*$. Thus σ_n is a final state.

Case 4. $a = f(z)$ for some $z \in A_2$.

Then the same arguments show that $A_0^* \cup A_1^* \cup A_2^*$ only contains the element z, and σ_n is a final state.

Feasibility of the second batch of questions

Lemma 8. *For each integer $n \geq 7$ there exists an integer $d \geq 3$ and an (n, m_n, d) code C_n such that $\mu(C_n) \geq 4$.*

Proof. If $7 \leq n < 11$ then by direct inspection in [4, Table I-A], one can obtain an $(n, e_n, 4)$ code $\mathcal{A}_n$ such that $e_n \geq m_n$ and $\mu(\mathcal{A}_n) \geq 4$. It follows that every set $C_n \subseteq \mathcal{A}_n$ such that $|\, C_n \,| = m_n$, is an $(n, m_n, 3)$ code with the additional property $\mu(C_n) = 4$, as required. For $n = 11, 12$, by inspection in [4, Table I-A], one can build an (n, e_1, d_1) code $\mathcal{D}_{n,1}$ and an (n, e_2, d_2) code $\mathcal{D}_{n,2}$, such that $e_1, e_2 \geq 2^{\frac{n+1}{2}}$, $d_1, d_2 \geq 3$, $\mu(\mathcal{D}_{n,1}), \mu(\mathcal{D}_{n,2}) \geq 4$ and $\Delta(\mathcal{D}_{n,1}, \mathcal{D}_{n,2}) \geq 1$. Then Lemma 4 assures that such conditions hold for any $n \geq 11$. Upon noticing that $2^{\frac{n+1}{2}} > m_n$ (Lemma 6), the desired conclusion follows by simply picking a subcode $C_n \subseteq \mathcal{D}_{n,1}$, such that $|\, C_n \,| = m_n$.

Corollary 9. *For $m = 5, 6, 7, \ldots$ let σ be a state of type $(1, m, \binom{m}{2})$. Then there exists a nonadaptive winning strategy S for σ which is* perfect.

Proof. Let $n = ch(\sigma)$. From the assumption $m \geq 5$ we get $n \geq 7$. By Definition 5, $m \leq m_n$. By Lemma 8 there exists an (n, m_n, d) code C_n with $d \geq 3$ and $\mu(C_n) \geq 4$. Picking now a subcode $C_n' \subseteq C_n$ such that $|C_n'| = m$, and applying Lemma 7 we have the desired conclusion.

The remaining cases

We shall extend Corollary 9 to the case $m = 1$ and $m = 3$. Further, for $m = 2$ and $m = 4$ we shall prove that the shortest nonadaptive winning strategy for a state of type $(1, m, \binom{m}{2})$ requires $ch(1, m, \binom{m}{2}) + 1$ questions.

Proposition 10. *For each $m = 1,2,3,4$, let $\lambda(m)$ be the length of the shortest nonadaptive winning strategy for some (equivalently, for every) state of type $(1, m, \binom{m}{2})$. Then $\lambda(1) = 4$, $\lambda(2) = 6$, $\lambda(3) = 6$, $\lambda(4) = 7$. For $m \in \{1,3\}$ the number $\lambda(m)$ satisfies the condition $\lambda(m) = ch(1, m, \binom{m}{2})$.*

Proof. For $m = 1$ we have $\lambda(1) \geq ch(1,1,0) = 4$. Conversely, by Lemma 7, using the singleton code $\{1111\}$, we also get $\lambda(1) \leq 4$.
For $m = 2$, by [6, page 75-76], we have $\lambda(1) \geq 6$. On the other hand, taking the code $C = \{111100, 001111\}$ and using Lemma 7, we obtain $\lambda(2) \leq 6$.
For $m = 3$ we have $\lambda(3) \geq ch(1,3,3) = 6$. Conversely, using the code $\{111100, 110011, 001111\}$ and Lemma 7 we get $\lambda(3) \leq 6$.
Finally, let us consider the case $m = 4$. We have $\lambda(4) \geq ch(1,4,6) = 6$. In fact it can be proved that no $(6,4,3)$ code exists with minimum Hamming weight ≥ 4.

Indeed, write such a code as $C = C^{(4)} \cup C^{(5)} \cup C^{(6)}$, where $C^{(i)} = \{x \in C \mid w_H(x) = i\}$, Thus direct inspection shows that $|C^{(5)} \cup C^{(6)}| \leq 1$ and $|C^{(4)}| \leq 3$. Further, if $|C^{(4)}| = 3$ then $|C^{(5)} \cup C^{(6)}| = 0$. In conclusion $|C| \leq 3$.

This and Lemmas 7 are to the effect that $\lambda(4) \geq 7$. On the other hand, taking the $(7,4,3)$ code $\{1111000, 0001111, 0110011, 1111111\}$ and again using Lemma 7, we obtain $\lambda(4) \leq 7$, and the proof is complete.

Combining Corollary 9 and Proposition 10 we have

Theorem 11. *For each integer $m = 1,3,5,6,7,\ldots$ there is a strategy to guess a number $x \in \{0,\ldots,2^m - 1\}$ with up to two lies in the answers, using a first batch of m nonadaptive questions and then, only depending on the answers to these questions, a second batch of $ch(2^m,0,0) - m$ questions. In case $m \in \{2,4\}$ the shortest such strategy requires precisely $ch(2^m,0,0) + 1$ questions.*

As shown in [6], in the fully adaptive Ulam-Rényi game a perfect winning strategy exists if, and only if, $m \neq 2$.

5 Open Problems

We conclude this paper with two open questions.

1. Is minimum adaptiveness compatible with shortest strategies also when more than two tests may be erroneous?

2. Suppose answers to be bits transmitted, e.g., by a satellite via a noisy channel. Then our results have the following equivalent counterpart: (i) the satellite sends us the m-tuple x of bits of the binary expansion of x; (ii) we feed the received m-tuple x' back to the satellite, without distortion—e.g., via a noiseless feedback channel; (iii) only depending on x' the satellite sends us a final tip z of $q_2(m) - m$ bits, which is received by us as z', in such a way that from the $q_2(m)$ bits $x'z'$ we are able to recover x, even if one or two of the bits of $x'z'$ are the result of distortion. Is it possible to reduce the m bit feedback in our model to a shorter feedback, while still allowing the satellite to send only $q_2(m)$ bits ?

References

[1] M. Aigner, *Searching with lies*, J. Combin. Theory, Ser. A, **74** (1995), pp. 43-56.

[2] R. S. Borgstrom and S. Rao Kosaraju, *Comparison-based search in the presence of errors*, In: Proceedings of the 25th Annual ACM Symposium on the Theory of Computing, San Diego, California, 16-18 May 1993, pp. 130-136.

[3] E. R. Berlekamp, *Block coding for the binary symmetric channel with noiseless, delayless feedback*, In: Error-correcting Codes, H.B. Mann (Editor), Wiley, New York (1968), pp. 61-88.

[4] A. E. Brouwer, J. B. Shearer, N.J.A. Sloane, W. D. Smith, *A New Table of Constant Weight Codes*, IEEE Transactions on Information Theory, **36** (1990), pp. 1334-1380.

[5] F. Cicalese, *Q-ary searching with lies*, in Proceedings of the Sixth Italian Conference on Theoretical Computer Science, Prato, Italy, (U. Vaccaro et al., Editors), World Scientific, Singapore, 1998, pp. 228-240.

[6] J. Czyzowicz, D. Mundici, A. Pelc, *Ulam's searching game with lies*, J. Combin. Theory, Ser. A, **52** (1989), pp. 62-76.

[7] R. Hill, *Searching with lies*, In: Surveys in Combinatorics, Rowlinson, P. (Editor), Cambridge University Press (1995), pp. 41–70.

[8] F.J. MacWilliams, N.J.A. Sloane, *The Theory of Error-Correcting Codes*, North-Holland, Amsterdam, 1977.

[9] D. Mundici, A. Trombetta, *Optimal comparison strategies in Ulam's searching game with two errors*, Theoretical Computer Science, **182** (1997), pp. 217-232.

[10] A. Negro, M. Sereno, *Ulam's Searching game with three lies*, Advances in Applied Mathematics, **13** (1992), pp. 404-428.

[11] A. Pelc, *Solution of Ulam's problem on searching with a lie*, J. Combin. Theory, Ser. A, **44** (1987), pp. 129-142.

[12] A. Pelc, *Searching with permanently faulty tests*, Ars Combinatoria, **38** (1994), pp. 65-76.

[13] A. Rényi, *Napló az információelméletről*, Gondolat, Budapest, 1976. (English translation: *A Diary on Information Theory*, J.Wiley and Sons, New York, 1984).

[14] J. Spencer, *Ulam's searching game with a fixed number of lies*, Theoretical Computer Science, **95** (1992), pp. 307-321.

[15] S.M. Ulam, *Adventures of a Mathematician*, Scribner's, New York, 1976.

Improving Mergesort for Linked Lists *

Salvador Roura

Departament de Llenguatges i Sistemes Informàtics,
Universitat Politècnica de Catalunya,
E-08028 Barcelona, Catalonia, Spain,
roura@lsi.upc.es

Abstract. We present a highly tuned mergesort algorithm that improves the cost bounds when used to sort linked lists of elements. We provide empirical comparisons of our algorithm with other mergesort algorithms. The paper also illustrates the sort of techniques that allow to speed a divide-and-conquer algorithm.

1 Introduction

The main goal of this paper is to improve the asymptotic average-case cost M_n of mergesort under the usual assumption that the list to be sorted includes a random permutation of n different keys. Recall that mergesort is the default algorithm to sort a linked list of keys. The meaning of "cost" above depends on the quantities of interest. For instance, for most of the versions of mergesort that we present in this paper, measuring M_n as the number of key comparisons yields $M_n = n \log_2 n + o(n \log n)$, which is optimal w.r.t. the leading term of the number of comparisons. However, as we shall see hereafter, we may also consider other operations, like reading keys from memory, comparing pointers, reading pointers from memory, writing pointers into memory, accessing keys in local variables, updating keys in local variables, accessing pointers in local variables and updating pointers in local variables. Taking into account and reducing the contribution to the cost of all these operations produces a mergesort algorithm asymptotically *twice* faster (in many computers) than the first version we will start with. The author believes that many of the results presented in this paper are original and did not appear previously. For the improvements already in the literature, we provide references to books where those improvements can be found [1], [2], [5].

On the other hand, the paper also exemplifies the kind of arguments that allow to speed a divide-and-conquer algorithm, in particular which factors do affect the leading term of the cost and which are asymptotically dismissable. We illustrate how we can analyse with little effort the asymptotic expected number of the elemental operations mentioned above, always in order to reduce their contribution to the cost. In other words, we will not restrict ourselves to the typical approach in the analysis of sorting methods, which only considers the number of comparisons. Finally, we will show how such a "common sense" rule as "dividing the problem into two halves is the best choice for a sorting method" is by no means an unquestionable truth.

* This research was supported by the ESPRIT LTR Project no. 20244 —ALCOM-IT— and the CICYT Project TIC97-1475-CE.

```
typedef struct node *link;
struct node { long key; link next; };
extern link DUMMY;

/* The list pointed to by c has at least one element. */
link mergesort(link c)
 { link a, b;
   if (c->next == DUMMY) return c;
   a = c; b = c->next;
   while ((b != DUMMY) && (b->next != DUMMY))
     { c = c->next; b = b->next->next; }
   b = c->next; c->next = DUMMY;
   return merge(mergesort(a), mergesort(b));
 }
```

Fig. 1. The first mergesort algorithm

2 The First Mergesort Algorithm

Our starting point is the version of mergesort for linked lists given in Sedgewick's classic book [5, pages 355 and 356] (that algorithm is presented in Figures 1 and 2 for completeness). It has been slightly adapted from the original one, in order to make it directly comparable with the rest of versions in this paper. In particular, we assume that the keys in the nodes are long integers, and that the last node of every list does not store the pointer `NULL`, but instead a pointer `DUMMY` to a global node whose key is larger than any of the keys in the list.

To analyse the cost of that mergesort algorithm (and most of the ones we will present later) we need to solve recurrences with the pattern

$$M_n = t_n + M_{\lceil n/2 \rceil} + M_{\lfloor n/2 \rfloor} \quad (1)$$

for $n \geq 2$, with some explicitly known value for M_1. Here t_n is the toll function, or non-recursive cost to sort a list with n nodes, which includes the cost to divide the list into two sublists plus the cost to merge the two recursively sorted sublists together. The toll function will turn out to be linear on n for any of the variants of mergesort introduced in this paper, in other words, we will always have $t_n = B \cdot n + o(n)$ for some constant B that will depend on the particular variant of mergesort that we consider at every moment.

There are several different mathematical techniques to solve recurrences following Equation (1). Since we are interested in analysing (and improving) the leading term of the asymptotic cost, we will use the so called *master theorem*, which is probably the easiest of the tools that provide this information. For instance, a simple application of the master theorem given in [4, page 452] yields the general solution

$$M_n = B \cdot n \log_2 n + o(n \log n) \; . \quad (2)$$

Therefore, improving the asymptotic cost of mergesort is equivalent to reducing the constant B, which depends on the cost to divide the original list and merge the two sorted sublists.

```
link merge(link a, link b)
 { struct node head; link c = &head;
   while ((a != DUMMY) && (b != DUMMY))
     if (a->key <= b->key) { c->next = a; c = a; a = a->next; }
                      else { c->next = b; c = b; b = b->next; }
   c->next = (a == DUMMY) ? b : a;
   return head.next;
 }
```

Fig. 2. The first merge algorithm

In the computation of B we will consider nine different elemental operations: Comparing two keys, reading a key —the key field— from memory given a pointer to the node where it is stored, comparing a pointer against DUMMY, reading (writing) a pointer —the next field— from (into) memory given a pointer to the node where it is (must be) stored, accessing (updating) a key stored in a local variable, and accessing (updating) a pointer stored in a local variable. The cost of these operations will be denoted $\mathcal{C}_k$, $\mathcal{R}_k$, $\mathcal{C}_p$, $\mathcal{R}_p$, $\mathcal{W}_p$, $\mathcal{A}_k$, $\mathcal{U}_k$, $\mathcal{A}_p$ and $\mathcal{U}_p$, respectively. Since all the algorithms presented in this paper are written in C, we have to make (reasonable) assumptions about how the C instructions are translated in terms of the elemental operations above. If necessary, those assumptions (which will become clear from the examples below) could be changed and the analyses in this paper recomputed accordingly.

Let us analyse the cost of the merge function in Figure 2. The body of the while is executed about n times on the average, because at each iteration we move one step forward in either a or b, and when the first of a or b reaches its end the other list is almost empty with high probability. Loosely speaking, this is what the following proposition states.

Proposition 1. *Let ℓ and m denote the number of keys in* a *and the number of keys in* b*, respectively. Then the expected number of keys in* b *larger than any of the keys in* a *is $m/(\ell+1)$, and the expected number of keys in* a *larger than any of the keys in* b *is $\ell/(m+1)$.*

Proof. Every of the keys in a or b has the same probability to appear in a than to appear in b. Choose a key from b at random. Then the probability that this key is larger than any of the ℓ keys in a is clearly $1/(\ell+1)$, since this event is equivalent to randomly pick, from a set with $\ell+1$ keys, a key which turns out to be the largest of the set. This argument can be extended to each of the m keys in b, concluding that the expected number of keys larger than all the ℓ keys in a is $m/(\ell+1)$. A similar argument proves the symmetrical case. ■

Setting $\ell = \lceil n/2 \rceil$ and $m = \lfloor n/2 \rfloor$ in the proposition above yields $n - o(n)$ (more precisely, $n - \Theta(1)$) as the expected number of iterations of the loop in Figure 2. Moreover, the rest of operations (passing parameters and allowing space for local variables, the first assignment to c, the final assignment to c->next and returning a pointer to the sorted list) only contributes $\Theta(1)$ time to the merging process. Therefore, we focus our

```
/* The list c has n elements, where 1 <= t <= n. */
link mergesort(link c, long t, link *s)
 { link m1, m2, s_rec; long t_rec;
   if (t == 1) { *s = c->next; c->next = DUMMY; return c; }
   t_rec = (t+1)/2;
   m1 = mergesort(c, t_rec, &s_rec);
   m2 = mergesort(s_rec, t - t_rec, s);
   return merge(m1, m2);
 }
```

Fig. 3. The second mergesort algorithm

attention on the cost of each iteration: we twice access a pointer and compare it against DUMMY; we then read two keys from memory and compare them; we write a `next` field, access two pointers, update two pointers, and read a `next` field. The constant of the linear term in the cost to merge the sorted sublists is thus

$$2(\mathcal{A}_p + C_p) + (2\mathcal{R}_k + C_k) + (\mathcal{W}_p + 2\mathcal{A}_p + 2\mathcal{U}_p + \mathcal{R}_p)\,. \tag{3}$$

On the other hand, the body of the `while` in Figure 1 is executed $\lfloor (n-1)/2 \rfloor$ times, so by a reasoning similar to the one above we can compute the constant of the linear term in the cost to split the list into two sublists as

$$((\mathcal{A}_p + C_p) + (\mathcal{R}_p + C_p) + (2\mathcal{U}_p + 3\mathcal{R}_p))/2\,. \tag{4}$$

Altogether, adding the constants in (3) and (4), we conclude that the cost of the mergesort algorithm in Figure 1 (with the merge function in Figure 2) is

$$M_n \sim (C_k + 2\mathcal{R}_k + 3C_p + 3\mathcal{R}_p + \mathcal{W}_p + 9\mathcal{A}_p/2 + 3\mathcal{U}_p)\, n \log_2 n\,. \tag{5}$$

3 Improving Mergesort

In order to improve the mergesort algorithm in the last section, we first realise that we should pass to it as an argument the number of nodes in the (sub)list to be sorted. This would allow us to reduce from linear to *a constant* the cost to divide the list into two parts, thus making this cost asymptotically irrelevant (see [1, page 173]). The trick consists in adding a parameter to the mergesort procedure. Then it would receive as input two parameters: a pointer `c` to the beginning of a list, and an integer `t`. This mergesort procedure only sorts the first `t` nodes of the list `c`, and returns two pointers: one to the beginning of the now sorted sublist with `t` nodes, the other to the beginning of another sublist with the nodes not sorted yet. As can be seen in Figure 3, after sorting the first half of the list we already have a pointer to the beginning of the second half of the list (because to sort a sublist we certainly must reach its end at least once), and therefore we do not need to explicitly traverse the original list to be sorted, nor to its end not even to its middle.

```
link merge(link a, link b)
 { struct node head; link c = &head;
   while (b != DUMMY)
     if (a->key <= b->key) { c->next = a; c = a; a = a->next; }
                      else { c->next = b; c = b; b = b->next; }
   c->next = a;
   return head.next;
 }
```

Fig. 4. The second merge algorithm

By this easy improvement, the process of dividing the list into two parts is no longer asymptotically relevant to the cost of mergesort. The cost of the procedure in Figure 3 (with `t` $= n$) only depends significantly on the constant of the cost of the merging phase (3), and hence is

$$M_n \sim \left(C_k + 2\mathcal{R}_k + 2C_p + \mathcal{R}_p + \mathcal{W}_p + 4\mathcal{A}_p + 2\mathcal{U}_p\right) n \log_2 n \,. \tag{6}$$

¿From now on, we will assume that our mergesort algorithm is the one given in Figure 3, and we will concentrate on finding faster merge functions. All the merge functions presented here have been devised to work correctly if there were repeated keys, and, moreover, to preserve the relative order of equal keys in the list to be sorted. Therefore, all the versions of mergesort in this paper are stable.

The first improved version of merge is presented in Figure 4. The unique difference with the one given in Figure 2 is that the `while` only checks for `b` to equal `DUMMY`. This merge function works even when the sublist `a` reaches its end before the sublist `b`, because at that moment `a` equals `DUMMY`, a pointer to a global node with key larger than any of the keys in the list. Therefore, the comparison `(a->key <= b->key)` always fails, and we move forward in `b` until we reach its end.

The reason for the variation above is clear: under the assumption that we are given a random permutation of keys, by Proposition 1 the expected number of iterations of the loop is still $n - o(n)$, while the cost of each iteration reduces in $\mathcal{A}_p + C_p$, the cost of the comparison `(a == DUMMY)`. The cost of mergesort with this variant of merge is thus

$$M_n \sim \left(C_k + 2\mathcal{R}_k + C_p + \mathcal{R}_p + \mathcal{W}_p + 3\mathcal{A}_p + 2\mathcal{U}_p\right) n \log_2 n \,. \tag{7}$$

The next improvement comes from the observation that we are updating as many pointers as we are traversing, but roughly half of them are already pointing to the right node (this happens when two consecutive nodes in `a` or `b` also appear consecutively in the final list). Therefore, we modify the merge function to traverse but not update those pointers (see [2, page 167, exercise 15]). As it is shown in Figure 5, the main loop keeps the invariant `a->key` $\le$ `b->key`, where `a` is the node that must be attached next to `c`. In that loop, we first link `a` to `c`, updating conveniently `a` and `c`, and afterwards we keep traversing pointers in the list `a` until we reach a node such that `a->key` $>$ `b->key`. At that moment, the same process starts again, but interchanging the roles of `a` and `b`. The rest of the merge function is designed to fit the main `while`.

```
link merge(link a, link b)
 { struct node head; link c;
   if (a->key <= b->key) c = &head;
    else { head.next = b; c = b; b = b->next;
           while (a->key > b->key) { c = b; b = b->next; }
         }
   while (b != DUMMY)
    { c->next = a; c = a; a = a->next;
      while (a->key <= b->key) { c = a; a = a->next; }
      c->next = b; c = b; b = b->next;
      while (a->key > b->key) { c = b; b = b->next; }
    }
   c->next = a;
   return head.next;
 }
```

Fig. 5. The third merge algorithm

```
link mergesed3(link a, link b)
 { struct node head; link c; long key_a, key_b;
   if (a->key <= b->key) { head.next = a; key_b = b->key; }
    else { head.next = b; c = b; b = b->next;
           while (a->key > b->key) { c = b; b = b->next; }
           c->next = a; key_b = b->key;
         }
   while (b != DUMMY)
    { for(;;)
       { c = a->next;
         if ((key_a = c->key) > key_b) { a->next = b; a = c; break; }
         a = c->next;
         if ((key_a = a->key) > key_b) { c->next = b; break; }
       }
      for(;;)
       { c = b->next;
         if ((key_b = c->key) >= key_a) { b->next = a; b = c; break; }
         b = c->next;
         if ((key_b = b->key) >= key_a) { c->next = a; break; }
       }
    }
   return head.next;
 }
```

Fig. 6. The fourth merge algorithm

The analysis of the cost of this variant of merge is similar to the ones done before. The cost of all the instructions out of the main loop is again $\Theta(1)$. The main loop is

	500 keys	5000 keys	50000 keys	500000 keys
Figures 1 and 2	0.014548 s (100)	0.20727 s (100)	2.8083 s (100)	43.004 s (100)
Figures 1 and 4	0.013595 s (93.45)	0.19195 s (92.61)	2.5864 s (92.10)	40.555 s (94.31)
Figures 1 and 5	0.012492 s (85.87)	0.17574 s (84.79)	2.3820 s (84.82)	38.593 s (89.74)
Figures 1 and 6	0.011921 s (81.94)	0.16718 s (80.66)	2.3149 s (82.43)	38.010 s (88.39)
Figures 3 and 2	0.012039 s (82.75)	0.16235 s (78.33)	1.9635 s (69.92)	26.579 s (61.81)
Figures 3 and 4	0.011154 s (76.67)	0.14747 s (71.49)	1.7631 s (62.78)	24.446 s (56.85)
Figures 3 and 5	0.010074 s (69.25)	0.13171 s (63.55)	1.5482 s (55.13)	22.098 s (51.39)
Figures 3 and 6	0.009453 s (64.98)	0.12230 s (59.01)	1.4733 s (52.46)	21.590 s (50.21)

Fig. 7. Empirical times for different algorithms and list sizes

executed once for every subsequence with keys from b that appear, in the final list, after some key from a. But this quantity is easy to compute.

Proposition 2. *Let ℓ and m denote the number of keys in* a *and the number of keys in* b*, respectively. Then the expected number of keys from* b *which appear immediately after a key from* a *in the final list is $\ell \cdot m/(\ell+m)$.*

Proof. Consider the list once it is sorted. Each position of the list has a key from b with probability $m/(\ell+m)$, because a and b are random permutations. If this happens, each position (except the first one) has a key from a before it with probability $\ell/(\ell+m-1)$. The expected contribution of every position of the list but the first one to the quantity we are computing is thus $\ell \cdot m/(\ell+m)(\ell+m-1)$, and there are $(\ell+m-1)$ of them. ■

Therefore, the main loop is executed $n/4-o(n)$ times on the average (just plug $\ell = \lceil n/2 \rceil$ and $m = \lfloor n/2 \rfloor$ in the last proposition), and the first and third lines in it too. The comparison of the `while` in the second line inside the main loop is evaluated once for every key in a (except maybe for a few keys at the very end), which means $n/2-o(n)$ times on the average. The body of that `while` is executed the same number of times, minus one per each iteration of the main loop, that is, $(n/2-o(n))-(n/4-o(n)) = n/4+o(n)$ times on the average. Finally, the asymptotic cost of the fourth line is the same as that of the second one. Altogether, the constant B associated to this merge function is

$$(\mathcal{A}_p + \mathcal{C}_p)/4 + 2(\mathcal{W}_p + 2\mathcal{A}_p + 2\mathcal{U}_p + \mathcal{R}_p)/4 + 2(2\mathcal{R}_k + \mathcal{C}_k)/2 + 2(2\mathcal{U}_p + \mathcal{A}_p + \mathcal{R}_p)/4\,,$$

and the cost of this variant of mergesort

$$M_n \sim (\mathcal{C}_k + 2\mathcal{R}_k + \mathcal{C}_p/4 + \mathcal{R}_p + \mathcal{W}_p/2 + 7\mathcal{A}_p/4 + 2\mathcal{U}_p)\, n \log_2 n\,. \tag{8}$$

Let us speed a little more the last merge function. On the one hand, every key in `a` and `b` is read from memory twice on the average, which can be reduced to just once by storing the key at the beginning of the current `a` and the key at the beginning of the current `b` in a couple of local variables. On the other hand, while traversing pointers in, say `a`, we do `{ c = a; a = a->next; }`, but it would be cheaper to do `{ c = a->next; }` and `{ a = c->next; }` alternatively, keeping track of which of `a` or `c` points to the node with the key to be compared next, and which to the previous node (this is sort of bootstrapping). These improvements can be found in Figure 6.

The analysis of this variant of merge follows the same lines as those presented before. The comparison of `b` against DUMMY is executed $n/4+o(n)$ times on the average. The first `for` is related to the keys in `a`. For each of those keys (except maybe for a small number of keys at the end of `a`) we either make the assignment in the first line and the comparison in the second line, or the assignment in the third line and the comparison in the fourth line. The same is true regarding the second `for` and the keys in `b` (except maybe for a small number of keys at the beginning of `b`). Therefore, we can add $n-o(n)$ times the cost of one of those assignments and one of those comparisons to the cost of merging. For the moment we have a term

$$(\mathcal{A}_p+\mathcal{C}_p)/4+(\mathcal{U}_p+\mathcal{R}_p)+(\mathcal{R}_k+\mathcal{U}_k+\mathcal{A}_k+\mathcal{C}_k) \tag{9}$$

contributing to the constant B.

We still have left computing the contribution of the sentences between brackets in the second and fourth lines of both `for`'s. For instance, the set of sentences `{ a->next = b; a = c; break; }` is executed once for every key k from `a` which lies immediately before a key from `b` in the final list, if the subsequence of keys from `a` where k appears has an odd number of keys. It is difficult to compute an exact expression for this quantity. However, under the assumption that the number of keys in `a` is $p\cdot n+O(1)$ for some $0<p<1$, the following proposition (we omit its proof) gives us a useful asymptotic approximation.

Proposition 3. *Let $\ell = p\cdot n+O(1)\geq 0$ with $0<p<1$ be the number of keys in* `a`. *Then the expected number of odd-length subsequences with keys from* `a` *which appear before some key from* `b` *in the final list is* $(1-p)p/(1+p)\cdot n+o(n)$.

Very similar arguments and asymptotic approximations produce the quantities $(1-p)p^2/(1+p)\cdot n+o(n)$ as the expected number of executions of the sentences between brackets in the fourth line of the first `for`, $(1-p)p/(2-p)\cdot n+o(n)$ for the second line of the second `for`, and $(1-p)^2p/(2-p)\cdot n+o(n)$ for the fourth line of the second `for`. Their total contribution to the constant B is thus

$$2\big(1/6(\mathcal{W}_p+2\mathcal{A}_p+\mathcal{U}_p)+1/12(\mathcal{W}_p+\mathcal{A}_p)\big)\ , \tag{10}$$

and adding (9) and (10) we deduce that this version of mergesort has cost

$$M_n \sim \big(\mathcal{C}_k+\mathcal{R}_k+\mathcal{C}_p/4+\mathcal{R}_p+\mathcal{W}_p/2+\mathcal{A}_k+\mathcal{U}_k+13\mathcal{A}_p/12+4\mathcal{U}_p/3\big)n\log_2 n\,. \tag{11}$$

In Figure 7 we show empirical average times (in seconds) for all the possible combinations of the mergesort algorithms and merge functions given in this paper. Those times were obtained by sorting several times (10000, 1000, 100 and 10 times, respectively) random lists with 500, 5000, 50000 and 500000 keys with a PC. Between parentheses we provide the relative times (in percentage) w.r.t. the first algorithm. It should be clear from Figure 7 that the theoretical improvements do reduce significantly the actual time to sort.

The last version of mergesort is optimal regarding the number of comparisons, and, at the merging phase, reads from memory each key and each pointer at most once, updates only the necessary pointers, and compares as few pointers against DUMMY as possible. So we could wonder if we can improve mergesort further. Most surprisingly, there is an affirmative answer to this question (at least from a theoretical point of view).

For the moment we have assumed that dividing the list to be sorted into two halves is the most reasonable choice. However, we can easily compute the expected cost of mergesort if we divide the original list into two sublists with $\ell = p \cdot n + O(1)$ and $m = n - \ell$ keys each, with $0 < p < 1$ and $p \neq 1/2$. Making use of Propositions 1, 2 and 3, we deduce that the constant B of the toll function in this case is

$$\begin{aligned} B(p) = {} & (1-p)p(\mathcal{A}_p + C_p) + (\mathcal{U}_p + \mathcal{R}_p) + (\mathcal{R}_k + \mathcal{U}_k + \mathcal{A}_k + C_k) \\ & + (1-p)p/(1+p)(\mathcal{W}_p + 2\mathcal{A}_p + \mathcal{U}_p) + (1-p)p^2/(1+p)(\mathcal{W}_p + \mathcal{A}_p) \\ & + (1-p)p/(2-p)(\mathcal{W}_p + 2\mathcal{A}_p + \mathcal{U}_p) + (1-p)^2 p/(2-p)(\mathcal{W}_p + \mathcal{A}_p) \,. \end{aligned} \tag{12}$$

Applying the discrete master theorem [4, page 452], we get the solution

$$M_n = \frac{B(p) \cdot n \log_2 n}{\mathcal{H}(p)} + o(n \log n) \,, \tag{13}$$

where $\mathcal{H}(p) = -(p \log_2 p + (1-p)\log_2(1-p))$.

The question is which is the value of p that minimises the function $f(p) = B(p)/\mathcal{H}(p)$. In this paper, we only consider the most significant operations, related to the costs C_k, $\mathcal{R}_k$, C_p, $\mathcal{R}_p$ and $\mathcal{W}_p$. In this case we have $B(p) = (C_k + \mathcal{R}_k + \mathcal{R}_p) + (1-p)p(C_p + 2\mathcal{W}_p)$, and the analysis of the behaviour of $f(p)$ produces the surprising result that $p = 1/2$ is not always the optimal choice. Indeed, when $(C_p + 2\mathcal{W}_p) > \tau \cdot (C_k + \mathcal{R}_k + \mathcal{R}_p)$ for the threshold value $\tau = 4/(2\ln 2 - 1) \simeq 10.35480$, the function $f(p)$ achieves a local *maximum* at $p = 1/2$, and two symmetrical absolute minima at p^* and $1 - p^*$, where p^* is the unique solution of the equation

$$\frac{C_p + 2\mathcal{W}_p}{C_k + \mathcal{R}_k + \mathcal{R}_p} = \frac{\log_2 p - \log_2(1-p)}{(1-p)^2 \log_2(1-p) - p^2 \log_2 p} \tag{14}$$

in the interval $(0, 1/2)$. The intuition is that choosing $p \neq 1/2$ increases the number of key comparisons, read keys and read pointers, but can reduce the number of pointer comparisons and updated pointers. A similar result (and the same threshold value) was shown for quicksort when used to sort an array of keys [3]. Note that from a practical standpoint, it is not likely (at least under the current technology) that $(C_p + 2\mathcal{W}_p) > \tau \cdot (C_k + \mathcal{R}_k + \mathcal{R}_p)$ for the usual times of elemental operations.

Acknowledgements

The author thanks Conrado Martínez for suggesting to consider this problem, and Rafel Cases for his useful comments to previous drafts of this paper.

References

[1] G.H. Gonnet and R. Baeza-Yates. *Handbook of Algorithms and Data Structures - In Pascal and C*. Addison-Wesley, Reading, MA, 2nd edition, 1991.

[2] D.E. Knuth. *The Art of Computer Programming: Sorting and Searching*, volume 3. Addison-Wesley, Reading, MA, 2nd edition, 1998.

[3] C. Martínez and S. Roura. Optimal sampling strategies in quicksort. In Kim G. Larsen, Sven Skyum, and Glynn Winskel, editors, *Proc. of the 25th International Colloquium (ICALP-98)*, volume 1443 of *LNCS*, pages 327–338. Springer, 1998.

[4] S. Roura. An improved master theorem for divide-and-conquer recurrences. In Pierpaolo Degano, Roberto Gorrieri, and Alberto Marchetti-Spaccamela, editors, *Proc. of the 24th International Colloquium (ICALP-97)*, volume 1256 of *LNCS*, pages 449–459. Springer, 1997.

[5] R. Sedgewick. *Algorithms in C*. Addison-Wesley, 3rd edition, 1998.

Efficient Algorithms for On-Line Symbol Ranking Compression (Extended Abstract)

Giovanni Manzini[1,2]

[1] Dipartimento di Scienze e Tecnologie Avanzate, Università del Piemonte Orientale "Amedeo Avogadro", I-15100 Alessandria, Italy.
[2] Istituto di Matematica Computazionale, CNR, I-56126 Pisa, Italy.

Abstract. Symbol ranking compression algorithms are known to achieve a very good compression ratio. Off-line symbol ranking algorithms (e.g., bzip, szip) are currently the state of the art for lossless data compression because of their excellent compression/time trade-off.

Some on-line symbol ranking algorithms have been proposed in the past. They compress well but their slowness make them impractical. In this paper we design some fast on-line symbol ranking algorithms by fine tuning two data structures (skip lists and ternary trees) which are well known for their simplicity and efficiency.

1 Introduction

The introduction of the Burrows-Wheeler transform (BWT) [5, 13] has set a new standard in data compression algorithms. Algorithms based on the BWT have speed comparable to the parsing-based algorithms (such as gzip and pkzip), and achieve a compression close to the much slower PPM-based algorithms (for the results of extensive testing see [1, 8]). Two highly optimized compressors (bzip and szip) which are based on the BWT are now available for many platforms [15, 17]. The main drawback of these algorithms is that they are not on-line, that is, they must process the whole input (or a large portion of it) before a single output bit can be produced. Since in many applications (e.g., data transmission) this is not acceptable, several efforts [9, 12, 20] have been done to design on-line counterparts of BWT-based algorithms. Since BWT-based algorithms can be seen as *symbol ranking* algorithms, the new algorithms are usually referred to as on-line symbol ranking algorithms.

Compared to bzip (the best known BWT based algorithm) on-line algorithms achieve a slightly worse compression ratio. This was to be expected since on-line algorithms make decisions knowing only a part of the input. Unfortunately, the on-line algorithms described in [9, 12, 20] are also slower than bzip (of a factor 10 or more) and this makes them impractical.

In this paper we use two well known data structures: skip lists [11, 16] and ternary trees [2, 16] to design several efficient on-line symbol ranking algorithms. We show that after some fine tuning our algorithms outperform gzip (in terms of compression/speed trade-off) and, for some kind of input files, they are as fast as bzip.

size	context	predictions
3.	iss	{ _ }
2.	ss	{ _ }
1.	s	{ s, _, w }
0.	*nil*	{ s, i, m, _, w }

Fig. 1. Construction of the next symbol candidate list in the Howard-Vitter symbol ranking algorithm with $c = 3$. Assume the input string is $s =$ **swiss_miss_is_missing**, and that we are coding the last i. For $i = 3,2,1,0$, we show the context of size i and the list of symbols following this context ordered by recency of occurrence. Note that $i = 0$ correspond to the empty context. In this case the prediction is done considering all symbols seen so far ordered by recency of occurrence. The list of candidate next symbol is obtained merging the predictions of each context removing duplicates. In our example we get {_, s, w, i, m} so the incoming symbol i is encoded with the rank 3.

2 Symbol ranking compression

Loosely speaking an on-line symbol ranking compression algorithm works as follows. Let s' denote the portion of the input string which has been already coded. On the basis of s' statistics, the algorithm builds a list of candidate next symbol ranked by likelihood of occurrence. That is, the most likely symbol is put in position 0, the second most likely in position 1, and so on. The incoming symbol, that is the symbol following s', is then encoded with its position in this list. These encoded values, which hopefully consist mainly of small integers, are then compressed using on-line (one pass) Huffman coding [18] or arithmetic coding [19].

The first symbol ranking algorithm proposed in the literature is the one by Howard and Vitter [10]. Given s', this algorithm considers its suffix of size c, $w = \beta_1 \cdots \beta_c$, where $c > 0$ is an assigned constant. The string w is called the the size c *context* of the incoming symbol. The list of candidate next symbol is built by looking at the previous occurrences of w inside s' and choosing the symbols following these occurrences. The process is repeated with the context of size $c-1$, $w' = \beta_2 \cdots \beta_c$, the context of size $c-2$, and so on (the details are given in Fig. 1).

Recently, other on-line symbol ranking algorithms have been proposed [9, 12, 20]. These algorithms build the candidate list using strategies similar to the one of Fig. 1. The main difference is that, instead of considering the context of a fixed size c, they look for the longest suffix of s' which appears elsewhere in s' and use it as the starting context. In other words, they use the longest context which can be used to get a prediction for the incoming symbol.

These algorithms are considered the on-line counterparts of the BWT-based algorithms. In fact, as observed for example in [6, 8], BWT-based algorithms use essentially the same principle, the only difference being that they make predictions on the basis of the whole input string. This is possible because they look at the whole input before the actual coding is done.

On-line symbol ranking algorithms can be seen also as simplified versions of PPM algorithms [6, 7]. For each context PPM algorithms maintain the list of candidate next symbol enriched with a frequency count for each symbol. The incoming symbol is encoded using roughly $-\log p$ bits, where p is the "empirical probability" of the incoming symbol determined on the basis of the frequency counts.

Not surprisingly, on-line symbol ranking algorithms achieve a compression ratio which is not as good as that of BWT-based algorithms or PPM algorithms, but they are generally superior to the dictionary based compressors (such as compress, and gzip). Unfortunately, on-line symbol ranking algorithms are not competitive in terms of running time. Howard and Vitter [10] develop fast procedures for encoding the ranks, but they do not provide enough data to assess the overall performance of their algorithms. Yokoo [20] reports that his algorithm is twenty times slower than gzip. For the algorithms in [12] and [9] we can only estimate, on the basis of the reported running times, that they are respectively fifteen times and seventy times slower than gzip.

In the next section we face the problem of designing efficient on-line symbol ranking algorithms. We concentrate our efforts in the design of fast procedures for determining the rank of the incoming symbol. In the past this was done using a variety of data structures: a binary tree with additional links in [20], suffix trees in [12], hashing in [9], and a multiply linked list in [10]. In our algorithms we use skip lists [11], and ternary trees [2], improved with some ad hoc modifications.

Note that our work is complementary to the one of Howard and Vitter [10]. They use a standard data structure for determining the ranks and develop fast procedures for encoding them (quasi-arithmetic coding and Rice coding). On the contrary, we develop new data structures for fast determination of the ranks and encode them using, as a black box, the CACM '87 arithmetic coding routines [19]. Finally, we stress that arithmetic coding has some latency, that is, it sometimes requires several input symbols before emitting some output. Hence, our algorithms are not strictly on-line[1]. If one wants a completely on-line algorithm the final encoding should be done using Huffman coding (possibly with multiple tables [17] to improve compression).

3 Design of efficient symbol ranking algorithms

In this section we describe several on-line symbol ranking algorithms. Our emphasis is on the design of efficient procedures for determining the rank of the incoming symbol. In all our algorithms we compress the sequence of ranks by means of arithmetic coding but we did not try to optimize this step.

Let s denote the whole input string and let s' denote the portion of the input which has been already coded. Our first design decision was to consider the suffixes of s' (which are the contexts of the incoming symbol) only up to a fixed maximum size $c > 0$. This is the approach of Howard and Vitter [10] (see Fig. 1). We decided to follow it, instead of the "unbounded context" approach of the more recent algorithms, for two reasons. The first one is that a bounded context leads to more efficient algorithms, and the second is that the results in [14] show that (for BWT-based algorithms) even a small context yields a very good compression.

Our basic strategy for determining the rank of the incoming symbol works as follows. We store all c-length contexts (that is, all c-length substrings) which have ap-

[1] Note, however, that the latency of arithmetic coding is negligible compared to the one of the Burrows-Wheeler algorithm.

peared in s'. For each context we maintain the list of the symbols following it ordered by recency of occurrence. This is achieved using the *Move-to-Front* strategy [4]: when we encounter the context w followed by the symbol α, we put α at the top of the list associated to w. Because of this update strategy we call this list the *MTF list* associated to w. MTF lists are used to determine the rank of the incoming symbol as follows. If the incoming symbol α is in the MTF list of the current context w its rank is simply its position in this list. This is by far the most common case. However, at the first occurrence of $w\alpha$ in s, α *will not* be in w's MTF list. In this case we say that α is "new" and its rank is determined by considering either a shorter context or those c-length contexts which are similar to w. We stress that the candidate next symbol list mentioned in the previous section was introduced only to illustrate the working of symbol ranking algorithms but we do not need to actually build it. Our algorithms build (implicitly) only its upper part, that is, the minimum portion required to determine the rank of the incoming symbol.

The algorithms described in the following sections differ in the strategy for encoding "new" symbols and in the data structure we use for maintaining the c-length substrings of s. Note that this data structure should perform efficiently a single operation: to locate a string which has been already inserted and to insert it if not present. We call this a `find/insert` operation.

3.1 Skip list based algorithms

In our first algorithm we maintain the ordered list of contexts (c-length substrings) using a skip list. The skip list [11] is a probabilistic data structure which has simple and efficient algorithms for the `find/insert` operation whose cost is, with high probability, logarithmic in the number of list elements. In order to use a skip list each context w must be converted to an integer: this is easily done by juxtaposing the bit representation of the symbols in w. Since the GNU C compiler supports 64 bit long integers, with a single value we can handle contexts of size up to 8. This is more than adequate since in our experiments we found that increasing the context beyond 6 increases significantly the memory usage without providing a noticeable improvement in the compression ratio (this is in accordance with the results in [14]).

The conversion from a string $w = \beta_1 \cdots \beta_c$ to an integer N_w is done "reversing" the symbols in w. That is, the most significant bits of N_w are those of β_c, while the less significant bits are those of β_1. As a result, in the skip list the context are sorted in right to left lexicographic order. This does not affect the performance of the `find/insert` operation, but ensures that the contexts close to w in the skip list, are (in general) similar to w (since they usually have a common suffix). Therefore, each time the incoming symbol α is "new", we compute its rank considering the MTF lists of the elements adjacent to w in the skip list. The details of the algorithm are best described with an example and are given in Fig. 2. Note that in [11] skip list elements have only forward pointers; since we need to access the contexts preceding and following the current context, we must add a backward pointer to each skip list element.

The MTF lists are implemented using an array of `char`'s (8-bit variables). This implementation requires very little memory, but has the drawback that we must examine all symbols in the MTF list until we find the desired one. In addition, when a new symbol must be inserted the whole list must be copied to a new (larger) array. Profiling

```
5. s_ ⟶ {m, i}
8. _i ⟶ {s}
7. mi ⟶ {s}
2. wi ⟶ {s}
6. _m ⟶ {i}
3. is ⟶ {_, s}
4. ss ⟶ {i, _}
1. sw ⟶ {i}
```

Fig. 2. Example of a symbol ranking algorithm based on a skip list. Assume that $c = 2$, the input string is $s =$ **swiss_miss_is_missing**, and that we are coding the letter n. Eight distinct contexts have been encountered so far; in the skip list they are ordered in right to left lexicographic order. The numbers on the left denote the order in which contexts have been inserted in the list. The rightmost column show the MTF list for each context. The current context is si; it will be inserted between mi and wi. Our algorithm will search the incoming symbol n in the MTF lists according to the following order: si, wi, mi, _m, _i, and so on. In this specific example the incoming symbol does not belong to any of these lists. The important point is that the contexts "similar" to si (those ending in i) are considered first. After the coding, the symbol n is added in the MTF list of the newly created context si; the other MTF lists are not modified.

```
5. s_ ⟶ {(m,6), (i,8)}
8. _i ⟶ {(s,3)}
7. mi ⟶ {(s,3)}
2. wi ⟶ {(s,3)}
6. _m ⟶ {(i,7)}
3. is ⟶ {(_,5), (s,4)}
4. ss ⟶ {(i,9) (_,5)}
1. sw ⟶ {(i,2)}
```

Fig. 3. The example of Fig. 2 with the MTF pointers added. If at any time during the coding we encounter the context is followed by the symbol s, we reach the skip list element corresponding to the next context (ss) following the pointer *4* which is associated to s in the is MTF list.

shows that these drawbacks have little impact in practice. This is probably due to the fact that the input of data compression algorithms are (usually) files with a strong structure. Experiments show that when we compress a (large) text file using a context of size 4, the incoming symbol α is the first symbol of the MTF list more than half of the times. For these reasons we did not implement the data structure for MTF lists described in [3] which is asymptotically faster but require a larger amount of memory.

The algorithm we have just described is already reasonably efficient. We can make it faster by storing some additional information in our data structure. Profiling shows that the find/insert operation takes up to the 35% of the total running time. We observe that the current context $w = \beta_1 \cdots \beta_c$ and the next symbol α completely determine the next context (which is in fact $\beta_2 \cdots \beta_c \alpha$). Hence, we can associate to each symbol in the MTF list a pointer to the skip list element corresponding to the next context. We call these additional pointers *MTF pointers* (see Fig. 3).

The use of MTF pointers increases the cost of updating the MTF list (symbols and pointers must be moved in lockstep), but this is (usually) more than compensated by the reduced number of `find/insert` operations. A big drawback of MTF pointers is that they increase significantly the space requirement of the algorithm. For this reason we have tested a variant in which we store the MTF pointer only for the first element of each MTF list (this is the rank 0 symbol according to the current context). It turns out that this new variant is as fast as the previous one and requires very little additional memory compared to the algorithm without MTF pointers (see Table 1).

We have called the variant which uses only rank 0 pointers algorithm Sr_sl and we have tested it using files of the Canterbury corpus. The results are reported in Section 4.

3.2 Ternary tree based algorithms

We have considered ternary trees as an alternative data structure for maintaining the set of contexts (the c-length substrings of the input string). We use a straightforward modification[2] of the algorithms described in [2]. The cost of the `find/insert` operation on ternary trees depends on the order in which the strings are inserted and is difficult to express analytically. In practice, the results in [2] show that this is a very efficient data structure.

The nodes of the ternary tree corresponding to c-length substrings have a pointer to a MTF list which is handled in the same way as in Section 3.1. The use of ternary trees instead of skip lists yields a substantial difference in the strategy for coding "new" symbols. Using ternary trees we no longer have an immediate access to the contexts which are adjacent (in lexicographic order) to the current context w. However, we can still access contexts which are "close" to w. Our idea consists in reversing each context before inserting it in the ternary tree (that is, if $w = \beta_1 \cdots \beta_c$ we start from the root searching the node corresponding to β_c, then we search for β_{c-1} and so on). In addition, each time we `find/insert` a context w we maintain the path $\pi_w = (n_0, n_1, \dots, n_k)$ going from n_0 (the root) to n_k (the node corresponding to w). If the incoming symbol α is not in the MTF list of w, we search it in the MTF lists of the contexts which can be found in the subtree rooted at n_{k-1} (excluding n_k). If α is not found there, we search it in the MTF lists of the contexts which can be found in the subtree rooted at n_{k-2} (excluding n_{k-1}), and so on.

MTF pointers can be also used in conjunction with ternary trees (MTF pointers now point to tree nodes corresponding to c-length substring). Again, we get the best performance by storing only the rank 0 pointers, that is the MTF pointers corresponding to the

[2] Bentley and Sedgewik consider null terminated strings of different lengths, whereas we need to store arbitrary strings of length c.

	# `find/insert`	Memory	Time
No MTF pointers	1,383,612	3.33	29.07
All MTF pointers	216,890	6.33	23.13
Rank 0 pointers	614,603	3.77	22.03

Table 1. Number of `find/insert` operations, memory usage (in megabytes) and running time (in seconds) for three different schemes of MTF pointers usage. The input file was *texbook.tex* (1,383,616 bytes) with context size $c = 4$.

first element of each MTF list. A new difficulty unfortunately arises when we use these pointers. Suppose that the symbol α is "new" for the context w. Our strategy requires that we search α in the MTF lists of context which are "close" to w. However, if the node corresponding to w has been obtained following an MTF pointer, rather than performing a tree search, we *do not* have the path $\pi_w = (n_0, n_1, \ldots, n_k)$ which we ordinarily use to find contexts similar to w. Initially, we solved this problem adding, to each node of the tree, a pointer to its parent node so that the path π_w could be reconstructed. However, since this causes a noticeable increase of memory usage, we decided to remove this pointer and to execute an additional tree search in the unfortunate case that a "new" symbol appears after a MTF pointers has been used. Profiling shows that this additional search are seldom necessary and that they do not affect significantly the running time.

The symbol ranking algorithm based on ternary trees and rank 0 pointers has been tested in Section 4 under the name Sr_tt.

3.3 Improved algorithms

Profiling shows that when we use skip lists or ternary trees combined with MTF pointers the most expensive operation of symbol ranking algorithms is the coding of "new" symbols. For most files, such symbols appear rather infrequently, however, since their coding requires the scanning of many MTF lists, their effect is noticeable. In this section we describe three new algorithms designed to cut down the cost of coding "new" symbols.

In the first algorithm we set an upper limit to the amount of search we perform for each "new" symbol. The algorithm receives an input parameter m, when a "new" symbol α is not among the first m symbols of the candidate next symbol list the algorithm outputs a `new_symbol` code followed by α. Clearly, this strategy reduces the running time at the expense of compression, and we must face the choice of an appropriate parameter m. After a few experiments we found that $m = 32$ represents a good compromise between speed and compression. We have applied this technique to Sr_tt and the resulting algorithm has been tested in Section 4 under the name Sr_tt_32.

We have also implemented a symbol ranking algorithm which closely follows the framework described in Section 2. In this algorithm "new" symbols are encoded using contexts of size $c-1, c-2, \ldots, 0$. To this end we maintain a MTF list for *every context* of size less or equal to c. If the symbol α has never appeared after the context $w = \beta_1 \cdots \beta_c$, we search it in the MTF list of the context $\beta_2 \cdots \beta_c$, and so on up to the MTF list of the empty context (note that these "lower order" contexts are precisely the suffixes of w). Since the algorithm examines at most $c+1$ lists the code of "new" symbols is significantly faster. However, we now have the problem of maintaining the MTF lists of the smaller contexts. Ideally, we should update these lists at each step, which means $c+1$ updates for input symbol. However, we found that updating only the lists which has been actually searched has a much better compression/time trade-off (in other words, if α is found in the MTF list of the context $\beta_2 \cdots \beta_c$, we do not update the list of the contexts of size $0, 1, \cdots, c-2$. This technique is commonly used in PPM compression algorithms).

The MTF lists of the "lower order" contexts can be easily accessed if we store the c-length contexts using a ternary tree. In fact, every node of the tree corresponds to a string of length l, $1 \leq l \leq c$, and the suffixes of a string w correspond to nodes which are in the path from the root to the node corresponding to w. When we `find/insert` the current context w, we keep track of these nodes so that if the incoming symbol turns out to be "new" we already have the relevant MTF lists at hand. We have implemented the above algorithm, with the addition of the rank 0 MTF pointers, and called it Sr_tt_fat (because of the extra memory it requires). We have tested it using files of the Canterbury corpus and the results are reported in Section 4.

To overcome the drawback of the extra memory required by Sr_tt_fat, we have devised a simple variant which is more space economical. In this variant we maintain MTF lists only for contexts of size 0, 1 and 2 (in addition to the MTF lists for the full c-length contexts). The elimination of the "middle-level" MTF lists clearly results in a saving of space. In addition, we can expect that the coding of "new" symbols should take less time since we search them directly in MTF lists of lower order in which they are more likely to be found. This variant can be implemented using either ternary trees or skip lists (since there is a small number of lower order MTF lists they are accessed directly by table look-up rather than through the data structure used for the contexts as in algorithm Sr_tt_fat). We decided to use ternary trees with an additional variant which further reduces the cost of `find/insert` operations. Since we no longer need to access the MTF lists of contexts which are similar to the current one, in the ternary tree we insert each context in its natural order (that is, if the context is $w = \beta_1 \cdots \beta_c$ we start from the root searching the node corresponding to β_1, then we search for β_2 and so on). At each leaf, in addition to the MTF list and to the rank 0 MTF pointer, we store a *suffix pointer*. The suffix pointer goes from the leaf corresponding to $\beta_1 \cdots \beta_c$ to the $(c-1)$-level node corresponding to $\beta_2 \cdots \beta_c$. Since the context following $\beta_1 \cdots \beta_c$ will be of the form $\beta_2 \cdots \beta_c \alpha$, the suffix pointer enables us to skip $c-1$ "levels" in the next search in the ternary tree. Note that, differently from the MTF pointers, once the suffix pointer is established it never changes. Since this algorithm uses the contexts in their natural order we have called it Sr_tt_nat. Its performance on the files of the Canterbury corpus are reported in Section 4.

4 Experimental results

Since our main interest is in the design of on-line symbol ranking algorithms which are time-efficient, we have tested our algorithms using the five largest files of the Canterbury corpus [1] (for small files the start-up overheads can dominate the compression time). In the full paper we will report the results for the whole Canterbury corpus in order to better assess the compression performance of our algorithms on different kinds of input files. The files considered here have the following characteristics.

fax (513,216 bytes). Black and white bitmap of an image belonging to the CCITT test set.
excl (1,029,744 bytes). Excel spreadsheet.
ecol (4,638,690 bytes). Complete genome of the E. Coli bacterium (a long sequence of A, C, T, G characters for the rest of us).
bibl (4,047,392 bytes). The King James version of the Bible.
wrld (2,473,400 bytes). The 1992 CIA world fact book.

Tables 2, 3, 4, report the performance of the algorithms described in the previous section. For a comparison we report data also for the algorithms gzip (with option -9 for maximum compression) and bzip2 (with option -1 for maximum speed, and option -9 for maximum compression).

Files:		fax	excl	ecol	bibl	wrld
Sr_sl	$c=4$	1.17	0.85	2.02	2.22	2.34
Sr_sl	$c=5$	1.16	0.82	2.02	2.11	2.13
Sr_sl	$c=6$	1.16	1.14	2.02	2.05	2.04
Sr_tt	$c=4$	1.16	0.82	2.02	2.21	2.33
Sr_tt	$c=5$	1.15	0.78	2.02	2.10	2.12
Sr_tt	$c=6$	1.15	0.98	2.02	2.04	2.02
Sr_tt_32	$c=4$	1.21	2.00	2.02	2.22	2.39
Sr_tt_32	$c=5$	1.20	1.98	2.02	2.12	2.18
Sr_tt_32	$c=6$	1.19	1.99	2.02	2.06	2.08
Sr_tt_fat	$c=4$	1.17	0.81	2.02	2.22	2.35
Sr_tt_fat	$c=5$	1.16	0.77	2.02	2.12	2.14
Sr_tt_fat	$c=6$	1.17	0.83	2.02	2.07	2.05
Sr_tt_nat	$c=4$	1.16	0.81	2.02	2.23	2.36
Sr_tt_nat	$c=5$	1.15	0.78	2.02	2.14	2.19
Sr_tt_nat	$c=6$	1.14	0.84	2.02	2.13	2.17
gzip -9		0.82	1.63	2.24	2.33	2.33
bzip2 -1		0.78	0.96	2.17	1.98	2.17
bzip2 -9		0.78	1.01	2.16	1.67	1.58

Table 2. Compression in bits per symbol (output bits over input bytes). A smaller number denotes better performance.

Due to the lack of space we cannot comment at length on the performance of the single algorithms. However, we can clearly see that the structure of the input file influences not only the compression ratio but also the running time of the algorithms. Consider for example the file *excl*. All algorithms, with the exception of Sr_tt_32, achieve a very good compression (less than 1 bit per symbol) but the running times for gzip and all algorithms using ternary trees are unusually high (note that the best compression is achieved by Sr_tt_fat). The results for *ecol* are also worth commenting. Since the file contains only four distinct symbols, one would expect a compression of at least two bits for input symbol. Our algorithms come very close to this value, but this is not true for either gzip (which also has a high running time) or bzip2. Finally, it must be noted that bzip2 preprocesses the input file using run-length encoding. This technique affects the compression ratio of the files containing long runs of identical symbols (*fax* and *excl* in our test set).

Overall, we can see that bzip2 achieves the best compression/speed trade-off. Its running time is also quite "stable", that is, it is little influenced by the structure of the input. Our symbol ranking algorithms usually compress better than gzip and are on a

Files:		fax	excl	ecol	bibl	wrld
Sr_sl	$c=4$	15.09	26.83	4.41	8.02	12.84
Sr_sl	$c=5$	19.80	33.77	5.45	11.04	18.53
Sr_sl	$c=6$	22.15	43.53	7.13	13.75	23.90
Sr_tt	$c=4$	16.05	228.16	3.93	5.43	10.71
Sr_tt	$c=5$	23.12	310.57	4.39	7.19	16.39
Sr_tt	$c=6$	29.43	395.44	4.79	10.32	23.59
Sr_tt_32	$c=4$	7.01	42.02	4.06	5.19	7.99
Sr_tt_32	$c=5$	9.03	57.24	4.44	6.75	11.26
Sr_tt_32	$c=6$	10.94	71.70	4.94	9.34	15.44
Sr_tt_fat	$c=4$	4.28	19.74	4.64	5.29	6.29
Sr_tt_fat	$c=5$	4.71	29.24	5.12	5.93	7.15
Sr_tt_fat	$c=6$	5.25	38.79	5.70	6.91	8.44
Sr_tt_nat	$c=4$	3.75	10.89	3.66	4.43	5.24
Sr_tt_nat	$c=5$	4.07	13.66	3.85	4.76	5.73
Sr_tt_nat	$c=6$	4.28	17.34	4.03	5.31	6.60
gzip -9		6.43	40.27	27.72	6.84	3.71
bzip2 -1		2.03	5.01	5.47	5.19	5.31
bzip2 -9		2.03	6.56	6.54	6.01	9.84

Table 3. Running time in milliseconds per input byte.

Files:		fax	excl	ecol	bibl	wrld
Sr_sl	$c=4$	2.45	5.57	0.01	0.36	1.56
Sr_sl	$c=5$	3.77	9.17	0.01	1.18	3.86
Sr_sl	$c=6$	4.36	12.33	0.04	2.64	6.15
Sr_tt	$c=4$	2.50	5.61	0.00	0.32	1.40
Sr_tt	$c=5$	4.36	10.17	0.01	1.02	3.57
Sr_tt	$c=6$	6.32	16.14	0.03	2.50	6.75
Sr_tt_fat	$c=4$	2.95	6.67	0.00	0.35	1.56
Sr_tt_fat	$c=5$	5.53	12.98	0.01	1.16	4.20
Sr_tt_fat	$c=6$	8.47	21.37	0.03	2.97	8.40
Sr_tt_nat	$c=4$	2.87	6.42	0.00	0.36	1.60
Sr_tt_nat	$c=5$	4.83	11.29	0.01	1.14	4.00
Sr_tt_nat	$c=6$	6.86	17.60	0.03	2.78	7.41
bzip2 -1		2.16	1.08	0.24	0.27	0.45
bzip2 -9		7.80	6.52	1.45	1.66	2.71

Table 4. Memory usage per input byte. The memory usage of Sr_tt_32 is the same of Sr_tt.

par with bzip2 -1 The improved algorithms described in Section 3.3 are usually faster than gzip, and are as fast as bzip2 for the text files (*bibl* and *wrld*) and the genome sequence (*ecol*).

5 Conclusions

The concept of symbol ranking is an important one in the field of data compression. The off-line symbol ranking algorithms (such as bzip and szip) constitute the state of the art in lossless data compression. In this paper we have shown that also on-line symbol ranking algorithms can have a very good compression/speed trade-off. Since symbol ranking algorithms are still in their infancy we can expect further advancements in the future.

More in general, our feeling is that data compression has recently seen the developments of new powerful compression techniques (see [13] for a recent review) but in many cases new ideas have been implemented with the wrong tools. We believe that the data structures described in this paper, as well as others well known in the algorithmic community, can be used to design more efficient compression algorithms.

References

[1] R. Arnold and T. Bell. The Canterbury corpus home page. `http://corpus.canterbury.ac.nz`.

[2] J. Bentley and R. Sedgewick. Fast algorithms for sorting and searching strings. In *Proceedings of the Eighth Annual ACM-SIAM Symposium on Discrete Algorithms*, pages 360–369, New Orleans, Louisiana, 1997.

[3] J. Bentley, D. Sleator, R. Tarjan, and V. Wei. A locally adaptive data compression scheme. *Communications of the ACM*, 29(4):320–330, April 1986.

[4] J. R. Bitner. Heuristics that dynamically organize data structures. *SIAM J. Comput.*, 8(1):82–110, 1979.

[5] M. Burrows and D. J. Wheeler. A block sorting lossless data compression algorithm. Technical Report 124, Digital Equipment Corporation, Palo Alto, California, 1994.

[6] J. G. Cleary and W. J. Teahan. Unbounded length contexts for PPM. *The Computer Journal*, 40(2/3):67–75, 1997.

[7] J. G. Cleary and I. H. Witten. Data compression using adaptive coding and partial string matching. *IEEE Transactions on Communications*, COM-32:306–315, 1984.

[8] P. Fenwick. The Burrows-Wheeler transform for block sorting text compression: principles and improvements. *The Computer Journal*, 39(9):731–740, 1996.

[9] P. Fenwick. Symbol ranking text compression with Shannon recoding. *J. UCS*, 3(2):70–85, 1997.

[10] P. Howard and J. Vitter. Design and analysis of fast text compression based on quasi-arithmetic coding. In *DCC: Data Compression Conference*. IEEE Computer Society TCC, 1993.

[11] W. Pugh. Skip lists: A probabilistic alternative to balanced trees. *Communications of the ACM*, 33(6):668–676, June 1990.

[12] K. Sadakane. Text compression using recency rank with context and relation to context sorting, block sorting and PPM*. In *Proc. Int. Conference on Compression and Complexity of Sequences (SEQUENCES '97)*. IEEE Computer Society TCC, 1997.

[13] D. Salomon. *Data Compression: the Complete Reference*. Springer Verlag, 1997.

[14] M. Schindler. A fast block-sorting algorithm for lossless data compression. In *Data Compression Conference*. IEEE Computer Society TCC, 1997. `http://eiunix.tuwien.ac.at/~michael/st/`.

[15] M. Schindler. The szip home page, 1997.
`http://www.compressconsult.com/szip/`.

[16] R. Sedgewick. *Algorithms in C*. Addison-Wesley, Reading, MA, USA, 3rd edition, 1997.

[17] J. Seward. The bzip2 home page, 1997.
`http://www.digistar.com/bzip2/index.html`.

[18] J. Vitter. Design and analysis of dynamic Huffman codes. *Journal of the ACM*, 34(4):825–845, October 1987.

[19] I. Witten, R. Neal, and J. Cleary. Arithmetic coding for data compression. *Communications of the ACM*, 30(6):520–540, June 1987.

[20] H. Yokoo. Data compression using a sort-based similarity measure. *The Computer Journal*, 40(2/3):94–102, 1997.

On List Update and Work Function Algorithms

Eric J. Anderson[1], Kris Hildrum[2], Anna R. Karlin[1], April Rasala[3], and Michael Saks[4]

[1] Dept. of Computer Science, Univ. of Wash., {eric,karlin}@cs.washington.edu
[2] Computer Science Div., Univ. of Calif., Berkeley, hildrum@cs.berkeley.edu
[3] Dept. of Computer Science, Dartmouth College, april.p.rasala@dartmouth.edu
[4] Dept. of Mathematics, Rutgers Univ., saks@math.rutgers.edu

Abstract. The *list update* problem, a well-studied problem in dynamic data structures, can be described abstractly as a metrical task system. In this paper, we prove that a generic metrical task system algorithm, called the *work function algorithm*, has constant competitive ratio for list update. In the process, we present a new formulation of the well-known "list factoring" technique in terms of a partial order on the elements of the list. This approach leads to a new simple proof that a large class of online algorithms, including Move-To-Front, is $(2-1/k)$-competitive.

1 Introduction

1.1 Motivation

The *list accessing* or *list update* problem is one of the most well-studied problems in competitive analysis [1],[2],[3],[4],[5]. The problem consists of maintaining a set S of items in an unsorted linked list, as a data structure for implementation of a dictionary. The data structure must support three types of requests: ACCESS(x), INSERT(x) and DELETE(x), where x is the name, or "key", of an item stored in the list. We associate a cost with each of these operations as follows: accessing or deleting the i-th item on the list costs i; inserting a new item costs $j+1$ where j is the number of items currently on the list before insertion. We also allow the list to be reorganized, at a cost measured in terms of the minimum number of transpositions of consecutive items needed for the reorganization. The standard model in the literature is that immediately after an access or an insertion, the requested item may be moved at no extra cost to a position closer to the front of the list. These exchanges are called *free exchanges*. Intuitively, using free exchanges, the algorithm can lower the cost on subsequent requests. In addition, at any time, two adjacent items in the list can be exchanged at a cost of 1. These exchanges are called *paid exchanges*.

The list update problem is to devise an algorithm for reorganizing the list, by performing free and/or paid exchanges, that minimizes search and reorganization costs. As usual, the algorithm will be evaluated in terms of its competitive ratio.

Many deterministic online algorithms have been proposed for the list update problem. Of these, perhaps the most well-known is the *Move-To-Front* algorithm: after accessing an item, move it to the front of the list, without changing the relative order of the other items. *Move-To-Front* is known to be $2-\frac{2}{k+1}$ competitive, and this is best possible [2],[7].

The static list update problem (where the list starts out with k elements in it, and all requests are accesses) can also be considered within the *metrical task system* framework introduced by Borodin, Linial and Saks [8].[1] Metrical task systems (MTS) are an abstract model for online computation that captures a wide variety of online problems (paging, list update and the k-server problem, to name a few) as special cases. A metrical task system is a system with n states, with a distance function d defined on the states: $d(i,j)$ is the distance between states i and j. The distances are assumed to form a metric. The MTS has a set $\mathcal{T}$ of allowable tasks; each task $\tau \in \mathcal{T}$ is a vector $(\tau(1), \tau(2), \ldots, \tau(n))$ where $\tau(i)$ is the cost of processing task τ in state i. An online algorithm is given a starting state and a sequence of tasks to be processed online, and must decide in which state to process each task. The goal of the algorithm is to minimize the total distance moved plus the total processing costs.

The list update problem can be viewed as a metrical task system as follows. The states of the list update MTS are the $k!$ possible orders the k elements in the list can be in. There are k tasks, one for each element x in the list; and $\tau_x(\pi)$, the cost of processing task τ_x in state π, is simply the depth of x in the list π. Finally, the distance between two states or permutations is just the number of inversions between the permutations.[2]

One of the initial results about metrical task systems was that the *work function algorithm* (WFA) has competitive ratio $2n-1$ for all MTS's, where n is the number of states in the metrical task system [8]. It was also shown that this is best possible, in the sense that there exist metrical task systems for which no online algorithm can achieve a competitive ratio lower than $2n-1$. However, for many MTS's the upper bound of $2n-1$ is significantly higher than the best achievable competitive ratio. For example, for list update with k elements in the list, $n = k!$, but we have constant competitive algorithms. Another example is the k-server problem on a finite metric space consisting of r points. For this problem, the metrical task system has $n = \binom{r}{k}$ states, but a recent celebrated result of Koutsoupias and Papadimitriou shows that in fact the *very same work function algorithm* is $2k-1$ competitive for this problem [9], nearly matching the known lower bound of k on the competitive ratio [10].

Unfortunately, our community understands very little at this point about how to design competitive algorithms that achieve close to the best possible competitive ratio for broad classes of metrical task systems. Indeed, one of the most intriguing open questions in this area is: *For what metrical task systems is the work function algorithm strongly competitive?* [3]

Burley and Irani have shown the existence of metrical task systems for which the work function algorithm is not strongly competitive [11]. However, these "bad" metrical

[1] As with much of the work on list accessing, we focus in this paper on the static case, where there are no insertions and deletions. The results described can be appropriately extended to the dynamic case.

[2] In this formulation, "free exchanges" are treated as made at unit cost immediately before the item is referenced. Because the cost of these exchanges is precisely offset by the lower reference cost, this model is identical to the standard model. See [6], Theorem 1. We continue to use the term "paid exchanges" to describe specifically those exchanges not involving the next-referenced element.

[3] We say an algorithm is *strongly competitive* if its competitive ratio is within a constant factor of the best competitive ratio achievable.

task systems seem to be rather contrived, and it is widely believed that the work function algorithm is in fact strongly competitive for large classes of natural metrical task systems. The desire to make progress towards answering this big question is the foremost motivation for the work described in this paper. We were specifically led to reconsider the list update problem when we observed the following curious fact (Proposition 5, Section 4): *The* Move-To-Front *algorithm for list update is a work function algorithm.*

This observation was intriguing for two reasons. First because it raised the question of whether work function algorithms generally (including those with tie-breaking rules different from that used in *Move-To-Front*) are strongly competitive for list update. This would provide an example of a substantially different type of metrical task system for which the work function algorithm is strongly competitive than those considered in the past.

The second and perhaps more exciting reason for studying work functions as they relate to list update is the tantalizing possibility that insight gained from that study could be helpful in the study of dynamic optimality for self-adjusting binary search trees [1],[12]. It is a long-standing open question whether or not there is a strongly competitive algorithm for dynamically rearranging a binary search tree using rotations, in response to a sequence of accesses. The similarity between *Move-To-Front* as an algorithm for dynamically rearranging linked lists, and the splay tree algorithm of Sleator and Tarjan [12] for dynamically rearranging binary search trees, long conjectured to be strongly competitive, is appealing. Our hope is that the use of work function-like algorithms might help to resolve this question for self-adjusting binary search trees.

1.2 Results

The main result of this paper is a proof that a class of work function algorithms is $O(1)$ competitive for the list update problem.[4] Proving this theorem requires getting a handle on the work function values, the optimal offline costs of ending up in each state. This is tricky, as the offline problem is very poorly understood. At present it is even unknown whether the problem of computing the optimal cost of executing a request sequence is NP-hard. The fastest optimal off-line algorithm currently known runs in time $O(2^k k!m)$, where k is the size of the list and m is the length of the request sequence [6].

Using the framework that we have developed for studying work functions and list update, we also present a new simple and illustrative proof that *Move-To-Front* and a large class of other online algorithms are $(2 - 1/k)$-competitive.

The rest of the paper is organized as follows. In Section 2, we present background material on work functions and on the work function algorithm. In Section 3, we present a formulation of the list update work functions in terms of a partial order on the elements of the list and use this formulation to prove that a large class of list update algorithms are $(2 - 1/k)$-competitive. Finally, in Section 4 we present our main result, that work function algorithms are strongly competitive for list update. The proof relies on an intricate construction; a number of technical details are omitted for lack of space.

[4] The proof does not achieve the best possible competitive ratio of 2.

2 Background

We begin with background material on work functions and work function algorithms.

Consider an arbitrary metrical task system, with states $s \in S$, and tasks $\tau \in T$. We define the *work function* $\omega_t(s)$ for any state s and index t to be the lowest cost of satisfying the first t requests of σ and ending up in state s [13],[8]. Suppose that σ_{t+1}, the $(t+1)$st request in σ, is the task τ. Because the states and task costs are time-independent, the work functions can be calculated through a dynamic programming formulation (this can be taken as the definition):

$$\omega_{t+1}(s') = \min_s \left(\omega_t(s) + \tau(s) + d(s,s')\right) \tag{1}$$

where $\tau(s)$ is the cost of executing task τ in state s. We note three elementary identities, which hold at all times t, and for all states s and s': (1) $\omega_{t+1}(s) \geq \omega_t(s)$, (2) $\omega_{t+1}(s) \leq \omega_t(s) + \sigma_{t+1}(s)$, and (3) $\omega_t(s) \leq \omega_t(s') + d(s,s')$.

The *Work Function Algorithm* (*WFA*), [13],[8], defined on an arbitrary metrical task system, is the following: when in state s_t, given a request $\sigma_{t+1} = \tau$, service τ in the state s_{t+1} such that

$$s_{t+1} = argmin_s(\omega_{t+1}(s) + d(s_t,s)) \tag{2}$$

where the minimum is taken over states s that are *fundamental at time* $t+1$, i.e., they must satisfy $\omega_{t+1}(s) = \omega_t(s) + \tau(s)$. Combining these two equations implies that s_{t+1} is chosen so that

$$s_{t+1} = argmin_s(\omega_t(s) + \tau(s) + d(s_t,s)). \tag{3}$$

This algorithm can be viewed as a compromise between two very natural algorithms: (1) A *greedy* algorithm which tries to minimize the cost spent on the current step, i.e., services the $(t+1)$st request τ in a state s that minimizes $d(s_t,s) + \tau(s)$. (2) A *retrospective* algorithm, which tries to match the optimal offline algorithm, i.e., chooses to service the $(t+1)$st request τ in a state s that minimizes $\omega_{t+1}(s)$.

Each of these algorithms is known to be noncompetitive for many natural problems. *WFA* combines these approaches and, interestingly, this results in an algorithm which is known to be strongly competitive for a number of problems for which both the *greedy* and *retrospective* algorithms are not competitive.[5]

A variant of this work function algorithm, which we'll call *WFA'*, is to service the request τ in the state s_{t+1} such that

$$s_{t+1} = argmin_s\left(\omega_{t+1}(s) + \tau(s) + d(s_t,s)\right). \tag{4}$$

The difference between *WFA* and *WFA'* is in the subscript of the work function. We actually feel that *WFA'* is a slightly more natural algorithm, in light of the discussion above about combining a greedy approach and a retrospective approach. It is this latter work function algorithm *WFA'* that we will focus on in this paper. Our proof that *WFA'*

[5] Varying the relative weighting of the greedy and retrospective components of the work function algorithm was explored in [14].

is $O(1)$ competitive for list update can be extended to handle WFA as well, though the proofs will be omitted in this extended abstract.[6]

3 A Different View on List Factoring

A technique which has been used in the past to analyze list update algorithms is the *list factoring* technique, which reduces the competitive analysis of list accessing algorithms to lists of size two [3],[7],[15],[4],[16]. For example, this technique, in conjunction with phase partitioning, was used to prove that an algorithm called *TimeStamp* is 2-competitive [4],[16]. In this section, we repeat the development of this technique, but present it in a somewhat different way, in terms of a partial order on elements.[7] This view leads us to a simple generalization of previous results and will assist us in our study of WFA'.

Consider the metrical task system corresponding to a list of length two. In this case there are two lists, (a,b) (a in front of b) and (b,a) (b in front of a), and the distance between these two states is 1. Since $\omega_t((a,b)) - 1 \leq \omega_t((b,a)) \leq \omega_t((a,b)) + 1$, for any t, the work functions at any time can be characterized by one of three distinct properties:

- $\omega_t((a,b)) < \omega_t((b,a))$, which we denote $a \succ b$,
- $\omega_t((a,b)) = \omega_t((b,a))$, which we denote $a \sim b$, or
- $\omega_t((a,b)) > \omega_t((b,a))$, which we denote $a \prec b$.

It is easy to verify directly from Equation 1 the transitions between these three properties as a result of references in the string σ.

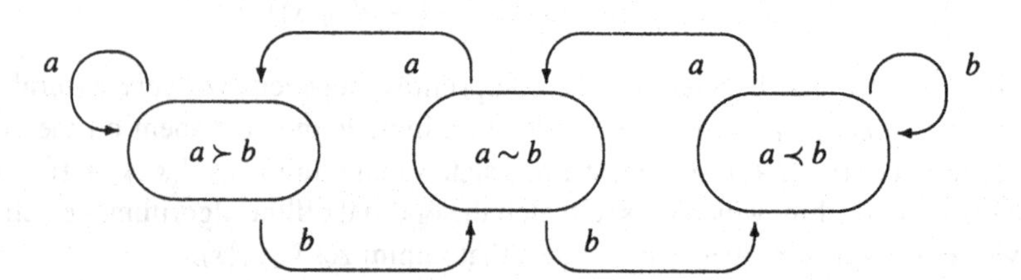

Fig. 1. The three-state DFA: the state $a \succ b$ corresponds to the case $\omega_t((a,b)) = \omega_t((b,a)) - 1$, the state $a \sim b$ corresponds to the case $\omega_t((a,b)) = \omega_t((b,a))$, and the state $a \prec b$ corresponds to the case $\omega_t((a,b)) = \omega_t((b,a)) + 1$

The resulting *three-state DFA* shown in Figure 1 can be used to completely characterize the work functions, the optimal offline list configuration, and the optimal cost to service a request sequence σ. The start state of the DFA is determined by the initial

[6] In addition, it is easy to show that prior results which hold for WFA also hold for WFA'. For example, WFA' is $2n-1$ competitive for any metrical task system with n states, and WFA' is $2k-1$ competitive for the k-server problem.

[7] This partial order has apparently been considered by Albers, von Stengel and Werchner in the context of randomized list update, and was used as a basis for an optimal randomized online algorithm for lists of length 4. [17]

order of the elements in the list: it is $a \succ b$ if the initial list is (a,b) and $a \prec b$ if the initial list is (b,a). Each successive request in σ results in a change of state in accordance with the transitions of the DFA, reflecting the work function values after serving that request. It is easily verified that the optimal cost of satisfying a sequence σ is precisely the number of references in the sequence plus the number of transitions *into* the middle DFA-state, i.e., the number of times a is referenced in the state $a \prec b$ plus the number of times b is referenced in the state $a \succ b$. The corresponding optimal offline strategy is: immediately before two or more references in a row to the same element, move that element to the front of the list.

Now consider list update for a list of length k. The cost of an optimal sequence can be written as the sum of (i) the distances between successive states (the number of exchanges performed)[8] and (ii) the reference costs at each state. The standard list factoring approach is to describe the cost of any optimal sequence for satisfying σ by decomposing it into $|\sigma|$ plus the sum over all pairs (a,b) of (i) the exchanges between a and b, and (ii) the pairwise incremental costs, i.e. the cost attributed to a when b is referenced but a is in front of b in the list and the cost attributed to b when a is referenced but b is in front of a in the list. But for any pair (a,b), the pairwise transpositions and the pairwise cost of references is a (perhaps suboptimal) solution to the list of length two problem for the subsequence of σ consisting of references only to a and b. Thus $|\sigma|$ plus the sum of the costs of the optimal length-two solutions over all pairs a,b is a lower bound for the optimal cost of satisfying σ.[9]

3.1 The partial order

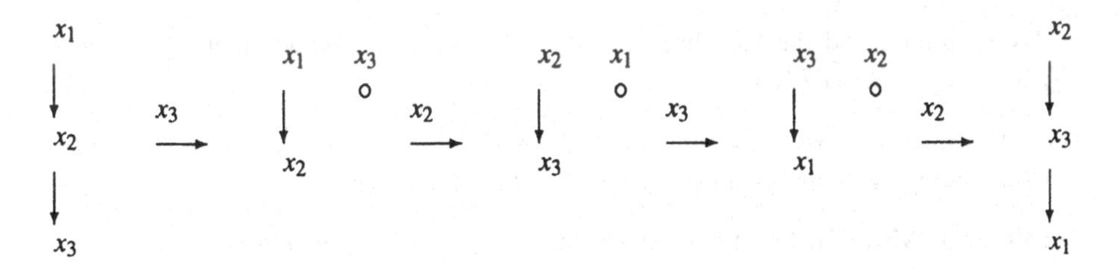

Fig. 2. Illustration of the evolution of the partial order on three elements in response to the request sequence $\sigma = x_3, x_2, x_3, x_2$ assuming the initial list is ordered x_1, x_2, x_3 from front to back. As usual, a directed edge from a to b indicates that $a \succ b$ in the partial order, whereas the absence of an edge indicates that $a \sim b$

We are thus led to consider the collection of $k(k-1)/2$ pairwise three-state DFAs, one for each pair a,b of elements in the list of length k. Consider the result of executing

[8] Recall that in our model we charge for each exchange, whether "paid" or "free"; the cost of free exchanges in the standard model precisely corresponds in our model to a reduced reference cost on the immediately following reference. See [6], Theorem 1.

[9] The lower bound is not tight; for a list of length five, initialized *abcde*, the sequence $\sigma =$ *ebddcceacde* provides one counterexample.

all these DFAs in parallel in response to requests in σ, starting from the states corresponding to the initial list. Figure 2 shows an example. It is easy to verify that the resulting states define a valid *partial order* on the k elements of the list. (For example, *Move-To-Front* is always consistent with this partial order.)

Define by G_t (respectively I_t) the number of elements greater than (respectively incomparable to) σ_t in this partial order immediately prior to its reference at time t. By the discussion above, the optimal cost of servicing a request sequence σ of length n and ending up in any state s is bounded below as follows: $\omega_n(s) \geq n + \sum_{1 \leq t \leq n} G_t$.

An easy counting argument shows that $\sum_t I_t \leq \sum_t G_t$: $\sum G_t$ is the cumulative number of transitions into middle states $a \sim b$ of the DFA's, $\sum I_t$ is the cumulative number of transitions out of middle states, and the starting state is always either $a \succ b$ or $a \prec b$ (not a middle state). Since you can't transition out of a middle state until you have transitioned into one, we have shown that

Lemma 1. *At all times T, $\sum_{t \leq T} I_t \leq \sum_{t \leq T} G_t$.*

This leads to a new, very simple proof that a collection of algorithms already known to be competitive, including *Move-To-Front*, *TimeStamp*, and many others, are all $2 - 1/k$ competitive.[10]

Theorem 2. *Any online list update algorithm that performs only free exchanges and maintains the invariant that the list order is consistent with the partial order is $(2 - 1/k)$-competitive.*

Proof. Any online algorithm A that maintains a list order consistent with the partial order and performs no paid exchanges has a total cost $A(\sigma)$ satisfying $A(\sigma) \leq n + \sum_t (I_t + G_t)$, where $|\sigma| = n$.

By Lemma 1 and the fact that $OPT(\sigma) \leq kn$, we can conclude that $A(\sigma) \leq n + 2\sum_t G_t \leq (2 - 1/k)OPT(\sigma)$. □

The following corollary is also immediate from Theorem 2, since both *Move-To-Front* and *TimeStamp* [4] maintain a list consistent with the partial order.

Corollary 3. Move-To-Front *and* TimeStamp *are $(2 - 1/k)$-competitive.*

4 On the Performance of Work Function Algorithms

4.1 Preliminaries

We begin with some definitions and facts. In what follows, the $(t+1)$st request σ_{t+1} is x, and the task cost $\tau_x(s)$ is denoted $x(s)$. We also define the $\uparrow_x$ binary relation on two states, $s \uparrow_x s'$, if s' can be derived from s by moving x forward (including $s = s'$) while leaving the relative positions of other elements undisturbed.

[10] Ran El-Yaniv has recently presented an different family of algorithms, all of which are $2 - 1/k$ competitive [5]. All algorithms in this family maintain lists consistent with the partial order and incur only free exchanges, and hence can also be proved $2 - 1/k$-competitive using this result.

We say that the state s is *wfa-eligible* at time t if it minimizes the expression $w_{t+1}(s)+x(s)+d(s_t,s)$. We say that the state s is *fundamental* at time t iff $\omega_{t+1}(s) = \omega_t(s)+x(s)$.[11]

We omit the easy proofs of the following facts.

Proposition 4. *Let s be an arbitrary state. Then:*

1. $\omega_{t+1}(s) = \omega_{t+1}(f)+d(f,s)$ for some state f that is fundamental at time t. (The state s is derived from some *fundamental state.)*

2. Suppose $\omega_{t+1}(s) = \omega_{t+1}(f)+d(f,s)$ where f is a fundamental state. Then $x(f) \leq x(s)$. (The depth of x in the fundamental state f is no greater than the depth of x in s.)

3. If s is wfa-eligible at time t, and $\omega_{t+1}(s) = \omega_{t+1}(f)+d(f,s)$, where f is a fundamental state at time t, then f is wfa-eligible at time t and $x(f) = x(s)$. (The fundamental state f is also wfa-eligible if s is.)

4. If $s \uparrow_x s'$, then $\omega_{t+1}(s) \geq \omega_{t+1}(s')$. (Moving x forward cannot increase the work function.)

We can now show (proof omitted in this extended abstract) that (a) there always exists a wfa-eligible state (the MTF state is one such) that requires no paid exchanges, and (b) that with such a restriction WFA' is equivalent to an algorithm we call *Move-To-Min-ω* (*Mtmw*) defined as follows: On a reference to x, move x forward to a state with lowest work function value immediately after the reference. In other words, if s_t is the state the algorithm is in immediately before servicing the $t+1$-st request σ_{t+1}, then *Mtmw* moves to a state s_{t+1} such that $s_{t+1} = argmin_{s\,:\,s_t\uparrow_x s}\omega_{t+1}(s)$ and satisfies σ_{t+1} there. Summarizing:

Proposition 5. *Mtmw is a special case of WFA' and* Move-To-Front *is a special case of Mtmw.*

4.2 WFA' is $O(1)$ competitive for list update.

The technically challenging part of the proof is the following lemma.

Lemma 6. *Consider $\sigma = \sigma_1, x, \sigma_2, x$, where in σ_2 there are no references to x, and $|\sigma| = t$. Let S be any fundamental state at the final time step t.*

Let N be the set of elements that are not referenced in σ_2 that are in front of x in S, and let R be the set of elements that are referenced in σ_2. Also, let $\hat{S}$ be S with x moved forward just in front of the element in N closest to the front of the list. Then

$$\omega_t(\hat{S}) \leq \omega_t(S)+|R|-|N|. \tag{5}$$

Proof. (sketch)

Suppose O is an optimal sequence ending in S after satisfying σ_1, x, σ_2, x, so that the cost of O is the work function value $\omega_t(S)$. Let T denote the state in which O satisfies the penultimate reference to x. We note that, at the point immediately prior to the penultimate reference to x (at time k, say), the cost of O is $\omega_{k-1}(T)$. In this

[11] These definitions, and the first three facts, are valid for all metrical task systems.

construction, we modify O between T and S to so as to obtain a state $\hat{S}$, with $S \uparrow_x \hat{S}$ and $\omega_t(\hat{S}) \le \omega_t(S) - |\mathcal{N}| + |\mathcal{R}|$.

Let N denote the total number of elements *not* referenced between $\sigma_k = x$ and $\sigma_t = x$. (This set specifically includes x, and is potentially much larger than $\mathcal{N}$, which is the number of such elements in front of x in S.) Order these non-referenced elements $p_1, \dots, p_N$ in the order they occur in the state T.

The construction of the lower-cost state $\hat{S}$ proceeds in three stages:

1. Rearrange the respective order of the non-referenced elements within T to obtain some state T'. In T', x will occupy the location of the front-most non-referenced element in T. All other non-referenced elements p in T' will satisfy the *non-decreasing depth property*, that $p(T) \le p(T')$.[12] All referenced elements remain at their original depths. (The specific definition of the state T' will emerge from the rest of the construction; the cost can be bounded by using only the non-decreasing depth property.) Evaluate $\sigma_k = x$ in this state T'.

Denoting by $I[X,Y]$ the number of interchanges of non-referenced elements other than x between states X and Y, it is straightforward to show (using the non-decreasing depth property) that $x(T') + d(T,T') \le x(T) + R + I[T,T']$.

2. Considering O as a sequence of transpositions and references transforming T to S, $O : T \to S$, apply a suitably chosen subsequence O', including all of the references and many of the transpositions, of O. This subsequence O' will transform T' to a state S'. In this state S', (i) each referenced element has the same depth as it does in S; (ii) the element x occupies the position of the front-most non-referenced element in S; and (iii) all other non-referenced elements in S' are in their same respective pairwise order as in S. Evaluate x in S'.

In Proposition 7, we show that a transformation from T' with the non-decreasing depth property to S' as so defined can be achieved by a suitably chosen subsequence of O. We also show that $I[T,T'] + |O'| \le |O|$, where $|O|$ denotes the cost of the sequence O.

3. Transform S' to the state $\hat{S}$, where $\hat{S}$ is defined by (i) $S \uparrow_x \hat{S}$, and (ii) the depth of x in $\hat{S}$ is the depth of the front-most non-referenced element in S (which is also its depth in S').

It is straightforward to show that $x(S') + d(S',\hat{S}) + |\mathcal{N}| \le x(S)$. The result now follows by comparing the cost of the modified sequence from and after $\omega_{k-1}(T)$ to the cost of the original sequence.

We now address the most intricate part of the construction:

Proposition 7. *Suppose S' is derived from S, such that (i) all referenced elements p have $p(S') = p(S)$, (ii) x occupies in S' the position of the front-most non-referenced element in S, and (iii) all other non-referenced elements are in their same respective order in S' as in S. Then there is a T' with the non-decreasing depth property, and a subsequence $O' \subseteq O$, such that (i) $O'(T') = S'$, and (ii) the cost of O is at least the cost of O' plus the cost of interchanges $I[T',T]$ of non-referenced elements (other than x) necessary to derive T' from T.*

[12] Recall that we denote the depth of an element p in the state X by $p(X)$.

Proof. (sketch) As above, we denote by p_i the non-referenced element occupying the i'th non-referenced position in T. For convenience, let z denote the location of x as a non-referenced element in T, $p_z = x$.[13]

We proceed by iteratively constructing T' from the end of the list, beginning with $O^{-1}(S')$. The location of referenced elements remains fixed throughout the construction. As a result, we consider only the N positions of non-referenced elements. For convenience, we describe the iteration as proceeding from $i = N$ to $i = 1$. (The "base case" is denoted by "$i = N+1$".) At each step, then, we define a map $O_i : T_i' \to S'$. The non-decreasing depth property is maintained for the elements (other than x) in $O_i^{-1}(S') = T_i'$ that occupy the locations i through N in T_i'. We show that any necessary interchanges of elements as we proceed from T_i' to T_{i-1}' correspond to transpositions in O_i'.

For each pair of elements $p, q \neq x$ at locations i and below in T_i', we can determine whether these two elements are in the same or in the opposite order in T. We denote by $I_i[T_i', T]$ the number of pairwise inversions of such elements (other than x). We denote by $|O|$ (respectively, $|O_i|$) the number of transpositions in the sequence O (respectively, O_i).

Formally, we can prove by induction (details omitted in this extended abstract) that for each i:

1. $O_i(T_i') = S'$ (and $O_i^{-1} : S' \to T_i'$)
2. $O_i \subseteq O$ in the sense of a subsequence of transpositions, and $|O| \geq |O_i| + I_i[T_i', T]$ (all swaps and inversions are accounted for)
3. $x(T_i') \leq p_i(T)$ (x is no deeper than position i)
4. $\forall p \neq x$ with $p(T) \geq p_i(T), p(T_i') \geq p(T)$ (all elements other than x at position i or below in T have the non-decreasing depth property)
5. $\forall p, q \neq x$ with $p(T), q(T) < p_i(T)$:
   (a) $p(S) < x(S) \iff p(T_i') \neq p(T)$, and $p(S) > x(S) \iff p(T_i') = p(T)$
   (b) $p(T) = q(T_i') \implies p(S) > q(S)$

We define $T' = T_1$, and note that the non-increasing depth property is satisfied for all $p_i \neq x$. We define $O' = O_1$, and note all of the inversions between non-referenced elements in T' have been accounted for, i.e., $I[T, T'] + |O'| \leq |O|$. Finally, we repeat that because the only transpositions removed from O are between non-referenced elements, the depths, and thus the reference costs, of all referenced elements remains identical between O and O'. □

We obtain the following corollary to Lemma 6.

Corollary 8. *Consider a request sequence σ where the last request (the t-th request in σ) is to x. If s is wfa-eligible after executing σ, then the depth of x in s is at most $2|R|$, where R is the set of elements that have been referenced since the penultimate reference to x.*

Proof. Let f be a fundamental state such that $\omega_{t+1}(s) = \omega_{t+1}(f) + d(f, s)$. By Proposition 4, f is also wfa-eligible and $x(f) = x(s)$. Suppose $x(s) > 2|R|$. Then $x(f) > 2|R|$. Elements in front of x in f either have or have not been referenced since the penultimate

[13] We use the terms "position" and "location" interchangeably to refer to the respective positions of non-referenced elements in T.

reference to x; so $x(f) > 2|R|$ implies $|N| > |R|$, where N is the set of elements in front of x in f that have not been referenced since the penultimate reference to x. Then by Lemma 6 there exists $\hat{f}$ with $\omega_t(\hat{f}) < \omega_t(f)$ and $f \uparrow_x \hat{f}$, contradicting the assumption that f is wfa-eligible. □

Finally, we use the lemma to obtain the main theorem.

Theorem 9. *WFA′ is O(1) competitive.*

Proof. (*sketch*) We consider only *Mtmw* here. Consider an arbitrary element x and let $\sigma = \sigma_0, x, \sigma_1, x, \sigma_2, x$, where in σ_1 and σ_2 there are no references to x. Then by Lemma 6 the *Mtmw* state, immediately before the final reference to x, is at depth at most $2r_1 + r_2$, where r_1 is the number of distinct elements referenced in σ_1 and r_2 is the number of distinct elements referenced in σ_2, not referenced in σ_1, that are moved in front of x at some point during the subsequence σ_2.

As usual, let G be the number of elements greater than x immediately before its final reference and let I be the number of elements incomparable to x immediately before its final reference. In addition, let $L(0)$ be the number of elements less than x immediately before its final reference *that were incomparable to x immediately before the penultimate reference to x*. We therefore have $r_1 + r_2 \leq G + I + L(0)$.

A simple counting argument similar to the proof of Lemma 1 proves that $\sum_t L(0)_t \leq \sum_t G_t$. Taken together with Lemma 1, we obtain the theorem. □

A fairly easy extension of the ideas presented here can be used to show that both WFA' and WFA are $O(1)$ competitive, even when the algorithms are allowed to perform paid exchanges.

It is fairly clear that our analyses of these algorithms are not tight. However, it is easy to show that WFA, even without paid exchanges, is no better than 3-competitive.

5 Acknowledgments

We gratefully acknowledge discussions with Susanne Albers, Ran El-Yaniv, and Sandy Irani, as well as the helpful comments of our anonymous reviewers.

This work was supported in part by NSF grant EIA-9870740 and BSF grant 96-00247 (Karlin), the CRA Distributed Mentor Project (Hildrum and Rasala), and an IBM Research Fellowship (Anderson).

References

[1] S. Albers and J. Westbrook. Self-organizing data structures. In *Online Algorithms: The State of the Art*, Fiat-Woeginger, Springer, 1998.

[2] D.D. Sleator and R.E. Tarjan. Amortized efficiency of list update and paging rules. *Communications of the ACM*, 28:202–208, 1985.

[3] J.L. Bentley and C. McGeoch. Amortized analysis of self-organizing sequential search heuristics. *Communications of the ACM*, 28(4):404–411, 1985.

[4] S. Albers. Improved randomized on-line algorithms for the list update problem. *SIAM Journal on Computing*, 27: 682–693, 1998.

[5] R. El-Yaniv. There are infinitely many competitive-optimal online list accessing algorithms. Discussion paper from The Center for Rationality and Interactive Decision Making. Hebrew University.

[6] N. Reingold and J. Westbrook. Off-line algorithms for the list update problem. *Information Processing Letters*, 60(2):75–80, 1996.

[7] S. Irani. Two results on the list update problem. *Information Processing Letters*, 38(6):301–306, 1991.

[8] A. Borodin, N. Linial, and M. Saks. An optimal online algorithm for metrical task systems. *Journal of the ACM*, 52:46–52, 1985.

[9] E. Koutsoupias and C. Papadimitriou. On the *k*-server conjecture. *Journal of the ACM*, 42(5): 971–983, September 1995.

[10] M. Manasse, L. McGeoch and D.D. Sleator. Competitive algorithms for server problems. *Journal of Algorithms*, 11:208–230, 1990.

[11] W. Burley and S. Irani. On algorithm design for metrical task systems. In *Proceedings of ACM-SIAM Symposium on Discrete Algorithms*, 1995.

[12] D.D. Sleator and R.E. Tarjan. Self-adjusting binary search trees. *Journal of the ACM*, 32: 652-686, 1985.

[13] M. Chrobak, L. Larmore. The server problem and on-line games. In *On-Line Algorithms, Proceedings of a DIMACS Workshop*, Vol 7 of *DIMACS Series in Discrete Mathematics and Computer Science*, pp. 11 – 64, 1991.

[14] W.R. Burley. Traversing layered graphs using the work function algorithm. *Journal of Algorithms*, 20(3):479–511, 1996.

[15] B. Teia. A lower bound for randomized list update algorithms. *Information Processing Letters*, 47:5–9, 1993.

[16] S. Albers, B. von Stengel and R. Werchner. A combined BIT and TIMESTAMP algorithm for the list update problem. *Information Processing Letters*; 56: 135– 139, 1995.

[17] S. Albers. Private communication.

The 3-Server Problem in the Plane
(Extended Abstract)

Wolfgang W. Bein[1], Marek Chrobak[2], and Lawrence L. Larmore[1]*

[1] Department of Computer Science, University of Nevada, Las Vegas, NV 89154.
{bein,lamore}@cs.unlv.edu.
[2] Department of Computer Science, University of California, Riverside, CA 92521.
marek@cs.ucr.edu.

Abstract. In the k-Server Problem we wish to minimize, in an online fashion, the movement cost of k servers in response to a sequence of requests. The request issued at each step is specified by a point r in a given metric space M. To serve this request, one of the k servers must move to r. (We assume that $k \geq 2$.)
It is known that if M has at least $k+1$ points then no online algorithm for the k-Server Problem in M has competitive ratio smaller than k. The best known upper bound on the competitive ratio in arbitrary metric spaces, by Koutsoupias and Papadimitriou [6], is $2k-1$. There is only a number of special cases for which k-competitive algorithms are known: for $k = 2$, when M is a tree, or when M has at most $k+2$ points.
The main result of this paper is that the Work Function Algorithm is 3-competitive for the 3-Server Problem in the Manhattan plane. As a corollary, we obtain a 4.243-competitive algorithm for 3 servers in the Euclidean plane. The best previously known competitive ratio for 3 servers in these spaces was 5.

1 Introduction

The *k-Server Problem* is defined as follows: we are given k mobile servers that reside in a metric space M. A sequence of requests is issued, where each request is specified by a point $r \in M$. To "satisfy" this request, one of the servers must be moved to r, at a cost equal to the distance from its current location to r. An algorithm $\mathcal{A}$ for the k-Server Problem decides which server should be moved at each step. $\mathcal{A}$ is said to be *online* if its decisions are made without the knowledge of future requests. Our goal is to minimize the total service cost.

We define an online algorithm $\mathcal{A}$ to be *C-competitive* if the cost incurred by $\mathcal{A}$ to service each request sequence ρ is at most C times the optimal (offline) service cost for ρ, plus possibly an additive constant independent of ρ. The *competitive ratio of* $\mathcal{A}$ is the smallest C for which $\mathcal{A}$ is C-competitive.

The k-Server Problem was introduced by Manasse, McGeoch and Sleator [8], who proved that no online algorithm can have a competitive ratio smaller than k if a metric space has at least $k+1$ points, and they presented an algorithm for the 2-Server Problem which is 2-competitive, and thus optimal, for any metric space. They also proposed the *k-Server Conjecture*, stating that, for each $k \geq 3$, there exists a k-competitive algorithm

* Research supported by NSF grant CCR-9503441.

that works in all metric spaces. So far, this conjecture has been settled only in a number of special cases, including trees and spaces with at most $k+2$ points [1, 2, 7]. Even some simple-looking special cases remain open, for example the 3-Server Problem on the circle, in the plane, or in 6-point spaces.

Recent research on the k-Server Conjecture has focussed on the *Work Function Algorithm* (WFA), as a possible candidate for a k-competitive algorithm. WFA is a tantalizingly simple algorithm that, at each step, chooses a server so as to minimize the sum of two quantities: the movement cost at this step, and the optimal cost of the new configuration. (More formally, the latter quantity is the optimal cost of serving past requests and ending in that configuration.) Thus one can think of WFA as a combination of two greedy strategies: one that minimizes the cost of the current move, and one that chooses the best configuration to be in.

Chrobak and Larmore [3, 4] proved that WFA is 2-competitive for $k = 2$. Their approach was based on a new technique, which involves introducing an algorithm-independent quantity called *pseudocost* that provides an upper bound on WFA's cost, and on estimating the pseudocost instead of the actual cost of WFA. The pseudocost approach can also be effectively used to prove that WFA is competitive for other problems, for example Task Systems [4].

For $k \geq 3$, a major breakthrough was achieved by Koutsoupias and Papadimitriou [5, 6], who proved that WFA is $(2k-1)$-competitive for k servers in arbitrary metric spaces. Their proof was based on the pseudocost method.

The main result of this paper is that WFA is 3-competitive for 3 servers in the Manhattan plane. Our research builds on the work from [3, 4, 6]. The proof uses the pseudocost method. The main difficulty in estimating the pseudocost is in finding an appropriate potential function. We construct a potential function Φ, and we formulate certain conditions on a metric space M under which Φ provides a certificate that WFA is 3-competitive in M. Then we show that the Manhattan plane satisfies these conditions, and we conclude that WFA is 3-competitive in the Manhattan plane for 3 servers. Since the Euclidean metric can be approximated by the city-block metric, this also gives a $3\sqrt{2}$-competitive algorithm for 3 servers in the Euclidean plane.

2 Preliminaries

Let M be a metric space. For points $x, y \in M$, we write xy to denote the distance between x and y. Unordered k-tuples of points in M will be called *configurations*, and they represent positions of our k servers. Configurations will be denoted by capital letters $X, Y, \ldots$ The configuration space is itself a metric space under the minimum-matching metric. We write XY to denote the minimum-matching distance between X and Y. For simplicity, we assume that the initial server configuration S^0 is fixed. Without loss of generality, we allow the algorithms to move any number of servers before or after each request, as long as between the times when two requests are issued, at least one of the servers visits the last request point. It is a simple exercise to verify that this additional freedom does not change the problem, but it makes the definitions associated with work functions easier to handle.

Work functions. Work functions provide information about the optimal cost of serving the past request sequence. For a request sequence ρ, by $\omega_\rho(X)$ we denote the minimum cost of serving ρ and ending in configuration X. We refer to ω_ρ as the *work function after* ρ. We use notation ω to denote any work function ω_ρ, for some request sequence ρ. Immediately from the definition of work functions we get that the optimal cost to service ρ is $opt(\rho) = \min_X \omega_\rho(X)$.

For given ρ, we can compute ω_ρ using simple dynamic programming. Initially, $\omega_\varepsilon(X) = S^0X$, for each configuration X (ε is the empty request sequence). For a non-empty request sequence ρ, if r is the last request in ρ, write $\rho = \sigma r$. Then ω_ρ can be computed recursively as $\omega_\rho = \omega_\sigma \wedge r$, where "$\wedge$" is the *update operator* defined as follows:

$$(\omega \wedge r)(X) = \min_{Y \ni r} \{\omega(Y) + YX\} \tag{1}$$

Note that $|\omega(X) - \omega(Y)| \leq XY$ for any work function ω and any configurations X and Y. This inequality we call the *Lipschitz property*. Koutsoupias and Papadimitriou [5, 6] proved that work functions also satisfy the following *quasiconvexity* property:

$$\omega(X) + \omega(Y) \geq \max_{x \in X-Y} \min_{y \in Y-X} \{\omega(X - x + y) + \omega(Y - y + x)\} \tag{2}$$

The Lipschitz property and quasiconvexity will be used extensively in our calculations.

The Work Function Algorithm. We define the *Work Function Algorithm* (WFA) to be an algorithm which chooses its service of the request sequence ρ as follows: Suppose that WFA is in configuration S, and that the current work function is ω. On request r, WFA chooses that $x \in S$ which minimizes $xr + \omega \wedge r(S - x + r)$, and moves the server from x to r.

WFA can be seen as a "linear combination" of two greedy strategies: one that minimizes the cost xr in the given step, and one that chooses the optimal configuration after r, that is, the configuration $S - x + r$ that minimizes $\omega \wedge r(S - x + r)$. Neither of these two greedy strategies is competitive.

Since $r \in S - x + r$, we have $\omega \wedge r(S - x + r) = \omega(S - x + r)$, so WFA can as well minimize $xr + \omega(S - x + r)$. Yet another possible formulation is to move to a new configuration S' that contains r and minimizes $SS' + \omega(S')$. It is not hard to show (by induction on the number of requests) that this is equivalent to our formulation, since WFA will only move one server at a time.

The pseudocost method. The pseudocost is a function that provides an upper bound on WFA's cost. More accurately, the pseudocost actually bounds the *sum* of WFA's and the optimal costs. Since the pseudocost is algorithm-independent, it is much easier to deal with than the actual cost of WFA.

For any work function ω and $r \in M$, we consider the maximum increase of the work function if r is requested:

$$\nabla_r(\omega) = \max_X \{\omega \wedge r(X) - \omega(X)\}$$

Suppose that $\rho = r^1 \ldots r^n$, and let ω^t denote the work function after requests $r^1 \ldots r^t$, that is $\omega^t = \chi_{S^0} \wedge r^1 \ldots r^t$. The *pseudocost* of ρ is defined as $\nabla_\rho = \sum_{t=1}^n \nabla_{r^t}(\omega^{t-1})$. The lemma below establishes the relationship between the cost of WFA and the pseudocost.

Lemma 1. [3, 6] $\nabla_\rho \geq cost_{\text{WFA}}(\rho) + opt(\rho)$.

According to the above lemma, in order to prove that WFA is C-competitive, it is sufficient to show that the pseudocost is $(C+1)$-competitive. To prove the latter fact, we use a standard potential argument, formalized in the lemma below.

Lemma 2. *Let $\Phi_{\omega,r} \in \mathbf{R}$ be defined for each work function ω with the last request r. Suppose that $\Phi_{\omega,r}$ satisfies the following properties:*
(OP) $\Phi_{\omega,r} + (C+1)\min(\omega) \geq 0$
(UP) *If $\mu = \omega \wedge s$ then $\Phi_{\mu,s} + \nabla_s(\omega) \leq \Phi_{\omega,r}$*
Then WFA *is C-competitive on M.*

Conditions (OP) and (UP) are referred to as, the *offset property* and the *update property*, respectively. Function $\Phi_{\omega,r}$ is called a *potential function*.

Proof. We use the notation from the previous lemma. Let also $\Phi^t = \Phi_{\omega^t,r^t}$. The proof is by amortized summation:

$$\begin{aligned}\nabla_\rho &= \sum_{t=1}^{n} \nabla_{r^t}(\omega^{t-1}) \\ &\leq \sum_{t=1}^{n} (\Phi^{t-1} - \Phi^t) \\ &= \Phi^0 - \Phi^n \\ &\leq (C+1)opt(\rho) + \Phi^0.\end{aligned}$$

In the last step we used the offset property (OP). Φ^0 is independent of ρ. By Lemma 1, it follows that WFA is C-competitive. ∎

A modified update property. Define the *shadow* of ω as

$$\tilde{\omega}(x) = \max_A \left\{ \sum_{a \in A} xa - \omega(A) \right\} \tag{3}$$

A configuration A is called an (ω,x)-*maximizer* if A maximizes the right-hand side in (3), that is, $\tilde{\omega}(x) = \sum_{a \in A} xa - \omega(A)$. If r is the last request in ω, then $\omega(A) = \omega(A - b + r) + br$, for some $b \in A$, and

$$\begin{aligned}\sum_{a \in A} xa - \omega(A) &= \sum_{a \in A} xa - br - \omega(A - b + r) \\ &\leq \sum_{a \in A-b+r} xa - \omega(A - b + r).\end{aligned}$$

We conclude that, without loss of generality, an (ω,x)-maximizer contains the last request.

Lemma 3. [6] *Suppose that A is an (ω,x)-maximizer. Then*

(a) *A is an* $(\omega\wedge x, x)$*-maximizer.*
(b) *A maximizes* $\omega\wedge x(A) - \omega(A)$.

Proof. The proofs for (a) and (b) use the quasiconvexity property and are quite similar. We give the proof for (b), and refer the reader to [6] for the proof of (a). (See also [4].)

To show (b), it is sufficient to prove that

$$\omega\wedge x(A) + \omega(B) \geq \omega(A) + \omega\wedge x(B) \tag{4}$$

for each configuration B. If $x \in B$ then $\omega(B) = \omega\wedge x(B)$ and (4) follows from $\omega\wedge x(A) \geq \omega(A)$. Suppose $x \notin B$. Since A is a (ω, x)-maximizer, $ax - \omega(A) \geq bx - \omega(A - a + b)$ for all $a \in A$ and $b \notin A$. Then, using quasi-convexity, we have

$$\begin{aligned}\omega\wedge x(A) + \omega(B) &= \min_{a\in A}\{\omega(A - a + x) + ax + \omega(B)\} \\ &\geq \min_{a\in A}\min_{b\in B-A}\{\omega(A - a + b) + ax + \omega(B - b + x)\} \\ &\geq \min_{b\in B-A}\{\omega(A) + bx + \omega(B - b + x)\} \\ &\geq \omega(A) + \omega\wedge x(B)\end{aligned}$$

and (4) follows. ■

We now introduce a *modified update property*, which we will use instead of the update property (UP) from Lemma 2. This will considerably simplify the calculations in the next section.

Corollary 4. *Let* $\Phi_{\omega,r}$ *satisfy the offset property* (OP) *and have the form* $\Phi_{\omega,r} = \tilde{\omega}(r) + \Psi_{\omega,r}$. *Suppose that* $\Phi_{\omega,r}$ *satisfies the following modified update property*
(MUP) *If* $\mu = \omega\wedge s$ *then* $\tilde{\omega}(s) + \Psi_{\mu,s} \leq \Phi_{\omega,r}$.
Then WFA is 3-competitive in M.

Proof. Let A be an (ω, s)-maximizer, and suppose that (MUP) holds. From Lemma 3, we have $\tilde{\mu}(s) = \sum_{a\in A} sa - \mu(A)$ and $\nabla_s(\omega) = \mu(A) - \omega(A)$. Then,

$$\begin{aligned}\Phi_{\mu,s} + \nabla_s(\omega) &= \sum_{a\in A} sa - \mu(A) + \Psi_{\mu,s} + \mu(A) - \omega(A) \\ &= \sum_{a\in A} sa - \omega(A) + \Psi_{\mu,s} \\ &\leq \Phi_{\omega,r},\end{aligned}$$

and, by Lemma 2, we conclude that WFA is 3-competitive. ■

3 The Potential Function

In this section and later throughout the paper we assume that $k = 3$. We now introduce our potential function Φ. We then show that Φ satisfies the offset property and that, under certain conditions, it also satisfies the update property.

Main idea. Finding an appropriate potential function is the main difficulty in proving competitiveness. One standard approach that applies to many online problems is to use the "lazy adversary" idea: Assume that the adversary is in some configuration X, and calculate the maximum cost of the algorithm on request sequences $\rho \in X^*$. Clearly, the potential has to be at least as large as the obtained quantity, so that the algorithm has enough "savings" to pay for serving such sequences ρ. Then subtract $(k+1)\cdot\omega(X)$, since the adversary will pay $\omega(X)$. From the analysis of spaces with $k+1$ points, it can be seen that we also need to add another quantity equal to the sum of all distances in X. Thus obtained value is referred to as the *lazy potential.*

As a better illustration, start with $k=2$. In that case, $X=\{x,y\}$, and, if we additionally assume that x is the last request, we only have one lazy request sequence in X to consider: $\rho=(xy)^*$. Thus the potential will be the maximum, over all choices of y, of WFA's cost on ρ, plus xy, and minus $3\cdot\omega(x,y)$. This value can be expressed by a closed-form expression and, indeed, this potential can be used to prove that WFA is 2-competitive for 2 servers [3, 4].

We now try to extend this idea to 3 servers. In this case, $X=\{x,y,z\}$ where, again, we assume that x is the last request. The main difficulty that arises for 3 servers is that now there are infinitely many possible request sequences on points in X, and it is not known whether the maximum cost of WFA on these sequences (or the pseudocost) can be expressed in closed form.

The general idea of our proof is to choose lazy sequences in which the adversary "reveals" his positions to the algorithm as late as possible. To this end, we only consider sequences of the form $\rho=(yx)^{i_1}z(yx)^{i_2}z\ldots$, where each i_j is large enough so that after requesting $(yx)^{i_j}$, requesting x or y does not change the work function. We call it a *procrastinating* potential. The derivation of the formula for the procrastinating potential is rather technical and involved, and since we do not need the derivation for our purpose, we state the formula without proof.

To simplify notation, throughout this section, we assume that ω is a work function with the last request r. Then, if no ambiguity arises, we will write $\omega(x,y)$, instead of $\omega(r,x,y)$. We will also write Φ_ω instead of $\Phi_{\omega,r}$, etc. Letters a,b,c,d,p,q,e,f, possibly with accents or subscripts, denote points in M. Let also

$$\ddot{\omega}(x)=\max_{a,a'}\{xa+xa'-\omega(a,a')\} \tag{5}$$

By the comments following the definition of the shadow $\tilde{\omega}(x)$ in the previous section, there is a maximizer of ω that contains r. Thus $\tilde{\omega}(x)$ and $\ddot{\omega}(x)$ are very closely related, namely $\tilde{\omega}(x)=\ddot{\omega}(x)+rx$. In particular, $\ddot{\omega}(r)=\tilde{\omega}(r)$. In this notation, the procrastinating potential is

$$\Phi_\omega=\ddot{\omega}(r)+\sup_{p,d,d'}\{\ddot{\omega}(p)+dd'-\omega(p,d)-\omega(p,d')\}$$

It is quite easy to show that Φ_ω satisfies the offset property (OP). The rest of this section focuses on the verification of the update property (UP).

Two other formulas. We now give two other formulas which we use to estimate our potential.

$$\Lambda_\omega = \sup_{p,q,e,e'} \{-rp + \ddot{\omega}(p) - rq + \ddot{\omega}(q) - \omega(p,q) + ee' - \omega(e,e')\}$$

$$\Gamma_\omega = \sup_{p,q,d,d',f} \{-rp + \ddot{\omega}(p) + rq + dd' - \omega(q,d) - \omega(q,d') + qf - \omega(p,f)\}$$

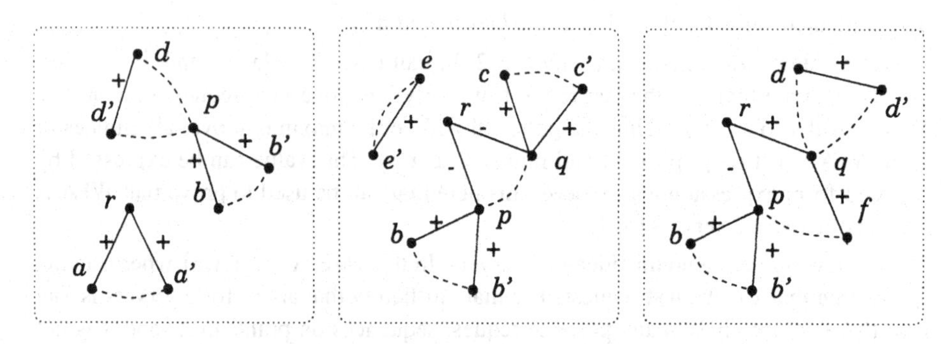

Fig. 1. A graphical representation of functions Φ, Λ and Γ.

The formulas for Φ, Λ and Γ are illustrated graphically in Figure 1. In this diagram, a solid line between x,y represents the distance xy, and a dashed line between x,y represents $\omega(x,y)$. The work function values are always negative. The "+" or "−" labels on solid lines show whether the corresponding distance is a positive or a negative term.

Now we are ready to state conditions under which WFA is 3-competitive in a given metric space.

Theorem 5. *Let M be any metric space. If $\Lambda_\omega \le \Phi_\omega$ and $\Gamma_\omega \le \Phi_\omega$ for any work function ω over M, then* WFA *for 3 servers on M is 3-competitive.*

Proof. We use Corollary 4. First, we need to verify the offset property (OP). Suppose that ω is minimized on configuration $\{r,a,b\}$. By choosing suitable points in the in the formula for Φ_ω, we get $\Phi_\omega \ge ra + rb - \omega(a,b) + aa + ab - \omega(a,b) + bb - \omega(a,b) - \omega(a,b) \ge -4\omega(a,b)$, and (OP) follows.

We now verify the update property. Recall that, without loss of generality, the maximizer contains the last request, that is, $\ddot{\omega}(r) = \tilde{\omega}(r)$. Therefore Φ_ω is of the form $\tilde{\omega}(r) + \Psi_\omega$. By Corollary 4, it is sufficient to show the following inequality:

$$\begin{aligned} sr + sa + sa' - \omega(a,a') + pb + pb' - \mu(s,b,b') \\ + dd' - \mu(s,p,d) - \mu(s,p,d') \le \Phi_\omega \end{aligned} \tag{6}$$

Before presenting the calculations for (6), we restate, in a more explicit form, the formulas (1) and (2). First, for $k = 3$, the update operation takes the form:

$$\mu(s,x,y) = \min \left\{ \begin{array}{l} \omega(x,y) + rs \\ \omega(s,x) + ry \\ \omega(s,y) + rx \end{array} \right\} \tag{7}$$

for all x, y. The quasiconvexity property implies that

$$\omega(x,y) + \omega(u,v) \geq \min \left\{ \begin{array}{l} \omega(x,u) + \omega(y,v) \\ \omega(x,v) + \omega(y,u) \end{array} \right\} \tag{8}$$

for any $x, y, u, v \in M$.

We are now ready to prove (6). The proof is by analysis of cases, depending on which of the three choices in equation (7), for each of $\mu(s,b,b')$, $\mu(s,p,d)$ and $\mu(s,p,d')$, realizes the minimum. Denote by LS the left side of equation (6).

Case 1: $\mu(s,b,b') = \omega(b,b') + rs$, $\mu(s,p,d) = \omega(p,d) + rs$, and $\mu(s,p,d') = \omega(p,d') + rs$. Then

$$\begin{aligned} \text{LS} &= sa - rs + sa' - rs - \omega(a,a') + pb + pb' - \omega(b,b') + dd' - \omega(p,d) - \omega(p,d') \\ &\leq ra + ra' - \omega(a,a') \ + \ pb + pb' - \omega(b,b') \ + \ dd' - \omega(p,d) - \omega(p,d') \\ &\leq \Phi_\omega \end{aligned}$$

Case 2: $\mu(s,b,b') = \omega(b,b') + rs$, $\mu(s,p,d) = \omega(p,d) + rs$, and $\mu(s,p,d') = \omega(s,p) + rd'$. By quasiconvexity and symmetry between a, a', we can assume that $\omega(a,a') + \omega(p,d) \geq \omega(a,d) + \omega(p,a')$. Then

$$\begin{aligned} \text{LS} &= sa - rs + dd' - rd' - \omega(a,a') + pb + pb' - \omega(b,b') + sa' - \omega(p,d) - \omega(p,s) \\ &\leq ra + rd - \omega(a,d) \ + \ pb + pb' - \omega(b,b') \ + \ sa' - \omega(p,s) - \omega(p,a') \\ &\leq \Phi_\omega \end{aligned}$$

Case 3: $\mu(s,b,b') = \omega(b,b') + rs$, $\mu(s,p,d) = \omega(p,d) + rs$, and $\mu(s,p,d') = \omega(s,d') + rp$. Then

$$\text{LS} = sa + sa' - \omega(a,a') - rs + pb + pb' - \omega(b,b') + dd' - \omega(s,d') - \omega(p,d)$$

If $\omega(s,d') + \omega(p,d) \geq \omega(s,p) + \omega(d,d')$, then

$$\begin{aligned} \text{LS} &\leq -rs + sa + sa' - \omega(a,a') - rp + pb + pb' - \omega(b,b') - \omega(s,p) \\ &\qquad + dd' - \omega(d,d') \\ &\leq \Lambda_\omega \end{aligned}$$

Otherwise, by quasiconvexity, $\omega(s,d') + \omega(p,d) \geq \omega(s,d) + \omega(p,d')$. Then

$$\begin{aligned} 2 \cdot \text{LS} &\leq sa + sa' - 2rs - \omega(a,a') + pb + pb' - \omega(b,b') \\ &\qquad + dd' - \omega(p,d) - \omega(p,d') + pb + pb' - 2rp - \omega(b,b') \\ &\qquad + sa + sa' - \omega(a,a') \ + \ dd' - \omega(s,d) - \omega(s,d') \\ &\leq 2 \cdot \Phi_\omega \end{aligned}$$

because $sa + sa' - 2rs \leq ra + ra'$ and $pb + pb' - 2rp \leq rb + rb'$.

Case 4: $\mu(s,b,b') = \omega(b,b') + rs$, $\mu(s,p,d) = \omega(s,p) + rd$, and $\mu(s,p,d') = \omega(s,p) + rd'$. By quasiconvexity and symmetry, we can assume that $\omega(a,a') + \omega(s,p) \geq \omega(s,a) + \omega(p,a')$. Then

$$\begin{aligned} \text{LS} &= sa + sa' - \omega(a,a') + pb + pb' - \omega(b,b') - 2\omega(s,p) \\ &\leq sa - \omega(s,a) + pb + pb' - \omega(b,b') + sa' - \omega(s,p) - \omega(p,a') \\ &\leq \Phi_\omega \end{aligned}$$

because $sa \leq rs + ra$.

Case 5: $\mu(s,b,b') = \omega(b,b') + rs$, $\mu(s,p,d) = \omega(s,p) + rd$, and $\mu(s,p,d') = \omega(s,d') + rp$. By quasiconvexity and symmetry, we can assume that $\omega(b,b') + \omega(s,d') \geq \omega(s,b) + \omega(b',d')$. Then

$$\begin{aligned} \text{LS} &= sa + sa' - \omega(a,a') + pb + pb' - rp - \omega(b,b') + dd' - rd - \omega(s,p) - \omega(s,'d) \\ &\leq rb' + rd' - \omega(b',d') + sa + sa' - \omega(a,a') + pb - \omega(s,p) - \omega(s,b) \\ &\leq \Phi_\omega \end{aligned}$$

The proof for the remaining cases will be given in the full version of the paper. ∎

4 WFA in the Manhattan Plane

In this section we assume that $k = 3$ and the given metric space is the Manhattan plane $\mathbf{R}^2_1$, that is $\mathbf{R}^2$ with the city-block metric: $xy = |x^1 - y^1| + |x^2 - y^2|$, where x^1, x^2 denote the coordinates of a point x.

We say that $(x_1, \dots, x_m)$ is a *linear m*-tuple if $x_i x_j + x_j x_\ell = x_i x_\ell$ for all $1 \leq i \leq j \leq \ell \leq m$. If B is a finite set of ordered pairs, we say that B is a *parallel bundle* if there exist points x, y such that (x,a,b,y) is a linear 4-tuple for all $(a,b) \in B$. We also say that the pairs in B are *parallel*. We use repeated double bars for a finite parallel bundle. For example we write $\begin{matrix} a \\ b \end{matrix} \Big\| \begin{matrix} c \\ d \end{matrix} \Big\| \begin{matrix} e \\ f \end{matrix}$ to indicate that $\{(a,b),(c,d),(e,f)\}$ is a parallel bundle. Similarly, if B is a set of unordered pairs, we say that B is a parallel bundle if each of the pairs can be ordered so that the resulting set of ordered pairs is a parallel bundle.

Lemma 6. *Let P be a finite set of pairs of points in the Manhattan plane. Then P is a union of two parallel bundles.*

Proof. Order each pair in P, so that $(a,b) \in P$ implies $a^1 \leq b^1$. Define B (resp. C) to be the set of all pairs in P, where $a^2 \leq b^2$ (resp. $b^2 \leq a^2$). Pick $N \in \mathbf{R}$ such that $N \geq |a^i|$ and $N \geq |b^i|$ for all $(a,b) \in B$ and $i = 1,2$. Define x, y by $x = (-N,-N)$ and $y = (N,N)$. Then, for any $(a,b) \in B$, (x,a,b,y) is a linear 4-tuple. Similarly define $u = (-N,N)$ and $v = (N,-N)$. Then, for any $(a,b) \in C$, (u,a,b,v) is a linear 4-tuple. ∎

Theorem 7. *WFA is 3-competitive for 3 servers in the Manhattan plane.*

Proof. By Theorem 5, it is sufficient to show that $\Lambda \leq \Phi$ and $\Gamma \leq \Phi$.

Proof of inequality $\Gamma \leq \Phi$. Pick points p,b,b',q,d,d',f such that

$$\Gamma_\omega = -rp + pb + pb' - \omega(b,b') + rq + dd' - \omega(q,d) - \omega(q,d') + qf - \omega(p,f)$$

The intuition behind the proof is that, by Lemma 6, the four pairs of points; $\{f,q\}$, $\{b,p\}$, $\{b',p\}$, and $\{d,d'\}$, must fall into at most two parallel bundles. Depending on how they fall, we take advantage of the parallel relationships between the pairs to obtain the result.

Case 1: $\begin{smallmatrix} b \\ p \end{smallmatrix}\Big\|\begin{smallmatrix} p \\ b' \end{smallmatrix}$. Pick x,y such that (x,b,p,y) and (x,p,b',y) are linear 4-tuples. Then (x,b,p,b',y) is a linear 5-tuple. Since $bp + b'p = bb' \leq rb + rb'$ and $rq - rp \leq pq$, we get

$$\begin{aligned}\Gamma_\omega &\leq rb + rb' - \omega(b,b') + qf + qp - \omega(f,p) + dd' - \omega(q,d) - \omega(q,d') \\ &\leq \Phi_\omega\end{aligned}$$

Case 2: $\begin{smallmatrix} d \\ d' \end{smallmatrix}\Big\|\begin{smallmatrix} f \\ q \end{smallmatrix}$. Pick a point x such that (x,d,d') and (x,f,q) are linear triples. Then $dd' = xd' - xd$ and $fq = xq - xf$. By quasiconvexity, without loss of generality, $\omega(b,b') + \omega(q,d) \geq \omega(b,d) + \omega(q,b')$. Therefore

$$\begin{aligned}\Gamma_\omega &\leq rq + pb' - rp - \omega(b',q) + xd' + xq - \omega(d',q) \\ &\qquad + bp - xd - \omega(b,d') - xf - \omega(f,p) \\ &\leq rq + rb' - \omega(b',q) + xd' + xq - \omega(d',q) + bp - \omega(x,b) - \omega(x,p) \\ &\leq \Phi_\omega\end{aligned}$$

The analysis of the remaining cases will be given in the full version of this paper.

Proof of inequality $\Lambda \leq \Phi$. Pick p,b,b',q,c,c',e,e' such that

$$\Lambda_\omega = -rp + pb + pb' - \omega(b,b') - rq + qc + qc' - \omega(c,c') + ee' - \omega(e,e')$$

The intuition behind the proof is that, by Lemma 6, the five pairs of points; $\{e,e'\}$, $\{b,p\}$, $\{b',p\}$, $\{q,c\}$, and $\{q,c'\}$, must fall into at most two parallel bundles. Depending on how they fall, we take advantage of the parallel relationships between the pairs to obtain the result.

Case 1: $\begin{smallmatrix} b \\ p \end{smallmatrix}\Big\|\begin{smallmatrix} p \\ b' \end{smallmatrix}$. Pick x,y such that (x,b,p,y) and (x,p,b',y) are linear 4-tuples. Then (x,b,p,b',y) is a linear 5-tuple, implying that $bp + pb' = bb'$. Without loss of generality, $\omega(b,b') + \omega(p,q) \geq \omega(b,p) + \omega(b',q)$. Therefore

$$\begin{aligned}\Lambda_\omega &\leq ee' - \omega(e,e') + qc + qc' - \omega(c,c') + bb' - \omega(b',q) - \omega(b,p) - rp - rq \\ &\leq re + re' - \omega(e,e') + qc + qc' - \omega(c,c') + bb' - \omega(b',q) - \omega(b,q) \\ &\leq \Phi_\omega\end{aligned}$$

Case 2: $\begin{smallmatrix} e \\ e' \end{smallmatrix} \Big\| \begin{smallmatrix} b \\ p \end{smallmatrix} \Big\| \begin{smallmatrix} c \\ q \end{smallmatrix}$. Pick x such that (x,e,e'), (x,b,p) and (x,c,q) are linear triples, implying that $ee' = xe' - xe$, $bp = xp - xb$ and $cq = xq - xc$. By quasiconvexity, without loss of generality, $\omega(e,e') + \omega(p,q) \geq \omega(e,p) + \omega(e',q)$. Therefore

$$\begin{aligned}\Lambda_\omega &\leq xp - rp + qc' - rq - xc - \omega(c,c') + xe' + xq - \omega(e',q) \\ &\qquad + b'p - ex - \omega(e,p) - xb - \omega(b,b') \\ &\leq rx + rc' - \omega(x,c') + xe' + xq - \omega(e',q) + b'p - \omega(x,p) - \omega(x,b') \\ &\leq \Phi_\omega\end{aligned}$$

The analysis of the remaining cases will be given in the full version of this paper.

Corollary 8. *There is a* $3\sqrt{2}$*-competitive algorithm for 3 servers in the Euclidean plane.*

Proof. Write $\|x,y\|_2$ and $\|x,y\|_1$ for the Euclidean and the city-block metric, respectively. Then $\frac{1}{\sqrt{2}}\|x,y\|_1 \leq \|x,y\|_2 \leq \|x,y\|_1$, for any two points $x,y \in \mathbf{R}^2$. For any request sequence in the plane, pretend that the metric is the city-block metric and follow WFA. The algorithm's cost in $\mathbf{R}^2_2$ is at most its cost in $\mathbf{R}^2_1$. The optimal cost in $\mathbf{R}^2_1$ is at most $\sqrt{2}$ times the optimal cost in $\mathbf{R}^2_2$. The result follows immediately from Theorem 7. ∎

5 Final Comments

We proved that WFA is 3-competitive for 3 servers in the Manhattan plane. This immediately implies that WFA is also 3-competitive in $\mathbf{R}^2_\infty$, the plane with the supnorm metric, because $\mathbf{R}^2_\infty$ and $\mathbf{R}^2_1$ are isometric.

We believe that our method can be used to prove that the WFA is 3-competitive in other metric spaces of interest. According to Theorem 5, in order to prove this result, one only needs to prove that $\Lambda, \Gamma \leq \Phi$. We conjecture that these inequalities are true on the circle. We also conjecture that the same method will work in $\mathbf{R}^2_2$ (improving the ratio from Corollary 8 to 3).

Theorem 5 does not apply to arbitrary metric spaces. We have an example of a metric space M and a work function ω for which Theorem 5 fails, namely $\Lambda_{\omega,r} > \Phi_{\omega,r}$. Nevertheless, by pursuing further our approach and studying lazy adversary sequences that generalize the procrastinating adversary, it may be possible to obtain even better potential functions that work in arbitrary metric spaces.

References

[1] Marek Chrobak, Howard Karloff, Tom H. Payne, and Sundar Vishwanathan. New results on server problems. *SIAM Journal on Discrete Mathematics*, 4:172–181, 1991.

[2] Marek Chrobak and Lawrence L. Larmore. An optimal online algorithm for k servers on trees. *SIAM Journal on Computing*, 20:144–148, 1991.

[3] Marek Chrobak and Lawrence L. Larmore. The server problem and on-line games. In *DIMACS Series in Discrete Mathematics and Theoretical Computer Science*, volume 7, pages 11–64, 1992.
[4] Marek Chrobak and Lawrence L. Larmore. Metrical task systems, the server problem, and the work function algorithm. In *Online Algorithms: State of the Art*, pages 74–94. Springer-Verlag, 1998.
[5] Elias Koutsoupias and Christos Papadimitriou. On the *k*-server conjecture. In *Proc. 26th Symp. Theory of Computing*, pages 507–511, 1994.
[6] Elias Koutsoupias and Christos Papadimitriou. On the *k*-server conjecture. *Journal of the ACM*, 42:971–983, 1995.
[7] Elias Koutsoupias and Christos Papadimitriou. The 2-evader problem. *Information Processing Letters*, 57:249–252, 1996.
[8] Mark Manasse, Lyle A. McGeoch, and Daniel Sleator. Competitive algorithms for server problems. *Journal of Algorithms*, 11:208–230, 1990.

Quartet Cleaning: Improved Algorithms and Simulations

Vincent Berry[1]*, Tao Jiang[2]**, Paul Kearney[3]***, Ming Li[3]†, and Todd Wareham[2]‡

[1] Département de Mathématiques, Université de Saint-Etienne
[2] Department of Computing and Software, McMaster University
[3] Department of Computer Science, University of Waterloo
[4] Department of Computing and Software, McMaster University

Abstract. A critical step in all quartet methods for constructing evolutionary trees is the inference of the topology for each set of four sequences (i.e. *quartet*). It is a well–known fact that all quartet topology inference methods make mistakes that result in the incorrect inference of quartet topology. These mistakes are called *quartet errors*. In this paper, two efficient algorithms for correcting bounded numbers of quartet errors are presented. These "quartet cleaning" algorithms are shown to be optimal in that no algorithm can correct more quartet errors. An extensive simulation study reveals that sets of quartet topologies inferred by three popular methods (Neighbor Joining [15], Ordinal Quartet [14] and Maximum Parsimony [10]) almost always contain quartet errors and that a large portion of these quartet errors are corrected by the quartet cleaning algorithms.

1 Introduction

The explosion in the amount of DNA sequence data now available [3] has made it possible for biologists to address important large scale evolutionary questions [11, 12, 19]. In the analysis of this data an evolutionary tree T that describes the evolutionary history of the set S of sequences involved is produced. T is modeled by an edge–weighted rooted tree where the leaves are labeled bijectively by sequences in S. Due to the large data sets involved, standard approaches for constructing evolutionary trees, such as maximum likelihood [9] and maximum parsimony [18], that exhaustively search the entire tree space are not feasible.

In recent years *quartet methods* for constructing evolutionary trees have received much attention in the computational biology community [1, 2, 4, 8, 14, 17]. Given a

* Supported in part by ESPRIT LTR Project no. 20244 — ALCOM-IT.
vberry@univ-st-etienne.fr
** Supported in part by NSERC Research Grant OGP0046613 and a CGAT grant.
jiang@cas.mcmaster.ca
*** Supported in part by NSERC Research Grant OGP217246-99 and a CITO grant.
pkearney@math.uwaterloo.ca
† Supported in part by NSERC Research Grant OGP0046506, CITO, a CGAT grant, and the Steacie Fellowship. mli@math.uwaterloo.ca
‡ Supported in part by a CGAT grant. harold@cas.mcmaster.ca

quartet of sequences $\{a,b,c,d\}$ and an evolutionary tree T, the *quartet topology* induced in T by $\{a,b,c,d\}$ is the path structure connecting a, b, c and d in T. Given a quartet $\{a,b,c,d\}$, if the path in T connecting sequences a and b is disjoint from the path in T connecting sequences c and d then the quartet is said to be *resolved* and is denoted $ab|cd$. Otherwise, the quartet is said to be *unresolved* and is denoted $(abcd)$. The four possible quartet topologies that can be induced by a quartet are depicted in Fig. 1.

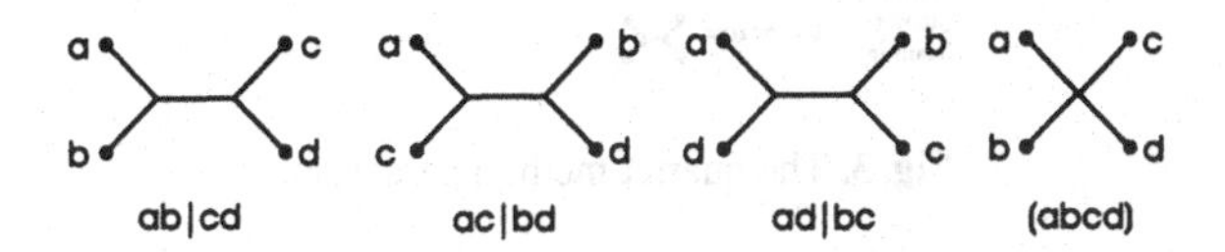

Fig. 1. The four quartet topologies for quartet $\{a,b,c,d\}$.

Quartet methods are based upon the fact that the topology of an evolutionary tree T (that is, T without it's edge weights) is uniquely characterized by its set Q_T of induced quartet topologies [5] (see Fig. 2). This suggests the following three step process, referred to as the *quartet method paradigm*, for estimating an unknown evolutionary tree T for a set S of sequences:

1. For each quartet $\{a,b,c,d\}$ of sequences in S, estimate the quartet topology induced by $\{a,b,c,d\}$ in T. The procedure for producing this estimate is called a *quartet topology inference method*. Let Q be the set of $\binom{n}{4}$ inferred quartet topologies.
2. The quartet topologies in Q are recombined to produce an estimate T' of T's topology. The procedure for producing T' is called a *quartet recombination method*.
3. T' is rooted and edge weights determined.

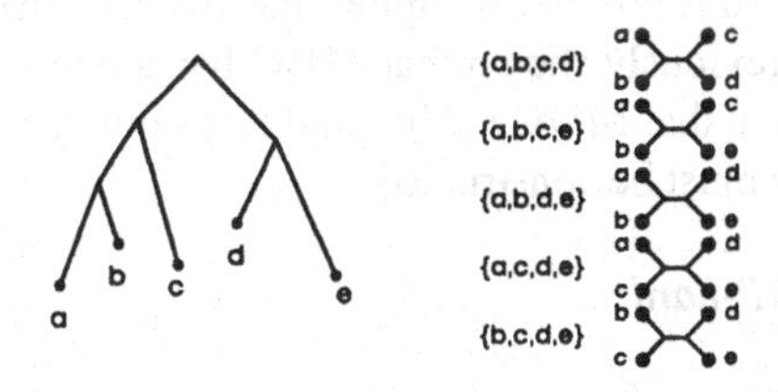

Fig. 2. An evolutionary tree T and its set Q_T of induced quartet topologies.

The quartet method paradigm is illustrated in Fig. 3. There are many quartet topology inference methods including maximum parsimony [10], maximum likelihood [9], neighbor joining [15] and the ordinal quartet method [14]. The reader is directed to [18] for a good overview of these methods. Notice that computationally intensive methods such as maximum likelihood and maximum parsimony can be used to infer quartet topology but are infeasible when used to infer the entire tree topology. Existing quartet recombination methods include the Q^* method [2], short quartet method [8], and quartet puzzling [17] among others [1, 13]. The third step in the quartet method is well–understood [4, 18].

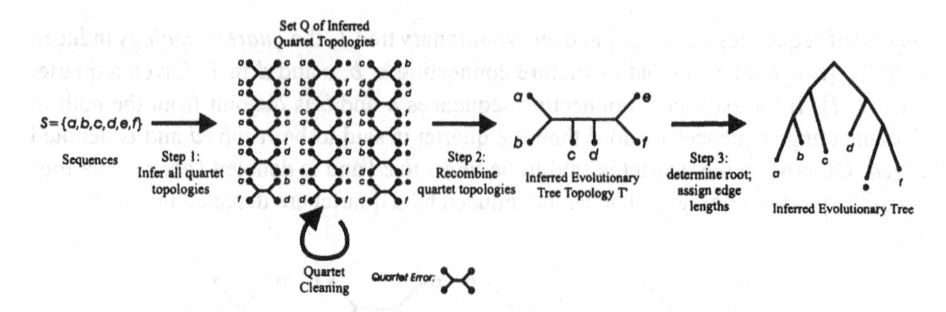

Fig. 3. The quartet method paradigm.

The algorithmic interest in the quartet method paradigm derives from the fact that quartet topology inference methods make mistakes, and so, the set Q of inferred quartet topologies contains *quartet errors*. The quartet $\{a,b,c,d\}$ is a quartet error if $ab|cd \in Q_T$ but $ab|cd \notin Q$. For example, in Fig. 3, $\{a,b,c,e\}$ is a quartet error since $ab|ce \in Q_T$ but $ac|be \in Q$. In this sense, Q is an estimate of Q_T. Consequently, the problem of recombining quartet topologies of Q to form an estimate T' of T is typically formulated as an optimization problem:

Maximum Quartet Consistency (MQC)
Instance: Set Q containing a quartet topology for each quartet of sequences in S and $k \in \mathcal{N}$.
Question: Is there an evolutionary tree T' labeled by S such that $|Q_{T'} \cap Q| \geq k$?

Problem **MQC** is NP–hard if the input Q is a partial set of quartets, i.e., quartet topologies can be missing [16] (this proof also implies that this version of **MQC** is MAX–SNP hard). Though it was previously shown that **MQC** has a polynomial time approximation scheme [13], proving that **MQC** is NP–hard turns out to be more difficult since a complete set of quartets must be constructed[1]:

Theorem 1. **MQC** *is NP-hard.*

Clearly, the accuracy of the estimate T' depends almost entirely on the accuracy of Q. In fact, many quartet recombination methods are very sensitive to quartet errors in Q. For example, the Q^* method [2] and the Short Quartet Method [8] can fail to recover the true evolutionary tree even if there is only one quartet error in Q. Hence, although much effort has been directed towards the development of quartet recombination methods, of prior importance is the development of methods that improve the accuracy of inferred quartet topologies.

This paper presents research on the detection and correction of quartet errors in Q. Methods for detecting and correcting quartet errors are called "quartet cleaning" methods (see Fig. 3). The principal contributions of the research presented here are two

[1] Several proofs have been omitted due to space constraints. Please contact the authors for details.

quartet cleaning methods, namely *global edge cleaning* and *local vertex cleaning*, and an extensive simulation study that establishes both the need for these quartet cleaning algorithms and their applicability.

These results are described in more detail in Sect. 1.1 and contrasted with previous results in Sect. 1.2.

1.1 Terminology and Results

Let S be a set of sequences, Q be a set of quartet topologies and T the true evolutionary tree for S. In this paper it can be assumed that Q contains a quartet topology for each quartet taken from S. Several concepts must be introduced so that quartet cleaning can be discussed in detail.

An edge e in tree T induces the bipartition (A,B) if $T-\{e\}$ consists of two trees where one is labeled by A and the other by B. This is denoted $e=(A,B)$. Let $Q(A,B)$ denote the set of quartet topologies of the form $aa'|bb'$ where $a,a' \in A$ and $b,b' \in B$. An internal vertex v in tree T induces the tripartition (A,B,C) if $T-\{v\}$ consists of three trees labeled by A, B and C, respectively. This is denoted $v=(A,B,C)$. Let $Q(A,B,C)$ denote the union of $Q(A,B)$, $Q(A,C)$ and $Q(B,C)$. Two bipartitions (A,B) and (C,D) are *compatible* if they can be induced in the same tree, *i.e.*, either $A \subseteq C$ or $C \subseteq A$. Similarly, two tripartitions are compatible if they can be induced in the same tree.

In order to assess the performance of quartet cleaning algorithms an understanding of the distribution of quartet errors in T is needed. Let $P_T(a,b)$ be the path between sequences a and b in tree T. If $ab|cd$ is induced in T then $P_T(a,c) \cap P_T(b,d)$ is called the *joining path* of $\{a,b,c,d\}$ in T. Notice that the joining path is necessarily non-empty (see Fig. 1). Define the quartet error $\{a,b,c,d\}$ to be *across* edge e if e is on the joining path of $\{a,b,c,d\}$ in T. Similarly, define the quartet error $\{a,b,c,d\}$ to be *across* vertex v if v is on the joining path of $\{a,b,c,d\}$ in T. These definitions permit the assignment of quartet errors in Q to edges/vertices of T.

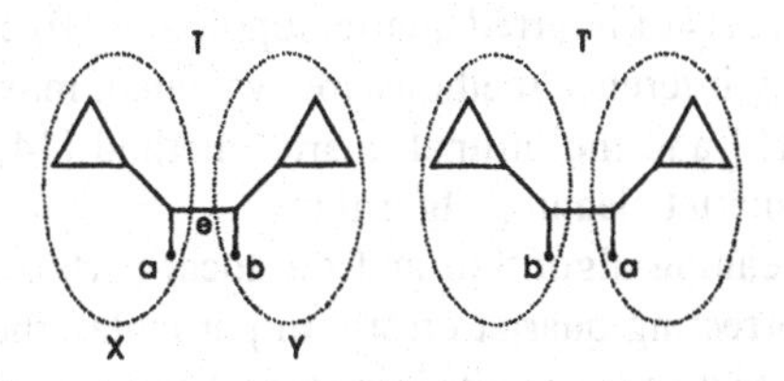

Fig. 4. Cleaning bounds.

Let $e=(X,Y)$ be the bipartition induced in T as depicted in Fig. 4. Observe that Q_T and $Q_{T'}$ differ by quartets of the form $ax|by$ where $x \in X$ and $y \in Y$. It follows that $|Q_T - Q_{T'}| = (|X|-1)(|Y|-1)$. If half of the quartets of the form $\{a,b,x,y\}$ with $x \in X$ and $y \in Y$ have quartet topology $ax|by$ in Q and the other half have quartet topology $bx|ay$ in Q then no quartet cleaning algorithm can guarantee that all quartet errors across e can be corrected under the **MQC** principle of optimality. This implies that $(|X|-1)(|Y|-1)/2$

is an upper bound on the number of quartet errors across an edge of T that can be corrected. This example motivates the following formulations of quartet cleaning:

Local Edge Cleaning A local edge cleaning algorithm with *edge cleaning bound b* corrects all quartet errors across any edge with fewer than b quartet errors across it.

Global Edge Cleaning A global edge cleaning algorithm with *edge cleaning bound b* corrects all quartet errors in Q if each edge of T has fewer than b quartet errors across it.

Analogous definitions apply for local and global vertex cleaning algorithms. Local edge/vertex cleaning is more robust that global edge/vertex cleaning since it can be applied to an edge/vertex independently of the number of quartet errors across other edges/vertices. This is a significant feature especially when some edges/vertices have a high number of quartet errors across them. In contrast, global cleaning algorithms are applicable only if all edges/vertices satisfy the cleaning bound.

The example from Fig. 4 illustrates that cleaning bounds should not be constant but vary with bipartition sizes. Hence, an edge $e = (X,Y)$ would have an edge cleaning bound that depends on $|X|$ and $|Y|$. In particular, the example demonstrates that the optimal edge cleaning bound is $(|X|-1)(|Y|-1)/2$.

The first contribution of the paper is an $O(n^4)$ time global edge cleaning algorithm with edge cleaning bound $(|X|-1)(|Y|-1)/2$. Following the above remarks, this algorithm is optimal in the number of quartet errors it can correct across an edge $e = (X,Y)$. The global edge cleaning algorithm is presented in Sect. 2.1.

The second contribution of the paper is an $O(n^7)$ time local vertex cleaning algorithm with vertex cleaning bound $(|X|-1)(|Y|-1)/4$. Although this algorithm has a smaller cleaning bound than the global edge cleaning algorithm, it is more robust since it is local. Hence, there are situations where the local vertex cleaning algorithm is superior to the global edge cleaning algorithm and vice–versa. The local vertex cleaning algorithm is presented in Sect. 2.3.

The third contribution of the paper is an extensive simulation study that assesses the utility of the above quartet cleaning algorithms. This study establishes the following:

- Regardless of the quartet topology inference method used to obtain Q, quartet errors are prevalent in sets of inferred quartet topologies. To establish this three popular quartet topology inference methods are evaluated: maximum parsimony [10], neighbor joining [15] and the ordinal quartet method [14]. This establishes that there is a need for quartet cleaning algorithms.
- The global edge cleaning algorithm and the local vertex cleaning algorithm are very effective at correcting quartet errors. In particular, the local vertex cleaning algorithm is more effective (due to its robustness) and both algorithms dramatically increase the accuracy of the inferred quartet topology set Q.

The simulation study is presented in Sect. 3.

1.2 Previous Results

The idea of quartet cleaning was introduced in [13]. This paper presented a polynomial time global edge quartet cleaning algorithm with cleaning bound $\alpha(|X|-1)(|Y|-1)/2$ where $\alpha > 0$ is a fixed constant. Hence, this algorithm is suboptimal. Although this algorithm has a polynomial time complexity, it is of very high degree, and so, of primarily theoretical interest. The algorithms presented here are more efficient, more effective (the global edge cleaning algorithm is optimal) and more robust.

2 Quartet Cleaning Algorithms

2.1 A Global Edge Quartet Cleaning Algorithm

Let S be a set of sequences that evolved on evolutionary tree T and let Q be a set of quartet topologies inferred from S. Assume that there are fewer than $(|A|-1)(|B|-1)/2$ quartet errors across each edge $e=(A,B)$ of T. In other words, assume that $|Q(A,B)-Q|<(|A|-1)(|B|-1)/2$. The following algorithm constructs T from Q by building T iteratively from the leaves up. Note that for the purposes of this algorithm, a leaf is in R is considered to be a rooted subtree.

Algorithm Global-Clean(Q)
1. Let $R:=S$.
2. For every pair of rooted subtrees T_1 and T_2 in R
3. Let A denote the leaf sequences of T_1 and T_2.
4. If $|Q(A,S-A)-Q|<(|A|-1)(|S-A|-1)/2$ then
5. Create a new tree T' that contains T_1 and T_2 as its subtrees.
6. Let $R:=R-\{T_1,T_2\}\cup\{T'\}$.
7. Repeat step 2 until $|R|=3$.
8. Connect the three subtrees in R at a new vertex and output the resulting unrooted tree.

Theorem 2. *Algorithm Global-Clean outputs the tree T correctly.*

Note that if not all edges of T satisfy the global cleaning bound of $(|A|-1)(|B|-1)/2$ then Global–Clean fails to return an evolutionary tree.

A straightforward implementation of Global–Clean results in a time complexity of $O(n^2\cdot n^4)=O(n^6)$, since each quartet can is checked at most $O(n^2)$ times. Section 2.2 presents a more careful implementation and analysis that yields a linear $O(n^4)$ global edge cleaning algorithm.

2.2 An Efficient Implementation of Global-Clean

The idea is as follows. First, observe that the most time consuming step in Global-Clean is step 4, *i.e.*, the identification of discrepancies between the set of quartet topologies induced by a candidate bipartition and those in the given set Q. Our idea is to avoid considering the same quartet repeatedly as much as possible. In order to achieve this, for each correct bipartition $(A,S-A)$, we record the set $Q(A,S-A)-Q$ using a linked list. Let A_1 and A_2 denote the leaf sequences of T_1 and T_2, respectively, considered in step 2 of Global-Clean, and let $A=A_1\cup A_2$, according to step 3. Then, in step 4, observe that $Q(A,S-A)-Q=(Q(A_1,S-A)-Q)\cup(Q(A_2,S-A)-Q)\cup(Q'(A_1,A_2,S-A)-Q)$, where $Q'(A_1,A_2,S-A)$ denotes the set of all quartet topologies $ab|cd$ with $a\in A_1,b\in A_2$, and $c,d\in S-A$. Since $Q(A_1,S-A)-Q\subseteq Q(A_1,S-A_1)-Q$, we can compute $Q(A_1,S-A)-Q$ from the list for $Q(A_1,S-A_1)$ in at most $|Q(A_1,S-A_1)|=O(n^2)$ time. Similarly, $Q(A_2,S-A)-Q$ can be computed in $O(n^2)$ time. Moreover, at each execution of step 2 (except for the first time), we need only check $O(n)$ new possibilities of joining subtrees: joining the subtree resulting from the previous iteration, say T_1, with each of the $O(n)$ other subtrees, say T_2. Since all the other possibilities have been examined in previous iterations, their outcomes can be easily retrieved. Because step 2 is repeated $n-2$ times, the overall time required for computing the sets $Q(A_1,S-A)-Q$ and $Q(A_2,S-A)-Q$ is $O(n\cdot n\cdot n^2)=O(n^4)$.

Next we show that computing the sets $Q'(A_1,A_2,S-A)-Q$ also globally requires $O(n^4)$ time, thanks to some delicate data structures. For each possible clustering $\{T_1,T_2\}$ of two subtrees, we split the set $Q'(A_1,A_2,S-A)$ into two lists, $Q'_-(A_1,A_2)$ and $Q'_+(A_1,A_2)$, depending on whether a topology contradicts or corresponds to the one in the set Q. That is, $Q'_-(A_1,A_2)=Q'(A_1,A_2,S-A)-Q$. When subtrees T_1 and T_2 are joined, the linked list $Q(A,S-A)$ associated to the new subtree $T'=\{T_1,T_2\}$ is simply obtained by first adding $Q(A_1,S-A)-Q$ and $Q(A_2,S-A)-Q$, which requires $O(n^2)$ time, and then adding the list $Q'_-(A_1,A_2)$, which requires $O(1)$ time if $Q'_-(A_1,A_2)$ (and $Q_+(A_1,A_2)$) is represented as a doubly-linked list. Note that $Q'_-(A_1,A_2)$ need not be scanned as its quartets are distinct from those already added to $Q(A,S-A)-Q$.

The Q'_- and Q'_+ lists of the new possible clusterings involving T' are easy to obtain. If A_i denotes the leaf sequences of a subtree $T_i\neq T_1,T_2$, then $Q'_-(A,A_i)=Q'_-(A_1,A_i)\cup Q'_-(A_2,A_i)-Q''(A_1,A_2,A_i)$ and $Q'_+(A,A_i)=Q'_+(A_1,A_i)\cup Q'_+(A_2,A_i)-Q''(A_1,A_2,A_i)$ where $Q''(A_1,A_2,A_i)$ denotes the members of $Q'(A_1,A_2,S-A)$ involving a leaf sequence in A_i. We proceed by scanning first $Q_-(A_1,A_2)$ and $Q_+(A_1,A_2)$ to remove the occurrences of the elements of $Q''(T_1,T_2,T_i)$ in the Q_- and Q_+ lists of $\{T_1,T_i\}$ and $\{T_2,T_i\}$. Then we merge these disjoint lists in $O(1)$ time to obtain those of $\{T',T_i\}$. Linking together all the occurrences of each quartet in the Q'_- and Q'_+ lists at the beginning of the algorithm enables us to remove each member of $Q''(A_1,A_2,A_i)$ from other Q'_- and Q'_+ lists in $O(1)$ time.

We remark that at any time of the algorithm, each quartet appears only a constant number of times in all Q'_- and Q'_+ lists. If $q=a,b,c,d$ and T_1,T_2,T_3,T_4 are the respective subtrees containing the leaf sequences a,b,c,d, then only the 6 clusterings $\{T_1,T_2\}$, $\{T_1,T_3\}$, $\{T_1,T_4\}$, $\{T_2,T_3\}$, $\{T_2,T_4\}$, and $\{T_3,T_4\}$ resolve q, and hence q belongs only to the Q'_- and Q'_+ lists of these 6 clusterings. As each occurrence of a quartet in the Q'_- and Q'_+ lists is checked or removed only once during the algorithm, the maintenance of the Q'_- and Q'_+ lists requires $O(n^4)$ time totally. Now, these lists (and their sizes) enable us to compute $|Q'(A_1,A_2,S-A)-Q|=|Q'_-(A_1,A_2)|$ in $O(1)$ time and thus $|Q(A,S-A)-Q|$ in $O(n^2)$ time, when needed.

As mentioned above, Global-Clean requires $n-2$ clusterings. Before each clustering we have to examine $O(n)$ new candidates, each requiring $O(n^2)$ time. Hence this implementation requires $O(n\cdot n\cdot n^2)=O(n^4)$ time.

2.3 A Local Vertex Quartet Cleaning Algorithm

Let S be a set of sequences that evolved on evolutionary tree T and let Q be a set of quartet topologies inferred from S. Define a bipartition (A,B) to be *good* if $|Q(A,B)-Q|\leq(|A|-1)(|B|-1)/4$. A tripartition (A,B,C) is good if each of its three induced bipartitions is good, *i.e.*, $|Q(A,B\cup C)-Q|\leq(|A|-1)(|B\cup C|-1)/4$, $|Q(B,A\cup C)-Q|\leq(|B|-1)(|A\cup C|-1)/4$ and $|Q(C,A\cup B)-Q|\leq(|C|-1)(|A\cup B|-1)/4$.

The following algorithm corrects all quartets errors across vertices of T that induce good tripartitions.

Algorithm Tripartition(Q)
1. Let $Tri(Q) := \emptyset$.
2. For every three sequences a, b and c
3. Let $A = \{a\}$; $B = \{b\}$; $C = \{c\}$.
4. For each $w \in S - \{a,b,c\}$.
5. If $aw|bc \in Q$ then $A = A \cup \{w\}$.
6. If $bw|ac \in Q$ then $B = B \cup \{w\}$.
7. If $cw|ab \in Q$ then $C = C \cup \{w\}$.
8. If (A,B,C) is good then let $Tri(Q) := Tri(Q) \cup \{(A,B,C)\}$.
9. Output $Tri(Q)$.

To show that the above algorithm is a local vertex cleaning algorithm with cleaning bound $(|A|-1)(|B|-1)/4$, it is first proven that $Tri(Q)$ contains all vertex induced tripartitions of T that are good.

Theorem 3. *Let v be a vertex of T that induces tripartition (A,B,C). If (A,B,C) is good then $(A,B,C) \in Tri(Q)$.*

Next we show that the tripartitions produced by the Tripartition algorithm are compatible. The following lemma will be useful.

Lemma 4. *Two tripartitions are compatible if and only if their induced bipartitions are all compatible with each other.*

Theorem 5. *If (A,B,C) and (X,Y,Z) are both in $Tri(Q)$ then they are compatible.*

Theorem 3 informs us that $Tri(Q)$ contains all good tripartitions of T and Theorem 5 informs us that $Tri(Q)$ contains only tripartitions compatible with the good tripartitions of T. Let T' be the tree obtained by combining the tripartitions of $Tri(Q)$. Then $Q_{T'}$ is a corrected version of Q. Note that unlike algorithm Global-Clean described in Sect. 2.1, algorithm Tripartition always produces a tree and this tree is either the same as T or is a contraction of T. A straightforward implementation of the Tripartition algorithm runs in $O(n^3 \cdot n^4) = O(n^7)$ time. An $O(n^5)$ implementation has been found recently by Della Vedova [7].

3 Simulation Study

In this section, a simulation study is presented that addresses the utility of the quartet cleaning algorithms. In particular, the simulation study addresses questions of need and applicability:

1. How common are quartet errors in sets of inferred quartet topologies?
2. How effective are the quartet cleaning algorithms at correcting quartet errors when they occur?

The approach of the simulation study was to simulate the evolution of sequences on an evolutionary tree T. From these sequences, a set of quartet topologies was inferred using a quartet inference method. This set of quartet topologies was then compared to the actual quartet topologies induced by T. Any quartet whose inferred and actual topologies did not match was considered a quartet error. By mapping these quartet errors back onto the vertices and edges of T, it was possible to establish the number of quartet errors across the vertices and edges of T. From this information it was determined which vertices and edges of T could be cleaned by the global edge and local vertex cleaning algorithms described in this paper.

The simulation study was extensive in that sequences were evolved under a broad range of parameters where the sequence length, evolutionary tree topology and evolutionary rates were varied. As well, three popular quartet inference methods, namely, Neighbor Joining [15], the Ordinal Quartet method [14] and Maximum Parsimony [10] were used to infer the sets of quartet topologies from the simulated sequences.

The details of the simulation study are as follows. For Neighbor Joining and the Ordinal Quartet method, distance matrices were generated from the sequences and corrected relative to the Kimura two–parameter (K2P) model of evolution [18, page 456]. For Maximum Parsimony, the sequences were corrected for transition/transversion bias by weighting the character–state transition matrix appropriately [18, pages 422–423]. Evolutionary trees and sequences were created in a manner similar to [6] using code adapted from program `listtree` in the PAML phylogeny analysis package [20]: Evolutionary tree topologies were generated by adding taxa at random, edge–lengths were assigned according to a specified mean edge–length, and sequences were evolved on these evolutionary trees using the K2P model with transition–transversion bias $\kappa = 0.5$. The simulation examined 1000 randomly–generated topologies on 10 sequences $\times$ 5 mean edge–lengths per topology $\times$ 100 edge–length sets per topology $\times$ 3 sequence lengths $= 1{,}500{,}000$ sequences datasets, where the set of mean edge–lengths considered was $\{0.025, 0.1, 0.25, 0.5, 0.75\}$ and the set of sequence lengths considered was $\{50, 200, 2000\}$. The simulation code consists of a number of C–language and shell–script programs that were compiled and run under the UNIX system on several of the SUN workstations owned by the computational biology group at McMaster University. The simulations took 5 days to complete.

The results of the simulation study are summarized in Table 1. Consider how the results address the two questions posed at the beginning of this section.

Firstly, the results clearly establish that quartet errors are prevalent, independent of the quartet topology inference method used. The number of quartet errors decreases as sequence length increases and mean edge–length decreases but remains significant even for sequence length 2000 and mean edge–length 0.025. Hence, there is a definite need for quartet cleaning algorithms.

Secondly, a comparison of the parenthesized and unparanthesized values in this table establishes the effectiveness of these algorithms. Both algorithms decrease the number of quartet errors significantly under a wide variety of conditions. For example, under the Ordinal Quartet Method with mean edge–length 0.1 and sequence length 200, the increase in accuracy is approximately 25%. When the mean edge–length is large and/or the sequence length is small, the increased accuracy is dramatic.

Table 1. Performance of the Quartet Cleaning Algorithms Under Simulation. The unparanthesized number is the average percent of all evolutionary trees (vertices per evolutionary tree) that have no quartet errors (no quartet errors across them) after global edge cleaning (local vertex cleaning) has been applied. The parenthesized number is the average percent of all evolutionary trees (vertices per evolutionary tree) that have no quartet errors (no quartet errors across them) before quartet cleaning is applied.

Quartet Inference Method	Mean Edge Length	Global Edge (% of Trees Cleanable)			Local Vertex (Average % of Vertices per Tree that are Cleanable)		
		Sequence Length			Sequence Length		
		50	200	2000	50	200	2000
Maximum Parsimony (Corrected)	0.025	10.48% (0.45%)	62.85% (10.85%)	95.02% (56.67%)	38.98% (29.20%)	76.15% (59.82%)	94.80% (87.27%)
	0.1	38.62% (0.22%)	79.60% (6.17%)	92.34% (27.06%)	55.91% (28.82%)	80.76% (55.57%)	89.90% (74.01%)
	0.25	26.88% (0.01%)	64.44% (0.71%)	78.48% (6.83%)	45.18% (20.09%)	67.96% (37.73%)	77.03% (53.93%)
	0.5	3.25% (0.00%)	23.46% (0.03%)	56.33% (0.87%)	23.50% (10.59%)	43.47% (22.38%)	61.71% (38.42%)
	0.75	0.30% (0.00%)	3.81% (0.00%)	27.46% (0.08%)	13.08% (6.03%)	25.46% (14.11%)	46.57% (28.48%)
Neighbor Joining (Corrected)	0.025	16.36% (0.73%)	70.47% (13.49%)	95.67% (57.18%)	43.79% (32.53%)	79.85% (62.96%)	95.06% (87.46%)
	0.1	47.10% (0.34%)	81.36% (6.92%)	91.72% (24.57%)	60.86% (32.04%)	81.57% (56.65%)	88.97% (72.31%)
	0.25	30.85% (0.02%)	64.18% (0.84%)	76.48% (6.16%)	47.69% (23.19%)	67.50% (38.44%)	75.48% (52.55%)
	0.5	3.91% (0.00%)	22.27% (0.07%)	51.80% (0.79%)	26.91% (16.37%)	43.25% (26.27%)	58.88% (38.08%)
	0.75	0.41% (0.00%)	3.63% (0.00%)	21.47% (0.06%)	16.78% (11.33%)	27.56% (19.41%)	43.48% (30.49%)
Ordinal Quartet Method (Corrected)	0.025	20.49% (1.21%)	74.33% (17.11%)	96.12% (59.25%)	47.79% (37.36%)	82.19% (66.38%)	95.45% (88.19%)
	0.1	64.15% (1.21%)	86.91% (12.38%)	93.84% (34.53%)	71.92% (40.32%)	86.32% (63.45%)	91.50% (78.00%)
	0.25	63.64% (0.16%)	81.72% (2.42%)	87.31% (13.82%)	68.85% (30.29%)	81.13% (47.25%)	85.44% (65.10%)
	0.5	39.03% (0.04%)	67.26% (0.19%)	81.75% (1.82%)	52.41% (23.67%)	69.18% (32.06%)	79.47% (45.22%)
	0.75	12.32% (0.00%)	35.18% (0.01%)	63.60% (0.16%)	34.28% (18.56%)	49.52% (24.86%)	66.47% (33.28%)

Acknowledgments: The authors would like to acknowledge John Tsang and Diane Miller for early discussions about the structure of the simulation and John Nakamura and Matt Kelly for assistance in debugging the initial version of the simulation code. They would especially like to acknowledge Chris Trendall for his work on speeding up and running the initial version of the simulation code and Mike Hu and Chris Trendall for their subsequent work on rewriting and further speeding up this code, which in large part made possible the results reported here.

References

[1] A. Ben-Dor, B. Chor, D. Graur, R. Ophir, and D. Pellep. From four–taxon trees to phylogenies: The case of mammalian evolution. *Proceedings of the Second Annual International Conference on Computational Molecular Biology*, pages 9–19, 1998.

[2] V. Berry and O. Gascuel. Inferring evolutionary trees with strong combinatorial evidence. *Proceedings of the Third Annual International Computing and Combinatorics Conference*, pages 111–123, 1997.

[3] M. Boguski. Bioinformatics – a new era. In S. Brenner, F. Lewitter, M. Patterson, and M. Handel, editors, *Trends Guide to Bioinformatics*, pages 1–3. Elsevier Science, 1998.

[4] D. Bryant. *Structures in Biological Classification.* Ph.D. Thesis, Department of Mathematics and Statistics, University of Canterbury, 1997.

[5] P. Buneman. The recovery of trees from measures of dissimilarity. In F.R. Hodson, D.G. Kendall, and P. Tautu, editors, *Mathematics in the Archaeological and Historical Sciences*, pages 387–395. Edinburgh University Press, Edinburgh, 1971.

[6] M. A. Charleston, M. D. Hendy, and D. Penny. The effects of sequence length, tree topology, and number of taxa on the performance of phylogenetic methods. *Journal of Computational Biology*, 1(2):133–152, 1994.

[7] G. Della Vedova. An improved algorithm for local vertex cleaning. *Manuscript*, 1998.

[8] P. Erdös, M. Steel, L. Székely, and T. Warnow. Constructing big trees from short sequences. *Proceedings of the 24th International Colloquium on Automata, Languages, and Programming*, 1997.

[9] J. Felsenstein. Evolutionary trees from DNA sequences: A maximum likelihood approach. *Journal of Molecular Evolution*, 17:368–376, 1981.

[10] W. M. Fitch. Toward defining the course of evolution: Minimal change for a specific tree topology. *Systematic Zoology*, 20:406–41, 1971.

[11] V. A. Funk and D. R. Brooks. *Phylogenetic Systematics as the Basis for Comparative Biology.* Smithsonian Institution Press, 1990.

[12] R. Gupta and B. Golding. Evolution of hsp70 gene and its implications regarding relationships between archaebacteria, eubacteria and eukaryotes. *Journal of Molecular Evolution*, 37:573–582, 1993.

[13] T. Jiang, P. E. Kearney, and M. Li. Orchestrating quartets: Approximation and data correction. *Proceedings of the 39th IEEE Symposium on Foundations of Computer Science*, pages 416–425, 1998.

[14] P. E. Kearney. The ordinal quartet method. *Proceedings of the Second Annual International Conference on Computational Molecular Biology*, pages 125–134, 1998.

[15] N. Saitou and M. Nei. The neighbor–joining method: A new method for reconstructing phylogenetic trees. *Molecular Biology and Evolution*, 4:406–425, 1987.

[16] M. Steel. The complexity of reconstructing trees from qualitative characters and subtrees. *Journal of Classification*, 9:91–116, 1992.

[17] K. Strimmer and A. von Haeseler. Quartet puzzling: A quartet maximum-likelihood method for reconstructing tree topologies. *Molecular Biology and Evolution*, 13(7):964–969, 1996.

[18] D. L. Swofford, G. J. Olsen, P. J. Waddell, and D. M Hillis. Phylogenetic inference. In D. M. Hillis, C. Moritz, and B. K. Mable, editors, *Molecular Systematics, 2nd Edition*, pages 407–514. Sinauer Associates, Sunderland, Massachusetts, 1996.

[19] L. Vigilant, M. Stoneking, H. Harpending, H. Hawkes, and A. C. Wilson. African populations and the evolution of human mitochondrial DNA. *Science*, 253:1503–1507, 1991.
[20] Z. Yang. PAML: a program package for phylogenetic analysis by maximum likelihood. *CABIOS*, 15:555–556, 1997.

Fast and Robust Smallest Enclosing Balls *

Bernd Gärtner

Institut für Theoretische Informatik, ETH Zürich, ETH-Zentrum, CH-8092 Zürich, Switzerland
(gaertner@inf.ethz.ch)

Abstract. I describe a C++ program for computing the smallest enclosing ball of a point set in d-dimensional space, using floating-point arithmetic only. The program is very fast for $d \leq 20$, robust and simple (about 300 lines of code, excluding prototype definitions). Its new features are a pivoting approach resembling the simplex method for linear programming, and a robust update scheme for intermediate solutions. The program with complete documentation following the *literate programming* paradigm [3] is available on the Web.[1]

1 Introduction

The smallest enclosing ball (or Euclidean 1-center) problem is a classical problem of computational geometry. It has a long history which dates back to 1857 when Sylvester formulated it for points in the plane [8]. The first optimal linear-time algorithm for fixed dimension was given by Megiddo in 1982 [4]. In 1991, Emo Welzl developed a simple randomized method to solve the problem in expected linear time [9]. In contrast to Megiddo's method, his algorithm is easy to implement and very fast in practice, for dimensions 2 and 3. In higher dimensions, a heuristic *move-to-front* variant considerably improves the performance.

The roots of the program I will describe go back to 1991 when I first implemented Emo Welzl's new method. Using the move-to-front variant I was able to solve problems on 5000 points up to dimension $d = 10$ (see [9] for all results). Back then, the program was written in MODULA-2 (the language I had learned in my undergraduate CS courses), and it was running on a 80386 PC with 20 MHz.

After the algorithm and the implementation results had been published, we constantly received requests for source code, from an amazingly wide range of application areas. To name some, there was environmental science (design and optimization of solvents), pattern recognition (finding reference points), biology (proteine analysis), political science (analysis of party spectra), mechanical engineering (stress optimization) and computer graphics (ray tracing, culling).

Soon it became clear that the MODULA-2 source was not of great help in serving these requests; people wanted C or C++ code. Independently, at least two persons ported the undocumented program to C. (One of them remarked that "although we don't understand the complete procedure, we are confident that the algorithm is perfect".) Vishwa Ranjan kindly made his carefully adapted C program accessible to me; subsequently,

* This work was supported by grants from the Swiss Federal Office for Education and Science (Projects ESPRIT IV LTR No. 21957 CGAL and No. 28155 GALIA).

[1] http://www.inf.ethz.ch/personal/gaertner/miniball.html

I was able to distribute a C version. (An independent implementation by David Eberly for the cases $d = 2,3$ based on [9] is available online. [2])

Another shortcoming of my program—persisting in the C code—was the lack of numerical stability, an issue not yet fashionable in computational geometry at that time. The main primitive of the code was to solve a linear system, and this was done with plain Gaussian elimination, with no special provisions to deal with ill-conditioned systems. In my defense I must say that the code was originally only made to get test results for Emo's paper, and in the tests with the usual random point sets I did for that, I discovered no problems, of course.

Others did. David White, who was developing code for the more general problem of finding the smallest ball enclosing balls, noticed unstable behavior of my program, especially in higher dimensions. In his code, he replaced the naive Gaussian elimination approach by a solution method based on *singular value decomposition* (SVD). This made his program pretty robust, and a C++ version (excluding the SVD routines which are taken from *Numerical Recipes in C* [6]) is available from David White's Web page. [3]

Meanwhile, I got involved in the CGAL project, a joint effort of seven European sites to build a C++ library of computational geometry algorithms.[4] To prepare my code for the library, I finally wrote a C++ version of it from scratch. As the main improvement, the code was no longer solving a complete linear system in every step, but was updating previous imformation instead. A "CGALized" version of this code is now contained in the library, and using it together with any exact (multiple precision) number type results in error-free computations.

Still, the numerical problems arising in floating-point computations were not solved. Stefan Gottschalk, one of the first users of my new C++ code, encountered singularities in the update routines, in particular if input points are (almost) cospherical, very close together or even equal. The effect is that center and squared radius of the ball maintained by the algorithm can become very large or even undefined due to exponent overflow, even though the smallest enclosing ball problem itself is well-behaved in the sense that small perturbations of the input points have only a small influence on the result.

As it turned out, previous implementations suffered from an inappropriate representation of intermediate balls that ignores the good-naturedness of the problem. The new representation scheme respects the underlying geometry—it actually resulted from a deeper understanding of the geometric situation— and solves most of the problems.

The second ingredient is a new high-level algorithm replacing the move-to-front method. Its goal is to decrease the overall number of intermediate solutions computed during the algorithm. This is achieved by reducing the problem to a small number of calls to the move-to-front method, with only a small point set in each call. These calls can then be interpreted as 'pivot steps' of the method. The advantages are a substantial improvement in runtime for dimensions $d \geq 10$, and much more robust behavior.

The result is a program which I think reaches a high level of efficiency and stability per lines of code. In simplicity, it is almost comparable to the popular approximation

[2] `http://www.cs.unc.edu/~eberly/`

[3] `http://vision.ucsd.edu/~dwhite`

[4] `http://www.cs.uu.nl/CGAL`

algorithms from the *Graphics Gems* collection [7, 10]; because the latter usually compute suboptimal balls, the authors stress their simplicity as the main feature. The code presented here shares this feature, while computing the optimal ball.

2 The Algorithms

Given an n-point set $P = \{p_1, \ldots, p_n\} \subseteq \mathbb{R}^d$, let $\text{MB}(P)$ denote the ball of smallest radius that contains P. $\text{MB}(P)$ exists and is unique. For $P, B \subseteq \mathbb{R}^d, P \cap B = \emptyset$, let $\text{MB}(P,B)$ be the smallest ball that contains P and has all points of B on its boundary. We have $\text{MB}(P) = \text{MB}(P, \emptyset)$, and if $\text{MB}(P,B)$ exists, it is unique. Finally, define $\overline{\text{MB}}(B) := \text{MB}(\emptyset, B)$ to be the smallest ball with all points of B on the boundary (if it exists).

A support set of (P,B) is an inclusion-minimal subset of P with $\text{MB}(P,B) = \text{MB}(S,B)$. If the points in B are affinely independent, there always exists a support set of size at most $d+1-|B|$, and we have $\text{MB}(S,B) = \overline{\text{MB}}(S \cup B)$.

If $p \notin \text{MB}(P,B)$, then p lies on the boundary of $\text{MB}(P \cup \{p\}, B)$, provided the latter exists—that means, $\text{MB}(P \cup \{p\}, B) = \text{MB}(P, B \cup \{p\})$. All this is well-known, see e.g. [9] and the references there.

The basis of our method is Emo Welzl's *move-to-front* heuristic to compute $\text{MB}(P,B)$ if it exists[9]. The method keeps the points in an ordered list L which gets reorganized as the algorithm runs. Let L_i denote the length-i prefix of the list, p^i the element at position i in L. Initially, $L = L_n$ stores the points of P in random order.

Algorithm 1.

```
mtf_mb(L_n, B):
  (* returns MB(L_n, B) *)
  mb := MB‾(B)
  IF |B| = d+1 THEN
     RETURN mb
  END
  FOR i = 1 TO n DO
     IF p^i ∉ mb THEN
        mb := mtf_mb(L_{i-1}, B ∪ {p^i})
        update L by moving p^i to the front
     END
  END
  RETURN mb
```

This algorithm computes $\text{MB}(P,B)$ incrementally, by adding one point after another from the list. One can prove that during the call to $\texttt{mtf\_mb}(L_n, \emptyset)$, all sets B that come up in recursive calls are affinely independent. Together with the above mentioned facts, this ensures the correctness of the method. By induction, one can also show that upon termination, a support set of (P,B) appears as a prefix L_s of the list L, and below we will assume that the algorithm returns the size s along with mb.

The practical efficiency comes from the fact that 'important' points (which for the purpose of the method are points outside the current ball) are moved to the front and will therefore be processed early in subsequent recursive calls. The effect is that the ball maintained by the algorithm gets large fast.

The second algorithm uses the move-to-front variant only as a subroutine for small point sets. Large-scale problems are handled by a *pivoting* variant which in every iteration adds the point which is most promising in the sense that it has largest distance from the current ball. Under this scheme, the ball gets large even faster, and the method usually terminates after very few iterations. (As the test results in Section 5 show, the move-to-front variant will still be faster for d not too large, but there are good reasons to prefer the pivoting variant in any case.)

Let $e(p,\mathrm{mb})$ denote the *excess* of p w.r.t. mb, defined as $||p-c||^2 - r^2$, c and r^2 the center and squared radius of mb.

Algorithm 2.

```
pivot_mb(L_n):
   (* returns MB(L_n) *)
   t := 1
   (mb, s) := mtf_mb(L_t, ∅)
   REPEAT
      (* Invariant: mb = MB(L_t) = MB‾(L_s), s ≤ t *)
      choose k > t with e := e(p^k, mb) maximal
      IF e > 0 THEN
         (mb, s') := mtf_mb(L_s, {p^k})
         update L by moving p^k to the front
         t := s + 1
         s := s' + 1
      END
   UNTIL e ≤ 0
   RETURN mb
```

Because mb gets larger in every iteration, the procedure eventually terminates. The computation of (mb, s') can be viewed as a 'pivot step' of the method, involving at most $d+2$ points. The choice of k is done according to a heuristic 'pivot rule', with the intention of keeping the overall number of pivot steps small. With this interpretation, the procedure pivot_mb is similar in spirit to the simplex method for linear programming [1], and it has in fact been designed with regard to the simplex method's efficiency in practice.

3 The Primitive Operation

During a call to algorithm pivot_mb, all nontrivial computations take place in the primitive operation of computing $\overline{\mathrm{MB}}(B)$ for a given set B in the subcalls to mtf_mb. The algorithm guarantees that B is always a set of affinely independent points, from which $|B| \leq d+1$ follows. In that case, $\overline{\mathrm{MB}}(B)$ is determined by the unique circumsphere of the points in B with center restricted to the affine hull of B. This means, the center c and squared radius r^2 satisfy the following system of equations, where $B = \{q_0, \dots, q_{m-1}\}, m \leq d+1$.

$$
\begin{aligned}
(q_i - c)^T (q_i - c) &= r^2, \quad i = 0, \ldots m-1, \\
\sum_{i=0}^{m-1} \lambda_i q_i &= c, \\
\sum_{i=0}^{m-1} \lambda_i &= 1.
\end{aligned}
$$

Defining $Q_i := q_i - q_0$, for $i = 0, \ldots, m-1$ and $C := c - q_0$, the system can be rewritten as

$$
\begin{aligned}
C^T C &= r^2, \\
(Q_i - C)^T (Q_i - C) &= r^2, \quad i = 1, \ldots, m-1, \\
\sum_{i=1}^{m-1} \lambda_i Q_i &= C.
\end{aligned}
\tag{1}
$$

Substituting C with $\sum_{i=1}^{m-1} \lambda_i Q_i$ in the equations (1), we deduce a linear system in the variables $\lambda_1, \ldots, \lambda_{m-1}$ which we can write as

$$
A_B \begin{pmatrix} \lambda_1 \\ \vdots \\ \lambda_{m-1} \end{pmatrix} = \begin{pmatrix} Q_1^T Q_1 \\ \vdots \\ Q_{m-1}^T Q_{m-1} \end{pmatrix}, \tag{2}
$$

where

$$
A_B := \begin{pmatrix} 2Q_1^T Q_1 & \cdots & 2Q_1^T Q_{m-1} \\ \vdots & & \vdots \\ 2Q_1^T Q_{m-1} & \cdots & 2Q_{m-1}^T Q_{m-1} \end{pmatrix}. \tag{3}
$$

Computing the values of $\lambda_1, \ldots, \lambda_{m-1}$ amounts to solving the linear system (2). C and r^2 are then easily obtained via

$$
C = \sum_{i=1}^{m-1} \lambda_i Q_i, \quad r^2 = C^T C. \tag{4}
$$

We refer to C as the *relative center* w.r.t. the (ordered) set B.

4 The Implementation

Algorithms 1 and 2 are implemented in a straightforward manner, following the pseudocode given above. In case of algorithm mtf_mb, the set B does not appear as a formal parameter but is updated before and after the recursive call by 'pushing' resp. 'popping' the point p^i. This stack-like behavior of B also makes it possible to implement the primitive operation in a simple, robust and efficient way. More precisely, the algorithm maintains a device for solving system (2) which can conveniently be updated if B changes. The update is easy when element p^i is removed from B—we just need to remember the status prior to the addition of p^i. In the course of this addition, however, some real work is necessary.

A possible device for solving system (2) is the explicit inverse A_B^{-1} of the matrix A_B defined in (3), along with the vector

$$v_B := \begin{pmatrix} Q_1^T Q_1 \\ \vdots \\ Q_{m-1}^T Q_{m-1} \end{pmatrix}.$$

Having this inverse available, it takes just a matrix-vector multiplication to obtain the values $\lambda_1, \ldots, \lambda_{m-1}$ that define C via (4).

Assume B is enlarged by pushing another point q_m. Define $B' = B \cup \{q_m\}$. Let's analyze how $A_{B'}^{-1}$ can be obtained from A_B^{-1}. We have

$$A_{B'} = \left(\begin{array}{ccc|c} & & & 2Q_1^T Q_m \\ & A_B & & \vdots \\ & & & 2Q_{m-1}^T Q_m \\ \hline 2Q_1^T Q_m & \cdots & 2Q_{m-1}^T Q_m & 2Q_m^T Q_m \end{array}\right),$$

and it is not hard to check that this equation can be written as

$$A_{B'} = L \left(\begin{array}{ccc|c} & & & 0 \\ & A_B & & \vdots \\ & & & 0 \\ \hline 0 & \cdots & 0 & z \end{array}\right) L^T,$$

where

$$L = \left(\begin{array}{ccc|c} 1 & & & 0 \\ & \ddots & & \vdots \\ & & 1 & 0 \\ \hline \mu_1 & \cdots & \mu_{m-1} & 1 \end{array}\right), \quad \mu := \begin{pmatrix} \mu_1 \\ \vdots \\ \mu_{m-1} \end{pmatrix} = A_B^{-1} \begin{pmatrix} 2Q_1^T Q_m \\ \vdots \\ 2Q_{m-1}^T Q_m \end{pmatrix} \tag{5}$$

and

$$z = 2Q_m^T Q_m - (2Q_1^T Q_m, \cdots, 2Q_{m-1}^T Q_m) A_B^{-1} \begin{pmatrix} 2Q_1^T Q_m \\ \vdots \\ 2Q_{m-1}^T Q_m \end{pmatrix}. \tag{6}$$

This implies

$$A_{B'}^{-1} = (L^T)^{-1} \left(\begin{array}{ccc|c} & & & 0 \\ & A_B^{-1} & & \vdots \\ & & & 0 \\ \hline 0 & \cdots & 0 & 1/z \end{array}\right) L^{-1}, \tag{7}$$

where

$$L^{-1} = \left(\begin{array}{ccc|c} 1 & & & 0 \\ & \ddots & & \vdots \\ & & 1 & 0 \\ \hline -\mu_1 & \cdots & -\mu_{m-1} & 1 \end{array} \right).$$

Expanding (7) then gives the desired update formula

$$A_{B'}^{-1} = \left(\begin{array}{c|c} A_B^{-1} + \mu\mu^T/z & -\mu/z \\ \hline -\mu^T/z & 1/z \end{array} \right), \tag{8}$$

with μ and z as defined in (5) and (6).

Equation (8) shows that A_B may become ill-conditioned (and the entries of A_B^{-1} very large and unreliable), if z evaluates to a very small number. The subsequent lemma develops a geometric interpretation of z from which we can see that this happens exactly if the new point q_m is very close to the affine hull of the previous ones. This can be the case e.g. if input points are very close together or even equal. To deal with such problems, we need a device that stays bounded in every update operation.

As it turns out, a suitable device is the $(d \times d)$-matrix

$$M_B := 2Q_B A_B^{-1} Q_B^T,$$

where

$$Q_B := (Q_1 \cdots Q_{m-1})$$

stores the points Q_i as columns. Lemma 1 below proves that the entries of M_B stay bounded, no matter what. We will also see how the new center is obtained from M_B, which is not clear anymore now.

Lemma 1.
(i) With μ as in (5), we have

$$\sum_{i=1}^{m-1} \mu_i Q_i = \bar{Q}_m,$$

where $\bar{Q}_m$ is the projection of Q_m onto the subspace spanned by the Q_i.
(ii) $M_B Q_m = \bar{Q}_m$.
(iii) $z = 2(Q_m - \bar{Q}_m)^T (Q_m - \bar{Q}_m)$, i.e. z is twice the distance from Q_m to its projection.
(iv) If C and r^2 are relative center and squared radius w.r.t. B, then the new relative center C' and squared radius r'^2 (w.r.t. B') satisfy

$$C' = C + \frac{e}{z}(Q_m - \bar{Q}_m), \tag{9}$$

$$r'^2 = r^2 + \frac{e^2}{2z}, \tag{10}$$

where

$$e = (Q_m - C)^T (Q_m - C) - r^2.$$

(v) M_B is updated according to

$$M_{B'} = M_B + \frac{2}{z}(Q_m - \bar{Q}_m)(Q_m - \bar{Q}_m)^T. \quad (11)$$

The proof involves only elementary calculations and is omitted here. Property (ii) gives M_B an interpretation as a linear function: M_B is the projection onto the linear subspace spanned by $Q_1, \ldots, Q_{m-1}$. Furthermore, property (v) implies that M_B stays bounded. This of course does not mean that no 'bad' errors can occur anymore. In (11), small errors in $Q_m - \bar{Q}_m$ can get hugely amplified if z is close to zero. Still, M_B degrades gracefully in this case, and the typical relative error in the final ball is by an order of magnitude smaller if the device M_B is used instead of A_B^{-1}.

The lemma naturally suggests an algorithm to obtain C', r'^2 and $M_{B'}$ from C, r^2 and M_B, using the values $\bar{Q}_m, e$ and z.

As already mentioned, even M_B may get inaccurate, in consequence of a very small value of z before. The strategy to deal with this is very simple: we ignore push operations leading to such dangerously small values! In the ambient algorithm `mtf_mb` this means that the point to be pushed is treated as if it were inside the current ball (in `pivot_mb` the push operation is never dangerous, because we push onto an empty set B). Under this scheme, it could happen that points end up outside the final ball computed by `mtf_mb`, but they will not be very far outside, if we choose the threshold for z appropriately.

The criterion is that we ignore a push operation if and only if the *relative* size of z is small, meaning that

$$\frac{z}{r_{curr}^2} < \varepsilon \quad (12)$$

for some constant ε, where r_{curr}^2 is the current squared radius. Now consider a subcall to `mtf_mb`$(L_s, \{p^k\})$ inside the algorithm `pivot_mb`, and assume that a point $p \in L_s$ ends up outside the ball mb_0 with support set S_0 and radius r_0 computed by this subcall.

One can check that after the last time the query '$p^i \notin$ mb ?' has been executed with p^i being equal to p in `mtf_mb`, no successful push operations have occured anymore. It follows that $\text{mb} = \text{mb}_0$ in this last query, the query had a positive answer (because p lies outside), and the subsequent push operation failed. This means, we had $z/r_0^2 < \varepsilon$ at that time.

Let r_{max} denote the radius of $\text{MB}(L_s, \{p^k\})$. Because of (10), we also had $e^2/2z \leq r_{max}^2$ at the time of the failing push operation, where e is the excess of p w.r.t. a ball $\overline{\text{MB}}(B \cup \{p^k\})$ with $B \subseteq S_0$. We then get

$$\left(\frac{e}{r_{max}^2}\right)^2 \leq \frac{2z}{r_{max}^2} \leq \frac{2z}{r_0^2} \leq 2\varepsilon.$$

Assuming that r_{max} is not much larger than r_0 (we expect push operations to fail rather at the end of the computation, when the ball is already large), we can argue that

$$\frac{e}{r_0^2} = O(\sqrt{\varepsilon}).$$

Moreover, because mb_0 contains the intersection of $\overline{\mathrm{MB}}(B \cup \{p^k\})$ with the affine hull of $B \cup \{p^k\}$, to which set p is quite close due to z being small, we also get

$$\frac{e_0}{r_0^2} = O(\sqrt{\varepsilon}), \tag{13}$$

where e_0 is the excess of p w.r.t. the final ball mb_0, as desired. This argument is not a strict proof for the correctness of our rejection criterion, but it explains why the latter works well in practice. In the code, ε is chosen as 10^{-32}. Because of (13), the relative error of a point w.r.t. the final ball is then expected to stay below 10^{-16} in magnitude. The latter value is the relative accuracy in the typical situations where the threshold criterion is not applied by the algorithm at all. Thus, ε is chosen in such a way that even when the criterion comes in, the resulting error does not go up.

Checking

While it is easy to verify that the computed ball is admissible in the sense that it contains all input points and has all support points on the boundary (approximately), its optimality does not yet follow from this; if there are less than $d+1$ support points, many balls are admissible with respect to this definition. The following lemma gives an optimality condition.

Lemma 2. *Let S be a set of affinely independent points. $\overline{\mathrm{MB}}(S)$ is the smallest enclosing ball of S if and only if its center lies in the convex hull of S.*

The statement seems to be folklore and can be proved e.g. by using the Karush-Kuhn-Tucker optimality conditions for constrained optimization [5], or by elementary methods.

From Section 2 we know that the algorithm should compute a support set S that behaves according to the lemma; still, we would like to have a way to check this in order to safeguard against numerical errors that may lead to admissible balls which are too large. Under the device A_B^{-1}, this is very simple—the coefficients λ_i we extract from system (2) in this case give us the desired information: exactly if they are all nonnegative, S defines the optimal ball.

The weakness in this argumentation is that due to (possibly substantial) errors in A_B^{-1}, the λ_i might appear positive, although they are not. One has to be aware that "checking" in this case only adds more plausibility to a seemingly correct result. Real checking would ultimately require the use of exact arithmetic, which is just not the point of this code.

Still, if the plausibility test fails (and some λ_i turn out to be far below zero), we *do* know that something went wrong, which is important information in evaluating the code.

Unfortunately, in using the improved device M_B during the computations, we do not have immediate access to the λ_i. To obtain them, we express C as well as the points $Q_1, \ldots, Q_{m-1}$ with respect to a different basis of the linear span of $Q_1, \ldots, Q_{m-1}$. In this representation, the linear combination of the Q_i that defines C will be easy to deduce.

The basis we use will be the set of (pairwise orthogonal) vectors $Q_i - \bar{Q}_i, i = 1, \ldots, m-1$. From the update formula for the center (9) we immediately deduce that

$$C = \sum_{i=1}^{m-1} f_i (Q_i - \bar{Q}_i),$$

where f_i is the value e/z that was computed according to Lemma 1(iv) when pushing q_i. This means, the coordinates of C in the new basis are $(f_1, \ldots, f_{m-1})$.

To get the representations of the Q_i, we start off by rewriting M_B as

$$M_B = \sum_{k=1}^{m-1} \frac{2}{z_k} (Q_k - \bar{Q}_k)(Q_k - \bar{Q}_k)^T,$$

which follows from Lemma 1(v). Here, z_k denotes the value z we got when pushing point q_k.

Now consider the point Q_i. We need to know the coefficient α_{ik} of $Q_k - \bar{Q}_k$ in the representation

$$Q_i = \sum_{k=1}^{m-1} \alpha_{ik} (Q_k - \bar{Q}_k).$$

With

$$M_{B^i} := \sum_{k=1}^{i} \frac{2}{z_k} (Q_k - \bar{Q}_k)(Q_k - \bar{Q}_k)^T \tag{14}$$

we get

$$M_{B^i} Q_i = Q_i$$

(after adding q_i to B, Q_i projects to itself). Via (14), this entails $\alpha_{i,i+1} = \cdots = \alpha_{i,m-1} = 0$ and

$$\alpha_{ik} = \frac{2}{z_k} (Q_k - \bar{Q}_k)^T Q_i, \quad k \leq i.$$

In particular we get $\alpha_{ii} = 1$. The coefficients λ_i in the equation

$$C = \sum_{i=1}^{m-1} \lambda_i Q_i$$

are now easy to compute in the new representations of C and the Q_i we have just developed. For this, we need to solve the linear system

$$\begin{pmatrix} a_{11} & \cdots & a_{m-1,1} \\ \vdots & & \vdots \\ a_{1,m-1} & \cdots & a_{m-1,m-1} \end{pmatrix} \begin{pmatrix} \lambda_1 \\ \vdots \\ \lambda_{m-1} \end{pmatrix} = \begin{pmatrix} f_1 \\ \vdots \\ f_{m-1} \end{pmatrix}$$

This system is triangular—everything below the diagonal is zero, and the entries on the diagonal are 1. So we can get the λ_i by a simple back substitution, according to

$$\lambda_i = f_i - \sum_{k=i+1}^{m-1} \alpha_{ki}\lambda_k.$$

Finally, we set

$$\lambda_0 = 1 - \sum_{k=1}^{m-1} \lambda_k,$$

and check whether all these values are nonnegative.

How much effort is necessary to determine the values α_{ik}? Here comes the punch line: if we actually represent the M_{B^i} according to (14) and during the push of q_i evaluate the product

$$M_{B^{i-1}}Q_i = \sum_{k=1}^{i-1} \frac{2}{z_k}(Q_k - \bar{Q}_k)(Q_k - \bar{Q}_k)^T Q_i = \sum_{k=1}^{i-1} \alpha_{ik}(Q_k - \bar{Q}_k)$$

according to this expansion, we have already computed α_{ik} by the time we need it for the checking!

Moreover, if we make representation (14) implicit by only storing the z_k and the vectors $Q_k - \bar{Q}_k$, we can even perform the multiplication $M_{B^{i-1}}Q_i$ with $\Theta(di)$ arithmetic operations, compared to $\Theta(d^2)$ operations when we really keep M_B as a matrix or a sum of matrices.

The resulting implementation of the 'push' routine is extremely simple and compact (about 50 lines of code), and it allows the checker to be implemented in 10 more lines of code.

5 Experimental Results

I have tested the algorithm on various point sets: random point sets (to evaluate the speed), vertices of a regular simplex (to determine the dimension limits) and (almost) cospherical points (to check the degeneracy handling). In further rounds, all these examples have been equipped with 'extra degeneracies' obtained by duplicating input points, replacing them by 'clouds' of points very close together, or embedding them into a higher dimensional space. This covers all inputs that have ever been reported as problematic to me. A test suite (distributed with the code) automatically generates all these scenarios from the master point sets and prints out the results.

In most cases, the correct ball is obtained by the pivoting method, while the move-to-front method frequently fails (results range from mildly wrong to wildly wrong on cospherical points, and under input point duplication resp. replacement by clouds). This means, although the move-to-front approach is still slightly faster than pivoting in low dimensions (see the results in the next paragraph), it is highly advisable to use the pivoting approach; it seems to work very well together with the robust update scheme based on the matrix M_B, as described in Section 4. The main drawbacks of the move-to-front method are its dependence on the order of the input points, and its higher number of push operations (the more you push, the more can go wrong). Of course, the input order can be randomly rearranged prior to computation (as originally suggested in [9]), but that eats up the gain in runtime over the pivoting method. On the other hand, if one does not rearrange, it is very easy to come up with bad input orders (try a set of points ordered along a line).

Random point sets. I have tested the algorithm on random point sets up to dimension 30 to evaluate the speed of the method, in particular with respect to the relation between the pivoting and the move-to-front variant. Table 1 (left) shows the respective runtimes for 100,000 points randomly chosen in the d-dimensional unit cube, in logarithmic scale (averaged over 100 runs). All runtimes (excluding the time for generating and storing the points) have been obtained on a SUN Ultra-Sparc II (248 MHz), compiling with the GNU `C++`-Compiler `g++` Version 2.8.1, and options `-O3 -funroll-loops`. The latter option advises the compiler to perform loop unrolling (and `g++` does this to quite some extent). This is possible because the dimension is fixed at compile time via a template argument. By this mechanism, one also gets rid of dynamic storage management.

As it turns out, the move-to-front method is faster than the pivoting approach up to dimension 8 but then loses dramatically. In dimension 20, pivoting is already more than ten times faster. Both methods are exponential in the dimension, but for applications in low dimensions (e.g. $d = 3$), even 1,000,000 points can be handled in about two seconds.

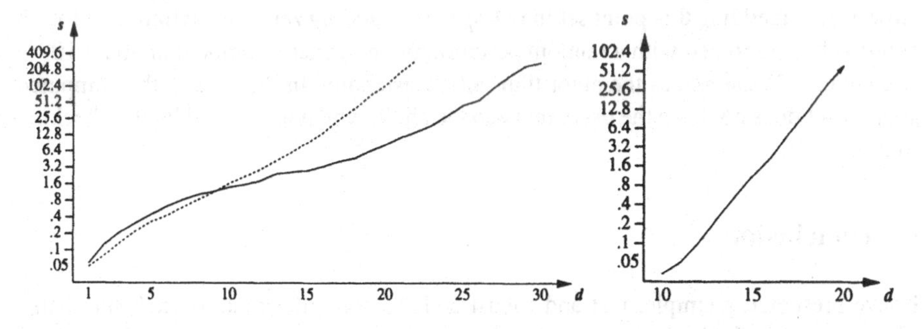

Table 1. Runtime in seconds for 100,000 random points in dimension d: pivoting (solid line) and move-to-front (dotted line) (left). Runtime in seconds on regular d-simplex in dimension d (right).

Vertices of a simplex. The results for random point sets suggest that dimension 30 is still feasible using the pivoting method. This, however, is not the case for all inputs. In high dimensions, the runtime is basically determined by the calls to the move-to-front method with point set $S \cup \{p^i\}$, S the current support set. We know that $|S| \leq d+1$, but if the input is random, $|S|$ will frequently be smaller (in dimension 20, for example, the average number of support points turns out to be around 17). In this case, a 'pivot step' and therefore the complete algorithm is much faster than in the worst case. To test this worst case, I have chosen as input the vertices of a regular d-simplex in dimension d, spanned by the unit vectors. In this case, the number of support points is d. Table 1 (right) shows the result (move-to-front and pivoting variant behave similarly). Note that to solve the problem on 20 points in dimension 20, one needs about as long as for $100,000$ random points in dimension 26!

As a conclusion, the method reaches its limits much earlier than in dimension 30, when it comes to the worst case. In dimension 20, however, you can still expect reasonable performance in any case.

Cospherical points. Here, the master point sets are *exactly* cocircular points in dimension 2, almost cospherical points in higher dimensions (obtained by scaling random vectors to unit length), a tesselation of the unit sphere in 3-space by longitude/latitude values, and vertices of a regular d-cube. While the pivoting method routinely handles most test scenarios, the move-to-front method mainly has problems with duplicated input points and slightly perturbed inputs. It may take very long and computes mildly wrong results in most cases. The slow behavior is induced by many failing push-operations due to the value z being too small, see Section 4. This causes many points which have mistakenly been treated as inside the current ball to reappear outside later.

The most difficult problems for the pivoting method arise from the set of 6144 integer points on the circle around the origin with squared radius $r^2 = 3728702916375125$. The set itself is handled without any rounding errors at all appearing in the result (this is only possible because r^2 still fits into a floating-point value of the `C++` type `double`). However, embedding this point set into 4-space (by adding zeros in the third and fourth coordinate), combined with a random perturbation by a relative amount of about 10^{-30} in each coordinate makes the algorithm fail occasionally. In these case, the computed support set does not have the orgin in its convex hull, which is detected by the checking routine.

6 Conclusion

I have presented a simple, fast and robust code to compute smallest enclosing balls. The program is the last step so far in a chain of improvements and simplifications of the original program written back in 1991. The distinguishing feature is the nice interplay between a new high-level algorithm (the pivoting method) and improved low-level primitives (the M_B-based update scheme).

For dimensions $d \leq 10$, the method is extremely fast, beyond that it slows down a bit, and for $d > 20$ it is not suitable anymore in some cases. This is because every 'pivot step' (a call to the move-to-front method with few points) takes time exponential in d. Even slight improvements here would considerably boost the performance of the whole algorithm. At this point, it is important to note that high dimension is *not* prohibitive for the smallest enclosing ball problem itself, only for the method presented. Interior point methods, or 'real' simplex-type methods in the sense that the pivot step is a polynomial-time operation (see e.g.[2]) might be able to handle very high dimensions in practice, but most likely at the cost of losing the simplicity and stability of the solution I gave here.

Acknowledgment

I would like to thank all the people that have in one way or the other contributed to this code over the last eight years, by using previous versions, making suggestions and reporting bugs. Special thanks go to Emo Welzl for acquainting me with his algorithm eight years ago, and for always keeping an eye on the progress.

References

[1] V. Chvátal. *Linear Programming.* W. H. Freeman, New York, NY, 1983.
[2] B. Gärtner. Geometric optimization. Lecture Notes for the *Equinoctial School on Geometric Computing*, ETH Zürich, 1997, `http://www.inf.ethz.ch/personal/gaertner/publications.html`
[3] Donald E. Knuth. Literate programming. *The Computer Journal*, 27(2):97–111, 1984.
[4] N. Megiddo. Linear-time algorithms for linear programming in R^3 and related problems. In *Proc. 23rd Annu. IEEE Sympos. Found. Comput. Sci.*, pages 329–338, 1982.
[5] A. L. Peressini, F. E. Sullivan, and J. J. Uhl. *The Mathematics of Nonlinear Programming.* Undergraduate Texts in Mathematics. Springer-Verlag, 1988.
[6] William H. Press, Brian P. Flannery, Saul A. Teukolsky, and William T. Vetterling. *Numerical Recipes in C.* Cambridge University Press, Cambridge, England, 2nd edition, 1993.
[7] J. Ritter. An efficient bounding sphere. In Andrew S. Glassner, editor, *Graphics Gems.* Academic Press, Boston, MA, 1990.
[8] J. J. Sylvester. A question on the geometry of situation. *Quart. J. Math.*, 1:79, 1857.
[9] Emo Welzl. Smallest enclosing disks (balls and ellipsoids). In H. Maurer, editor, *New Results and New Trends in Computer Science*, volume 555 of *Lecture Notes Comput. Sci.*, pages 359–370. Springer-Verlag, 1991.
[10] X. Wu. A linear-time simple bounding volume algorithm. In D. Kirk, editor, *Graphics Gems III.* Academic Press, Boston, MA, 1992.

Efficient Searching for Multi-dimensional Data Made Simple (Extended Abstract)

Enrico Nardelli[1], Maurizio Talamo[2], and Paola Vocca[3]

[1] Dipartimento di Matematica Pura ed Applicata, Università di L'Aquila, Via Vetoio, Coppito I-67010 L'Aquila, Italy, nardelli@univaq.it

[2] Dipartimento di Informatica e Sistemistica, Università di Roma "La Sapienza", Via Salaria 113, I–00198 Rome, Italy, talamo@dis.uniroma1.it

[3] Dipartimento di Matematica Università di Roma "Tor Vergata", Via della Ricerca Scientifica, I–00133 Rome, Italy, vocca@mat.uniroma2.it

Abstract. We introduce an innovative decomposition technique which reduces a multi–dimensional searching problem to a sequence of one–dimensional problems, each one easily manageable in optimal time×space complexity using traditional searching strategies. The reduction has no additional storage requirement and the time complexity to reconstruct the result of the original multi–dimensional query is linear in the dimension.

More precisely, we show how to preprocess a set of $S \subseteq \mathbb{N}^d$ of *multi–dimensional* objects into a data structure requiring $O(m \log n)$ space, where $m = |S|$ and n is the maximum number of different values for each coordinate. The obtained data structure is *implicit*, i.e. does not use pointers, and is able to answer the *exact match* query in $7(d-1)$ steps. Additionally, the model of computation required for querying the data structure is very simple; the only arithmetic operation needed is the addition and no shift operation is used.

The technique introduced, overcoming the multi–dimensional bottleneck, can be also applied to non traditional models of computation as external memory, distributed, and hierarchical environments. Additionally, we will show how the proposed technique permits the effective realizability of the well known perfect hashing techniques on real data.

The algorithms for building the data structure are easy to implement and run in polynomial time.

1 Introduction

The efficient representation of multi–dimensional points set plays a central role in many large–scale computations, including, for instance, object management in distributed environments (CORBA, DCOM); object–oriented and deductive databases management [2, 5, 25, 10, 19], and spatial and temporal data manipulation [20, 24]. All these applications manage very large amounts of multi–attribute data. Such data can be considered as points in a d–dimensional space. Hence, the key research issue, in order to provide "good" implementations of these applications, is the design of an efficient data structure for searching in the d–dimensional space. A fundamental search operation is the *exact match* query, that is, test the presence of a point in the multi–dimensional set when all its coordinates are specified. Another important operation is the *prefix–partial match* query which looks for a set of points, possibly empty, for whom only the first $k \leq d$ coordinates are specified.

We deal with the exact match query by using an innovative decomposition technique which reduces a multi–dimensional searching problem to a sequence of one–dimensional problems, each one easily manageable in optimal time×space complexity using traditional searching strategies. The reduction requires no additional storage besides that one required for data and the time complexity to reconstruct the result of the original multi–dimensional query is linear in the dimension. The technique introduced, overcoming the multi–dimensional bottleneck, can be applied in more general contexts, such as distributed and hierarchical environments. Additionally, it can be positively used, jointly with perfect hashing techniques, when dealing with real data.

The technique is based on two main steps. In the first step, we reduce the d–dimensional searching problem to a sequence of d one–dimensional searching problems. In the second step, the multi–dimensional data is reconstructed using a set of $(d-1)$ 2–place functions. Each function is represented using a new data structure derived from a decomposition of the 2–place functions into a set of "sparse" 2–place functions easily representable. The decomposition technique of 2–place functions is an application of a more general technique introduced in [22] and successively refined in [23] for testing reachability in general directed graphs. The same technique has been successfully applied in [21] to the problem of implicitly representing a general graph.

The data structure we present has the following characteristics:

- *general and deterministic*: We represent any multi–dimensional point set and our space and time bounds are worst-case deterministic;
- *space and time efficient*: Exact match query requires $7(d-1)$ steps and prefix–partial match $7(k-1)+t$ steps, where t is the number of points reported, using $O(m \log n)$ space, where m is the size of the point set and n is the maximum number of values a coordinate can receive;
- *easy to implement*: The algorithms used to build the data structure, although somewhat tricky to analyze, are very simple and run in $O(n^3)$ time; no operations are needed for searching other than one–dimensional array accesses;
- *simple computation model*: The only arithmetic operation required for querying the data structure is the addition and no shift operation is used.

Due to its relevance, the multi–dimensional searching problem has been deeply investigated. In computational geometry and for spatial databases, the problem has been solved only for small values of the dimension [18, 20] and the solutions proposed grow exponentially with d. The same problem has been studied for temporal databases [24]. In this case, we have empirical results, only, and the worst case is unbounded. In a general setting, there are two major techniques for implementing the multi–dimensional searching problem: trees and hashing. For the first, several data structure have been developed as d-dimensional version of data structures for the one–dimensional problem (e.g. B-trees [3], compacted tries [17], digital search trees [14]). In this case, even though the space complexity is optimal exact match queries require a logarithmic number of steps in the worst case. Hashing and perfect hashing techniques have the drawback that, for each search performed, it may be required the evaluation of computationally complex functions [8, 7, 6]. Hence, numerically robust implementations are required. With our technique, each search only requires a constant number of table accesses, and addresses to be accessed are computed with only a constant number of additions.

Concerning the comparison of our technique with the less powerful computational models considered in the so-called word-RAM approach [9], namely the RISC model, the are two issues to be considered. First, our technique does not need to use a shift operation, which may require at least $\log m$ additions to be simulated, where m is the problem size. Second, the overall space needed for computations in the word-RAM model is $O(2^w)$ bits, where w is the word size, and for the model to be of interest this quantity has to be considerably larger than the problem size m, namely $2^w \gg m$ ([9], pag. 371). Contrast this with the overall space needed in our approach that, expressed in terms of m, can be written as $O(m \log^2 m)$.

The paper is structured as follows: In Section 2 we describe the representation of the multi-dimensional problem by means of a sequence of 2-place functions; In Section 3 we give some definitions and notations, and present some decomposition theorems; Using these theorems, in Section 4 we describe the data structure for representing a 2-place function and, hence, a multi-dimensional points set; then, in Section 5 we present some application of our technique. Finally, in Section 6 we outline some open problems and future research directions.

2 Problem Representation

In this section, we show first how to reduce a multi-dimensional problem to a set of one-dimensional problems, and then how to reconstruct the original problem.

Given $S \subseteq \mathbb{N}^d$, with $m = |S|$. Let $x = x_1, \ldots, x_d \in S$, then $n_i \doteq |\{x_i : x \in S\}|$. The reduction is defined by the following set of functions:

$$g_i : \mathbb{N} \longmapsto \{1, \ldots n_i\} \qquad 1 \leq i \leq d. \tag{1}$$

Each function g_i maps the values of a coordinate to a set of integers of bounded size. This mapping can be easily represented with data structures for one-dimensional searching, such as B-trees or perfect hashing tables. Without loss of generality, from now on we assume $S \subseteq U^d$, where $U = \{1, \ldots n\}$, being $n = max_i\{n_i\}$.

Hence, let $x = x_1, \ldots, x_d \in S \subseteq U^d$ be a generic key of S, where x_i denotes the value of the i-th coordinate. Let $a = a_1, a_2, \ldots, a_d$ be a value in U^d. We denote with $a(i)$ the subsequence of its first i coordinates, namely $a(i) = a_1, a_2, \ldots, a_i$, called a *partial value* or the *prefix* (*of length* i) of a. We write $b(j) \subset a(i)$ when $j < i$ and $b_k = a_k$, for $k = 1, 2, \ldots, j$. In the same way, we define the prefix for a key in S.

Let $S(a(i))$ be the subset of S containing all keys that are coincident on the prefix $a(i)$. Note that $|S(a(d-1))| \leq n$ and $|S(a(d))| \leq 1$. For any $a(i)$ such that $|S(a(i))| > 1$ and $\nexists b(j) \subset a(i)$ such that $S \supseteq S(b(j)) \supset S(a(i))$ we say that $a(i)$ is the *maximal shortest common prefix* of $S(a(i))$ *with respect to* S. We assume that it does not exist a maximal shortest common prefix $a(i)$ such that $S(a(i)) = S$, since otherwise we can consider a reduced dimension universe, by simply deleting the maximal shortest common prefix from every key.

The representation mechanism we use for keys is based on a suitable coding of subsets of keys with common prefixes of increasing length, starting from the maximal shortest common prefixes. We denote with f_i a 2-place function such that $f_i : U^i \times U \mapsto S$. We code keys using these functions in an incremental way.

Given a set T of keys, we denote with s_l^T, $1 \leq l \leq k_T$, the l-th key in a fixed, but arbitrarily chosen, total ordering of the k_T keys in T. The choice of the order is immaterial: we use it only to make the description clearer.

Let us now assume $a(i)$ is a maximal shortest common prefix with respect to S. For reasons that will be clearer in the following, we only take into account maximal shortest common prefixes longer than 1. We then represent $S(a(i)), i > 1$, with the following technique.

First we represent the $i-1$ smallest elements in $S(a(i))$ as it follows:

$$\begin{cases} f_1(a_1, a_2) = s_1^{S(a(i))} \\ \dots \\ f_{i-1}(a_1 \dots a_{i-1}, a_i) = s_{i-1}^{S(a(i))} \end{cases}$$

Now, if $k_{S(a(i))} \leq i-1$, we have represented all elements in $S(a(i))$ and we are done. Otherwise we still have to represent the $k_{S(a(i))} - (i-1)$ remaining elements in $S' = S(a(i)) \setminus \bigcup_{l=1}^{i-1} s_l^{S(a(i))}$.

All keys in S' can then be partitioned in subsets, possibly just one, each containing keys with a common prefix $a(i+j) \supset a(i)$, and such that, for each subset S'_r, $a(i+j_r)$ is the maximal shortest common prefix of $S'_r = S(a(i+j_r)) \cap S'$ with respect to S'.

We now represent the $k_{S'_r}$ keys in S'_r by recursively applying the same approach. Namely, we first represent the j_r smallest keys in S'_r as it follows:

$$\begin{cases} f_i(a_1 \dots a_i, a_{i+1}) = s_1^{S'_r} \\ \dots \\ f_{i+j_r-1}(a_1 \dots a_{i+j_r-1}, a_{i+j_r}) = s_{j_r}^{S'_r} \end{cases}$$

Now, if $k_{S'_r} \leq j_r - 1$, we have represented all elements in S'_r and we are done. Otherwise, we still have to represent the $k_{S'_r} - (i + j_r - 1)$ remaining elements in $S''_r = S'_r \setminus \bigcup_{l=1}^{j_r} s_l^{S'_r}$.

All keys in S''_r can then be partitioned in subsets, possibly just one, each containing keys with a common prefix $a(i+j_r+h) \supset a(i+j_r)$, and such that, for each subset $S''_{r,q}$, $a(i+j_r+h_{r,q})$ is the maximal shortest common prefix of $S''_{r,q} = S(a(i+j_r+h_{r,q})) \cap S''_r$ with respect to S''_r. And now the representation process goes on recursively.

We now show an example of the application of the definitions introduced above.

Example 1. Assume $d = 6$ and $n = 9$. Consider a set $S = \{233121, 233133, 233135, 233146, 234566, 234577, 234621, 234622, 234623, 343456\}$. Then there are only two maximal shortest common prefixes with respect to S, namely 23 of length 2 and 343456 of length 6.

We then set $f_1(2,3) = 233121$ and $f_1(3,4) = 343456$: since $k_{S(23)} \not\leq 2-1$ while $k_{S(343456)} \leq 6-1$ then $S(343456)$ has been completely represented, while for keys remaining in $S' = S(23) \setminus \{233121\}$ we have to recursively apply the same technique.

The maximal shortest common prefixes in $S' = \{233133, 233135, 233146, 234566, 234577, 234621, 234622, 234623\}$ are 2331 of length $2+2$ and 234 of length $2+1$. It is $S'_1 = S(2331) \cap S' = \{233133, 233135, 233146\}$ and $S'_2 = S(234) \cap S' = \{234566, 234577, 234621, 234622, 234623\}$.

We then set $f_2(23,3) = 233133$ and $f_3(233,1) = 233135$; we also set $f_2(23,4) = 234566$. Since $k_{S'_1} \not\leq 2$ and $k_{S'_2} \not\leq 1$ then both for keys remaining in $S''_1 = S'_1 \setminus \{233133, 233135\}$ and for those in $S''_2 = S'_2 \setminus \{234566\}$ we have to recursively apply the same technique. We obtain the following sets: $f_4(2331,4) = 233146$, $f_3(234,5) = 234577$, $f_3(234,6) = 234621$, $f_4(2346,2) = 234622$, and $f_5(23462,3) = 234623$.

Given a d–dimensional set of points $S \subseteq U^d$ a point $x = a_1, a_2, \ldots, a_d$ can be searched by incrementally evaluating the 2–place functions f_i. At each step i, with $i = \{1, \ldots, d-1\}$, two cases are possible: $f_i(a_1 \ldots a_i, a_{i+1}) = x$ and we are done. Otherwise, the search continues with the evaluation of f_{i+1}. It is trivial to verify that the search ends reporting x if and only if $x \in S$. In the next section we show how to efficiently represent 2–place functions so that the above search strategy can be executed in a constant number of steps.

3 2–place Functions Representation

In order to state the main result of this section we need to recall some definitions and give new notations.

3.1 Definitions

A *bipartite graph* $G = (A \cup B, E)$ is a graph with $A \cap B = \emptyset$ and edge set $E \subseteq A \times B$.

Given a 2–place function $f : A \times B \longmapsto \mathbb{N}$, a unique labeled bipartite graph $G = (A \cup B, E)$ can be built, such that the label of $(x,y) \in E$ is equal to z if and only if $x \in A$, $y \in B$, and $f(x,y) = z \in Z$. Hence, the representation of a 2–place function is equivalent to test adjacency in the bipartite graph and lookup the label associated to the edge, if it exists. For ease of exposition, in the following, we will deal with labeled bipartite graphs instead of 2–place functions. Moreover, from now on, n_A and n_B denote the number of vertices in A and B, respectively, and m is the number of edges of the bipartite graph.

Given a bipartite graph $G = (A \cup B, E)$, $x \in A \cup B$ is *adjacent* to $y \in A \cup B$ if $(x,y) \in E$. Given a vertex x, the set of its adjacent vertices is denoted by $\alpha(x)$; $\delta(x) \doteq |\alpha(x)|$ is the *degree* x. The notation is extended to a set S of vertices as $\alpha(S) \doteq \cup_{x \in S} \alpha(x)$ and $\delta(S) \doteq \sum_{x \in S} \delta(x)$. The maximum degree among vertices in S is denoted by Δ_S. In particular, Δ_A and Δ_B denote the maximum degree among vertices in A and B, respectively. A bipartite graph is *regular* if all vertices have the same degree $\Delta = \Delta_A = \Delta_B$. A bipartite graph $G = (A \cup B, E)$ is *bi-regular* if all vertices in A have the same degree Δ_A and all vertices in B have the same degree Δ_B.

Given a set of vertices $S \in A$ or $S \in B$, $\delta_S(x) \doteq |\alpha(x) \cap S|$ denotes the number of vertices in S adjacent to x. Furthermore, $\alpha_j(S) \doteq \{x \in \alpha(S) : \delta_S(x) = j\}$ denotes the set of vertices in $\alpha(S)$ incident to S with exactly j edges. Given a set of vertices $S \in A \cup B$, the *sub-bipartite induced by* S is the sub-bipartite $G' \doteq (S, E_S)$, with $E_S = E \cap (S \times S)$.

A *h-cluster* S is a set of vertices, either in A or in B, s.t. $\delta_S(x) \leq h$, $x \in \alpha(S)$. A 1-cluster is simply called *cluster*.

3.2 Partitioning into h-Clusters

We present an algorithm which, given a bipartite graph $G = (A \cup B, E)$, computes a h-cluster $C \subset A$, with $h = \lceil \log n_B \rceil$; hence, the sub-bipartite induced by $C \cup B$ has the property $\Delta_{\alpha(C)} \leq h$. Of course, this can be done trivially if C consists of at most h vertices. Somewhat surprisingly, it turns out that a clever selection of vertices of the h-cluster, we can find a h-cluster of $\Omega\left(\frac{n_A}{\Delta_B}\right)$ vertices, hence a significant fraction of all vertices in A.

The idea behind the algorithm derives from the following observation: when we add a new vertex x to the h-cluster, then for each vertex y in $\alpha(x)$, its degree $\delta_C(y)$ with respect to C increases by one. A trivial approach would be to just check that for each vertex $y \in \alpha(C) \cap \alpha(x)$, $\delta_C(y) \leq h-1$ holds; this guarantees $\Delta_{\alpha(C)} \leq h$ after the insertion. Unfortunately, on the long run this strategy does not work. A smarter strategy must look forward, to guarantee that not only the current choice is correct, but that it does not restrict too much successive choices. A new vertex x is added to the cluster in h successive steps, at each step j observing how x increases the number $|\alpha_{j-1}(C)|$ of vertices adjacent to C having degree $j-1$ with respect to C. At each step the selection is passed by those vertices which do not increase too much the number $|\alpha_{j-1}(C)|$ of vertices adjacent to C having degree $j-1$ with respect to C, where "too much" means no more than t times the average value over all candidates at step j, for some suitable choice of t.

We will prove that this strategy causes the number $|\alpha_h(C)|$ of vertices adjacent to C having degree h with respect to C to increase very slowly, thus ensuring that this number is less than 1 until at least $\frac{n_A}{\beta \Delta_B}$ vertices have been chosen, for a fixed constant β.

The algorithm is presented in Figure 1; from now on, C_i denotes the h-cluster at the end of step i, and $S_{i,j}$ the set of vertices, to be added to C_{i-1}, that passed the selection step j. Furthermore, the notation $\alpha_0(C_i)$ is extended to denote the set $B - \alpha(C_i)$ of all vertices in B not adjacent to C_i.

Lemma 2. $|S_{i,j}| \geq (n_A - i + 1)\left(1 - \frac{1}{t}\right)^j$.

Proof. At each step j we select those vertices $x \in S_{i,j-1}$ such that $|\alpha_{j-1}(C_{i-1}) \cap \alpha(x)|$ is no more than t times the average value $\mu_{i,j-1}$ over all vertices in $S_{i,j-1}$. If a set of n non-negative integers with average value μ then at most n/t elements have value greater than $t\mu$ and, hence, at least $n(1-1/t)$ elements have value at most $t\mu$, thus $|S_{i,j}| \geq (1-1/t)|S_{i,j-1}|$, with $|S_{i,0}| = n_A - i + 1$; the Lemma follows.

Lemma 3. *Let $n_{i,j} \doteq |\alpha_j(C_i)|$, that is the number of vertices y in $\alpha(C_i)$ s.t. $\delta_{C_i}(y) = j$. Then*

$$n_{i,j} \leq \left[\frac{t\Delta_B(i-1)}{(n_A - i + 1)\left(1 - \frac{1}{t}\right)^{\frac{j-1}{2}}}\right]^j \frac{n_B}{j!}$$

Proof. The proof is by induction on the step j.
Base step: $j = 1$. At step $(i,1)$, $\mu_{i,0}$ is the average degree of the $n_A - i + 1$ vertices in $A - C_{i-1}$ with respect to vertices not connected to $\alpha(C_{i-1})$. Thus, $\mu_{i,0} \leq \frac{m}{n_A - i + 1}$, and a vertex x that is added to $S_{i,1}$ verifies $\delta_{\alpha_0(C_{i-1})}(x) \leq \frac{tm}{n_A - i + 1}$.

$C_0 \leftarrow \emptyset$;
$i \leftarrow 0$;
repeat
  $i \leftarrow i+1$;
  $S_{i,0} \leftarrow A - C_{i-1}$;
  for $j \leftarrow 1$ **to** h **do begin**
    $\mu_{i,j-1} = \frac{\sum_{x \in S_{i,j-1}} |\alpha_{j-1}(C_{i-1}) \cap \alpha(x)|}{|S_{i,j-1}|}$;
    $S_{i,j} \leftarrow \{x \in S_{i,j-1} : |\alpha_{j-1}(C_{i-1}) \cap \alpha(x)| \leq t\mu_{i,j-1}\}$;
  end ;
  pick a vertex $x \in S_{i,j}$;
  $C_i \leftarrow C_{i-1} \cup \{x\}$;
until $S_{i,j} = \emptyset$;
end .

Fig. 1. Algorithm Select.

If x is added to C_{i-1}, $n_{i-1,1}$ is increased by at most $\frac{tm}{n_A - i + 1}$ new vertices. Hence,

$$n_{i,1} \leq n_{i-1,1} + \frac{tm}{n_A - i + 1} \leq \sum_{k=1}^{i-1} \frac{tm}{n_A - k} \leq \frac{tm(i-1)}{n_A - i + 1} .$$

Since $m \leq \Delta_B n_B$, $n_{i,1} \leq \frac{t\Delta_B n_B (i-1)}{n_A - i + 1}$, and the base step is proved.

Induction step: $j-1 \rightarrow j$. At step (i,j), $\mu_{i,j-1}$ is the average degree of candidate vertices in $S_{i,j-1}$ with respect to vertices in $\alpha_{j-1}(C_{i-1})$. By Lemma 2 and since the total number of edges outgoing from $\alpha_{j-1}(C_{i-1})$ is at most $\Delta_B n_{i-1,j-1}$, we have $\mu_{i,j-1} \leq \frac{\Delta_B n_{i-1,j-1}}{(n_A - i + 1)\left(1 - \frac{1}{t}\right)^{j-1}}$.

A vertex x is added to $S_{i,j-1}$ if it verifies $\delta_{\alpha_{j-1}(C_{i-1})}(x) \le t\mu_{i,j-1}$. Hence, if x is added to C_{i-1}, $n_{i-1,j}$ is increased by at most $t\mu_{i,j-1}$ new vertices. Thus,

$$n_{i,j} \le n_{i-1,j} + \frac{t\Delta_B n_{i-1,j-1}}{(n_A - i + 1)\left(1 - \frac{1}{t}\right)^{j-1}} \le \sum_{k=1}^{i-1} \frac{t\Delta_B n_{k,j-1}}{(n_A - k)\left(1 - \frac{1}{t}\right)^{j-1}}$$

$$\le \frac{t\Delta_B}{(n_A - i + 1)\left(1 - \frac{1}{t}\right)^{j-1}} \sum_{k=1}^{i-1} n_{k,j-1}$$

$$\le \frac{t\Delta_B}{(n_A - i + 1)\left(1 - \frac{1}{t}\right)^{j-1}} \sum_{k=1}^{i-1} \left[\frac{t\Delta_B(k-1)}{(n_A - k + 1)\left(1 - \frac{1}{t}\right)^{\frac{j-2}{2}}}\right]^{j-1} \frac{n_B}{(j-1)!}$$

$$\le \left[\frac{t\Delta_B}{(n - i + 1)\left(1 - \frac{1}{t}\right)^{\frac{j-1}{2}}}\right]^{j} \frac{n_B}{(j-1)!} \sum_{k=1}^{i-1} (k-1)^{j-1}$$

$$\le \left[\frac{t\Delta_B}{(n - i + 1)\left(1 - \frac{1}{t}\right)^{\frac{j-1}{2}}}\right]^{j} \frac{n_B}{j!} (i-1)^j .$$

This concludes the induction step.

Theorem 4. *Let $G = (A \cup B, E)$ be a bipartite graph. For $h \ge \lfloor \log n_B \rfloor$, Algorithm Select finds a h-cluster C of $\left\lceil \frac{n_A}{(2e^{\frac{3}{2}}+1)\Delta_B} \right\rceil$ vertices in time $O(|C| n_A \Delta_A)$.*

Proof. If $t = h \ge 2$, then $\left(1 - \frac{1}{t}\right)^{\frac{h-1}{2}} \ge \frac{1}{\sqrt{e}}$. Let $i_{\max}$ be the value of index i at the end of the execution of Algorithm Select. Considering that $h! \ge \left(\frac{h}{e}\right)^h$, Lemma 3 implies:

$$n_{i_{\max},h+1} \le \left[\frac{e^{\frac{3}{2}} \Delta_B (i_{\max} - 1)}{n_A - i_{\max} + 1}\right]^{h+1} n_B \; .$$

If $i_{\max} < \left\lceil \frac{n_A}{2e^{\frac{3}{2}}\Delta_B + 1} \right\rceil$ then $\frac{e^{\frac{3}{2}}\Delta_B(i_{\max}-1)}{n_A - i_{\max}+1} < \frac{1}{2}$, hence, $n_{i_{\max},h+1} \le \frac{n_B}{2^{h+1}} < 1$ for $h \ge \lfloor \log n_B \rfloor \ge \lceil \log n_B \rceil - 1$, and $C_{i_{\max}}$ is a h-cluster.

From now on β denotes the constant $2e^{\frac{3}{2}} + 1 < 10$. The following theorem will be used to derive the space complexity of the proposed data structure.

Theorem 4 leads to the following

Corollary 5. *Let $G = (A \cup B, E)$ be a bipartite graph. For $h \ge \lfloor \log n_B \rfloor$, A can be partitioned into $\lceil 2\beta\Delta_B \rceil \cdot \lceil \log n_A \rceil$ h-clusters. The time complexity is $n_A^2 \Delta_A$.*

Proof. The sequence of clusters is computed by repeatedly selecting a h-cluster and removing its vertices from A. Let us suppose that after k iterations the number n'_A of vertices remained in A is greater than $n_A/2$, but after $k+1$ iterations is less than or equal

$n_A/2$. By Theorems 4, during the first k iterations, algorithm Select finds h-clusters of at least $\left\lceil \frac{n_A/2}{\beta\Delta_B} \right\rceil$ vertices. Hence, in k iterations at least $k\left\lceil \frac{n_A/2}{\beta\Delta_B} \right\rceil$ vertices have been removed from A, so $k \leq 2\beta\Delta_B$.

We can repeat the same argument to the remaining vertices, each time halving the number of vertices still in A; this can obviously repeated no more than $\lceil \log n_A \rceil$ times.

3.3 Partitioning into Clusters

The following lemma characterizes the complexity of partitioning a bipartite graph $G = (A \cup B, E)$ into clusters (1-clusters). Clusters will be used to build the ground data structure upon which the others are based.

Lemma 6. *Let $G = (A \cup B, E)$ be a bipartite graph. B can be partitioned in $1 + \Delta_B(\Delta_A - 1)$ clusters. The time required is $O(n_B\Delta_A\Delta_B)$*

Proof. Let $B_1, \ldots, B_k$ be a partition of B into clusters so that B_i is a maximal cluster for $B - \bigcup_{j=1}^{i-1} B_j$. Each vertex $y \in B_i$ has at most Δ_B adjacent vertices, and each of them has at most $\Delta_A - 1$ adjacent vertices different from y. Hence, a vertex $y \in B_i$ prevents at most $\Delta_B(\Delta_A - 1)$ vertices to be included in the same cluster. Since the cluster is maximal, each vertex in B either has been chosen in B_i or has been excluded from it, so $n_B \leq |B_i|(1 + \Delta_B(\Delta_A - 1))$. Hence $|B_i| \geq \frac{n_B}{1+\Delta_B(\Delta_A-1)}$. The lemma follows.

Note that the bound given by Lemma 6 is tight, since there exists an infinite class of regular bipartite graphs that cannot be decomposed in less than $1 + \Delta(\Delta - 1)$ clusters [11].

4 The Data Structure

In this Section we present the data structure for the multi–dimensional searching problem.

Based upon the decomposition theorems given in Section 3.2, we previously present a data structure for labeled bipartite that allows us to represent a bipartite graph $G = (A \cup B, E)$ in $O(n + m \log n)$ space, and to test if two vertices are adjacent with a constant number of steps. For sake of clarity, we first describe a simpler data structure that represent bi-regular bipartite graphs, then extend the result to represent all bipartite graphs.

4.1 Representing Bi-Regular Bipartite Graphs

Given a bi-regular bipartite graph $G = (A \cup B, E)$, we partition A in h-clusters according to Corollary 5; hence, we obtain a sequence of at most $\lceil 2\beta\Delta_B h \rceil$ bipartite graphs $G_i = (A_i, B, E_i)$, where A_i is the i-th h-cluster and $E_i \doteq E \cap (A_i \times B)$. Then we partition the vertex set B of each bipartite $G_i = (A_i \cup B, E_i)$ into clusters. Lemma 6 ensures that each bipartite graph is decomposed into at most $1 + h(\Delta_A - 1)$ clusters.

We define the following arrays:

- `hclus` of size n_A; $i = $ `hclus`$[x]$ is the index of the unique h-cluster A_i to which $x \in A$ belongs;

- clus of size $n_B \times \lceil 2\beta\Delta_B h \rceil$; $j =$ clus$[y,i]$ is the index of the unique cluster $B_{i,j}$ in G_i to which $y \in B$ belongs;
- join$_i$ of size $n_A \times (1 + h(\Delta_A - 1))$; $y =$ join$[x,j]$ is the unique possible vertex $y \in B$ adjacent to x in the j-th cluster in the unique i-th h-cluster to which x belongs.

Adjacency on the bipartite graph can be tested in 3 steps since $(x,y) \in E$ if and only if, given $i \doteq$ hclus$[x]$ and $j \doteq$ clus$[y,i]$, $y =$ join$[x,j]$ holds. The total space required is

$$O(n_A + n_B \lceil 2\beta\Delta_B h \rceil + n_A(1 + h(\Delta_A - 1))) = O((n+m)\log n) \ .$$

Note that if $m \leq n$ then isolated vertices can be trivially represented, so the space complexity becomes $O(n + m\log n)$.

4.2 Representing Bipartite Graphs and 2–place Functions

We now show how to obtain for general bipartite graphs the same results as for bi-regular graphs. Given a bipartite graph $G = (A \cup B, E)$, we first partition B into maximal subsets B_i, s.t. $\forall y \in B_i$, $2^i \leq \delta(y) < 2^{i+1}$. We obtain a sequence of at most $h = \lceil \log n \rceil$ bipartite graphs $G_i = (A \cup B_i, E_i)$, where B_i is the i-th subset of B and $E_i \doteq E \cap (A \times B_i)$.

Then, according to Corollary 5, for each such bipartite graph G_i we partition A into h-clusters $A_{i,j}$, obtaining a sequence of at most $\lceil 2^{i+1}\beta \rceil\, h$ bipartite graphs $G_{i,j}$, and further partition each h-cluster into at most h subsets $A_{i,j,k}$ s.t. $\forall x \in A_{i,j,k}$, $2^k \leq \delta_{B_i}(x) < 2^{k+1}$, obtaining a sequence of bipartite graphs $G_{i,j,k}$.

Finally, for each bipartite graph $G_{i,j,k}$, we partition the set B_i into clusters; Lemma 6 ensures that each bipartite graph $G_{i,j,k}$ is decomposed into at most $1 + h(\Delta_{A_{i,j,k}} - 1)$ clusters.

We define the following arrays:

- range of size n_B; $i =$ range$[y]$ is the index of the unique subset B_i to which y belongs;
- hclus of size $n_A \times h$; $j =$ hclus$[x,i]$ is the index of the unique h-cluster $A_{i,j}$ to which $x \in A$ belongs in G_i.
- subs of size $n_A \times h$; $k =$ subs$[x,i]$ is the index of the unique subset $A_{i,j,k}$ in the unique h-cluster to which $x \in A$ belongs in G_i.
- For each vertex $y \in B_i$, we define an array ranges$_y$ of size $\lceil 2^{i+1}\beta \rceil\, h$; ranges$_y[j]$ is a reference to the array clus, which contains the cluster indices of y in all subsets $A_{i,j,k}$; it is empty if y is not adjacent to any vertex in $A_{i,j}$. The total space needed for array ranges$_y[j]$ for all $y \in B$ is

$$O(\sum_{B_i} \sum_{y \in B_i} \lceil 2^{i+1}\beta \rceil\, h) = O(\sum_{B_i} \sum_{y \in B_i} \delta(y)\beta h) = O(mh) \ .$$

  Reading ranges$_y[j]$ requires 2 steps, one to read the initial address of the array given y, and one to access its j-th element.
- For each vertex $y \in B_i$, and each h-cluster $A_{i,j}$ connected to y, we define an array clus; clus$[k]$ is the index of the unique cluster in $G_{i,j,k}$ to which $y \in B_i$ belongs; it is empty if y is not adjacent to any vertex in $A_{i,j,k}$. For each vertex $y \in B_i$, since

$2^i \leq \delta(y) < 2^{i+1}$, at most 2^{i+1} such arrays are defined, each of them having size h. Hence, the total space needed for all arrays `clus` is

$$O(\sum_{B_i}\sum_{y\in B_i} 2^i h) = O(\sum_{B_i}\sum_{y\in B_i} \delta(y)h) = O(mh) \ .$$

- `joins` of size $n_A \times h$; `joins`$[x,i]$ is a reference to the array `join`, which contains all vertices in B_i adjacent to x. It is empty if x is not adjacent to any vertex in B_i.
- For each vertex $x \in A_{i,j,k}$, and each set B_i connected to x, we define an array `join` of size $(1+h(2^{k+1}-1))$; `join`$[l]$ is the (unique) possible vertex adjacent to x in the l-th cluster of $G_{i,j,k}$; it is empty if x is not adjacent to any vertex the l-th cluster of $G_{i,j,k}$. For each vertex $x \in A$, the space needed for all its related arrays `join` is $O(\sum_{B_i} 2\delta_{B_i}(x)) = O(h\delta(x)$, so the total space for arrays `join` for all $x \in A$ is $O(mh)$.

Adjacency on the bipartite graph can be tested in constant time since $(x,y) \in E$ if and only if, given $i \doteq$ `range`$[y]$, $j \doteq$ `hclus`$[x,i]$, $k \doteq$ `subs`$[x,i]$, `clus` $\doteq$ `ranges`$_y[j]$, $l \doteq$ `clus`$[k]$ and `join` $\doteq$ `joins`$[x,i]$, $y =$ `join`$[l]$ holds. The test requires 7 steps. The total space required is

$$O(n_B + n_A h + mh) = O((n+m)\log n) \ .$$

Also in this case, if $m \leq n$ then isolated vertices can be trivially represented, so the space complexity becomes $O(n + m\log n)$.

¿From the above discussion, we have the following theorem:

Theorem 7. *There exists a data structure that represents a bipartite graph with n vertices and m edges in space $O(n + m\log n)$. Vertex adjacency can be tested in 7 steps. Preprocessing time is $O(n^2\Delta)$, where Δ is the maximum vertex degree.*

The representation of a 2–place function and the lookup operation which given two objects, return a value associated to the pair, is equivalent to the following: given a bipartite graph $G = (A \cup B, E)$, and a *labeling function* $\mathcal{L} : E \to \mathbb{N}$, and $x \in A$, $y \in B$, if $(x,y) \in E$ return $\mathcal{L}(x,y)$. This can be easily accomplished with the previously described data structure and, whenever $(x,y) \in E$, extending `join`$[l]$ to contain both the (unique) possible vertex y adjacent to x in the l-th cluster of $G_{i,j,k}$ and the value $\mathcal{L}(x,y)$. This leads to the following theorem:

Theorem 8. *There exists a data structure that represents a 2–place function of size m between objects from a domain of size n in space $O(n + m\log n)$. The lookup operation requires 7 steps. Preprocessing time is $O(n^3)$.*

4.3 Representing Multi-dimensional Data

Given a point set $S \subseteq U^d$, with $m = |S|$ and $n = |U|$. Let $\langle f_1, \ldots, f_{d-1}\rangle$ be the sequence of 2–place functions representing S, as described in Section 2. Additionally, let $m = |S|$ and m_i be the size of the 2–place function f_i, for $1 \leq i \leq d-1$. By the definition, we have $m = \sum_{i=1}^{d-1} m_i$. Moreover, for any i, $n_i \leq m_i$, n_i being the size of the domain set of f_i. By Theorem 8, we can state the following theorem:

Theorem 9. *There exists an implicit data structure that represents a set $S \subseteq U^d$ of multi–dimensional points in space $O(m\log n)$. The exact match and prefix–partial match queries can be performed in $7(d-1)$ and $7(d-1)+t$ steps, respectively, where t is the number of points reported. The preprocessing time is $O(dn^3)$.*

5 Extensions

5.1 External Memory Data Structure

Due to its nature, the above described data structure can be efficiently applied to secondary storage. In this paper, we consider the standard two-level I/O model introduced by Aggarwal and Vitter in [1]. In this case, we can devise a powerful compression technique leading to a space optimal data structure.

In the data structure described in Section 4.2, the critical arrays are `ranges`, `clus`, and `join`, that is those requiring a total space $O(mh)$, which in terms of external memory storage implies $O(mh/B)$ blocks. The following lemma counts the number of non-empty entries in these arrays:

Lemma 10. *Let k' be the total number of non-empty entries in arrays* $\mathtt{ranges}_y$ *for all $y \in B$; k'' be the total number of non-empty entries in all arrays* `clus`*; and k''' be the total number of non-empty entries in arrays* `join`*. Then $k' \leq m$, $k'' \leq m$, $k''' \leq m$.*

Proof. If $\mathtt{ranges}_y[j]$ is not empty, then some edge (x,y) belongs to $G_{i,j}$; on the other hand, there is a unique bipartite graph $G_{i,j}$ containing such edge. Hence, the total number of non-empty entries in arrays $\mathtt{ranges}_y$ for all $y \in B$ is at most m.

If $\mathtt{clus}[k]$ is not empty for some vertex $y \in B_i$ and some h-cluster $A_{i,j}$ connected to y, then some edge (x,y) belongs to $G_{i,j,k}$; since there is a unique bipartite graph $G_{i,j,k}$ containing such edge, the total number of non-empty entries in all arrays `clus` is at most m.

If $\mathtt{join}[l]$ is not empty, then some edge (x,y) belongs to $G_{i,j,k}$; there is a unique bipartite graph $G_{i,j,k}$ containing such edge; the total number of non-empty entries in arrays `join` for all $x \in A$ is at most m.

Let a be an array of size k. We partition a into intervals of B elements, and represent each interval by a reference to the block containing the non empty entries in that interval. It is easy to see that an array a of size k with k' empty entries can be represented in $O(\frac{k}{B} + k')$ space, thus in $O(\frac{k}{B^2} + \frac{k'}{B})$ blocks; furthermore, one access to $\mathtt{a}[i]$ maps to 2 memory accesses. hence the external memory version of Theorem 8 and Theorem 9 can be stated as follows:

Theorem 11. *There exists an external–memory data structure that represents a 2–place function of size m between objects from a domain of size n with $O(\frac{n\log n+m}{B})$ blocks. The lookup operation requires 10 I/Os.*

Theorem 12. *There exists an external memory implicit data structure that represents a set $S \subseteq \mathbb{N}^d$ of multi–dimensional points with $O(m/B)$ blocks. The exact match and prefix–partial match queries require $10(d-1)$ and $10(d-1)+t/B$ I/Os, respectively, where t is the number of points reported by the partial match query.*

5.2 Incremental Exact Match Queries

The representation we propose for a multi–dimensional point set S allows to efficiently perform the exact match operation in a more general context. In fact we can define the *incremental exact match* query, where the coordinates are specified incrementally, that is, the search starts when the first coordinate is given, and proceeds refining the searching space as soon as the other coordinates are specified. This definition of exact search is particularly useful in distributed environments where the request for a query is expressed by sending messages along communication links [15, 16, 4, 12, 13] and not all coordinates reside on the same machine.

Another field of application of the incremental exact match query is for the *interactive exploratory search* on Web. In this case the user can specify the searching keys one by one so as to obtain intermediate results.

Also, the incremental exact match query is particularly practical when dealing with a point sets from a very high multi–dimensional space (order of thousands of keys). In this case we can manage the query in a distributed environment by specifying only $k \ll d$ keys a time in order to prevent network congestion and to obtain a more reliable answer.

5.3 Improving Conventional Searching Data Structures

Our decomposition technique can be positively applied to one-dimensional hashing and perfect hashing when dealing with real keys. Let w be the machine word length, and $K \gg w$ the key length. We can divide each key in K/w sub-keys, and reduce the original one–dimensional searching problem to a multi–dimensional searching problem, which can be solved with our technique with no additional storage and with a constant number of I/Os.

Another important application it to the trie data structure. With a technique similar to the one above described, we can consider larger node sizes.

6 Open Problems

One important open problem is that of dynamizing the data structure; even an incremental-only version data structure would be a useful improvement. Another important research direction is to extend the operation set to include other operations useful for the management of a multi-dimensional data set (e.g. range queries, retrieve maximal elements, orthogonal convex-hull, *etc.*)

We are currently carrying out an extensive experimentation on secondary memory, based on a data sets derived from a business application. This experimentation activity is still at its beginning, the main purpose being primarily to test the effective speedup in the lookup operation and the overall size of the representation on these data sets. Preliminary experimentation results show that the behavior of our data structure is very fast and works very well in the average case.

References

[1] A. Aggarwal and J. S. Vitter. The input/output complexity of sorting and related problems. *Communications of the ACM*, 31(9):1116–1127, September 1988.

[2] R. Agrawal, A. Borgida, and H. V. Jagadish. Efficient management of transitive relationship in large data and knowledge bases. In *Proceedings of the International Conference on the Management of Data*, pages 253–262, Portland, OR, 1989.
[3] R. Bayer and C. McCreight. Organization and maintenance of large ordered indexes. *Acta Informatica*, 1(3):173–179, 1972.
[4] R. Devine. Design and implementation of DDH: A distributed dynamic hashing algorithm. In *4th Int. Conf. on Foundations of Data Organization and Algorithms (FODO)*, Chicago, 1993.
[5] D.Mayer and B.Vance. A call to order. In *Proceedings of the International Conference on Principle of Database Systems*, 1993.
[6] A. Fiat and M. Naor. Implicit $O(1)$ probe search. In *Proceedings of the Twenty First Annual ACM Symposium on Theory of Computing*, pages 336–344, Seattle, Washington, 1989.
[7] A. Fiat, M. Naor, J. P. Schmidt, and A. Siegel. Non-oblivious hashing. In *Proceedings of the Twentieth Annual ACM Symposium on Theory of Computing: Chicago, Illinois, May 2–4, 1988*, pages 367–376, New York, NY 10036, USA, 1988. ACM Press.
[8] M. L. Fredman, J. Komlós, and E. Szemeredi. Sorting a sparse table with $O(1)$ worst case access time. In *Proc. 23rd Ann. IEEE Symp. on Foundations of Computer Science*, pages 165–169, 1982.
[9] T. Hagerup. Sorting and searching on the word RAM. In M. Morvan, C. Meinel, and D. Krob, editors, *STACS: Annual Symposium on Theoretical Aspects of Computer Science*, pages 366–398. LNCS 1373, Springer-Verlag, 1998.
[10] P. C. Kanellakis, S. Ramaswamy, D. E. Vengroff, and J. S. Vitter. Indexing for data models with constraints and classes. *Journal of Computer and System Science*, 52:589–612, 1996.
[11] D. J. Kleitman and K. J. Winston. The asymptotic number of lattices. *Annuals of Discrete Matemathics*, 6:243–249, 1980.
[12] B. Kröll and P. Widmayer. Distributing a search tree among a growing number of processors. In *ACM SIGMOD Int. Conf. on Management of Data*, pages 265–276, Minneapolis,MN, 1994.
[13] B. Kröll and P. Widmayer. Balanced distributed search trees do not exists. In S. Akl et al., editor, *4th Int. Workshop on Algorithms and Data Structures (WADS'95)*, pages 50–61, Kingston, Canada, 1995. LNCS 955, Springer-Verlag.
[14] Douglas Lea. Digital and Hilbert *K-D* trees. *Information Processing Letters*, 27(1):35–41, 1988.
[15] W. Litwin, M. A. Neimat, and D. A. Schneider. LH^*–linear hashing for distributed files. In *ACM SIGMOD Int. Conf. on Management of Data*, Washington, D.C., 1993.
[16] W. Litwin, M. A. Neimat, and D. A. Schneider. LH^*–a scalable distributed data structure. *ACM Trans. Database Systems*, 21(4):480–525, 1996.
[17] D. Morrison and R. Patricia. Practical algorithm to retrieve information coded in alphanumeric. *Journal of the ACM*, 15:514–534, 1968.
[18] F. P. Preparata and M. I. Shamos. *Computational Geometry*. Springer-Verlag, Berlin, New York, 1985.
[19] S. Ramaswamy and S. Subramanian. Path caching: A technique for optimal external searching. In *Proc. ACM Symp. Principles of Database System*, pages 25–35, 1994.
[20] H. Samet. *The Design and Analysis of Spatial Data Structures*. Addison-Wesley, Reading, MA, 1990.
[21] M. Talamo and P.Vocca. Compact implicit representation of graphs. In J. Hromkovič and O. Sýkora, editors, *Proceedings of 24th International Workshop on Graph–Theoretic Concepts in Computer Science WG'98*, pages 164–176. LNCS 1517, Springer-Verlag, 1998.
[22] M. Talamo and P. Vocca. A data structure for lattice representation. *Theoretical Computer Science*, 175(2):373–392, April 1997.

[23] M. Talamo and P. Vocca. A time optimal digraph browsing on a sparse representation. Technical Report 8, Matemathics Department, University of Rome "Tor Vergata", 1997.

[24] A. Tansel, J. Clifford, S. Gadia, S. Jajodia, A. Segev, and R. Snodgrass, editors. *Temporal Databases: Theory, Design, and Implementation.* Benjamin/Cummings, Redwood City, CA, 1993.

[25] M. Yannakakis. Graph-theoretic methods in database theory. In ACM, editor, *PODS '90. Proceedings of the Ninth ACM SIGACT-SIGMOD-SIGART Symposium on Principles of Database Systems: April 2–4, 1990, Nashville, Tennessee*, volume 51(1), New York, NY 10036, USA, 1990. ACM Press.

Geometric Searching over the Rationals*

Bernard Chazelle

Department of Computer Science,
Princeton University, and NEC Research Institute
chazelle@cs.princeton.edu

Abstract. We revisit classical geometric search problems under the assumption of rational coordinates. Our main result is a tight bound for point separation, ie, to determine whether n given points lie on one side of a query line. We show that with polynomial storage the query time is $\Theta(\log b/\log\log b)$, where b is the bit length of the rationals used in specifying the line and the points. The lower bound holds in Yao's cell probe model with storage in $n^{O(1)}$ and word size in $b^{O(1)}$. By duality, this provides a tight lower bound on the complexity on the polygon point enclosure problem: given a polygon in the plane, is a query point in it?

1 Introduction

Preprocess n points in the plane, using $n^{O(1)}$ storage, so that one can quickly tell whether a query line passes entirely below or above the points. This *point separation* problem is dual to deciding whether a query point lies inside a convex polygon. As is well known, this can be done in $O(\log n)$ query time and $O(n)$ storage, which is optimal in the algebraic decision tree model [8, 9]. This is suitable for infinite-precision computations [3, 4, 20], but it does not allow for bucketing or any form of hashing. Unfortunately, these happen to be essential devices in practice. In fact, the computational geometry literature is rife with examples of speed-ups derived from finite-precision encodings of point coordinates, eg, range searching on a grid [17], nearest neighbor searching [11, 12], segment intersection [13], point location [16].

To prove lower bounds is usually difficult; even more so when hashing is allowed. Algebraic models are inadequate and one must turn to more general frameworks such as the cell probe model [18] or, in the case of range searching, the arithmetic model [7, 19]. As a searching (rather than computing) problem, point separation lends itself naturally to the cell probe model and this is where we confine our discussion. Our main interest is in pinpointing what sort of query time can or cannot be achieved with polynomial storage. Note that some restriction on storage is essential since constant query time is trivially achieved with exponential space.

Let P be a set of n points in the plane, whose coordinates are rationals of the form p/q, where p and q are b-bit integers. A cell probe algorithm for point separation consists of a table of size n^c, with each cell holding up to b^d bits, for some arbitrarily large constants c,d. A query is answered by looking up a certain number of cells and outputting yes or no, depending on the information gathered. For lower bound purposes,

* This work was supported in part by NSF Grant CCR-96-23768, ARO Grant DAAH04-96-1-0181, NEC Research Institute, Ecole Polytechnique, and INRIA.

the query time counts only the number of cells that are looked up during the computation.

Theorem 1. *Given any cell-probe algorithm for point separation, there exist an input of n points and a query line that require $\Omega(\log b/\log\log b)$ time. The lower bound is tight.*

The upper bound can be achieved on a standard unit-cost RAM. Take the convex hull of the points and, given the query line, search for the edges whose slopes are nearest that of the line. Following local examination of the relative heights of the line and edge endpoints, conclude whether there is point separation or not. This is elementary computational geometry and details can be skipped. The main point is that the problem reduces to predecessor searching with respect to slopes (rational numbers over $O(b)$ bits), which can be done optimally using a recent algorithm of Beame and Fich [2]. Their algorithm preprocesses n integers in $[0,N]$, so that the predecessor of any query integer can be found in $O(\log\log N/\log\log\log N)$ time, using $n^{O(1)}$ storage. By appropriate scaling and truncation, their scheme can be used for predecessor searching over the rationals, with the query time becoming $O(\log b/\log\log b)$, for rationals with $O(b)$-bit numerators and denominators.

2 The Complexity of Point Separation

The input consists of a set P of n points in $\mathbf{R}^2$, which is encoded in a table T of size n^c, where c is an arbitrarily large constant. To simplify the notation we can replace c by $\max\{c,d\}$, and require that each cell should hold at most $w = b^c$ bits. A cell probe algorithm is characterized by a table assignment procedure (ie, a function mapping any P to an assignment of the table T to actual values) together with an infinite sequence of functions f_1, f_2, etc. Given a query ℓ (ie, a certain line in $\mathbf{R}^2$), we evaluate the index $f_1(\ell)$ and look up the table entry $T[f_1(\ell)]$. If $T[f_1(\ell)]$ encodes whether ℓ separates the point set or not, we answer the query and terminate. Otherwise, we evaluate $f_2(\ell, T[f_1(\ell)])$ and look up the entry $T[f_2(\ell, T[f_1(\ell)])]$, and we iterate in this fashion until a cell probe finally reveals the desired answer. Note that such a framework is so general it easily encompasses every known solution to the point separation problem.

We use Miltersen's reformulation [15] of a cell probe algorithm as a communication complexity game between two players [14]. Alice chooses a set $\mathcal{L}_1$ of candidate queries (ie, a set of lines in the plane), while Bob decides on a collection $\mathcal{P}_1$ of n-point sets. Note that each pair $(\ell, P) \in \mathcal{L}_1 \times \mathcal{P}_1$ specifies a problem instance. Alice and Bob's task is then to exhibit a problem instance $(\ell, P) \in \mathcal{L}_1 \times \mathcal{P}_1$ that requires $\Omega(\log b/\log\log b)$ probes in T to answer. They do that by simulating each probe by a round in a communication complexity game.

The n^c possible values of the index $f_1(\ell)$ partition $\mathcal{L}_1$ into equivalence classes. Alice chooses one of them and sends to Bob the corresponding value of $f_1(\ell)$. Of all the possible 2^w assignments of the entry $T[f_1(\ell)]$ Bob chooses one of them and narrows down his candidate set $\mathcal{P}_1$ to the set $\mathcal{P}_2$ of point sets leading to that chosen value of $T[f_1(\ell)]$. Bob sends back to Alice his choice of $T[f_1(\ell)]$. Knowing ℓ and $T[f_1(\ell)]$, Alice chooses a value for $f_2(\ell, T[f_1(\ell)])$ and communicates it to Bob, etc. Each round

k produces a new pair $(\mathcal{L}_{k+1}, \mathcal{P}_{k+1})$ with the property that, for all queries in $\mathcal{L}_{k+1}$ and all point sets in $\mathcal{P}_{k+1}$, Bob and Alice exchange the same information during the first k rounds, which are thus unable to distinguish among any of the problem instances in $\mathcal{L}_{k+1} \times \mathcal{P}_{k+1}$.

We say a query line (resp. point set) is *active* at the beginning of round k if it belongs to $\mathcal{L}_k$ (resp. $\mathcal{P}_k$). The set $\mathcal{L}_k \times \mathcal{P}_k$ is called *unresolved* if it contains at least two problem instances (ℓ, P) and (ℓ', P') with different yes/no outcomes: in such a case, Bob and Alice need to proceed with round k, and the cost of the protocol (ie, the minimum number of rounds necessary) is at least k. We show that for some suitable $n = n(b)$, given any cell probe table assignment procedure, there exist a starting set $\mathcal{L}_1$ of query lines and a starting collection $\mathcal{P}_1$ of n-point sets in the plane that allow Bob and Alice to produce a nested sequence of unresolved sets

$$\mathcal{L}_1 \times \mathcal{P}_1 \supseteq \cdots \supseteq \mathcal{L}_t \times \mathcal{P}_t,$$

where $t = \Theta(\log b / \log\log b)$.

The protocol between Bob and Alice builds on our earlier work on approximate searching over the Hamming cube [5], which itself borrows ideas from the work of Ajtai [1] on predecessor searching. A protocol for predecessor queries of a similar flavor was recently devised independently by Beame and Fich [2].

2.1 Points and Lines

Let p_i denote the point (i, i^2), and given $i < j$, let $a_{ij} = \frac{1}{2}(i+j, i^2+j^2)$ and $b_{ij} = ((i+j)/2, ij)$. Any of Bob's n-point sets P is of the form

$$P = \left\{ p_{i_1}, X_{i_1 i_2}, p_{i_2}, X_{i_2 i_3}, \ldots, p_{i_{s-1}}, X_{i_{s-1} i_s}, p_{i_s} \right\},$$

for some $i_1 < \cdots < i_s$, where $n = 2s - 1$ and X denotes the symbol a or b (not necessarily the same one throughout the sequence). Thus, P can be specified by an *index set* $I = I(P) = \{i_1, \ldots, i_s\}$ consisting of s distinct b-bit integers and a bit vector $\sigma = \sigma(P)$ of length $s-1$ specifying the X's. For technical reasons, we require that all the integers of the index set I be even.

The starting query set $\mathcal{L}_1$ consists of the lines of the form, $y = 2kx - k^2$, for all odd b-bit integers k. Note that this is the equation of the line through p_k tangent to the parabola $y = x^2$. The number of bits needed to encode any point coordinate or line coefficient is $2b$ (and not b, a minor technicality). Note that the problem does not become suddenly easier with other representations such as $\alpha x + \beta y = 1$, and that for the purposes of our lower bound, all such representations are essentially equivalent. The following is immediate.

Lemma 2. *Let p_{i_j} and $p_{i_{j+1}}$ be two points of P and let ℓ be the line $y = 2kx - k^2$, where $i_j < k < i_{j+1}$. The line ℓ separates the point set P if and only if the symbol X in $X_{i_j i_{j+1}}$ is of type b.*

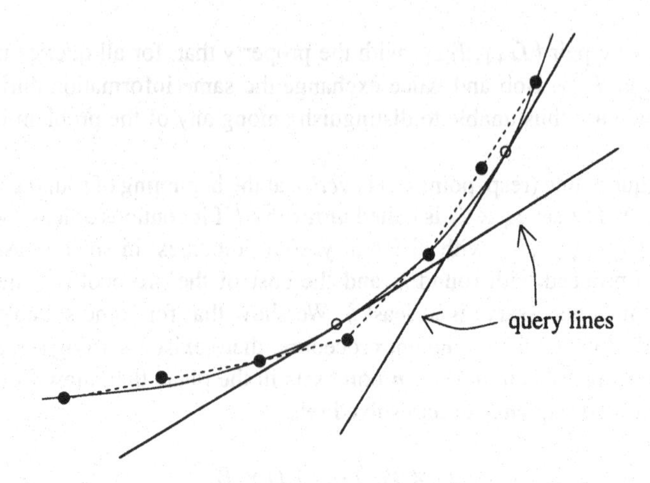

Fig. 1. A set P with $n = 7$ points and two queries with different answers.

2.2 A Hierarchy of Tree Contractions

Keeping control of Alice query lines is quite simple. The same cannot be said of Bob's point sets. Not only Bob's collections of point sets must be kept large but they must include point sets of all shape (but not size; remember that their size n is fixed). This variety is meant to make the algorithm's task more difficult. Some point sets must stretch widely with big gaps between consecutive points, while others must be confined to narrow intervals. For this reason, we cannot define point sets by picking points at random uniformly. Instead, we use a tree and a hierarchy of contractions of subtrees to define intervals from which we can specify the point sets.

Consider the perfect binary tree whose leaves (nodes of depth b) correspond to the integers 0 through $2^b - 1$, and let $\mathcal{T}_1$ denote its subtree of depth d^t sharing its root, where[1]

$$t = \left\lfloor \frac{\log b}{2 \log\log b} \right\rfloor \quad \text{and} \quad d = \lfloor c^2 \log b \rfloor \tag{1}$$

We assume throughout that the bit size b and the constant c are both suitably large. Note that b greatly exceeds d^t and so the tree $\mathcal{T}_1$ is well defined. Given a node v of the tree $\mathcal{T}_1$, let $\mathcal{T}_1(v)$ denote its subtree of depth d^{t-1} rooted at v. Contract *all* the edges of $\mathcal{T}_1$ except those whose (lower) incident node happens to be a leaf of $\mathcal{T}_1(v)$, for some node v of depth at most $d^t - d^{t-1}$ and divisible by d^{t-1}. This transforms the tree $\mathcal{T}_1$ into a smaller one, denoted $\mathcal{U}_1$, of depth d. Note that the depth-one subtree formed by an internal node v of $\mathcal{U}_1$ and its $2^{d^{t-1}}$ children forms a contraction of the tree $\mathcal{T}_1(v)$ (Fig.2).

Repeating this process leads to the construction of $\mathcal{U}_k$ for $1 < k \leq t$. Given an internal node v of $\mathcal{U}_{k-1}$, the depth-one tree formed by v and its children is associated with the subtree $\mathcal{T}_{k-1}(v)$, which now plays the role of $\mathcal{T}_1$ earlier, and is renamed $\mathcal{T}_k$. For

[1] All logarithms are to the base two.

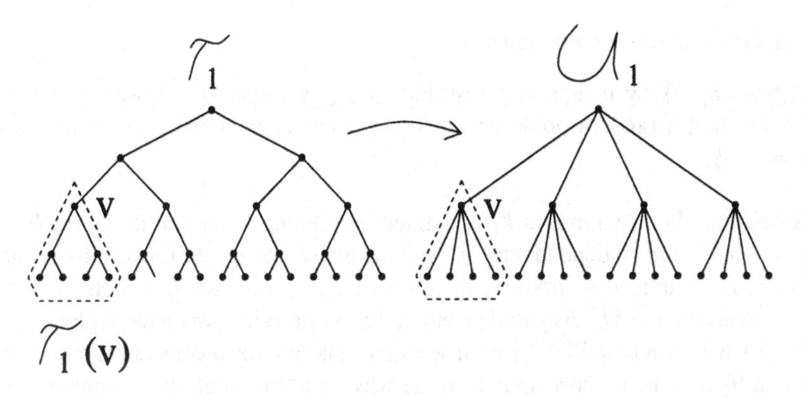

Fig. 2. The tree $\mathcal{T}_1$ and its contraction into $\mathcal{U}_1$.

any node $u \in \mathcal{T}_k$ of depth at most $d^{t-k+1} - d^{t-k}$ and divisible by d^{t-k}, let $\mathcal{T}_k(u)$ denote the subtree of $\mathcal{T}_k$ of depth d^{t-k} rooted at u: as before, turn the leaves of $\mathcal{T}_k(u)$ into the children of u by contracting the relevant edges. This transforms $\mathcal{T}_k$ into the desired tree $\mathcal{U}_k$ of depth d.

The contraction process is the same for all $k < t$, but not for $k = t$. We simply make all the leaves of $\mathcal{T}_t$ into the children of the root and remove the other internal nodes, which produces a depth-one tree $\mathcal{T}_t$ with 2^d leaves. Although $\mathcal{T}_k$ is defined nondeterministically, it is always a perfectly balanced binary tree of depth d^{t-k+1}.

Lemma 3. *Any internal node of any $\mathcal{U}_k$ has exactly $2^{d^{t-k}}$ children if $k < t$, and 2^d children if $k = t$.*

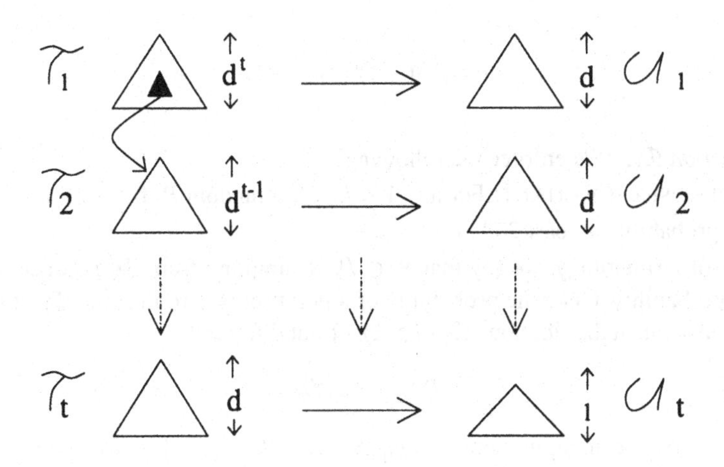

Fig. 3. The hierarchy of trees.

2.3 A Product Space Construction

We define any $\mathcal{P}_k$ by means of a distribution $\mathcal{D}_k$. We specify a lower bound on the probability that a random point set P_k drawn from $\mathcal{D}_k$ is active prior to round k, ie, belongs to $\mathcal{P}_k$.

• *Distribution* $\mathcal{D}_1$: A random P_1 is defined by picking a random index set I_1 (more on this below) and, independently, a random bit vector σ_1 uniformly distributed in $\{0,1\}^{s-1}$: I_1 is defined recursively in terms of $I_2,\dots,I_t$. Each I_k is defined with respect to a certain tree $\mathcal{U}_k$. Any node v in any $\mathcal{U}_k$ is naturally associated with an interval of integers between 0 and 2^b-1 of size larger than any fixed constant (go back to the node v of $\mathcal{T}_1$ to which it corresponds to see why): call the smallest even integer in that interval the *mark point* of v. We define a random index set I_1 by setting $k=1$ in the procedure below:

- For $k=t$, a random I_k (within some $\mathcal{T}_k$) is formed by the mark points of w^5 nodes selected at random, uniformly without replacement, among the leaves of the depth-one tree $\mathcal{U}_k$.
- For $k<t$, a random I_k (within some $\mathcal{T}_k$) is defined in two stages:
  [1] For each $j=1,2,\dots,d-1$, choose w^5 nodes of $\mathcal{U}_k$ of depth j at random, uniformly without replacement, among the nodes of depth j that are not descendants of nodes chosen at lower depth ($<j$). The $(d-1)w^5$ nodes selected are said to be *picked by* I_k.
  [2] For each node v picked by I_k, recursively choose a random I_{k+1} within $\mathcal{T}_{k+1}=\mathcal{T}_k(v)$. The union of these $(d-1)w^5$ sets I_{k+1} forms a random I_k within $\mathcal{T}_k$.

Note that a random P_1 (drawn) from $\mathcal{D}_1$ is active with probability 1 since no information has been exchanged yet between Bob and Alice. We see by induction that a random I_k consists of $s=(d-1)^{t-k}w^{5(t-k+1)}$ integers. Setting $k=1$, and using the fact that $n=2s-1$, we have the identity

$$n=2(d-1)^{t-1}w^{5t}-1. \tag{2}$$

• *Distribution* $\mathcal{D}_k$: We enforce the following

• POINT SET INVARIANT: For any $1\le k\le t$, a random P_k from $\mathcal{D}_k$ is active with probability at least 2^{-w^2}.

By abuse of terminology, we say that $P_k\in\mathcal{D}_k$ if sampling from $\mathcal{D}_k$ produces P_k with nonzero probability. Once the probability of a point set is zero in some $\mathcal{D}_k$, it remains so in all subsequent distributions $\mathcal{D}_j$ ($j>k$), or put differently,

$$\mathcal{D}_1\supseteq\cdots\supseteq\mathcal{D}_t.$$

Let $P_1=(I_1,\sigma_1)$ be an input point set $\{p_{i_1},X_{i_1i_2},\dots,X_{i_{s-1}i_s},p_{i_s}\}$ in $\mathcal{D}_1$. In the recursive construction of I_1, if v is a node of $\mathcal{U}_k$ picked by I_k in step [1], let $\{i_a,\dots,i_b\}$ be the I_{k+1} defined recursively within $\mathcal{T}_{k+1}=\mathcal{T}_k(v)$. The set

$$P_{|v}\stackrel{\text{def}}{=}\left\{p_{i_a},X_{i_ai_{a+1}},\dots,X_{i_{b-1}i_b},p_{i_b}\right\}$$

is called the *v-projection* of P_1. Similarly, one may also refer to the v-projection of any P_j ($j \leq k$), which might be empty. Obviously, it is possible to speak of a random $P_{|v}$ (with v fixed), independently of any P_1, as the point set formed by a random I_k and a uniformly distributed random bit vector σ_k of size $|I_k| - 1$. It is this distribution that will be understood in any further reference to a random $P_{|v}$.

Assume that we have already defined $\mathcal{D}_k$, for $k < t$. A distribution $\mathcal{D}_k$ is associated with a specific tree $\mathcal{T}_k$. To define $\mathcal{D}_{k+1}$, we must first choose a node v in $\mathcal{U}_k$ and make $\mathcal{T}_{k+1} = \mathcal{T}_k(v)$ our reference tree for $\mathcal{D}_{k+1}$. Any n-point set of $\mathcal{D}_k$ whose probability is not explicitly set below is assigned probability zero under $\mathcal{D}_{k+1}$. Consider each possible point set $P_{|v}$ in turn (for v fixed), and apply the following rule:

- If $P_{|v}$ is the v-projection of some P_k in $\mathcal{P}_k$, then take one[2] such P_k, and set its probability under $\mathcal{D}_{k+1}$ to be that of picking $P_{|v}$ randomly.
- Otherwise, take one $P_k \in \mathcal{D}_k$ whose v-projection is $P_{|v}$, and again set its probability under $\mathcal{D}_{k+1}$ to be that of picking $P_{|v}$ randomly.

During that round k, Bob reduces the collection of active point sets in $\mathcal{D}_{k+1}$ to form $\mathcal{P}_{k+1}$. To summarize, a random P_k is defined with reference to a specific tree $\mathcal{T}_k$. Note that the distribution $\mathcal{D}_k$ is isomorphic to that of a random $P_{|v}$, for fixed $v \in \mathcal{U}_{k-1}$, or equivalently, a random (I_k, σ_k), where σ_k is a uniformly distributed random bit vector of size $|I_k| - 1$.

2.4 Alice's Query Lines

As the game progresses, $\mathcal{L}_1$ decreases in size to produce the nested sequence $\mathcal{L}_1 \supseteq \cdots \supseteq \mathcal{L}_t$. Prior to round k, the currently active query set $\mathcal{L}_k$ is associated with the same reference tree $\mathcal{T}_k$ used to define a random P_k. As we observed in the last section, each node of $\mathcal{U}_k$ corresponds to a unique interval of integers in $[0, 2^b)$. By abuse of notation, we also let $\mathcal{L}_k$ designate the set of integers j defining the lines $y = 2jx - j^2$ in the set. We maintain the following:

- QUERY INVARIANT: For any $1 \leq k \leq t$, the fraction of the leaves in $\mathcal{U}_k$ whose intervals intersect $\mathcal{L}_k$ is at least $1/b$.

Lemma 4. *If $\mathcal{L}_t$ and $\mathcal{P}_t$ satisfy their respective invariant, then $\mathcal{L}_t \times \mathcal{P}_t$ is unresolved.*

Proof. Suppose that $\mathcal{L}_t$ satisfies the query invariant and that $\mathcal{L}_t \times \mathcal{P}_t$ is not unresolved: we show that $\mathcal{P}_t$ must then violate the point set invariant. For each leaf of $\mathcal{U}_t$ whose interval intersects $\mathcal{L}_t$, pick one $j_i \in \mathcal{L}_t$ in that interval. By Lemma 3 and the query invariant, this gives us a sequence $j_1 < \cdots < j_m$ of length

$$m \geq \frac{2^d}{b}. \tag{3}$$

Given $P_t \in \mathcal{D}_t$, we define the *spread* of P_t, denoted spread(P_t), as the number of intervals of the form $[j_i, j_{i+1}]$ $(0 \leq i \leq m)$ that intersect the index set $I(P_t)$ (Fig.4); for consistency we write $j_0 = 0$ and $j_{m+1} = 2^b - 1$. Suppose that the spread $|S|$ is defined by some fixed

[2] It does not matter which one, but it has to be unique.

set S of size less than w^4. Of the $m+1$ candidate intervals $[j_i, j_{i+1}]$, a random I_t must then avoid $m+1-|S|$ of them. Although such an interval may not always enclose a whole leaf interval, it does contain at least one mark point, and so the choice of I_t is confined to at most $2^d - m - 1 + |S|$ leaves of $\mathcal{U}_t$. Thus, the probability that the spread is defined by S is bounded by

$$\binom{2^d + |S| - m - 1}{w^5} \Big/ \binom{2^d}{w^5} \leq \left(1 - \frac{m - |S|}{2^d}\right)^{w^5}.$$

Summing over all S's of size less than w^4, it follows from (3) that

$$\text{Prob}\left[\text{spread}(P_t) < w^4\right] \leq \sum_{k<w^4} \binom{m+1}{k} \left(1 - \frac{1}{2b}\right)^{w^5} \leq 2^{-w^4}. \tag{4}$$

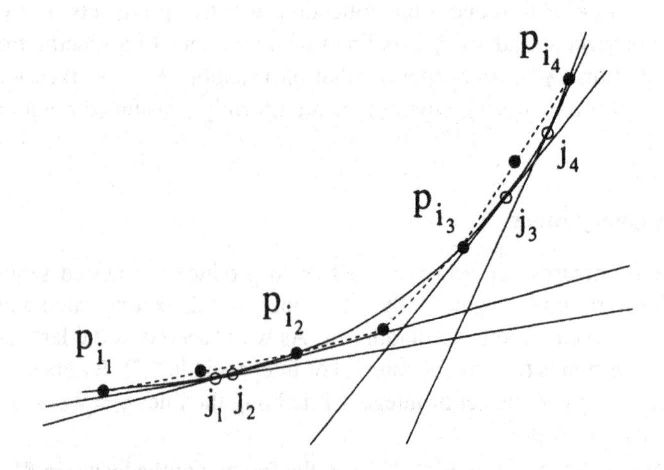

Fig. 4. A spread of 3 determined by $[j_0, j_1], [j_2, j_3], [j_4, j_5]$.

Suppose now that the spread is at least w^4. Then

$$P_t = \left\{ p_{i_1}, X_{i_1 i_2}, p_{i_2}, X_{i_2 i_3}, \ldots, p_{i_{s-1}}, X_{i_{s-1} i_s}, p_{i_s} \right\}$$

includes a subset P^* of at least $w^4 - 1$ points p_{i_j}, every one of which can be paired with a line $y = 2kx - k^2$ of $\mathcal{L}_t$, where $i_j < k < i_{j+1}$. Pick a random P_t from $\mathcal{D}_t$, and let Ξ denote the event: "all queries from $\mathcal{L}_t$ give the same answer yes/no with respect to point set P_t." By Lemma 2, the $X_{i_j, i_{j+1}}$'s are all of the form $a_{i_j, i_{j+1}}$ or all of the form $b_{i_j, i_{j+1}}$ (no mix). As we observed earlier, $\mathcal{D}_t$ is isomorphic to the distribution of a random (I_t, σ_t), where σ_t is a string of $w^5 - 1$ bits (drawn uniformly, independently). The constraint on the X's reduces the choice of a random P_t by a factor of at least 2^{w^4-2}, and hence,

$$\text{Prob}\left[\Xi \mid \text{spread}(P_t) \geq w^4\right] \leq 2^{2-w^4}. \tag{5}$$

Putting together (4,5), we find

$$\begin{aligned}\mathrm{Prob}[\Xi] &= \mathrm{Prob}[\Xi \mid \mathrm{spread}(P_t) < w^4]\cdot \mathrm{Prob}[\mathrm{spread}(P_t) < w^4] \\ &\quad + \mathrm{Prob}[\Xi \mid \mathrm{spread}(P_t) \geq w^4]\cdot \mathrm{Prob}[\mathrm{spread}(P_t) \geq w^4] \\ &\leq 2^{-w^4} + 2^{2-w^4} < 2^{-w^2},\end{aligned}$$

which violates the point set invariant. □

During the k-th round, Alice chooses an index in Bob's table. As we discussed earlier, the set of n^c possible choices partitions her current query set $\mathcal{L}_k$ into as many equivalence classes. An internal node v of $\mathcal{U}_k$ is called *heavy* if one (or more) of these classes intersects the intervals associated with a fraction at least $1/b$ of the children of v. The following is a variant of a result of Ajtai [1].

Lemma 5. *The union of the intervals associated with the heavy nodes of $\mathcal{U}_k$ contains at least a fraction $1/2b$ of the leaves' intervals.*

Proof. Fix an equivalence class and color the nodes of $\mathcal{U}_k$ whose intervals intersect it. Mark every non-root colored node that is heavy with respect to the equivalence class. Then, mark every descendant in $\mathcal{U}_k$ of a marked node. Let N be the number of leaves in $\mathcal{U}_k$ and let N_j be the number of leaves of $\mathcal{U}_k$ whose depth-j ancestor in $\mathcal{U}_k$ is colored and unmarked (we include v as one of its ancestors). For $j > 1$, an unmarked, colored, depth-j node is the child of an unmarked, colored, depth-$(j-1)$ node that is not heavy for the chosen class, and so $N_j < N_{j-1}/b$. We have $N_1 \leq N$ and, for any $j > 0$,

$$N_j \leq \frac{N}{b^{j-1}}.$$

Repeating this argument for all the other equivalence classes, we find that all the unmarked, colored nodes (at a fixed depth $j > 0$) are ancestors of at most $n^c N/b^{j-1}$ leaves. This implies that the number of unmarked, colored leaves is at most $n^c N/b^{d-1} < N/2b$. (This follows from (1, 2).) The query invariant guarantees that at least N/b leaves of $\mathcal{U}_k$ are colored and so at least $N/2b$ are both colored and marked. It follows that the marked nodes whose parents are unmarked are themselves are ancestors of at least $N/2b$ leaves: all these nodes are heavy. □

Alice's strategy is to keep her active query sets as "entangled" as possible with Bob's point sets. Put differently, ideally the two should form a low-discrepancy set system [6] (at least in the one-way sense). The next result says that this is true on at least one level of $\mathcal{U}_k$, where many heavy nodes end up being picked by a random I_k.

Lemma 6. *For any $0 < k < t$, there is a depth j $(0 < j < d)$ such that, with probability at least 2^{-w^2-1}, a random P_k from $\mathcal{D}_k$ is active and its index set I_k picks at least w^3 heavy depth-j nodes in its associated $\mathcal{U}_k$.*

Proof. Recall that $\mathcal{D}_k$ is isomorphic to a random (I_k, σ_k). Fix σ_k once and for all. The heavy nodes of $\mathcal{U}_k$ are ancestors of at least a fraction $1/2b$ of the leaves (Lemma 5). It follows that, for some $0 < j < d$, at least a fraction $1/2bd$ of the nodes of depth j are heavy. Among these, I_k may pick only those that are not picked further up in the tree: this caveat rules out fewer than dw^5 candidate nodes, which by Lemma 3, represents a fraction at most $dw^5/2^d$ of all the nodes of depth j. So, it appears that among the set of depth-j nodes that may be picked by I_k, the fraction α of heavy ones satisfies

$$\alpha \geq \left(\frac{1}{2db} - dw^5 2^{-d}\right) \Big/ \left(1 - dw^5 2^{-d}\right) > \frac{1}{3db}.$$

The index set I_k picks w^5 depth-j nodes of $\mathcal{U}_k$ at random with no replacement. By Hoeffding's classical bounds [10], the probability that the number of heavy ones picked exceeds the lemma's target of w^3 is at least

$$1 - e^{-2w^5(\alpha - 1/w^2)^2} > 1 - 2^{-w^3}.$$

It follows from the point set invariant and the independence of I_k and σ_k that, with probability at least $2^{-w^2} - 2^{-w^3}$, a random P_k is active and its index set I_k picks at least w^3 heavy depth-j nodes in its associated $\mathcal{U}_k$. □

2.5 Probability Amplification

During the k-th round, Bob sends to Alice the contents of the cell $T[f_k(\ell, T[f_1(\ell)], \ldots)]$. The 2^w possible values partition the current collection $\mathcal{P}_k$ of active point sets into as many equivalence classes. We exploit the product nature of the distribution $\mathcal{D}_k$ to amplify the probability of being active by projecting the distribution on one of its factors.

Lemma 7. *For any $0 < k < t$, there exists a heavy node v of $\mathcal{U}_k$ such that, with probability at least $1/2$, a random P_{k+1} drawn from the distribution $\mathcal{D}_{k+1}$ associated with $\mathcal{T}_{k+1} = \mathcal{T}_k(v)$ belongs to $\mathcal{P}_k$.*

Proof. We refer to the depth j in Lemma 6. Let $p_{|S}$ denote the conditional probability that a random P_k from $\mathcal{D}_k$ belongs to $\mathcal{P}_k$, given that S is exactly the set of heavy nodes of depth j picked by I_k. Summing over all subsets S of heavy depth-j nodes of size at least b^3,

$$\sum_S \text{Prob}[S = \text{set of heavy depth-}j\,\text{nodes picked by}\,I_k] \cdot p_{|S}$$

is the sum, over all S, of the probability that $P_k \in \mathcal{P}_k$ and that S is precisely the set of heavy nodes of depth j picked by its index set I_k. By Lemma 6, this sum is at least 2^{-w^2-1}, and therefore $p_{|S^*} \geq 2^{-w^2-1}$, for some set S^* of at least w^3 heavy nodes of depth j.

Because a random P_k whose I_k picks v consists of a random (I_{k+1}, σ_{k+1}) drawn at node v independently of the rest of (I_k, σ_k), its v-projection has a distribution isomorphic to that of (I_{k+1}, σ_{k+1}), which is also $\mathcal{D}_{k+1}$. The same is true even if the distribution on

P_k is conditioned upon having S as the set of heavy depth-j nodes picked by I_k. If P_k belongs to $\mathcal{P}_k$ then its v-projection maps to a unique set $P_{k+1} \in \mathcal{D}_{k+1}$ also in $\mathcal{P}_k$.

Let $p_{|v}$ denote the probability that a random P_{k+1} drawn from the distribution $\mathcal{D}_{k+1}$ associated with $\mathcal{T}_{k+1} = \mathcal{T}_k(v)$ belongs to $\mathcal{P}_k$. It follows that

$$p_{|S^*} \leq \prod_{v \in S^*} p_{|v}.$$

Since $|S^*| \geq w^3$, it follows that

$$p_{|v} \geq \left(2^{-w^2-1}\right)^{1/|S^*|} \geq \frac{1}{2},$$

for some $v \in S^*$. □

Both query and point set invariants are trivially satisfied before round 1. Assume now that they hold at the opening of round $k < t$. Let v denote the node of $\mathcal{U}_k$ in Lemma 7. The n^c possible ways of indexing into the table T partition Alice's query set $\mathcal{L}_k$ into as many equivalence classes. Because v is heavy, the intervals associated with a fraction at least $1/b$ of its children intersect a particular equivalence class. Alice chooses such a class and the query lines in it as her new query set $\mathcal{L}_{k+1}$. The tree $\mathcal{U}_{k+1}$ is naturally derived from $\mathcal{T}_{k+1} = \mathcal{T}_k(v)$, and the query invariant is satisfied at the beginning of round $k+1$.

Upon receiving the index from Alice, Bob must choose the contents of the table entry while staying consistent with past choices. By Lemma 7, a random P_{k+1} from $\mathcal{D}_{k+1}$ (distribution associated with $\mathcal{T}_{k+1}$) is active at the beginning of round k with probability at least a half. There are 2^w choices for the table entry, and so for at least one of them, with probability at least $(1/2)2^{-w} > 2^{-w^2}$, a random point set from $\mathcal{D}_{k+1}$ is active at the beginning of round k and produces a table with that specific entry value. These point sets constitute the newly active collection $\mathcal{P}_{k+1}$, and the point set invariant still holds at the beginning of round $k+1$.

To show that t rounds are needed, we must prove that $\mathcal{L}_k \times \mathcal{P}_k$ is unresolved, for any $k \leq t$. In fact, because of the nesting structure of these products, it suffices to show that $\mathcal{L}_t \times \mathcal{P}_t$ is unresolved, which follows from Lemma 4. This proves the lower bound of Theorem 1. □

Acknowledgments

I wish to thank the anonymous reviewer for useful suggestions. I also thank Faith Fich for informing me of her recent work with Paul Beame, and in particular, their upper bound for predecessor searching in a bounded universe.

References

[1] Ajtai, M. *A lower bound for finding predecessors in Yao's cell probe model*, Combinatorica, 8 (1988), 235–247.

[2] Beame, P., Fich, F. *Optimal bounds for the predecessor problem*, Proc. 31st Annu. ACM Symp. Theory Comput. (1999), to appear.

[3] Ben-Or, M. *Lower bounds for algebraic computation trees*, Proc. 15th Annu. ACM Symp. Theory Comput. (1983), 80–86.

[4] Björner, A., Lovász, L., Yao, A.C. *Linear decision trees: Volume estimates and topological bounds*, Proc. 24th Annu. ACM Symp. Theory Comput. (1992), 170–177.

[5] Chakrabarti, A., Chazelle, B., Gum, B., Lvov, A. *A good neighbor is hard to find*, Proc. 31st Annu. ACM Symp. Theory Comput. (1999), to appear.

[6] Chazelle, B. *The Discrepancy Method: Randomness and Complexity*, Cambridge University Press, to appear.

[7] Fredman, M.L. *A lower bound on the complexity of orthogonal range queries*, J. ACM, 28 (1981), 696–705.

[8] Grigoriev, D., Karpinksi, M., Meyer auf der Heide, F., Smolensky, R. *A lower bound for randomized algebraic decision trees*, Computational Complexity, 6 (1997), 357–375.

[9] Grigoriev, D., Karpinksi, M., Vorobjov, N. *Improved lower bound on testing membership to a polyhedron by algebraic decision trees*, Proc. 36th Annu. IEEE Symp. Foundat. Comput. Sci. (1995), 258–265.

[10] Hoeffding, W. *Probability inequalities for sums of bounded random variables*, J. Amer. Stat. Assoc., 58 (1963), 13–30.

[11] Karlsson, R.G. *Algorithms in a restricted universe*, Tech Report CS-84-50, Univ. Waterloo, Waterloo, ON, 1984.

[12] Karlsson, R.G., Munro, J.I. *Proximity on a grid*, Proc. 2nd Symp. Theoret. Aspects of Comput. Sci., LNCS Springer, vol.182 (1985), 187–196.

[13] Karlsson, R.G., Overmars, M.H. *Scanline algorithms on a grid*, BIT, 28 (1988), 227–241.

[14] Kushilevitz, E., Nisan, N. *Communication Complexity*, Cambridge University Press, 1997

[15] Miltersen, P.B. *Lower bounds for union-split-find related problems on random access machines*, Proc. 26th Annu. ACM Symp. Theory Comput. (1994), 625–634.

[16] Müller, H. *Rasterized point location*, Proc. Workshop on Graph-Theoretic Concepts in Comput. Sci., Trauner Verlag, Linz (1985), 281–293.

[17] Overmars, M.H. *Computational geometry on a grid: an overview*, ed. R. A. Earnshaw, Theoretical Foundations of Computer Graphics and CAD, NATO ASI, vol.F40, Springer-Verlag (1988), 167–184.

[18] Yao, A.C. *Should tables be sorted?*, J. ACM, 28 (1981), 615–628.

[19] Yao, A.C. *On the complexity of maintaining partial sums*, SIAM J. Comput. 14 (1985), 277–288.

[20] Yao, A.C. *Decision tree complexity and Betti numbers*, Proc. 26th Annu. ACM Symp. Theory Comput. (1994), 615–624.

On Computing the Diameter of a Point Set in High Dimensional Euclidean Space

Daniele V. Finocchiaro[1] and Marco Pellegrini[2]

[1] Scuola Normale Superiore in Pisa
Piazza dei Cavalieri 7, I-56126 Pisa, Italy
fino@cibs.sns.it

[2] Institute for Computational Mathematics of CNR
via S. Maria 46, I-56126 Pisa, Italy
pellegrini@imc.pi.cnr.it

Abstract. We consider the problem of computing the diameter of a set of n points in d-dimensional Euclidean space under Euclidean distance function. We describe an algorithm that in time $O(dn\log n + n^2)$ finds with high probability an arbitrarily close approximation of the diameter. For large values of d the complexity bound of our algorithm is a substantial improvement over the complexity bounds of previously known exact algorithms. Computing and approximating the diameter are fundamental primitives in high dimensional computational geometry and find practical application, for example, in clustering operations for image databases.

1 Introduction

We consider the following problem: given a set of n points in d-dimensional Euclidean space, compute the maximum pairwise Euclidean distance between two points in the set. This problem is known as the *diameter problem* or the *furthest pair* problem. There are several efficient algorithms for the case when $d = 2$ [26] and $d = 3$ [2, 27, 6, 5, 23] which, however, do not extend to higher dimensional spaces. The diameter problem is one of the basic problems in high dimensional computational geometry [15, 16, 17].[1] In this paper we consider a setting in which the number of points n and the dimension of the space d are equally important in the complexity analysis.

The exact solution to the diameter problem in arbitrary dimension can be found using the trivial algorithm that generates all possible $\Theta(n^2)$ inter-point distances and determines the maximum value. This algorithm runs in time $O(dn^2)$. As noted in [26] substantial improvements of the asymptotic complexity in terms of n and d must overcome the fact that for $d \geq 4$ the number of diametral pairs of points can be $\Omega(n^2)$ [12],[2] while computing a single inter-point distance takes time $\Omega(d)$. Yao in [33] gives an algorithm to compute the diameter in time $O(n^{2-a(d)}\log^{1-a(d)} n)$ where $a(d) = 2^{-(d+1)}$ for $d \geq 3$ and fixed. The technique in [33] can be extended to an $o(n^2)$ algorithm in non-fixed dimension for $d \leq 1/2 \log\log n$.

[1] Gritzmann and Klee have proposed the term *Computational Convexity* to denote the study of combinatorial and algorithmic aspects of polytopes in high dimensional spaces.

[2] Instead, in dimensions 2 and 3, the number of diametral pairs of points is $O(n)$ [11, 28].

A result of Yao [32] (cited in [33]) shows that all inter-point distances can be computed in time $O(M(n,d) + nd + n^2)$, where $M(n,d)$ is the time to multiply an $n \times d$ matrix by a $d \times n$ matrix. Using the asymptotically fastest known square matrix multiplication algorithm [7] we have $M(n,d) \leq O(\min\{nd^{s-1}, dn^{s-1}\})$, where $s \approx 2.376$. The furthest pair is then found trivially with $O(n^2)$ extra time.

In order to obtain faster algorithms for large values of d, we relax the requirement by considering algorithms that approximate the diameter up to multiplicative factors arbitrarily close to one. The main result in this paper (Theorem 12) is the following: the diameter of a point-set of size n in dimension d is approximated in time $O(dn \log n + n^2)$ within a factor $1 + \varepsilon$, for any real value $\varepsilon > 0$, with probability $1 - \delta$. The constants hidden in the big-Oh notation depend on user controlled parameters ε and δ, but not on d. Our result matches or improves asymptotically the trivial exact algorithm in the range $d \geq 4$. Our algorithm matches asymptotically Yao's algorithms for d in the range: $\frac{1}{2} \log\log n \leq d \leq n^{1/(s-1)} \approx n^{0.726}$, and attains better performance for $d \geq n^{1/(s-1)} \approx n^{0.726}$.

Another approximation algorithm in literature, [10], attains a fixed approximation factor $c = \sqrt{5 - 2\sqrt{3}}$, which is not arbitrarily close to one.

Subsequent to the acceptance of this paper we got news of a forthcoming paper of Borodin, Ostrowsky and Rabani [4] where, amongst other important results, the approximate furthest pair problem is solved in time roughly $O(nd \log n + n^{2-\theta(\varepsilon^2)})$ with high probability. Their result improves asymptotically over our bound for $d < n/\log n$, however, for reasonable values of ε, say $\varepsilon = 0.1$, the asymptotic improvement is of the order of $n^{1/100}$, and such gain should be weighted against a more complex coding effort.

1.1 Applications

Among the applications of computing (or approximating) the diameter in high dimensional spaces we mention those in data clustering for images database [18]. An image is mapped into a point in an high dimensional space by associating a dimension to each term of a wavelet expansion of the image considered as a two-dimensional piecewise constant function [13, 19]. Thus similarity clustering of images can be translated in the problem of determining clusters of such points. One of the most natural measure of quality for a cluster is its diameter. Note that in such applications the number of dimensions is very high thus the complexity due to the dimension must be taken into consideration in the design of efficient algorithms.

1.2 The Algorithm

Our algorithm has been inspired by a recent technique for nearest neighbor search described by Kleinberg [21]. Although the formulation of the closest pair problem seems not too different from that of the diameter (searching for the smallest inter-point distance instead of the maximum), the mathematical properties of the two problems are quite different so that in general not any efficient algorithm for the closest pair problem yields an efficient one for the farthest pair problem.

Intuitively, the method of Kleinberg is based on the idea that if a vector $x \in \mathbb{R}^d$ is longer than a vector $y \in \mathbb{R}^d$ then this relation is preserved with probability greater than

$1/2$ in a projection of x and y over a random line. Thus using several projections of the same set of vectors and a majority voting scheme we can retrieve, with probability close to one, the "actual" relative length relation between x and y. The theory of range spaces of bounded VC-dimension is invoked to determine how many random lines are needed to satisfy the constraints imposed by the error parameter ε and the confidence parameter δ.

1.3 Organization of the Paper

In Section 2 we review some basic properties of projections of random vectors and of range spaces with bounded VC-dimension. In Section 3 we give the preprocessing and query algorithms for furthest point queries. Finally in Section 4 we apply the data structures of Section 3 to the problem of determining the diameter, thus establishing the main result.

2 Basic Results

We denote with S^{d-1} the set of directions in $\mathbb{R}^d$; it can be identified with the set of unit vectors, or with the set of points on the unit sphere. For any two vectors $x, y \in \mathbb{R}^d$, we define the set of directions over which the length of the projection of x is longer or equal than that of y:

$$Z_{x;y} = \{ v \in S^{d-1} : |v \cdot x| \geq |v \cdot y| \} .$$

We denote with $z_{x;y}$ the relative measure of $Z_{x;y}$ with respect to the entire set of directions; it represents the probability that the projection of x on a random direction is longer than the projection of y:

$$z_{x;y} = \frac{|Z_{x;y}|}{|S^{d-1}|} = \Pr[|v \cdot x| \geq |v \cdot y|] .$$

Note that for $x \neq y$ the probability that $|v \cdot x| = |v \cdot y|$ is null.

The idea of this work is that if x is 'significantly' longer than y, then the set $Z_{x;y}$ is large, and it is very probable that a random direction belongs to it. The following lemma captures this idea:

Lemma 1. *If* $(1-\gamma)\|x\| \geq \|y\|$, *with* $0 \leq \gamma \leq 1$, *then* $z_{x;y} \geq 1/2 + \gamma/\pi$.

Proof Sketch.. Consider the plane $\mathcal{A}$ containing the vectors x, y, and the projection $\pi_{\mathcal{A}}(v)$ of v on this plane. Observe that $v \cdot x = \pi_{\mathcal{A}}(v) \cdot x$ and $v \cdot y = \pi_{\mathcal{A}}(v) \cdot y$. So the relation $v \in Z_{x;y}$ depends only on $\pi_{\mathcal{A}}(v)$, or equivalently on the angle ϕ between $\pi_{\mathcal{A}}(v)$ and the vector x (or y).

Note that ϕ is uniformly distributed in $[0, 2\pi)$. Let θ be the angle formed by x and y: we find that $v \in Z_{x;y}$ whenever $\cos^2(\theta - \phi) \leq \cos^2\phi \cdot \|x\|^2/\|y\|^2$, and the quantity $z_{x;y}$ is minimum when x and y are orthogonal. In this case

$$z_{x;y} = \frac{2}{\pi} \tan^{-1} \frac{\|x\|}{\|y\|} \geq \frac{2}{\pi} \tan^{-1} \frac{1}{1-\gamma} .$$

In order to bound to the latter quantity, we consider that $\tan^{-1}(1-\gamma)^{-1}$ is a convex function, and its derivative in $\gamma = 0$ is $1/2$. So we obtain

$$\tan^{-1}\frac{1}{1-\gamma} \geq \tan^{-1} 1 + \frac{1}{2}\gamma = \frac{\pi}{4} + \frac{1}{2}\gamma \,,$$

from which the lemma follows. □

Intuitively, we want to fix a set V of directions, and compare the lengths of vectors by comparing the lengths of their respective projections over elements of V, and making a majority vote. For two vectors x, y we will write that

$$x \rhd_V y \quad \Longleftrightarrow \quad |V \cap Z_{x;y}| \geq 1/2|V| \,. \tag{1}$$

This means that the projection of x is longer than the projection of y over at least half of the vectors in V. Note that $x \rhd_V y$ and $y \rhd_V x$ can simultaneously hold with respect to the same set V.

For a fixed set of directions V, consider two arbitrary vectors x and y such that x is significantly longer than y. Lemma 1 says that the set $Z_{x;y}$ is large, so we can hope that in V there are enough vectors of $Z_{x;y}$ so that $x \rhd_V y$. However, we want this to hold for *any* vectors x, y, with respect to the *same* (fixed) set V. For this reason we cannot use directly Lemma 1, but we will use some VC-dimension techniques. For a detailed treatment of this topic see [31, 1] or the seminal paper [29]. Here we will present only the definitions needed.

A *range space* is a pair $(\mathcal{P}, \mathcal{R})$, where $\mathcal{P} = (\Omega, \mathcal{F}, \mu)$ is a probability space, and $\mathcal{R} \subseteq \mathcal{F}$ is a collection of measurable (w.r.t. the measure μ) subsets of Ω. A finite set $A \subseteq \Omega$ is said to be *shattered* by $\mathcal{R}$ if every subset of A can be expressed in the form $A \cap R$ for some $R \in \mathcal{R}$. The *VC-dimension VC-dim*$(\mathcal{R})$ of the range space $(\mathcal{P}, \mathcal{R})$ is the maximum size of a set that can be shattered by $\mathcal{R}$ (hence no set of size $>$ *VC-dim*$(\mathcal{R})$ can be shattered).

There is a natural identification between collections of sets and families of binary-valued function: to each subset $R \in \mathcal{R}$ corresponds its indicator function $f_R(x) : \Omega \to \{0,1\}$ such that $f_R(x) = 1 \iff x \in R$. This identification is useful to express combinations of range spaces in the following manner.

For an integer $k \geq 2$, let $u : \{0,1\}^k \to \{0,1\}$ and $f_1, \ldots, f_k : \mathcal{R} \to \{0,1\}$ be binary-valued functions (u represents the combination operator and the f_i's are indicator functions). Define $u(f_1, \ldots, f_k) : \mathcal{R} \to \{0,1\}$ as the binary-valued function $x \mapsto u(f_1(x), \ldots, f_k(x))$. Finally, if $\mathcal{A}_1, \ldots, \mathcal{A}_k$ are families of binary-valued functions, we define $\mathcal{U}(\mathcal{A}_1, \ldots, \mathcal{A}_k)$ to be the family of binary-valued functions $\{u(f_1, \ldots, f_k) : f_i \in \mathcal{A}_i \quad \forall i\}$. In this way we can obtain $\mathcal{A} \oplus \mathcal{B} = \{A \cup B : A \in \mathcal{A}, B \in \mathcal{B}\}$ by choosing $u(a,b) = \max\{a,b\}$, and $\mathcal{A} \odot \mathcal{B} = \{A \cap B : A \in \mathcal{A}, B \in \mathcal{B}\}$ by choosing $u(a,b) = a \cdot b$.

We will use the following theorem by Vidyasagar [30, Th. 4.3], which improves on a result by Dudley [8, 9]:

Theorem 2. *If VC-dim*$(\mathcal{A}_i)$ *is finite for each* i, *then* $\mathcal{U} = \mathcal{U}(\mathcal{A}_1, \ldots, \mathcal{A}_k)$ *also has finite VC-dimension, and VC-dim*$(\mathcal{U}) < \alpha_k \cdot d$, *where* $d = \max_i$ *VC-dim*$(\mathcal{A}_i)$, *and* α_k *is the smallest integer such that* $k < \alpha_k / \log_2(e\alpha_k)$.

We consider the probability space $\mathcal{P}$ made by the set of directions S^{d-1}, with the uniform distribution, where $\mu(X) = |X|/|S^{d-1}|$; we take as $\mathcal{R}$ the collection of all the sets $Z_{x;y}$. The following lemma bounds the VC-dimension of this range space:

Lemma 3. *The VC-dimension of the range space $(\mathcal{P}, \mathcal{R})$ defined above is strictly less than $25(d+1)$.*

Proof. For a vector $u \in \mathbb{R}^d$, let H_u denote the closed hemisphere $\{ v \in S^{d-1} : u \cdot v \geq 0 \}$, and let $\mathcal{H}$ be the collection of all such hemispheres. Then any set $Z_{x;y}$ can be expressed as the Boolean combination $(H_{x-y} \cap H_{x+y}) \cup (H_{-x-y} \cap H_{-x+y})$. Thus we have $\mathcal{R} \subseteq \mathcal{U}$, where $\mathcal{U} = (\mathcal{H} \odot \mathcal{H}) \oplus (\mathcal{H} \odot \mathcal{H})$. A theorem by Radon [3] implies that $VC\text{-}dim(\mathcal{H}) \leq d+1$. We then apply Theorem 2, with $u(a,b,c,d) = \max\{a \cdot b, c \cdot d\}$, and the fact that $\alpha_4 = 25$ to obtain that $VC\text{-}dim(\mathcal{R}) \leq VC\text{-}dim(\mathcal{U}) < 25(d+1)$. □

A subset A of Ω is called a γ-approximation (or γ-sample) for a range space $(\mathcal{P}, \mathcal{R})$ is for every $R \in \mathcal{R}$ it holds

$$\left| \frac{|R \cap A|}{|A|} - \mu(R) \right| \leq \gamma \,. \tag{2}$$

This means that A can be used to obtain a good estimate of the measure of any set $R \in \mathcal{R}$. The main result, which follows from Lemma 3 and from a fundamental theorem by Vapnik and Chervonenkis, is the following.

Lemma 4. *With probability at least $1-\delta$ a set V of cardinality*

$$f(\gamma,\delta) = \frac{16}{\gamma^2}\left(25(d+1)\log\frac{400(d+1)}{\gamma^2} + \log\frac{4}{\delta}\right) = \Theta(d \log d)$$

of vectors chosen uniformly at random from S^{d-1} is a γ-approximation for the range space defined above.

The lemma above permits to make comparisons in the following way: choose a set V of $f(\varepsilon/\pi, \delta)$ random directions. With probability $1-\delta$ it is a (ε/π)-approximation. For any x, y with $(1-\varepsilon)||x|| \geq ||y||$, we have that $\mu(Z_{x;y}) \geq 1/2 + \varepsilon/\pi$, by Lemma 1. From the definition of γ-approximation it follows that $|Z_{x;y} \cap V| \geq 1/2|V|$, so $x \triangleright_V y$ by definition (1). This is the idea which we are going to use in the following section. However, in order to save computations, we will make comparisons using random subsets of the fixed set V.

3 An Algorithm for Farthest-Point Queries

We will present in this section a (ε, δ)-approximation scheme for computing the farthest site of a query point q. This means that with probability $1-\delta$ the algorithm gives an answer which is within a factor $1-\varepsilon$ from the optimal one. The parameters ε and δ are chosen by the user before the algorithm starts.

Let $P = \{p_1, \ldots, p_n\}$ the set of given sites, and $q \in \mathbb{R}^d$ the query point. For simplicity we will assume that n is a power of 2. Let p^* be the farthest-site from q, i.e., the point such that

$$d(p^*, q) = \max_{p_i \in P} d(p_i, q) \ ,$$

where $d(p,q) = \|p - q\|$ is the standard Eucledian metric in $\mathbb{R}^d$. Let Z_ε be the set of sites that are far from q 'almost' like p^*:

$$Z_\varepsilon = \{ p_i \in P : d(p_i, q) \geq (1 - \varepsilon) d(p^*, q) \} \ .$$

The purpose of the algorithm is to give, with probability at least $1 - \delta$, an element of Z_ε.

3.1 Building the Data Structure

The preprocessing stage is the following. Let $\varepsilon_0 = \log(1+\varepsilon)/\log n$. We choose randomly L vectors from S^{d-1}, where $L = f(\varepsilon_0/12, \delta) = \Theta(d \log d \log^2 n \log\log n)$. These vectors can be obtained for example by the method described by Knuth [22, p. 130]. Let $V = \{v_1, \ldots, v_L\}$ the set of L directions generated above. The data structure is a matrix M, of dimension $L \times n$, where $M[i,j] = v_i \cdot p_j$.

3.2 Processing a Query

We first define the following relation between sites of P: for a set of directions $\Gamma \subseteq V$, we say that $p_i \succeq_\Gamma p_j$ if the projection of p_i is farther than the projection of p_j from the projection of q, for at least half of the directions in Γ. Formally:

$$p_i \succeq_\Gamma p_j \quad \Longleftrightarrow \quad (p_i - q) \triangleright_\Gamma (p_j - q) \ .$$

We will use this relation to build a 'tournament' between sites, by making comparisons with respect to a fixed set Γ: if $p_i \succeq_\Gamma p_j$ and $p_j \not\succeq_\Gamma p_i$, the winner of the comparison is p_i; if $p_i \not\succeq_\Gamma p_j$ and $p_j \succeq_\Gamma p_i$, the winner is p_j; finally, if both $p_i \succeq_\Gamma p_j$ and $p_j \succeq_\Gamma p_i$, the winner is chosen arbitrarily. Note that the above description is complete, i.e., it cannot occur that $p_i \not\succeq_\Gamma p_j$ and $p_j \not\succeq_\Gamma p_i$.

Phase A. We first extract a random subset $\Gamma \subseteq V$ of cardinality $c_1 \log^3 n$ (the value of c_1 will be given in Lemma 8). We make extractions with replacement, so we permit Γ to be a multiset. Let $b = \log|\Gamma| = \Theta(\log\log n)$. We assume for simplicity that b has an integer value.

We compute the values $v \cdot q$ for all $v \in \Gamma$. Then we build a complete binary tree T of depth $\log n$. We associate randomly the sites in P to the leafs of T. To every internal node x at height $1 \leq h \leq b$, we associate a random subset $\Gamma_x \subseteq \Gamma$ of size $c_2' + c_2 h$ (the appropriate values for c_2 and c_2' will be given in Lemma 9). To higher nodes, with height $h > b$, we associate the entire set Γ.

Now we make a tournament between the sites, proceeding from the leafs towards the root of T: to each internal node x we associate the winner of the comparison between the sites associated to its children, with respect to the set Γ_x. Let $\tilde{p}_A$ be the winner of this tournament, i.e., the site which at the end of Phase A will be associated to the root of T.

Phase B. Independently from all the above, we randomly choose a subset $P_0 \subseteq P$ of size $c_3 \log^3 n$, where c_3 will be defined in Lemma 7. We compute the distances between q and the sites in P_0. Let $\tilde{p}_B$ be the site of P_0 farthest from q.

The algorithm finishes returning the site, between $\tilde{p}_A$ and $\tilde{p}_B$, which is farthest from q.

3.3 Correctness

We now prove that the algorithm correctly computes an element of Z_ε. First of all, observe that, as a consequence of Lemma 4, the set V created during the preprocessing is an $(\varepsilon_0/12)$-approximation of the range space $(\mathcal{P}, \mathcal{R})$, with probability $1-\delta$. In the following, we will assume that this event occurred.

Lemma 5. *Let V be an $(\varepsilon_0/12)$-approximation, and v a random element of V. Let $q \in \mathbb{R}^d$, $p_i, p_j \in P$, and suppose that $(1-\gamma)d(p_i,q) \geq d(p_j,q)$, with $\varepsilon_0 \leq \gamma \leq 1/2$. The probability that $|v \cdot (p_i - q)| \geq |v \cdot (p_j - q)|$ is at least $1/2 + \gamma/5$.*

Proof. Let $x = p_i - q$ and $y = p_j - q$. The measure $z_{x;y}$ of $Z_{x;y}$ is at least $1/2 + \gamma/\pi$, by Lemma 1. As V is an $(\varepsilon_0/12)$-approximation, the probability that an element of V belongs to $Z_{x;y}$ is at most $\varepsilon_0/12$ less than that value. So this probability is at least $z_{x;y} - \varepsilon_0/12 \geq 1/2 + \gamma/5$, and the lemma follows. □

Lemma 6. *Suppose $(1-\gamma)d(p_i,q) \geq d(p_j,q)$, and let Γ' be a random subset of Γ, of cardinality k. The probability that $p_j \succeq_{\Gamma'} p_i$ is at most $e^{-\gamma^2 k/36}$.*

Proof Sketch.. Let $x = p_i - q$ and $y = p_j - q$. The relation $p_j \succeq_{\Gamma'} p_i$ is equivalent to $y \triangleright_{\Gamma'} x$, that means $|\Gamma' \cap Z_{y;x}| \geq 1/2 k$. As $S^{d-1} \setminus Z_{y;x} \subseteq Z_{x;y}$, we need only to bound the probability that $|\Gamma' \cap Z_{x;y}| \leq 1/2 k$. Define k Poisson random variables $X_1, \ldots, X_k$, where X_r is 1 if the r-th vector in Γ' belongs to $Z_{x;y}$, and 0 otherwise. From Lemma 5 it follows that the probability of success is $\Pr[X_r = 1] \geq 1/2 + \gamma/5$. Defining the random variable $X = \sum_r X_r = |\Gamma' \cap Z_{x;y}|$, the event we are interested in is $X \leq 1/2 k$. The lemma follows by some calculations and by application of the Chernoff bound (see e.g. [24]). □

The motivation for Phase B is that if in the set Z_ε there are too many sites, one of them could eliminate p^* during the early stage of the tournament, and then be eliminated by some site not in Z_ε. On the other side, if Z_ε is sufficiently small, there is a high probability that p^* will reach at least level b of the tournament tree: in this case, we can show that the winner $\tilde{p}_A$ is an element of Z_ε. Intuitively, at each comparison the winner can be slightly closer to q than the loser: if p^* is eliminated too early in the competition, all this small errors can take us too close to q; if instead p^* reaches at least level b, the final error is small with high probability.

We begin by proving that if Z_ε is large, then Phase B succeeds in finding an approximate farthest-point of q with high probability:

Lemma 7. *If $|Z_\varepsilon| \geq \gamma_1 n/\log^3 n$, then with probability at least $1-\delta$ it holds that $\tilde{p}_B \in Z_\varepsilon$.*

Proof. When we randomly select the elements of P_0 from P, the probability of taking an element not in Z_ε is less than $1-\gamma_1/\log^3 n$. The probability that no element of P_0 belongs to Z_ε is less than that quantity raised to power $|P_0| = c_3 \log^3 n$. We can choose $c_3 \equiv c_3(\gamma_1, \delta, n)$ such that $(1-\gamma_1/\log^3 n)^{c_3 \log^3 n} \leq \delta$. If we choose such a c_3, with probability at least $1-\delta$ there is at least one element of Z_ε in P_0, so $\tilde{p}_B$ certainly belongs to Z_ε. □

Let us see now what happens if Z_ε is small. Define the constant $\gamma_2 \equiv \gamma_2(\varepsilon)$ such that $e^{-\gamma_2} \geq 1-\varepsilon$. We denote by $\mathcal{E}_1$ the following event:

$$\mathcal{E}_1 \equiv \left\{ \exists\, p_i, p_j \in P, \quad i \neq j: \quad p_j \succeq_\Gamma p_i, \quad \left(1-\frac{\gamma_2}{\log n}\right) d(p_i,q) \geq d(p_j,q) \right\} ,$$

that is, the event that there exist two sites, with distances significantly different from q, for which the comparison based on projections on the vectors of Γ gives the wrong answer. We can give a bound to the probability that this event occurs:

Lemma 8. *The probability of event $\mathcal{E}_1$ is less or equal to $\delta/3$.*

Proof. There are $n(n-1) \leq n^2$ ordered pairs of sites to consider: for each pair we apply Lemma 6, with $\gamma = \gamma_2/\log n$ and $k = |\Gamma| = c_1 \log^3 n$, to bound the probability of error. The probability of $\mathcal{E}_1$ is bounded by the sum of these probabilities. The lemma follows by defining suitably the value of $c_1 \equiv c_1(\gamma_2, \delta, n)$ in such a way that $n^2 \cdot e^{-c_1\gamma_2^2 \log n/36} \leq \delta/3$. □

Let us denote by $\mathcal{E}_2$ the event "p^* does not reach level b in the tournament tree", that is, it is not assigned, during the tournament in Phase A, to any node at level b.

Lemma 9. *If $|Z_\varepsilon| < \gamma_1 n/\log^3 n$, the probability of $\mathcal{E}_2$ is less or equal to $2\delta/3$.*

Proof. Consider the leaf of T to which p^* was assigned, the node x at level b along the path from this leaf to the root (x is the node to which p^* should be assigned, if $\mathcal{E}_2$ does not occur), and the subtree T^x rooted at x.

We first of all exclude the possibility that in T^x there be sites of Z_ε other than p^*. The number of leafs in T^x is $2^b = |\Gamma| = c_1 \log^3 n$, so the probability that $|T^x \cap Z_\varepsilon| = 1$ is

$$\binom{n-|Z_\varepsilon|}{2^b-1}\binom{n}{2^b-1}^{-1} .$$

By defining suitably $\gamma_1 \equiv \gamma_1(c_1, \delta, n)$ we can make this probability greater than $1-\delta/3$. So we can assume that p^* is the only site of Z_ε in the subtree T^x. We now show that in this case it reaches level b with probability $1-\delta/3$.

If p^* reached level $h-1$, it is compared, at level h, with an element not in Z_ε, made using $|\Gamma_x| = c_2' + c_2 h$ vectors of Γ. From Lemma 6, with $\gamma = \varepsilon$, it follows that the probability that p^* loses the comparison is bounded by

$$e^{-\varepsilon^2(c_2'+c_2 h)/36} = e^{-\varepsilon^2 c_2'/36}(e^{-\varepsilon^2 c_2/36})^h .$$

We can define the constants $c_2' \equiv c_2'(\varepsilon,\delta)$ and $c_2 = c_2(\varepsilon)$ such that the left factor is less than $\delta/3$, and the right factor is less than 2^{-h}.

Summing these quantities for $h = 1,\dots,b$ we finally obtain that the probability that p^* is eliminated is bounded by $\delta/3$. Together with the fact that the probability that more than one site of Z_ε belongs to T^x is less than $\delta/3$, this implies the lemma. □

We can now prove that if Z_ε is small, Phase A succeeds in finding a good answer:

Lemma 10. *If* $|Z_\varepsilon| < \gamma_1 n/\log^3 n$, *then with probability at least* $1-\delta$ *it holds that* $\tilde{p}_A \in Z_\varepsilon$.

Proof. With probability $1-\delta$ neither $\mathcal{E}_1$ nor $\mathcal{E}_2$ occur. In this case we show by induction on $h \geq b$ that there exists a site $p_{(h)}$, assigned to a node at level h, such that

$$d(p_{(h)},q) \geq \left(1-\frac{\gamma_2}{\log n}\right)^{h-b} d(p^*,q) \ , \tag{3}$$

and this implies that $\tilde{p}_A \in Z_\varepsilon$.

The fact that $\mathcal{E}_2$ does not occur guarantees that p^* reached level b, so we can start the induction, for $h = b$, with $p_{(h)} = p^*$. Suppose now that (3) holds at level h, and let $p'_{(h)}$ be the adversary of $p_{(h)}$. If $d(p'_{(h)},q) \geq (1-\gamma_2/\log n)d(p_{(h)},q)$, then both $p_{(h)}$ and $p'_{(h)}$ satisfy (3) for level $h+1$, so relation (3) holds in any case. If instead $d(p'_{(h)},q) < (1-\gamma_2/\log n)d(p_{(h)},q)$, as we are assuming that $\mathcal{E}_1$ does not occur, certainly $p_{(h)}$ wins the comparison, so $p_{(h+1)} = p_{(h)}$ and relation (3) is valid for level $h+1$.

From the inductive argument it follows that (3) holds for $h = \log n$, where $p_{(\log n)} = \tilde{p}_A$ is the site assigned to the root of T. Being $b \geq 1$, from the fact that $(1-x/m)^{m-1} \geq e^{-x}$, and from the definition of γ_2, we obtain

$$d(\tilde{p}_A,q) \geq \left(1-\frac{\gamma_2}{\log n}\right)^{\log n-1} d(p^*,q) \geq e^{-\gamma_2} d(p^*,q) \geq (1-\varepsilon)d(p^*,q) \ ,$$

that is what we wanted to prove. □

The correctness of the algorithm follows from both Lemma 7 and Lemma 10.

3.4 Complexity

The preprocessing consists in the computation of a distance, which requires time $O(d)$, for each element of the matrix M. The time needed is thus $O(dLn) = O(d^2 n \log d \log^2 n)$. The space required to store M is $O(Ln) = O(dn \log d \log^2 n)$.

To answer a query, we first compute $v \cdot q$ for $v \in \Gamma$. The time required for this is $O(|\Gamma| \cdot d) = O(d \log^3 n)$. Each comparison of type $p_i \succeq_{\Gamma_x} p_j$ requires time $O(|\Gamma_x|)$, because all values $v \cdot p_i$ are already stored in M. The number of nodes at level h is $n2^{-h}$, so the total number of operations at levels $1 \leq h \leq b$ is

$$\sum_{h=0}^{b} (c_2' + c_2 h) n 2^{-h} = c_2' n \sum 2^{-h} + c_2 n \sum h 2^{-h} \leq 2(c_2' + c_2)n = O(n) \ .$$

For levels $h > b$ each comparison is made using the entire set Γ, where $|\Gamma| = 2^b$. We have

$$\sum_{h=b+1}^{\log n} |\Gamma| \cdot n2^{-h} = \sum_{h=b+1}^{\log n} |\Gamma| \cdot n2^{b-h}2^{-b} = n \sum_{h=b+1}^{\log n} 2^{b-h} \leq n .$$

We thus obtain that the total cost of Phase A is $O(n + d\log^3 n)$. Phase B requires time $O(d\log^3 n)$. We conclude that the time required to answer a query is $O(n + d\log^3 n)$. We summarize the properties of the farthest-point algorithm in the following theorem:

Theorem 11. *With probability* $1-\delta$ *the set* V *created in the preprocessing is an* $(\varepsilon_0/12)$-*approximation. In this case, the probability that the algorithm returns an element of* Z_ε *is, for each query, at least* $1-\delta$. *The complexity of the algorithm is* $O(d^2 \log d \cdot n \log^2 n)$ *for the preprocessing and* $O(n + d\log^3 n)$ *for answering a query.*

4 Computing the Diameter of a Point Set

We now show how to use the algorithm for farthest-point queries to find the diameter d_P of a set of points $P = \{p_1, \ldots, p_n\}$ in $\mathbb{R}^d$.

First of all, we use a dimension-reduction technique, by projecting all the points on a random subspace of dimension $k = O(\varepsilon^{-2} \log n)$. Let P' the resulting set of points in $\mathbf{R}^k$ and $d_{P'}$ its diameter. The Johnson–Lindenstrauss Lemma (see [20, 14]) affirms that with high probability all inter-point distances are preserved, so that $(1+\varepsilon/2)d_P \geq d_{P'} \geq (1-\varepsilon/2)d_P$.

Next, we compute the set V and the matrix M as in the preprocessing stage of the farthest-point algorithm, using $\varepsilon/2$ as approximation parameter.

For each $p'_i \in P'$, we perform a farthest-point query where $q = p'_i$ and the sites are the remaining points in P'; let $F_\varepsilon(p'_i)$ the point returned by the algorithm described in Section 3. We compute all the distances $d(p'_i, F_\varepsilon(p'_i))$. Let $\tilde{d}_{P'}$ the maximum distance computed in this way. Our main result is the following:

Theorem 12. *Let* d_P *be the diameter of the point set* P, *and* $\tilde{d}_{P'}$ *the value computed by the algorithm above. Then, with probability* $1-\delta$, *it holds that* $(1+\varepsilon/2)d_P \geq \tilde{d}_{P'} \geq (1-\varepsilon)d_P$. *The complexity of the algorithm is* $O(nd\log n + n^2)$.

Proof. Let (p^*, q^*) be a pair of points in P' such that $d(p^*, q^*) = d_{P'}$. When we make the farthest-point query with $q = p^*$, the point $F_\varepsilon(p^*)$ is, with probability $1-\delta$, such that $d(p^*, F_\varepsilon(p^*)) \geq (1-\varepsilon/2)d_{P'}$. The value $\tilde{d}_{P'}$ computed in the last step is $\tilde{d}_{P'} \geq (1-\varepsilon/2)d_{P'} \geq (1-\varepsilon/2)^2 d_P \geq (1-\varepsilon)d_P$. Moreover, the distances computed by this algorithm are certainly less than $d_{P'}$.

The time to perform the projection at the first step is $O(nd\log n)$. Then we perform the preprocessing and n queries of the algorithm in Section 3, in dimension $O(\log n)$. As a consequence of Theorem 11, we obtain that the overall complexity of the algorithm is $O(nd\log n + n^2)$. □

Acknowledgements. We thank Piotr Indyk and Jon Kleinberg for pointing out the paper [4], and Yuval Rabani for sending us a preliminary version of it.

References

[1] N. Alon and J. Spencer. *The Probabilistic Method.* Wiley, 1992.

[2] S. N. Bespamyatnikh. An efficient algorithm for the three-dimensional diameter problem. In *Proc. 9th ACM-SIAM SODA*, pages 137–146, 1998.

[3] A. Blumer, A. Ehrenfeucht, D. Haussler, and M. Warmuth. Learnability and the Vapnik-Chervonenkis dimension. *J. ACM*, 4(36):929–965, 1989.

[4] A. Borodin, R. Ostrovsky, and Y. Rabani. Subquadratic approximation algorithms for clustering problems in high dimensional space. To appear in *Proc. STOC*, 1999.

[5] B. Chazelle, H. Edelsbrunner, L. Guibas, and M. Sharir. Diameter, width, closest line pair, and parametric searching. *Discrete Comput. Geom.*, 10:183–196, 1993.

[6] K. L. Clarkson and P. W. Shor. Applications of random sampling in computational geometry, II. *Discrete Comput. Geom.*, 4:387–421, 1989.

[7] D. Coppersmith and S. Winograd. Matrix multiplication via arithmetic progression. In *Proc. 19th ACM STOC*, pages 1–6, 1987.

[8] R. M. Dudley. Central limit theorems for empirical measures. *Annals of Probability*, 6:899–929, 1978.

[9] R. M. Dudley. *A Course on Empirical Processes.* In *Lecture Notes in Mathematics*, No. 1097, pages 1–142, 1984.

[10] Ö. Eğecioğlu and B. Kalantari. Approximating the diameter of a set of points in the Euclidean space. *Information Processing Letters*, 32:205–211, 1989.

[11] P. Erdös. On sets of distances of *n* points. *Amer. Math. Monthly*, 53:248–250, 1946.

[12] P. Erdös. On sets of distances of points in Euclidean spaces. *Magyar Tud. Akad. Mat. Kutató Int. Közl.*, 5:165–169, 1960.

[13] M. Flickner, H. Sawhney, W. Niblack, J. Ashley, Q. Huang, B. Dom, M. Gorkani, J. Hafner, D. Lee, D. Petkovic, D. Steele, and P. Yanker. Query by image and video content: The QBIC system. *Computer*, 28(9):23–32, 1995.

[14] P. Frankl and H. Maehara. The Johnson-Lindenstrauss lemma and the sphericity of some graphs. *Journal of Combinatorial Theory*, Series B 44:355–362, 1988.

[15] P. Gritzmann and V. Klee. Inner and outer j-radii of convex bodies in finite dimensional normed spaces. *Discrete Computat. Geom.*, 7:255–280, 1992.

[16] P. Gritzmann and V. Klee. Computational complexity of inner and outer j-radii of polytopes in finite-dimensional normed spaces. *Mathematical programming*, 59:163–213, 1993.

[17] P. Gritzmann and V. Klee. On the complexity of some basic problems in computational convexity: I. Containment problems. *Discrete Math.*, 136:129–174, 1994.

[18] V. N. Gudivada and V. V. Raghavan. Guest editors' introduction: Content-based image retrieval systems. *Computer*, 28(9):18–22, 1995.

[19] C. E. Jacobs, A. Finkelstein, and D. H. Salesin. Fast multiresolution image querying. In Robert Cook, editor, *SIGGRAPH 95 Conference Proceedings*, pages 277–286. Addison Wesley, August 1995.

[20] W. B. Johnson and J. Lindenstrauss. Extensions of Lipschitz mappings into a Hilbert space. In *Conference in Modern Analysis and Probability*, volume 26 of *Contemporary Mathematics*, pages 189–206, 1984.

[21] J. M. Kleinberg. Two algorithms for nearest-neighbor search in high dimensions. In *29th ACM STOC*, pages 599–608, 1997.

[22] D. E. Knuth. *Seminumerical Algorithms*, volume 2 of *The Art of Computer Programming.* Addison-Wesley, 2nd edition, 1981.

[23] J. Matoušek and O. Schwarzkopf. A deterministic algorithm for the three-dimensional diameter problem. In *Proc. 25th STOC*, pages 478–484, 1993.

[24] R. Motwani and P. Raghavan. *Randomized Algorithms.* Cambridge University Press, 1995.

[25] K. Mulmuley. *Computational Geometry. An Introduction Through Randomized Algorithms*. Prentice Hall, 1994.

[26] F. P. Preparata and M. I. Shamos. *Computational Geometry: an Introduction*. Springer Verlag, 1985.

[27] E. A. Ramos. Construction of 1-d lower envelopes and applications. In *Proc. 13th Annual Symp. on Computational Geometry*, pages 57–66, 1997.

[28] S. Straszewicz. Sur un problème géometrique de P. Erdös. *Bull. Acad. Polon. Sci. Cl. III*, 5:39–40, 1957.

[29] V. N. Vapnik and A. Y. Chervonenkis. On the uniform convergence of relative frequencies of events to their probabilities. *Theory Prob. Appl.*, 16:264–280, 1971.

[30] M. Vidyasagar. Learning and generalization with applications to neural networks. Preliminary version of [31], November 1995.

[31] M. Vidyasagar. *A Theory of Learning and Generalization: with Applications to Neural Networks and Control System*. Springer-Verlag, London, 1997.

[32] A. C. Yao. Personal communication. Also in "On computing the distance matrix of *n* points in *k* dimensions", TR IBM San Jose Research Center, about 1982.

[33] A. C. Yao. On constructing minimum spanning trees in *k*-dimensional spaces and related problems. *SIAM Journal of Computing*, 11(4):721–736, 1982.

A Nearly Linear-Time Approximation Scheme for the Euclidean k-median Problem

Stavros G. Kolliopoulos[1] and Satish Rao[1]

No Institute Given

Abstract. In the *k-median* problem we are given a set S of n points in a metric space and a positive integer k. The objective is to locate k medians among the points so that the sum of the distances from each point in S to its closest median is minimized. The k-median problem is a well-studied, NP-hard, basic clustering problem which is closely related to facility location. We examine the version of the problem in Euclidean space. Obtaining approximations of good quality had long been an elusive goal and only recently Arora, Raghavan and Rao gave a randomized polynomial-time approximation scheme for the Euclidean plane by extending techniques introduced originally by Arora for Euclidean TSP. For any fixed $\varepsilon > 0$, their algorithm outputs a $(1+\varepsilon)$-approximation in $O(nkn^{O(1/\varepsilon)}\log n)$ time.

In this paper we provide a randomized approximation scheme for points in d-dimensional Euclidean space, with running time $O(2^{1/\varepsilon^d} n \log n \log k)$, which is nearly linear for any fixed ε and d. Our algorithm provides the first polynomial-time approximation scheme for k-median instances in d-dimensional Euclidean space for any fixed $d > 2$. To obtain the drastic running time improvement we develop a structure theorem to describe hierarchical decomposition of solutions. The theorem is based on a novel *adaptive decomposition* scheme, which guesses at every level of the hierarchy the structure of the optimal solution and modifies accordingly the parameters of the decomposition. We believe that our methodology is of independent interest and can find applications to further geometric problems.

1 Introduction

In the *k-median* problem we are given a set S of n points in a metric space and a positive integer k. The objective is to locate k *medians* (facilities) among the points so that the sum of the distances from each point in S to its closest median is minimized. The k-median problem is a well-studied, NP-hard problem which falls into the general class of *clustering* problems: partition a set of points into clusters so that the points within a cluster are close to each other with respect to some appropriate measure. Moreover k-median is closely related to *uncapacitated facility location,* a basic problem in the operations research literature (see e.g. [9]). In the latter problem except for the set S of points we are given also a cost c_i for *opening* a facility at point i. The objective is to open an unspecified number of facilities at a subset of S so as to minimize the sum of the cost to open the facilities (*facility cost*) plus the cost of assigning each point to the nearest open facility (*service cost*).

1.1 Previous Work

The succession of results for k-median is as follows. Lin and Vitter [15] used their filtering technique to obtain a solution of cost at most $(1+\varepsilon)$ times the optimum but using $(1+1/\varepsilon)(\ln n+1)k$ medians. They later refined their technique to obtain a solution of cost $2(1+\varepsilon)$ while using at most $(1+1/\varepsilon)k$ medians [14]. The first non-trivial approximation algorithm that achieves feasibility as well, i.e. uses k medians, combines the powerful randomized algorithm by Bartal for approximation of metric spaces by trees [4, 5] with an approximation algorithm by Hochbaum for k-median on trees [13]. The ratio thus achieved is $O(\log n \log\log n)$. This algorithm was subsequently refined and derandomized by Charikar et al. [6] to obtain a guarantee of $O(\log k \log\log k)$. Only very recently, Charikar and Guha and independently Tardos and and Shmoys reported the first constant-factor approximations [7]. In contrast, the uncapacitated facility location problem, in which there is no a priori constraint on the number of facilities, seems to be better understood. Shmoys, Tardos and Aardal [18] gave a 3.16 approximation algorithm. This was later improved by Guha and Khuller [12] to 2.408 and more recently to 1.736 by Chudak [8].

For the problem of interest in this paper, i.e. k-median on the Euclidean plane, a randomized polynomial-time approximation scheme was given by Arora, Raghavan and Rao in [3]. For any fixed $\varepsilon > 0$, their algorithm outputs a $(1+\varepsilon)$-approximation with probability $1-o(1)$ and runs in $O(nkn^{O(1/\varepsilon)}\log n)$ time, worst case. This development followed the breakthrough approximation scheme of Arora [1, 2] for the Traveling Salesman Problem and other geometric problems. While the work in [3] used techniques from the TSP approximation scheme, the different structure of the optimal solutions for k-median and TSP necessitated the development of a new structure theorem to hierarchically decompose solutions. We elaborate further on this issue during the exposition of our results in the next paragraph. The dependence of the running time achieved by the methods of Arora, Raghavan and Rao on $1/\varepsilon$ is particularly high. For example, the approximation scheme can be extended to higher-dimension instances but runs in quasi-polynomial time $O(n^{(\log n/\varepsilon)^{d-2}})$ for a set of points in R^d for fixed $d > 2$.

1.2 Results and Techniques

Results. We provide a randomized approximation scheme which, for any fixed $\varepsilon > 0$, outputs in expectation a $(1+\varepsilon)$ approximately optimal solution, in time $O(2^{1/\varepsilon^d} n \log n \log k)$ worst case. This time bound represents a drastic improvement on the result in [3]. For any fixed accuracy ε desired, the dependence of the running time on $1/\varepsilon$ translates to a (large) constant hidden in the near-linear asymptotic bound $O(n \log n \log k)$ compared to the exponent of a term polynomial in n in the bound of Arora, Raghavan and Rao. Moreover, for inputs in R^d, our algorithm extends to yield a running time of $O(2^{1/\varepsilon^d} n \log n \log k)$, which yields for the first time a polynomial-time approximation scheme for fixed $d > 2$. The ideas behind the new k-median algorithm yield also improved, nearly linear-time, approximation schemes for uncapacitated facility location. We now elaborate on the techniques we use to obtain our results.

Techniques. Perhaps the most important contribution of our paper lies in the new ideas we introduce to overcome the limitations of the approach employed by Arora, Raghavan and Rao in [3]. To introduce these ideas we sketch first some previous developments, starting with the breakthrough results in [1, 2] (see also [16] for a different approximation scheme for Euclidean TSP).

A basic building block for Arora's [1, 2] results on TSP was a structure theorem providing insight into how much the cost of an optimal tour could be affected in the following situation. Roughly speaking, the plane is recursively dissected into a collection of rectangles of geometrically decreasing area, represented by a quadtree data structure. For every box in the dissection one places a fixed number, dependent on the desired accuracy ε, of equidistant *portals* on the boundary of the box. The optimal TSP tour can cross between adjacent rectangles any number of times; a *portal-respecting* tour is allowed to cross only at portals. How bad can the cost of a deflected, portal-respecting, tour be compared to the optimum? Implicitly, Arora used a charging argument on the edges in an optimal solution to show that the edges could be made to be portal respecting. We sketch now his approach which was made explicit and applied to k-median in [3].

The original solution is assumed to consist of a set of edges and is assumed to be surrounded by a rectangle with sidelength polynomial in n (cf. Section 2). At level i of the dissection, the rectangles at this level with sidelength 2^i are cut by vertical and horizontal lines into rectangles of sidelength 2^{i-1}. The x- and y-coordinates of the dissection are randomly shifted at the beginning, so that the probability that an edge s in a solution is cut at level i is $O(\text{length}(s)/2^i)$. Let m denote the number of portals along the dissection lines. If s is cut at level i it must be deflected through a portal, paying additional cost $O(2^i/m)$. Summing over all the $O(\log n)$ levels of the decomposition, the expected deflection cost of edge s in a portal-respecting solution is at most

$$\sum_{i=1}^{O(\log n)} O\left(\frac{\text{length}(s)}{2^i}(2^i/m)\right) \quad (1)$$

Selecting $m = \Theta(\log n/\varepsilon)$ demonstrates the existence of a portal-respecting solution of cost $(1+\varepsilon)OPT$. The running time of the dynamic programming contains a $k2^{O(m)}$ term, hence the $kn^{1/\varepsilon}$ term in the overall running time.

Arora additionally used a "patching lemma" argument to show that the TSP could be made to cross each box boundary $O(1/\varepsilon)$ times. This yielded an $O(n(\log n)^{O(1/\varepsilon)})$ time algorithm for TSP. (This running time was subsequently improved by Rao and Smith to $O(2^{O(1/\varepsilon^2)} + n\log n)$ [17] while still using $\Theta(\log n/\varepsilon)$ portals). The k-median method did not, however, succumb to a patching lemma argument, thus the running time for the algorithms in [3] remained $O(kn^{O(1/\varepsilon)})$.

Our method reduces the number of portals m to $\Theta(1/\varepsilon)$. That is, we remove the $\log n$ factor in the number of portals that appear to be inherent in Arora, Raghavan, and Rao's charging based methods and even in Arora's charging plus patching based methods.

Adaptive dissection. We outline some of the ideas behind the reduced value for m. The computation in (1) exploits linearity of expectation by showing that the "average" dissection cut is good enough. The complicated dependencies between the dissection lines across all $O(\log n)$ levels seem prohibitive to reason about directly. On the other hand, when summing the expectations across all levels an $O(\log n)$ factor creeps in, which apparently has to be offset by setting m to $\log n/\varepsilon$. We provide a new structure theorem to characterize the structure of near-optimal solutions. In contrast to previous approaches, given a rectangle at some level in the decomposition, it seems a good idea to choose more than one possible "cuts" hoping that one of them will hit a small number of segments from the optimum solution. This approach gives rise to the *adaptive dissection* idea, in which the algorithm "guesses" the structure of the part of the solution contained in a given rectangle and tunes accordingly the generation of the subrectangles created for the next level of the dissection. In the k-median problem the guess consists of the area of the rectangle which is empty of facilities. Let L be the maximum sideline of the subrectangle containing facilities. Cutting close to the middle of this subrectangle with a line of length L should, in a probabilistic sense, mostly dissect segments from the optimal solution of length $\Omega(L)$, forcing them to deflect by $L/m = \varepsilon L$. A number of complications arise by the fact that a segment may be cut by both horizontal and vertical dissection lines. We note that the cost of "guessing" the empty area is incorporated into the size of the dynamic programming lookup table by trying all possible configurations. Given the preeminence of recursive dissection in approximation schemes for Euclidean problems ([1, 2, 3, 17]) we believe that the adaptive dissection technique is of independent interest and can prove useful in other geometric problems as well.

Although the adaptive dissection technique succeeds in reducing the required number of portals to $\Theta(1/\varepsilon)$ and thus drastically improve the dependence of the running time on $1/\varepsilon$, the dynamic program has still to enumerate all possible rectangles. Compared to the algorithm in [3] we apparently have to enumerate even more rectangles due to the "guess" for the areas without facilities. We further reduce the size of the lookup table by showing that the boundaries of the possible rectangles can be appropriately spaced and still capture the structure of a near-optimal solution.

2 Preliminaries

An *edge* (u,v) is a line segment connecting input points u and v. In the context of a solution, i.e. a selection of k medians, an *assignment edge* is an edge (u,v) such that one of the points is the median closest to the other in the solution. Unless otherwise specified, the *sidelength* of a rectangle with sides parallel to the axes is the length of its largest side.

We assume that the input points are on a unit grid of size polynomial in the number of input points. This assumption is enforced by a preprocessing phase described in [3] incurring an additive error of $O(1/n^c)$ times the optimal value, for some constant $c > 0$. The preprocessing phase can be implemented by a plane sweep in conjunction with the minmax diameter algorithm of Gonzalez [11]. The latter can be implemented to run in $O(n\log k)$ time [10]. The total preprocessing time is $O(n\log n)$.

3 The Structure Theorem

In this section we prove our basic Structure Theorem that shows the existence of approximately optimal solutions with a simple structure. Our exposition focuses on 2-

dimensional Euclidean instances. It is easy to generalize to d dimensions. Given a set of points S and a set of facilities $F \subset S$, we define the *greedy cost* of the solution to be the cost of assigning each point to its closest facility. We proceed to define a recursive randomized decomposition. The decomposition has two processes: the Sub-Rectangle process, and the Cut-Rectangle process.

Sub-Rectangle:
Input: a rectangle B containing at least one facility.
Process: Find the minimal rectangle B' containing all the facilities. Let its maximum sidelength be S. Grow the rectangle by $S/3$ in each dimension. We call the grown rectangle B''.
Output: $B_S = B'' \cap B$.

Notice that $B - B_S$ contains no facilities.

Cut-Rectangle:
Input: a rectangle B containing at least one facility.
Process: Randomly cut the rectangle into two rectangles with a line that is orthogonal to the maximal side in the middle third of the rectangle.
Output: The two rectangles.

The recursive method applies alternatingly the Sub-Rectangle and Cut-Rectangle processes to produce a decomposition of the original rectangle containing the input. We remark that the original rectangle is not necessarily covered by leaf rectangles in the decomposition, due to the sub-rectangle steps.

We place $m+1$ evenly spaced points on each side of each rectangle in the decomposition, where m will be defined later and depends on the accuracy of the sought approximation. We call these points *portals.* We define a *portal-respecting path* between the two points to be a path between the two points that only crosses rectangles that enclose one of the points at portals. We define the *portal-respecting distance* between two points to be the length of the shortest portal respecting path between the points. We begin by giving three technical lemmata which will be of use in the main Structure Theorem.

Lemma 1. *If the maximal rectangle R separating points v and w has sidelength D the difference between the portal respecting distance and the geometric distance between v and w is $O(D/m)$.*

We define a *cutting* line segment in the decomposition to be (i) either a line segment l that is used in the Cut-Rectangle procedure to divide a rectangle R into two rectangles or (ii) a line segment l used to form the boundary of a sub-rectangle R in the Sub-Rectangle procedure. In both cases we say that *l cuts R*. We define the *sidelength* of a cutting line l as the sidelength of the rectangle cut by l. Observe that the length of a cutting line is upperbounded by its sidelength.

Lemma 2. *If any two parallel cutting line segments produced by the application of Cut-Rectangle are within distance L, one of the line segments has sidelength at most $3L$.*

Proof. Let l_1, l_2 be the two cutting segments at distance L. Assume wlog that they are both vertical, l_1 is the longer of the two lines, l_1 is on the left of l_2 and cuts a rectangle R of sidelength greater than $3L$ into R_1 and R_2. Then l_1 is produced first in the decomposition. Thus l_2 is contained within R_2 (see Fig. 1a), and since it comes second can only cut a rectangle R_2' contained within R_2. By the definition of Cut-Rectangle, if s is the sidelength of R_2', l_2 is drawn at least $s/3$ away from the left boundary of R_2' which implies $s/3 \leq L$. Thus $s < 3L$. □

The next lemma relates the length of a cutting line segment produced by Sub-Rectangle to the length of assignment edges it intersects.

Lemma 3. *If a cutting line segment X produced by Sub-Rectangle intersects an assignment edge (v,f) of length D, X has sidelength at most $5D$.*

Proof. Let R be the rectangle cut by X. Observe that f must be contained in R. Let s be the sidelength of the rectangle R. Wlog assume that the horizontal dimension of R is maximal. By the definition of Sub-Rectangle there is $y < s$ such that $s = (5/3)y$ and the two vertical strips of width $y/3$ at the sides of R are empty of facilities. Therefore $y/3 < D$ which implies that $s < 5D$. □

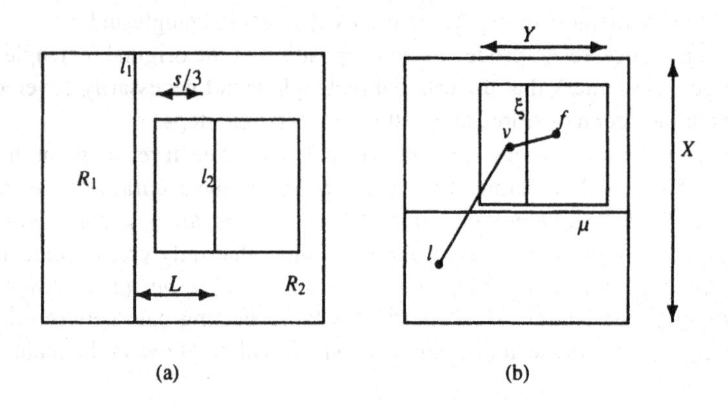

Fig. 1. (a) Demonstration of Lemma 2. (b) Case A2 in the proof of Theorem 4.

We assign each point to the closest facility under the portal respecting distance function. The *modified cost* of the assignment is the sum over all the points of the portal-respecting distances to their respective assigned facility.

Theorem 4. **(Structure Theorem)** *The expected difference between the modified cost and the greedy cost, C, of the facility location problem is $O(C/m)$.*

Proof of theorem: By linearity of expectation, it suffices to bound the expected cost increase for a given assignment edge. For a point v we define f as v's closest facility and assume f to be to the right of v (without loss of generality.) We define l as the closest facility to the left of v. We denote the distance from v to f by D, and the distance from v to l by L. The idea behind the analysis of the portal-respecting solution is that assigning v to either f or l (based on the amount by which the decomposition distorts each distance) will be enough to show near-optimal modified cost.

We assume without loss of generality that v and f are first separated by a vertical cutting line. (We turn the configuration on its side and do the same argument if this condition does not hold.)

The semicircle of diameter $2L$ centered at v and lying entirely to the left of the vertical line containing v is empty in its interior. Therefore we obtain:

Lemma 5. *In the decomposition, v and f are not separated for the first time by a cutting line of sidelength in the interval* $(8D, L/2)$.

Proof of lemma: We know that v and l are separated by the time the sidelength of any enclosing rectangle is at most L. Observe that by Lemma 3, a cutting line of sidelength $> 8D$ separating v and f can only be produced by Cut-Rectangle. For any rectangle of sidelength $L/2$ containing v, there are no facilities to the left of v. Thus, by the sub-rectangle process, for boxes of sidelength $s < L/2$, the left boundary of any rectangle containing v of sidelength $s < L/2$ is within distance $s/5$ of v. By the Cut-rectangle process, any cutting line for a rectangle box is at least $s/3$ to the right of its left boundary which implies that the cutting line is at least $2s/15$ to the right of v. Thus, the length D line segment cannot be cut until the box size is at most $(15/2)D$. □

We proceed to a case analysis based on which of the two edges (v,l) or (v,f) is separated first by the decomposition. Observe that (v,l) can be separated for the first time by either a vertical or a horizontal line.

CASE A. *Edge (v,l) is separated first.* If v and f are first separated by a line produced by Sub-Rectangle, the increase in cost is $O(1/m)5D$ by Lemma 3. Therefore we can assume in the examination of CASE A that v and f are first separated by a vertical cutting line produced by Cut-Rectangle. The following calculation is straightforward and will be of use. Let $E(\Delta, Z)$ denote the event that an edge of length Δ is separated by a cutting line produced by Cut-Rectangle of sidelength Z. Then $Pr[E(\Delta, Z)] \leq 3\Delta/Z$.

We will now calculate the expectation of the cost increase for the two possible sub-cases. Let μ, ξ denote the lines separating for the first time (v,l) and (v,f) respectively.

CASE A1. Edge (v,l) is separated for the first time by a vertical cutting line. With some probability p, ξ has sidelength $L/2$ or more. By Lemma 5, ξ has sidelength at most $8D$ with probability $(1-p)$. Therefore, by Lemma 1, the cost increase is $O(D/m)$ with probability $(1-p)$. Now we turn to the case in which ξ has length $L/2$ or more. If μ is produced by Sub-Rectangle, by Lemma 3, μ has sidelength at most $5L$. If μ is produced by Cut-Rectangle, by Lemma 2 either ξ or μ has length at most $3L$. By Lemma 1 the cost increase is $O(L/m)$ regardless of the operation producing μ. Probability p is at most $c(2D/L)(1+1/2+1/4+\ldots) = O(D/L)$ for appropriate constant c. Therefore the expected cost increase is at most $(1-p)O(D/m) + pO(L/m) = O(D/m) + O((D/L)(L/m)) = O(D/m)$.

Remark: The probability calculation for Case A1 depended only on the choice of which vertical line first cuts (v,f). This will be useful in Section 4 when we restrict our choices for vertical cutting lines.

CASE A2. Edge (v,l) is separated for the first time by a horizontal cutting line. We compute first the expectation of the cost increase conditioned upon the sidelength X of line

μ. See Figure 1b. Edge (v,f) is cut by a line of sidelength Y with probability at most $3D/Y$. Observe that this is true regardless of the value of X. Moreover $L/2 \leq Y \leq X$ or by Lemma 5, $Y \leq 8D$. The upper bound of X holds since (v,f) is contained in the rectangle of sidelength X that contains v and f and l. A consequence of the latter fact is that μ cannot have been produced by a Sub-Rectangle operation. The conditional expected cost increase is bounded by $\sum_{L/2\leq Y\leq X|\exists i,Y\in[2^{i-1},2^{i}]}(D/Y)(Y/m) = O((D/m)\log(X/L))$. We now remove the conditioning on X. Line μ is produced by Cut-Rectangle thus it has sidelength X with probability at most $3L/X$. The expectation of the cost increase is $\sum_{X\geq L|\exists i,X\in[2^{i-1},2^{i}]}(L/X)\log(X/L)O(D/m) = O(D/m)$. **CASE B.** *Edge (v,f) is separated first.* The analysis is similar to Case A above with two Cases B1, B2 based on whether (v,l) is separated by a vertical or a horizontal line. □

4 Modifying the Structure Theorem

The Structure Theorem in the previous section demonstrates that a portal-respecting $(1+\varepsilon)$-optimal solution exists while only placing $O(1/\varepsilon)$ portals on the boundary of the decomposition rectangles. Using ideas from [3], Theorem 4 by itself would suffice to construct a dynamic programming algorithm running in $O(2^{1/\varepsilon}kn^4)$ time. In this section we show how to effectively bound the number of rectangles to be enumerated by the dynamic program and thus obtain a nearly linear-time algorithm.

We give first some definitions. Consider the rectangle of sidelength N that surrounds the original input. We assume that $N = m2^{\rho}$ for some integer ρ. Then, we call the vertical (horizontal) lines that are numbered 0 mod 2^i, $1 \leq i \leq \rho$, starting from the left (top) *i-allowable*. Note that all the lines are 1-allowable, the top and leftmost lines are ρ-allowable, and that any j-allowable line is i-allowable for all $i < j$. A rectangle R of sidelength s is *allowable* if the boundaries lie on t-allowable lines where t is the maximal value such that $2^t < s/m$.

We modify the Sub-rectangle and Cut-rectangle processes as follows.

Sub-rectangle:
Input: An allowable rectangle containing at least one facility.
Process: Perform the sub-rectangle process of the previous section.
Output: the minimal allowable rectangle that contains the rectangle computed in the Process.

Cut-rectangle:
Input: An allowable rectangle containing at least one facility.
Process: Choose a cutting line in the middle third of the rectangle uniformly among all lines that produce two allowable subrectangles.
Output: The two allowable subrectangles.

Notice that the decomposition is essentially the same as before. The primary difference is that the randomization in the cut-rectangle process is diminished. If we add back some randomization up front by shifting the original rectangle surrounding the input, we can get the same result as in the Structure Theorem on the expected increased cost of a portal respecting solution.

Theorem 6. *If the original rectangle is randomly shifted by (a,b) where a and b are chosen independently and so that each line is equally likely to be l-allowable, then the expected difference between the modified cost and the greedy cost, C, of the k-median problem is $O(C/m)$.*

Proof sketch: All the deterministic lemmata from Section 3 continue to hold in the restricted version of the decomposition. The randomized portion in the proof of Theorem 4 only reasons about two types of events. Moreover, it reasons about each event in isolation (cf. Cases A1 and A2 in the proof). Thus, we need only be concerned with the probabilities of each event. The random shift up front along with the randomization inside the process will ensure that these two types of events occur in our allowable decomposition procedure with approximately the same probability as in the previous decomposition.

The first event (cf. Case A1) is that a cut-rectangle line of length X cuts a line segment of length D. The probability of this event is required to be at most $3D/X$ in the proof of the Structure Theorem. We show that this continues to hold albeit the constant is larger than 3.

We assume for simplicity, that the line segment is in a rectangle R of size exactly $X = m2^i$ at some point. (This will be true to within a constant factor.) If $D < 2^i$, we know that the segment is cut by at most one i-allowable line. The probability of this event is at most $D/2^i$ due to the random shift. The cut-rectangle chooses from $m/3$ i-allowable lines uniformly at random. Thus, the line segment is cut with probability $1/(m/3)$ times $D/2^i$, which is $3D/X$ as required. If $D > 2^i$, we notice that D intersects at most $\lceil D/2^i \rceil < 2D/2^i$ lines. By the union bound the probability that this line segment is cut during Cut-Rectangle on R is uppebounded by $2D/2^i$ times $3/m$, i.e., $6D/X$.

The second event (cf. Case A2) is the intersection of two events; a horizontal line of length X cuts a segment of length L and a vertical line of length Y cuts a segment of length D. In the proof of the Structure Theorem, the probability was assumed to be upper bounded by the product of the probability bounds of the two events, i.e., $3L/X$ times $3D/Y$.

For our restricted decomposition, the probability of each event in isolation can be bounded by $6L/X$ and $6D/Y$ as argued above. Moreover, we chose the horizontal and vertical shifts independently and we choose the horizontal and vertical cut-lines in different sub-rectangle processes. Thus, we can also argue that the probability of the two events is at most $6L/X$ times $6D/Y$. □

We now prove a lemma bounding the number of allowable rectangles.

Lemma 7. *The number of allowable rectangles that contain l or more points is $O(m^4(n/l)\log n)$.*

Proof. Our proof uses a charging argument. Let R_l be a rectangle on the plane that has minimum sidelength, say L and contains l points. We bound the cardinality of the set S_l of allowable rectangles, which are distinct from R_l, contain at least l points and have at least one point in common with R_l. Let R_a be such a rectangle. Then R_a has sidelength at least L, otherwise it would have been chosen instead of R_l.

We bound the number of allowable rectangles in S_l with sidelength $X \in [2^{i-1}, 2^i]m$ by $O(m^4)$ as follows. The corners must fall on the intersection of two i-allowable lines that are within distance X of R_l. The number of i-allowable lines that are within distance X of R_l is $O(X/2^{i-1})$ since $X \geq L$. Thus, the number of corner choices is $O(m^2)$. Two corners must be chosen, so the number of rectangles in S_l of sidelength $X \in [2^{i-1}, 2^i]m$ is $O(m^4)$. Since there are $O(\log n)$ values of i, $|S_l| = O(m^4 \log n)$.

Now, we remove R_l and its points from the decomposition, and repeat the argument on the remaining $n-l$ points. The number of repetitions until no points are left is $O(n/l)$ therefore by induction, we get a bound of $O(m^4(n/l)\log n)$ on the number of allowable rectangles that contain at least l points. □

5 The Dynamic Program

We have structural theorems relative to a particular decomposition. Unfortunately, the decompositions are defined with respect to the facility locations. However, in reality they only use the facility locations in the Sub-rectangle steps. Moreover, the number of subrectangles is at most polynomial in the size of the original rectangle. Indeed, the number of allowable subrectangles is polynomial in m and n. Thus, we can perform dynamic programming to find the optimal solution. The structure of the lookup table is similar to the one used in [3]. We exploit our Structure Theorem and the analysis on the total number of allowable rectangles to obtain a smaller number of entries.

The table will consist of a set of entries for each allowable rectangle that contains at least one point. For each allowable rectangle, we will also enumerate the following

- the number of facilities in the rectangle,
- a distance for each portal to the closest facility in the rectangle and
- a distance for each portal to the closest facility outside the rectangle.

Actually, we will only approximate the distances to the nearest facility inside and out of the rectangle to a precision of s/m for a rectangle of sidelength s. Moreover, we do not consider distances of more than $10s$. Finally, the distance value at a portal only changes by a constant from the distance value at an adjacent portal. This, will allow us to bound the total number of table entries by $k2^{O(m)}$ for each allowable rectangle. See [3] for further details on the table construction.

We can compute the entries for a rectangle of sidelength s, by looking at either all the subrectangles from the Sub-rectangle process or by looking at all ways of cutting the rectangle into allowable smaller rectangles. This is bounded by $O(2^{O(m)})$ time per table entry. We bound the table size by noting that the total number of allowable rectangles is, by Lemma 7, at most $O(m^4 n \log n)$. Thus, we can bound the total number of entries in the table by $O(k2^{O(m)} n \log n)$.

We can improve this to $O(2^{O(m)} n \log k \log n)$ as follows. If an allowable rectangle contains fewer than $l < k$ nodes, we only need to keep $l2^O(m)$ entries for it, since at most l facilities can be placed inside it. Moreover, the number of allowable rectangles containing between l and $2l$ points is shown by Lemma 7 to be $O(m^4(n/l)\log n)$. We can now bound the total number of entries by

$$k2^{O(m)}O(m^4(n/k)\log n) + \sum_{l=2^i}^{l<k} 2^{O(m)}O(m^4(n/l)\log n)l = O(2^{O(m)} n \log k \log n).$$

We are now ready to state the main result of the paper.

Theorem 8. *Given an instance of the k-median problem in the 2-dimensional Euclidean space, and any fixed $m > 0$, there is a randomized algorithm that computes a $(1+1/m)$-approximation, in expectation, with worst case running time $O(2^{O(m)} n \log k \log n)$.*

Repeating the algorithm $O(\log n)$ times gives a $(1+O(1/m))$ approximation guarantee with probability $1-o(1)$. The algorithm can easily be extended to instances in the d-dimensional Euclidean space. We omit the details.

Theorem 9. *Given an instance of the k-median problem in the d-dimensional Euclidean space, and any fixed $m > 0$, there is a randomized algorithm that computes a $(1+1/m)$-approximation, in expectation, with worst case running time $O(2^{O(m^d)} n \log k \log n)$.*

For the uncapacitated facility location problem we do not need to keep track of the number of facilities open for subrentagles, hence we obtain an approximation scheme with running time $O(2^{O(m^d)} n \log n)$. We omit the details.

References

[1] S. Arora. Polynomial time approximation schemes for Euclidean TSP and other geometric problems. In *Proc. 37th FOCS*, 2–11, 1996.
[2] S. Arora. Nearly linear time approximation schemes for Euclidean TSP and other geometric problems. In *Proc. 38th FOCS*, 554–563, 1997.
[3] S. Arora, P. Raghavan, and S. Rao. Polynomial time approximation schemes for the euclidean *k*-medians problem. In *Proc. 30th STOC*, 106–113, 1998.
[4] Y. Bartal. Probabilistic approximation of metric spaces and its algorithmic applications. In *Proc. 37th FOCS*, 184–193, 1996.
[5] Y. Bartal. On approximating arbitrary metrics by tree metrics. In *Proc. 30th STOC*, 161–168, 1998.
[6] M. Charikar, C. Chekuri, A. Goel, and S. Guha. Rounding via trees: deterministic approximation algorithms for group Steiner tree and *k*-median. In *Proc. 39th FOCS*, 114–123, 1998.
[7] M. Charikar, S. Guha, E. Tardos, and D. Shmoys. A constant factor approximation algorithm for the k-median problem. To appear in Proc. STOC 99.
[8] F. A. Chudak. Improved approximation algorithms for uncapacitated facility location. In R. E. Bixby, E. A. Boyd, and R. Z. Ríos-Mercado, editors, *Proc. 6th IPCO*, volume 1412 of *LNCS*, 180–194. Springer-Verlag, Berlin, 1998.
[9] G. Cornuéjols, G. L. Nemhauser, and L. A. Wolsey. The uncapacitated facility location problem. In P. Mirchandani and R. Francis, editors, *Discrete Location Theory*. John Wiley and Sons, Inc., New York, 1990.
[10] T. Feder and D. H. Greene. Optimal algorithms for approximate clustering. In *Proc. 20th STOC*, 434–444, 1988.
[11] T. Gonzalez. Clustering to minimize the maximum intercluster distance. *Theoretical Computer Science*, 38:293–306, 1985.
[12] S. Guha and S. Khuller. Greedy strikes back: improved facility location algorithms. In *Proc. 9th SODA*, 649–657, 1998.
[13] D. S. Hochbaum. Heuristics for the fixed cost median problem. *Mathematical Programming*, 22:148–162, 1982.

[14] J. H. Lin and J. S. Vitter. Approximation algorithms for geometric median problems. *Information Processing Letters*, 44:245–249, 1992.
[15] J. H. Lin and J. S. Vitter. ε-approximations with minimum packing constraint violation. In *Proc. 24th STOC*, 771–782, 1992.
[16] G. Mitchell. Guillotine subdvisions approximate polygonal subdivisions: Part II - A simple PTAS for geometric *k*-MST, TSP and related problems. Preliminary manuscript, April 30 1996. To appear in *SIAM J. Computing*.
[17] S. Rao and W. D. Smith. Improved approximation schemes for geometrical graphs via *spanners* and *banyans*. In *Proc. STOC*, 540–550, 1998.
[18] D. B. Shmoys, E. Tardos, and K. I. Aardal. Approximation algorithms for facility location problems. In *Proc. STOC*, 265–274, 1997.

Sum Multi-coloring of Graphs

Amotz Bar-Noy[1], Magnús M. Halldórsson[2], Guy Kortsarz[3], Ravit Salman[4], and Hadas Shachnai[5]

[1] Dept. of EE, Tel-Aviv University, Tel-Aviv, Israel. amotz@eng.tau.ac.il.
[2] Science Institute, University of Iceland, Reykjavik, Iceland. mmh@hi.is.
[3] Dept. of CS, Open University, Tel Aviv, Israel. guyk@tavor.openu.ac.il.
[4] Dept. of Mathematics, Technion, Haifa, Israel. maravit@tx.technion.ac.il.
[5] Dept. of Computer Science, Technion, Haifa, Israel. hadas@cs.technion.ac.il.

Abstract. Scheduling dependent jobs on multiple machines is modeled by the graph *multi-coloring* problem. In this paper we consider the problem of minimizing the average completion time of all jobs. This is formalized as the *sum multi-coloring (SMC)* problem: Given a graph and the number of colors required by each vertex, find a multi-coloring which minimizes the sum of the largest colors assigned to the vertices. It reduces to the known *sum coloring* (SC) problem in the special case of unit execution times.

This paper reports a comprehensive study of the SMC problem, treating three models: with and without preemption allowed, as well as co-scheduling where tasks cannot start while others are running. We establish a linear relation between the approximability of the maximum independent set (IS) and SMC in all three models, via a link to the SC problem. Thus, for classes of graphs where IS is ρ-approximable, we obtain $O(\rho)$-approximations for preemptive and co-scheduling SMC, and $O(\rho \cdot \log n)$ for non-preemptive SMC. In addition, we give constant-approximation algorithms for SMC under different models, on a number of fundamental classes of graphs, including bipartite, line, bounded degree, and planar graphs.

1 Introduction

Any multi-processor system has certain resources, which can be made available to one job at a time. A fundamental problem in distributed computing is to efficiently schedule jobs that are competing on such resources. The scheduler has to satisfy the following two conditions: (*i*) *mutual exclusion*: no two conflicting jobs are executed simultaneously. (*ii*) *no starvation*: the request of any job to run is eventually granted. The problem is well-known in its abstracted form as the *dining/drinking philosophers* problem (see, e.g., [D68, L81]).

Scheduling dependent jobs on multiple machines is modeled as a graph *coloring* problem, when all jobs have the same (unit) execution times, and as graph *multi-coloring* for arbitrary execution times. The vertices of the graph represent the jobs and an edge in the graph between two vertices represents a dependency between the two corresponding jobs, that forbids scheduling these jobs at the same time. More formally, for a weighted undirected graph $G = (V, E)$ with n vertices, let the *length* of a vertex v be a positive integer denoted by $x(v)$, also called the *color requirement* of v. A multi-coloring of the vertices of G is a mapping into the power set of the positive integers, $\Psi : V \mapsto 2^N$. Each vertex v is assigned a set of $x(v)$ distinct numbers (colors), and adjacent vertices are assigned disjoint sets of colors.

The traditional optimization goal is to minimize the total number of colors assigned to G. In the setting of a job system, this is equivalent to finding a schedule, in which the time until *all* the jobs complete running is minimized. Another important goal is to minimize the *average* completion time of the jobs, or equivalently, to minimize the sum of the completion times. In the *sum multi-coloring* (SMC) problem, we look for a multi-coloring Ψ that minimizes $\sum_{v \in V} f_\Psi(v)$, where the completion time $f_\Psi(v)$ is the maximum color assigned to v by Ψ. We study the sum multi-coloring problem in three models:

- In the PREEMPTION model (p-SMC), each vertex may get any set of colors.
- In the NO-PREEMPTION (np-SMC) model, the set of colors assigned to each vertex has to be contiguous.
- In the CO-SCHEDULING model (co-SMC), the vertices are colored in rounds: in each round the scheduler completely colors an independent set in the graph.

The PREEMPTION model corresponds to the scheduling approach commonly used in modern operating systems [SG98]: jobs may be interrupted during their execution and resumed at later time. The NO-PREEMPTION model captures the execution model adopted in real-time systems, where scheduled jobs must run to completion. The CO-SCHEDULING approach is used in some distributed operating systems [T95]. In such systems, the scheduler identifies subsets of *cooperating processes*, that can benefit from running at the same time interval (e.g., since the processes in the set communicate frequently with each other); then, each subset is executed simultaneously on several processors, until *all* the processes in the subset complete.

The SMC problem has many other applications, including traffic intersection control, session scheduling in local-area networks (see, e.g., in [IL97]), compiler design and VLSI routing [NSS94].

Related work

When all the color requirements are equal to 1, the problem, in all three models, reduces to the previously studied *sum coloring* (SC) problem (A detailed survey appears in [BBH+98]). A good sum coloring would tend to color many vertices early; the following natural heuristic was proposed in [BBH+98].

MAXIS: Choose a maximum independent set in the graph, color all of its vertices with the next available color; iterate until all vertices are colored.

This procedure was shown to yield a 4-approximation, and additionally, in the case that IS can only be approximated within a factor of ρ, it gives a 4ρ-approximation.

On the other hand, SC also shares the inapproximability of IS, which immediately carries over to SMC: SC cannot be approximated in general within $n^{1-\varepsilon}$, for any $\varepsilon > 0$, unless $NP = ZPP$ [FK96, BBH+98]. It is also NP-hard to approximate within some factor $c > 1$ on bipartite graphs [BK98].

Our Results

This paper reports a comprehensive study of the SMC problem. We detail below our main results, and summarize further results, that will appear in the full version of this paper.

General graphs: Our central results are in establishing a linear relation between the approximability of the maximum independent set (IS) and SMC in all three models, via a link to the SC problem. For classes of graphs where IS is solvable, we obtain a 16-approximation to both pSMC and coSMC, as well as $O(\log \min(n, p))$-approximation to npSMC. For classes of graphs where IS is ρ-approximable, these ratios translate to $O(\rho)$-approximations for pSMC and coSMC, and $O(\rho \cdot \log n)$ for npSMC.

Important special classes of graphs: We also study special classes of graphs. For pSMC, we describe a 1.5-approximation algorithm for bipartite graphs. We generalize this to a $(k+1)/2-$ratio approximation algorithm for $k-$colorable graphs, when the coloring is given. Also, we present a $(\Delta+2)/3$ approximation algorithm for graphs of maximum degree Δ, and a 2-approximation for line graphs.

For npSMC, we describe a 2.796-approximation algorithm for bipartite graphs, and a $1.55k+1$-approximation algorithm when a k-coloring of the graph is given. These bounds are absolute, i.e. independent of the cost of the optimal solution, and also have the advantage of using few colors.

Comparison of multi-coloring models: We explore the relationship among the three models and give a construction, which indicates why the np-SMC model is "harder". Namely, while finding independent sets iteratively suffices to approximate both the p-SMC and the co-SMC problems, any such solution must be $\Omega(\log p)$ off for the np-SMC problem.

Performance bounds for the MAXIS algorithm: An immediate application of the MAXIS algorithm for the SC problem appears in the $O(1)-$approximation algorithm for the p-SMC problem. Furthermore, most of our algorithms reduce to MAXIS for the SC problem, if the color requirements are uniform. It is therefore natural to find the exact performance of MAXIS for the SC problem. In [BBH+98] it was shown that this algorithm yields a 4-approximation for the sum coloring problem. We give here a construction which shows, that MAXIS cannot achieve an approximation factor better than 4.

As for the SMC problem, we note that it is usually preferable to color many vertices early, even if it means that more colors will be needed in the end. Thus, the MAXIS algorithm is a natural candidate heuristic also for the SMC problems. However, we show that MAXIS is only an $\Omega(\sqrt{p})$ approximation algorithm for p-SMC, where p is the largest color requirement in the graph. For the np-SMC problem, its performance cannot even be bounded in terms of p.

Further results: Our results for the SMC problem can be extended to apply also to (i) the *weighted SMC* problem, where each vertex v is associated with a weight $w(v)$ and the goal is to minimize $\sum_{v \in V} w(v) \cdot f_\Psi(v)$, and (ii) *on-line scheduling* of dependent jobs. A summary of our results for these problems will be given in the full version of the paper.

Organization of the Extended Abstract:

Due to space limitations, we include in this abstract only the main results of our study. Detailed proofs are in the full version of the paper [BHK+98]. In Section 2 we introduce some notation and definitions. Section 3 presents approximation algorithms for the p-SMC problem. Section 4 describes the results for the np-SMC problem, and Section 5 discusses the co-SMC problem. Finally, in Section 6 we briefly present our lower bound of 4, for the MAXIS algorithm.

2 Definitions and notation

For a given undirected graph $G = (V,E)$ with n vertices, and the mapping $x : V \rightarrow N$, we denote by $S(G) = \sum_v x(v)$ the sum of the color requirements of the vertices in G. We denote by p the maximum color requirement in G, that is $p = \max_{v \in V} x(v)$. An *independent set* in G is a subset I of V such that any two vertices in I are non-adjacent. Given a *multi-coloring* Ψ of G, denote by C_i the independent set that consists of vertices with $i \in \Psi(v)$, by $c_1^{\Psi}(v), \ldots, c_{x(v)}^{\Psi}(v)$ the collection of $x(v)$ colors assigned to v, and by $f_{\Psi}(v) = c_{x(v)}^{\Psi}(v)$ the largest color assigned to v. The *multi-color sum* of G with respect to Ψ is $\mathrm{SMC}(G,\Psi) = \sum_{v \in V} f_{\Psi}(v)$.

A multi-coloring Ψ is *contiguous* (non-preemptive), if for any $v \in V$, the colors assigned to v satisfy $c_{i+1}^{\Psi}(v) = c_i^{\Psi}(v) + 1$ for $1 \leq i < x(v)$. In the context of scheduling, this means that all the jobs are processed without interruption. A multi-coloring Ψ solves the co-scheduling problem, if the set of vertices can be partitioned into k disjoint independent sets $V = I_1 \cup \cdots \cup I_k$ with the following two properties: (i) $c_1^{\Psi}(v) = c_1^{\Psi}(v')$ for any $v, v' \in I_j$, for $1 \leq j \leq k$. (ii) $c_{x(v)}^{\Psi}(v) < c_1^{\Psi}(v')$ for all $v \in I_j$ and $v' \in I_{j+1}$ for $1 \leq j < k$. In the context of scheduling, this means scheduling to completion all the jobs corresponding to I_j, and only then starting to process the jobs in I_{j+1}, $\forall\ 1 \leq j < k$.

The *minimum multi-color sum* of a graph G, denoted by $\mathrm{pSMC}(G)$, is the minimum $\mathrm{SMC}(G,\Psi)$ over all multi-colorings Ψ. We denote the minimum contiguous multi-color sum of G by $\mathrm{npSMC}(G)$. The minimum multi-color sum of G for the co-scheduling problem is denoted by $\mathrm{coSMC}(G)$. Indeed, for any graph G:

$$S(G) \leq \mathrm{pSMC}(G) \leq \mathrm{npSMC}(G) \leq \mathrm{coSMC}(G) .$$

3 The preemptive sum multi-coloring problem

In the preemptive version of the sum multi-coloring problem, a vertex may get *any* set of $x(v)$ colors. Our first result is an $O(1)$-approximation for the p-SMC problem on general graphs. This is an extension of the result for the sum-coloring problem in the sense that it establishes a connection between sum multi-coloring and maximum weighted independent sets. Then we address several families of graphs: bipartite graphs, bounded degree graphs, and line graphs.

General graphs

En route to approximating pSMC, we consider another measure of a multi-coloring, the sum of the *average* color value assigned to a vertex. We approximate it by reducing it to the *sum coloring* problem of a derived *weighted* graph. We then transform a multi-coloring with small sum of averages to one with small multi-color sum.

In the weighted sum coloring problem, each vertex v has a weight $w(v)$, and we need to assign to each vertex v a (single) color $\Psi(v)$, so as to minimize $\sum_{v \in V} w(v)\Psi(v)$. The weighted MAXIS algorithm, selecting an independent set with maximum *weight* in each round, gives a $4-$approximation [BBH+98].

We denote by $\mathrm{AV}_\Psi(v)$ the average color assigned to v by Ψ, namely, $\mathrm{AV}_\Psi(v) = (\sum_{i\in\Psi(v)} i)/x(v)$. Let $\mathrm{SA}_\Psi(G) = \sum_v \mathrm{AV}_\Psi(v)$ denote the sum of averages of Ψ, and let $\mathrm{SA}^*(G)$ be the minimum possible average sum. Clearly, $SA^*(G) \le \mathrm{pSMC}(G)$.

Given a multi-coloring instance (G,x), we construct a weighted graph (G',w) as follows. The graph has $x(v)$ copies of each vertex v connected into a clique, with each copy of v adjacent to all copies of neighbors of v in G. The weight $w(v_i)$ of each copy v_i of v will be $1/x(v)$.

There is a one-one correspondence between multi-colorings Ψ of (G,x) and colorings of G', as the $x(v)$ copies of a vertex v in G all receive different colors. Let Ψ also refer to the corresponding coloring of G'. Observe that

$$\mathrm{SC}_\Psi(G',w) = \sum_{v\in V(G)} \sum_{i=1}^{x(v)} w(v_i)\cdot c_i^\Psi(v) = \sum_{v\in V(G)} \frac{\sum_{i=1}^{x(v)} c_i^\Psi(v)}{x(v)} = SA_\Psi(G,x).$$

Lemma 1. *Define the weight of a vertex to be* $1/x(v)$. *Suppose that a multi-coloring* Ψ *of a graph G has the property that each color i is an independent set of weight within a* ρ *factor from optimal on the subgraph induced by yet-to-be fully colored vertices. Then,* $SA_\Psi(G) \le 4\rho\cdot\mathrm{pSMC}(G)$. *Further, if* Ψ *is contiguous, then* $\mathrm{SMC}(G,\Psi) \le 4\rho\cdot\mathrm{pSMC}(G)$.

Proof. A coloring Ψ that satisfies the hypothesis, also implicitly satisfies the property of MAXIS on (G',w). Hence, by the result of [BBH+98],

$$\mathrm{SA}_\Psi(G,x) = \mathrm{SC}_\Psi(G',w) \le 4\rho\cdot\mathrm{SC}^*(G',w) = 4\rho\cdot\mathrm{SA}^*(G,x) \le 4\rho\cdot\mathrm{pSMC}(G).$$

Note, that for any coloring, the final color of a vertex differs from the average color by at least exactly half the color requirement of the vertex, i.e., $f_\Psi(v) \ge AV_\Psi(v) + (x(v)-1)/2$, with equality holding when Ψ is contiguous. Thus, $\mathrm{SMC}(G,\Psi) = \mathrm{SA}_\Psi(G) + \frac{S(G)-n}{2}$; also $\mathrm{pSMC}(G) \ge \mathrm{SA}^*(G) + \frac{S(G)-n}{2}$. We conclude that if Ψ is contiguous, then

$$\mathrm{SMC}(G,\Psi) = SA_\Psi(G) + \frac{S(G)-n}{2} \le 4\rho\cdot SA^*(G) + \frac{S(G)-n}{2} \le 4\rho\cdot\mathrm{pSMC}(G). \square$$

Theorem 2. pSMC *can be approximated within a factor of* 16ρ.

Proof. Given an instance (G,x), obtain a multi-coloring Ψ by applying weighted MAXIS on the derived graph G'. Then, form the multi-coloring Ψ' that "doubles" each independent set of Ψ:

$$\Psi'(v) = \{2i, 2i'+1 : i = c_t^\Psi(v), t \le \lceil x(v)/2\rceil \text{ and } i' = c_{t'}^\Psi(v), t' \le \lfloor x(v)/2\rfloor\}.$$

Observe that Ψ' assigns each vertex v $x(v)$ colors.

Let $m_v = c^\Psi_{\lceil x(v)/2\rceil}$ be the median color assigned to v by Ψ. The largest color used by Ψ' is $f_\Psi(v) = c^\Psi_{\lceil x(v)/2\rceil} + c^\Psi_{\lfloor x(v)/2\rfloor} \le 2m_v$. Also $m_v \le 2\cdot AV_\Psi(v)$, since less than half the elements of a set of natural numbers can be larger than twice its average. Thus $f_{\Psi'}(v) \le 4\cdot AV_\Psi(v)$ and $\mathrm{SMC}(G,\Psi') \le 4\cdot SA_\Psi(G)$. Thus, by Lemma 1, Ψ' is a 16ρ–approximation of $\mathrm{pSMC}(G)$. □

Bipartite Graphs, and k-colorable graphs

Consider graphs that can be colored with k colors, i.e., the set of vertices V can be partitioned into k disjoint independent sets $V = V_1 \cup \cdots \cup V_k$. Consider the following **Round-Robin** algorithm: For $1 \leq i \leq k$ and $h \geq 0$, at round $t = k \cdot h + i$ give the color t to all the vertices of V_i that still need a color. It is not hard to see that $f(v) \leq k \cdot x(v)$ for all $v \in V$. Hence, the Round-Robin algorithm is a k-approximation algorithm. In this subsection we give a non-trivial algorithm for $k-$colorable graphs, with the coloring given, whose approximation factor is at most $(k+1)/2$. This gives a $2.5-$ratio approximation for planar graphs, a $2-$ratio approximation for outerplanar and series-parallel graphs, and a $k/2+1$ approximation for graphs with tree-width bounded by k. In particular, for bipartite graphs the approximation ratio is bounded by $1.5 - 1/(2n)$. For simplicity, the result is described for bipartite graphs only. The result for general k is derived similarly.

We need the following definitions and notations. Let G be a bipartite graph $G(V_1, V_2, E)$ with n vertices, such that edges connect vertices in V_1 with vertices in V_2. We denote by $\alpha(G)$ the size of a maximum independent set in G. We use the term *processing* an independent set $W \subseteq V$ to mean assigning the next available color to all the vertices of W. Suppose that the first i colors assigned by a multi-coloring Ψ are distributed among the vertices. The *reduced graph* of G is the graph for which the $x(v)$ values are decreased accordingly with the colors assigned so far, deleting vertices that were fully colored. Finally, let $\gamma(n) = (2n^2)/(3n-1)$.

Informally, the algorithm distinguishes the following two cases: If the size of the maximum independent set in the current reduced graph is "large," the algorithm chooses to process a maximum independent set. Otherwise, if the maximum independent set is "small," the algorithm works in a fashion similar to Round-Robin. Once a vertex (or a collection of vertices) is (are) assigned their required number of colors, the algorithm re-evaluates the situation.

Algorithm 1 BC

While *some vertices remain* **do**

1. *Let $\tilde{G}$ be the current reduced graph, and let $\tilde{n}$ be its number of vertices.*
2. *If $\alpha(\tilde{G}) \leq \gamma(\tilde{n})$* **do**
   (a) *Let m be the minimum $x(v)$ in $\tilde{G}$. Assume without loss of generality that V_1 contains at least as many vertices v with $x(v) = m$ as V_2.*
   (b) *Give the next m colors to the remaining vertices in V_1, and the following m colors those in V_2.*
3. *else $(\alpha(\tilde{G}) > \gamma(\tilde{n}))$*
   (a) *Choose a maximum independent set $I \subset V$ of size $\alpha(\tilde{G})$. Let m be the minimum $x(v)$ value in I.*
   (b) *Give the next m colors to all the vertices in I.*

The algorithm runs in polynomial time, since finding a maximum independent set in a bipartite graph can be performed in polynomial time using flow techniques (cf., [GJ79]) and since in each iteration at least one vertex is deleted.

Theorem 3. BC *approximates* pSMC *on bipartite graphs within a factor of* 1.5.

Bounded degree graphs and line graphs

A natural algorithm for the multi-coloring problem is the Greedy (First-Fit) algorithm. It processes the vertices in an arbitrary order, assigning to each vertex $v \in V$ the set of the smallest $x(v)$ colors, with which none of its preceding neighbors have been colored. This method has the advantage of being *on-line*, processing requests as they arrive. Let Δ denote the maximum degree of G.

We can show that Greedy is exactly $\Delta + 1$-approximate. Instead, we consider the modified version, Sorted Greedy (SG), which orders the vertices in a non-decreasing order of color requirements, before applying Greedy. This slight change improves the approximation ratio by a factor of nearly 3. The proof is omitted.

Theorem 4. SG *provides a* $(\Delta+2)/3$*-approximation to* pSMC(G)*, and that is tight.*

SG also has good approximation ratio for line graphs and intersection graphs of k-uniform hypergraphs.

Theorem 5. SG *provides a* $2-4/(\Delta+4)$*-approximation to* pSMC(G) *on line-graphs. More generally, it provides a* $k(1-2(k-1)/(\Delta+2k))$ *approximation ratio on intersection graphs of k-uniform hyper-graphs.*

Proof. Given a line graph G, form a graph H that is a disjoint collection $\{C_1, C_2, \ldots, C_{|V(H)|}\}$ of the maximal cliques in G. Add a singleton clique for each vertex that appears only once.

The minimum contiguous multi-coloring sum of H is given by ordering the vertices of each C_i in a non-decreasing order of color requirements, for

$$\text{pSMC}(H) = \mathcal{S}(H) + \sum_{(u,v)\in E(H)} \min(x(u), x(v)).$$

Observe that since each vertex in G appears at most twice in H, $\mathcal{S}(H) = 2\mathcal{S}(G)$, and any multi-coloring of G corresponds to a multi-coloring of H of at most double the weight, pSMC$(H) \leq 2 \cdot$ pSMC(G). Further, there is a one-one correspondence between the edges of G and H. Thus, we have

$$\text{pSMC}(G) \geq \mathcal{S}(G) + \frac{1}{2} \sum_{(u,v)\in E(G)} \min(x(u), x(v)).$$

Letting $d = 2(\sum_{(u,v)\in E} \min(x(u), x(v)))/\mathcal{S}(G)$, we bound the performance ratio by

$$\frac{\text{SMC}(G, \text{SG})}{\text{pSMC}(G)} \leq f(d) \equiv \frac{1+d/2}{1+d/4}.$$

Since $f(d)$ is monotone increasing, and $d \leq \Delta$, we have $f(d) \leq f(\Delta) = 2 - 4/(\Delta+4)$. This matches the bound proved for sum coloring for regular edge graphs. □

4 The non-preemptive sum multi-coloring problem

In the non-preemptive version of the sum multi-coloring problem, the set of colors assigned to any vertex must be contiguous. This makes it harder to obtain algorithms with good approximation ratios. We give here an algorithm for general graphs with a logarithmic factor (times the ratio of the independent set algorithm used), as well as constant factor algorithms for bipartite and k-colorable graphs.

General graphs

Let r be the number of different color-requirements (lengths) in the graph. The following is a $\rho \cdot \min\{O(\log p), O(\log n), r\}$-approximation algorithm for the npSMC problem.

Algorithm 2 *SameLengthIS*

Let $Small \leftarrow \{v \mid x(v) \leq p/n^2\}$.
Color first the vertices of $Small$, arbitrarily but fully and non-preemptively.
Let V' denote $V \setminus Small$.
$G' \leftarrow (V', E', x', w)$, where
$x'(v) \leftarrow 2^{\lceil \log x(v) \rceil}$, for each $v \in V'$,
$w(v) \leftarrow 1/x'(v)$, the weight of each $v \in V'$, and
$E' \leftarrow E \cup \{(u,v) : x'(u) \neq x'(v)\}$
Apply weighted MAXIS to G', coloring the vertices fully in each step.

Theorem 6. SameLengthIS *approximates* npSMC *within a factor of* $\min\{8\rho \log p, 16\rho \log n + 1.5\}$, *assuming IS can be approximated within* ρ.

Proof. The color sum of the vertices in *Small* is at most $p/n^2(1+2+\ldots+n) < p/2$. Also, they use at most p/n colors in total, and yield an added cost of at most p to the coloring of V'. Thus, coloring *Small* contributes at most an additive 1.5 to the performance ratio, when $p \geq n^2$.

Observe that G' is a super-graph of G, that the algorithm produces a valid multi-coloring, and that it is contiguous since jobs are executed to completion. Since the weights of G' are upper bounds on the weights of G, the cost of the coloring on G' is an upper bound of its cost on G.

Observe that each color class is an independent set of weight at least ρ factor from maximum among independent sets in the current remaining graph. Thus, by Lemma 1, $\mathrm{SMC}(G, \mathrm{SameLengthIS}) \leq 4\rho \cdot \mathrm{pSMC}(G')$.

Any independent set of G is partitioned into at most $z = \log(\min\{p, n^2\})$ independent sets in G', one for each length class. In addition, the rounding up of the lengths at most doubles the optimum cost of the instance. Thus, $\mathrm{pSMC}(G') \leq 2z \cdot \mathrm{pSMC}(G)$. □

Also, the factor $\log p$ can be replaced by $\log q$, where q is the ratio between the largest to the smallest color requirement.

Notice that our analysis rates our algorithm (that in fact produces a co-schedule, as detailed in the next section) in terms of a stronger adversary, that finds the best possible *preemptive* schedule.

Bipartite graphs and k-colorable graphs

Let G be a graph whose k-coloring $C_1, C_2, \ldots C_k$ is given. Let a be a constant to be optimized, and let $d = a^k$. Let $C_i[x,y]$ denote the set of vertices in C_i of lengths in the interval $[x,y]$. Our algorithm is as follows:

Steps(G,a)
Let X be a random number uniformly chosen from $[0,1]$.
$d \leftarrow a^k$
$Y \leftarrow d^X$.
for $i \leftarrow 0$ to $\log_d p$ do
for $j \leftarrow 1$ to k do
$A_{ij} \leftarrow d^{i-1} a^j Y$
Color vertices of $C_j[A_{ij}/d, A_{ij}]$ using the next $\lfloor A_{ij} \rfloor$ colors

Steps can be derandomized, by examining a set of evenly spaced candidates for the random number X. The additive error term will be inversely proportional to the number of schedules evaluated. This yields the following results.

Theorem 7. *If G is bipartite, then we can approximate* npSMC(G) *within a* 2.796 *factor. If a k-coloring of a graph G is given, then we can approximate* npSMC(G) *within a* $1.55k+1$ *factor.*

We note that the ratios obtained are *absolute* in terms of $S(G)$, instead of being relative to the optimal solution. We strongly suspect that these are the best possible absolute ratios.

We illustrate our approach by an example. Given a bipartite graph G, we color the vertices into sets C_1 and C_2. Let $C_i[a,b]$ denote the set of vertices in C_i whose length is between a and b, inclusive. The idea is as follows:

Color all vertices in $C_1[1,1]$ with the first color, followed by $C_2[1,2]$ with the next 2 colors, $C_1[2,4]$ with the next 4, $C_2[3,8]$ with the next 8, etc.
In general, color $C_1[2^{2i-2}+1, 2^{2i}]$, followed by $C_2[2^{2i-1}+1, 2^{2i+1}]$, for $i = 0, 1, \ldots$.

For a given vertex v in say C_1, the worst case occurs when $x(v) = 2^{2i} + 1$, for some i. Then, v is finished in step $x(v) + \sum_{j=0}^{i} (2^{2j} + 2^{2j+1}) = x(v) + 2^{2i+2} - 1 = 5x(v) - 5$. The same can be argued for any $u \in C_2$. Thus, we have bounded the *worst case* completion time of *any* vertex by a factor of 5, giving a guarantee on the schedule completion to the system and to each vertex.

5 The co-scheduling problem

Recall the definition of the co-scheduling problem. In this version, the multi-coloring first chooses an independent set I_1. It assigns colors to all the vertices in I_1 using the first x_1 colors, where $x_1 = \max_{v \in I_1} x(v)$. Then, the multi-coloring cannot use the first x_1 colors for other vertices and therefore uses colors $x_1 + 1, x_1 + 2$ etc. to color the rest of the vertices. The goal is to minimize the sum of all $f_\Psi(v)$.

We show that Algorithm SameLengthIS of the previous section approximates the co-SMC problem within a constant factor, comparing favorably with the logarithmic factor for np-SMC. We then construct a graph H for which coSMC$(H) = \Omega(\log p) \cdot$ npSMC(H), indicating that this discrepancy in the performance is inherent.

General graphs

Theorem 8. SameLengthIS *approximates* coSMC *within a factor of* 16ρ, *assuming IS can be approximated within a factor of* ρ.

Proof. Let Ψ be the coloring produced by SameLengthIS on G'. As the algorithm colors in rounds, it produces a valid co-schedule. Each color class is an independent set of weight within ρ factor of maximum, among independent sets in the current remaining graph. Thus, by Lemma 1, $\mathrm{SMC}(G,\Psi) \leq 4\rho \cdot \mathrm{coSMC}(G')$.

We now relate the optimal schedules of G' and G. Let X be the length of the longest job in a given round of an optimal schedule of G. Rounding the length to a power of two results in a length $X' = 2^{\lceil \lg X \rceil}$. Consider the schedule that breaks the round into a sequence of rounds of length $1,2,4,\dots,X'$, for a total length of $2X'-1 < 4X$. This is a valid schedule of G', where each vertex is delayed by at most a factor of 4. Hence, $\mathrm{coSMC}(G') \leq 4 \cdot \mathrm{coSMC}(G)$, and the theorem follows. □

A construction separating co-SMC and np-SMC

We have given $O(1)$-approximations to the p-SMC problem and to the coSMC problem, while only a $\min\{O(\log p), O(\log n)\}$-approximation for the npSMC problem. It would be curious to know the precise relationships among these three models. We give a partial answer by constructing an instance H for which $\mathrm{coSMC}(H) = \Omega(\log p) \cdot \mathrm{npSMC}(H)$.

For $i = 0,\dots,\log p$, $j = 1,\dots,2^{(\log p)-i+1}$, and $k = 1,\dots,2^i$, the graph H has vertex set $V = \{v_{i,j,k}\}$ where the color requirements of $v_{i,j,k}$ is 2^i, and edge set $E = \{(v_{i,j,k}, v_{i',j',k'}) : i = i' \text{ and } k \neq k'\}$. In words, the graph has p vertices of each color requirement $\ell = 2^i$, arranged in completely connected independent sets of size p/ℓ with vertices of different requirements non-adjacent.

Consider the straightforward non-preemptive coloring where the different color requirements are processed independently and concurrently. Then, the makespan of the schedule is p, and since the graph has $p \lg p$ vertices, the multi-coloring sum is $O(p^2 \log p)$.

On the other hand, any independent set contains at most 2^i vertices from each color requirement group i. Hence, in any independent set of length ℓ in a co-schedule, there are at most 2ℓ vertices. Thus, at most $2t$ vertices are completed by step t, for each $t = 1,2,\dots,p\log p/2$. In particular, at most half of the vertices are completed by step $p \log p/4$. Thus, the coloring sum of the remaining vertices is $\Omega(p^2 \log^2 p)$. Hence, a $\Omega(\log p)$ separation between these models. As $n = p \cdot \log p$, this result also implies a $\Omega(\log n)$−separation.

6 The MAXIS heuristic

We describe in this section a construction showing that the performance ratio of MAXIS for the sum coloring problem is exactly 4, up to low order terms. Given the central use of (weighted) sum coloring in our algorithms for multi-coloring general graphs, this yields a good complementary bound for our heuristics.

We represent a coloring of a graph G by k colors as a tuple of length k: $\langle c_1,\dots,c_k\rangle$. The size of the set C_i of the vertices colored by i is c_i. By definition, for a given pattern $P = \langle c_1,\dots,c_k\rangle$ the sum coloring is $SC(P) = \sum_{i=1}^{k} i \cdot c_i$.

Given a pattern $P = \langle c_1, \ldots, c_k \rangle$, we describe a *chopping* procedure which constructs a graph G_P with $n = \sum_{i=1}^{k} c_i$ vertices. In G_P, there are k independent sets $C_1, \ldots, C_k$ that cover all the vertices of the graph, such that $|C_i| = c_i$. We now place the vertices of the graph in a matrix of size $c_1 \times k$. The ith column contains c_i ones at the bottom and $c_1 - c_i$ zeros at the top. Each vertex is now associated with a one entry in the matrix.

The chopping procedure first constructs an independent set I_1 of size c_1. It collects the vertices from the matrix line after line, from the top line to the bottom line. In each line, it collects the vertices from right to left. Each one entry that is collected becomes a zero entry. Then it adds edges from I_1 to all the other vertices, as long as these edges do not connect two vertices from the same column. In a same manner, the procedure constructs I_2. The size of I_2 is the number of ones in the first column after the first step. At the beginning of the ith step, the procedure has already constructed $I_1, \ldots, I_{i-1}$, defined all the edges incident to these vertices, and replaced all the one entries associated with the vertices of the sets $I_1, \ldots, I_{i-1}$ by zeros. During the ith step, the chopping procedure constructs in a similar manner the independent set I_i, the size of which is the number of ones in the first column of the matrix at the beginning of the step. Again, each one entry that is collected becomes a zero entry. Then the procedure connects the vertices of I_i with the remaining vertices in the matrix, as long as these edges do not connect two vertices from the same row. The procedure terminates after h steps when the matrix contains only zeros, leaving the graph with another coloring of $G, I_1, \ldots, I_h$, which corresponds to a possible coloring by MAXIS.

To achieve the desired bound, we build a pattern as follows. We let the first two entries be equal, and after two steps of chopping they should be equal to the third entry. After two additional chopping steps we would like to have the first four entries to be equal, and so on. Small examples of such patterns are $\langle 4,4,1 \rangle$ for $k = 3$ and $\langle 36,36,9,4 \rangle$ for $k = 4$. For $k = 4$, MAXIS produces the pattern $\langle 36,18,9,6,4,3,3,2,1,1,1,1 \rangle$ and then the ratio is $240/151 > 1.589$. More formally, for $x > 1$, consider the pattern $LB = \left\langle x, x, \frac{x}{4}, \frac{x}{9}, \ldots, \frac{x}{(k-1)^2}, \frac{x}{k^2} \right\rangle$. We choose x such that LB contains only integral numbers (e.g., $x = (k!)^2$). In the chopping procedure, once we arrive at a pattern of equal size, we just take these $k+1$ columns as the next $k+1$ entries. Thus, it yields the pattern $A(LB) = \left\langle x, \frac{x}{2}, \frac{x}{4}, \frac{x}{6}, \frac{x}{9}, \frac{x}{12}, \ldots, \frac{x}{(k-1)^2}, \frac{x}{(k-1)k}, \frac{x}{k^2}, \ldots, \frac{x}{k^2} \right\rangle$ where $\frac{x}{k^2}$ appears $k+1$ times.

It can be shown, that $SC(LB) < (H_k + 2.65)x$ and $SC(A(LB)) > (4H_k - 0.65)x$, where $H_k = \sum_{i=1}^{k} 1/i$. The approximation ratio of MAXIS is $4 - O(1/\ln k)$.

References

[BBH+98] A. Bar-Noy, M. Bellare, M. M. Halldórsson, H. Shachnai, and T. Tamir. On chromatic sums and distributed resource allocation. *Information and Computation*, 140:183–202, 1998.

[BHK+98] A. Bar-Noy, M. M. Halldórsson, G. Kortsarz, R. Salman and H. Shachnai. Sum Multi-Coloring of Graphs. TR #CS0949, Technion IIT, CS Dept., Nov. 1998. `http://www.raunvis.hi.is/~mmh/Papers/smc.html`

[BK98] A. Bar-Noy and G. Kortsarz. The minimum color-sum of bipartite graphs. *J. Algorithms*, 28:339–365, 1998.

[D68] E. W. Dijkstra. Cooperating Sequential Processes, in *Programming Languages*, ed. by F. Genuys, Academic Press, 43–112, 1968.

[FK96] U. Feige and J. Kilian. Zero Knowledge and the Chromatic number. *J. Computer Sys. Sci.*, 57(2):187-199, Oct. 1998.

[GJ79] M. R. Garey and D. S. Johnson. Computers and Intractability: A Guide to the Theory of NP-Completeness. W. H. Freeman, 1979.

[IL97] S. Irani and V. Leung, Probabilistic Analysis for Scheduling with Conflicts, *SODA '97*, 286–295.

[L81] N. Lynch. Upper Bounds for Static Resource Allocation in a Distributed System. *J. of Computer and System Sciences*, 23:254–278, 1981.

[NSS94] S. Nicoloso, M. Sarrafzadeh, and X. Song. On the Sum Coloring Problem on Interval Graphs. *Algorithmica*, 23:109–126, 1999.

[SG98] A. Silberschatz and P. Galvin. Operating System Concepts. Addison-Wesley, 5th Edition, 1998.

[T95] A. S. Tannenbaum, *Distributed Operating Systems.* Prentice-Hall, 1995.

Efficient Approximation Algorithms for the Achromatic Number

Piotr Krysta[1*] and Krzysztof Loryś[2**]

[1] Max-Planck-Institut für Informatik, Im Stadtwald, D-66123 Saarbrücken, Germany.
krysta@mpi-sb.mpg.de

[2] Institute of Computer Science, University of Wrocław, Przesmyckiego 20, PL-51151 Wrocław, Poland. lorys@tcs.uni.wroc.pl

Abstract. The achromatic number problem is as follows: given a graph $G = (V, E)$, find the greatest number of colors in a coloring of the vertices of G such that adjacent vertices get distinct colors and for every pair of colors some vertex of the first color and some vertex of the second color are adjacent. This problem is NP-complete even for trees. We present improved polynomial time approximation algorithms for the problem on graphs with large girth and for trees, and linear time approximation algorithms for trees with bounded maximum degree. We also improve the lower bound of Farber *et al.* for the achromatic number of trees with maximum degree bounded by three.

1 Introduction

The *achromatic number* of a graph is the maximum size k of a vertex coloring of the graph, where every pair of the k colors is assigned to some two adjacent vertices and adjacent vertices are colored with different colors. The *achromatic number problem* is to compute the achromatic number of a given graph. This concept was first introduced in 1967 by Harary *et al.* [8] in a context of graph *homomorphism* (sec [7]). A related well known problem is the *chromatic* number problem, that is computing the minimum size of a vertex coloring of a graph, where adjacent vertices are colored with different colors. These two problems are different in nature. For example, removing edges from a graph cannot increase its chromatic number but it can increase the achromatic number [10].

Previous Literature. Yannakakis and Gavril [11] showed that the achromatic number problem is NP-complete. It is NP-complete also for bipartite graphs as proved by Farber *et al.* [7]. Furthermore, Bodlaender [1] proved that the problem remains NP-complete when restricted to connected graphs that are simultaneously cographs and interval graphs. Cairnie and Edwards [3] show that the problem is NP-complete even for trees. (However, the chromatic number of a tree can be computed in polynomial time. For a tree that is not a single node, the chromatic number is always two.)

Let l be a fixed positive integer. There are known exact polynomial time algorithms for two very restricted classes of trees: for trees with not more than l leaves, and for

* The author is supported by Deutsche Forschungsgemeinschaft (DFG) as a member of the Graduiertenkolleg Informatik, Universität des Saarlandes, Saarbrücken.

** Partially supported by Komitet Badań Naukowych, grant 8 T11C 032 15.

trees with $\binom{l}{2}$ edges and with at least $\binom{l-1}{2}+1$ leaves [7]. Cairnie and Edwards [4] have proved that the achromatic number problem for trees with constant maximum degree can also be solved in polynomial time. Some phases of their algorithm are based on enumeration, and the algorithm has running time $\Omega(m^{126})$, where m is the number of edges of the tree.

An *α-approximation* algorithm [9] for a maximization problem Π is a polynomial time algorithm, that always computes a solution to the problem Π, whose value is at least a factor $\frac{1}{\alpha}$ of the optimum. We call α the *approximation ratio*. For a formal definition of *asymptotic* approximation ratio see [9], but it is clarified below what we mean by asymptotic.

Chaudhary and Vishwanathan [5] presented a 7-approximation algorithm for the achromatic number problem on trees. They also give an $O(\sqrt{n})$-approximation algorithm for graphs with *girth* (i.e. length of the shortest cycle) at least six, where n is the number of vertices of the graph.

Our Results. Our first result is a 2-approximation algorithm for trees, which improves the 7-approximation algorithm in [5]. Our algorithm is based on a different idea from that of [5]. The algorithm is presented in Section 5.

Let $d(n)$ be some (possibly increasing) function and T be a tree with n vertices. Assuming that the maximum degree of T is bounded by $d(n)$, we developed an alternative, to that of Section 5, combinatorial approach to the problem. This let us reduce the approximation ratio of 2 to 1.582. Additional result is a 1.155-approximation algorithm for binary trees, i.e. with maximum degree at most 3. The ratios 1.582 and 1.155 are proved to hold asymptotically as the achromatic number grows. For example, the first algorithm produces an achromatic coloring with at least $\frac{1}{1.582}\Psi(T) - O(d(n))$ colors, where $\Psi(T)$ is the achromatic number of T. We show that the algorithms for bounded degree trees can be implemented in linear time in the unit cost RAM model. Although, our algorithms for bounded degree trees are approximate and the algorithm of [4] for constant degree trees is an exact one, our algorithms have linear running time (running time of the algorithm in [4] is $\Omega(m^{126})$) and they also work on trees with larger maximum degree (e.g. $\log(n)$ or even $(n-1)^{1/4}$, where n is number of vertices of the tree). We also improve a result of Farber *et al.* [7] giving a better lower bound for the achromatic number of trees with maximum degree bounded by three. These results appear in Section 4.

Our next result presented in Section 6 is an $O(n^{3/8})$-approximation algorithm for graphs with girth at least six, which improves the $O(n^{1/2})$-approximation in [5]. This algorithm is a consequence of our 2-approximation algorithm for trees.

Let $\Psi = \Psi(G)$ be the achromatic number of a graph G. Chaudhary and Vishwanathan [5] show an $O(\Psi(G)^{2/3})$-approximation algorithm for graphs G with girth ≥ 6, and also state a theorem that for graphs with girth ≥ 7, there is an $O(\sqrt{\Psi})$-approximation algorithm for the problem. They also conjectured that for any graph G, there is a $\sqrt{\Psi(G)}$-approximation algorithm for the achromatic number problem. We prove their conjecture for graphs of girth ≥ 6, showing an $O(\sqrt{\Psi})$-approximation algorithm for such graphs. This result is also described in Section 6.

Our approximation algorithms are based on a tree partitioning technique. With this technique we prove some combinatorial results for trees, that can be of independent interest.

2 Preliminaries

In this paper we consider only undirected finite graphs. For a graph G, let $E(G)$ and $V(G)$ denote the set of edges and the set of vertices of G, respectively. Given $v_1, v_2 \in V(G)$, the *distance* between v_1 and v_2 is the number of edges in the shortest path between v_1 and v_2. Let $e_1, e_2 \in E(G)$. The distance between e_1 and e_2 is the minimum of the distances between an end-vertex of e_1 and an end-vertex of e_2. We say that e_1, e_2 are *adjacent* if the distance between e_1 and e_2 is 0. Moreover, we say that a vertex $v \in V(G)$ and an edge $e \in E(G)$ are *adjacent* if v is an end-vertex of e.

A *coloring* of a graph $G = (V, E)$ with k colors is a partition of the vertex set V into k disjoint sets called *color classes*, such that each color class is an independent set. A coloring is *complete* if for every pair C_1, C_2 of different color classes there is an edge $\{v_1, v_2\} \in E$ such that $v_1 \in C_1$, $v_2 \in C_2$ (C_1 and C_2 are adjacent). The *achromatic number* $\Psi(G)$ of the graph G is the greatest integer k, such that there exists a complete coloring of G with k color classes. A *partial complete coloring* of G is a coloring in which only some of the vertices have been colored but every two different color classes are adjacent. We also consider coloring as a mapping $c : V \longrightarrow \mathcal{N}$, where $c(v) > 0$ denotes the color (color class) assigned to v. It is obvious that any partial complete coloring with k colors can be extended to a complete coloring of the entire graph with at least k colors. Thus, the number of colors of a partial complete coloring is a lower bound for the achromatic number, and we can restrict the attention to subgraphs in order to approximate the achromatic number of the whole graph.

A *star* is a tree whose all edges are adjacent to a common vertex, called *center* of the star. The *size* of a star is the number of its edges. A *path with trees* is a path with trees hanging from some internal vertices of the path. These trees are called *path trees* and the path is called *spine*. A *path with stars* is a path with stars hanging from some internal vertices (being stars' centers) of the path.

For a given tree T we call a *leaf edge* an edge which is adjacent to a leaf of T. A *system of paths with trees* for T is a family $\{T_1, \ldots, T_k\}$ of subtrees of T such that: (1) each subtree T_i is a path with trees, (2) any two subtrees T_i, T_j ($i \neq j$) are vertex disjoint, and (3) the family is maximal, i.e. if we add to the family any edge of the tree which does not belong to the family, then we violate condition (1) or (2). The edges of the tree which do not belong to any subtree T_i of the given system of paths and are not adjacent to any leaf, are called *links*. If a given system of paths with trees consists of paths with empty trees, then it is called *system of paths*. We say that there is a *conflict* via an edge (or link) if there are two vertices colored with the same color and joined by the edge (or link). We proceed with an easy fact.

Lemma 1. *Let $T = (V, E)$ be any tree and $\mathcal{S}$ be any system of paths for T. Then $\mathcal{S}$ contains at least $|E| - \#($ leaf edges of $T) + 1$ edges.*

3 Coloring Paths

Here, we optimally color any path. This is a starting point for Section 4.

Lemma 2. *There is a linear time algorithm, that finds a complete coloring of any path P with $\Psi(P)$ colors.*

Proof. Let $f(l) = \binom{l}{2}$ if l is odd, and $f(l) = \binom{l}{2} + \frac{l-2}{2}$ if l is even. Let P consists of $f(l)$ edges. We show how to color P with l colors. Let l be odd. We proceed by induction on l. For $l = 1$ and $l = 3$ the appropriate colorings are respectively: (1) and $(1)-(2)-(3)-(1)$. Note that the last color is 1. Now let us suppose we can color any path with $f(l)$ edges using l colors, such that the last vertex is colored with 1. Then, we extend this coloring appending to the end, the following sequence of colored vertices: $(1)-(l+1)-(2)-(l+2)-(3)-(l+1)-(4)-(l+2)-\ldots-(l-1)-(l+2)-(l)-(l+1)-(l+2)-(1)$. A similar proof for an even l and a simple proof of optimality are omitted. To color optimally P, we first compute $f(l)$ (l is equal to $\Psi(P)$). It is straightforward to check that the above algorithm runs in linear time. □

The above lemma can be extended to any system $\mathcal{S}$ of paths in which each link is adjacent to the beginning of some path (this property will be used later). We first define a partial order $\prec$ on paths of the system. Let us consider a path p of our system, such that there is a vertex u on p with links: $(u,v_1),\ldots,(u,v_j)$. Let $p(u,v_i) \neq p$ denote the path of the system with vertex v_i, for $i = 1,\ldots,j$. For any path p of the system and any vertex u on p, if $j > 1$, then we require that the number of paths *preceding* p among the paths $p(u,v_1),\ldots,p(u,v_j)$ is at most one, i.e. $\#\{p(u,v_i) : i = 1,\ldots,j, p(u,v_i) \prec p\} \leq 1$. This partial order on paths of the system can be extended to a linear order, which we call *path order*. Then we form one big path P by connecting subsequent paths by the path order. Now we can color P as in the proof of Lemma 2. However, this could cause some conflicts via links. Let s be the first position in P where such a conflict appears. We can remove it by replacing colors of all vertices of P starting from this position, with the colors of their right neighbors. We continue this procedure until there are no conflicts.

The partial order can be found in linear time and extended to a linear order using the topological sort in time $O(|\mathcal{S}|)$ [3]. Computing P thus takes linear time as well. After storing P in an array, the new colors on P can be computed in linear time by pointers manipulation (the details are omitted). Hence, we have:

Lemma 3. *There is a linear time algorithm which finds an optimal complete coloring of any system of paths in which each link is adjacent to the beginning of some path.*

4 Coloring Bounded Degree Trees

We first show how to find in a bounded degree tree a large system of paths with stars. A lemma below can be proved by an induction on the height of the tree. This also gives a simple algorithm to compute a system of paths: given a tree with root r, compute recursively the systems of paths for each subtree of r and output the union of these systems and one of the edges adjacent to r.

Lemma 4. *In any tree $T = (V,E)$ with maximum degree d, there exists a system $\mathcal{S}$ of paths with stars consisting of at least $\frac{d}{2.5\cdot(d-1)} \cdot |E|$ edges from E. Moreover each link is adjacent to the first vertex of some path.*

Let T be a tree with maximum degree d and let $k = max\{p : \binom{p}{2} \leq \frac{d}{2.5(d-1)}|E|\}$. So $k \geq \sqrt{\frac{d}{2.5(d-1)}}\sqrt{2|E|}$. Let $h : \mathcal{N} \longrightarrow \mathcal{R}_+$ be a function such that $h(k) = O(d) < k$. We design an algorithm that finds a complete coloring of T with at least $\frac{1}{1.582} \cdot \Psi(T) - h(k)$ colors. The algorithm is a generalization of the algorithm from Lemma 2.

First, we give a description of some ideas behind the algorithm. We compute in T a system $\mathcal{S}$ of paths with stars by Lemma 4. We have many paths with stars and some of them are connected by links. We will be taking dynamically consecutive paths with stars from $\mathcal{S}$ according to the path order defined in Section 3, and concatenating them, forming finally a big path P with stars.

The subpath of the spine of P (together with its stars) used to add two new colors $l+1, l+2$ (for an odd l – see the proof of Lemma 2) is called a *segment* $(l+1, l+2)$. The colors $l+1$ and $l+2$ are called *segment colors*, while the colors $1, 2, \ldots, l$ are called *non-segment colors* for segment $(l+1, l+2)$. The vertices of the segment $(l+1, l+2)$ with segment (resp. non-segment) colors are called segment (resp. non-segment) vertices. During the coloring of segment $(l+1, l+2)$, we want to guarantee connections between the segment colors and non-segment ones. As in the proof of Lemma 2, all the segments begin and end with color 1. We assume that: the end-vertex of a segment is the beginning vertex of the next segment, each segment begins and ends with a vertex with color 1, the consecutive positions (vertices of the spine) of the segment are numbered by numbers $1, 2, \ldots$.

We will color P segment by segment. For each $i \in \{1, \ldots, k-2\}$, define the set $P(i) = \{i+1, i+2, \ldots, k\}$ if i is odd, and $P(i) = \{i+2, i+3, \ldots, k\}$ if i is even. $P(i)$ is intended to contain all the segment colors of the segments that should contain i as a non-segment color. The sets $P(i)$ will be changing as the algorithm runs. We define a *sack* S to be a set of some pairs (x, y) of colors such that $x \neq y$. Intuitively, S will contain the connections between some pairs of colors such that we have to guarantee these connections. These connections will be realized in the last phase of the algorithm. We need also a variable *waste* to count the number of edges that we lose (do not use effectively) in the coloring.

We now give a description of main steps of the algorithm. Using Lemma 4 find in T a system of paths $\mathcal{S}$ with stars with $\geq \frac{d}{2.5\cdot(d-1)} \cdot |E|$ edges. Set $S \leftarrow \emptyset$ and $waste \leftarrow 0$.

We show how to color segment by segment, interleaving the phases of coloring a segment and taking dynamically next paths with stars from $\mathcal{S}$ w.r.t. the path order. Assume that we have colored all the previous segments $(i, i+1)$ of P for $i = 2, 4, 6, \ldots, l-2$, and thus we have taken into account all the connections between colors $1, 2, \ldots, l-1$ (l is even). (For an odd l the proof is almost identical.) From now on we describe how to color the next segment $(l, l+1)$. We first take some next paths from $\mathcal{S}$ that follow the path order, for coloring of the whole segment $(l, l+1)$. We concatenate these paths one by one. In some cases this concatenation will be *improved* later. We call a *segment* the portion of the big path composed with these concatenated paths. These paths are called *segment paths*.

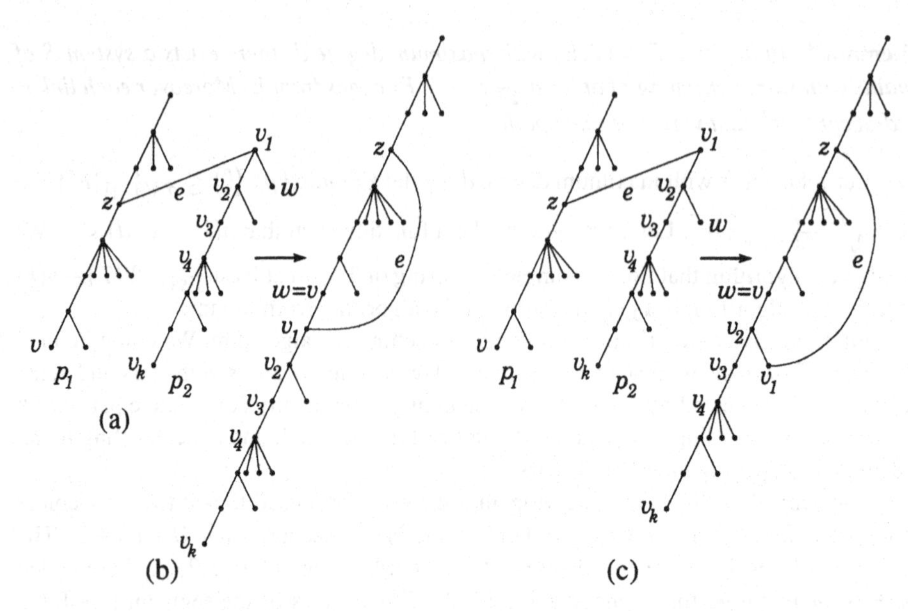

Fig. 1. Improving the concatenation in *Fixing positions* phase, subcases.

Fixing positions. We establish the segment and non-segment positions (vertices), such that if there is a link between one segment path and the first vertex v of a second segment path, then v is assigned a non-segment position.

Establish the (preliminary) segment and non-segment positions (only) on the spine of the segment exactly as in the proof of Lemma 2. Let p_1 and p_2 be any two segment paths, such that p_1 has a link e from a vertex z of p_1 to the beginning vertex v_1 of p_2. Let $v_1, v_2, \ldots, v_k$ be the consecutive vertices of p_2, and v be the last vertex on p_1 ((a) in Figure 1).

If v_1 is assigned a segment position, then if p_2 has an odd length, then improve the concatenation of the segment paths, reversing p_2: glue the vertex v with v_k. If p_2 is of even length, and v_1 is assigned a segment position, then proceed as follows. If there is a star of size ≥ 1 centered at v_1, then make an end-vertex (say w) of this star, the first vertex of path p_2: glue v with w (improving the concatenation) ((b) in Figure 1). If there is no star centered at v_1, and there is a star of size ≥ 1 centered at v_2, then let (v_2, w) be its arbitrary edge and:

1. If p_1 and p_2 are located within segment $(l, l+1)$, then if z has a segment position (color l or $l+1$), then glue v with w (improving the concatenation), and treat the edge (v_2, v_1) as a new star of size one ((c) in Figure 1).
2. If p_1 is located in some previous segment than $(l, l+1)$, then z has been colored so far, and if $c(z) \in \{l, l+1\}$, then perform the steps from the previous step 1 with "improving the concatenation" (as (c) in Figure 1).

If there are no stars with centers at v_1 and at v_2, then glue v_2 with v (improving the concatenation) and edge (v_2, v_1) is treated as a new star of size one. Thus, v_2 is assigned a segment position and v_1 – a non-segment position.

We will color all the segment positions with colors l and $l+1$. It follows that there will be no conflicts, since for each link at least one of its end-vertices has a non-segment position. Throughout the algorithm we use the following easy observation (applicable also for coloring in the proof of Lemma 2): if we permute arbitrarily non-segment colors on their positions, then we will still obtain a valid coloring of segment $(l,l+1)$; and secondly: we can exchange the two segment colors $l,l+1$ with each other, obtaining still a valid coloring.

Balancing. We want to color segment $(l,l+1)$. We first calculate an end-position of this segment (position on the spine). Let $N(j) = \{i : j \in P(i)\}$ for $j \in \{l,l+1\}$ and let $W(l,l+1) = N(l) \cap N(l+1)$. Let $\sigma_0(p,q)$, for some segment positions $p < q$, denote sum of the sizes of all stars centered at positions $p' \in \{p,p+1,\ldots,q\}$ of segment $(l,l+1)$ such that $p' \equiv 0 \; mod \; 4$, and let $\sigma_2(p,q)$ denote an analogous sum for positions $p' \in \{p,p+1,\ldots,q\}$ such that $p' \equiv 2 \; mod \; 4$. Given any position p' of segment $(l,l+1)$ let $\delta(p') = \sigma_0(1,p') - \sigma_2(1,p')$. Given a position p' of segment $(l,l+1)$, we define a $p'' = \gamma(p') \leq p'$ to be the smallest position index such that $\delta(p'') \geq \frac{\delta(p')}{2}$ if $\delta(p') \geq 0$, and such that $\delta(p'') \leq \frac{\delta(p')}{2}$ if $\delta(p') < 0$. As the end of the segment we take minimum position index q such that $2|W(l,l+1)| + 2d - 4 \leq q - 1 + \sigma_0(1,q-1) + \sigma_2(1,q-1)$. We set $S \leftarrow S \cup \{(a,l),(b,l+1) : a \in N(l) \setminus W(l,l+1),\; b \in N(l+1) \setminus W(l,l+1)\}$. Now we perform balancing: Exchange segment colors l and $l+1$ with each other on all segment positions to the right of the position $\gamma(q-1)$. Let $l' \in \{l,l+1\}$ be the color of the first position to the right of $\gamma(q-1)$. Set $S \leftarrow S \cup \{(l', c(\gamma(q-1)))\}$. During coloring segment $(l,l+1)$, we take into account only colors from the intersection of $N(l)$ and $N(l+1)$. The remaining colors $a \in N(l) \setminus W(l,l+1)$ and $b \in N(l+1) \setminus W(l,l+1)$ are added to S.

Coloring. We first give intuitions behind the coloring steps. The stars with centers in segment vertices will be used to shorten the segment. Namely we can skip a colored fragment $(l) - (x) - (l+1) - (y) - (l)$ of the spine if we find some segment vertices with stars and such that some of them has color l and some – color $l+1$. We just assign colors $l,l+1$ to centers of the stars, and the colors x,y to appropriate end-vertices of the stars. To use economically stars with centers in segment vertices, the balancing above guarantees roughly the same number $(\pm(2d-4))$ of edges of the stars for the segment vertices with color l, and for the segment vertices with color $l+1$.

The stars with centers in non-segment vertices will be used to reduce sizes of sets $P(i)$. A set $P(i)$ can be considered as the set of segment colors: $j \in P(i)$ means that the color i will be to appear (as a non-segment color) in a segment, in which j is a segment color. But possibly we can realize the connection (i,j) before j becomes a segment color: before we start coloring segment $(j,j+1)$, if we find a star with a center colored by i, we will color end-vertices of the star with the *greatest* colors from $P(i)$ and delete them from $P(i)$. This let us reduce the number of non-segment colors we have to consider during coloring of segment $(j,j+1)$ and let us estimate $|S|$ and the value of *waste*.

Now, we describe the coloring in more detail. Color all the segment positions with colors l, $l+1$. Now, color all non-segment positions on the spine of segment $(l,l+1)$ and on end-vertices of the stars with colors from $W(l,l+1)$. The coloring is performed dynamically to avoid conflicts: take the consecutive non-segment colors from

$W(l, l+1)$ and put them on arbitrary non-segment spine positions or end-vertices of the stars (called non-segment positions as well). If $c \in W(l, l+1)$ causes a conflict on a non-segment position u via a link, then color u with any other color from $W(l, l+1)$, avoiding the conflict. If c gives conflicts on each non-segment position, then put all other non-segment colors on that positions, and add (c, l), $(c, l+1)$ to S, and set $waste \leftarrow waste + 3$.

If during the improving of concatenation one glued two vertices, each adjacent to a link, into a vertex x, then this may cause a conflict for two non-segment colors c and c': one cannot color x with neither c nor c'. There may be several such conflict positions x within segment $(l, l+1)$. In this case, proceed by analogy to the above: if there is a conflict for c, c' on each such position x, then put on that positions other non-segment colors, and add (c, l), $(c, l+1)$, (c', l), $(c', l+1)$ to sack S, and set $waste \leftarrow waste + 6$.

Put segment colors from the sets $P(i)$ on end-vertices of the non-segment stars: if i is a non-segment color on the vertex-center of a star, then assign to the end-vertices v of the star the greatest colors from $P(i)$, and delete these colors from $P(i)$. If some color $c \in P(i)$ gives a conflict on some v via a link, then put c on some other end-vertex of the star. If c gives conflicts on all end-vertices of the star, then continue assigning with the next greatest color in $P(i)$, delete it from $P(i)$ and if $|P(i)| = 1$, then set $waste \leftarrow waste + d - 1$. This completes description of the coloring of segment $(l, l+1)$.

Last step. (This step is performed after all the segments have been colored.) Realize connections from S: greedily use every third edge from the spine to the right of the end-position of the last colored segment. This completes description of the algorithm.

In the algorithm we use the greatest colors from $P(i)$. Note, that this trick assures that $N(l+1) \subseteq N(l)$, so no pair $(b, l+1)$ will be added to S in balancing. Moreover, this also gives that for each color $a \in \{1, \dots, k\}$, there exists at most one segment color l, such that (a, l) will be added to S, and let us estimate $|S|$ after the algorithm stops: $|S| \leq 4k$ (we skip the details). Similar ideas give: the total number of wasted edges $= waste + (3d-1) \cdot |S| \leq (15d - 3.5)k$.

Our algorithm could fail only when it would attempt to use more than $\binom{k}{2}$ edges. However, this is not the case, since with $h(k) = 15d - 3.5$, it is easy to show that $\binom{k-h(k)}{2} + h(k) \cdot (k - h(k)) \leq \binom{k}{2}$, for every k. Thus the number of colors we have used is at least $k - h(k)$. Now, recall that $k \geq \sqrt{\frac{d}{2.5(d-1)}}\sqrt{2|E|}$, and thus $k - h(k) \geq \sqrt{\frac{d}{2.5(d-1)}}\sqrt{2|E|} - h(k) \geq \sqrt{\frac{d}{2.5(d-1)}}\Psi(T) - h(k)$.

The linear time implementation of the algorithm is quite natural and is omitted due to lack of space. The above ideas let us prove the following theorem.

Theorem 5. *Let $T = (V, E)$ be any tree with maximum degree $d = d(|V|)$. Given T, the above algorithm produces in $O(|E|)$-time a complete coloring of T with at least $\sqrt{\frac{d}{2.5 \cdot (d-1)}} \cdot \Psi(T) - h(d)$ colors, where $h(d) = 15 \cdot d - 3.5$.*

Since $\sqrt{\frac{2.5 \cdot (d-1)}{d}} < \sqrt{2.5} < 1.582$, Theorem 5 gives asymptotic 1.582-approximation algorithm for the achromatic number problem on bounded degree trees. We can assume, e.g., that $d(|V|) = O(\log(|V|))$, or even $d(|V|) = (4 \cdot (|V| - 1))^{1/4}$ (see also below). We now improve the algorithm for binary trees. A binary tree is a tree with maximum degree 3. Let a 2-*tendril* be a path of length at most 2. We present an analogous lemma to Lemma 4 and a theorem based on it.

Lemma 6. *In any binary tree $T = (V,E)$ there exists a system of paths with 2-tendrils, with at least $\frac{3}{4} \cdot |E|$ edges from E. Moreover, each link is adjacent to the first vertex of some path.*

Theorem 7. *There is an asymptotic* 1.155*-approximation linear time algorithm for the achromatic number problem on binary trees. Given a binary tree T, it produces a complete coloring of T with at least $\frac{1}{1.155} \cdot \Psi(T) - O(1)$ colors.*

Farber *et al.* [7] prove, that the achromatic number of a tree with m edges and maximum degree $(4m)^{1/4}$, is at least $\sqrt{m}$. An obvious upper bound is $\sqrt{2}\sqrt{m}$. (Therefore, we may assume in Theorem 5 that $d \leq (4 \cdot (|V|-1))^{1/4}$.) Using the coloring algorithm from Theorem 7 we obtain the following improvement.

Theorem 8. *Let T be any tree with m edges and maximum degree at most* 3. *Then we have that $\Psi(T) \geq 1.224 \cdot \sqrt{m} - c$, for some fixed positive constant c.*

5 Coloring Arbitrary Trees

In this section we give a 2-approximation algorithm for trees. Let $T = (V,E)$ be a given tree, and $|E| = m$. For an internal node $v \in V$ of T, let $star(v)$ be the set of all leaf edges adjacent to v. Let $\Psi(T) = k$, and σ be any complete coloring of T with k colors. By $E' \subseteq E$ we denote a set of *essential* edges for σ, which for each pair of different colors $i, j \in \{1, \dots, k\}$ contains one edge linking a vertex colored with i and a vertex colored with j. Notice that $|E'| = \binom{k}{2}$.

Lemma 9. *For each $r \leq k$ and any internal vertices $v_1, \dots, v_r \in V$ the following holds $|\bigcup_{i=1}^{r} star(v_i) \cap E'| \leq \frac{(k-1)+(k-r)}{2} \cdot r$ for any E' as above.*

The above lemma follows from the fact that centers of stars $star(v_i)$ can be colored with at most r colors, so each of the edges in $\bigcup_{i=1}^{r} star(v_i)$ has one of its endpoints colored with one of these r colors.

We say that a color l is *saturated* if during a course of the algorithm, l has edge connections with each other color, that we use in the partial complete coloring. Condition **Cond (2)** appearing below is defined in the proof of Theorem 10.

The algorithm is as follows. Set $\overline{m} \leftarrow m$, and for each $e \in E$, mark e "good". **Step (1)**: Set $k \leftarrow \max\{i : \binom{i}{2} \leq \overline{m}\}$. Select a maximum size star $star(v)$ $(v \in V)$ and unmark "good" for one of its edges. Continue this marking process until **Cond (1)**: for each set of r $(r \leq k)$ stars in T the number of "good" edges is $\leq \frac{(k-1)+(k-r)}{2} \cdot r$. Set $\overline{m} \leftarrow$ #(all edges of E marked "good"). End **Step (1)**. If $\binom{k}{2} > \overline{m}$, then go to **Step (1)**. Replace tree T with T with only "good" edges.

We have now two cases. If #(leaf edges of T) $\leq \frac{3}{4}\overline{m}$, then compute a system of paths $\mathcal{S}$ in T, and optimally color $\mathcal{S}$ by Lemma 3.

In the other case, if #(leaf edges of T) $> \frac{3}{4}\overline{m}$, proceed as follows. Sort all stars $\{star(v) : v \in V\}$ of T w.r.t. their sizes. Let $s_1 \geq s_2 \geq \ldots \geq s_x$ be these sizes. Find the smallest integer i such that $s_i < \frac{\sqrt{m}}{\sqrt{2}} - i$. We consider two cases. If $i \geq \frac{\sqrt{m}}{\sqrt{2}} - 1$, color the first i largest stars with $\frac{\sqrt{m}}{\sqrt{2}}$ colors: center of j-th star ($j = 1,2,\ldots,i$) is colored with color j, and its $\frac{\sqrt{m}}{\sqrt{2}} - j$ sons are colored with the consecutive colors $j+1, j+2, \ldots, \frac{\sqrt{m}}{\sqrt{2}}$. Else, if $i < \frac{\sqrt{m}}{\sqrt{2}} - 1$, use the procedure from the case of "$i \geq \frac{\sqrt{m}}{\sqrt{2}} - 1$" for the first $i-1$ largest stars, saturating colors $1,2,\ldots,i-1$: center of j-th star is colored with color j and its sons with colors $j+1, j+2, \ldots, \frac{\sqrt{m}}{\sqrt{2}}$, for $j = 1,2,\ldots,i-1$. The remaining colors $i, i+1, \ldots, \frac{\sqrt{m}}{\sqrt{2}}$ will be saturated using the fact that now we have many stars with small sizes ($\leq \frac{\sqrt{m}}{\sqrt{2}}$). We namely perform the following steps, called the *final saturation steps*.

Partition all the stars into two sets B_1 and B_2 of all stars with centers at odd distance from the root of T and resp. at even distance from the root. Set $j \leftarrow i$. Now, we try to saturate color j. While **Cond (2)** do: take for B_p one of the sets B_1, B_2 with greater number of edges; keep on picking the consecutive maximum size stars from B_p until color j is saturated; delete these stars from B_p. Set $j \leftarrow j+1$. End While. This completes description of the algorithm.

Theorem 10. *The above algorithm is a 2-approximation algorithm for the achromatic number problem on a tree.*

Proof. Note, that if condition **Cond (1)** holds for any set of r ($r \leq k$) stars of the greatest size, then it also holds for any set of r stars. Thus it is possible to perform procedure **Step (1)** in polynomial time. To prove the approximation ratio we show that the algorithm finds a complete coloring with at least $\frac{\sqrt{m}}{\sqrt{2}}$ colors, which is half of optimal number of colors. If #(leaf edges of T) $\leq \frac{3}{4}m$, then by Lemma 1, S has at least $\frac{1}{4}m$ edges, so the resulting partial complete coloring uses $\geq \frac{\sqrt{m}}{\sqrt{2}}$ colors. We prove this also for case of #(leaf edges of T) $> \frac{3}{4}m$. We have many edges in the stars. We show that the algorithm colors the stars with $\geq \frac{\sqrt{m}}{\sqrt{2}}$ colors. By procedure **Step (1)** and Lemma 9, the number of edges in $i-1$ stars we have used before the final saturation steps is at most $A = \frac{(\sqrt{2}\sqrt{m}-1)+(\sqrt{2}\sqrt{m}-i+1)}{2} \cdot (i-1)$. Defining $A' = \frac{\frac{\sqrt{m}}{\sqrt{2}}-1+\frac{\sqrt{m}}{\sqrt{2}}-i+1}{2} \cdot (i-1) + \frac{\sqrt{m}}{\sqrt{2}} \cdot (i-1)$, we have $A = A'$. The first element of A' can be considered as the number of edges we have used effectively in the coloring, while the second element of this sum can be considered as the number of edges we have lost (before final saturation steps). We prove that the final saturation steps will saturate all colors $i, i+1, \ldots, \frac{\sqrt{m}}{\sqrt{2}}$. **Cond (2)** is defined to be true iff: $e_1 =$ #(edges in B_1) $\geq \frac{\sqrt{m}}{\sqrt{2}} - j$ or $e_2 =$ #(edges in B_2) $\geq \frac{\sqrt{m}}{\sqrt{2}} - j$. So the procedure stops if $e_1 + e_2 \leq 2 \cdot (\frac{\sqrt{m}}{\sqrt{2}} - j - 1) = \delta$. The formula for A together with A' can be generalized also to the case of coloring in case "$i < \frac{\sqrt{m}}{\sqrt{2}} - 1$" together with "While **Cond (2)**" procedure. (We just glue all centers of the stars used to saturate color j in "While **Cond (2)**" into one star, for $j = i, i+1, \ldots$.) So after the above procedure, when **Cond (2)** is not true any more, we have used so far at most $\Delta = \frac{(\sqrt{2}\sqrt{m}-1)+(\sqrt{2}\sqrt{m}-j+1)}{2} \cdot (j-1)$ edges. Thus, if **Cond (2)** will not be true, then

$\frac{3}{4}\overline{m} - \Delta \leq \delta$. It is easy to show that the smallest j, such that $\frac{3}{4}\overline{m} - \Delta \leq \delta$, is greater than the number of colors $\frac{\sqrt{\overline{m}}}{\sqrt{2}}$ to saturate. Obviously, the algorithm has polynomial running time. □

6 Coloring Large Girth Graphs

In this section we prove that for graphs with girth at least six, the achromatic number can be approximated in polynomial time with ratio $O(n^{3/8})$.

We define in a given graph $G = (V, E)$ a subset $M \subseteq E$ to be an *independent matching* if no two edges in M have a common vertex and there is no edge in $E \setminus M$ adjacent to more than one edge in M. Let $G = (V, E)$ be a given graph, with $|V| = n$, $|E| = m$, and with girth at least six. For a given edge $e \in E$, let $N(e)$ denote the set of all edges at distance at most one from e. We now describe the algorithm of Chaudhary and Vishwanathan [5] with our modification. The steps below are performed for all parameters $f = 1, 2, \ldots, m$.

1. Set $I \leftarrow \emptyset$, $i \leftarrow 1$.
2. Choose any edge $e_i \in E$ and set $I \leftarrow I \cup \{e_i\}$, $E \leftarrow E \setminus N(e_i)$.
3. If $E \neq \emptyset$ then set $i \leftarrow i + 1$, go to step 2.
4. If $|I| > f$ then output a partial complete coloring using edges in I, else partition each $N(e_i)$ into two trees by removing the edge e_i. Then use the algorithm of Section 5 to produce a partial complete coloring for each such tree and output the largest size coloring.

Theorem 11. *Let G be a graph with n vertices and with girth at least six. The above algorithm is a $(\sqrt{2}+\varepsilon)\sqrt{\Psi(G)}$-approximation algorithm for the achromatic number problem on G (for any $\varepsilon > 0$). Moreover $(\sqrt{2}+\varepsilon)\sqrt{\Psi(G)} = O(n^{3/8})$.*

Proof. I is an independent matching and having any independent matching of size $\binom{l}{2}$, we can generate a partial complete coloring of size l. In this case we have a coloring of size $\geq \sqrt{2f}$. In the other case, since girth ≥ 6, removing the edge e_i from $N(e_i)$, vertex set of $N(e_i)$ can be partitioned into two trees. Consider a maximum coloring σ of G and let E' denote a set of essential edges for σ. If $|I| \leq f$, then at least one of the sets $N(e_i)$, say $N(e_{i_0})$, contains $\geq (|E'|/f)$ essential edges. $N(e_{i_0})$ consists of two trees, so one of them contains $\geq \frac{(|E'|/f)-1}{2}$ essential edges. So we can set $\overline{m} = \frac{(|E'|/f)-1}{2}$ in the proof of Theorem 10 (see **Step (1)** of algorithm in Section 5), and from this proof the tree can be colored with at least $c = \sqrt{\frac{\overline{m}}{2}}$ colors. Thus the number of colors is $c = \frac{1}{\sqrt{8f}}\sqrt{\Psi(\Psi-1) - 2f}$, where $\Psi = \Psi(G)$, and $|E'| = \binom{\Psi}{2}$. For f, such that $c = \sqrt{2f}$, $\frac{1}{\sqrt{8f}}\sqrt{\Psi(\Psi-1)-2f} = \sqrt{2f}$. So, $c = \frac{1}{\sqrt{8}}\sqrt{\sqrt{16\Psi^2 - 16\Psi + 1} - 1}$. It can be shown that $c \geq c' \cdot \sqrt{\Psi}$, for a constant $c' = \frac{1}{\sqrt{2}+\varepsilon}$ and $\varepsilon > 0$. Since $c \geq c'\sqrt{\Psi} = \frac{1}{(\sqrt{2}+\varepsilon)\sqrt{\Psi}}\Psi$, the approximation ratio is $(\sqrt{2}+\varepsilon)\sqrt{\Psi}$. Any n-vertex graph with girth at least g has at most $n\lceil n^{\frac{2}{g-2}} \rceil$ edges [2], and $\Psi = O(\sqrt{|E|})$, so the approximation ratio is $O(n^{3/8})$. □

7 Open Problems

We suggest the following open problems: (1) Improving the approximation ratios. (2) Is there an $O(\sqrt{\Psi})$-approximation algorithm for the problem on other classes of graphs? (3) Is the achromatic number problem on trees Max SNP-hard?

References

[1] H.L. Bodlaender, *Achromatic Number is NP-complete for Cographs and Interval Graphs*, Information Processing Letters, **31**: 135–138, 1989.

[2] B. Bollobás, *Extremal Graph Theory*, Academic Press, London, 1978.

[3] N. Cairnie and K.J. Edwards, *Some Results on the Achromatic Number*, Journal of Graph Theory, **26**: 129–136, 1997.

[4] N. Cairnie and K.J. Edwards, *The Achromatic Number of Bounded Degree Trees*, Discrete Mathematics, **188**: 87–97, 1998.

[5] A. Chaudhary and S. Vishwanathan, *Approximation Algorithms for the Achromatic Number*, Proc. 8th ACM-SIAM Symposium on Discrete Algorithms, 558–563, 1997.

[6] T.H. Cormen, C.E. Leiserson and R.L. Rivest, *Introduction to Algorithms*, MIT Press, Cambridge, MA, 1990.

[7] M. Farber, G. Hahn, P. Hell and D. Miller, *Concerning the Achromatic Number of Graphs*, Journal of Combinatorial Theory, Series B, **40**: 21–39, 1986.

[8] F. Harary, S. Hedetniemi and G. Prins, *An Interpolation Theorem for Graphical Homomorphisms*, Portugaliae Mathematica, **26**: 453–462, 1967.

[9] R. Motwani, *Lecture Notes on Approximation Algorithms*, Department of Computer Science, Stanford University, Stanford, 1992.

[10] T.L. Saaty and P.C. Kainen, *The Four-Color Problem: assaults and conquest*, Dover Publishers, New York, 1986.

[11] M. Yannakakis and F. Gavril, *Edge Dominating Sets in Graphs*, SIAM Journal on Applied Mathematics, **38**: 364–372, 1980.

Augmenting a $(k - 1)$-Vertex-Connected Multigraph to an ℓ-Edge-Connected and k-Vertex-Connected Multigraph

Toshimasa Ishii, Hiroshi Nagamochi, and Toshihide Ibaraki

Department of Applied Mathematics and Physics,
Kyoto University, Kyoto 606-8501, Japan.
{ishii, naga, ibaraki}@kuamp.kyoto-u.ac.jp

Abstract. Given an undirected multigraph $G = (V, E)$ and two positive integers ℓ and k, we consider the problem of augmenting G by the smallest number of new edges to obtain an ℓ-edge-connected and k-vertex-connected multigraph. In this paper, we show that an $(k-1)$-vertex-connected multigraph G $(k \geq 4)$ can be made ℓ-edge-connected and k-vertex-connected by adding at most 2ℓ surplus edges over the optimum, in $O(\min\{k, \sqrt{n}\}kn^3 + n^4)$ time, where $n = |V|$.

1 Introduction

The problem of augmenting a graph by adding the smallest number of new edges to meet edge-connectivity or vertex-connectivity requirement has been extensively studied as an important subject in network design, and many efficient algorithms have been developed so far. However, it was only very recent to have algorithms for augmenting both edge-connectivity and vertex-connectivity simultaneously (see [9, 10, 11, 12] for those results).

Let $G = (V, E)$ stand for an undirected multigraph with a set V of vertices and a set E of edges. We denote the number of vertices by n, and the number of pairs of adjacent vertices by m. The local edge-connectivity $\lambda_G(x, y)$ (resp., the local vertex-connectivity $\kappa_G(x, y)$) is defined to be the maximum ℓ (resp., k) such that there are ℓ edge disjoint (resp., k internally vertex disjoint) paths between x and y in G (where at most one edge between x and y is allowed to be in the set of internally vertex disjoint paths). The *edge-connectivity* and *vertex-connectivity* of G are defined by $\lambda(G) = \min\{\lambda_G(x,y) \mid x, y \in V, x \neq y\}$ and $\kappa(G) = \min\{\kappa_G(x,y) \mid x, y \in V, x \neq y\}$. We call a multigraph G *ℓ-edge-connected* if $\lambda(G) \geq \ell$. Analogously, G is *k-vertex-connected* if $\kappa_G(x, y) \geq k$ for all $x, y \in V$. The *edge-connectivity augmentation problem* (resp., the *vertex-connectivity augmentation problem*) asks to augment G by the smallest number of new edges so that the resulting multigraph G' becomes ℓ-edge-connected (resp., k-vertex-connected).

As to the edge-connectivity augmentation problem, Watanabe and Nakamura [20] first proved that the problem can be solved in polynomial time for any given integer ℓ. However, some special cases of this problem are NP-hard; for example, augmenting G to attain ℓ-edge-connectivity while preserving simplicity of the given graph [1, 15].

As to the vertex-connectivity augmentation problem, the problem of making a $(k-1)$-vertex-connected multigraph k-vertex-connected was shown to be polynomially solvable for $k = 2$ [3] and for $k = 3$ [21]. It was later found out that, for $k \in \{2, 3, 4\}$, the vertex-connectivity augmentation problem can be solved in polynomial time in [3, 7] (for $k = 2$), [6, 21] (for $k = 3$), and [8] (for $k = 4$), even if the input multigraph G is not necessarily $(k-1)$-vertex-connected. However, whether there is a polynomial time algorithm for an arbitrary constant k was still an open question (even if G is $(k-1)$-vertex-connected). Recently Jordán presented an $O(n^5)$ time approximation algorithm for $k \geq 4$ [13, 14] in which the difference between the number of new edges added by the algorithm and the optimal value is at most $(k-2)/2$.

The problem of augmenting both edge- and vertex-connectivities has been studied in [9, 10, 11, 12]. For two integers ℓ and k, we say that G is *(ℓ,k)-connected* if G is ℓ-edge-connected and k-vertex-connected. The *edge-and-vertex-connectivity augmentation problem*, denoted by EVAP(ℓ,k), asks to augment G by adding the smallest number of new edges so that the resulting multigraph G' becomes (ℓ,k)-connected, where $\ell \geq k$ is assumed without loss of generality. Recently, the authors proved that EVAP($\ell,2$) can be solved in $O((nm+n^2\log n)\log n)$ time [10], and that EVAP($\ell,3$) can be solved in polynomial time (in particular, $O(n^4)$ time if an input graph is 2-vertex-connected) for any fixed integer ℓ [11, 12]. However, for arbitrary integers ℓ and k, whether there is a polynomial time algorithm for EVAP(ℓ,k) in which the difference between the number of new edges added by the algorithm and the optimal value is $O(\ell)$ was still an open question.

In this paper, we consider problem EVAP(ℓ,k) for a $(k-1)$-vertex-connected graph. One may consider an algorithm that first meets one of the edge-connectivity and vertex-connectivity requirements, and then meets the other requirement. It is easily seen that this sequential algorithm does not lead to an optimal solution. In this paper, we prepare an algorithm that augments the graph by considering both edge-connectivity and vertex-connectivity in each step. However, still in this case, the resulting multigraph may not be optimally augmented from the original multigraph. To evaluate approximation error, we first present a lower bound on the number of edges that is necessary to make a given multigraph G (ℓ,k)-connected, and then show that the lower bound plus 2ℓ edges suffice if the input graph is $(k-1)$-vertex-connected with $k \geq 4$. The task of constructing such a set of new edges can be done in $O(\min\{k,\sqrt{n}\}kn^3+n^4)$ time.

2 Preliminaries

2.1 Definitions

For a multigraph $G=(V,E)$, an edge with end vertices u and v is denoted by (u,v). Given two disjoint subsets of vertices X, $Y \subset V$, we denote by $E_G(X,Y)$ the set of edges connecting a vertex in X and a vertex in Y, and denote $c_G(X,Y)=|E_G(X,Y)|$. A singleton set $\{x\}$ is also denoted x. In particular, $E_G(u,v)$ is the set of multiple edges with end vertices u and v and $c_G(u,v)=|E_G(u,v)|$ denotes its multiplicity. Given a multigraph $G=(V,E)$, its vertex set V and edge set E may be denoted by $V(G)$ and $E(G)$, respectively. For a subset $V' \subseteq V$ (resp., $E' \subseteq E$) in G, $G[V']$ (resp., $G[E']$) denotes the subgraph induced by V' (resp., $G[E']=(V,E')$). For $V' \subset V$ (resp., $E' \subseteq E$), we denote subgraph $G[V-V']$ (resp., $G[E-E']$) also by $G-V'$ (resp., $G-E'$). For $E' \subset E$, we denote $V(G[E'])$ by $V[E']$. For an edge set F with $F \cap E = \emptyset$, we denote the augmented graph $G=(V,E \cup F)$ by $G+F$. A *partition* $X_1,\cdots,X_t$ of a vertex set V is a family of nonempty disjoint subsets X_i of V whose union is V, and a *subpartition* of V is a partition of a subset of V. A *cut* is defined to be a subset X of V with $\emptyset \neq X \neq V$, and the *size* of a cut X is defined by $c_G(X,V-X)$, which may also be written as $c_G(X)$. A subset X *intersects* another subset Y if none of subsets $X \cap Y$, $X-Y$ and $Y-X$ is empty. A family $\mathcal{X}$ of subsets $X_1,\cdots,X_p$ is called *laminar* if no two subsets in $\mathcal{X}$ intersect each other (however, possibly $X_i \subseteq X_j$ for some $X_i,X_j \in \mathcal{X}$).

For a subset X of V, a vertex $v \in V-X$ is called a *neighbor* of X if it is adjacent to some vertex $u \in X$, and the set of all neighbors of X is denoted by $\Gamma_G(X)$. A maximal connected subgraph G' in a multigraph G is called a *component* of G, and the number of components in G is denoted by $p(G)$. A *disconnecting set* of G is defined as a cut S of V such that $p(G-S)>p(G)$ holds and no $S' \subset S$ has this property. Let $\hat{G}$ denote the simple graph obtained from G by replacing multiple edges in $E_G(u,v)$ by a single edge (u,v) for all $u,v \in V$. A component G' of G with

$|V(G')| \geq 3$ always has a disconnecting set unless $\hat{G}'$ is a complete graph. If G is connected and contains a disconnecting set, then a disconnecting set of the minimum size is called a *minimum disconnecting set*, whose size is equal to $\kappa(G)$. A cut $T \subset V$ is called *tight* if $\Gamma_G(T)$ is a minimum disconnecting set in G. A tight set D is called *minimal* if no proper subset D' of D is tight (hence, the induced subgraph $G[D]$ is connected). We denote a family of all minimal tight sets in G by $\mathcal{D}(G)$, and denote the maximum number of pairwise disjoint minimal tight sets by $t(G)$. For a vertex set S in G, we call the components in $G-S$ *the S-components*, and denote the family of all S-components by $C(G-S)$. Note that the vertex set S is a disconnecting set in a connected multigraph G if and only if $|C(G-S)| \geq 2$. Clearly, for a minimum disconnecting set S, every S-component is tight, and the union of two or more (but not all) S-components is also tight.

Lemma 1. [13] *Let G be k-vertex-connected. If $t(G) \geq k+1$, then any two cuts $X, Y \in \mathcal{D}(G)$ are pairwise disjoint (i.e., $t(G) = |\mathcal{D}(G)|$).* □

Lemma 2. [18] *Let $C \subseteq E$ be a cycle in a k-vertex-connected graph $G = (V,E)$ such that $\kappa(G-e) = k-1$ holds for every $e \in C$. Then there exists a vertex $v \in V[C]$ with $|\Gamma_G(v)| = k$.* □

We call a disconnecting set S a *shredder* if $|C(G-S)| \geq 3$. A tight set T is called a *superleaf*, if T contains exactly one cut in $\mathcal{D}(G)$ and no $T' \supset T$ satisfies this property. The following lemmas summarize some properties of superleaves.

Lemma 3. [2] *Let $G = (V,E)$ be a connected multigraph with $t(G) \geq \kappa(G)+3$. Then every two superleaves are pairwise disjoint. Hence, a superleaf is disjoint from all other cuts in $\mathcal{D}(G)$ except for the cut in $\mathcal{D}(G)$ contained in it.* □

Lemma 4. *Let S be a minimum shredder and $D \in \mathcal{D}(G)$ be a minimal tight set in a connected graph $G = (V,E)$. If D is contained in a cut $T \in C(G-S)$ and no cut in $\mathcal{D}(G)$ other than D is contained in T, then T is the superleaf with $T \supseteq D$.* □

Lemma 5. *Let S be a minimum shredder in a connected graph $G = (V,E)$. If $p(G-S) \geq \kappa(G)+1$ holds, then every superleaf Q in G satisfies $Q \cap S = \emptyset$.* □

Lemma 6. *Let G be a connected multigraph with $t(G) \geq \kappa(G)+3$. Then $t(G+e) \leq t(G)-1$ holds for an edge e which connects two pairwise disjoint cuts $D_1, D_2 \in \mathcal{D}(G)$ with $\Gamma_G(D_2) \cap Q_1 = \emptyset$, where Q_1 is the superleaf containing D_1.* □

2.2 Preserving Edge-Connectivity

Given a multigraph $G = (V,E)$, a designated vertex $s \in V$, and a vertex $u \in \Gamma_G(s)$, we construct a graph $G' = (V,E')$ by deleting one edge (s,u) from $E_G(s,u)$, and adding a new edge to $E_G(s,v)$ with $v \in V-s$. We say that G' is obtained from G by *shifting* (s,u) to (s,v).

Let $G = (V,E)$ satisfy $\lambda_G(x,y) \geq \ell$ for all pairs $x,y \in V-s$. For an edge $e = (s,v)$ with $v \in V-s$, if there is a pair $x,y \in V-s$ satisfying $\lambda_{G-e}(x,y) < \ell$, then there is a unique cut $X \subset V-s$ with $c_G(X) = \ell$ and $v \in X$ such that all cuts $X' \subset X$ with $v \in X'$ satisfies $c_G(X') > \ell$. We call such X *λ-critical* with respect to $v \in \Gamma_G(s)$.

Theorem 7. *Let $G = (V,E)$ be a multigraph with a designated vertex $s \in V$ such that $\lambda_G(x,y) \geq \ell$ for all $x,y \in V-s$. Let X be a λ-critical cut with respect to $v \in \Gamma_G(s)$ if any; $X = V$ otherwise. Then for any $v' \in X \subset V-s$, the shifted multigraph $G-(s,v)+(s,v')$ satisfies $\lambda_{G-(s,v)+(s,v')}(x,y) \geq \ell$ for all $x,y \in V-s$.* □

Given a multigraph $G=(V,E)$, a designated vertex $s \in V$, vertices $u,v \in \Gamma_G(s)$ (possibly $u=v$) and a nonnegative integer $\delta \le \min\{c_G(s,u), c_G(s,v)\}$, we construct graph $G'=(V,E')$ by deleting δ edges from $E_G(s,u)$ and $E_G(s,v)$, respectively, and adding new δ edges to $E_G(u,v)$. We say that G' is obtained from G by *splitting* δ pairs of edges (s,u) and (s,v). A sequence of splittings is *complete* if the resulting graph G' does not have any neighbor of s.

Let $G=(V,E)$ satisfy $\lambda_G(x,y) \ge \ell$ for all pairs $x,y \in V-s$. A pair $\{(s,u),(s,v)\}$ of two edges in $E_G(s)$ is called *λ-splittable*, if the multigraph G' resulting from splitting edges (s,u) and (s,v) satisfies $\lambda_{G'}(x,y) \ge \ell$ for all pairs $x,y \in V-s$. The following theorem is proven by Lovász [17, Problem 6.53].

Theorem 8. [5, 17] *Let $G=(V,E)$ be a multigraph with a designated vertex $s \in V$ with even $c_G(s)$, and $\ell \ge 2$ be an integer such that $\lambda_G(x,y) \ge \ell$ for all $x,y \in V-s$. Then for each $u \in \Gamma_G(s)$ there is a vertex $v \in \Gamma_G(s)$ such that $\{(s,u),(s,v)\}$ is λ-splittable.* □

Repeating the splitting in this theorem, we see that, if $c_G(s)$ is even, there always exists a complete splitting at s such that the resulting graph G' satisfies $\lambda_{G'-s}(x,y) \ge \ell$ for every pair of $x,y \in V-s$. It is shown in [19] that such a complete splitting at s can be computed in $O((m+n\log n)n\log n)$ time.

We call a cut $X \subset V-s$ *dangerous* if $c_G(X) \le \ell+1$ holds. Note that $\{(s,u),(s,v)\}$ is not λ-splittable if and only if there is a dangerous cut $X \subset V-s$ with $\{u,v\} \subseteq X$.

Theorem 9. [5] *Let $G=(V,E)$ be a multigraph with a designated vertex $s \in V$, and $\ell \ge 2$ be an integer such that $\lambda_G(x,y) \ge \ell$ for all pairs $x,y \in V-s$. Let $u \in \Gamma_G(s)$. Then there are at most two maximal dangerous cuts X with $u \in X$; i.e., no cut $X' \supset X$ with $u \in X'$ is dangerous. In particular, if there are exactly two maximal dangerous cuts X_1 and X_2 with $u \in X_1 \cap X_2$, then $c_G(X_1 \cup X_2)=\ell+2$, $c_G(X_1 \cap X_2)=\ell$, $c_G(X_1-X_2)=c_G(X_2-X_1)=\ell$, and $c_G(s, X_1 \cap X_2)=1$ hold.* □

Corollary 10. *Let $G=(V,E)$ be a multigraph satisfying the assumption of Theorem 9 and $\lambda(G-s) \ge k$. For each vertex $u \in \Gamma_G(s)$, $|\{v \in \Gamma_G(s) | \{(s,u),(s,v)\}$ is not λ-splittable$\}| \le |\{e \in E_G(s) | \{(s,u),e\}$ is not λ-splittable$\}| \le \ell+2-k$.* □

Conversely, we say that G' is obtained from G by *hooking up* an edge (u,v), if we construct G' by replacing (u,v) with two edges (s,u) and (s,v) in G.

2.3 Preserving Vertex-Connectivity

Let $G=(V,E)$ denote a multigraph with $s \in V$ and $|V| \ge k+2$ such that $\kappa_G(x,y) \ge k$ holds for all pairs $x,y \in V-s$. A pair $\{(s,u),(s,v)\}$ of two edges in $E_G(s)$ is called *κ-splittable*, if the multigraph G' resulting from splitting edges (s,u) and (s,v) satisfies $\kappa_{G'}(x,y) \ge k$ for all pairs $x,y \in V-s$. If $G-s$ is not k-vertex-connected, then $\kappa(G-s)=k-1$ holds, since G satisfies $\kappa_{G-s}(x,y) \ge \kappa_G(x,y)-1 \ge k-1$ for all pairs $x,y \in V-s$. Hence if $\{(s,u),(s,v)\}$ is not κ-splittable, then the resulting graph G' has a disconnecting set $S \subset V-s$ with $|S|=k-1$ and $p(G'-S)=2$, and a cut $T \in C(G'-S)$ with $T \subset V-s$, $\{u,v\} \cap T \ne \emptyset$, $\{u,v\} \subseteq T \cup S$ and $E_{G'}(s,T)=\emptyset$. The following theorem for a shredder is given by [2, Lemma 5.6].

Theorem 11. [2] *Let $G=(V,E)$ be a multigraph with a designated vertex $s \in V$, and $k \ge 2$ be an integer such that $\kappa_G(x,y) \ge k$ for all pairs $x,y \in V-s$, $\kappa(G-s)=k-1$, and $t(G-s) \ge k+2$. Let Q_1, Q_2, and Q_3 be three distinct superleaves in $G-s$ such that $\Gamma_{G-s}(Q_1) \cap Q_2 = \emptyset = \Gamma_{G-s}(Q_1) \cap Q_3$ and $\Gamma_{G-s}(Q_1)$ is not a shredder in $G-s$. Then, for $x \in \Gamma_G(s) \cap D_1$, $y \in \Gamma_G(s) \cap D_2$, and $z \in \Gamma_G(s) \cap D_3$, where $D_i \subseteq Q_i$, $i=1,2,3$, denote the cuts in $\mathcal{D}(G-s)$, at least one of $\{(s,x),(s,y)\}$,*

$\{(s,y),(s,z)\}$, and $\{(s,z),(s,x)\}$ is κ-splittable. Moreover, $t(G') = t(G) - 2$ holds for the resulting graph G' (note that $\Gamma_G(s) \cap D_i \neq \emptyset$ holds for $i = 1,2,3$ since otherwise $\kappa_G(x,y) \geq k$ cannot hold for all pairs $x,y \in V - s$). □

The following result is a slight generalization of [2, Lemma 5.7] in which a graph G is assumed to satisfy $\kappa_G(x,y) \geq k(\geq 2)$ for all pairs $x,y \in V - s$, but removal of any edge incident to s from G violates this property.

Theorem 12. *Let $G = (V,E)$ be a multigraph with a designated vertex $s \in V$, and $k \geq 2$ be an integer such that $\kappa_G(x,y) \geq k$ for all pairs $x,y \in V - s$, $\kappa(G-s) = k-1$ and $t(G-s) \geq \max\{2k-2, k+2\}$. Let $Q_1 \subset V - s$ be an arbitrary superleaf such that $S = \Gamma_{G-s}(Q_1)$ is a shredder in $G-s$. If $G-s$ has a cut $T \in C((G-s)-S) - \{Q_1\}$ with $c_G(s,T) \geq 2$, then $\{(s,x),(s,y)\}$ is κ-splittable for any pair $\{x,y\}$ such that $x \in \Gamma_G(s) \cap Q_1$ and $y \in \Gamma_G(s) \cap T$.* □

2.4 Lower Bound

For a multigraph G and a fixed integer $\ell \geq k \geq 4$, let $opt(G)$ denote the optimal value of EVAP(ℓ,k) in G, i.e., the minimum size $|F|$ of a set F of new edges to obtain an (ℓ,k)-connected graph $G+F$. In this section, we derive two types of lower bounds, $\alpha(G)$ and $\beta(G)$, on $opt(G)$.

Let X be a cut in G. To make G ℓ-edge-connected and k-vertex-connected, it is necessary to add at least $\max\{\ell - c_G(X), 0\}$ edges between X and $V-X$, or at least $\max\{k - |\Gamma_G(X)|, 0\}$ edges between X and $V - X - \Gamma_G(X)$ if $V - X - \Gamma_G(X) \neq \emptyset$. Given a subpartition $\mathcal{X} = \{X_1, \cdots, X_p, X_{p+1}, \cdots, X_q\}$ of V, where $V - X_i - \Gamma_G(X_i) \neq \emptyset$ holds for $i = p+1, \cdots, q$, we can sum up "deficiencies" $\max\{\ell - c_G(X_i), 0\}$, $i = 1, \cdots, p$, and $\max\{k - |\Gamma_G(X_i)|, 0\}$, $i = p+1, \cdots, q$. As adding one edge to G contributes to the deficiency of at most two cuts in $\mathcal{X}$, we need at least $\lceil \alpha(G)/2 \rceil$ new edges to make G (ℓ,k)-connected, where

$$\alpha(G) = \max_{\text{all subpartitions } \mathcal{X}} \left\{ \sum_{i=1}^{p} (\ell - c_G(X_i)) + \sum_{i=p+1}^{q} (k - |\Gamma_G(X_i)|) \right\}, \tag{2.1}$$

and the maximum is taken over all subpartitions $\mathcal{X} = \{X_1, \cdots, X_p, X_{p+1}, \cdots, X_q\}$ of V with $V - X_i - \Gamma_G(X_i) \neq \emptyset$, $i = p+1, \cdots, q$.

We now consider another case in which different type of new edges become necessary. For a disconnecting set S of G with $|S| \leq k-1$, let $T_1, \cdots, T_q$ denote all the components in $G-S$, where $q = p(G-S)$. To make G k-vertex-connected, a new edge set F must be added to G so that all T_i form a single connected component in $(G+F)-S$. For this, it is necessary to add at least $p(G-S)-1$ edges to connect all components in $G-S$. Define

$$\beta(G) = \max_{\substack{\text{all disconnecting sets} \\ S \text{ in } G \text{ with } |S| < k}} \left\{ p(G-S) \right\}. \tag{2.2}$$

Thus at least $\beta(G)-1$ new edges are necessary to make G (ℓ,k)-connected.

Lemma 13. (Lower Bound) *For a given multigraph G, let*

$$\gamma(G) = \max\{\lceil \alpha(G)/2 \rceil, \beta(G) - 1\}.$$

Then $\gamma(G) \leq opt(G)$ holds, where $opt(G)$ denotes the minimum number of edges augmented to make G (ℓ,k)-connected. □

Remark: Both of $\alpha(G)$ and $\beta(G)$ can be computed in polynomial time. The algorithm of computing $\alpha(G)$ will be mentioned in Section 4. It can be seen that $\beta(G)$ is computable in polynomial time, since finding all minimum shredders in the given graph can be done in polynomial time [2]. □

Based on this, we shall prove the next result in this paper.

Theorem 14. *Let G be a $(k-1)$-vertex-connected multigraph G with $k \geq 4$. Then, for any integer $\ell \geq k$, $\gamma(G) \leq opt(G) \leq \gamma(G)+2\ell$ holds, and a feasible solution F of* EVAP(ℓ,k) *with $\gamma(G) \leq |F| \leq \gamma(G)+2\ell$ can be found in $O(\min\{k,\sqrt{n}\}kn^3+n^4)$ time, where n is the number of vertices in G.* □

3 An Algorithm for EVAP(ℓ, k)

In this section, we present a polynomial time algorithm, called EV-AUG, for finding a nearly optimal solution to EVAP(ℓ,k) for a given $(k-1)$-vertex-connected graph and a given integer $\ell \geq k \geq 4$. The algorithm EV-AUG consists of the following three major steps. In each step, we also give some properties to verify its correctness. The proof for these properties will be given in the subsequent sections.

Algorithm EV-AUG

Input: An undirected multigraph $G=(V,E)$ with $|V| \geq k+1$, $\kappa(G)=k-1$, and an integer $\ell \geq k \geq 4$.

Output: A set of new edges F with $|F| \leq opt(G)+2\ell$ such that $G^* = G+F$ satisfies $\lambda(G^*) \geq \ell$ and $\kappa(G^*) \geq k$.

Step I (Addition of vertex s and associated edges): If $t(G) \leq 2\ell+1$ holds, then we can see from Lemma 2 that G can be made k-vertex-connected by adding a new edge set F_0' with $|F_0'| \leq 2\ell$. Moreover, G can be made ℓ-edge-connected by adding a new edge set F_0'' with $|F_0''| \leq \lceil \alpha(G)/2 \rceil$, by using the algorithm AUGMENT in [19]. Output $F_0' \cup F_0''$ as an solution, which satisfies $|F_0'| + |F_0''| \leq opt(G)+2\ell$.

If $t(G) \geq 2\ell+2$ holds, then Lemma 1 says that every two cuts in $\mathcal{D}(G)$ are pairwise disjoint. Then add a new vertex s together with a set F_1 of edges between s and V so that the resulting graph $G_1 = (V \cup \{s\}, E \cup F_1)$ satisfies

$$c_{G_1}(X) \geq \ell \quad \text{for all cuts } X \text{ with } \emptyset \neq X \subset V, \tag{3.1}$$

$$c_{G_1}(s,D) \geq 1 \quad \text{for all minimal tight sets } D \in \mathcal{D}(G), \tag{3.2}$$

where $|F_1|$ is *minimum* subject to (3.1) and (3.2) (Section 4 describes how to find such a minimum F_1).

Property 15. The above set of edges F_1 satisfies $|F_1| = \alpha(G)$. □

If $c_{G_1}(s)$ is odd, then add one edge $\hat{e} = (s,w)$ to F_1 for a vertex w arbitrarily chosen from V. Denote the resulting graph again by G_1. After setting $G' := G_1$ and $\overline{G'} := G_1 - s$, we go to Step II.

Step II (Edge-splitting): While $t(\overline{G'}) \geq 2\ell+2$ holds (hence superleaves in $\overline{G'}$ are pairwise disjoint by Lemma 3), repeat the following procedure (A) or (B); if the condition

$$\beta(\overline{G'}) - 1 \geq \lceil t(\overline{G'})/2 \rceil \ (\geq \ell+1) \tag{3.3}$$

holds, then procedure (A) else procedure (B) (If $t(\overline{G'}) \leq 2\ell+1$, then go to Step III).

Procedure (A)

Choose a minimum disconnecting set S^* in $\overline{G'}$ satisfying $p(\overline{G'}-S^*)=\beta(\overline{G'})$.

(Case-1) $\overline{G'}$ has a cut $T^* \in C(\overline{G'}-S^*)$ with $c_{G'}(s,T^*) \geq 2$. Then the next property holds.

Property 16. In Case-1, there is a pair of two edges $\{(s,u),(s,v)\}$ such that (*i*) u and v are contained in distinct cuts in $C(\overline{G'}-S^*)$, (*ii*) $\{(s,u),(s,v)\}$ is λ-splittable and κ-splittable, and (*iii*) the multigraph $G'' := G' - \{(s,u),(s,v)\} + (u,v)$ resulting from splitting edges (s,u) and (s,v) in G' satisfies $\beta(G''-s)=\beta(\overline{G'})-1$. □

Then split the two edges $\{(s,u),(s,v)\}$ in Property 16. Set $G' := G' - \{(s,u),(s,v)\} + \{(u,v)\}$, $\overline{G'} := G'-s$, and go to Step II.

(Case-2) Every cut $T^* \in C(\overline{G'}-S^*)$ satisfies $c_{G'}(s,T^*)=1$.

Property 17. In Case-2, if one of the following cases (1) – (4) holds, then there is a λ-splittable and κ-splittable pair of two edges incident to s after hooking up at most one split edge and shifting at most one edge incident to s. Otherwise we can make G (ℓ,k)-connected by adding $\beta(G)-1$ edges.
(1) There is an edge $e=(s,v)$ with $v \in S^*$.
(2) There is a split edge $e=(u,v) \notin E$ with $u \in S^*$ and $v \in T_1 \in C(\overline{G'}-S^*)$.
(3) There is a split edge $e=(u,v) \notin E$ with $\{u,v\} \subseteq T_1 \in C(\overline{G'}-S^*)$ such that $p(\overline{G'}-S^*)=p((\overline{G'}-e)-S^*)$ holds.
(4) There is a split edge $e=(u,v) \notin E$ with $u,v \in S^*$. □

If G' satisfies one of the conditions (1) – (4) in Property 17, then split the two edges $\{(s,x),(s,y)\}$ defined in Property 17: set $G' := G' - \{(s,x),(s,y)\} + \{(x,y)\}$, $\overline{G'} := G'-s$, and return to Step II. Otherwise make G (ℓ,k)-connected by augmenting $\beta(G)-1$ edges, according to Property 17, and halt after outputting the set of edges added to G as an optimal solution.

Property 18. Each iteration of procedure (A) decreases $\beta(\overline{G'})$ at least by one, and does not increase $c_{G'}(s)$. □

Procedure (B)

Choose an arbitrary superleaf Q_1 in $\overline{G'}$. Then there is an edge (s,x_1) with $x_1 \in D_1 \in \mathcal{D}(\overline{G'})$ such that $D_1 \subseteq Q_1$. Let $S=\Gamma_{\overline{G'}}(Q_1)$.

(Case-3) S is a shredder in $\overline{G'}$.

Property 19. In Case-3, there is an edge (s,x_2) with $x_2 \in D_2 \in \mathcal{D}(\overline{G'})$ such that $D_2 \subseteq T_2 \in C(\overline{G'}-S)-\{Q_1\}$ holds and $\{(s,x_1),(s,x_2)\}$ is λ-splittable and κ-splittable. □

After splitting two edges (s,x_1) and (s,x_2) in Property 19, set $G' := G' - \{(s,x_1),(s,x_2)\} + \{(x_1,x_2)\}$ and $\overline{G'} := G'-s$, and return to Step II.

(Case-4) S is not a shredder in $\overline{G'}$.

Property 20. In Case-4, in addition to Q_1, there are two distinct superleaves Q_2 and Q_3 in $\overline{G'}$ such that $Q_2 \cap S = \emptyset = Q_3 \cap S$ holds, and each of $\{(s,x_1),(s,x_2)\}$, $\{(s,x_2),(s,x_3)\}$, and $\{(s,x_3),(s,x_1)\}$ is λ-splittable, where $x_2 \in D_2 \cap \Gamma_{G'}(s)$ and $x_3 \in D_3 \cap \Gamma_{G'}(s)$ for cuts D_i with $D_i \in \mathcal{D}(\overline{G'})$ and $D_i \subseteq Q_i$ for $i=2,3$. □

Take the two edges (s,x_2) and (s,x_3) in Property 20. Then Theorem 11 says that at least one of $\{(s,x_1),(s,x_2)\}$, $\{(s,x_2),(s,x_3)\}$, and $\{(s,x_3),(s,x_1)\}$ (say, $\{(s,x_1),(s,x_2)\}$) is κ-splittable. After splitting the two edges (s,x_1) and (s,x_2), set $G' := G' - \{(s,x_1),(s,x_2)\} + \{(x_1,x_2)\}$ and $\overline{G'} := G' - s$, and return to Step II.

Property 21. Each iteration of procedure (B) decreases $t(\overline{G'})$ and $c_{G'}(s)$ at least by one, respectively. □

Step III (Edge augmentation): Let $G_3 = (V \cup s, E \cup F_3 \cup F_3')$ denote the current multigraph, where F_3 denotes the set of split edges, and $F_3' := E_{G_3}(s,V)$. Let $\overline{G_3} := G_3 - s$.

Property 22. In Step III, $t(\overline{G_3}) \leq 2\ell + 1$ and $|F_3| + |F_3'|/2 = \lceil \alpha(G)/2 \rceil$ hold, and all cuts $X \subset V$ satisfies $c_{G_3}(X) \geq \ell$. □

Then find a complete edge-splitting at s in G_3 according to Theorem 8 to obtain $G_3^* = (V, E \cup F_3^*)$ (ignoring the isolated vertex s) with $\lambda(G_3^*) \geq \ell$. Note that $|F_3^*| = \lceil \alpha(G)/2 \rceil$ holds from Property 22. Moreover, by Lemma 2 and Property 22, G_3^* can be made k-vertex-connected by adding a set F_4^* of new edges with $|F_4^*| \leq 2\ell$. Output $F_3^* \cup F_4^*$ as an solution, where its size satisfies $|F_3^*| + |F_4^*| \leq \lceil \alpha(G)/2 \rceil + 2\ell \leq opt(G) + 2\ell$. □

4 Justification of Step I

Proof of Property 15: We show that a set F_1 of edges satisfying Property 15 can be found by applying the algorithm ADD-EDGE [11], as follows.

Let $G_1' = (V \cup s, E \cup F_1')$ be a multigraph such that F_1' is *minimal* subject to (3.1) and (3.2). Note that $|F_1'| \geq \alpha(G)$ holds, since otherwise G_1' violates (3.1) or (3.2).

If $G_1' - e$ violates (3.1) for an edge $(s,u) \in E_{G_1'}(s)$, then there is a λ-critical cut $X_u \subset V$ with respect to u. If $G_1' - e$ violates (3.2) for an edge $(s,u) \in E_{G_1'}(s)$, then there is a cut $D \in \mathcal{D}(G)$ satisfying $c_{G_1'}(s,D) = 1$. We call such a cut D *κ-critical.* Note that every λ-critical cut X satisfies $c_{G_1'}(s,X) = \ell - c_G(X)$ and every κ-critical cut X satisfies $c_{G_1'}(s,X) = k - |\Gamma_G(X)|\ (=1)$. Hence if there is a subpartition $\mathcal{X}$ of V satisfying $\Gamma_{G_1'}(s) \subseteq \cup_{X \in \mathcal{X}} X$ such that every cut $X \in \mathcal{X}$ is λ-critical or κ-critical, then $|F_1'| = \alpha(G)$ holds, since $|F_1'| = \sum_{i=1}^{p}(\ell - c_G(X_i)) + \sum_{i=p+1}^{q}(k - |\Gamma_G(X_i)|) \leq \alpha(G)$ holds for $\mathcal{X} = \{X_1, \cdots, X_q\}$ from the maximality of $\alpha(G)$.

Otherwise there is a pair of two cuts X and Y with $X \cap Y \neq \emptyset$ such that X is λ-critical, Y is κ-critical, and $(Y - \sum\{X_i | X_i \text{ is λ-critical}\}) \cap \Gamma_{G_1'}(s) \neq \emptyset$ holds, because it can be easily seen that a family of all λ-critical cuts is laminar and every two cuts in $\mathcal{D}(G)$ are pairwise disjoint. Then we can replace and remove some edges in $E_{G_1'}(s,V)$ to decrease the number of such pairs to zero by applying the algorithm ADD-EDGE [11], since the algorithm ADD-EDGE depends on the property that every two cuts in $\mathcal{D}(G)$ are pairwise disjoint and every κ-critical cut D satisfies $c_{G_1'}(s,D) = 1$. Thus, we can find a set F_1 of edges satisfying Property 15. □

5 Justification of Step II

In Step II, $t(\overline{G'}) \geq 2\ell + 2$ holds and G' satisfies (3.1) and (3.2). First we consider the correctness of Procedure (A). For this, we prove Properties 16 – 18. Since condition (3.3) implies $p(\overline{G'} - S^*) \geq \ell + 1 \geq k + 1$, Lemma 5 tells that

$$\text{every superleaf } Q \text{ in } \overline{G'} \text{ satisfies } Q \cap S^* = \emptyset. \tag{5.1}$$

Proof of Property 16: Let Q_1 be a superleaf in $\overline{G'}$ satisfying $\Gamma_{\overline{G'}}(Q_1) = S^*$, and choose a vertex $x_1 \in Q_1$. Note that such Q_1 exists, since otherwise Lemma 4 and (5.1) imply that every cut in $C(\overline{G'} - S^*)$ contains at least two cuts in $\mathcal{D}(\overline{G'})$ and hence $\beta(\overline{G'}) \leq \lceil t(\overline{G'})/2 \rceil$ holds, contradicting condition (3.3). Let N_1 be a set of vertices $v \in \Gamma_{G'}(s) \cap (V - S^*)$ such that $\{(s,x_1),(s,v)\}$ is λ-splittable. Since $t(\overline{G'}) \geq 2\ell + 2$ and (5.1) imply $|\Gamma_{G'}(s) \cap (V - S^*)| \geq 2\ell + 2$, we have $|N_1| \geq 2\ell + 2 - \{\ell + 2 - (k-1)\} = \ell + k - 1$ by Corollary 10. Let $C^*(\overline{G'} - S^*)$ be a family of cuts $T \in C(\overline{G'} - S^*) - Q_1$ which contains a vertex in N_1.

(1) If there is a cut $T \in C^*(\overline{G'} - S^*)$ that satisfies $c_{G'}(s,T) \geq 2$, then $\{(s,x_1),(s,v)\}$ with $v \in T \cap N_1$ is λ-splittable (from the definition of N_1) and κ-splittable (from Theorem 12). If $\overline{G'}$ has another disconnecting set $S' \neq S^*$ satisfying $p(\overline{G'} - S') = \beta(\overline{G'})$, then it is not difficult to see from condition (3.3) that $\overline{G'}$ has exactly one disconnecting set $S' \neq S^*$ satisfying $p(\overline{G'} - S') = \beta(\overline{G'})$, and that there is a cut $T_1 \in C(\overline{G'} - S^*)$ which contains $(\beta(\overline{G'}) - 1)$ S'-components, and moreover, by Corollary 10 and $\beta(\overline{G'}) - 1 \geq \ell + 1$, T_1 contains a cut $T \in C(\overline{G'} - S')$ containing a vertex $v_1 \in N_1$.

(2) Otherwise every $T \in C^*(\overline{G'} - S^*)$ satisfies $c_{G'}(s,T) = 1$. By applying Lemma 4 to shredder S^* in $\overline{G'}$, we see that every $T \in C^*(\overline{G'} - S^*)$ is a superleaf in $\overline{G'}$, which contains exactly one vertex in N_1. Now, by assumption, $\overline{G'}$ has a cut $T^* \in C(\overline{G'} - S^*)$ satisfying $c_{G'}(s,T^*) \geq 2$, and $T^* \notin C^*(\overline{G'} - S^*)$ and $T^* \cap N_1 = \emptyset$ hold. We choose one edge (s,x_2) such that $x_2 \in T^*$ (clearly, $x_2 \notin N_1$). Then there is an edge (s,v) with $v \in N_1$ such that $\{(s,x_2),(s,v)\}$ is λ-splittable, because $|N_1| \geq \ell + k - 1$ holds and Corollary 10 says that there are at most $(\ell+2) - (k-1)$ neighbors v of s such that $\{(s,x_2),(s,v)\}$ is not λ-splittable. By $T^* \cap N_1 = \emptyset$, vertices $x_2 \in T^*$ and $v \in N_1$ are contained in different S^*-components in $\overline{G'}$. Thus, $\{(s,x_2),(s,v)\}$ is κ-splittable by Theorem 12.

Finally, we see that $\beta(G'' - s) = \beta(\overline{G'}) - 1$ holds in $G'' := G' - \{(s,u),(s,v)\} + \{(u,v)\}$ for a pair of edges $\{(s,u),(s,v)\}$ chosen in the above (1) and (2), since the edge (u,v) connects two distinct S-components for a disconnecting set S in $\overline{G'}$ that satisfies $p(\overline{G'} - S) = \beta(\overline{G'})$. □

Proof of Property 17: Every cut $T \in C(\overline{G'} - S^*)$ contains exactly one vertex in $\Gamma_{G'}(s)$, and hence is a superleaf in $\overline{G'}$ from Lemma 4. Hence by (5.1), $t(\overline{G'}) = \beta(\overline{G'}) = |C(\overline{G'} - S^*)|$ holds. Now from $p(\overline{G'} - S^*) \geq \ell + 1$, $c_{G'}(X) \geq p(\overline{G'} - S^*)|X| \geq (\ell + 1)$ holds for every cut $X \subseteq S^*$. If one of the conditions (1) – (4) holds, then we hook up at most one split edge and/or shift at most one edge incident to s, to obtain a multigraph G'' in which we can apply Property 16 to find λ-splittable and κ-splittable pair of two edges incident to s.

(1) Let X be a λ-critical cut with respect to v (if any). Now $\overline{G'}$ has a cut $T_1 \in C(\overline{G'} - S^*)$ satisfying $T_1 \cap X \neq \emptyset$, since every cut $Y \subseteq S^*$ satisfies $c_{G'}(Y) \geq \ell + 1$. The graph $G'' := G' - e + e'$ resulting from shifting e to an edge $e' = (s,v_1)$ with $v_1 \in X \cap T_1$ satisfies (3.1) and (3.2) by Lemma 7 and by (5.1), respectively. Now $c_{G''}(s,T_1) = 2$ holds.

(2),(3) The graph $G'' := G' - e + \{(s,u),(s,v)\}$ resulting from hooking up e also satisfies condition (3.3), because $t(\overline{G'}) = \beta(\overline{G'}) \geq 2\ell + 2$ and $\ell \geq 4$ hold. Now $c_{G''}(s,T_1) \geq 2$ holds.

(4) The graph $G' - e + \{(s,u),(s,v)\}$ resulting from hooking up e also satisfies condition (3.3), because $t(\overline{G'}) = \beta(\overline{G'}) \geq 2\ell + 2$ and $\ell \geq 4$ hold. Then applying the edge-shifting in (1) to the graph $G' - e + \{(s,u),(s,v)\}$, we obtain a multigraph G'' for which $c_{G''}(s,T_1) = 2$ holds for some $T_1 \in C(\overline{G''} - S^*)$.

Finally if none of (1) – (4) holds, we show that G can be made (ℓ,k)-connected by adding $\beta(G) - 1$ new edges to G. In this case, every $T \in C(\overline{G'} - S^*)$ satisfies $c_{G'}(s,T) = 1$, $E_{G'}(s,S^*) = \emptyset$ holds, and every edge $e = (u,v) \in F'$ satisfies $\{u,v\} \subset V - S^*$ and $p((\overline{G'} - e) - S^*) = p(\overline{G'} - S^*) + 1$, where F' is the set of all split edges (note $F' \cap E = \emptyset$). We easily see that $\kappa(\overline{G'} + F_2) \geq k$ holds for any new edge set F_2 such that (N^*, F_2) is a spanning tree for $N^* = \Gamma_{G'}(s)$. We choose an F_2 as follows. Since G' satisfies (3.1), by a complete edge-splitting at s in G' as described in Theorem 8, we can obtain $G'' = (V, E \cup F'')$ (ignoring the isolated vertex s) with $\lambda(G'') \geq \ell$. Note that in G'', every split edge $e=(u,v)\in F''$ satisfies $\{u,v\}\subset V-S^*$ and $p((G''-e)-S^*)=p(G''-S^*)+1$, because every newly split edge connects some T_i and T_j satisfying $T_i,T_j\in C(\overline{G'}-S^*)$ and $T_i\neq T_j$. Therefore $p(G-S^*)=p((G''-F'')-S^*)=p(G''-S^*)+|F''|$ holds. Let $C(G''-S^*)=\{T_1',T_2',\cdots,T_b'\}$, where $b=p(G''-S^*)$. Then $\kappa(G''+F^*)\geq k$ holds for $F^*=\{(x_i,x_{i+1})\mid i=1,\cdots,b-1\}$, where x_i is a vertex in $T_i'\cap N^*$, since $(N^*,F_2=(F''-F')\cup F^*)$ is a spanning tree. Note that $|F''|+|F^*|=|F''|+p(G''-S^*)-1=p(G-S^*)-1\leq\beta(G)-1$ implies that $F''\cup F^*$ is an optimal solution since $|F''\cup F^*|$ attains a lower bound $\beta(G)-1$. □

Proof of Property 18: Each iteration of procedure (A) decreases $\beta(\overline{G'})$ at least by one from Property 16, and does not increase $c_{G'}(s)$, since hooking up does not increase $\beta(\overline{G'})$ and at most one edge is hooked up in each case. □

Next we prove Properties 19 – 21 for the correctness of Procedure (B). Now $\beta(\overline{G'})\leq\lceil t(\overline{G'})/2\rceil$ holds.

Proof of Property 19: Assume that the property does not hold. Let $\mathcal{A}_1$ denote the family of superleaves Q in $\overline{G'}$ satisfying $Q\cap S=\emptyset$. Then, by Theorem 12 and Corollary 10, every $Q\in\mathcal{A}_1$ satisfies $Q\in C(\overline{G'}-S)$ satisfying $c_{G'}(s,Q)=1$ or $\Gamma_{G'}(s)\cap D\subseteq X$ for some $D\in\mathcal{D}(\overline{G'})$ with $D\subseteq Q$ and a maximal cut $X\subset V$ with satisfying $c_{G'}(s,X)\leq\ell+2-(k-1)$ and containing $x_1\in X$ and all vertices $v\in\Gamma_{G'}(s)$ such that $\{(s,x_1),(s,v)\}$ is not λ-splittable. Therefore $|\mathcal{A}_1|\leq\ell+3-k+p(\overline{G'}-S)$ holds.

If $p(\overline{G'}-S)\geq k$ holds, then Lemma 5 implies that every superleaf Q in $\overline{G'}$ satisfies $Q\in\mathcal{A}_1$, and hence $|\mathcal{A}_1|=t(\overline{G'})$ holds. This implies $t(\overline{G'})=|\mathcal{A}_1|\leq\ell+3-k+p(\overline{G'}-S)\leq\ell+3-k+\beta(\overline{G'})$. ¿From this and $\beta(\overline{G'})\leq\lceil t(\overline{G'})/2\rceil$, we have $\lceil t(\overline{G'})/2\rceil\leq\ell+4-k$ $(k\geq 4)$, contradicting $t(\overline{G'})\geq 2\ell+2$.

If $p(\overline{G'}-S)\leq k-1$, then $|\mathcal{A}_1|\leq\ell+3-k+p(\overline{G'}-S)\leq\ell+2$ holds. Since $|S|=k-1$ holds and every two superleaves in $\overline{G'}$ are pairwise disjoint, we have $t(\overline{G'})\leq|\mathcal{A}_1|+k-1\leq\ell+k+1\leq 2\ell+1$, contradicting $t(\overline{G'})\geq 2\ell+2$. □

Proof of Property 20: We can see this property from Corollary 10 and $|\{Q\subset V\mid Q$ is a superleaf in $\overline{G'}$ with $Q\cap S\neq\emptyset\}|\geq 2\ell-k+2$. Here we omit the details. □

Proof of Property 21: For the resulting graph G'', $c_{G''}(s)=c_{G'}(s)-2$ clearly holds. Moreover, $t(\overline{G'})\leq t(\overline{G''})-1$ follows from Lemma 6 and Theorem 12. □

6 Complexity

First the family $\mathcal{D}(G)$ of all minimal tight sets and superleaves in G can be computed in $O(\min\{k-1,\sqrt{n}\}mn)$ time by using the standard network flow technique n times [4]. If $t(G)\leq 2\ell+1$, then F_0' can be computed in $O(\min\{k,\sqrt{n}\}kn^3)$ time by applying Phase 5 of Jordán's algorithm in [13], and F_0'' can be computed in $O((m+n\log n)n\log n)$ time, by using the algorithm AUGMENT in [19]. If $t(G)\geq 2\ell+2$, G_1 can be computed in $O(n^2m+n^3\log n)$ time, since the procedure is based on the algorithm ADD-EDGE* [11], which takes $O(n^2m+n^3\log n)$ time.

In Step II, $\max\{|\Gamma_{\overline{G'}}(Q)|\ |Q$ is a superleaf in $\overline{G'}\}=\beta(\overline{G'})$ if (3.3) holds, as mentioned in the proof of Property 16. So whether (3.3) holds or not can be checked and a disconnecting set S^* with $p(\overline{G'}-S^*)=\beta(\overline{G'})$ can be found if (3.3) holds, by computing $\max\{|\Gamma_{\overline{G'}}(Q)|\ |Q$ is a superleaf in $\overline{G'}\}$. In the last case of procedure (A), it is known in [19] that a complete splitting can be found in $O(n(m+n\log n)\log n)$ time. The remaining procedure of Step II is executed in $O(n^4)$ time as follows. ¿From Property 18 (resp., Property 21), the number of iterations of procedure (A) (resp., (B)) is at most n by $\beta(G)\leq n$ (resp., $t(G)\leq n$). In each iteration, we can find a κ-splittable pair in $O(\min\{k-1,\sqrt{n}\}m)$ time by using the standard network flow technique [4]. For the vertex $x_1 \in \Gamma_{G'}(s)$, a maximal dangerous cut $X \subset V$ with $\{x_1,y\} \subseteq X$ can be found in $O(n^3)$ time by computing a maximum flow between two vertices x_1 and s in $G' - \{(s,x_1),(s,y)\} + (x_1,y)$ [16] (if any). Hence we can find a λ-splittable pair in $O(n^3)$ time. Note that, in procedure (A), a λ-critical cut $X \subset V$ can be found in $O(n(m+n\log n))$ time by using the algorithm AUGMENT in [19].

Summarizing the argument given so far, Theorem 14 is now established. □

7 Concluding Remarks

In this paper, we gave a polynomial time algorithm for augmenting a given $(k-1)$-vertex-connected multigraph G, $k \geq 4$, to an ℓ-edge-connected and k-vertex-connected graph by adding at most 2ℓ surplus edges over the optimum. However, if $\ell = k \geq 4$, there is an algorithm [13, 14] that produces at most $(k-2)/2$ surplus edges over the optimum. Therefore, it is a future work to close the gap between this and our bound.

References

[1] J. Bang-Jensen and T. Jordán, *Edge-connectivity augmentation preserving simplicity*, Proc. 38nd IEEE Symp. Found. Comp. Sci., 1997, pp. 486–495.
[2] J. Cherian and R. Thurimella, *Fast algorithms for k-shredders and k-node connectivity augmentation*, Proceedings 28th ACM Symposium on Theory of Computing, 1996, pp. 37–46.
[3] K. P. Eswaran and R. E. Tarjan, *Augmentation problems*, SIAM J. Computing, Vol.5, 1976, pp. 653–665.
[4] S. Even and R. E. Tarjan, *Network flow and testing graph connectivity*, SIAM J. Comput., Vol. 4, 1975, pp. 507–518.
[5] A. Frank, *Augmenting graphs to meet edge-connectivity requirements*, SIAM J. Discrete Mathematics, Vol.5, 1992, pp. 25–53.
[6] T. Hsu and V. Ramachandran, *A linear time algorithm for triconnectivity augmentation*, Proc. 32nd IEEE Symp. Found. Comp. Sci., 1991, pp.548–559.
[7] T. Hsu and V. Ramachandran, *Finding a smallest augmentation to biconnect a graph*, SIAM J. Computing, Vol.22, 1993, pp.889–912.
[8] T. Hsu, *Undirected vertex-connectivity structure and smallest four-vertex-connectivity augmentation*, Lecture Notes in Computer Science 1004, Springer-Verlag, 6th International Symp. on Algorithms and Computation, 1995, pp. 274-283.
[9] T. Hsu and M. Kao, *Optimal bi-level augmentation for selectively enhancing graph connectivity with applications*, Lecture Notes in Computer Science 1090, Springer-Verlag, 2nd International Symp. on Computing and Combinatorics, 1996, pp. 169-178.
[10] T. Ishii, H. Nagamochi, and T. Ibaraki, *Augmenting edge-connectivity and vertex-connectivity simultaneously*, Lecture Notes in Computer Science 1350, Springer-Verlag, 8th ISAAC, 1997, pp. 102-111.
[11] T. Ishii, H. Nagamochi and T. Ibaraki, *Optimal augmentation of a biconnected graph to a k-edge-connected and triconnected graph*, Proceedings of 9th Annual ACM-SIAM Symposium on Discrete Algorithms, San Francisco, California January 1998, pp. 280–289.
[12] T. Ishii, H. Nagamochi and T. Ibaraki, *k-edge and 3-vertex connectivity augmentation in an arbitrary multigraph*, Lecture Notes in Computer Science 1533, Springer-Verlag, 9th ISAAC, 1998, pp. 159–168.
[13] T. Jordán, *On the optimal vertex-connectivity augmentation*, J. Combinatorial Theory, Series B, Vol.63, 1995, pp.8–20.

[14] T. Jordán, *A note on the vertex-connectivity augmentation problem*, J. Combinatorial Theory, Series B., Vol.71, 1997, pp. 294-301.
[15] T. Jordán, *Two NP-complete augmentation problems*, Preprint no. 8, Department of Mathematics and Computer Science, Odense University, 1997.
[16] A. V. Karzanov, *Determining the maximal flow in a network by the method of preflows*, Soviet Math. Doklady, Vol.15, 1974, pp. 434–437.
[17] L. Lovász, Combinatorial Problems and Exercises, North-Holland, 1979.
[18] W. Mader, *Ecken vom Grad n in Minimalen n-fach zusammenhängenden Graphen*, Arch. Math. Vol.23, 1972, pp. 219–224.
[19] H. Nagamochi and T. Ibaraki, *Deterministic $\tilde{O}(nm)$ time edge-splitting in undirected graphs*, J. Combinatorial Optimization, 1, (1997), pp. 5–46.
[20] T. Watanabe and A. Nakamura, *Edge-connectivity augmentation problems*, J. Comp. System Sci., Vol.35, 1987, pp.96–144.
[21] T. Watanabe and A. Nakamura, *A smallest augmentation to 3-connect a graph*, Discrete Appl. Math., Vol.28, 1990, pp.183–186.

An Optimisation Algorithm for Maximum Independent Set with Applications in Map Labelling*

Bram Verweij and Karen Aardal

Deptartment of Computer Science, Utrecht University, The Netherlands
{bram,aardal}@cs.uu.nl

Abstract. We consider the following map labelling problem: given distinct points $p_1, p_2, \ldots, p_n$ in the plane, find a set of pairwise disjoint axis-parallel squares $Q_1, Q_2, \ldots, Q_n$ where p_i is a corner of Q_i. This problem reduces to that of finding a maximum independent set in a graph.
We present a branch and cut algorithm for finding maximum independent sets and apply it to independent set instances arising from map labelling. The algorithm uses a new technique for setting variables in the branch and bound tree that implicitly exploits the Euclidean nature of the independent set problems arising from map labelling. Computational experiments show that this technique contributes to controlling the size of the branch and bound tree. We also present a novel variant of the algorithm for generating violated odd-hole inequalities. Using our algorithm we can find provably optimal solutions for map labelling instances with up to 950 cities within modest computing time, a considerable improvement over the results reported on in the literature.

1 Introduction

When designing maps an important question is how to place the names of the cities on the map such that each city name appears close to the corresponding city, and such that no two names overlap. Various problems related to this question are referred to as *map labelling problems*.

A basic map labelling problem is described as follows: given a set $P = \{p_1, p_2, \ldots, p_n\}$ of n distinct points in $\mathbb{R}^2$, determine the supremum σ^* of all reals σ, for which there are n pairwise disjoint, axis-parallel $\sigma \times \sigma$ squares $Q_1, Q_2, \ldots, Q_n$, where p_i is a corner of Q_i for all $i = 1, \ldots, n$. By "pairwise disjoint squares" we mean that no overlap between any two squares is allowed. Once the squares are known they define the boundaries of the area where the labels can be placed. The decision variant (DP) of this problem is for fixed σ to decide whether there exists a set of squares $Q_1, \ldots, Q_n$ as described above. Formann and Wagner [13] showed that problem DP is $\mathcal{NP}$-complete. Kučera *et al.* [18] observed that there are only $O(n^2)$ possible values that σ^* can take. Optimising over those can be done by solving only $O(\log n)$ problems DP with different σ using binary search. Moreover, they present an algorithm that solves the map labelling problem DP.

* This research was (partially) supported by ESPRIT Long Term Research Project 20244 (project ALCOM IT: *Algorithms and Complexity in Information Technology*).

We study the following generalisation (ODP) of problem DP: given σ, find as many pairwise disjoint squares as possible. Clearly, a solution to problem DP exists if and only if a solution to problem ODP exists in which n squares are found. An advantage of studying ODP instead of DP is that a feasible solution to ODP actually represents a partial map labelling. Our experience shows that it is relatively easy to find feasible solutions to ODP of good quality. Since DP is $\mathcal{NP}$-complete, problem ODP is $\mathcal{NP}$-hard, and therefore we do not expect that there exists a polynomial time algorithm for solving ODP. So, we have to resort to enumerative methods such as branch and bound if we want to solve ODP to optimality, unless $\mathcal{P} = \mathcal{NP}$. Formann and Wagner [13] developed a $\frac{1}{2}$-approximation algorithm for ODP. Different heuristic algorithms (including simulated annealing) are discussed by Christensen *et al.* [8]. Van Dijk *et al.* [10] considered genetic algorithms, Wagner and Wolff [26] propose a hybrid heuristic. Cromly [9] proposed a semi-automatic LP based approach for finding feasible solutions to ODP. Zoraster [29, 30] used Lagrangean relaxation to make a heuristic algorithm for ODP. We formulate ODP as an *independent set problem*, see Section 1.1 and develop a *branch and cut algorithm* for solving ODP to optimality. Branch and cut is a branch and bound algorithm in which a so-called *cutting plane algorithm* may be called in every node of the branch and bound tree. For readers unfamiliar with branch and cut algorithms, we give a brief description in Section 2.1.

1.1 The Independent Set Formulation

An *independent set* S in a graph G is a set of nodes such that no two nodes in S are adjacent in G. Problem ODP can be formulated as a maximum cardinality independent set problem on the *conflict graph* $G_{P,\sigma} = (V_P, E_{P,\sigma})$ associated with the map labelling problem: for each Q_i, $i = 1, \ldots, n$, we add four nodes to V_P, corresponding to the possible placements of square Q_i. We add edge $\{u,v\}$ to $E_{P,\sigma}$ if u and v correspond to placements of Q_i and Q_j for some $i \neq j$ with $Q_i \cap Q_j \neq \emptyset$, or to distinct placements of Q_i for some i. Using this construction, any valid placement of the squares $Q_1, \ldots, Q_n$ corresponds to an independent set in $G_{P,\sigma}$ of size n, and vice versa. This relation to the independent set problem has been used to derive polynomial time approximation algorithms [1, 4] for problem ODP. Kakoulis and Tollis [17] show how to approach more general map labelling problems than problem ODP using essentially an independent set formulation.

The complement of a graph $G = (V,E)$ has the same node set as G and is denoted by $\bar{G} = (V, \bar{E})$, where $\{u,v\} \in \bar{E}$ if and only if $\{u,v\} \notin E$. A set of nodes $S \subset V$ is a maximum independent set in G if and only if S is a maximum clique in $\bar{G}$. Several optimisation algorithms for finding maximum independent sets [24, 25, 20, 22, 28, 14] and maximum cliques [5, 2, 3, 7, 6] have been reported on in the literature. The independent set formulation and the clique formulation are equally suitable for algorithmic design. In our study we use the independent set formulation.

Only labels that are positioned closely together intersect each other, which means that the independent set instances that arise from map labelling are very sparse. Altough sparse graphs are notoriously hard for the independent set codes reported on in the literature (sparse problems with only 200 nodes are considered hard), the graphs arising from map labelling are different because they have a nice topological structure, namely, the conflict graph can easily be embedded in the Euclidean plane such that the edges of the graph only connect nodes that are close together (e.g., by placing a node in the center of the square it represents).

1.2 Contribution and Outline

Our main contribution is that we demonstrate that it is possible to solve map labelling instances with up to 950 cities (or 3800 nodes) quickly using our branch and cut algorithm for the independent set problem. The optimisation algorithms reported on so far are useless for map labelling instances with more than 50 cities [8, page 219]. We show that it is possible to enhance the standard branch and cut algorithm with a recursive technique that is applicable for the Euclidean instances arising from map labelling. Finally, we present a novel variant of lifting odd hole inequalities based on path decomposition. In Section 2 we describe our branch and cut algorithm. Computational experiments on standard test problems from the map labelling literature are reported on in Section 3.

2 An Algorithm for Maximum Independent Sets

We begin by giving a brief description of a branch and cut algorithm in Section 2.1 directed to readers less familiar with this topic. In Section 2.2 we specify the details of our implementation of the branch and cut algorithm. This includes how we set values of variables in the branch and cut algorithm in order to speed up the algorithm, and a new technique that makes use of the possibility to decompose a problem in a node of the search tree yielding small easy-to-solve integer programs that we can solve to optimality. This enhancement proved very useful for the map labelling instances as they, due to their sparsity, quite often give rise to decomposable subproblems once a number of variables have been set. We conclude this section by giving the separation algorithms for two families of inequalities, namely clique and odd hole inequalities. In Section 2.3 we describe a local search algorithm for finding good feasible solutions, and thereby lower bounds on the optimal value of our problem. In this section we use $n = |V|$ and $m = |E|$ to denote the number of vertices and edges, respectively, of the graph used in the independent set formulation of ODP.

2.1 Branch and Cut

A branch and cut algorithm is a branch and bound algorithm where we may call a *cutting plane algorithm* in each node of the search tree. Here we give a short description of a basic version of branch and bound and of a cutting plane algorithm. For further details we refer to the book by Wolsey [27].

Consider the problem to determine $z_{\mathrm{OPT}} = \max\{z(x) : x \in P, x \text{ integer}\}$, where z is a linear function in x, and where P is a polyhedron. We refer to this problem as problem Π. Branch and bound makes use of the *linear programming relaxation* (LP-relaxation) $\bar{\Pi}$: $\bar{z} = \max\{z(x) : x \in P\}$. It is easy to see that $\bar{z} \geq z_{\mathrm{OPT}}$. At the top level, or *root node*, of the branch and bound tree we have problem $\bar{\Pi}$. At level k of the tree we have a collection of problems, say $\bar{\Pi}_1, \ldots, \bar{\Pi}_l$ such that the corresponding polyhedra $P_1, \ldots, P_l$ are pairwise disjoint, and such that all integral vectors in P are contained in $P_1 \cup \cdots \cup P_l$.

The algorithm works as follows. We maintain a set of *open problems*, the best known value of an integer solution $z^* = z(x^*)$, and the corresponding integer solution x^*. At first, problem $\bar{\Pi}$ is the only open problem. In iteration i, we select an open problem $\bar{\Pi}^i$ and solve it. If problem $\bar{\Pi}^i$ is infeasible we remove $\bar{\Pi}^i$ from the list of open problems and continue with the next iteration. If the optimal solution to $\bar{\Pi}^i$, $\bar{x}^i$, is integral, i.e., $\bar{x}^i$ is a solution to problem Π, we set $x^* := \bar{x}^i$ if $z(\bar{x}^i) > z^*$, we remove $\bar{\Pi}^i$ from the list of open problems and we proceed to the next iteration. Otherwise, we identify a component j of the vector $\bar{x}^i$ such that $\bar{x}^i_j$ is not integral, and "branch" on x_j, i.e., we formulate two new open problems, say $\bar{\Pi}^i_1$ and $\bar{\Pi}^i_2$ by adding constraints $x_j \leq \lfloor \bar{x}^i_j \rfloor$ and $x_j \geq \lceil \bar{x}^i_j \rceil$ to P^i. If x_j is a 0-1 variable we add constraints $x_j = 0$ and $x_j = 1$. Note that $\bar{x}^i$ neither belongs to P^i_1 nor to P^i_2. The value of $z(\bar{x}^i)$ is an upper bound on the value of any solution to the problems $\bar{\Pi}^i_1$ and $\bar{\Pi}^i_2$. We replace $\bar{\Pi}^i$ by $\bar{\Pi}^i_1$ and $\bar{\Pi}^i_2$ in the set of open problems and proceed to the next iteration. The algorithm stops after the set of open problems becomes empty.

When using branch and bound to solve integer programming problems it is crucial that we obtain good lower and upper bounds on the optimal value as the bounds are used to prune the search tree. In order to obtain good upper bounds we strengthen the LP-relaxation by adding *valid inequalities*. Let $X = P \cap \mathbb{Z}^n$, i.e., X is the set of feasible solutions to Π, and let $\text{conv}(X)$ denote the convex hull of feasible solutions. Observe that $z_{\text{OPT}} = \max\{z(x) : x \in \text{conv}(X)\}$. An inequality $\pi^T x \leq \pi_0$ is *valid* for $\text{conv}(X)$ if $\pi^T x \leq \pi_0$ for all $x \in \text{conv}(X)$. For any valid inequality $\pi^T x \leq \pi_0$, the set $\{x \in \text{conv}(X) : \pi^T x = \pi_0\}$ is called a *face* of $\text{conv}(X)$ if it is nonempty and not equal to $\text{conv}(X)$. A face is called a *facet* of $\text{conv}(X)$ if it is not contained in any other face of $\text{conv}(X)$. The facets of $\text{conv}(X)$ are precisely the inequalities that are necessary in the description of $\text{conv}(X)$. If the problem of optimising over X is $\mathcal{NP}$-hard, then we cannot expect to find an explicit description of $\text{conv}(X)$ unless $\mathcal{NP} = \text{co-}\mathcal{NP}$. In practice we therefore limit ourselves to certain families of facet-defining, or at least high-dimensional, valid inequalities. Given a family $\mathcal{F}$ of valid inequalities and a vector $\bar{x}$, the problem of determining whether there exists an inequality belonging to $\mathcal{F}$ that is violated by $\bar{x}$ is called the *separation problem* based on $\mathcal{F}$. Even if it is $\mathcal{NP}$-hard to optimise over X, the separation problem based on a specific family of valid inequalities for $\text{conv}(X)$ might be polynomially solvable.

In a *cutting plane algorithm* we maintain an LP-relaxation of a problem Π. We start with the formulation P. Solve $\bar{\Pi}$ given P. If the optimal solution $\bar{x}$ is integral, then we stop. Otherwise, we call the *separation algorithms* based on all families of valid inequalities that we consider. If any violated inequalities are identified we add them to P. If no violated inequalities are found we stop.

2.2 Branch and Cut for the Independent Set Formulation

Here we describe the parts of the branch and cut algorithm that are specific for our problem and for our implementation. We use the following notation. For any set S and finite discrete set I, we use $x \in S^I$ to denote a vector x of dimension $|I|$ whose components are indexed by the elements of I. For any $I' \subseteq I$ we use the implicit sum notation $x(I')$ to denote $\sum_{i \in I'} x_i$.

Given a graph $G = (V,E)$, the set of all incidence vectors of independent sets in G is given by

$$X_{\text{IS}} = \{x \in \{0,1\}^V \mid \forall \{v,w\} \in E : x(\{v,w\}) \leq 1\}.$$

We consider the following integer programming version of the maximum cardinality independent set formulation.

$$\max\{z(x) = x(V) : x \in X_{\text{IS}}\}.$$

In a certain node i of the branch and cut tree, all the variables that we have been branching on in order to create node i have been set to either zero or one. In order to trigger pruning of the search tree, either through integrality or through infeasibility, it is important to try to set other variables equal to zero or one as well. We have implemented three variable setting algorithms: setting by *reduced costs*, by *logical implications*, and the new scheme *variable setting by recursion and substitution*. We will return to these algorithms later in this section.

We use two families of facet-defining valid inequalities in the cutting plane algorithm: *clique inequalities* and *lifted odd hole inequalities*. Let $C \subseteq V$ be a subset of the nodes that induces a clique, i.e., a complete subgraph, in G. Padberg [23] showed that the clique inequality $x(C) \leq 1$ is valid for $\text{conv}(X_{\text{IS}})$, and that it defines a facet of $\text{conv}(X_{\text{IS}})$ if and only if C is maximal. We call an odd length chordless cycle in G an *odd hole* of G. Let $H \subseteq V$ be a subset of the nodes that induces an odd hole in G. It was shown by Padberg that the odd hole inequality $x(H) \leq \lfloor |H|/2 \rfloor$ is valid for $\text{conv}(X_{\text{IS}})$ and that it defines a facet for $\text{conv}(X_{\text{IS}}) \cap \{x \in \mathbb{R}^V \mid x_v = 0 \text{ for all } v \in V \setminus H\}$. Hence, in general odd hole inequalities do not define facets of $\text{conv}(X_{\text{IS}})$. Padberg suggested a way of increasing the dimension of the faces induced by the odd hole inequalities, called *lifting*. The separation algorithms based on the two families of inequalities together with our variant of Padberg's lifting scheme for odd hole inequalities are described at the end of this section.

In our branch and cut algorithm we also incorporated a local search algorithm for finding feasible solutions of good quality, and thereby good lower bounds on the optimal value. This local search algorithm is described in Section 2.3.

At the initialisation of the branch and cut algorithm we consider a stronger linear relaxation than the linear relaxation obtained from X_{IS}. We observe that every pair of vertices connected by an edge in the conflict graph is a clique on two nodes. This clique is, however, in general not maximal. Therefore we substitute each constraint $x(\{u,v\}) \leq 1$ for all $\{u,v\} \in E$ by a constraint based on a maximal clique containing the edge $\{u,v\}$. Let $\mathcal{C}$ be a collection of maximal cliques in G such that each edge of G is contained in at least one clique in $\mathcal{C}$. Define $P_{\mathcal{C}} = \{x \in \mathbb{R}^V : x(C) \leq 1 \text{ for all } C \in \mathcal{C}, x \geq 0\}$, and let $\Pi_{\mathcal{C}}$ be defined as $\max\{z(x) : x \in P_{\mathcal{C}}, x \text{ integer}\}$. As in Section 2.1 we define the LP-relaxation of problem $\Pi_{\mathcal{C}}$ to be the problem $\bar{\Pi}_{\mathcal{C}}$: $\max\{z(x) : x \in P_{\mathcal{C}}\}$. As mentioned in Section 2.1 the branch and cut algorithm maintains a collection $\mathcal{Q}$ of *open problems*. We initialise $\mathcal{Q} := \{\bar{\Pi}_{\mathcal{C}}\}$.

The only algorithmic details that remain to be defined are how to select a branching variable and how to select the next open problem. We select a variable to branch on based on pseudo costs as suggested by Linderoth and Savelsbergh [19]. The open problem $\bar{\Pi}^i$ to remove from Q is chosen using the best value node selection rule. Furthermore, we give precedence to node i if we found a better heuristic solution in node $p(i)$.

Variable Setting and SRS. Let (x^i, π) be an optimal primal-dual pair for an LP relaxation in node i and let c^π denote the reduced cost vector. If for some $v \in V$ we have $x^i_v = 0$ (or $x^i_v = 1$) and $c^\pi_v < z(x^*) - z(x^i)$ (or $c^\pi_v > -(z(x^*) - z(x^i))$), we can set $x^j_v := 0$ (or $x^j_v := 1$, respectively) in all nodes j in the part of the branch and bound tree rooted at node i. We denote the neighbourhood of a node v in G by $N(v)$, i.e., $N(v) = \{u \in V \mid \{u, v\} \in E\}$. If, in a node j of the branch and bound tree, we have that x^j_v is set to 1 for some v (either by branching or by reduced cost), we can set $x^j_u := 0$ for all $u \in N(v)$. If on the other hand, there exists v such that x^j_u is set to 0 for all $u \in N(v)$, we can set $x^j_v := 1$.

Observe that deciding whether to set a variable based on reduced cost depends on the gap $z(x) - z(x^*)$. This gap decreases every time we find a new x^*. We try to set variables *in every node* of the branch and bound tree whenever this occurs. This is done in a "lazy" fashion. For this purpose, we store a tree T that mirrors the branch and bound tree. For each node in the branch and bound tree, we store the variables that we can set in its corresponding node in T, together with the value $z(x^*)$ for which we last applied the variable setting procedure in that node. Suppose we decide to solve problem $\bar{\Pi}^i$ for some node $i \in T$. For each node j on the path from the root of T to node i, if $z(x^*)$ has improved since the last time we applied variable setting in node j, then we reconstruct the final LP reduced cost of node j and re-apply the variable setting procedure. Reconstructing the final LP reduced cost can be done using the final LP basis of node j. Storing a basis in each node of T would require $\Theta(n+m)$ memory per node, which may become prohibiting if the tree size becomes large. In each node j, however, it suffices to store the difference between its final basis and the final basis of its parent node $p(j)$.

Another, even more important enhancement that we developed is referred to as *variable setting by recursion and substitution*, or SRS. We call a variable that is not set to 0 or 1 *free*. Focus on node i after the cutting plane algorithm is done. Denote the subset of nodes in G corresponding to free variables in node i of the branch and bound tree by F^i. We check whether the graph $G(F^i) = (F^i, E(F^i))$ is disconnected. If this is the case, denote the connected components of $G(F^i)$ by $G(F^i_1), \ldots, G(F^i_k)$ such that $|F^i_j| \leq |F^i_{j+1}|$ for $j = 1, \ldots, k-1$. We identify the largest component, $G(F^i_k)$, as the main component of the problem in node i, which we leave aside for the moment. Our goal is to exploit the fact that we can find maximum independent sets on the smaller components easily. Denote x^i restricted to $W \subseteq V$ by x^W, i.e. $x^W_v = x^i_v$ if $v \in W$ and $x^W_v = 0$ if $v \notin W$. For all $j = 1, \ldots, k-1$, if $x^{F^i_j}$ is integer we set $\tilde{x}^{F^i_j} := x^{F^i_j}$. Otherwise we recursively find a maximum independent set with incidence vector $\tilde{x}^{F^i_j}$ on $G(F^i_j)$. Furthermore, we substitute these partial solutions back into x^i to give us $\tilde{x}^i$ as follows:

$$\tilde{x}^i := x^i - \sum_{j=1}^{k-1} x^{F^i_j} + \sum_{j=1}^{k-1} \tilde{x}^{F^i_j}.$$

Proposition 1. *There exists an optimal solution x to Π^i with $x_v = \tilde{x}^i_v$ for all $v \in \cup_{j=1}^{k-1} F^i_j$.*

For the proof, see the full version of the paper. It follows that we can set $x^i := \tilde{x}^i$, thereby tightening $z(x^i)$. Finally, for any node j descendant of node i, we set the variables x_v for $v \in \bigcup_{j=1}^{k-1} F^i_j$ to $\tilde{x}^i_v$.

Separation and Lifting. For the separation of clique inequalities we do follow the ideas of Nemhauser and Sigismondi [22]. We refer to the full paper for the technical details regarding our precise implementation.

Let x denote a fractional solution. We start by describing the separation algorithm for the basic odd hole inequalities $x(H) \leq \lfloor |H|/2 \rfloor$ given the vector x. We first find an odd cycle starting from some node $v \in V$ using the construction described by Grötschel, Lovász, and Schrijver [15]. To find a shortest odd cycle containing node v, Grötschel, Lovász, and Schrijver construct an auxiliary bipartite graph $\tilde{G} = ((V^1, V^2), \tilde{E})$ and cost vectors $c \in [0,1]^E$ and $\tilde{c} \in [0,1]^{\tilde{E}}$ as follows. Each node $v \in V$ is split into two nodes v^1 and v^2, v^i is included in V^i $(i = 1,2)$. For each edge $\{u,v\} \in E$, we add the edges $\{u^1,v^2\}$ and $\{u^2,v^1\}$ to $\tilde{E}$, and set $c_{\{u,v\}} = \tilde{c}_{\{u^1,v^2\}} = \tilde{c}_{\{u^2,v^1\}} = 1 - x_u - x_v$. Observe that a path from $u^1 \in V^1$ to $v^2 \in V^2$ in $\tilde{G}$ corresponds to a walk of odd length in G from u to v.

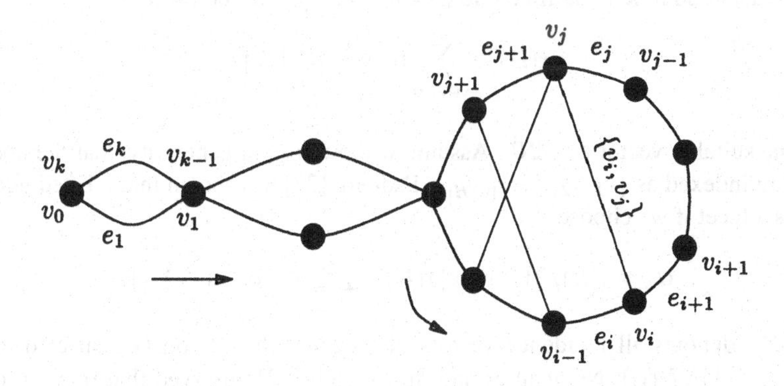

Fig. 1. Identifying an odd hole in a closed walk.

A shortest path from v^1 to v^2 in $\tilde{G}$ corresponds to a shortest odd length closed walk in G containing v. The reason that we are looking for a shortest path is that a subset of nodes H that induces an odd hole that corresponds to a violated odd hole inequality will have $c(E(H)) < 1$, and a short closed walk in $\tilde{G}$ is more likely to lead to a violated lifted odd hole inequality than a long one. Shortest paths in a graph with non-negative edge lengths can be found using Dijkstra's algorithm [11]. Hence, we can find a closed walk

$$C := (v = v_0, e_1, v_1, e_2, v_2, \ldots, v_{k-1}, e_k, v_k = v)$$

in G with odd k that is minimal with respect to c and $|C|$ by using Dijkstra's algorithm to find a shortest path (with respect to $\tilde{c}$) of minimal cardinality from v^1 to v^2 in $\tilde{G}$. Some of the v_i may occur more than once, and the walk may have chords. Let $j \in 2, \ldots, k-1$ be the smallest index such that there exists an edge $\{v_i, v_j\} \in E$ for some $i \in 0, \ldots, j-2$ (such i, j exists because $\{v_0, v_{k-1}\} \in E$). Let $i \in 0, \ldots, j-2$ be the largest index such that $\{v_i, v_j\} \in E$. Let $H := \{v_i, v_{i+1}, \ldots, v_j\}$. We claim that H induces an odd hole in G (see Figure 1). Clearly H induces a cycle. If $|H| = 3$ then H is a clique and we ignore it. Otherwise, by choice of i and j, H does not contain chords. Now suppose, seeking contradiction, that $|H| = j - i + 1$ is even. Then,

$$C' := (v = v_0, e_1, \ldots, v_{i-1}, e_i, v_i, \{v_i, v_j\}, v_j, e_{j+1}, v_{j+1}, \ldots, e_k, v_k = v)$$

is an odd length closed walk in G containing v. Moreover, $c(\{e_{i+1}, \ldots, e_j\}) = (j-i) - (2\sum_{p=i+1}^{j-1} x_{v_p}) - x_{v_i} - x_{v_j}$. It follows from $x(\{v_p, v_{p+1}\}) \leq 1$ that $\sum_{p=i+1}^{j-1} x_{v_p} \leq (j-i-1)/2$. Therefore,

$$c(\{e_{i+1}, \ldots, e_j\}) \geq (j-i) - (j-i-1) - x_{v_i} - x_{v_j} = c_{\{v_i, v_j\}},$$

so C' is not longer than C with respect to c. However, C' is of smaller cardinality, which contradicts to our choice of C. Hence H induces an odd hole in G.

Given an odd hole H, a lifted odd hole inequality is of the form

$$x(H) + \sum_{v \in V \setminus H} \alpha_v x_v \leq \lfloor |H|/2 \rfloor$$

for some suitable vector $\alpha \in \mathbb{N}^V$. Assume without loss of generality, that the nodes in $V \setminus H$ are indexed as $\{v_1, v_2, \ldots, v_{|V \setminus H|}\}$. Padberg [23] has shown that a lifted odd hole induces a facet if we choose

$$\alpha_{v_i} = \lfloor |H|/2 \rfloor - \max\{x(H) + \textstyle\sum_{j=1}^{i-1} \alpha_{v_j} x_{v_j} : x \in X_{\mathrm{IS}}^i\},$$

where X_{IS}^i denotes all incidence vectors of independent sets on G restricted to $H \cup \{v_1, \ldots, v_{i-1}\}) \setminus N(v_i)$. Nemhauser and Sigismondi [22] observed that $\alpha_v = 0$ for $v \in V \setminus H$ if $|N(v) \cap H| \leq 2$. This implies that the independent set problems that have to be solved in order to compute the lifting coefficients α are relatively small. In our computational experiments on map labelling problems we have observed independent set problems with up to approximately 40 nodes. We lift the variables in non-decreasing lexicographic order of $(|\frac{1}{2} - x_v|, -|N(v) \cap H|)$, where ties are broken at random.

Moreover, we can use the fact that a hole has small path width [12]. Therefore we can build an initial path decomposition of the hole that we can extend in a straightforward greedy fashion each time we find a variable with a nonzero lifting coefficient. In our experiments on map labelling problems, it rarely occurred that the path width of the resulting path decomposition exceeded 20. This enables us to compute the lifting coefficients very efficiently. The advantage of using a path decomposition algorithm is that the size of a maximal independent set can be found using the path decomposition in time that is exponential only in the path width, not in the size of the graph.

2.3 A Local Search Heuristic for Independent Sets

We proceed by describing the primal heuristics we use to improve x^*. First we apply a simple rounding heuristic that gives a feasible solution. We then use this solution as a starting point for a combination of local search heuristics.

Suppose we have a fractional solution x^{LP} to problem $\bar{\Pi}$ and we want to construct an integer solution x^{R} to problem Π. The rounding procedure works as follows. For each $v \in V$, if $x_v > \frac{1}{2}$ we set $x_v^{\mathrm{R}} := 1$, otherwise we set $x_v^{\mathrm{R}} := 0$. The feasibility of x^{R} follows from the observation that $x_v^{\mathrm{LP}} > \frac{1}{2}$ implies $x_u^{\mathrm{LP}} \leq \frac{1}{2}$ for all $u \in N(v)$, which together gives $x^{\mathrm{R}}(\{u,v\}) \leq 1$ for each $\{u,v\} \in E$.

After the rounding procedure, we first apply a 1-opt procedure. We start by taking $x^{1\text{-opt}} := x^{\mathrm{R}}$. For each $v \in V$, we check if all $u \in N(v)$ have $x_u^{1\text{-opt}} = 0$ and if so, we set $x_v^{1\text{-opt}} := 1$. Clearly, $x^{1\text{-opt}}$ is a feasible solution if x^{R} is a feasible solution. The combined time complexity of the rounding and the 1-opt procedures is $O(n)$.

Finally, we apply a 2-opt procedure. We start by taking $x^{2\text{-opt}} := x^{1\text{-opt}}$. For each $v \in V$, we check whether there exists $u,w \in N(v)$ such that $\tilde{x} := x^{2\text{-opt}} - e_v + e_u + e_w$ is feasible. As soon as we find such u,w, we replace $x^{2\text{-opt}}$ by $\tilde{x}$. We continue until no such u,v,w exists. The 2-opt algorithm can be implemented to work in $O((z(x^{2\text{-opt}}) - z(x^{1\text{-opt}}) + 1)n^3)$ time.

When applying the above heuristics from a node in the branch and bound tree, we make sure that all variables that are set by branching, reduced cost, logical implication, and SRS remain at their value in x^{LP}.

3 Computational Results

In order to evaluate the behaviour of our algorithm on independent set problems that come from map labelling, we implemented our algorithm in C++, using our own framework for solving MIP's based on the CPLEX 6.0.1 linear program solver [16]. We use the LEDA graph data structure [21] to represent graphs. We tested our algorithm on the same class of map labelling problems as used by Christensen [8] and van Dijk [10]. These problems are generated by placing n (integer) points on a standard map of size 792 by 612. The points have to be labelled using labels of size 30×7. For each $n \in \{100, 150, \ldots, 750, 800\}$ we randomly generated 50 maps. Figure 2 shows the average branch and bound tree sizes and running times of our algorithm on these problem instances. The reported running times were observed on a 168 MHz Sun Ultra Enterprise 2.

To evaluate the influence of SRS on the performance of our algorithm, we conducted a second set of experiments. In these experiments, the density of the problems was kept

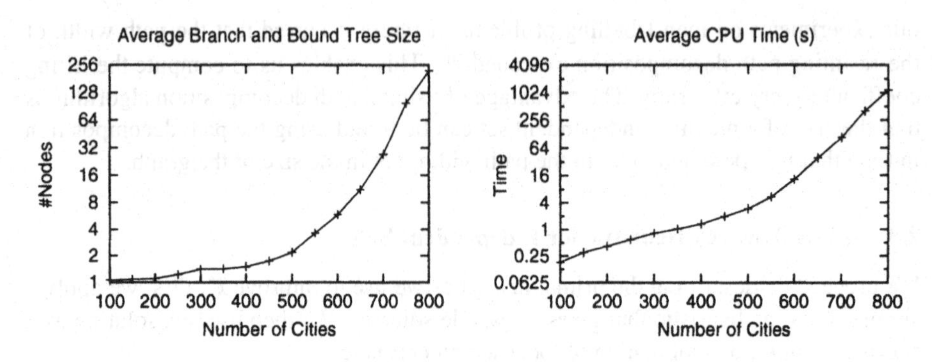

Fig. 2. Performance of our algorithm on random map labelling problems.

the same by making the map size a function of the number of cities. For a problem with n cities, we use a map of size $\lfloor 792/\sqrt{750/n} \rfloor \times \lfloor 612/\sqrt{750/n} \rfloor$. For each $n \in \{600, 650, \ldots, 900, 950\}$ we randomly generated 50 maps. ¿From each generated map, we selected its largest connected component and used that as the input for our algorithm both with and without SRS. Figure 3 shows the average branch and bound tree sizes and running times for these experiments. The reported branch and bound tree sizes for the case with SRS includes the nodes in branch and bound trees of recursive calls.

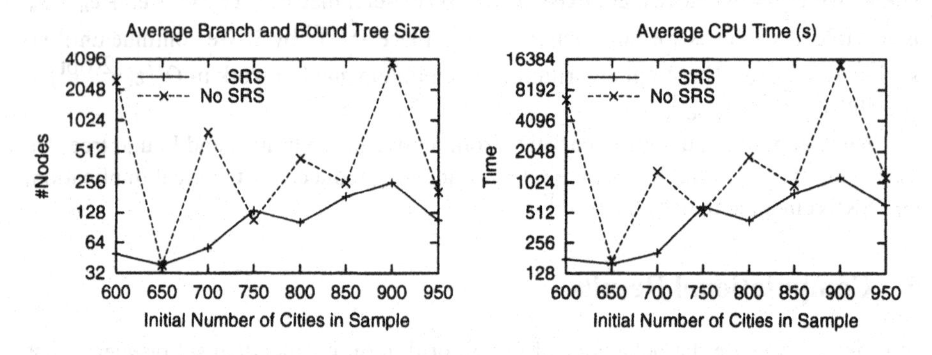

Fig. 3. The impact of SRS.

The experiments show that map labelling instances with up to 950 cities can be solved to optimality using our branch and cut algorithm with reasonable computational effort. It turns out to be crucial to exploit the sparsity of the resulting independent set problems, and the new variable setting procedure, SRS, is very efficient in doing so.

Christensen [8] reports on experiments with different heuristic algorithms on problem instances with up to 1500 cities. The running times of different heuristics as mentioned by Christensen fall in the range of tenth of seconds (for random solutions) to approximately 1000 seconds (for the heuristic by Zoraster [30]). We have observed in our experiments that local search, starting from fractional LP solutions in the branch and bound tree, finds an optimal solution already after a couple of minutes of computing time. Moreover, due to the strong relaxation resulting from the maximal clique and lifted odd hole inequalities, we can prove our solutions to be at most 1 or 2 labels away from optimal at an early stage of the computation. Most of the time is spent by our algorithm in decreasing the upper bound to prove optimality. We conclude that we can turn our branch and cut algorithm into a very robust heuristic for map labelling by adding heuristic pruning of the search tree.

References

[1] P. K. Agarwal, M. van Kreveld, and S. Suri. Label placement by maximum independent set in rectangles. *Comp. Geom. Theory and Appl.*, 11:209–218, 1998.

[2] L. Babel. Finding maximum cliques in arbitrary and special graphs. *Computing*, 46(4):321–341, 1991.

[3] L. Babel and G. Tinhofer. A branch and bound algorithm for the maximum clique problem. *ZOR – Methods and Models of Opns. Res.*, 34:207–217, 1990.

[4] V. Bafna, B. O. Narayanan, and R. Ravi. Non-overlapping local alignments (weighted independent sets of axis parallel rectangles). In Selim G. Akl, Frank K. H. A. Dehne, Jörg-Rüdiger Sack, and Nicola Santoro, editors, *Proc. 4th Int. Workshop on Alg. and Data Structures*, volume 955 of *Lec. Notes Comp. Sci.*, pages 506–517. Springer-Verlag, 1995.

[5] E. Balas and J. Xue. Weighted and unweighted maximum clique algorithms with upper bounds from fractional coloring. *Algorithmica*, 15:397–412, 1996.

[6] E. Balas and C. S. Yu. Finding a maximum clique in an arbitrary graph. *SIAM J. on Comp.*, 15:1054–1068, 1986.

[7] R. Carraghan and P. M. Pardalos. An exact algorithm for the maximum clique problem. *Opnl. Res. Letters*, 9:375–382, 1990.

[8] J. Christensen, J. Marks, and S. Shieber. An empirical study of algorithms for point-feature label placement. *ACM Trans. on Graphics*, 14(3):203–232, 1995.

[9] R. G. Cromley. An LP relaxation procedure for annotating point features using interactive graphics. In *Proc. AUTO-CARTO VII*, pages 127–132, 1985.

[10] S. van Dijk, D. Thierens, and M. de Berg. On the design of genetic algorithms for geographical applications. In W. Banzhaf, J. Daida, A. E. Eiben, M. H. Garzon, V. Honavar, M. Jakiela, and R. E. Smith, editors, *Proc. of the Genetic and Evolutionary Comp. Conf. (GECCO-99)*. Morgan Kaufmann, Jul 1999.

[11] E. Dijkstra. A note on two problems in connexion with graphs. *Numeriche Mathematics*, 1:269–271, 1959.

[12] B. de Fluiter. *Algorithms for Graphs of Small Treewidth.* PhD thesis, Utrecht University, Utrecht, 1997.

[13] M. Formann and F. Wagner. A packing problem with applications to lettering of maps. In *Proc. 7th Ann. ACM Symp. Comp. Geom.*, pages 281–288, 1991.

[14] C. Friden, A. Hertz, and D. de Werra. An exact algorithm based on tabu search for finding a maximum independent set in graph. *Comp. Opns. Res.*, 17(5):375–382, 1990.

[15] M. Grötschel, L. Lovász, and A. Schrijver. *Geometric Algorithms and Combinatorial Optimization.* Springer-Verlag, Berlin, 1988.

[16] ILOG, Inc. CPLEX Devision. *Using the CPLEX Callable Library*, 1997.

[17] K. G. Kakoulis and I. G. Tollis. A unified approach to labeling graphical features. In *Proc. 14th Ann. ACM Symp. Comp. Geom.*, pages 347–356, 1998.

[18] L. Kučera, K. Mehlhorn, B. Preis, and E. Schwarzenecker. Exact algorithms for a geometric packing problem. In *Proc. 10th Symp. Theo. Aspects Comp. Sci.*, volume 665 of *Lec. Notes Comp. Sci.*, pages 317–322. Springer-Verlag, 1993.
[19] J. T. Linderoth and M. W. P. Savelsbergh. A computational study of search strategies for mixed integer programming. *INFORMS J. on Comp.*, to appear.
[20] C. Mannino and A. Sassano. An exact algorithm for the maximum stable set problem. *Comp. Optn. and Appl.*, 3:243–258, 1994.
[21] K. Mehlhorn, S. Näher, and C. Uhrig. *The LEDA User Manual Version 3.7.* Max-Planck-Institut für Informatik, Saarbrücken, 1998.
[22] G. L. Nemhauser and G. Sigismondi. A strong cutting plane/branch-and-bound algorithm for node packing. *J. of the Opns. Res. Soc.*, 43(5):443–457, 1992.
[23] M. W. Padberg. On the facial structure of set packing polyhedra. *Mathematical Programming*, 5:199–215, 1973.
[24] F. Rossi and S. Smriglio. A branch-and-cut algorithm for the maximum cardinality stable set problem. Technical Report 353, "Centro Vito Volterra"-Università di Roma Tor Vergata, Jan 1999.
[25] E. C. Sewell. A branch and bound algorithm for the stability number of a sparse graph. *INFORMS J. on Comp.*, 10(4):438–447, 1998.
[26] F. Wagner and A. Wolff. An efficient and effective approximation algorithm for the map labeling problem. In P. Spirakis, editor, *Proc. 3rd Ann. Eur. Symp. on Alg.*, volume 979 of *Lec. Notes Comp. Sci.*, pages 420–433. Springer-Verlag, 1995.
[27] L. Wolsey. *Integer Programming.* John Wiley & Sons, Inc., New York, 1998.
[28] J. Xue. *Fast algorithms for the vertex packing problem.* PhD thesis, Carnegie Mellon University, Pittsburgh, PA, 1991.
[29] S. Zoraster. Integer programming applied to the map label placement problem. *Cartographica*, 23(3):16–27, 1986.
[30] S. Zoraster. The solution of large 0-1 integer programming problems encountered in automated cartography. *Opns. Res.*, 38(5):752–759, 1990.

A Decomposition Theorem for Maximum Weight Bipartite Matchings with Applications to Evolutionary Trees

Ming-Yang Kao[1], Tak-Wah Lam[2], Wing-Kin Sung[2], and Hing-Fung Ting[2]

[1] Department of Computer Science, Yale University, New Haven, CT 06520, U.S.A., kao-ming-yang@cs.yale.edu‡

[2] Department of Computer Science and Information Systems, University of Hong Kong, Hong Kong, {twlam, wksung, hfting}@csis.hku.hk§

Abstract. Let G be a bipartite graph with positive integer weights on the edges and without isolated nodes. Let n and W be the node count and the total weight of G. We present a new decomposition theorem for maximum weight bipartite matchings and use it to design an $O(\sqrt{n}W)$-time algorithm for computing a maximum weight matching of G. This algorithm bridges a long-standing gap between the best known time complexity of computing a maximum weight matching and that of computing a maximum cardinality matching. Given G and a maximum weight matching of G, we can further compute the weight of a maximum weight matching of $G-\{u\}$ for all nodes u in $O(W)$ time. As immediate applications of these algorithms, the best known time complexity of computing a maximum agreement subtree of two ℓ-leaf rooted or unrooted evolutionary trees is reduced from $O(\ell^{1.5}\log \ell)$ to $O(\ell^{1.5})$.

1 Introduction

Let $G=(X,Y,E)$ be a bipartite graph with positive integer weights on the edges. A *matching* of G is a subset of node-disjoint edges of G. Let $\text{mwm}(G)$ (respectively, $\text{mm}(G)$) denote the maximum weight (respectively, cardinality) of any matching of G. A *maximum weight* matching is one whose weight is $\text{mwm}(G)$. Let N be the largest weight of any edge. Let W be the total weight of G. Let n and m be the numbers of nodes and edges of G; to avoid triviality, we maintain $m=\Omega(n)$ throughout the paper.

The problem of finding a maximum weight matching of a given G has a rich history. The first known polynomial-time algorithm is the $O(n^3)$-time Hungarian method [23]. Fredman and Tarjan [12] used Fibonacci heaps to improve the time to $O(n(m+n\log n))$. Gabow [13] introduced scaling to solve the problem in $O(n^{3/4}m\log N)$ time by taking advantage of the integrality of edge weights. Gabow and Tarjan [14] improved the scaling method to further reduce the time to $O(\sqrt{n}m\log(nN))$. For the case where the edges all have weight 1, i.e., $N=1$ and $W=m$, Hopcroft and Karp [18] gave an $O(\sqrt{n}W)$-

‡ Research supported in part by NSF Grant 9531028.

§ Research supported in part by Hong Kong RGC Grant HKU-7027/98E.

time algorithm.[1] It has remained open since [14] whether the gap between the running times of the latter two algorithms can be closed for the case $W = o(m\log(nN))$.

This paper resolves the open problem in the affirmative by giving an $O(\sqrt{n}W)$-time algorithm for general W. The algorithm does not use scaling but instead employs a novel decomposition theorem for weighted bipartite matchings. We also use the theorem to solve the *all-cavity maximum weight matching* problem which, given G and a maximum weight matching of G, asks for $\text{mwm}(G - \{u\})$ for all nodes u in G. The case where $N = 1$ has been studied by Chung [4]. Recently, Kao, Lam, Sung, and Ting [21] gave an $O(\sqrt{n}m\log N)$-time algorithm for general N. This paper presents a new algorithm that runs in $O(W)$ time.

As immediate applications, we use the new matching algorithms to speed up two of the best known algorithms for comparing *evolutionary* trees, which are trees with leaves labeled by distinct species [1, 2, 17, 26]. Different models of the evolutionary relationship of the same species may result in different evolutionary trees. An *agreement subtree* of two evolutionary trees is an evolutionary tree which is also a topological subtree of the two given trees. A *maximum* agreement subtree is one with the largest possible number of leaves. A basic problem in computational biology is to extract the maximum amount of evolutionary information shared by two given models of evolution. To a useful extent, this problem can be solved by computing a maximum agreement subtree of the corresponding evolutionary trees [11].

Algorithms for computing a maximum agreement subtree for unrooted or rooted evolutionary trees have been studied intensively in the past few years. The unrooted case is technically more difficult than the rooted case. Steel and Warnow [25] gave the first polynomial-time algorithm for two unrooted trees with no restriction on the degrees. Let ℓ be the number of leaves in the input trees. Their algorithm runs in $O(\ell^{4.5}\log\ell)$ time. Farach and Thorup [8] reduced the time to $O(\ell^{2+o(1)})$ for unrooted trees [8] and $O(\ell^{1.5}\log\ell)$ for rooted trees [9]. For the unrooted case, the time complexity was improved by Lam, Sung, and Ting [24] to $O(\ell^{1.75+o(1)})$ and later by Kao, Lam, Przytycka, Sung, and Ting [20, 22] to $O(\ell^{1.5}\log\ell)$. Faster algorithms for trees with degrees bounded by a constant have also been discovered [5, 7, 19, 20, 22]. In this paper, we use the new algorithm for computing a single maximum weight matching to reduce the time for general rooted trees to $O(\ell^{1.5})$. We use both new matching algorithms to reduce the time for general unrooted trees to $O(\ell^{1.5})$ time as well.

Section 2 presents the decomposition theorem and uses it to compute the weight of a maximum weight matching. Section 3 gives an algorithm to construct a maximum weight matching. Section 4 solves the all-cavity matching problem. Section 5 applies the matching algorithms to maximum agreement subtrees.

[1] For $N = 1$, Feder and Motwani [10] gave another matching algorithm which is more efficient for dense graphs; the time complexity is $O(\sqrt{n}W/(\log n/\log(n^2/m))$, which remains $O(\sqrt{n}W)$ whenever the graph has at most $n^{2-\varepsilon}$ edges for any $\varepsilon > 0$.

2 The decomposition theorem

In §2.1, we state the decomposition theorem and use it to compute the weight $\text{mwm}(G)$ in $O(\sqrt{n}W)$ time. In §2.2 and §2.3, we prove the theorem. In §3, we further construct a maximum weight matching itself within the same time bound.

2.1 An algorithm for computing mwm(G)

Let $w(u,v)$ denote the weight of an edge $uv \in G$; if u is not adjacent to v, let $w(u,v) = 0$. A *cover* of G is a function $C : X \cup Y \to \{0,1,2,\ldots\}$ such that $C(x) + C(y) \geq w(x,y)$ for all $x \in X$ and $y \in Y$. C is a *minimum weight* cover if $\sum_{z \in X \cup Y} C(z)$ is the smallest possible. A minimum weight cover is a dual of a maximum weight matching as stated in the next fact.

Fact 1 (see [3]) *Let C be a cover and M be a matching of G. The following statements are equivalent.*

1. *C is a minimum weight cover and M is a maximum weight matching of G.*
2. $\sum_{uv \in M} w(u,v) = \sum_{u \in X \cup Y} C(u)$.
3. *Every node in* $\{u \mid C(u) > 0\}$ *is matched by some edge in M, and* $C(u) + C(v) = w(u,v)$ *for all* $uv \in M$.

For an integer $h \in [1,N]$, we divide G into two lighter bipartite graphs G_h and G_h^Δ as follows. Note that the total weight of G_h and G_h^Δ is at most W.

- G_h is formed by the edges uv of G with $w(u,v) \in [N-h+1,N]$. Each edge uv in G_h has weight $w(u,v) - (N-h)$. For example, G_1 is formed by the heaviest edges of G, and the weight of each edge is exactly one.
- Let C_h be a minimum weight cover of G_h. G_h^Δ is formed by the edges uv of G with $w(u,v) - C_h(u) - C_h(v) > 0$. The weight of uv is $w(u,v) - C_h(u) - C_h(v)$.

The next theorem is the decomposition theorem.

Theorem 1. $\text{mwm}(G) = \text{mwm}(G_h) + \text{mwm}(G_h^\Delta)$; *particularly,* $\text{mwm}(G) = \text{mm}(G_1) + \text{mwm}(G_1^\Delta)$.

Proof. See §2.3.

Theorem 1 suggests the following recursive algorithm to compute $\text{mwm}(G)$.

Procedure Compute-MWM(G)

1. Construct G_1 from G.
2. Compute $\text{mm}(G_1)$ and find a minimum weight cover C_1 of G_1.
3. Compute G_1^Δ from G and C_1.
4. If G_1^Δ is empty, then return $\text{mm}(G_1)$; otherwise, return $\text{mm}(G_1)$+Compute-MWM(G_1^Δ).

Lemma 2. Compute-MWM(G) *correctly finds* $\text{mwm}(G)$ *in* $O(\sqrt{n}W)$ *time.*

Proof. The correctness of Compute-MWM follows from Theorem 1. Below we analyze its running time. We initialize a maximum heap [6] in $O(m)$ time to store the edges of G according to their weights. Let $T(n,W)$ be the running time of Compute-MWM excluding this initialization. Let L be the set of the heaviest edges in G. Then, Step 1 takes $O(|L|\log m)$ time. In Step 2, a maximum cardinality matching of G_1 can be found in $O(\sqrt{n}|L|)$ time [10, 18]. From this matching, C_1 can be found in $O(|L|)$ time [3]. Let L' be the set of the edges of G adjacent to some node u with $C_1(u) > 0$; i.e., L' consists of the edges of G whose weights are reduced in G_1^{Δ}. Step 3 updates every edge of L' in the heap in $O(|L'|\log m)$ time. Since the total weight of G_1^{Δ} is at most $W - |L'|$, Step 4 requires at most $T(n, W - |L'|)$ time. In summary, as $L \subseteq L'$, $T(n,W) \leq O(\sqrt{n}|L'|) + T(n, W - |L'|) = O(\sqrt{n}W)$. Thus, the running time of Compute-MWM is $T(n,W) + O(m) = O(\sqrt{n}W)$ as stated.

2.2 Unfolded graphs

The proof of Theorem 1 makes use of the *unfolded* graph $\phi(G)$ of G defined as follows.

- For each node u of G, $\phi(G)$ has α images of u, denoted as $u^1, u^2, \ldots, u^{\alpha}$, where α is the weight of the heaviest edge incident to u.
- For each edge uv of G, $\phi(G)$ has the edges $u^1v^{\beta}, u^2v^{\beta-1}, \ldots, u^{\beta}v^1$, where $\beta = w(u,v)$.

The next lemma relates G and $\phi(G)$. Let M be a matching of G. Then, $\phi(M) = \bigcup_{uv \in M}\{u^1v^{\beta}, \ldots, u^{\beta}v^1 \mid \beta = w(u,v)\}$. A path in G is *alternating* for M if (1) its edges alternate between being in M and being not and (2) in case the first (respectively, last) edge of the path is not in M, the first (respectively, last) node of the path is not matched by M.

Lemma 3. *If M is a maximum weight matching of G, then $\phi(M)$ is a maximum cardinality matching of $\phi(G)$. Consequently,* $\mathrm{mwm}(G) = \mathrm{mm}(\phi(G))$.

Proof. Since $\phi(M)$ is a matching of $\phi(G)$ and $\mathrm{mwm}(G) = \sum_{uv \in M} w(u,v) = |\phi(M)|$, it suffices to prove $|\phi(M)| = \mathrm{mm}(\phi(G))$. From basics of bipartite matchings [15, 16], to prove this by contradiction, we may assume that $\phi(G)$ has an alternating path $a_1^{i_1}, a_2^{i_2}, \ldots, a_p^{i_p}$ for $\phi(M)$ such that p is odd and the first and the last edge of the path are not in $\phi(M)$. Let P be the corresponding path $a_1, a_2, \ldots, a_p$ in G. The *net change* of a path in G is the total weight of its edges in M minus that of its edges not in M. To contradict the maximality of M, it suffices to use P to construct an alternating path Q of G for M with a positive net change.

Note that the edges of P alternate between being in M and being not and that $a_1a_2, a_{p-1}a_p \notin M$. Since $w(a_k, a_{k+1}) = i_k + i_{k+1} - 1$, the net change of P is $w(a_1,a_2) - w(a_2,a_3) + \cdots + w(a_{p-1},a_p) = i_1 + i_p - 1 \geq 1$. If neither a_1 nor a_p is matched by M, then $Q = P$ is as desired. Otherwise, there are two cases: (1) exactly one of a_1 and a_p is matched by M; (2) both a_1 and a_p are matched by M. By symmetry, Cases 1 and 2 are similar, and we only detail the proof of Case 1 while further assuming that a_1 is matched by M. Since $a_1a_2 \notin M$, $a_0a_1 \in M$ for some a_0. Let $Q = a_0, a_1, \ldots, a_p$. Then, Q is alternating for M. The net change of Q is $-w(a_0,a_1) + i_1 + i_p - 1$. Since $a_0a_1 \in M$, $\phi(M)$ has the edges $a_0^1a_1^{\beta}, \ldots, a_0^{\beta}a_1^1$, where $\beta = w(a_0,a_1)$. Then, since $a_1^{i_1}$ is not matched

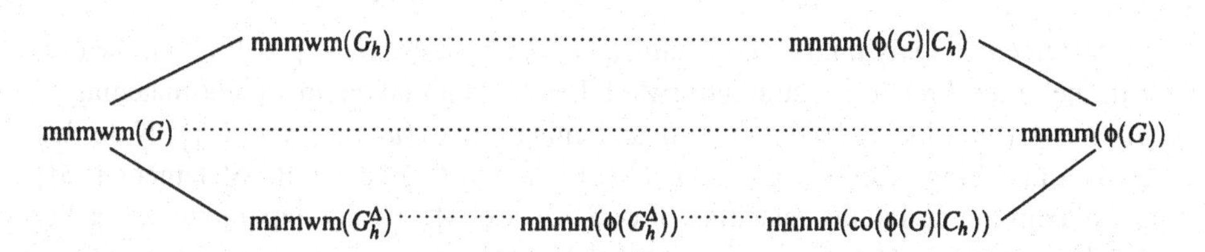

Fig. 1. The relationship between the graphs involved in the proof of Theorem 1. Matchings that have equal value are connected by a dotted line.

by $\phi(M)$, $i_1 \geq w(a_0,a_1)+1$. Therefore, the net change of Q is positive, and Q is as desired.

2.3 Proof of Theorem 1

Let $\phi(G)|C_h$ be the subgraph of $\phi(G)$ induced by the edges incident to nodes u^j with $j \leq C_h(u)$. Lemma 4 below shows the relationship among $G, G_h, G_h^\Delta, \phi(G)$, $\phi(G)|C_h, \phi(G)-\phi(G)|C_h, \phi(G_h^\Delta)$, as depicted in Figure 1. Theorem 1 follows immediately from Lemmas 3 and 4.

Let $V(H)$ denote the node set of a graph H.

Lemma 4. *1.* $\mathrm{mm}(\phi(G)|C_h) = \mathrm{mwm}(G_h)$.
 2. $\mathrm{mm}(\phi(G)-\phi(G)|C_h) = \mathrm{mm}(\phi(G_h^\Delta))$.
 3. $\mathrm{mm}(\phi(G)) = \mathrm{mm}(\phi(G)|C_h) + \mathrm{mm}(\phi(G)-\phi(G)|C_h)$.

Proof. The statements are proved as follows.

Statement 1. Let D be a minimum weight cover of $\phi(G)|C_h$. Since C_h is a minimum weight cover of G_h, by Fact 1, it suffices to show $\sum_{u^i} D(u^i) = \sum_u C_h(u)$. For $\sum_{u^i} D(u^i) \leq \sum_u C_h(u)$, consider $D' : V(\phi(G)|C_h) \to \{0,1,2,\ldots\}$ where for every $u^i \in V(\phi(G)|C_h)$, $D'(u^i) = 1$ if $i \leq C_h(u)$ and 0 otherwise. Since all the edges in $\phi(G)|C_h$ must be of weight 1 and must be attached to some u^i where $i \leq C_h(u)$, D' is a weighted cover of $\phi(G)|C_h$. Thus, $\sum_{u^i} D(u^i) \leq \sum_{u^i} D'(u^i) \leq \sum_u C_h(u)$. For $\sum_{u^i} D(u^i) \geq \sum_u C_h(u)$, consider $D'' : V(G_h) \to \{0,1,2,\ldots\}$ where for every $u \in V(G_h)$, $D''(u) = \sum_i D(u^i)$. We claim that D'' is a weighted cover of G_h and thus, $\sum_{u^i} D(u^i) = \sum_u D''(u) \geq \sum_u C_h(u)$. To show that D'' is a weighted cover of G_h, note that for every edge uv in G_h of weight z, according to the construction, $\phi(G)|C_h$ has at least z edges of the form u^iv^j. As D is a node cover of $\phi(G)|C_h$, $D''(u)+D''(v) = \sum_i D(u^i) + \sum_j D(v^j) \geq z$. Hence, D'' is a weighted cover of G_h.

Statement 2. To show $\mathrm{mm}(\phi(G)-\phi(G)|C_h) \leq \mathrm{mm}(\phi(G_h^\Delta))$, let M be a maximum cardinality matching of $\phi(G)-\phi(G)|C_h$. Note that for any edge $u^iv^j \in M$, $i > C_h(u)$ and $j > C_h(v)$. Let $M' = \left\{u^{i-C_h(u)}v^{j-C_h(v)} \mid u^iv^j \in M\right\}$. Note that M' is a matching in $\phi(G_h^\Delta)$ and $|M| = |M'|$. Thus, $\mathrm{mm}(\phi(G)-\phi(G)|C_h) = |M| = |M'| \leq \mathrm{mm}(\phi(G_h^\Delta))$. To show $\mathrm{mm}(\phi(G_h^\Delta)) \leq \mathrm{mm}(\phi(G)-\phi(G)|C_h)$, let M'' be a maximum cardinality matching of $\phi(G_h^\Delta)$. Similarly, we can verify that $\left\{u^{i+C_h(u)}v^{j+C_h(v)} \mid u^iv^j \in M''\right\}$ is a matching of $\phi(G)-\phi(G)|C_h$. Thus, $\mathrm{mm}(\phi(G_h^\Delta)) = |M''| \leq \mathrm{mm}(\phi(G)-\phi(G)|C_h)$.

Statement 3. Only $\text{mm}(\phi(G)) \geq \text{mm}(\phi(G)|C_h) + \text{mm}(\phi(G) - \phi(G)|C_h)$ will be proved; the other direction is straightforward. Let M be a maximum weight matching of G_h. Let $M_1 = \{u^i v^j \in \phi(G) \mid uv \in M \text{ and either } i \leq C_h(u) \text{ or } j \leq C_h(v)\}$. Note that M_1 is a matching of $\phi(G)|C_h$ with $\sum_u C_h(u) = \text{mwm}(G_h)$ edges. By Statement 1, M_1 is a maximum cardinality matching of $\phi(G)|C_h$. Let M_2 be any maximum cardinality matching of $\phi(G) - \phi(G)|C_h$. We claim that $M_1 \cup M_2$ forms a matching of $\phi(G)$. Then, $\text{mm}(\phi(G)|C_h) + \text{mm}(\phi(G) - \phi(G)|C_h) = |M_1| + |M_2| \leq \text{mm}(\phi(G))$.

Suppose $M_1 \cup M_2$ is not a matching. Then there exist two edges $e_1 \in M_1$ and $e_2 \in M_2$ that share an endpoint. Since $e_1 \in M_1$, it is of the form $u^i v^j$ where $uv \in M, i + j = w(u,v) + 1$, with $i \leq C_h(u)$ or $j \leq C_h(v)$. Without loss of generality, assume $i \leq C_h(u)$. With respect to G_h, M and C_h satisfy Fact 1 and $C_h(u) + C_h(v)$ equals the weight of uv in G_h, i.e., $w(u,v) - (N - h)$. Putting all relations together, $j \geq (N - h) + 1 + C_h(v)$.

As $i \leq C_h(u)$, u^i is not adjacent to any edge in $\phi(G) - \phi(G)|C_h$. Thus, the endpoint shared by e_1 and e_2 must be v^j. Let e_2 be $t^k v^j$. As $e_2 \in \phi(G) - \phi(G)|C_h$, $k > C_h(t)$, $j > C_h(v)$ and $j + k = w(t,v) + 1$. Therefore, $j < w(t,v) + 1 - C_h(t)$. Since $j \geq (N - h) + 1 + C_h(v)$, $C_h(t) + C_h(v) < w(t,v) - (N - h)$. However, C_h is a weighted cover of G_h and thus, $C_h(t) + C_h(v) \geq w(t,v) - (N - h)$, reaching a contradiction.

3 Construct a maximum weight matching

The algorithm in §2 only computes the value of $\text{mwm}(G)$. To report the edges involved, we first construct a minimum weight cover of G in $O(\sqrt{n}W)$ time and then use this cover to construct a maximum weight matching in $O(\sqrt{n}m)$ time.

Lemma 5. *Assume that h, G_h, C_h, and G_h^Δ are defined as in §2. In addition, let C_h^Δ be any minimum weight cover of G_h^Δ. If D is a function on $V(G)$ such that for every $u \in V(G)$, $D(u) = C_h(u) + C_h^\Delta(u)$. Then D is a minimum weight cover of G.*

Proof. Note that for any edge uv of G, its weight in G_h^Δ is $w(u,v) - C_h(u) - C_h(v)$. Since C_h^Δ is a weighted cover, $C_h^\Delta(u) + C_h^\Delta(v) \geq w(u,v) - C_h(u) - C_h(v)$. Thus, $D(u) + D(v) = C_h(u) + C_h^\Delta(u) + C_h(v) + C_h^\Delta(v) \geq w(u,v)$. It follows that D is a weighted cover of G. To show that D is minimum, we observe that

$$\begin{aligned}
\textstyle\sum_{u \in V(G)} D(u) &= \textstyle\sum_{u \in V(G)} C_h(u) + C_h^\Delta(u) \\
&= \textstyle\sum_{u \in V(G)} C_h(u) + \sum_{u \in V(G)} C_h^\Delta(u) \\
&= \text{mwm}(G_h) + \text{mwm}(G_h^\Delta) && \text{by Fact 1} \\
&= \text{mwm}(G). && \text{by Theorem 1}
\end{aligned}$$

By Fact 1, D is minimum.

By Lemma 5, a minimum weight cover of G can be computed using a recursive procedure similar to Compute-MWM.

Procedure Compute-Min-Cover(G)

1. Construct G_1 from G.
2. Find a minimum weight cover C_1 of G_1.
3. Compute G_1^Δ from G and C_1.

4. If G_1^Δ is empty, then return C_1; otherwise, let C_1^Δ = Compute-Min-Cover(G_1^Δ) and return D where for all nodes u in G, $D(u) = C_1(u) + C_1^\Delta(u)$.

Lemma 6. Compute-Min-Cover(G) *correctly finds a minimum weight cover of G in $O(\sqrt{n}W)$ time.*

Proof. The correctness of Compute-Min-Cover(G) follows from Lemma 5. For the time complexity, the analysis is similar to that of Lemma 2.

Given a minimum weight cover D of G, a maximum weight matching of G is constructed from D as follows. Let H be a subgraph of G which contain all edges uv with $w(uv) = D(u) + D(v)$. We make two copies of H. Call them H^a and H^b. For every node u of H, let u^a and u^b denote the corresponding nodes in H^a and H^b, respectively. We union H^a and H^b to form H^{ab}, and add to H^{ab} the set of edges $\{u^a u^b \mid u \in V(H),\ D(u) = 0\}$. Note that H^{ab} has at most $2n$ nodes and at most $3m$ edges. We find a maximum cardinality matching K of H^{ab} using the matching algorithm in [10, 18]. By Lemma 7, the matching $\{uv \mid u^a v^a \in K\}$ is a maximum weight matching of G. The time complexity of the construction is dominated by the computation of the maximum cardinality matching K, which is $O(\sqrt{n}m)$ [10, 18].

Lemma 7. *Let K be a maximum cardinality matching of H^{ab}. Then, $K^a = \{uv \mid u^a v^a \in K\}$ is a maximum weight matching of G.*

Proof. First, we show that H^{ab} has a perfect matching. Let M be a maximum weight matching of G. Since $D(u) + D(v) = w(u,v)$ for every edge $uv \in M$, M is also a matching of H. Let U be the set of nodes in H unmatched by M. By Fact 1, $D(u) = 0$ for all $u \in U$. Let Q be $\{u^a u^b \mid u \in U\}$. Let $M^a = \{u^a v^a \mid uv \in M\}$ and $M^b = \{u^b v^b \mid uv \in M\}$. Note that $Q \cup M^a \cup M^b$ forms a matching in H^{ab} and every node in H^{ab} is matched by either Q, M^a or M^b. Thus, H^{ab} has a perfect matching.

Since K is a perfect matching, for every node u with $D(u) > 0$, u^a must be matched by K. Since there is no edge between u^a and any x^b in H^{ab}, there exists some v^a with $u^a v^a \in K$. Thus, every node u with $D(u) > 0$ must be matched by some edge in K^a. Therefore, $\Sigma_{uv \in K^a} w(u,v) = \Sigma_{u \in X \cup Y, D(u) > 0} D(u) = \Sigma_{u \in X \cup Y} D(u) = \text{mwm}(G)$.

4 All-cavity maximum weight matchings

This section shows that given G and a maximum weight matching M of G, we can compute $\text{mwm}(G - \{u\})$ for all node u in G using $O(W)$ time.

Recall that $\phi(M)$ is a maximum cardinality matching of $\phi(G)$ (see Lemma 3). For each u^i in $\phi(G)$, let $\Delta(u^i) = 0$ if there is an even-length alternating path for $\phi(M)$ starting from u^i; otherwise, $\Delta(u^i) = 1$. Consider any node u in G, let $u^1, u^2, \ldots, u^\beta$ be its corresponding nodes in $\phi(G)$. The following lemma states the relationship among all $\Delta(u^i)$.

Lemma 8. *If $\Delta(u^i) = 0$, then for all $i \le j \le \beta$, $\Delta(u^j) = 0$. Furthermore, we can construct $\beta - i + 1$ node-disjoint even-length alternating paths $P_i, P_{i+1}, \ldots P_\beta$ for $\phi(M)$, where each P_j starts from u^j.*

Proof. As $\Delta(u^i) = 0$, let $P_i = u_0^{i_0}, v_0^{j_0}, u_1^{i_1}, v_1^{j_1}, \ldots, u_{p-1}^{i_{p-1}}, v_{p-1}^{j_{p-1}}, u_p^{i_p}$ of $\phi(G)$ be a shortest even-length alternating path for $\phi(M)$ where $u_0^{i_0} = u^i$. Note that P_i must be simple and $u_p^{i_p}$ is not matched by $\phi(M)$.

Based on P_i, we can construct an even-length alternating path for $\phi(M)$ starting from u^{i+1}. Let $h = \min\{q \mid u_q^{i_q+1}$ is not matched by $\phi(M)\}$, which must exist according to the definition of P_i and $\phi(M)$. Then, $P_{i+1} = u^{i+1}, v_0^{j_0-1}, u_1^{i_1+1}, v_1^{j_1-1}, \ldots, u_h^{i_h+1}$ is an even-length alternating path for $\phi(M)$.

Similarly, even-length alternating path P_j for $\phi(M)$ starting from u^j can be found for $j = i+2, \cdots, \beta$. Also, it can be verified that $P_i, P_{i+1}, \cdots P_\beta$ are node-disjoint.

The next lemma shows that, given $\text{mwm}(G)$, we can compute $\text{mwm}(G - \{u\})$ from the values $\Delta(u^i)$.

Lemma 9. $\Sigma_{1 \leq i \leq \beta} \Delta(u^i) = \text{mwm}(G) - \text{mwm}(G - \{u\})$.

Proof. Let k be the largest integer such that $\Delta(u^k) = 1$. By Lemma 8, $\Delta(u^i) = 1$ for all $1 \leq i \leq k$, and 0 otherwise. Thus, $\Sigma_{1 \leq i \leq \beta} \Delta(u^i) = k$.

Below, we prove the following two equalities:
(1) $\text{mm}(\phi(G) - \{u^1, \ldots, u^k\}) = \text{mm}(\phi(G)) - k$.
(2) $\text{mm}(\phi(G) - \{u^1, \ldots, u^\beta\}) = \text{mm}(\phi(G) - \{u^1, \ldots, u^k\})$.
Then, by Lemma 3, $\text{mwm}(G) = \text{mm}(\phi(G))$ and $\text{mwm}(G - \{u\}) = \text{mm}(\phi(G) - \{u^1, \ldots, u^\beta\})$. This implies $\text{mwm}(G) - \text{mwm}(G - \{u\}) = k$ and the lemma follows.

First, we show Equality (1). Let H be the set of edges of $\phi(M)$ incident to u^i with $1 \leq i \leq k$. Let $M' = \phi(M) - H$. Then, $|M'| = |\phi(M)| - k$. We claim that M' is a maximum cardinality matching of $\phi(G) - \{u^1, ..., u^k\}$. Hence, $\text{mwm}(\phi(G) - \{u^1, ..., u^k\}) = |\phi(M)| - k$; Equality (1) follows. Suppose M' is not a maximum cardinality matching of $\phi(G) - \{u^1, \ldots, u^k\}$. Then, there exists an odd-length alternating path P for M' in $\phi(G) - \{u^1, \ldots, u^k\}$ whose both ends are not matched by M' [15, 16]. P must start from some node v^j with $u^i v^j \in \phi(M)$ and $i < k$. Otherwise, P is alternating for $\phi(M)$ in G and $\phi(M)$ cannot be a maximum cardinality matching of $\phi(G)$. Let Q be a path formed by joining $u^i v^j$ with P. Q is an even-length alternating path for $\phi(M)$ starting from u^i in $\phi(G)$. This contradicts the fact that there is no even-length alternating path for $\phi(M)$ starting from u^i for $i < k$.

To show Equality (2), by Lemma 8, we construct $\beta - k$ node-disjoint alternating paths $P_{k+1}, \ldots, P_\beta$ for $\phi(M)$ where P_j starts at u^j. Let M'' be $\phi(M) \oplus P_{k+1} \oplus \ldots \oplus P_\beta$. Note that $|M''| = |\phi(M)|$ and there are no edges in M'' incident to any u^i with $k+1 \leq i \leq \beta$. Then, $M'' - H$ is a matching of $\phi(G) - \{u^1, \ldots, u^\beta\}$ with size at least $|M''| - k = |\phi(M)| - k$. Since $\text{mm}(\phi(G) - \{u^1, \ldots, u^k\}) = |\phi(M)| - k$ by Equality (1) and $\text{mm}(\phi(G) - \{u^1, \ldots, u^\beta\}) \leq \text{mm}(\phi(G) - \{u^1, \ldots, u^k\})$, Equality (2) follows.

Thus, to find $\text{mwm}(G - \{u\})$ for all nodes u in G, it suffices to find $\Delta(u^i)$ for all u^i in $\phi(G)$. This can be done in $O(W)$ time by Lemma 10.

Lemma 10. *For all $u^i \in \phi(G)$, $\Delta(u^i)$ can be computed in $O(W)$ time.*

Proof. We want to find out, for all u^i, whether $\Delta(u^i) = 0$, i.e., whether there is an even-length alternating path $u^i, v^j, \ldots$ for $\phi(M)$. Let $A_j = \{u^i \in \phi(G) \mid$ the shortest even-length alternating path for $\phi(M)$ starting from u^i is of length $2j\}$. Note that a node is in $\bigcup_j A_j$ if and only if there is an even-length alternating path for $\phi(M)$ starting from that node. As A_0 is the set of nodes in $\phi(G)$ that is unmatched by $\phi(M)$, A_0 can be found in $O(W)$ time. Observe that a node x is in A_{j+1} if and only if $x \notin A_0 \cup A_1 \cup \ldots \cup A_j$ and there is a length-2 path between x and a node in A_j such that the edge incident to x belongs to $\phi(M)$. By examining all such length-2 paths, A_{j+1} can be computed in $O(t_j)$ time where t_j is the sum of the degree of the nodes in A_j. Therefore, all A_j can be found inductively in $O(\sum_j t_j)$ time. Since the sets A_i are disjoint and there are W edges in $\phi(G)$, $\sum_j t_j \leq 2W$ and hence all A_i can be found in $O(W)$ time.

5 Maximum agreement subtrees

An evolutionary tree T is a tree whose leaves labeled by distinct symbols. Let $\mathcal{L}(T)$ be the set of labels used in T. For $L \subseteq \mathcal{L}(T)$, L induces a subtree T whose nodes are either the leaves labeled with L or the least common ancestors of any two nodes labeled with L, and whose edges preserve the ancestor-descendant relationship of T. A maximum agreement subtree of T_1 and T_2 is defined as follows: Let T_1' be a subtree of T_1 induced by some subset of $\mathcal{L}(T_1) \cap \mathcal{L}(T_2)$. T_2' is defined similarly for T_2. T_1' and T_2' are each called an *agreement subtree* of T_1 and T_2 if they have a leaf-label preserving isomorphism. A *maximum agreement subtree* of T_1 and T_2 is an agreement subtree that contains the largest possible number of labels. Denote $\mathrm{mast}(T_1, T_2)$ as the number of labels in a maximum agreement subtree of T_1 and T_2.

In the following subsections, we present two algorithms for computing a maximum agreement subtree of T_1 and T_2, depending on whether T_1 and T_2 are rooted and unrooted. Both algorithms run in $O(\ell^{1.5})$ time where $\ell = \max\{|T_1|, |T_2|\}$. To simplify our discussion, we focus on computing the value of $\mathrm{mast}(T_1, T_2)$.

5.1 Rooted maximum agreement subtrees

Let T_1 and T_2 be any rooted evolutionary trees. This section shows that with the new matching algorithm, the algorithm of Farach and Thorup [9] for computing $\mathrm{mast}(T_1, T_2)$ can be improved to run in $O(\ell^{1.5})$ time.

Fact 2 (see [9]) *Let $t_{n,W}$ be the required time to compute a maximum weight matching of a bipartite graph with node count n and total weight W. Then, the rooted maximum agreement subtree of T_1 and T_2 can be found in $O(\ell^{1+o(1)} + t_{\ell,O(\ell)})$ time.*

To utilize Fact 2, Farach and Thorup [9] applied the Gabow-Tarjan matching algorithm [14] and showed that $\mathrm{mast}(T_1, T_2)$ can be computed in $O(\ell^{1+o(1)} + \sqrt{\ell}\ell \log \ell) = O(\ell^{1.5} \log \ell)$ time. This is the fastest known algorithm in the literature. We replace the matching algorithm and give an improvement as shown in the following theorem.

Theorem 11. *The rooted maximum agreement subtree of T_1 and T_2 can be computed in $O(\ell^{1.5})$ time.*

Proof. We replace the Gabow-Tarjan matching algorithm with the matching algorithm described in §2. Then $t_{\ell,O(\ell)}$ is equal to $O(\ell^{1.5})$ instead of $O(\ell^{1.5}\log\ell)$. Thus, from Fact 2, mast(T_1,T_2) can be computed in $O(\ell^{1.5})$ time.

5.2 Unrooted maximum agreement subtrees

Consider two unrooted evolutionary trees T_1 and T_2. This section shows that mast(T_1,T_2) can be computed in $O(\ell^{1.5})$ time.

Similar to §5.1, computing unrooted maximum agreement subtree is based on bipartite matching algorithms. However, to improve the time complexity, in addition to apply our new matching algorithms, we need to take advantage of the structure of the bipartite graphs involved. Lemma 12 shows that our new matching algorithms can be further adapted to take advantage of the structure of the bipartite graphs.

Lemma 12. *Consider a bipartite graph G. Let u, v be any two nodes of G and let w be the total weight of all the edges of G that are not attached to u and v. Then* mwm(G) *can be computed in* $O(\sqrt{n}w)$ *time. Furthermore,* mwm$(G-\{u\})$ *for all* $u \in G$ *can be computed in the same time complexity.*

Proof. To be shown in the full paper.

We are ready to show the computation of mast(T_1,T_2). Let match(w,w') and cavity(w,w') be the time required to solve the maximum weight matching problem and the all-cavity matching problem of a bipartite graph G whose total weight is at most w' and whose total weight after excluding the edges adjacent to two particular nodes is at most w. The work of Kao et al. [20, 22] can be interpreted as follows:

Fact 3 (see [20, 22]) *If* match(w,w') *and* cavity(w,w') *are* $\Omega(w^{1+o(1)})$, *then* mast(T_1,T_2) *can be computed in* $T(\ell,\ell)$ *time where*

$$T(\ell,\ell') = \ell^{1+o(1)} + \sum_{\sum x_i=\ell} (\text{match}(x_i,\ell') + \text{cavity}(x_i,\ell')) + \sum_{\ell_i<\frac{\ell}{2},\sum\ell_i=\ell} T(\ell_i,\ell')$$

Based on the Gabow-Tarjan matching algorithm [14] and the all-cavity matching algorithm in [21], both match(w,w') and cavity(w,w') equal $O(w^{1.5}\log(ww'))$. Therefore, by Fact 3, mast(T_1,T_2) can be computed in $O(\ell^{1.5}\log\ell)$ time. If we apply Lemma 12, both match(w,w') and cavity(w,w') equal $O(w^{1.5})$ and the time for finding mast(T_1,T_2) is reduced to $O(\ell^{1.5})$.

References

[1] R. Agarwala and D. Fernández-Baca, *A polynomial-time algorithm for the perfect phylogeny problem when the number of character states is fixed*, SIAM Journal on Computing, 23 (1994), pp. 1216–1224.

[2] H. L. Bodlaender, M. R. Fellows, and T. J. Warnow, *Two strikes against perfect phylogeny*, in Lecture Notes in Computer Science 623: Proceedings of the 19th International Colloquium on Automata, Languages, and Programming, Springer-Verlag, New York, NY, 1992, pp. 273–283.

[3] J. Bondy and U. Murty, *Graph Theory with Applications*, North-Holland, New York, NY, 1976.
[4] M. J. Chung, *$O(n^{2.5})$ time algorithms for the subgraph homeomorphism problem on trees*, Journal of Algorithms, 8 (1987), pp. 106–112.
[5] R. Cole and R. Hariharan, *An $O(n \log n)$ algorithm for the maximum agreement subtree problem for binary trees*, in Proceedings of the 7th Annual ACM-SIAM Symposium on Discrete Algorithms, 1996, pp. 323–332.
[6] T. H. Cormen, C. L. Leiserson, and R. L. Rivest, *Introduction to Algorithms*, MIT Press, Cambridge, MA, 1990.
[7] M. Farach, T. M. Przytycka, and M. Thorup, *Computing the agreement of trees with bounded degrees*, in Lecture Notes in Computer Science 979: Proceedings of the 3rd Annual European Symposium on Algorithms, P. Spirakis, ed., Springer-Verlag, New York, NY, 1995, pp. 381–393.
[8] M. Farach and M. Thorup, *Fast comparison of evolutionary trees*, Information and Computation, 123 (1995), pp. 29–37.
[9] M. Farach and M. Thorup, *Sparse dynamic programming for evolutionary-tree comparison*, SIAM Journal on Computing, 26 (1997), pp. 210–230.
[10] T. Feder and R. Motwani, *Clique partitions, graph compression and speeding-up algorithms*, Journal of Computer and System Sciences, 51 (1995), pp. 261–272.
[11] C. R. Finden and A. D. Gordon, *Obtaining common pruned trees*, Journal of Classification, 2 (1985), pp. 255–276.
[12] M. L. Fredman and R. E. Tarjan, *Fibonacci heaps and their uses in improved network optimization algorithms*, Journal of the ACM, 34 (1987), pp. 596–615.
[13] H. N. Gabow, *Scaling algorithms for network problems*, Journal of Computer and System Sciences, 31 (1985), pp. 148–168.
[14] H. N. Gabow and R. E. Tarjan, *Faster scaling algorithms for network problems*, SIAM Journal on Computing, 18 (1989), pp. 1013–1036.
[15] Z. Galil, *Efficient algorithms for finding maximum matching in graphs*, ACM Computing Surveys, 18 (1986), pp. 23–38.
[16] A. M. H. Gerards, *Matching*, in Handbooks in Operations Reserach and Management Science, volume 7, M. O. Ball, T. L. Magnanti, C. L. Monma, and G. L. Nemhauser, eds., Elsevier Science, 1995, pp. 135–224.
[17] D. Gusfield, *Efficient algorithms for inferring evolutionary trees*, Networks, 21 (1991), pp. 19–28.
[18] J. E. Hopcroft and R. M. Karp, *An $n^{5/2}$ algorithm for maximum matching in bipartite graphs*, SIAM Journal on Computing, 2 (1973), pp. 225–231.
[19] M. Y. Kao, *Tree contractions and evolutionary trees*, SIAM Journal on Computing, 27 (1998), pp. 1592–1616.
[20] M. Y. Kao, T. W. Lam, T. M. Przytycka, W. K. Sung, and H. F. Ting, *General techniques for comparing unrooted evolutionary trees*, in Proceedings of the Twenty-Ninth Annual ACM Symposium on Theory of Computing, El Paso, Texas, 4–6 May 1997, pp. 54–65.
[21] M. Y. Kao, T. W. Lam, W. K. Sung, and H. F. Ting, *All-cavity maximum matchings*, in Lecture Notes in Computer Science 1350: Proceedings of the 8th Annual International Symposium on Algorithms and Computation, 1997, pp. 364–373.
[22] M. Y. Kao, T. W. Lam, W. K. Sung, and H. F. Ting, *Cavity matchings, label compressions, and unrooted evolutionary trees*, 1997. Submitted for journal publication.
[23] H. W. Kuhn, *The Hungarian method for the assignment problem*, Naval Research Logistics Quarterly, 2 (1955), pp. 83–97.
[24] T. W. Lam, W. K. Sung, and H. F. Ting, *Computing the unrooted maximum agreement subtree in sub-quadratic time*, Nordic Journal of Computing, 3 (1996), pp. 295–322.

[25] M. STEEL AND T. WARNOW, *Kaikoura tree theorems: Computing the maximum agreement subtree*, Information Processing Letters, 48 (1993), pp. 77–82.
[26] L. WANG, T. JIANG, AND E. LAWLER, *Approximation algorithms for tree alignment with a given phylogeny*, Algorithmica, 16 (1996), pp. 302–315.

Faster Exact Solutions for Some NP-Hard Problems (Extended Abstract)

Limor Drori[1] and David Peleg[2] *

[1] Applied Materials, Rehovot, Israel, limor_drori@amat.com
[2] Department of Computer Science and Applied Mathematics, The Weizmann Institute of Science, Rehovot, 76100 Israel, peleg@wisdom.weizmann.ac.il

Abstract. This paper considers a number of NP-complete problems, and provides faster algorithms for solving them exactly. The solutions are based on a recursive partitioning of the problem domain, and careful elimination of some of the branches along the search without actually checking them. The time complexity of the proposed algorithms is of the form $O(2^{\varepsilon n})$ for constant $0 < \varepsilon < 1$, where n is the output size of the problem. In particular, such algorithms are presented for the *Exact SAT* and *Exact Hitting Set* problems (with $\varepsilon = 0.3212$), and for the *Exact 3SAT* problem (with $\varepsilon = 0.2072$). Both algorithms improve on previous ones proposed in the literature.

1 Introduction

One of the main avenues in our struggle with NP-complete problems involves attempting to develop faster exhaustive-search algorithms for their solution. In particular, certain problems can be solved by algorithms whose complexity depends on specific parameters of the problem at hand, which may be considerably smaller than the input size.

One natural parameter arises in problems Π whose solution space consists of all n bit vectors, where a vector $\bar{x}$ is a legal solution of the problem if it satisfies a certain n-ry predicate $\mathtt{Pred}_\Pi(\bar{x})$. Clearly, every problem of this type can be solved optimally by a naive exhaustive search algorithm, cycling through all 2^n possible solutions and testing each of them individually. But for certain problems it may be possible to reduce the search cost significantly below 2^n by applying clever ways of eliminating some of the cases without actually checking them directly.

One approach for achieving that is the *recursion with elimination* technique. This method is based on a recursive partitioning of the solution space, carefully eliminating some of the branches along the search by using special transformation rules fitted for the problem at hand. Algorithms based on this approach were developed for a number of problems, including an $O(2^{0.276n})$ time algorithm for *Maximum Independent Set (MIS)* [12, 5, 9] and a number of variants of *Satisfiability (SAT)*. In particular, a bound of $O(1.5^n)$ (or $O(2^{0.585n})$) was proved for *3SAT*, the variant of SAT in which each clause contains at most three literals [6, 10]. For the general SAT problem there is an $O(2^{0.773n})$ time and space solution [9], but the best known time bound using polynomial space is

* Supported in part by a grant from the Israel Ministry of Science and Art.

still $O(2^n)$. The complexity of certain algorithms developed for SAT can be bounded in terms of two other parameters, namely L, the length of the input formula and K, the number of clauses it contains. Specifically, SAT can be solved in time $O(2^{0.308K})$ or $O(2^{0.105L})$ [8, 7, 4].

A different approach to the efficient solution of NP-complete problems is the *2-table* method introduced in [11]. This method is based on splitting the n output variables into two sets of $n/2$ variables each, creating all possible $2^{n/2}$ partial solutions on each set, and then scanning all possible combinations of one partial solution from each set in a sorted cost order, ensuring that overall, only $O(2^{n/2})$ cases need to be tested. The method thus yields an $O(2^{n/2})$-time, $O(2^{n/2})$-space algorithm. The class of problems for which this method is applicable is given an axiomatic characterization in [11], and is shown to include, in particular, the *Knapsack, Exact Hitting Set (XHS)* and *Exact 3SAT (X3SAT)* problems.

A generalization of this method to 4 tables, also presented in [11], succeeds in reducing the space requirements of the algorithm to $O(2^{n/4})$, but does not improve the time requirements. In fact, an open problem posed in [11] (which is still open to date, to the best of our knowledge), is whether variants of the k-tables approach can yield $o(2^{n/2})$-time algorithms for any of the problems mentioned above (cf. [2]).

While the current paper does not answer those questions, it does demonstrate that if the answer for the above questions is negative, then the difficulty is not inherent to the problems under consideration but rather to the k-table method itself. This is established by providing faster-than-$2^{n/2}$ algorithms for solving a number of the problems handled in [11]. The solutions are based on the recursion with elimination approach, and their time complexity is $O(2^{\varepsilon n})$ for constant $0 < \varepsilon < 1/2$. Their space complexity is polynomial in n. Hence the complexity of the 4-tables algorithm for these problems can be improved upon, albeit perhaps not by a variant of the k-tables technique.

In particular, the following results are presented in this abstract. We first derive an $O(2^{0.3212n})$ time algorithm for *Exact SAT (XSAT)*, the variant of SAT in which the solution assignment must satisfy that each clause in the formula has exactly one true literal. We also give an $O(2^{0.2072n})$ time algorithm for X3SAT. By a straightforward reduction to XSAT, we also get an $O(2^{0.3212n})$ time algorithm for XHS.

Finally let us remark that very recently we have managed to improve the results for XSAT and XHS, obtaining an $O(2^{0.2985n})$ time algorithm for both problems [1].

2 An algorithm for Exact 3SAT

2.1 Terminology

Let $X = \{x_1, \ldots, x_n\}$ be a set of Boolean variables. A *truth assignment* for X is a function $\tau : X \mapsto \{0,1\}$; we say that u is "true" under τ if $\tau(u) = 1$, and "false" otherwise. With each variable u we associate two *literals*, u and $\bar{u}$. The truth assignment τ is expanded to literals by setting $\tau(\bar{u}) = 1$ if $\tau(u) = 0$, and $\tau(\bar{u}) = 0$ if $\tau(u) = 1$.

A *clause* over X is a set of literals from X. A clause is satisfied by a truth assignment τ iff exactly one of its literals is true under τ. A collection C of clauses over X is satisfiable iff there exists some truth assignment that simultaneously satisfies all the clauses in C.

The *Exact 3 Satisfiability (X3SAT)* problem (called *one-in-three 3SAT* in [3]) is defined as follows. A clause C is called a k-clause, for integer $k \geq 1$, if it consists of k literals. Given a set $X = \{x_1, \ldots, x_n\}$ of variables and a collection $C = \{C_1, \ldots, C_m\}$ of 3-clauses over X, decide whether there exists a truth assignment for X, such that each clause in C has exactly one true literal. Note that a clause may contain two or more literals with the same variable. For example, the following are legal 3-clauses: $\{x_1, x_1, x_1\}$, $\{x_1, x_1, x_2\}$, $\{x_1, \bar{x}_1, x_2\}$, etc. However, note that the first of those can never be satisfied; in the second, the only satisfying assignment is $\tau(x_1) = 0$ and $\tau(x_2) = 1$; and in the third, every satisfying assignment must have $\tau(x_2) = 0$.

Let us introduce the following terminology. For a clause C, let $X(C)$ denote its set of variables. A variable occurring in a single clause is called a *singleton*. A variable x appearing in the same affinity in all clauses in C (i.e, always as x, or always as $\bar{x}$) is hereafter referred to as a *constant variable* (it is sometimes called a *pure literal* as well). For a literal ℓ, let $x(\ell)$ denote the corresponding variable, and let $V(\ell)$ denote its affinity, namely, $V(\ell) = 1$ if $\ell = x$ and $V(\ell) = 0$ if $\ell = \bar{x}$. The opposite affinity is denoted by $\bar{V}(\ell) = 1 - V(\ell)$. It is also convenient to use the notation $\bar{\ell}$, for a literal ℓ, to signify the opposite literal, i.e, $\bar{\ell} = \bar{x}$ if $\ell = x$, and $\bar{\ell} = x$ otherwise.

2.2 Cannonical instances

Our general strategy is based on simplifying the instance at hand via either reducing the number of clauses or reducing the number of variables occurring in them. This is done by using two basic operations, namely, fixing the truth assignment of certain variables, or identifying certain variable pairs with each other. More formally, for a variable x and a bit $b \in \{0,1\}$, we denote by $\text{FIX}(x,b)$ the restriction of the problem instance at hand to truth assignments τ in which $\tau(x) = b$. Applying this operation allows us to eliminate x from the instance entirely, as follows. First, eliminate x from every clause where it occurs as a literal ℓ with affinity $V(\ell) = 1 - b$. (This makes the clause smaller, and hence may help us to eliminate the clause as well, as is discussed shortly.) Next, observe that every clause C where x occurs as a literal with affinity $V(\ell) = b$, is satisfied by x, hence it can be discarded. However, note also that as ℓ is satisfied, all other literals in C must be falsified, which immediately forces us to apply FIX to those literals as well. This chain reaction may proceed until no additional variables can be fixed.

Two specific cases of the FIX operation deserve special notation. For a literal ℓ, we write $\text{FIXTRUE}(\ell)$ to mean $\text{FIX}(x(\ell), V(\ell))$, and $\text{FIXFALSE}(\ell)$ to mean $\text{FIX}(x(\ell), \bar{V}(\ell))$.

Also, for two literals ℓ_1 and ℓ_2 (of different variables) we denote by $\text{IDENTIFY}(\ell_2, \ell_1)$ the restriction of the instance to truth assignments τ in which $\tau(\ell_2) = \tau(\ell_1)$. Applying this operation allows us to discard of the variable $x(\ell_2)$ in our instance, by replacing any occurrence of ℓ_2 in a clause with ℓ_1, and any occurrence of $\bar{\ell}_2$ with $\bar{\ell}_1$. As a result of this operation, it might happen that in some clause C (not necessarily the one that caused us to apply the operation, but some other clause), there are now two copies of the variable x_1. For example, this will happen if we applied $\text{IDENTIFY}(x_1, x_2)$ and had some other clause $C = (x_1, x_2, x_3)$. So after the identification C becomes (x_1, x_1, x_3). Note that these are two distinct occurrences of x_1, so $V(x_1)$ must be set to 0, otherwise this clause is satisfied twice. Similarly, if we had a clause $C = (x_1, \bar{x}_2, x_3)$, then after the identification it becomes $(x_1, \bar{x}_1, x_3)$, and necessarily we must fix $V(x_3)$ to 0. Again, such simplification may lead to a chain reaction, continuing until no further simplifications can be performed.

Note that the FIX and IDENTIFY operations are both linear in the input size.

We would like to identify the class of instances that cannot be simplified automatically. Towards that end, a *cannonical instance* of X3SAT is defined as an instance enjoying the following properties:

(P1) Any two clauses in the instance share at most one common variable.
(P2) There is at most one singleton in every clause.
(P3) In every clause there are exactly three different variables.

The above definition is justified by the following simple observations, which indicate how non-cannonical instances can be easily transformed into cannonical ones. (Some of the proofs are omitted from the extended abstract).

Claim 1 *If the instance contains a* 1*-clause,* $C = \{\ell\}$*, then the clause can be discarded along with the variable* $x(\ell)$.

Claim 2 *If the instance contains a* 2*-clause,* $C = \{\ell_1, \ell_2\}$*, then the clause can be discarded along with one of the variables.*

Claim 3 *Whenever one variable in a* 3*-clause* C *is fixed, the instance can be simplified by discarding at least one more variable and clause.*

Claim 4 *If the instance contains two clauses with literals based on the same variables, then the instance can be either decided (by an* $O(1)$*-step test) or simplified by discarding one clause and possibly some of the variables.*

Proof. Let the two clauses be $C = \{\ell_1, \ell_2, \ell_3\}$ and $C' = \{\ell'_1, \ell'_2, \ell'_3\}$, with $x(\ell_i) = x(\ell'_i)$ for $i = 1,2,3$. Not all three affinities are identical, since the two clauses are different, hence there are seven cases to consider.

If $\ell'_j \neq \ell_j$ for every $j = 1,2,3$ then the instance cannot be satisfied.

Now suppose exactly one variable occurs with the same affinity in the two clauses, without loss of generality $\ell'_3 = \ell_3$. In this case, the truth assignment may not satisfy ℓ_3, since this will force all other literals in C and C' to be falsified, which cannot be done. Hence any satisfying truth assignment τ is forced to set $\tau(x(\ell_3)) = \bar{V}(\ell_3)$, and we must apply FIXFALSE(ℓ_3). Hence by Claim 3, it is possible to discard the two clauses and the variable.

Finally suppose that exactly one variable occurs with opposite affinities in the two clauses, without loss of generality $\ell'_3 = \bar{\ell}_3$. In this case the instance cannot be satisfied. ∎

Claim 5 *If the instance contains two clauses with two common variables, then the instance can be either decided (by an* $O(1)$*-step test) or simplified by discarding one clause and possibly some of the variables.*

Proof. Let the two clauses be $C = \{\ell_1, \ell_2, \ell_3\}$ and $C' = \{\ell'_1, \ell'_2, \ell_4\}$, where $x(\ell_i) = x(\ell'_i)$ for $i = 1,2$ and $x(\ell_3) \neq x(\ell_4)$. The following three cases are possible:

1. $\ell'_1 = \ell_1$ and $\ell'_2 = \ell_2$. In this case any satisfying truth assignment may satisfy at most one of ℓ_1 and ℓ_2. If ℓ_1 or ℓ_2 is satisfied, necessarily ℓ_3 and ℓ_4 must both be falsified. Conversely, if neither ℓ_1 nor ℓ_2 is satisfied, necessarily ℓ_3 and ℓ_4 must both be satisfied. Hence ℓ_3 and ℓ_4 must be given the same truth assignment, and we may apply IDENTIFY(ℓ_4, ℓ_3) and discard $x(\ell_4)$ and the clause C'.

2. $\ell'_1 = \ell_1$ and $\ell'_2 \neq \ell_2$. In this case no satisfying truth assignment may satisfy ℓ_1, because then one of the clauses will have two satisfied variables. Thus we may apply FIXFALSE(ℓ_1). Hence by Claim 3, it is possible to discard the variables $x(\ell_3), x(\ell_4)$ and the two clauses.

3. $\ell'_1 \neq \ell_1$ and $\ell'_2 \neq \ell_2$. If both ℓ_1 and ℓ_2 are satisfied then C is not satisfied, and if both ℓ_1 and ℓ_2 are not satisfied then C' is not satisfied. Hence any satisfying truth assignment τ must satisfy exactly one of the pair $\{\ell_1, \ell_2\}$. This means that we must apply IDENTIFY$(\ell_2, \bar{\ell}_1)$. Also, this forces any satisfying truth assignment to falsify both ℓ_3 and ℓ_4, hence we may apply FIXFALSE(ℓ_3) and FIXFALSE(ℓ_4). Consequently, we may discard the two clauses. ▮

Claim 6 *If the instance contains a clause with two singletons or more, then the instance can be simplified by discarding that clause and two variables.*

Claim 7 *If every clause in an instance contains a singleton, and all variables are constant, then the instance is satisfiable (and a satisfying truth assignment can be found in time linear in the input size).*

Now we present a procedure of time complexity $O(mn)$, which given a non-cannonical instance C of the problem transforms it to a cannonical instance C'.

Procedure CANNONIZEX3SAT(C)

While the instance C is not cannonical *repeat:*

(a) For every 1-clause C containing literal ℓ do:
Apply FIXTRUE(ℓ) and discard the variable $x(\ell)$. Discard any other clause C' which contains $x(\ell)$ as follows:
If the variable is in the same affinity in C' as in C (and thus satisfies the clause C'), the other variables of C' must not be satisfied, so the FIXFALSE operation can be applied to each of them. If the variable is not in the same affinity in C', then it can be removed from C', and a 2-clause would be left.

(b) For every 2-clause C do:
Simplify C by Claim 2.

(c) For every clause containing a repeated variable x_1 do:
If x_1 is repeated three times with the same affinity (i.e, $\{x_1, x_1, x_1\}$), a contradiction occurs, and C is decided. Else discard the repetition as follows:

1. If x_1 appears twice with the same affinity b (i.e, $\{x_1, x_1, x_2\}$ or $\{\bar{x}_1, \bar{x}_1, x_1\}$): x_1 must be set to $\bar{b}$ (in order to avoid the clause from being satisfied twice). Furthermore, the third variable must be set to 1 or 0 for satisfying the clause.
2. If x_1 appears twice, and in both affinities (i.e, $\{x_1, \bar{x}_1, x_2\}$): The variable not repeated must be falsified.

(d) For each clause in C with more than one singleton, simplify C by Claim 6.
(e) For every two clauses with two or more common variables do:
Decide or simplify C by Claims 4 and 5.

End_While

The output of Procedure CANNONIZEX3SAT(C) is of the form (b, C') where b is:

$$b = \begin{cases} 0, & \text{a contradiction occured,} \\ 1, & \text{the cannonization ended successfuly.} \end{cases}$$

and C' is the resulting cannonical instance if $b = 1$.

Each iteration of the algorithm takes $O(m)$ time and discards at least one variable, and so the algorithm terminates in time $O(mn)$.

Lemma 8. *A non-cannonical instance C can be transformed into a cannonical instance C' in time $O(mn)$, with C' containing (strictly) fewer variables than C.*

2.3 A recursive algorithm

Let C be an instance of the X3SAT problem. The recursive algorithm X3SAT(C) for finding a satisfying truth assignment τ or a contradiction operates as follows. Let us first describe the basic recursive procedure TEST(C,x,v), where C is an instance, x is a variable and $v \in \{0,1\}$.

Procedure TEST(C,x,v)

1. Apply FIX(x,v).
2. Transform C into a cannonical C' by invoking Procedure
   $(b, C') \leftarrow$ CANNONIZEX3SAT(C).
3. If $b = 0$ (a contradiction occurs) then return 0.
4. Else if $C' = \emptyset$ (all variables are discarded) then return 1.
5. Otherwise, recursively invoke $b' \leftarrow$ X3SAT(C') and return b'.

Main Algorithm X3SAT(C)

1. If every clause contains a singleton variable and all variables are constant then return 1.
2. Else choose a variable x_i according to the following steps.
   (a) If there is a non-constant variable x_i in C, choose it.
   (b) Else pick a clause with no singleton in it, and choose one of its variables.
3. Invoke the recursive procedure $b' \leftarrow$ TEST($C,x_i,1$).
4. If $b' = 1$, then return 1 and halt. Otherwise invoke $b' \leftarrow$ TEST($C,x_i,0$) and return b'.

2.4 Analysis

In the full paper we prove the following theorem.

Theorem 9. *The recursive algorithm* X3SAT(C) *solves the X3SAT problem in time complexity* $O(m \cdot 2^{0.2072n})$*, where n and m are the number of variables and clauses of C, respectively.*

In order to illustrate the flavor of the proof, let us provide a simpler analysis for a weaker bound, namely, $O(m \cdot 2^{0.2557n})$.

Suppose that the algorithm is applied to an instance C containing n variables. First, note that if the condition of step 1 holds, then the procedure terminates correctly by Claim 7. Hence from now on we restrict our attention to the case where this condition is not met. In this case, the algorithm picks a variable x_i, and tests both possible ways of fixing its truth value, namely 0 and 1. In both cases, fixing the value of x_i yields a non-cannonical instance, which can be simplified further, by identifying some variables and fixing the truth values of some others. Once a cannonical instance C' is obtained, it is handled recursively by the algorithm. The crucial observation is that in each case, the resulting instance C' has fewer variables than the original C. Analyzing the number of variables discarded in each step is the central component of our analysis, and it yields the recurrence equations governing the complexity of the algorithm.

Let $f(m,n)$ denote the worst-case time complexity of Algorithm X3SAT on m-clause, n-variable instances. Note that in the case of 3-clauses, the maximal number of clauses in an instance is $\binom{2n}{3} = O(n^3)$, hence the input length is $m \log n = O(n^3 \log n)$.

The number of variables discarded at each step of the recursion is analyzed as follows. Recall that step 2 chooses some *test variable* x_i. It is chosen by step 2a or step 2b. We shall next examine these two cases separately.

Test variable chosen in Step 2a. Let us first provide a straightforward analysis for the case where x_i was chosen by step 2a, namely, it is a non-constant variable. Given the linear dependence of $f(m,n)$ on m, we shall henceforth ignore m, and analyze only the dependence of the complexity on n, denoted by the function $f(n)$. Let C_1 and C_2 be two clauses in which x_i appears in opposing affinities. Without loss of generality let $C_1 = \{x_1, \ell_2, \ell_3\}$ and $C_2 = \{\bar{x}_1, \ell_4, \ell_5\}$.

When TEST(C,x_1,1) is invoked, x_1 is fixed to 1. Hence C_1 is satisfied, so we may apply FIXFALSE(ℓ_2) and FIXFALSE(ℓ_3). Moreover, by Claim 2 we apply IDENTIFY($\ell_5, \bar{\ell}_4$). Thus x_1, $x(\ell_2)$, $x(\ell_3)$ and $x(\ell_5)$ are discarded.

Likewise when TEST(C,x_1,0) is invoked, x_1 is fixed to 0. Hence C_2 is satisfied, so we may apply FIXFALSE(ℓ_4) and FIXFALSE(ℓ_5), and again, by Claim 2 we apply IDENTIFY($\ell_3, \bar{\ell}_2$). Thus x_1, $x(\ell_4)$, $x(\ell_5)$ and $x(\ell_3)$ are discarded.

It follows that the recurrence equations governing the complexity of Algorithm X3SAT are the following. By the above analysis we get that in case the variable x_i is always chosen by step 2a, then $f(m,n)$ satisfies

$$f(m,n) \le \begin{cases} O(1), & \text{step 1 applies,} \\ O(mn) + 2f(m,n-4), & \text{otherwise.} \end{cases}$$

This recurrence solves to give $f(m,n) = O(m \cdot 2^{n/4})$. The improved result of Theorem 9 is based on refining this analysis by breaking the discussion further into subcases, and deriving an improved recurrence equation for each case. Given the linear dependence of $f(m,n)$ on m, we shall henceforth ignore m, and analyze only the dependence of the complexity on n, denoted by the function $f(n)$. Hence the corresponding recurrence is

$$f(n) \leq 2f(n-4) \tag{1}$$

Test variable chosen in Step 2b. Now let us analyze the case where x_i was chosen by step 2b of the algorithm X3SAT. In this case, each variable appears in all of its clauses with the same affinity. Without loss of generality let us assume all variables appear in the positive form. Furthermore, x_i was chosen from a clause which contained no singleton. Let $C_1 = \{x_1, x_2, x_3\}$ be the clause and x_1 be the variable chosen, and let C_2 be some other clause containing x_1. Since C is cannonical, two clauses may not have two or more common variables, by property (P1). Hence let $C_2 = \{x_1, x_4, x_5\}$. When TEST$(C, x_1, 1)$ is invoked, x_2, x_3, x_4 and x_5 are falsified, resulting in five fixed variables.

When TEST$(C, x_1, 0)$ is invoked, IDENTIFY$(x_3, \bar{x}_2)$ is imposed by C_1, and IDENTIFY $(x_5, \bar{x}_4)$ by C_2. Therefore three variables are discarded.

By the above analysis we get that in case the variable x_i is always chosen by step 2b, then $f(n)$ satisfies

$$f(n) \leq f(n-3) + f(n-5) \tag{2}$$

It follows that the time complexity of Algorithm X3SAT is bounded by a function $f(n)$ which obeys inequalities (1) and (2). These inequalities both solve to give $f(n) = 2^{\varepsilon n}$ for some constant $0 < \varepsilon < 1$, whose value is determined by the specific inequalities. In particular, the constraints imposed by the specific inequalities at hand can be calculated to be the following:

$f(n) \leq 2f(n-4) \Longrightarrow \varepsilon \geq 0.25$,

$f(n) \leq f(n-3) + f(n-5) \Longrightarrow \varepsilon \geq 0.2557$.

Hence the bound achieved on the worst-case time complexity of our algorithm is $O(m \cdot 2^{0.2557n})$.

3 An algorithm for Exact SAT

3.1 Terminology

The terminology of XSAT is similar to that of X3SAT as explained in Section 2.1. The only difference is in the definition of the collection C, which in X3SAT is restricted to have 3-clauses; in XSAT this restriction no longer holds.

3.2 Cannonical instances

Our general strategy is the same as that of X3SAT as presented in Section 2.2. We shall again use the FIX, FIXTRUE, FIXFALSE and IDENTIFY operations. We shall also use

ℓ'_i to denote a literal on the same variable as ℓ_i (i.e, $x(\ell'_i) = x(\ell_i)$). Furthermore, Claims 1, 2, 3 and 7 still hold.

A *cannonical instance* of XSAT is defined as an instance enjoying the following properties:

(Q1) Any two 3-clauses share at most one common variable.
(Q2) No clause contains fewer than three variables.
(Q3) No clause contains the same variable more than once.
(Q4) There is at most one singleton in every clause.
(Q5)
There are no two constant variables such that each clause either contains both or contains neither.
(Q6) There are no two clauses with the same number of variables r in which $r-1$ variables appear in both clauses and with the same affinity.
(Q7) There are no two clauses such that all variables of one appear in the other.

This definition is again motivated by a sequence of claims, including claims 1, 2, 3, 7 of the previous section, and the simple observations presented next.

Claim 10 *If the instance contains two constant variables such that each clause either contains both variables or contains neither, then the instance can be simplified by discarding one of the variables.*

Claim 11 *If the instance contains two clauses with the same number of variables $r \geq 3$, in which $r-1$ variables appear in both clauses and with the same affinity, then the instance can be simplified by discarding a variable and one of the clauses.*

Proof. Let the two clauses be $C_1 = \{\ell_1, \ldots, \ell_{r-1}, \ell_r\}$ and $C_2 = \{\ell_1, \ldots, \ell_{r-1}, \ell_t\}$ where $x(\ell_r) \neq x(\ell_t)$, $r \geq 3$ and $|C_1| = |C_2|$. Any satisfying truth assignment must either satisfy exactly one of the literals $\ell_1, \ldots, \ell_{r-1}$ and falsify ℓ_r and ℓ_t or falsify all the literals $\ell_1, \ldots, \ell_{r-1}$ and satisfy both ℓ_r and ℓ_t. Hence ℓ_r and ℓ_t must have the same truth assignment, and it is possible to apply IDENTIFY(ℓ_r, ℓ_t) and discard $x(\ell_t)$ and one of the clauses. ∎

Claim 12 *If the instance contains a clause with two singletons or more, then the instance can be simplified by discarding at least one variable.*

Lemma 13. [Subsets lemma] *If the instance contains two clauses C_1 and C_2 such that all variables of C_1 appear in C_2 (i.e, $X(C_1) \subseteq X(C_2)$), then the instance can be either decided (by an $O(1)$-step test) or simplified by discarding one clause and some of the variables.*

Proof. Let the two clauses be $C_1 = \{\ell_1, \ell_2, \ldots, \ell_m\}$ and $C_2 = \{\ell'_1, \ell'_2, \ldots, \ell'_k\}$ where $m \leq k$. The affinity of the common variables can be classified according to the following cases:

- There are three or more variables with opposite affinities in C_1 and C_2. Then in every truth assignment some of these variables satisfy C_1 and the rest satisfy C_2. Since there are at least three such variables overall, at least one of the clauses is satisfied twice. Consequently, a contradiction occurs.

- There are exactly two variables with opposite affinities in C_1 and C_2. Without loss of generality let $x(\ell_1)$ and $x(\ell_2)$ be the variables. IDENTIFY$(\ell_2,\bar{\ell}_1)$ may be applied, because otherwise one of the clauses is satisfied twice. Furthermore, all other variables in C_1 and C_2 must be falsified. Hence $x(\ell_2),\dots,x(\ell_k)$ and the two clauses are discarded.
- There is exactly one variable with opposite affinities in C_1 and C_2. Without loss of generality let this variable be $x(\ell_1)$. If $k=m$, then $x(\ell_1)$ satisfies one of the clauses, let it be C_1. Thus any other variable satisfying C_2 causes C_1 to be satisfied twice. Hence a contradiction occurs. If $k>m$, then any satisfying truth assignment must FIXTRUE(ℓ_1) otherwise C_2 is satisfied twice. Hence it is possible to falsify $\ell_2,\dots,\ell_m$ and C_2 must be satisfied by one of the remaining variables $\ell'_{m+1},\dots,\ell'_k$.
- $\ell_i=\ell'_i$ for each $1\le i\le m$. In this case, the variable satisfying C_1 will necessarily satisfy C_2 as well. Thus $\ell'_{m+1},\dots,\ell'_k$ must be falsified, and C_2 may be discarded.

∎

Now we shall present a procedure of time complexity $O(mn)$, which given a non-cannonical instance C transforms it to a cannonical instance C'.

Procedure CANNONIZEXSAT(C)

While the instance C is not cannonical *repeat:*

(a) For every 1-clause C containing literal ℓ do:
Apply FIXTRUE(ℓ) and discard the variable $x(\ell)$. Discard or simplify any other clause C' which contains $x(\ell)$ as follows:
If the variable is in the same affinity in C' as in C (and thus satisfies the clause C'), the other variables of C' must not be satisfied, so the FIXFALSE operation can be applied to each of them. If the variable is not in the same affinity in C', then it can be removed from C'.

(b) For every 2-clause C do:
Simplify C by Claim 2.

(c) For every clause containing a repeated variable x_1 do:
If x_1 is repeated at least twice in each affinity (i.e, $\{x_1,x_1,\bar{x}_1,\bar{x}_1,\dots\}$), a contradiction occurs, and C is decided. Else discard the repetition as follows:
1. If x_1 appears at least twice in one affinity b and does not appear at all in the other affinity, $\bar{b}$, then there are two cases to consider. If there are no other variables in the clause, a contradiction occurs. Else set x_1 to $\bar{b}$ and remove it from the clause.
2. If x_1 appears at least twice in one affinity b and exactly once in the other affinity, $\bar{b}$, then again set x_1 to $\bar{b}$. Furthermore, apply FIXFALSE to all other literals in the clause and discard the clause.
3. If x_1 appears exactly once in each affinity (i.e, $\{x_1,\bar{x}_1,x_2\}$), apply FIXFALSE to all other literals in the clause and discard the clause.

(d) For every two clauses which contain three variables each and share two or more common variables do:
Decide or simplify C by Claims 4 and 5.

(e) For each clause in C which has more than one singleton in it do:
Simplify C by Claim 12.

(f) For every two constant variables such that each clause in C either contains both variables or contains neither do:
Simplify C by Claim 10.
(g) For every two clauses with the same number of variables r in which $r-1$ variables appear in both clauses with the same affinity do:
Simplify C by Claim 11.
(h) For every two clauses where all variables of one appear in the other do:
Decide or simplify C by Lemma 13.

End_While

The output of CANNONIZEXSAT(C) is of the form (b, C') with the same values as defined in X3SAT. Each iteration of the algorithm takes $O(m)$ time and discards at least one variable, and so the algorithm terminates in time $O(mn)$, hence Lemma 8 still applies.

3.3 A recursive algorithm

Let C be an instance. The recursive algorithm XSAT(C) for finding a satisfying truth assignment τ or a contradiction is identical to that of Section 2.3, except that it uses the more general cannonization procedure CANNONIZEXSAT instead of procedure CANNONIZEX3SAT of Section 2.3 and step 2b for choosing a variable is changed to picking the *smallest* clause with no singleton in it, instead of just any clause of Section 2.3.

In the full paper we prove the following theorem.

Theorem 14. *The recursive algorithm* XSAT(C) *has time complexity* $O(m \cdot 2^{0.3212n})$, *where n and m are the number of variables and clauses of C, respectively.*

4 An algorithm for Exact Hitting Set

The *Exact Hitting Set (XHS)* problem is defined as follows. Given a Boolean matrix M with n rows and m columns, decide whether there exists a subset of the rows, such that each column j in M has exactly one row i such that $M_{ij} = 1$.

Theorem 15. *There exists a recursive algorithm of time complexity* $O(m \cdot 2^{0.3212n})$ *for XHS.*

Proof. The proof is by a simple reduction from XHS to XSAT. Given an instance M of the XHS problem, construct an instance C of XSAT as follows. Define a variable x_i for each row $1 \leq i \leq n$, and for each column j create a clause $C_j = \{x_i \mid M_{ij} = 1\}$. Note that the number of variables in the instance C equals the number of rows in the given matrix M. It is straightforward to verify that $XHS(M) = True$ iff $XSAT(C) = True$. ∎

Acknowledgement We are grateful to two anonymous referees for their suggestions, which helped to improve the presentation of the paper.

References

[1] L. Drori and D. Peleg. *Improved Exact Solutions for Exact Hitting Set and Related Problems*. Unpublished manuscript, March. 1999.

[2] A. Ferreira. On space-efficient algorithms for certain NP-complete problems. *Theoret. Computer Sci.*, 120:311–315, 1993.

[3] M.R. Garey and D.S. Johnson. *Computers and Intractability: a Guide to the Theory of NP-Completeness. W.H. Freeman and Co.*, San Francisco, CA, 1979.

[4] Edward A. Hirsch. Two new upper bounds for SAT. In *Proc. 9th ACM-SIAM Symp. on Discrete Algorithms*, pages 521–530, January 1998.

[5] Tang Jian. An $O(2^{0.304n})$ algorithm for solving maximum independent set problem. *IEEE Trans. on Communication*, 35:847–851, September 1986.

[6] O. Kullmann. Worst-case analysis, 3-sat decision and lower bounds: approaches for improved sat algorithms. In *DIMACS SAT Workshop*, 1996.

[7] O. Kullmann and H. Luckhardt. Deciding propositional tautologies: Algorithms and their complexity. Unpublished manuscript, 1997.

[8] B. Monien and E. Speckenmeyer. Solving satisfiability in less than 2^n steps. *Discrete Applied Mathematics*, 10:287–295, April 1985.

[9] J.M. Robson. Algorithms for Maximum Independent Sets. *Journal of Algorithms* , 7:425-440, 1986.

[10] I. Schiermeyer. Pure literal look ahead: An $O(1.497^n)$ 3-satisfiability algorithm. In *Workshop on the Satisfiability Problem*, 1996.

[11] R. Schroeppel and A. Shamir. A $T = O(2^{n/2})$, $S = O(2^{n/4})$ algorithm for certain NP-complete problems. *SIAM J. on Computing*, 10:456–464, 1981.

[12] Robert E. Tarjan and Anthony E. Trojanowski. Finding a maximum independent set. *SIAM J. on Computing*, 6:537–546, September 1977.

A Polyhedral Algorithm for Packings and Designs

Lucia Moura

Department of Computer Science, University of Toronto
Toronto, Canada, M5S 3G4

Abstract. We propose a new algorithmic technique for constructing combinatorial designs such as t-designs and packings. The algorithm is based on polyhedral theory and employs the well-known branch-and-cut approach. Several properties of the designs are studied and used in the design of our algorithm. A polynomial-time separation algorithm for clique facets is developed for a class of designs, and an isomorph rejection algorithm is employed in pruning tree branches. Our implementation is described and experimental results are analysed.

1 Introduction

Computational methods have been important in combinatorial design theory. They are useful for constructing "starter" designs for recursive constructions for infinite families [17] and they also play a central role in applications [2]. Techniques that have been widely used include backtracking, several local search methods (such as hill-climbing, simulated annealing, genetic algorithms) as well as several algorithms using tk-matrices (see [8, 17]). In [11], we propose the general approach of employing polyhedral algorithms to combinatorial design problems. In the present article, we design and implement the first branch-and-cut algorithm for constructing t-designs and packings.

A t–(v,k,λ) *design* is a pair $(V,\mathcal{B})$ where V is a v-set and $\mathcal{B}$ is a collection of k-subsets of V called *blocks* such that every t-subset of V is contained in exactly λ blocks of $\mathcal{B}$. A t–(v,k,λ) *packing design* is defined by replacing the condition "in exactly λ blocks" in the above definition by "in at most λ blocks". The *packing number*, denoted by $D_\lambda(v,k,t)$, is the *maximum* number of blocks in a t–(v,k,λ) packing design. Important classes of t-designs are Steiner systems (t–$(v,k,1)$ designs) and balanced incomplete block designs (2–(v,k,λ) designs); designs with $k=t+1$ are also of special interest, for instance Steiner triple systems and Steiner quadruple systems. Central questions in combinatorial design theory are the existence of t-designs and the determination of packing numbers. It is well known that if a t–(v,k,λ) design exists, it is a maximum packing. Thus, our algorithm constructs designs by searching for maximal packings. In [18], a polyhedral algorithm specific for 2-designs is proposed, using a different integer programming formulation.

The main contributions of this paper is the design and implementation of a new branch-and-cut algorithm for packings and t-designs. We incorporate into the algorithm some aspects specific to combinatorial design problems such as a new clique separation based on design properties and an isomorph rejection scheme to remove isomorphic subproblems from the branch-and-cut tree. The effects of various parameters on performance are analyzed through experiments. Our method is competitive with other techniques for generating designs including backtracking and randomized search. Our algorithm produced new maximal cyclic t-$(v,k,1)$ packings for $t = 2,3,4,5$, $k = t+1, t+2$ and small v. We hope this article will be a starting point for further research in the application of similar techniques to combinatorial design problems.

2 Polyhedra for Designs and Packings

In this section, we present integer programming formulations for t-designs and their extensions to packings. Similar formulations for designs with prescribed automorphism groups can be employed (see [11]).

We concentrate on designs without repeated blocks, known as *simple* designs. Such a design can be represented by an *incidence vector*, that is a 0-1 vector $x \in \mathbb{R}^{\binom{v}{k}}$ indexed by the k-subsets of a v-set and such that $x_S = 1$ if and only if S is a block of the design. The polyhedron associated with a design is defined as the convex hull of the incidence vectors of all designs of that kind. Let us denote by $T_{t,v,k,\lambda}$ and $P_{t,v,k,\lambda}$ the polyhedra associated with the t–(v,k,λ) designs and packing designs, respectively.

Let $W^v_{t,k}$ be the $\binom{v}{t} \times \binom{v}{k}$ matrix with rows indexed by the t-subsets and columns by k-subsets of a v-set, and such that $[W^v_{t,k}]_{T,K} = 1$ if $T \subseteq K$ and $[W^v_{t,k}]_{T,K} = 0$, otherwise. For a detailed study of these matrix and their role in design theory see [4].
It is easy to see that t–(v,k,λ) designs correspond to the solutions $x \in \mathbb{R}^{\binom{v}{k}}$ of

$$\text{(DP)} \quad \begin{cases} W^v_{t,k}\, x = \lambda \mathbf{1}, \\ x \in \{0,1\}^{\binom{v}{k}}. \end{cases}$$

The maximum packings correspond to solutions $x \in \mathbb{R}^{\binom{v}{k}}$ of

$$\text{(PDP)} \quad \begin{cases} \text{maximize} & \mathbf{1}^T x \\ \text{subject to} & W^v_{t,k}\, x \le \lambda \mathbf{1}, \\ & x \in \{0,1\}^{\binom{v}{k}}. \end{cases}$$

Based on these integer programming formulations, we rewrite the design polytopes as $T_{t,v,k,\lambda} = conv\{x \in \{0,1\}^{\binom{v}{k}} : W^v_{t,k}\, x = \lambda \mathbf{1}\}$ and $P_{t,v,k,\lambda} = conv\{x \in \{0,1\}^{\binom{v}{k}} : W^v_{t,k}\, x \le \lambda \mathbf{1}\}$. For $\lambda = 1$ these are special cases of set partitioning and set packing, respectively. For $\lambda > 1$, the polytope $P_{t,v,k,\lambda}$ is the polytope for independent sets of an independence system. The general problems of set partitioning, set packing and maximum independence systems are known to be NP-hard [7].

An independence system is a pair $(I, \mathcal{I})$ where I is an n-set and $\mathcal{I}$ is a family of subsets of I, with the property that $I_1 \subseteq I_2 \in \mathcal{I}$ implies $I_1 \in \mathcal{I}$; the individual members of $\mathcal{I}$ are called independent sets. Sets $J \subseteq I$ such that $J \notin \mathcal{I}$ are said to be *dependent* and minimal such sets are called *circuits*. An independence system is characterized by its family of circuits. We observe that the circuits of our specific independence system are the sets of $(\lambda+1)$ k-subsets of $[1,v]$ sharing a common t-subset, as we state in the following proposition.

Proposition 1. *(Packings and independence systems)*
A t–(v,k,λ) packing is an independent set of the independence system given by the circuits

$$\left\{\{K_1,K_2,\ldots,K_{\lambda+1}\} : K_i \in \binom{[1,v]}{k} \text{ and } |K_1 \cap K_2 \cap \ldots \cap K_{\lambda+1}| \geq t\}\right\}.$$

The following proposition gives basic properties of the design polytopes.

Proposition 2. *(Basic properties - see [11])*
i. If $k \leq v-t$ then $dim T_{t,v,k,\lambda} \leq \binom{v}{k} - \binom{v}{t}$.
ii. $P_{t,v,k,\lambda}$ is full dimensional.
iii. The inequalities $x_i \geq 0$, $i = 1,\ldots,\binom{v}{k}$, are facet inducing for $P_{t,v,k,\lambda}$.

Let $(I,\mathcal{I})$ be a p-regular independence system, i.e. one with all circuits with same cardinality p. A subset $I' \subseteq I$ is said to be a clique if $|I'| \geq p$ and all p-subsets of I' are circuits of $(I,\mathcal{I})$.

Theorem 3. *(Clique inequalities for independence systems - Nemhauser and Trotter [15])*
Suppose $I' \subseteq I$ is a maximal (with respect to set-inclusion) clique in a p-regular independence system $S = (I,\mathcal{I})$. Then, $\sum_{i \in I'} x_i \leq p-1$ defines a facet of $P(S)$.

2.1 Characterization of Clique Inequalities of $P_{t,v,k,\lambda}$ and Efficient Separation

A clique in such a $(\lambda+1)$-regular independence system corresponds to a group of k-subsets of $[1,v]$ with the following intersecting properties.

Definition 4. *(s-wise t-intersecting set systems)*
Given $s \geq 2$ and $v,t \geq 1$, a family $\mathcal{A}$ of subsets of $[1,v]$ is said to be s-wise t-intersecting, *if any s members $A_1,\ldots,A_s$ of $\mathcal{A}$ are such that $|A_1 \cap \ldots \cap A_s| \geq t$. A family $\mathcal{A}$ is said to be* k-uniform *if every member of $\mathcal{A}$ has cardinality k. Let $I^s(v,k,t)$ denote the set of all k-uniform s-wise t-intersecting families of subsets of $[1,v]$. Let $MI^s(v,k,t)$ denote the set of all families in $I^s(v,k,t)$ that are maximal with respect to set inclusion (i.e. $\mathcal{A} \in I^s(v,k,t)$ such that for any $\mathcal{B} \in I^s(v,k,t)$, if $\mathcal{B} \supseteq \mathcal{A}$ then $\mathcal{B} = \mathcal{A}$).*

Proposition 5. *(Characterization of generalized cliques)*
Let $\mathcal{A}$ be a family of k-subsets of $[1,v]$. Then $\mathcal{A}$ is a generalized clique for the independence system associated with a t–(v,k,λ) packing if and only if $\mathcal{A} \in I^{(\lambda+1)}(v,k,t)$. Moreover, a clique $\mathcal{A}$ is maximal if and only if $\mathcal{A} \in MI^{(\lambda+1)}(v,k,t)$.

Proof. By the definition above of clique in an independence system, $\mathcal{A}$ corresponds to a clique for the set packing independence system if and only if $\mathcal{A}$ is a family of k-subsets of $[1,v]$ such that all subfamilies of $\mathcal{A}$ with $(\lambda+1)$ elements are circuits. By Proposition 1, this is equivalent to $\mathcal{A} \in I^{(\lambda+1)}(v,k,t)$. Clearly, the clique is maximal if and only if $\mathcal{A} \in MI^{(\lambda+1)}(v,k,t)$.

The following corollary characterizes the clique inequalities for the polytope $P_{t,v,k,\lambda}$.

Corollary 6. *(Clique inequalities for $P_{t,v,k,\lambda}$)*
Let $\mathcal{A} \in I^{(\lambda+1)}(v,k,t)$. Then, the inequality

$$\sum_{K \in \mathcal{A}} x_K \leq \lambda$$

is valid for the polytope $P_{t,v,k,\lambda}$. Moreover, the above inequality induces a facet of $P_{t,v,k,\lambda}$ if and only if $\mathcal{A} \in MI^{(\lambda+1)}(v,k,t)$.

Proof. It follows from Proposition 5 and Theorem 3.

In [13], we classify all pairwise t-intersecting families of k-sets of a v-set for $k-t \leq 2$ and all t and v. Some of these results can be translated in terms of clique facets as follows:

1. For any $t \geq 1$, $v \geq t+3$, $\lambda = 1$ and $k = t+1$, there exists exactly two distinct (up to isomorphism) clique facets for $P_{t,v,t+1,1}$, namely

$$\sum_{K \in \binom{[1,v]}{t+1}: K \supseteq T} x_K \leq 1, \quad \text{for all } T \in \binom{[1,v]}{t}, \tag{1}$$

$$\sum_{K \in \binom{L}{t+1}} x_K \leq 1, \quad \text{for all } L \in \binom{[1,v]}{t+2}. \tag{2}$$

2. For any $t \geq 1$, $v \geq t+6$, $\lambda = 1$ and $k = t+2$, there exists exactly 15 distinct types of clique facets for $t = 1$, and 17 distinct types of clique facets for $t \geq 2$. These cliques are given explicitly for $t = 1$ and $t = 2$, and by a construction for $t \geq 3$ (see [13] for their forms).
3. For any $t \geq 1$, $v \geq t+2$, $\lambda \geq 2$ and $k = t+1$, there exists exactly one distinct (up to isomorphism) generalized clique, namely $\sum_{K \in \binom{[1,v]}{t+1}: K \supseteq T} x_K \leq 1$.
4. For arbitrary $k > t$ and λ, there exists a $v_0 = v_0(t,\lambda,k)$ such that all clique facets for $v \geq v_o$ are determined by those for v_0.

The knowledge of the clique structure can help us designing separation algorithms. For example, for the case of $k = t+1$ and $\lambda = 1$ the separation of clique inequalities turns out to be quite efficient, as shown in the following algorithm.

Let $G^v_{t,k} = \left(\binom{[1,v]}{k}, E\right)$ be the intersection graph of $W^v_{t,k}$, that is, the graph such that $K_1, K_2 \in \binom{[1,v]}{k}$ are linked by an edge if and only if $|K_1 \cap K_2| \geq t$.

Algorithm: Separation of clique inequalities for $P_{t,v,t+1,1}$
Input: a fractional solution $\bar{x}$ to (PDP)
Output: a violated clique inequality or "There are no violated cliques"
  for every edge $\{K_1, K_2\} \in E$
    take $L = K_1 \cup K_2$
    if $\sum_{K \in \binom{L}{t+1}} \bar{x}_K > 1$ then
      return "Violated clique: ", L
  return "There are no violated cliques."

Let us specify our measure of complexity. The size of a design problem is measured by the number of bits needed for its integer programming formulation. For t–(v,k,λ) designs, the problem size is exactly $\binom{v}{t} \times \binom{v}{k} + \binom{v}{t} \log\lambda$. Our measure of complexity is the number of basic operations such as arithmetic operations and comparisons.

Corollary 7. *(Separation of cliques in $P_{t,v,t+1,1}$)*
The clique facets in $P_{t,v,t+1,1}$ can be separated in polynomial time.

Proof. The statement is implied by the correctness of the previous algorithm. Any fractional solution $\bar{x}$ to (PDP) must satisfy $W^v_{t,t+1}\bar{x} \leq 1$. So, inequalities (1) are satisfied and the only possible violated cliques are the ones in (2). Note that there is exactly one clique of type (2) passing through each edge $\{K_1, K_2\}$ in the graph $G^v_{t,t+1}$, for $|K_1 \cup K_2| = t+2$. This shows the correctness of the algorithm. The polynomiality can be checked by noticing that every iteration takes polynomial number of steps and the number of iterations is at most the square of the number of variables in the problem.

3 A Branch-and-Cut Algorithm for Packings and Designs

Besides being an alternative to tackle combinatorial design problems, the branch-and-cut approach offers other advantages. We can adapt the general framework in order to deal with design specific issues such as: fixing subdesigns, extending designs, forbidding sub-configurations, proving non-existence of designs and assuming the action of an automorphism group (see [12] for details).

Our implementation handles t–$(v,k,1)$ designs and packings, both ordinary ones and admitting cyclic automorphism groups. In the following, we describe subalgorithms and other issues specific to our implementation. The reader is referred to [1] for the general branch-and-cut approach.

3.1 Initialization

Some variable fixing can be done, in the original problem of finding t-designs, before running the branch-and-cut algorithm. For any t–$(v,k,1)$ design and any given $(t-1)$-subset S of $[1,v]$, the blocks of the design that contain S are unique up to permutations. Indeed, we can assume w.l.o.g. that the following blocks are present in the design $B_i = [1,t-1] \cup [t+(i-1)(k-t+1), t+i(k-t+1)-1]$, $1 \leq i \leq \frac{v-t+1}{k-t+1}$,, $E = [1,t-2] \cup \bigcup_{1 \leq i \leq k-(t-2)} \{t+(i-1)(k-t+1)\}$, and that all the other subsets of $[1,v]$ containing $[1,t]$ are not present in the design.

3.2 Separation Algorithms

The fractional intersecting graph: The separation of clique facets relies on finding violated cliques in the intersection graph of the original matrix. It is well-known [5] that in any set packing problem, we can restrict our attention to the *fractional intersection graph*, i.e., the subgraph of the intersection graph induced by the fractional variables. In our experiments, this reduces the size of the graph we have to deal with from several thousand to a few hundred nodes.

Separation of clique facets: Given $\bar{x}$, we must find violated cliques in the fractional graph $G^v_{k,t}$ (corresponding to $\bar{x}$). A clique C is considered violated if $\sum_{i \in C} \bar{x}_i - 1 >$ VIOLATION-TOLERANCE. Previously generated inequalities are stored in a pool of cuts in order to reuse them in other nodes of the tree. This is a common feature in a branch-and-cut algorithm (see [1]).

GENERAL-CLIQUE-SEPARATION: The general clique detection employed by our algorithm works for a general graph. We borrowed several ideas from Hoffman and Padberg [5] and Nemhauser and Sigismondi [14]. For every node v, we search for a violated clique containing v. If the neighborhood of v is small, say under 16 nodes, we enumerate every clique in the neighborhood; otherwise, we use two greedy heuristics proposed in [14].

SPECIAL-CLIQUE-SEPARATION (for k=t+1): We implemented the special clique separation algorithm for designs with $k = t + 1$, given on page 465. Recall that this algorithm uses the knowledge of the clique structure for these problems and examines each edge of the graph exactly once, since there is at most one violated clique passing through each edge. Note that we use the fractional intersection graph in place of $G^v_{k,t}$, as mentioned previously.

Criteria for abandoning the cutting-plane algorithm: Our algorithm stops cutting-plane iterations if any one of the following conditions is satisfied: the optimal solution to the LP relaxation is integral (i.e. the subproblem rooted at the node was solved), the subproblem is infeasible, or the addition of the cuts is "not producing much improvement". The last condition is measured by the number of cuts and the "quality" of cuts at the previous iteration. The cutting-plane algorithm is abandoned whenever the number of cuts in the previous iteration is smaller than a parameter MIN-NUMBER-OF-CUTS or the maximum violation is smaller than a parameter MIN-WORTHWHILE-VIOLATION.

3.3 Partial Isomorph Rejection

For combinatorial design problems, several subproblems in the branch-and-cut tree may be equivalent. Recall that a node in the branch-and-cut tree corresponds to the subproblem in which the variables in the path from the root to the node have their values fixed either to zero or one. Let N be any node of the tree and denote by $\mathcal{F}_0(N)$ and $\mathcal{F}_1(N)$ the collection of blocks corresponding to the variables fixed to 0 and 1, respectively, in the path from the root of the tree to N. If N and M are nodes in the tree with $(\mathcal{F}_0(N), \mathcal{F}_1(N))$ isomorphic to $(\mathcal{F}_0(M), \mathcal{F}_1(M))$ then equivalent problems are going to be unnecessarily solved. The partial isomorph rejection we describe in this section aims at reducing the number of such equivalent subproblems.

Let N be a node of the branch-and-cut tree with two children N_0 and N_1. The original branching scheme would make N_0 correspond to $x_K = 0$ and N_1 correspond to $x_K = 1$, for some variable K. This branching scheme is modified so that the number of nodes in the tree is reduced by avoiding some subproblems that are equivalent to others already considered, as we discuss next. Let A be the permutation group acting on $[1, v]$ that fixes $\mathcal{F}_0(N)$ and $\mathcal{F}_1(N)$, and let $A(K)$ be the orbit of K under A. The new branching scheme for partial isomorph rejection lets N_0 correspond to "$x_L = 0$ for all $L \in A(K)$" and N_1 correspond to "$x_L = 1$ for some $L \in A(K)$". The tree reduction comes from letting N_1

correspond, w.l.o.g., to "$x_K = 1$" instead. Thus the new branching scheme implies

$$\mathcal{F}_0(N_0) = \mathcal{F}_0(N) \cup A(K), \quad \mathcal{F}_1(N_0) = \mathcal{F}_1(N),$$
$$\mathcal{F}_0(N_1) = \mathcal{F}_0(N), \quad \mathcal{F}_1(N_1) = \mathcal{F}_1(N) \cup \{K\}.$$

This new branch-and-cut tree has at most as many problems as the regular one, since when $|A(K)| > 1$ not only K but other variables in $A(K)$ are being simultaneously fixed at 0 in N_0. All we need is an algorithm that, given collections $\mathcal{F}_0$ and $\mathcal{F}_1$ of k-subsets of $[1,v]$ and a k-subset K of $[1,v]$, computes: (1) the permutation group A acting on $[1,v]$ that fixes $\mathcal{F}_0$ and $\mathcal{F}_1$; and (2) the orbit $A(K)$ of K under A.

The first problem is equivalent to finding the permutation group acting on the vertices of a special graph that fixes some subsets of the vertices. Consider the bipartite graph $G_{[1,v],\mathcal{F}_0\cup\mathcal{F}_1}$ whose vertex partition corresponds to points in $[1,v]$ and sets in $\mathcal{F}_0 \cup \mathcal{F}_1$, and such that $p \in [1,v]$ is connected to $F \in \mathcal{F}_0 \cup \mathcal{F}_1$ if and only if $p \in F$. Thus, our problem is equivalent to finding the permutation group acting on the vertices of the graph that fixes vertices in $[1,v]$, in $\mathcal{F}_0$ and in $\mathcal{F}_1$. This can be computed using the package *Nauty* [9], by Brendan McKay, the "most powerful general purpose graph isomorphism program currently available" [6].

The second problem can be solved by a simple algorithm that we describe now. A collection $\mathcal{S}$ of k-sets is initialized with K. At every step, a different set L in $\mathcal{S}$ is considered and, for all $\pi \in A$, the set $\pi(L)$ is added to $\mathcal{S}$. The algorithm halts when all sets in $\mathcal{S}$ have been considered, and thus, $A(K) = \mathcal{S}$.

A small variation of the two previous methods can dramatically improve efficiency when $|\mathcal{F}_0 \cup \mathcal{F}_1| \ll v$. Let $R = \cup_{\mathcal{B}\in\mathcal{F}_0\cup\mathcal{F}_1} \mathcal{B}$. Consider the graph $G_{R,\mathcal{F}_0\cup\mathcal{F}_1}$ instead of $G_{[1,v],\mathcal{F}_0\cup\mathcal{F}_1}$ and apply the method described above to compute an automorphism group A'. The points in $[1,v] \setminus R$ are isolated vertices in $G_{[1,v],\mathcal{F}_0\cup\mathcal{F}_1}$, and therefore form a cycle in any permutation in A. Thus A' is the restriction of A to points in R. In order to compute $A(K)$, we use the method described above and compute $A'(K \cap R)$, and then compute $A(K)$ by taking all the k-subsets of $[1,v]$ that contain some set in $A'(K \cap R)$. Two kinds of improvements in efficiency are observed. In the first part, the original graph gets reduced by $[1,v] \setminus R$ nodes. In the second part, if $|K \cap R| < k$ the size of the set $\mathcal{S}$ in the second problem is reduced by a factor of $\binom{[1,v]\setminus R}{k-|K\cap R|}$.

3.4 Branch-and-Cut Tree Processing

We implemented two strategies for the selection of the branching variable in a node. The first one selects the variable with largest fractional value, and the second one, the variable closest to 0.5. Both strategies turned out to be equivalent, since in our problems most of the fractional variables are smaller than 0.5. Our selection of the next node to process is done as a depth-first search, giving priority to nodes with variables fixed to 1. To help in the detection of a globally optimal solution, we use the general upper bound on the size of a packing given by the Schönheim bounds (see [10]).

4 Computational Results

In this section, we report on computational experiments with the branch-and-cut implementation described in the previous section. The tests are run on a Sun Ultra 2 Model

2170 workstation with 245 MB main memory and 1.2 GB virtual memory, operating system SunOS 5.5.1. Our branch-and-cut consists of about 10,000 lines of code written in C++ language and compiled with g++ compiler. The following packages are linked with our code: LEDA Library version 3.2.3 [16] for basic data structures such as lists and graphs, CPLEX package version 4.0.8 [3] for solving linear programming subproblems, and Nauty package version 2.0 [9] for finding automorphism groups of graphs.

Our main conclusions are outlined as follows. It is advantageous to run the isomorph rejection algorithm. In Table 2 we report on isomorph rejection statistics. We compare the same instances of packings with and without isomorph rejection (specified by column (IRej)). Only the packings with $v = 5, 11, 12, 14$ require a call to the algorithm. From these parameters, only $v = 5, 11$ benefit from nontrivial orbits, and only $v = 11$ profits from the isomorph rejection. We observe that the time spent in the isomorph rejection algorithm is very small compared to the total time. The packing for $v = 11$ could not be found without the isomorph rejection. The difficulty encountered for $v = 11$ is that the Schönheim upper bound is not met by the packing size. Therefore, the program might find a solution of size 17, but has to go over most of the branches to conclude it is optimal. The isomorph rejection reduces the amount of branches to be searched for. We conclude that the isomorph rejection algorithm is effective since it spends little extra time and adds the benefit of tackling the hardest problems.

The impact of cutting is analysed through a comparison between branch-and-cut with a straight forward branch-and-bound. We observed in our experiments that for designs there was no clear winner in terms of total time. For packings, the total time using cuts is either comparable or substantially smaller than without cuts, especially for the larger instances (Table 3). In all cases, using cuts reduces the number of explored tree nodes and the number of times the algorithm backtracks. This is reflected in the often smaller number of solved linear programming problems and time spend on solving them. However, the time spent on cut separation makes the cutting version worst for some instances. Larger problems should benefit from the cutting version, since linear programming tends to dominate the running time.

The specialized separation is also more efficient in practice than the general separation heuristic for clique facets. Table 5 summarizes the results of 8 runs corresponding to a combination of parameters described in Table 4. Even and odd numbered runs correspond to specialized and general separation algorithms, respectively. From these tables we observe that the specialized separation is done much faster than the general one (see column (ST)). In all runs the specialized separation produced savings in the total running time of up to 50%.

Parameters that affect the trade-off between branching and cutting, are also analysed in Table 5. This table shows the influence of the parameters MIN-WORTHWHILE-VIOLATION and VIOLATION-TOLERANCE. Let us denote them by MWV and V, respectively. An interval $[v_{min}, v_{max}]$ is assigned to each of these parameters. The algorithm initially sets the value of the parameter to v_{max}; as the number of fractional variables decreases, the parameter is continuously reduced towards v_{min}. The 4 combinations shown in Table 4 are tried. The best runs involve (MWV, V) = $[(0.3, 0.3), (0.3, 0.3)]$ and $[(0.3, 0.6), (0.3, 0.6)]$. The main conclusion is that the performance is positively affected by requiring stronger cuts. In all other tables, these parameters are set as MWV = V = [(0.3,0.6)].

We also did a comparison of our branch-and-cut with a leading general purpose branch-and-bound software (Cplex Mixed Integer Programming Optimizer), as a reference (see Table 5, column run#=Cplex). As problem sizes grow, our algorithm runs much faster than CPLEX (2 to 10 times faster in the two largest problems).

Some difficult problems given by Steiner quadruple systems are shown in Table 6. The case $v = 14$ is already a hard instance. In earlier versions of this implementation it took about 6 hs to solve this problem. Currently, it takes about 40 minutes to solve it. The next instance $v = 16$ is still a challenge for this implementation. Although the design is known to exist, other computational methods also fail to find this design (Mathon, personal communication).

Finally, we summarize our findings on cyclic packings. Recall that these problems are solved by assuming a cyclic automorphism group action on the design, which produces reductions of the problem size. Tables 7 and 8 compare the size of maximal t–$(v,k,1)$ packings to the size of regular packings for $t = 2,3,4,5$, $k = t+1, t+2$, and small v. Columns (Bl) and (C) indicate the number of base blocks and the total number of blocks in the cyclic packings, respectively. To the best of our knowledge, this is the first time this quantities are computed. In columns (D) and (S) we include known values for the size of a maximal ordinary packing and Schönheim upper bounds, respectively. Thus, we must have $C \leq D \leq S$. Observe that in most cases C is not much smaller than D (or not much smaller than S in the cases that D is unknown, see Table 8). Our experiments show that the sizes of cyclic packings are close to maximal ordinary packings (compare (C) and (D)), and they are much easier to compute and compact to store. Therefore, these objects should be very attractive for applications. The size of a maximal cyclic packing also offers a good lower bound for the packing number.

References

[1] Caprara, A., and Fischetti, M.: Branch-and-Cut Algorithms. In *Annotated Bibliographies in Combinatorial Optimization,* M. Dell'Amico, et al, Eds., John Wiley & Sons, 1997, ch. 4.

[2] Colbourn, C., and van Oorschot, P.: Applications of Combinatorial Designs in Computer Science. *ACM Computing Surveys 21* (1989), 223–250.

[3] Using the CPLEX Callable Library, version 4.0, 1995. http://www.cplex.com.

[4] Godsil, C.: Linear Algebra and Designs. Manuscript (229 p.), 1995.

[5] Hoffman, K., and Padberg, M.: Solving Airline Crew-Scheduling Problems by Branch-and-Cut. *Manag. Sci. 39* (1993), 657–682.

[6] Kocay, W.: On Writing Isomorphism Programs. In Wallis [17], 1996, pp. 135–175.

[7] Lenstra, J., and Rinnooy Kan, A.: Complexity of Packing, Covering and Partitioning. In *Packing and Covering in Combinatorics*, A. Schrijver, Ed. Mathematical Centre Trats 106, 1979, pp. 275–291.

[8] Mathon, R.: Computational Methods in Design Theory. In Wallis [17], 1996, pp. 29–48.

[9] McKay, B.: Nauty User's Guide (version 1.5). Tech. Rep. TR-CS-90-02, Dept. Computer Science, Austral. Nat. Univ., 1990.

[10] Mills, W., and Mullin, R.: Coverings and Packings. In *Contemporary Design Theory: a Collection of Surveys*, J. Dinitz and D. Stinson, Eds., Wiley, 1992, pp. 371–399.

[11] Moura, L.: Polyhedral Methods in Design Theory. In Wallis [17], 1996, pp. 227–254.

[12] Moura, L.: *Polyhedral Aspects of Combinatorial Designs*. PhD thesis, University of Toronto, 1999.

[13] Moura, L.: Maximal s-Wise t-Intersecting Families of Sets: Kernels, Generating Sets, and Enumeration. To appear in *J. Combin. Theory. Ser. A* (July 1999), 22 pages.

[14] Nemhauser, G., and Sigismondi, G.: A Strong Cutting Plane/Branch-and-Bound Algorithm for Node Packing. *J. Opl. Res. Soc. 43* (1992), 443–457.

[15] Nemhauser, G., and Trotter Jr., L.: Properties of Vertex Packing and Independence System Polyhedra. *Math. Programming 6* (1974), 48–61.

[16] Stefan, N., and Mehlhorn, K.: LEDA-Manual Version 3.0. Tech. Rep. MPI-I-93109, Max-Planck-Institute für Informatik, 1993.

[17] Wallis, W. (Ed.): *Computational and Constructive Design Theory*. Kluwer Academic Publishers, 1996.

[18] Wengrzik, D.: Schnittebenenverfahren für Blockdesign-Probleme. Master's thesis, Universität Berlin, 1995.

Appendix

Table 1. Problem sizes and statistics

t-(v,k,1) packings	ordinary rows	ordinary cols	cyclic packings rows	cyclic packings cols
2-(5,3,1)	10	10	2	0
2-(6,3,1)	15	20	3	1
2-(7,3,1)	21	35	3	2
2-(8,3,1)	28	56	4	2
2-(9,3,1)	36	84	4	7
2-(10,3,1)	45	120	5	4
2-(11,3,1)	55	165	5	10
2-(12,3,1)	66	220	6	11
2-(13,3,1)	78	286	6	16
2-(14,3,1)	91	364	7	14
3-(6,4,1)	20	15	4	1
3-(7,4,1)	35	35	5	2
3-(8,4,1)	58	70	7	8
3-(9,4,1)	84	126	10	9
3-(10,4,1)	120	210	12	19
3-(11,4,1)	165	330	15	25
3-(12,4,1)	220	495	19	37
3-(13,4,1)	286	715	22	49
3-(14,4,1)	364	1001	26	67
3-(15,4,1)	455	1365	31	82
3-(16,4,1)	560	1820	35	110
3-(17,4,1)	680	2380	40	132
3-(18,4,1)	816	3060	46	160
3-(19,4,1)	969	3876	51	195
3-(20,4,1)	1140	4845	57	238
3-(21,4,1)	1330	5985	64	270
3-(22,4,1)	1540	7315	70	325
4-(8,5,1)	70	56	10	2
4-(9,5,1)	126	126	14	9
4-(10,5,1)	210	252	22	18
4-(11,5,1)	330	462	30	37
4-(12,5,1)	495	792	43	52
4-(13,5,1)	715	1287	55	93
4-(14,5,1)	1001	2002	73	122

t-$(v,k,1)$		ordinary				cyclic	
		before fixing		after fixing			
design	b	rows	cols	rows	cols	rows	cols
2-(7,3,1)	7	21	35	15	21	3	2
2-(9,3,1)	12	36	84	28	55	4	7
2-(13,3,1)	26	78	286	66	219	6	16
2-(15,3,1)	35	105	445	91	363	7	25
2-(19,3,1)	57	171	969	153	815	9	42
2-(21,3,1)	70	210	1330	190	1139	10	55
2-(25,3,1)	100	300	2300	286	2023	12	80
2-(27,3,1)	117	351	2925	325	2599	13	97
2-(31,3,1)	155	495	4495	465	4059	15	130
2-(33,3,1)	176	528	5456	496	4959	16	151
3-(8,4,1)	14	56	70	50	54	7	8
3-(10,4,1)	30	120	210	112	181	12	19
3-(14,4,1)	91	364	1001	352	934	26	67
3-(16,4,1)	140	560	1820	546	1728	35	110

- b: number of blocks
- Sch: Schönheim bound
- IRej: isomorph rejection algorithm
- IE: number of calls to IRej
- IR: number of successful IRej's
- Mxl: Max. depth of node in successful IRej
- IT: Total time in IRej
- BB: number of explored B&B nodes
- BT: number of backtracks
- LPT: total time solving LP problems
- LPs: number of LP problems
- ST: total time in separation alg.
- Cl: total number of added cliques
- TotT: total time

Table 2. The effect of partial isomorph rejection on 2-$(v,3,1)$ packings

v	b	Sch	IRej	BB	BT	IE	IR	Mxl	IT	LPT	LPs	ST	Cl	TotT
5	2	3	yes	3	1	1	1	0	0	0	3	0	0	0.01
5	2	3	no	7	3	0	0	-	0	0.02	8	0	1	0.02
6	4	4	yes	3	0	0	0	-	0	0	3	0.01	0	0.01
6	4	4	no	3	0	0	0	-	0	0.01	3	0	0	0.01
7	7	7	yes	1	0	0	0	-	0	0.01	1	0	0	0.02
7	7	7	no	1	0	0	0	-	0	0.01	1	0	0	0.02
8	8	8	yes	5	0	0	0	-	0	0.02	8	0.02	6	0.06
8	8	8	no	5	0	0	0	-	0	0.02	8	0.02	6	0.05
9	12	12	yes	2	0	0	0	-	0	0.02	3	0	1	0.04
9	12	12	no	2	0	0	0	-	0	0.02	3	0.02	1	0.05
10	13	13	yes	11	0	0	0	-	0	0.08	14	0.05	4	0.14
10	13	13	no	11	0	0	0	-	0	0.05	14	0.08	4	0.13
11	17	18	yes	649	324	324	11	10	0.26	6.41	829	2.46	319	9.94
11	17	18	no	-	-	-	-	-	-	-	-	-	-	(*)
12	20	20	yes	17	1	1	0	-	0	0.22	24	0.25	11	0.53
12	20	20	no	17	1	0	0	-	0	0.24	24	0.27	11	0.53
13	26	26	yes	13	0	0	0	-	0	0.12	14	0.34	1	0.48
13	26	26	no	13	0	0	0	-	0	0.13	14	0.33	1	0.49
14	27	28	yes	77	27	27	0	-	0.05	1.82	121	0.85	86	2.91
14	27	28	no	77	27	0	0	-	0	1.81	121	0.88	86	2.93

(*) the algorithm failed to find the designs even after exploring 400,000 branches.

Table 3. Branch-and-cut versus branch-and-bound for 2-$(v,3,1)$ packings

v	b	Cuts	BB	BT	IE	IR	Mxl	IT	LPT	LPs	ST	Cl	TotT
5	2	yes	3	1	1	1	0	0	0	3	0	0	0.01
5	2	no	3	1	1	1	0	0	0	3	0	0	0.01
6	4	yes	3	0	0	0	-	0	0	3	0.01	0	0.01
6	4	no	3	0	0	0	-	0	0.01	3	0	0	0.01
7	7	yes	1	0	0	0	-	0	0.01	1	0	0	0.02
7	7	no	1	0	0	0	-	0	0.01	1	0	0	0.02
8	8	yes	5	0	0	0	-	0	0.02	8	0.02	6	0.06
8	8	no	8	1	1	0	-	0	0.04	8	0	0	0.05
9	12	yes	2	0	0	0	-	0	0.02	3	0	1	0.04
9	12	no	4	0	0	0	-	0	0.02	4	0	0	0.02
10	13	yes	11	0	0	0	-	0	0.08	14	0.05	4	0.14
10	13	no	13	0	0	0	-	0	0.06	13	0	0	0.08
11	17	yes	649	324	324	11	10	0.26	6.41	829	2.46	319	9.94
11	17	no	1063	531	531	17	10	0.33	7.91	1063	0	0	9.72
12	20	yes	17	1	1	0	-	0	0.22	24	0.25	11	0.53
12	20	no	848	414	414	0	-	1.75	5.69	848	0	0	8.97
13	26	yes	13	0	0	0	-	0	0.12	14	0.34	1	0.48
13	26	no	44	14	14	0	-	0	0.34	44	0	0	0.46
14	28	yes	77	27	27	0	-	0.05	1.82	121	0.85	86	2.91
14	28	no	314	145	145	0	-	0.15	3.57	314	0	0	4.71

Table 4. Parameter combination for several runs.

V	MWV	clique separation algorithm general	specialized
[0.1,0.1]	[0.1,0.3]	run# 1	run# 2
[0.1,0.1]	[0.3,0.3]	run# 3	run# 4
[0.3,0.3]	[0.3,0.3]	run# 5	run# 6
[0.3,0.6]	[0.3,0.6]	run# 7	run# 8

Table 5. Clique separation and parameter variations for 2–$(v,3,1)$ designs

v	b	run#	BB	BT	LPT	LPs	ST	Cl	TotT
25	100	1	71	4	23.1	110	92.21	91	115.76
25	100	3	71	4	23.5	110	92.21	93	116.28
25	100	5	64	0	11.62	81	79.51	30	91.38
25	100	7	57	0	12.49	65	72.7	9	85.42
25	100	2	60	1	26.37	98	20.28	97	46.94
25	100	4	60	1	25.75	98	19.96	97	45.99
25	100	6	66	0	17.11	91	18.49	42	35.78
25	100	8	60	0	9.966	69	17.27	12	27.44
25	100	Cplex	97					1	63.83
27	117	1	75	3	50.46	124	156.84	121	207.89
27	117	3	74	3	42.84	121	153.92	118	197.35
27	117	5	102	18	74.5	144	146.66	81	222.79
27	117	7	76	4	32.89	89	132.96	16	166.52
27	117	2	71	2	41.69	112	33.66	109	75.86
27	117	4	71	2	43.31	112	33.75	109	77.55
27	117	6	99	14	48.23	137	32.58	63	82.29
27	117	8	76	4	32.61	89	29.89	16	63.09
27	117	Cplex	230					6	340.48
31	155	1	135	13	165.04	200	368.27	155	536.09
31	155	3	135	13	171.42	200	375.35	155	549.59
31	155	5	119	1	98.6	160	329.61	56	429.22
31	155	7	117	9	101.22	132	318.37	24	421.60
31	155	2	112	0	122.77	166	85.07	119	208.57
31	155	4	112	0	126.88	166	82.92	119	210.55
31	155	6	112	0	70.42	145	79.29	44	150.42
31	155	8	113	7	87.61	127	76.35	25	165.56
31	155	Cplex	2538					34	3879.00
33	176	1	141	11	381.24	215	526.76	181	911.19
33	176	3	141	11	382.21	215	537.29	181	922.63
33	176	5	219	52	768.50	295	508.96	170	1288.51
33	176	7	257	67	599.14	295	509	64	1122.38
33	176	2	118	5	255.11	182	122.8	152	379.71
33	176	4	118	5	252.65	182	133.36	152	387.85
33	176	6	123	0	106.33	155	112.80	42	220.02
33	176	8	174	26	309.52	198	114.61	53	430.36
33	176	Cplex	1267					14	2651.92

Table 6. Steiner Quadruple Systems: 3-$(v,4,1)$ designs

v	b	BB	BT	IE	IR	Mxl	IT	LPT	LPs	ST	Cl	TotT
8	14	1	0	0	0	-	0	0	1	0	0	0.01
10	30	1	0	0	0	-	0	0.07	1	0	0	0.07
14	91	1700	838	0	0	-	0	1974.75	1755	35.59	57	2488.7

Table 7. Cyclic 2-$(v,k,1)$/3-$(v,k,1)$ packings

$t=2$

	$k=3$				$k=4$			
v	Bl	C	D	S	Bl	C	D	S
8	1	8	8	8	1	2	2	4
9	1	9	12	12	0	0	3	4
10	1	10	13	13	0	0	5	7
11	1	11	17	18	0	0	6	8
12	2	16	20	20	1	3	6	9
13	2	26	26	26	1	13	13	13
14	1	14	28	28	1	14	14	14
15	3	35	35	35	1	15	15	15
16	2	32	37	37	1	16	20	20
17	2	34	44	45	1	17	20	21
18	3	42	48	48	1	18	22	22
19	3	57	57	57	1	19	25	28
20	2	40	60	60	2	25	30	30
21	4	70	70	70	1	21	31	31
22	3	66	73	73	1	22	37	38
23	3	69	83	84	1	23	40	40
24	4	80	88	88	2	30	42	42

$t=3$

	$k=4$				$k=5$			
v	Bl	C	D	S	Bl	C	D	S
8	1	10	14	14	0	0	2	4
9	1	9	18	18	0	0	3	7
10	3	30	30	30	1	2	6	8
11	3	33	35	35	1	11	11	15
12	5	45	51	54	1	12	12	19
13	4	52	65	65	1	13	18	23
14	6	84	91	91	1	14	28	36
15	7	105	105	105	3	33	42	42
16	10	132	140	140	2	32	48	48
17	9	153	156	157	4	68	68	68
18	11	198	198	202	-	-	?	75
19	12	228	228	228	-	-	?	83
20	15	285	285	285	-	-	?	112
21	15	315	315	315	-	-	?	126

Table 8. Cyclic 4-$(v,k,1)$/5-$(v,k,1)$ packings

$t=4$

	$k=5$				$k=6$			
v	Bl	C	D	S	Bl	C	D	S
8	1	8	8	11	0	0	1	1
9	1	9	18	25	1	3	3	6
10	3	30	36	36	0	0	5	11
11	6	66	66	66	1	11	11	14
12	6	72	?	84	2	14	22	30
13	9	117	?	140	2	26	26	41
14	11	154	?	182	3	42	42	53

$t=5$

	$k=6$				$k=7$			
v	Bl	C	D	S	Bl	C	D	S
8	1	4	4	6	0	0	1	1
9	1	9	12	16	0	0	1	1
10	3	30	30	41	0	0	3	8
11	6	66	66	66	0	0	6	17
12	10	110	132	132	1	12	12	24
13	12	156	?	182	2	26	26	55
14	(*)	≥ 273	?	326	3	30	42	82
15	(*)	≥ 370	?	455	-	-	?	113

Bl: base blocks in the cyclic packing C: size of maximal cyclic packing
D: size of maximal packing S: Schönheim upper bound

- indicates our algorithm did not find the cyclic packings
? indicates that the regular packing number is unknown
(*) indicates that a (not necessarily optimal) cyclic packing was found by the algorithm

Threshold Phenomena in Random Lattices and Efficient Reduction Algorithms

Ali Akhavi

GREYC - Université de Caen, F-14032 Caen Cedex, France
ali.akhavi@info.unicaen.fr

Abstract. Two new lattice reduction algorithms are presented and analyzed. These algorithms, called the Schmidt reduction and the Gram reduction, are obtained by relaxing some of the constraints of the classical LLL algorithm. By analyzing the worst case behavior and the average case behavior in a tractable model, we prove that the new algorithms still produce "good" reduced basis while requiring fewer iterations on average. In addition, we provide empirical tests on random lattices coming from applications, that confirm our theoretical results about the *relative behavior* of the different reduction algorithms.

1 Introduction

A Euclidean lattice is the set of all integer linear combinations of p linearly independent vectors in $\mathbb{R}^n$. The vector space $\mathbb{R}^n$ is then called *the ambient space.* Any lattice can be generated by many bases (all of them of cardinality p). The lattice basis reduction problem aims to find bases with good Euclidean properties, that is sufficiently short vectors and almost orthogonal. The problem is old and there exist numerous notions of reduction; the most natural ones are due to Minkowski or to Korkhine–Zolotarev. For a general survey, see for example [7, 18]. Both of these reduction processes are "strong", since they build reduced bases with somehow best Euclidean properties. However, they are also computationally hard to find, since they demand that the first vector of the basis should be a shortest one in the lattice. It appears that finding such an element in a lattice is likely to be NP-hard [1, 20].
Fortunately, even approximate answers to the reduction problem have numerous theoretical and practical applications in computational number theory and cryptography: Factoring polynomials with rational coefficients [11], finding linear Diophantine approximations [9], breaking various cryptosystems [10, 14],[19] and integer linear programming [6, 12]. In 1982, Lenstra, Lenstra and Lovász [11] gave a powerful approximation reduction algorithm. It depends on a real approximation parameter $t \in]1,2[$ and is called LLL(t). It begins with the Gram-Schmidt orthogonalizing process, then it aims to ensure, for each index i, $1 \le i \le p-1$, a lower bound on the ratio between the lengths ℓ_i and ℓ_{i+1} of two successive orthogonalized vectors,

$$\frac{\ell_{i+1}}{\ell_i} \geq s \qquad \text{with } s^2 = \frac{1}{t^2} - \frac{1}{4} \leq \frac{3}{4}. \tag{1}$$

So, for reducing an n-dimensional lattice, it performs at least $n-1$ iterations. This celebrated algorithm seems difficult to analyze precisely, both in the worst–case and in average–case. The original paper [11] gives an upper bound for the number of iterations of LLL(t), which is polynomial in the data size. When given p input vectors of $\mathbb{R}^n$ of

length at most M, the data size is $O(np\log M)$ and the upper bound is $p^2 \log_t M + p$. Daudé and Vallée [4] exhibited an upper bound for the average number of iterations (in a simple natural model) which asymptotically equals $(p^2/2)\log_t n + p$.
There is already a wide number of variations around the LLL algorithm (due for instance to Kannan or Schnorr [6, 13]) whose goal is to find lattice bases with sharper Euclidean properties than the original LLL algorithm.
Here, we choose the other direction, and we present two new variations around the LLL–reduction that are a priori weaker than the usual LLL reduction. They are called Schmidt–reduction and Gram–reduction. As for the LLL-reduction, they depend both on the parameter s. The Gram reduction also depends on another parameter γ. When $\gamma = 0$, the Gram–reduction coincides with the LLL reduction. Our algorithms are modifications of the LLL algorithm; they have exactly the same structure but they are based on different and weaker tests on the ratio between the lengths of orthogonalized vectors. Our purpose is twofold. On one hand, we propose more time–efficient reductions for lattices of high dimension. Although the new reduced bases are less sharp, they can play the same rôle as the Lovász-reduced ones in most of the applications, and they are obtained faster. On the other hand, the new algorithms are easier to analyze precisely so that the randomness and efficiency issues are much better understood.
Plan of the paper. In Section 2, we define the new reductions and we compare them to the LLL reduction: We give the Euclidean qualities of any reduced basis and the worst-case complexity of the reduction algorithms. Section 3 presents the main tools of the average–analysis in a tractable model. In Section 4, we show a general threshold phenomenon for the ratios of lengths of two different orthogonalized vectors and we compare the different reduction processes on random lattices. In Section 5, we report empirical tests on random lattices coming from applications: the new reduction algorithms remain more time–efficient and their outputs are still strong enough to be useful in applications.
Summary of results.

(a) First, we show that our reduced bases have always Euclidean properties similar to the LLL reduced ones. In particular, the shortest vector is at most $(1/s)^{n-1}$ times longer than a shortest element in the lattice. Observe that most of applications use only the first vector of the reduced basis.

(b) For the worst–case number of iterations, we show for all the reduction algorithms, the same upper bound as for the LLL algorithm. So, we cannot distinguish between these algorithms by a worst–case analysis.

Then, we compare these reductions by means of average–case analysis. We adopt a tractable probabilistic model, where the p vectors of the input basis are chosen uniformly and independently in the unit ball of $\mathbb{R}^n$. For the average analysis, we use various tools: We begin with the result due to Daudé and Vallée [4] about the distribution function of the length ℓ_a of the a-th orthogonalized vector associated with random bases. Then, we generalize a method due to Laplace to the two–dimensional–case and we apply this machinery to study the distribution function of the ratio ℓ_b/ℓ_a between the lengths of two different orthogonalized vectors. More precisely, we choose for a and b two affine functions of the dimension n of the ambient space:

Definition 1. *For θ any real constant in $[0,1]$, and r any integer constant, the quantity $f(n) := \theta n + r$ is called* an affine index *(or simply* an index*) iff it is an element of $\{1,...,n\}$. Moreover, such an index is called* a beginning index *iff the slope θ satisfies $\theta < 1$. It is called* an ending index *iff the slope θ satisfies $\theta = 1$.*

We consider the asymptotics of the distribution function of the ratio ℓ_b/ℓ_a for $n \to \infty$ and when $a := \alpha n + i$, $b := \beta n + j$ are two indexes. By "almost surely", we mean that the probability tends exponentially to 1 with the dimension n. We exhibit some quite different phenomena according to the position of the pair (a,b) with respect to the dimension n of the ambient space $\mathbb{R}^n$.

(c) For a pair (a,b) of beginning indexes, the distribution function of ℓ_b/ℓ_a presents a threshold phenomenon that is of independent interest. More precisely, given two real constants α and β in $[0,1[$, and two integer constants i and j, the probability

$$\Pr\{\ell_{\beta n+j}/\ell_{\alpha n+i} < v\}$$

follows a 0–1 law when n tends to infinity and the jump happens when v equals $\sqrt{1-\beta}/\sqrt{1-\alpha}$. Then, for any fixed s, we exhibit $\omega_0(s) < 1$ such that, when the ambient space is of sufficiently high dimension n, any random input of dimension $p := \omega n$ with $\omega < \omega_0(s)$ is almost surely reduced after $p-1$ iterations (in the sense of all the previous reductions). Furthermore, we show that the new algorithms, are quite efficient, even in the most difficult case of the full dimensional lattices ($p = n$), since the numbers K_S and K_G of iterations of the Schmidt and the Gram algorithms are almost surely $n-1$: For any $\varepsilon > 0$, there exists N such that for any $n > N$,

$$\Pr\{K_S = n-1\} \geq 1 - s^{(1-\varepsilon)n} \quad ; \quad \Pr\{K_G = n-1\} \geq 1 - s^{(1-\varepsilon)n^\gamma}.$$

(d) On the contrary, for a pair (a,b) of ending indexes the distribution function of ℓ_b/ℓ_a does not present a threshold phenomenon anymore. More precisely, given two positive integer constants i and j, the probability

$$\Pr\{\ell_{n-j}/\ell_{n-i} < v\}$$

admits a limit that is a continuous function of v. Thus, the LLL algorithm is much less time–efficient, since we show that the number K_L of iterations of the LLL algorithm is strictly greater then $n-1$, with a non–negligeable probability,

$$\Pr\{K_L > n-1\} > 1/\sqrt{1+(1/s)^2}.$$

For the average number of iterations of the LLL algorithm, the only known upper–bound remains $n^2 \log_t n + n$ [4].

2 New reductions and worst-case analysis

First, we recall how the Euclidean properties of a basis are usually evaluated in lattice theory. Then, we define two new reductions: For s a real parameter defined by 1, and $\gamma \in]0,1]$ a fixed real, we introduce the (s,γ)-Gram reduction and the s-Schmidt reduction. We compare all these reductions from two points of view, the Euclidean properties of the output basis, and the worst-case computational complexity of the algorithms. We obtain the results $(a),(b)$ of the introduction.

2.1 Two measures of quality

Let $\mathbb{R}^n$ be endowed with the usual scalar product $\langle \, , \rangle$ and Euclidean length $|u| = \langle u,u \rangle^{1/2}$. A lattice of $\mathbb{R}^n$ is the set of all integer linear combinations of a set of linearly independent vectors. Generally it is given by one of its basis $(b_1, b_2, \ldots, b_p)$ and the number p is the dimension of the lattice. So, if M is the maximum length of the vectors b_i, the data-size is $(np \log M)$. In a lattice, there exist some invariant quantities that does not depend on the choice of a basis. Among these invariants, the *n successive minima* Λ_i are defined as follows: Λ_i is the smallest positive number t so that there exist in the lattice, i independent vectors of lengths at most t. So, Λ_1 is the length of a shortest vector.

Intuitively, a reduced basis of a lattice consists of short vectors or equivalently it is nearly orthogonal. The shortness of the vectors is measured by *the length defects*. The i-th length defect $\mu_i(b)$ compares $|b_i|$ to the i-th minimum Λ_i,

$$\mu_i(b) = |b_i|/\Lambda_i.$$

All the reduction algorithms begin with the usual Gram-Schmidt orthogonalization process, which associates to a basis $b = (b_1, b_2, \ldots, b_p)$ an orthogonal basis $b^* = (b_1^*, b_2^*, \ldots, b_p^*)$ and a triangular matrix $(m) = (m_{ij})$ that expresses system b into system b^*. The vector b_1^* equals b_1 and for $i \geq 2$, b_i^* is the component of b_i orthogonal to the vector subspace spanned by $b_1, \ldots, b_{i-1}$:

$b_1^* = b_1$, and $b_i^* = b_i - \sum_{j<i} m_{ij} b_j^*$, where $m_{ij} = \langle b_i, b_j^* \rangle / |b_j^*|^2$, for $j < i$.

It is clear that $m_{ij} = 0$ for $i < j$ and $m_{ii} = 1$.

The length ℓ_i of the vector b_i^* does play an important role in the sequel. The ratio $\ell_i/|b_i|$ is the sinus of the angle between b_i and the vector space spanned by $b_1, \ldots, b_{i-1}$. So, a nearly orthogonal basis, has all its ratios $\ell_i/|b_i|$ near to 1 and the *orthogonality defect* $\rho(b)$ measures the "nearly orthogonality" for b,

$$\rho(b) = \prod_{i=1}^{p} |b_i|/\ell_i. \tag{2}$$

A basis b is called *size-reduced* if $|m_{ij}| \leq 1/2$, for $1 \leq j < i \leq p$. (3)

Size-reduction is an easy tool to shorten a basis, since there is a simple algorithm that obtains a size-reduced basis from any basis, by integral translations of each b_i parallely to the previous b_j $(j < i)$. But size-reduction alone does not guarantee the usual quality needed for a reduced basis.

2.2 Concepts of reduction

Definition 2. *Let t, s be two real parameters related by (1). Given a basis $b = (b_1, b_2, \ldots, b_p)$ of a lattice L and for an index i, $1 \leq i \leq p-1$, we consider*

$$\text{the } t\text{-Lovász condition}: \quad (\ell_{i+1}^2 + m^2{}_{i+1,i}\ell_i^2) / \ell_i^2 \geq 1/t^2 \tag{4}$$

$$\text{the } s\text{-Siegel condition}: \quad \ell_{i+1}/\ell_i \geq s \tag{5}$$

$$\text{the } s\text{-Schmidt condition}: \quad \ell_{i+1}/\ell_1 \geq s^i \tag{6}$$

$$\text{the } (s,\gamma)\text{-Gram condition}: \quad (6) \text{ and } \ell_{i+1}/\ell_i \geq s^{i^\gamma}, \text{ with } 0 < \gamma \leq 1 \tag{7}$$

Let C_i be one of the above conditions, for a fixed index i. The basis b is called C-reduced if it is size-reduced and if it satisfies the C_i condition, for all indexes i, $1 \leq i \leq p-1$.

The condition (4) is introduced by Lovász [11] and it is used in the original LLL algorithm. The s-Siegel condition is an immediate consequence of together (4) and size-reduction (3); it is in fact, always used rather than the t-Lovász condition and in the sequel we will often do so. An average study of ratios (ℓ_{i+1}/ℓ_i) shows that they have all 1 as mean values, but their variances increase with the index i. When i is closed to n, the ratios are very dispersed. So that, the s-Siegel conditions are more and more difficult to satisfy and it is reasonable to fix a lower bound for (ℓ_{i+1}/ℓ_i) that decreases with the index i. We introduce here the (s,γ)-Gram condition that takes in consideration the previous remark and is exactly the classical s-Siegel condition, as $\gamma = 0$. The s-Schmidt condition[1] is the less sharp introduced here. The s-Schmidt and s-Siegel conditions can not be compared, for a fixed index i. However, if the whole basis is $s-$Siegel reduced, then it is also $s-$Schmidt reduced. The next lemma compares more *locally* the above conditions; it is useful to study the computational complexity of the reduction algorithms.

Lemma 3. *Let t,s two parameters related by (1). Let $(b_1,\ldots,b_p)$ be a basis and $i \in \{1,\ldots,p\}$ a fixed index.*

1. *If the s-Siegel condition is not satisfied and $|m_{i+1,i}| \leq 1/2$, then the t-Lovász condition is not satisfied either.*
2. *If the (s,γ)-Gram condition is not satisfied, neither is the s-Siegel condition. Conversely, the choice of $\gamma = 0$ in the (s,γ)-Gram reduction leads exactly to the s-Siegel reduction.*
3. *If the s-Schmidt condition is satisfied for the index i, but not for the index $i+1$, then the s-Siegel is not satisfied for the index $i+1$ either.*

2.3 Comparing the quality of reduced bases

The next theorem shows that the s-Schmidt reduction provides only a short vector of the lattice. For the three other reductions, all vectors of the reduced bases are short and the basis is nearly orthogonal. For the proof, see [2].

Theorem 4. *Let $b = (b_1,\ldots,b_p)$ be a basis of a lattice.*

1. *If b is s-Schmidt reduced, then its first length defect is bounded from above:*
$$\mu_1(b) \leq (1/s)^{(p-1)}. \tag{8}$$
2. *If b is (s,γ)-Gram reduced, then its first length defect is upper bounded as in (8). All the other lengths defects and the orthogonality defect are also bounded from above:*
$$\mu_i(b) \leq (1/s)^{(p-1)^{1+\gamma}}, \textit{ for all } i \in \{2,\ldots,p\} \quad \textit{and} \quad \rho(b) \leq (1/s)^{(p^{2+\gamma})/(2+\gamma)}.$$

2.4 Comparing reduction algorithms

Let us consider the generic C-reduction algorithm, where C is one of the conditions introduced in Definition 2.

[1] The Schönhage's semi–reduction [16] is not too far from our Schmidt reduction.

The C-reduction algorithm:

Input: A basis $(b_1,\ldots,b_p)$ of a lattice L.
Output: A C-reduced basis b of the lattice L.
Initialization: Compute the orthogonalized system b^* and the matrix m.
i := 1;
While i < p do
  $b_{i+1} := b_{i+1} - \lfloor m_{i+1,i} \rceil b_i$ ($\lfloor x \rceil$ is the integer nearest to x).
  Test the C_i condition.
    If true, make $(b_1,\ldots,b_{i+1})$ size-reduced by translations; **set i := i + 1;**
    If false, swap b_i and b_{i+1}; update b^* and m; if $i \neq 1$ then **set i := i − 1;**

During an execution, the index i varies in $\{1,\ldots,n\}$. When i equals some $k \in \{1,\ldots,p-1\}$, the beginning lattice generated by $(b_1,\ldots,b_k)$ is already reduced. Then, the C_k condition is tested. If the test is positive, size-reduction is performed and the beginning lattice generated by $(b_1,\ldots,b_{k+1})$ is reduced. So, i is incremented. Otherwise, the vectors b_k and b_{k+1} are swapped. At this moment, nothing guarantees that $(b_1,\ldots,b_k)$ "remains" reduced. So, i is decremented. The algorithm updates b^* and m, translates the new b_k in the direction of b_{k-1} and tests the C_{k-1} condition. Thus, the index i may fall down to 1. Finally, when i equals p, the whole basis is reduced and the algorithm terminates.
The following Theorem shows that the C-reduction algorithm terminates always and performs a polynomial number of iterations. However, this worst–case analysis does not distinguish between the four previous reduction algorithms.

Theorem 5. *Let s and t be real parameters related by (1), C one of the four conditions of Definition 2, and $(b_1, b_2, \ldots, b_p)$ any integer input basis. The maximum number K of iterations of the generic C-reduction satisfies*

$$K \leq p(p-1)\log_t M + p - 1, \quad \textit{where} \quad M := \max_{i\in\{1,\ldots,p\}} |b_i|.$$

Sketch of proof. When C denotes the t-Lovász condition, the original proof of [11] is based on the decrease of the integer quantity

$$D := \prod_{i=1}^{p-1}\prod_{j=1}^{i} \ell_j^2,$$

by the factor $(1/t^2)$, whenever a test is negative. When C is another condition, a negative test for an index i means that $(b_1,\ldots,b_{i-1})$ is C-reduced, whereas $(b_1,\ldots,b_i)$ is not. So, by Lemma 3, the t-Lovász test would be negative either.

3 The lengths ℓ_i in the probabilistic model and the Laplace method

First, we define the probabilistic model. Then we give various tools that we use on the average–analysis of the next Section. We begin with the result due to Daudé and Vallée [4] about the distribution function of the length ℓ_a of the $a-th$ orthogonalized vector associated with random bases. Then, we generalize a method due to Laplace for evaluating asymptotics for integrals, to the two–dimensional–case.

3.1 The probabilistic model

All the previous reduction algorithms act in the same way on a basis and on its transformed by a homothety. On the other hand, choosing a continuous model makes it possible to use powerful mathematical tools. All these reasons legitimate working with bases of length less than 1. The most natural and simplest model is thus the uniform model over all legal inputs to the reduction algorithms. So, in our analysis, the input vectors $b_1, b_2, \ldots, b_p$, $(p \leq n)$ are chosen independently and uniformly inside the unit ball $\mathcal{B}_n$ of $\mathbb{R}^n$. Clearly, they form almost surely an independent system, called a random basis (p-dimensional). The lattice that it generates is called a random lattice. It is full–dimensional if $p = n$. Classical methods (see Section 3 in [4]) generalize our results on a discrete model .

3.2 The distribution of variables ℓ_i

Under the uniform model, Daudé and Vallée showed that the squares of the lengths of ℓ_i's follow a Beta law (Corollary of the next Theorem). Before describing the result, we recall some usual definitions. For $u \in [0,1]$ and two reals p and q, the random variable X of the interval $[0,1]$ follows the Beta law of parameters p and q if its distribution function satisfies

$$\Pr\{X \leq u\} = \frac{\int_0^u x^{p-1}\,(1-x)^{q-1}\,dx}{\int_0^1 x^{p-1}\,(1-x)^{q-1}\,dx}.$$

The numerator is called an *incomplete Beta integral* and denoted $B(p,q,u)$. The normalization coefficient (the denominator) is simply a *Beta integral* $B(p,q)$ [21].
The next theorem describes the density of the random variables ℓ_i. It plays a central rôle in our probabilistic analysis.

Theorem 6. *[4] If the vectors $b_1, b_2, \ldots, b_p$ are independently and uniformly distributed inside the unit ball $\mathcal{B}_n$ of $\mathbb{R}^n$, then the lengths ℓ_i of their orthogonalized vectors are independent variables. The density $f_{i,n}(u)$ of ℓ_i on the interval $[0,1]$ is given by*

$$f_{i,n}(u) = \frac{2}{B(\frac{n-i+1}{2}, \frac{i+1}{2})} u^{n-i}(1-u^2)^{\frac{i-1}{2}}.$$

All previous reduction algorithms deal with ratios of the form $\frac{\ell_b}{\ell_a}$. In the sequel,

$$a(n) := \alpha n + i \quad \text{and} \quad b(n) := \beta n + j, \tag{9}$$

are always affine indexes (Definition 1). We look for asymptotic equivalents for the distribution function

$$G_{n,a,b}(v) := \Pr\{\ell_{b(n)}/\ell_{a(n)} < v\}$$

of ratios ℓ_b/ℓ_a. When $a(n) \neq b(n)$, the previous Theorem shows that $\ell_{b(n)}$ and $\ell_{a(n)}$ are independent variables and gives their density. So, the density $f_{n,a,b}(x,y)$ of the couple $(\ell_{a(n)}, \ell_{b(n)})$ satisfies $f_{n,a,b}(x,y) = f_{n,a}(x) f_{n,b}(y)$. The exact expression of $G_{n,a,b}(v)$ involves the polygon $\Delta(v)$ and $\Delta(\infty)$

$$\Delta(v) = \{(x,y) \in [0,1]^2 : y/x < v\} \quad \text{and} \quad \Delta(\infty) = [0,1]^2, \tag{10}$$

and is expressed as a ratio of two generic integrals,

$$G(v) = \frac{I_n(v)}{I_n(\infty)}, \quad \text{where} \quad I_n(v) := \int_{\Delta(v)} g(x,y) e^{nh(x,y)}\,dxdy, \tag{11}$$

$$\text{with} \quad g(x,y) = (1-x^2)^{\frac{i-1}{2}} x^{-i} (1-y^2)^{\frac{j-1}{2}} y^{-j}, \quad \text{and} \tag{12}$$

$$h(x,y) = (\alpha/2)\ln(1-x^2) + (\beta/2)\ln(1-y^2) + (1-\alpha)\ln(x) + (1-\beta)\ln(y). \tag{13}$$

An appropriate way to study the asymptotic behaviors of integrals such $I(v)$ and $I(\infty)$ is the Laplace method (see for example [3, 5, 17]). The next subsection first explains the Laplace's idea. Then we generalize the Laplace method for g and h functions of two variables, and the integration domain, a polygon.

3.3 The Laplace method for integrals

Let us consider the convergent integral $I_n = \int_\Delta F(x,n)dx$.
It often happens that the graph of $F(x,n)$, considered as a function of x, has somewhere a sharp peak, and that the contribution of a neighborhood of the peak is almost equal to the whole integral, when n is large. Then we can try to approximate F in that neighborhood by simpler functions, for which the integral can be evaluated. The advantage is that we need only a local approximation for F. This idea is due to Laplace and it is often used to find the asymptotic behavior of simple integrals. For our needs, we have to generalize Laplace's idea for double integrals that appear in (11): the integration domain Δ is always a convex plan polygon and the functions g and h satisfy some strong assumptions. For a more precise version and for the proof, see [2].

Proposition 7. *Let Δ be a polygon and I_n a sequence of absolutely convergent integrals*

$$I_n = \int_\Delta g(x,y)e^{nh(x,y)}\,dxdy, \quad \textit{where}$$

(a) the function h has an absolute and strict maximum on Δ, say at (x_0,y_0),
(b) there are two linear forms X and Y of variables $(x-x_0, y-y_0)$ such that near (x_0,y_0) and inside Δ, the functions h and g are approximated as follows

$$h(x,y) - h(x_0,y_0) = \Theta(H_1(X) + H_2(Y)), \qquad \textit{when } (x,y) \longrightarrow (x_0,y_0), \tag{14}$$

where H_1, H_2 are one–variable polynomials of low degrees[2] $\rho \geq 1$ and $\sigma \geq 1$.

$$g(x,y) \sim CX^\lambda Y^\mu, \qquad \textit{when } (x,y) \longrightarrow (x_0,y_0). \tag{15}$$

where $C \neq 0$, $\lambda > -1$ and $\mu > -1$ are real constants.

Then when n is large, $\quad I_n = \Theta\left(e^{nh(x_0,y_0)}\, n^{-\frac{\lambda+1}{\rho}-\frac{\mu+1}{\sigma}}\right).$

Remarks **1.** If the maximum (x_0,y_0) is inside the polygon Δ and not on its boundary, then λ and μ are positive even integers. If the maximum (x_0,y_0) is on the boundary but not at a vertex of the polygon Δ, then $Y=0$, introduced in (b), is the equation of the polygon side that contains (x_0,y_0) and λ is a positive and even integer. If the maximum (x_0,y_0) is at a vertex of the polygon $\Delta(v)$, then $X=0, Y=0$ are the equations of the two polygon sides containing (x_0,y_0).

[2] The low degree of a polynomial $\sum_{i=0}^{r} a_i X^i$ is the minimal index i such that $a_i \neq 0$.

2. We need the following quite particular case in Theorem 8: If
(*i*) the maximum (x_0, y_0) is at a vertex of the polygon Δ and ν denotes the tangent of the angle of Δ at its vertex (x_0, y_0) which is acute,
(*ii*) the linear forms X and Y equal $X := x - x_0$, $Y =: y - y_0$ and they may not correspond to the polygon sides' directions,
(*iii*) in (14), $H_1(X) = H_2(X) = -AX^2$, with $A < 0$, then when n is large,

$$I_n \sim \frac{C}{4} B\left(\frac{\lambda+1}{2}, \frac{\mu+1}{2}, \frac{\nu^2}{1+\nu^2}\right) \Gamma(\frac{\lambda+\mu+2}{2}) e^{nh(x_0,y_0)} \left(\frac{1}{An}\right)^{\frac{\lambda+\mu+2}{2}}. \tag{16}$$

4 From ratios ℓ_b/ℓ_a to the reduction algorithms

We apply here the machinery developed in the last Section for studying the asymptotic distribution function of the ratio ℓ_b/ℓ_a between the lengths of two different orthogonalized vectors. More precisely, when n denotes the dimension n of the ambient space $\mathbb{R}^n$, $a(n)$ and $b(n)$ are two affine indexes , as defined by (9). Theorem 8 exhibits some quite different phenomena according to the position of the pair (a, b) with respect to n.
Then we differentiate the behaviors of the previous reduction algorithms (Theorem 9). The reduction algorithms operate always on a random basis $(b_1, \ldots, b_p)$, $(p \leq n)$. When using the term "almost surely", we mean that the probability tends exponentially to 1 with the dimension n. In this section, we prove the points $(c), (d)$ of the Introduction.

4.1 Asymptotic behavior of $G_{n,a,b}(\nu)$

Proposition 7 shows that the asymptotic behaviors of integrals $I_n(\nu)$ and $I_n(\infty)$ that are involved in $G_{n,a,b}(\nu)$'s expression (11) depend strongly on the maximum of the function h (13) on the integration domain. The function h has a strict and global maximum on $\Delta(\infty) = [0,1]^2$, at $(\sqrt{1-\alpha}, \sqrt{1-\beta})$. Three different behaviors arise according to the relative positions of this point and $\Delta(\nu)$:
1. The point $(\sqrt{1-\alpha}, \sqrt{1-\beta})$ is not in $\Delta(\nu)$. Then, one shows easily that the function h has a global and strict maximum on $\Delta(\nu)$, which is on the boundary $y = \nu x$, say at $(\xi, \nu\xi)$. Since $\Delta(\infty)$ contains strictly $\Delta(\nu)$, clearly, $D_{\alpha,\beta,\nu} := \exp\left(h(\xi, \nu\xi) - h(\sqrt{1-\alpha}, \sqrt{1-\beta})\right)$ < 1. Then, Proposition 7 gives equivalents for $I(\nu)$ and $I(\infty)$. Thus, the distribution function tends exponentially to 0 with n. More precisely, there are two constants $D_{\alpha,\beta,\nu} < 1$ and $f_{(i,j)}$, such that

$$G_{n,a,b}(\nu) \sim \Theta\left(\left(D_{\alpha,\beta,\nu}\right)^n \, n^{f_{(i,j)}}\right).$$

2. The point $(\sqrt{1-\alpha}, \sqrt{1-\beta})$ is inside $\Delta(\nu)$, but not in its boundary $y = \nu x$. So, the function h has the same maximum on $\Delta(\nu)$ and on $\Delta(\infty)$. Moreover, the maximum $(\sqrt{1-\alpha}, \sqrt{1-\beta})$ has *the same neighborhood* as an element of $\Delta(\nu)$ and as an element of $\Delta(\infty)$. By Proposition 7, $I(\nu)$ and $I(\infty)$ have the same equivalents and the distribution function tends to 1 with n. Further, since $G_{n,a,b}(\nu) = 1 - G_{n,b,a}(1/\nu)$, the convergence is exponential,

$$1 - G_{n,a,b}(\nu) \sim \Theta\left(\left(D_{\beta,\alpha,\frac{1}{\nu}}\right)^n \, n^{f_{(j,i)}}\right).$$

3. The point $(\sqrt{1-\alpha}, \sqrt{1-\beta})$ is on the boundary $y = \nu x$ of $\Delta(\nu)$. The function h has the same maximum on $\Delta(\nu)$ and on $\Delta(\infty)$. But the point $(\sqrt{1-\alpha}, \sqrt{1-\beta})$ has not the *the* *same neighborhood* as an element of $\Delta(v)$ and as an element of $\Delta(\infty)$. Then, it arrises from (16) that for any fixed v, the distribution function $G(v)$ tends to a constant, strictly in $]0,1[$.

Theorem 8. *Let the vectors $b_1, b_2, \ldots, b_n$ be independently and uniformly distributed inside the unit ball $\mathcal{B}_n$ of $\mathbb{R}^n$, and let ℓ_a denote the length of the a-th orthogonalized vector.*

(i) If $a(n) := \alpha n + i$, $b(n) := \beta n + j$ are beginning indexes, i.e. $(\alpha, \beta) \in [0,1[^2$, then the distribution function $G(v)$ of the ratio ℓ_b/ℓ_a follows asymptotically and exponentially with n, a 0–1 law. The jump is at $(\sqrt{1-\alpha}, \sqrt{1-\beta})$.

(ii) If $a(n) := n-i$, $b(n) := n-j$, i.e. $\alpha = \beta = 1$ and i, j are two positive integer constants, then the asymptotic distribution function of ℓ_b/ℓ_a does not follow a 0–1 law, but variates continuously *with i, j and v, in* $]0,1[$:

$$\Pr\{\frac{\ell_{n-j}}{\ell_{n-i}} < v\} \rightarrow \frac{B\left(\frac{j+1}{2}, \frac{i+1}{2}, \frac{v^2}{1+v^2}\right)}{B\left(\frac{j+1}{2}, \frac{i+1}{2}\right)}, \quad \text{when } n \rightarrow \infty. \tag{17}$$

4.2 Satisfying the C condition for a fixed index

Theorem 8 shows that for any beginning index, a random lattice satisfies any C condition of Definition 2. In short, for the random bases of the uniform model the "serious" reduction problems occur for the ending indexes. Considering such indexes, we classify the previous reductions in two groups: First, the s-Schmidt and (s,γ)-Gram conditions are almost surely satisfied even for an ending index, by *(i)* of Theorem 8. Second, the s-Siegel and t-Lovász conditions are not, by *(ii)* of Theorem 8.

4.3 Full–dimensional random lattices in $\mathbb{R}^n$

Here, we confirm the separation in two groups, that we made in (4.2) for the behaviors of different reduction algorithms previously introduced.

First, let C denote the s-Schmidt or the (s,γ)-Gram condition. The previous paragraph showed that for any index $i \in \{1, \ldots, n\}$, C_i is almost surely satisfied. The next Theorem makes precise that the conditions are almost surely satisfied *all together*. So, all the tests C_i are positive and the number of iterations equals $n-1$. In other words, almost surely, the C–reduction algorithm just size-reduce the input basis and verify that the tests are fulfilled. The proof [2] is technical (particularly for the Gram reduction) and is based once more on the Laplace's idea for evaluating asymptotic behavior of *sums*.

Second, the s-Siegel (or the t-Lovász) reduction. If a random basis is reduced, in particular the last Siegel condition is satisfied. By the relation (17), we find the asymptotic probability that last s-Siegel condition is not satisfied and thus we give a lower bound for the probability that a random lattice is not Siegel reduced. Equivalently during an execution of the Siegel reduction algorithm, the index i is decremented at least once, with a non–negligible probability.

Theorem 9. *Let K_L, K_G, K_S denote the numbers of iterations of the s-Siegel, (s,γ)-Gram and s-Schmidt reduction, when they operate on a random full-dimensional basis.*

For any $0 < \varepsilon < 1$, there exists N (which depends also on γ in (s,γ)-Gram reduction), such that for $n > N$,

$$\Pr\{K_L > n-1\} \geq 1/\sqrt{1+(1/s)^2} \tag{18}$$

$$\Pr\{K_G = n-1\} \geq 1 - s^{(1-\varepsilon)n^\gamma} \tag{19}$$

$$\Pr\{K_S = n-1\} \geq 1 - s^{(1-\varepsilon)n} \tag{20}$$

4.4 Lattices of low dimension in $\mathbb{R}^n$.

For any real parameter s, $0 < s < 1$, let $\omega_0(s)$ be the greatest real satisfying

$$s^2 < (1-\omega)\omega^{\frac{\omega}{1-\omega}}$$

By using similar methods (and in a simpler way) as in the proof of (19) and (20), we show that the number of iterations of the s−Siegel algorithm when it works on random input of dimension $p := \omega n$ with $\omega < \omega_0(s)$ is almost surely $p-1$. Comparing this result with (18) shows, in particular, the importance of the ratio between the dimensions of the input basis and the ambient space on the average behavior of the reduction process.

5 Statistical evaluations

The previous analysis shows that for random inputs of the uniform model, the new reduced bases are obtained faster. Several questions remain in practice.
(1) Is the output basis of the LLL algorithm of much better quality?
(2) We showed that, very probably, the number of iterations of the classical LLL algorithm is strictly greater than $n-1$ (the minimum possible). On the other hand, the only bound established for the average number of iterations of the LLL algorithm is $n^2 \log n$ [4]. Is, in practice, the t-Lovász reduced basis much more slowly to be obtained, in comparison with the new reduced bases?
(3) The uniform model is a tractable one for a mathematical average analysis. But in usual applications of lattice reduction, the lattices are not really of the uniform model. Are the new reduced bases obtained still faster with random inputs coming from applications?
Our next experimental study gives an insight to the answers of these questions. For every model considered, we report experimental mean values obtained with 20 random inputs. For each model, we provide three tables. The first table shows the average number of iterations of the different reduction algorithms. The second table gives the n-th root of the orthogonality defect of different output bases. (the first line corresponds to the input basis). The third table describes the ratio between the lengths of the shortest output vector of the new reduction algorithms and the classical LLL algorithm. For the approximation parameters (s or t), we generally use the usual values $t^2 = 4/3$ ($s^2 = 1/2$), unless for the last row of each Table, where the optimal Schmidt algorithm is considered ($s^2 = 3/4$).

5.1 Experimentations on the uniform model

Table 1 first largely confirms our average analysis. To generate such random integer inputs we use ideas of Knuth and Brent[8]. Then, the number of iterations of the new reduction algorithms is always the minimum possible for lattices of dimension greater than 20. Moreover, the output bases of the LLL algorithm is not of better quality than the new reduced bases, while it requires much more iterations to be obtained. In other words, on random inputs of the uniform model, the LLL algorithm demands a non negligeable number of iterations to build a basis of similar quality than its input. Contrarily, the new algorithms, immediately detect, the acceptable quality of such an input.

dim	10	20	30	40	10	20	30	40	10	20	30	40
LLL	59.8	132.7	149.8	191.4	1.20	1.54	1.71	1.82	1	1	1	1
Gr/Sch	9.4	19	29	39	1.42	1.69	1.71	1.77	1.04	1.07	1.00	1.00
Sch opt	22.9	19.8	29	39	1.33	1.69	1.71	1.77	1.04	1.07	1.00	1.00

Table 1. Comparison between the number of iterations and the quality of the output bases on random inputs of *the uniform model*. The input vectors are of length at most 2^{dim}, where *dim* is the dimension of the input. On the right , we report the number of iterations, on the middle the n-th root of the orthogonality defect, and on the left the ratio of lengths of the shortest output vector (reference: LLL)

5.2 Experimentations in the "SubsetSum" model

Given $a_1,\ldots,a_n,M$, consider the basis $(b_1,\ldots,b_{n+1})$ of $\mathbb{Z}^{n+2}$ formed with the rows of the following matrix:

$$\begin{pmatrix} 2\,0\,0\,.\,.\,.\; na_1\;0 \\ 0\,2\,0\,.\,.\,.\; na_2\;0 \\ .\;.\;.\;.\;.\;.\;.\;.\;. \\ .\;.\;.\;.\;.\;2\;na_n\;0 \\ 1\,1\,.\,.\,.\,1\;nM\;1 \end{pmatrix}.$$

Lattices generated by such a basis are used by Schnorr and Euchner [15] to solve almost all subset sum problems. Moreover, lattices of similar shape are used in many other applications [19]. Here, our results are obtained from 20 random inputs of the "SubsetSum" model, generated as follows: Pick random numbers $a_1,\ldots,a_n$ in the interval $[1,2^n]$, pick a random set $I\subset\{1,\ldots,n\}$ and put $M=\sum_{i\in I}a_i$. (The choice of a_i in the interval $[1,2^n]$ lead us to deal with lattices of density 1, which are the most difficult in cryptographical applications.)
Table 2 shows that the new reduced basis are obtained steel quite faster, while on average, they remain of a similar quality. Let us point out in particular, the optimal Schmidt algorithm which obtains an output vector at most twice longer than the the shortest output vector of the classical LLL algorithm, after a number of iterations that is on average three times less.

6 Conclusion

We have presented and analyzed two efficient variations of the LLL algorithm.

Number of iterations										
dim	10	20	30	40	50	60	80	90	100	110
LLL	140	449	782	1131	1775	1763	2163	2266	2363	2600
Sch	29.9	75.1	137	208	279	340	462	483	500	596
Gram	36	87	148	218	286	372	813	951	1115	1363
Sch opt	62	144	254	393	542	614	737	758	779	810
n-th root of the orthogonality defect										
rand	$1.1e3$	$2e6$	$0.2e9$	$4.3e12$	$5e15$	$6.6e18$	$8.9e24$	$1e28$	$1.2e31$	$1.4e34$
LLL	1.2	1.6	2.1	2.8	3	3.3	3.8	3.8	4.2	4.1
Sch	2.3	5.9	10	18	31	54	64	48	38	35
Gram	2	4.3	8.6	16	28	31	19	19	20	19
Sch opt	1.5	2.6	4	6.3	8.3	7.3	9	8.3	10	9.9
Ratio between the lenghts of the shortest output vectors (reference LLL)										
Sch	1.4	2.4	2.7	4.2	6.5	5.4	5.7	6.4	5.4	3.8
Gram	1.4	2	2.6	3.9	5.9	5.4	2.6	3.8	3.5	3.4
Sch opt	1.1	1.4	1.6	1.6	1.9	1.7	1.8	1.7	1.7	2.1

Table 2. Comparison between the number of iterations and the quality of the output bases on random inputs of *the Subset Sum model* of density 1.

From a theoretical point of view, we have exhibited several threshold phenomena in random lattices of the uniform model: (1) The distribution function of ℓ_b/ℓ_a, when a and b are beginning indexes, which follows asymptotically a 0–1 law (2) the gap that occurs in the behaviors of these distribution functions as (a,b) becomes a pair of ending indexes, (3) the gap between the reduction probabilities of a random input for (s,γ)-Gram reduction ($\gamma > 0$) and s-Siegel reduction (which coincides with (s,γ)-Gram, with $\gamma = 0$).

From a practical point of view, empirical tests of Section 5 show that our new reductions are quite interesting. They provide some new tools for lattice reduction and can be very useful to build up an algorithmic strategy of reduction.

Acknowledgments. This work is part of my Ph.D. thesis [2]. I thank Brigitte Vallée, who is my thesis advisor, for regular helpful discussions. Many thanks to her and to Philippe Flajolet for their encouragement and for their help to improve the paper.

References

[1] AJTAI, M. The shortest vector problem in L_2 is NP-hard for randomized reduction. *Elect. Colloq. on Comput. Compl.* (1997). (http://www.eccc.unitrier.de/eccc).

[2] AKHAVI, A. *Analyse comprative d'algorithmes de réduction sur les réseaux aléatoires.* PhD thesis, Université de Caen, 1999.

[3] BENDER, C., AND ORZAG, S. *Advanced Mathematical Methods for Scientists and Engineers.* MacGraw-Hill, New York, 1978.

[4] DAUDÉ, H., AND VALLÉE, B. An upper bound on the average number of iterations of the LLL algorithm. *Theoretical Computer Science 123* (1994), 95–115.

[5] DE BRUIJN, N. G. *Asymptotic methods in Analysis.* Dover, New York, 1981.

[6] KANNAN, R. Improved algorithm for integer programming and related lattice problems. In *15th ACM Symp. on Theory of Computing* (1983), pp. 193–206.

[7] KANNAN, R. Algorithmic geometry of numbers. *Ann. Rev. Comput. Sci. 2* (1987), 231–267.

[8] KNUTH, D. E. *The Art of Computer Programming*, 2nd ed., vol. 2: Seminumerical Algorithms. Addison-Wesley, 1981.

[9] LAGARIAS, J. C. The computational complexity of simultaneous diophantine approximation problems. In *23rd IEEE Symp. on Found. of Comput. Sci.* (1982), pp. 32–39.

[10] LAGARIAS, J. C. Solving low-density subset problems. In *IEEE Symp. on Found. of Comput. Sci.* (1983).

[11] LENSTRA, A. K., LENSTRA, H. W., AND LOVÁSZ, L. Factoring polynomials with rational coefficients. *Math. Ann. 261* (1982), 513–534.

[12] LENSTRA, H. Integer programming with a fixed number of variables. *Math. Oper. Res. 8* (1983), 538–548.

[13] SCHNORR, C. P. A hierarchy of polynomial time lattice basis reduction algorithm. *Theoretical Computer Science 53* (1987), 201–224.

[14] SCHNORR, C. P. Attacking the Chor-Rivest cryptosystem by improved lattice reduction. In *Eurocrypt* (1995).

[15] SCHNORR, C. P., AND EUCHNER, M. Lattice basis reduction: Improved practical algorithms and solving subset sum problems. In *Proceedings of the FCT'91 (Altenhof, Germany), LNCS 529* (1991), Springer, pp. 68–85.

[16] SCHÖNHAGE, A. Factorization of univariate integer polynomials by diophantine approximation and an improved basis reduction algorithm. In *Lect. Notes Comput. Sci.* (1984), vol. 172, pp. 436–447.

[17] SEDGEWICK, R., AND FLAJOLET, P. *An Introduction to the Analysis of Algorithms.* Addison-Wesley Publishing Company, 1996.

[18] VALLÉE, B. Un problème central en géométrie algorithmique des nombres: la réduction des réseaux. Autour de l'algorithme LLL. *Informatique Théorique et Applications 3* (1989), 345–376.

[19] VALLÉE, B., GIRAULT, M., AND TOFFIN, P. How to break Okamoto's cryptosystem by reducing lattice bases. In *Proceedings of Eurocrypt* (1988).

[20] VAN EMDE BOAS, P. Another NP-complete problem and the complexity of finding short vectors in a lattice. *Rep. 81-04 Math. Inst. Univ. Amsterdam* (1981).

[21] WHITTAKER, E., AND WATSON, G. *A course of Modern Analysis*, 4th ed. Cambridge University Press, Cambridge (England), 1927. reprinted 1973.

On Finding the Maximum Number of Disjoint Cuts in Seymour Graphs*

Alexander A. Ageev

Sobolev Institute of Mathematics
pr. Koptyuga 4, 630090, Novosibirsk, Russia
ageev@math.nsc.ru

Abstract. In the CUT PACKING problem, given an undirected connected graph G, it is required to find the maximum number of pairwise edge disjoint cuts in G. It is an open question if CUT PACKING is NP-hard on general graphs. In this paper we prove that the problem is polynomially solvable on Seymour graphs which include both all bipartite and all series-parallel graphs. We also consider the weighted version of the problem in which each edge of the graph G has a nonnegative weight and the weight of a cut D is equal to the maximum weight of edges in D. We show that the weighted version is NP-hard even on cubic planar graphs.

1 Introduction

In the CUT PACKING problem, given an undirected connected graph G, it is required to find the maximum number of pairwise edge disjoint cuts in G. This problem looks natural and has various connections with many well-known optimization problems on graphs. E. g., one can observe that INDEPENDENT SET can be treated as a constrained version of CUT PACKING whose collection of feasible cuts consists of all stars. The directed counterpart of CUT PACKING — DICUT PACKING — is well known to be polynomially solvable on general graphs (see Lucchesi [14], Lucchesi and Younger [15], Frank [8] or Grötschel et al. [11, p. 252]). By contrast, the complexity status of CUT PACKING on general graphs still remains an open problem. In this paper we use a characterization in [1] and an algorithmic result on joins due to Frank [9] to prove that CUT PACKING is polynomially solvable when restricted to the family of Seymour graphs. To present a rigorous definition of Seymour graph we need to introduce a few more notions. At this point we only notice that the family includes both all bipartite and all series-parallel graphs (Seymour [18][19]). It should be also noted that the complexity status of the recognition problem for Seymour graphs remains unclear; it is known only that the problem belongs to co-NP [1]. All we have to say for ourselves is that our result also holds for recognizable in polynomial time subfamilies of Seymour graphs such as, e. g., above mentioned bipartite[1] and series-parallel graphs.

---

* This research was partially supported by the Russian Foundation for Basic Research, grants 97-01-00890, 99-01-00601.

[1] András Frank pointed out to me that the polynomial-time solvability in the case of bipartite graphs was shown earlier and in a different way by D. H. Younger [20].

Graphs in the paper are undirected with possible multiple edges and loops. Given a graph G and $v \in V(G)$, $N_G(v)$ will denote the set of vertices adjacent to v in G. Any inclusion-wise minimal cut can be represented in the form $\delta_G(X)$, i. e., as the set of edges leaving a set $X \subseteq V(G)$. If the context allows we shall omit subscripts and write $N(v)$, $\delta(X)$.

A *join* in a graph G is a set of edges $J \subseteq E(G)$ such that each circuit of G has at most half of edges in J. Matchings and shortest paths between any two vertices are simple examples of joins. The term "join" originates from the classical notion of T-join introduced by Edmonds and Johnson in their seminal paper [7]. Given a graph G and a vertex subset $T \subseteq V(G)$ of even cardinality, a subset J of edges of G is called a *T-join* if the set of vertices having odd cardinality in the subgraph of G spanned by J coincides with T. It can be shown that any minimum cardinality T-join is a join and, moreover, any join is a minimum cardinality T-join where T is the set of vertices having odd cardinality in the subgraph of G spanned by J (Guan's lemma [12]). This establishes a one-to-one correspondence between joins and minimum cardinality T-joins of a graph.

Let $\{D_1, \dots, D_k\}$ be a collection of disjoint cuts in G. Pick an edge e_i in each cut D_i and set $J = \{e_1, \dots, e_k\}$. Since each cut and each circuit in a graph may have in common only an even number of edges , J is a join. If a join J in G can be represented in this way, we call $\{D_1, \dots, D_k\}$ a *complete packing* of J and say that J *admits a complete packing* .

There are graphs in which not every join admits a complete packing. A simple example is K_4. Any two cuts of K_4 intersect and therefore, every join admitting a complete packing in K_4 consists of at most one edge whereas any perfect matching of K_4 constitutes a join of cardinality 2.

A graph G is called a *Seymour graph* if every join of G admits a complete packing.

A co-NP characterization of Seymour graphs is provided by the following theorem which is an easy corollary of a stronger result in [1] (a bit shorter alternative proof can be found in [2]).

Let G be a graph and J be a join of G. We say that a circuit C of G is *J-saturated* if exactly half of the edges of C lie in J. A subgraph H of G is called *J-critical* if it is non-bipartite and constitutes the union of two J-saturated circuits. E.g., if $G = K_4$ and J is a perfect matching, then G itself is J-critical.

Theorem 1 (Ageev, Kostochka & Szigeti [1]). *A graph G is not a Seymour graph if and only if G has a join J and a J-critical subgraph having maximum degree* 3.

The "if part" of this theorem is easy and holds even when no bound on the maximum degree is assumed (see [1]).

To prove that CUT PACKING is polynomially solvable on Seymour graphs we shall use the above characterization and the following algorithmic result.

Theorem 2 (Frank [9]). *Given an undirected graph G, a join of maximum cardinality in G can be found in polynomial time.*

The running time of Frank's algorithm is bounded by $O(mn)$, where $n = |V(G)|$, $m = |E(G)|$ [9]. Let G be a Seymour graph and J be a join of G with maximum cardinality returned by Frank's algorithm. By the definition of Seymour graph J admits a complete packing $\{D_1, \ldots, D_k\}$ where $k = |J|$. Since the size of any collection of disjoint cuts in G does not exceed k, the collection $\{D_1, \ldots, D_k\}$ provides an optimal solution for CUT PACKING on G. The question remains how to find such a collection in polynomial time.

We shall prove that even a bit more can be done.

Theorem 3. *Given a Seymour graph G and a join J in G, a complete packing of J can be found in polynomial time.*

This contrasts with the well-known result of Middendorf and Pfeiffer [17] that, given a planar cubic graph G and a join J of G, it is NP-complete to determine whether J admits a complete packing.

We notice that alternative and a bit more sophisticated proofs of Theorem 3 can be extracted directly from the arguments in [1] and [2].

Our proof of Theorem 3 relies upon three basic facts. The first fact — Lemma 4 — states that the family of Seymour graphs is closed under the operation of star contraction of a graph. To prove this we use Theorem 1. Lemma 4 is of apparent interest irrelatively the scope of this paper. In [3] we develop the related subject to elaborate an alternative co-NP characterization of Seymour graphs. The second fact — Lemma 5 — is well known among "T-joins" experts and reveals a connection between the same operation and joins admitting complete packings. Unfortunately, no published proof of this important statement appeared. To make the paper self-contained we include such a proof in Section 3. The third fact is the classical result due to Edmonds and Johnson [7] that, given a graph G and an even vertex subset T, there exists a polynomial-time algorithm to find a T-join of G with minimum cardinality. We shall use the straightforward corollary of it and Guan's lemma: given a graph G and an edge subset J, one can decide in polynomial time if J is a join of G.

Together with the original setting it seems natural to test for NP-hardness a bit more general, weighted version of the problem (WEIGHTED CUT PACKING): given a graph G with nonnegative edge weights $w(e)$, find a collection of disjoint cuts $\{D_1, \ldots, D_k\}$ in G maximizing

$$\sum_{i=1}^{k} \max\{w(e) : e \in D_i\}.$$

CUT PACKING is equivalent to that special case of WEIGHTED CUT PACKING in which all weights $w(e)$ are equal to 1. We shall demonstrate in the last section of this paper that WEIGHTED CUT PACKING is NP-hard even on cubic planar graphs.

2 Seymour graphs and star contractions

Since every bipartite graph is a Seymour graph and every other graph can be obtained by contraction of a bipartite graph, the family of Seymour graphs is not closed under the operation of contracting an edge. The operation that truly preserves the Seymour property is star contraction.

Let G be an undirected graph. We say that a graph H is obtained from G by *contracting a star* if H is a result of contracting edges of some star in G.

If X is a vertex subset of a graph G then G/X will denote a graph which is obtained from G by identifying (shrinking) all vertices in X (after this shrinking new multiple edges and loops may appear). Notice that no edge is deleted from G and so G/X has the same set of edges as G.

Lemma 4. *Every graph obtained from a Seymour graph by contracting a star is a Seymour graph.*

Proof. Suppose to the contrary that a Seymour graph G has a vertex u such that contracting the star at u results in a non-Seymour graph H. We can equivalently think of H as the subgraph $G/X - u$ of G/X where $X = N_G(u)$ (for an illustration see Fig. 1). Let ξ denote the vertex of G/X that is the result of shrinking the set X in G. Notice that by Theorem 1, G does not have any J'-critical subgraph for any join J' of G. Since H is not a Seymour graph, by Theorem 1 it has a join J and a J-critical subgraph $S = C_1 \cup C_2$ of maximum degree 3, where C_1 and C_2 are J-saturated circuits. Note that J is also a join of G. Note first that ξ lies in S; otherwise S would be a J-critical subgraph of G, which by Theorem 1 would contradict the assumption that G is a Seymour graph. Let S', C_1', and C_2' denote the subgraphs of G spanned by the edges of G lying in S, C_1, and C_2 respectively. Let $\{x_1, x_2, \dots, x_k\}$ denote the subset of vertices of S' lying in X. Recall that S has maximum degree 3 and hence, by the definition of J-critical subgraph, each vertex of S has degree either 2 or 3. It follows that $1 \le k \le 3$. Note first that in fact $2 \le k \le 3$, since $k = 1$ means that S' is isomorphic to S which implies that S' is a J-critical subgraph of G, i. e. G is not a Seymour graph. Then at least one of the graphs C_1' and C_2' is a path. Recall that x_i are neighbours of u in G. For each i, let e_i denote an edge connecting x_i and u. Denote by T the subgraph of G obtained from S' by adding the vertex u and the edges e_i. Our goal in the remaining part of the proof is to show that T is a J^*-critical subgraph of G for some join J^*. By the remark just after Theorem 1 this contradicts the assumption that G is a Seymour graph and thus proves the lemma. Observe first that S can be obtained from T by contracting the star at u. Thus, since S is non-bipartite, T is non-bipartite as well.

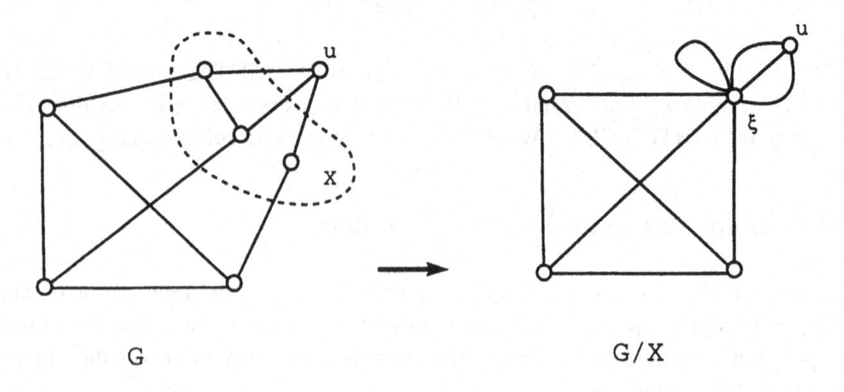

Fig. 1. A graph G and the graph G/X obtained from G by shrinking the vertex subset $X = N_G(u)$ to a new vertex ξ.

CASE 1: $k = 2$.
Let $J^* = J \cup \{e_1\}$. Of the circuits C_1 and C_2 at least one passes through ξ. Assume first that, say, C_1 does, whereas C_2 not. Then C_1' is a path, whereas C_2' is a circuit in G. Set $C_1'' = C_1' \cup \{e_1\} \cup \{e_2\}$. Then, by construction, $T = C_1'' \cup C_2'$ and, moreover, C_1'' and C_2' are both J^*-saturated circuits. Assume now that C_1' and C_2' are both paths. Then set $C_i'' = C_i' \cup \{e_1\} \cup \{e_2\}$, $i = 1,2$. Again, by construction, $T = C_1'' \cup C_2''$, and moreover, C_1'' and C_2'' are both J^*-saturated circuits. Thus, in either case T is a J^*-critical subgraph of G.

CASE 2: $k = 3$ (for an illustration refer to Fig. 1).
In this case ξ has degree 3 in S' and, consequently, C_1 and C_2 both pass through ξ. Therefore C_1' and C_2' are both paths in G. Since $S = C_1 \cup C_2$, it follows that of the three edges incident with ξ in S, exactly one lies in both C_1 and C_2. W.l.o.g. we may assume that this edge is incident with x_1 in G. Our assumption implies that x_1 is a common endpoint of C_1' and C_2' and we may assume further that the other endpoint of C_1' is x_2 whereas that of C_2' is x_3. Now set $J^* = J \cup \{e_1\}$, $C_1'' = C_1' \cup \{e_1\} \cup \{e_2\}$, and $C_2'' = C_2' \cup \{e_1\} \cup \{e_3\}$. Then, by construction, $T = C_1'' \cup C_2''$ and moreover, C_1'' and C_2'' are both J^*-saturated circuits. It means that, again, T is a J^*-critical subgraph of G. □

3 Joins admitting complete packings and star contractions

In this section to make the paper self-contained we give a proof for a folklore lemma which is one of the crucial points in our proof of Theorem 3.

Let G be a graph. For $v \in V(G)$, denote by $G \star v$ the graph that is obtained from G by contracting the star at v. Let J be a join of G. A vertex $v \in V(G)$ is called *J-marginal* if v is incident with exactly one edge e in J and $J \setminus \{e\}$ is a join of $G \star v$.

Lemma 5. *Let G be a graph and J be a join of G. If J admits a complete packing, then G has a J-marginal vertex.*

Proof. Since J admits a complete packing, G has a collection of edge disjoint cuts such that each cut contains exactly one edge in J. Among all such collections choose a collection $C = \{\delta(X_1), \delta(X_2), \ldots, \delta(X_k)\}$ with minimum $|X_1|$. For each i, let $J \cap X_i = \{e_i\}$. Let v be the end of e_1 that lies in X_1. We claim that $X_1 = \{v\}$. Assume to the contrary that $X_1 \setminus \{v\} \neq \emptyset$. We show first that for any i, $X_1 \cap X_i$ is either $\emptyset$ or X_1. Assume not, that is $0 < |X_1 \cap X_i| < |X_1|$ for some i. Since $e_1 \notin \delta(X_i)$, either both ends of e_1 lie in X_i or both not; for the similar reason the same holds with respect to e_i and X_1. Note that we may assume that both ends of e_1 lie in X_i, since the replacement of the set X_i by its complement does not alter the collection C. Assume that both ends of e_i do not lie in X_1. Set $X_1' = X_1 \cap X_i$ and $X_i' = X_1 \cup X_i$. It is easy to check that $\delta(X_1') \cap \delta(X_i') = \emptyset$, $\delta(X_1') \cup \delta(X_i') = \delta(X_1) \cup \delta(X_i)$, and $e_1 \in \delta(X_1')$, $e_i \in \delta(X_i')$. Replace in the collection C the cuts $\delta(X_1)$ and $\delta(X_i)$ by $\delta(X_1')$ and $\delta(X_i')$ respectively. This yields a new collection $C' = \{\delta(X_1'), \delta(X_2), \ldots, \delta(X_i'), \ldots, \delta(X_k)\}$ with the same properties except that $|X_1'| < |X_1|$, which contradicts the choice of C. Assume now that both ends of e_i lie in X_1. Then we set $X_1' = X_1 \cap \overline{X}_i$ and $X_i' = X_1 \cup \overline{X}_i$. Similarly, $\delta(X_1') \cap \delta(X_i') = \emptyset$, $\delta(X_1') \cup \delta(X_i') = \delta(X_1) \cup \delta(X_i)$, and, unlike the above, $e_1 \in \delta(X_i')$, $e_i \in \delta(X_1')$. This implies that, again, the collection C can be replaced by another collection having the same properties except that it includes the cut $\delta(X_1')$ with $|X_1'| < |X_1|$, which is a contradiction to the choice of C. Thus we have proved that for each $i \neq 1$, either $X_1 \cap X_i = \emptyset$ or $X_1 \subset X_i$. It immediately follows that $X_1 = \{v\}$, since otherwise we could replace in C the cut $\delta(X_1)$ by the star cut $\delta(v)$. Set $S = N_G(v) \cup \{v\}$. Now consider the graph $G \star v$ and denote by ω the vertex of $G \star v$ that is the result of contracting the star at v in G. Since the cuts in C are edge disjoint, we have either $S \cap X_i = \emptyset$ or $S \subset X_i$ for each $i \neq 1$. Now for each $i \neq 1$, set $X_i^* = (X_i \setminus S) \cup \{\omega\}$ if $S \subset X_i$ and $X_i^* = X_i$ otherwise. It follows that $C^* = \{\delta(X_2^*), \delta(X_3^*), \ldots, \delta(X_k^*)\}$ is a collection of disjoint cuts of $G \star v$ with $e_i \in \delta(X_i^*)$ for each $i \neq 1$. This implies that $J \setminus \{e_1\}$ is a join in $G \star v$ and C^* is a complete packing of $J \setminus \{e_1\}$, as desired. □

4 Proof of Theorem 3

We describe a polynomial-time algorithm that, given a Seymour graph G and a join J of G, finds a complete packing of J.

Let G be a graph and J be a join of G. If v is a J-marginal vertex of G, then $e(v)$ will denote the (unique) edge in J that is incident with v. The algorithm runs through k similar steps where $k = |J|$, and outputs a complete packing $\{D_1, D_2, \ldots, D_k\}$.
Step 0. Set $G_1 := G$, $J_1 := J$.
Step i, $1 \leq i \leq k$. Look successively through the vertices of G_i to find a J_i-marginal vertex v_i of G_i. To check if a vertex v of G_i is J_i-marginal, call one of the known polynomial-time algorithms that can test if $J_i \setminus \{e(v)\}$ is a join of $G_i \star v$. Set D_i to be equal to the set of edges of the star at v_i in G_i. Set $G_{i+1} := G_i \star v_i$, $J_{i+1} := J_i \setminus \{e(v_i)\}$. If $i < k$ go to step $i+1$, otherwise end up.

We now establish the correctness of the algorithm. By Lemma 4 each graph G_i is a Seymour graph, since each was obtained from the Seymour graph G by a sequence of star contractions. Using this, Lemma 5 and a simple induction on i we further obtain that, for each $i = 1, \ldots, k$, J_i is a join admitting a complete packing in G_i. Moreover, by Lemma 5, for each $i = 1, \ldots, k$, G_i does contain a J_i-marginal vertex and thus the algorithm never terminates before step k. Next, it is clear from the description that D_i and D_j are disjoint whenever $i \neq j$. And finally, since every cut of G_i is a cut of G, each D_i is a cut of G. □

It is clear that the running time of the algorithm is $O(n^2 C(m,n))$ where $n = |V(G)|$, $m = |E(G)|$ and $C(m,n)$ is the running time of the chosen procedure to test if a given set of edges is a join. As such, one can use any of the known algorithms for finding a minimum cardinality T-join: the original $O(n^3)$-algorithm of Edmonds and Johnson ([7]), the $O(mn \log n)$-algorithm of Barahona et al. ([4],[6]) or an $O(n^{3/2} \log n)$-algorithm for the special case of planar Seymour graphs ([16],[10],[5]).

As was already mentioned above, the running time of Frank's algorithm for finding a join of maximum cardinality is $O(mn)$ [9]. Thus, if we use, e. g., the algorithm of Barahona et al. ([4],[6]) for finding a minimum cardinality T-join, the overall running time of our algorithm for solving CUT PACKING on Seymour graphs will be bounded by $O(mn^3 \log n)$.

5 NP-hardness result

Theorem 6. *WEIGHTED CUT PACKING is NP-hard on the cubic planar graphs.*

Proof. We use a reduction from the following NP-complete decision problem [17]: given a graph G and a join J of G, to decide if a join J admits a complete packing in G. Assign $(0,1)$-weights w to the edges of G in the following way: $w(e) = 1$ if $e \in J$ and $w(e) = 0$ otherwise. Let $\{D_1, D_2, \dots, D_k\}$ be a collection of edge disjoint cuts in G of maximum total w-weight. We may assume that each D_i in the collection has weight 1 and thus the total weight of the collection is k. We claim then that $k = |J|$ if and only if J admits a complete packing. Indeed, if $k = |J|$ then $\{D_1, D_2, \dots, D_k\}$ is a complete packing of J. On the other hand, if J admits a complete packing $\{D'_1, D'_2, \dots, D'_{|J|}\}$ then this packing has weight $|J|$. However, any collection of disjoint cuts in G has weight at most $|J|$ and so $k = |J|$. □

Acknowledgements. The author is grateful to András Frank for helpful comments and an anonymous referee for pointing out the related results on DICUT PACKING.

References

[1] Ageev, A., Kostochka, A., Szigeti, Z.: A characterization of Seymour graphs. J. Graph Theory **24** (1997) 357–364.
[2] Ageev, A.A., Kostochka, A.V.: Vertex set partitions preserving conservativeness. Preprint 97-029, Bielefeld University, 1997. Submitted for journal publication.
[3] Ageev, A., Sebő, A.: An alternative co-NP characterization of Seymour graphs. In preparation.
[4] Barahona, F.: Applicationde l'optimisationcombinatuare a certainesmodeles de verres de spins: complexite et simulation. Master's thesis, Universite de Grenoble, France, 1980.
[5] Barahona, F.: Planar multicommodity flows, max cut and the Chinese postman problem. In: Polyhedral Combinatorics, DIMACS Series in Discrete Mathematics and Computer Science **1** (1990) 189–202.
[6] Barahona, F., Maynard, R., Rammal, R., Uhry, J.P.: Morphology of ground states of a 2-dimensional frustration model. Journal of Physics A: Mathematical and General **15** (1982) 673–699.
[7] Edmonds, J., Johnson, E.L.: Matchings, Euler tours and the Chinese postman problem, Math. Programming **5** (1973) 88–124.
[8] Frank, A: How to make a digraph strongly connected. Combinatorica **1** (1981) 145–153.
[9] Frank, A: Conservative weightings and ear-decompositions of graphs. Combinatorica **13** (1993) 65–81.
[10] Gabow, H.N.: A scaling algorithm for weighting matching in general graphs. In: 26th Annual Symposium on Foundations of Computer Science, IEEE, N.Y., 1985, 90–100.
[11] Grötschel, M., Lovász, L., Schrijver, A.: Geometric algorithms and combinatorial optimization, Springer-Verlag, 1988.
[12] Guan, M.G.: Graphic programming using odd and even points. Chinese Mathematics **1** (1962) 273–277.
[13] Lovász, L., Plummer, M.D.: Matching Theory. Akadémiai Kiadó, Budapest, 1986.
[14] Lucchesi, C.L.: A minimax equality for directed graphs, Ph.D. Thesis, University of Waterloo, Waterloo, Ontario, 1976.

[15] Lucchesi, C.L., Younger, D.H.: A minimax theorem for directed graphs. J. of London Math. Soc. (2) **17** (1978) 369–374.

[16] Matsumoto, K., Nishizeki, T., Saito, N.: Planar multicommodity flows, maximum matchings and negative cycles. SIAM J. on Computing **15** (1986) 495–510.

[17] Middendorf, M., Pfeiffer, F.: On the complexity of the disjoint paths problem. Combinatorica **13** (1993) 97–107.

[18] Seymour, P.D.: The matroids with the max-flow min-cut property. J. Comb. Theory B**23** (1977) 189–222.

[19] Seymour, P.D.: On odd cuts and plane multicommodity flows. Proc. London Math. Soc. Ser. (3) **42** (1981) 178–192.

[20] Younger, D.H.: Oral communication (1993).

Dilworth's Theorem and Its Application for Path Systems of a Cycle – Implementation and Analysis

András A. Benczúr[1,2]*, Jörg Förster[3]**, and Zoltán Király[2]***

[1] Computer and Automation Institute, Hungarian Academy of Sciences
[2] Eötvös University, Budapest
http://www.cs.elte.hu/~{benczur,joergf,kiraly}
[3] Hewlett Packard GmbH, Germany

Abstract. Given a set $\mathcal{P}$ of m subpaths of a length n path P, Győri's theorem gives a min-max formula for the smallest set $\mathcal{G}$ of subpaths, the so-called generators, such that every element of $\mathcal{P}$ arises as the union of some members of $\mathcal{G}$. We present an implementation of Frank's algorithm for a generalized version of Győri's problem that applies to subpaths of cycles and not just paths. The heart of this algorithm is Dilworth's theorem applied for a specially prepared poset.

- We give an $O(n^{2.5}\sqrt{\log n} + m\log n)$ running time bound for Frank's algorithm by deriving non-trivial bounds for the size of the poset passed to Dilworth's theorem. Thus we give the first practical running time analysis for an algorithm that applies to subpaths of a cycle.
- We compare our algorithm to Knuth's $O((n+m)^2)$ time implementation of an earlier algorithm that applies to subpaths of a path only. We apply a reduction to the input subpath set that reduces Knuth's running time to $O(n^2\log^2 n + m\log n)$. We note that derivatives of Knuth's algorithm seem unlikely to be able to handle subpaths of cycles.
- We introduce a new "cover edge" heuristic in the bipartite matching algorithm for Dilworth's problem. Tests with random input indicate that this heuristic makes our algorithm (specialized to subpaths of a path) outperform Knuth's one for all except the extremely sparse ($m \approx n/2$) inputs. Notice that Knuth's algorithm (with our reduction applied) is better by a factor of approximately $\sqrt{n}$ in theory.

1 Introduction

Győri's theorem [9] gives a min-max formula for the following problem. Let P be a path (or a cycle, as in an extension of the theorem due to Frank and Jordán [6]); let a collection $\mathcal{P}$ of its subpaths be given. Then we want to find another collection $\mathcal{G}$ of subpaths, the so-called *generators* such that each element of $\mathcal{P}$ arises as the union of some paths in $\mathcal{G}$. Note that the problem for subpaths of a tree is known to be NP-complete [14]. The theorem is often stated in the terminology of horizontally convex rectilinear bodies [9] or interval systems [13] that can easily be transformed to our terminology.

---

* Supported from grants OTKA T-30132, T-17580, T-29772 and FKFP 0206/1997
** Work was done while visiting Eötvös University, Budapest
*** Supported from grants OTKA F-14919, T-17580, T-30059, T-29772 and FKFP 0607/1999

Frank and Jordán [6] find a remarkable connection of Győri's problem to a large collection of so-called connectivity augmentation problems. Some of these problems bear algorithms very similar to the one presented in the paper; most of them are based on Dilworth's min-max theorem for posets. A poset (*partially ordered set*) is a set with a transitive comparison relation $<$. The relation is not necessarily complete: *incomparable* elements x,y satisfy neither $x<y$ nor $y<x$. Dilworth's theorem states that the maximum number of pairwise incomparable elements of a poset is equal to the minimum number of *chains* (some k elements with $x_1 < x_2 < \cdots < x_k$) such that each element of the poset belongs to at least one chain. Dilworth's theorem lies in the heart of Frank's algorithm [5]; as it will turn out, it is the computational bottleneck both in theory and in practice. We implement and analyze the Frank–Jordán extension of Győri's theorem and in particular give efficient algorithms for Dilworth's theorem.

1.1 Previous results

Franzblau and Kleitman [7] turned Győri's proof to a polynomial time algorithm for finding the minimum and maximum in question. This algorithm was simplified by Knuth [13] who implemented the algorithm and derived an $O((n+m)^2)$ running time bound for m subpaths of a path of n edges. Knuth's algorithm has no simple extension for subpaths of cycles. Frank and Jordán gave the first (non-combinatorial) algorithm [6] that applies to subpaths of a cycle as well; that algorithm was then turned to a combinatorial one by Frank [4, 5]. In the center of both algorithms we find Dilworth's theorem for a certain poset derived from the input set of subpaths.

The only known algorithm to find the minimum number of chains and the maximum incomparable subset in Dilworth's theorem uses a reduction to a bipartite matching problem (Ford and Fulkerson [8]). For a poset of p elements where the number of comparable pairs is q, the resulting bipartite graph has $2p$ vertices and q edges; the maximum matching can then be found in $O(p\sqrt{q})$ time by the Hopcroft–Karp matching algorithm [10]. The hunt for a better bipartite matching algorithm remains open [12].

1.2 Our results

We describe an implementation of Frank's algorithm [5] for Győri's theorem and improve Frank's $O(n^4+m)$ running time bound to $O(n^{2.5}\sqrt{\log n}+m\log n)$. Our algorithm, unlike Knuth's one [13], uses the extension of Győri's theorem to cycles [6] and applies to subpaths of a cycle instead of a path.

Our new algorithms for Dilworth's theorem are of own interest. Instead of taking the q-element set of all comparable elements, our algorithms take the typically much smaller $\tilde{q}$ "cover" or immediate successor pairs typically used to describe a poset. In the usual poset terminology an element (I,e) *covers* another (J,f), denoted by $(I,e) \prec (J,f)$, if $(I,e)<(J,f)$ and no element (I',e') satisfies $(I,e)<(I',e')<(J,f)$; the cover relation gives the so-called Hasse diagram—the typical description of the poset.

Our running time analysis consists of non-trivial combinatorial proofs. We give tight bounds for the poset size derived from the input subpath system. We prove that an initial reduction may discard all except $O(n\log n)$ subpaths from the input without altering the output. This reduction also applies in Knuth's algorithm [13] and improves its running time to $O(n^2 \log^2 n + m \log n)$; this becomes the theoretically fastest algorithm if subpaths of a path are considered.

We test our implementation with random input. We specialize our algorithm to subpaths of a path and compare its performance with Knuth's algorithm. In practice our algorithm turns out better apart from extreme sparse problems with $n \approx 2m$ (note that we test Knuth's algorithm with our initial subpath set reduction). We believe this is due to our improved way to perform Dilworth's theorem: while in theory we were not able to show $\tilde{q} \ll q$ in the posets arising during the algorithm, our tests indicate this may be the case—at least for average or random sets of subpaths.

1.3 The Győri–Frank–Jordán theorem

Frank and Jordán [6] prove an extended version of Győri's theorem for subpaths of a cycle C. They introduce the following reformulation of the theorem that we also use in the bulk of the paper. Consider the set of *subpath-edge pairs* (I,e) where $I \in \mathcal{P}$ and $e \in I$ is an edge of cycle C. Notice that a set of subpaths $\mathcal{G}$ generates $\mathcal{P}$ iff for all (I,e) there is an element $G \in \mathcal{G}$ with $e \in G \subseteq I$; we say that G *generates* the pair (I,e). We may restrict attention to "minimal copies" of subset-edge pairs: if $I \subset J$, then it suffices to generate (I,e) since then (J,e) is also generated. In Frank's [5, 4] terminology, (I,e) is *essential* if there is no other subpath $J \in \mathcal{S}$ with $e \in J \subset I$.

Let $\mathcal{E}$ denote the set of essential subset-edge pairs. If (I,e) and $(J,f) \in \mathcal{E}$ cannot be generated by the same subpath G, then we call them *independent*. Independence implies one of the following cases: either $e \in I - J$, or $f \in J - I$, or $I \cap J$ consists of two disconnected components and e and f lie in different components (this may happen if $I \cup J = C$). By further investigation we find two mutually exclusive cases for non-independent elements (I,e) and $(J,f) \in \mathcal{E}$ (see Fig. 1):

Comparable. We say that $(I,e) < (J,f)$ if e and f belong to the same connected component of $I \cap J$ and the following four edges (possibly some of which are equal) follow in clockwise order: the first edge of J, edge f, edge e, and the last edge of I. The elements of $\mathcal{E}$ form a poset with the above comparison relation.

Crossing. If two non-independent elements (I,e) and (J,f) of $\mathcal{E}$ are non-comparable, then we say they are *crossing*; crossing is *clockwise* if the first edge of I, the first edge of J, f and e follow in clockwise order.

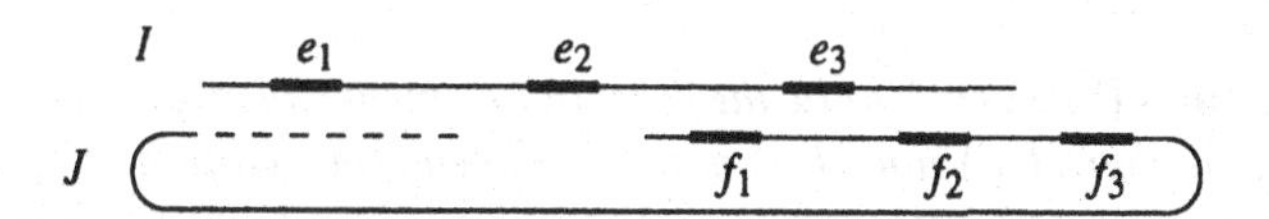

Fig. 1. Two subpaths $I,J \subset C$. J may or may not wrap around and consequently $I \cap J$ may or may not consist of two connected components. The subpath-edge pairs (I,e_1) and (I,e_2) are independent of all (J,f_i) for $i=1,2,3$; similarly (J,f_3) is independent of all (I,e_i) for $i=1,2,3$. As for the remaining arrangements, $(J,f_1) < (I,e_3)$; (I,e_3) clockwise cross (J,f_2); and (J,f_2) anti-clockwise cross (I,e_3).

The size of a set of pairwise independent subpath-edge pairs is a clear lower bound on the size of a generator subpath set. The Győri–Frank–Jordán min-max theorem claims the opposite is true as well:

Theorem 1 (Győri [9], Frank and Jordán [6]). *The minimum number of subpaths that generate a given set of subpaths $\mathcal{P}$ of a cycle C is equal to the maximum number of pairwise independent subpath-edge pairs (I,e) with $e \in I \in \mathcal{P}$.*

1.4 Frank's algorithm

Frank's algorithm [5] is a four-phase procedure (see Fig. 2) that finds a minimum generator set for subpaths of a cycle C. The algorithm is based on the following observation [6]: a given chain $(I_1,e_1) < (I_2,e_2) < \dots < (I_k,e_k)$ can be generated by a single subpath between e_k and e_1. Now one might want to applying Dilworth's theorem over the essential set $\mathcal{E}$ to obtain an optimum chain decomposition that defines an optimum generator system. Unfortunately however, the notion of incomparability under the partial order $<$ is weaker than the notion of independence: two crossing elements are incomparable but they always possess a common generator subpath.

The main idea now in Frank's [5, 4] algorithm is to remove crossing elements and thus make independent and incomparable mean the same. After some initialization steps described in Section 2, Phase 2 of Frank's algorithm (Fig. 2) constructs a cross-free set $\mathcal{K}$ by considering pairs in $\mathcal{E}$ and removing one of them if they cross. Then Phase 3 applies Dilworth's theorem to $\mathcal{K}$ to construct generators $\mathcal{G}$ as well as a pairwise independent subset I of $\mathcal{K}$ such that $\mathcal{G}$ and I have the same size.

Phase 3 (Fig. 2) thus yields a set $\mathcal{G}$ that generates $\mathcal{K}$ but not all of $\mathcal{E}$. In the final Phase 4 we identify all (J,f) not generated by $\mathcal{G}$. As the main result, Frank [5] gives a simple procedure to adjust $\mathcal{G}$ so that it will generate (J,f) without increasing its size or "un-generating" previously generated elements.

In order for Frank's theorem to hold, we must follow a specific rule to construct the "cross-free" $\mathcal{K}$: We consider each $(I,e) \in \mathcal{E}$ and discard all subpath-edge pairs that cross (I,e). For now the order of the (I,e) is arbitrary; later we choose a certain order that makes the algorithm efficient. Finally for the last "correction" phase to work, we must remember the order of deletions and consider discarded elements in reverse order; for each $(J,f) \in \mathcal{E}$ that we discard, we also need to record the element (I,e) causing its deletion:

Lemma 2 (Frank [5]). *Let (I,e) be the essential element causing the removal of (J,f) from $\mathcal{K}$.* ***(i)*** *Neither (I,f) nor (J,e) may have been removed from $\mathcal{K}$ prior to (J,f).* ***(ii)*** *If G_1 generates (I,f) and G_2 generates (J,e) and neither G_1 nor G_2 generates (J,f), then they satisfy $G_2 \subset G_1$.* ***(iii)*** *For G_1 and G_2 as above, if we define the subpaths G_1' and G_2' by exchanging the endpoints of G_1 and G_2, then one of G_1' and G_2' generates (J,f).* ***(iv)*** *If (K,g) is a subset-edge pair generated by G_1 or G_2, then (K,g) is generated by G_2' if (a) g is clockwise after e or (b) $g = e$ but K precedes I and (c) by G_1' in all other cases (see Fig. 3).* □

Phase 1 **preprocess the input**
$\mathcal{K}, \mathcal{E} \leftarrow$ essential subpath-edge pairs
Phase 2 **for** all $(I,e) \in \mathcal{K}$ **do**
  for all $(J,f) \in \mathcal{K}$ **do**
    if (I,e) and (J,f) cross
      then mark (J,f) by $(I,f),(J,e)$ and delete (J,f) from $\mathcal{K}$
Phase 3 **call** DILWORTH($\mathcal{K}$) to find a pairwise independent set I
  and subpaths $\mathcal{G}$ generating $\mathcal{K}$
Phase 4 **for** all $(J,f) \in \mathcal{E} - \mathcal{K}$ in the reverse order of deletion **do**
  $(I,f),(J,e) \leftarrow$ marks of (J,f)
  find $G_1 \in \mathcal{G}$ generating (I,f) and $G_2 \in \mathcal{G}$ generating (J,e)
  if neither G_1 nor G_2 generate (J,f)
    then exchange the endpoints of G_1 and G_2 in $\mathcal{G}$

Fig. 2. Frank's algorithm

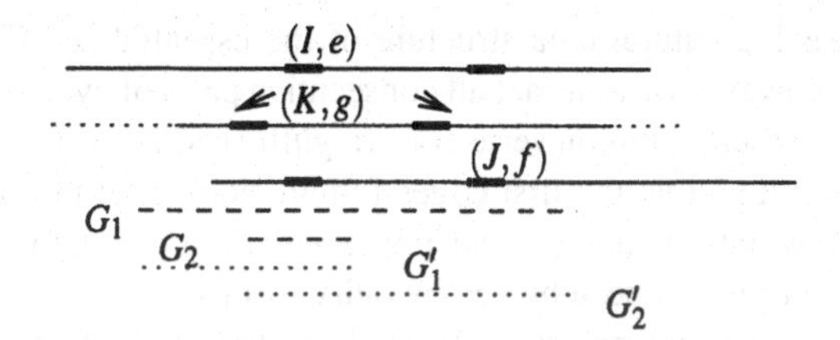

Fig. 3. The scenario of Lemma 2.

2 Our implementation

Unfortunately the straight implementation of Algorithm 2 has a poor time complexity. The set of all subset-edge pairs (even after removing duplicates) may have size $\Omega(n^3)$ and even $|\mathcal{E}| = \Omega(n^2)$. Thus the best runtime bound Frank [5] achieves is $O(n^3)$ for Phase 1 and $O(n^4)$ for Phase 2. Next we sketch our implementation where the computational bottleneck is an algorithm for Dilworth's theorem on a poset. All additional work will be bounded by a term $O(nm + n^2 \log^2 n)$. Due to space limitations we only sketch our implementation; for details see the full paper [2].

Initial reduction. We save a large amount of work by a simple preprocessing step that, in $O(m \log n)$ time, deletes all but $O(n \log n)$ subpaths from the input without altering the optima in question. Observe that if $I = \bigcup I_i$ for $I_i \in \mathcal{P}$, then the generators for I_i will generate I and thus I may be removed from the input. We quickly identify and discard a vast amount of such subpaths I.

For simplicity we explain the reduction for subpaths of paths only. We use a divide-and-conquer approach: We consider the first $n/2$ edges $P_1 \subset P$ and recursively reduce the set of subpaths $\{I : I \in \mathcal{P} \text{ and } I \subset P_1\}$. We do the same for the last $n/2$ edges P_2.

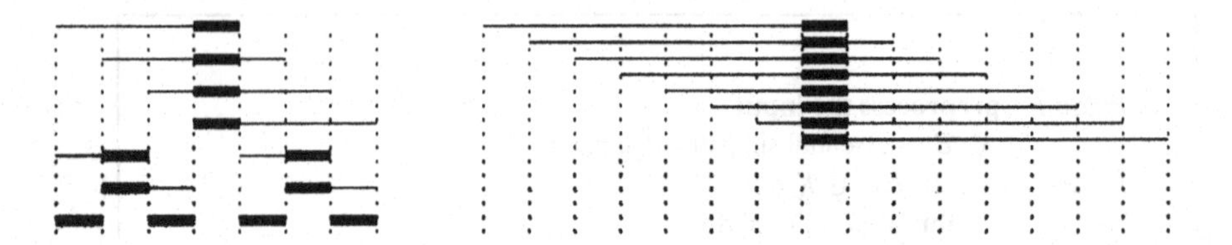

Fig. 4. Left: the instance $\mathcal{P}_3$; a sub-instance $\mathcal{P}_2$ is within the shaded area. Right: the recursive way to build the instances. On both sides of the figure thick subset-edge pairs are all essential and no two of them cross.

Now take the subset $\mathcal{P}' \subset \mathcal{P}$ of paths that are the subpaths of neither P_1 nor P_2. Notice that I can be removed unless it the shortest element of $\mathcal{P}_1$ either starting or ending at a certain edge of P. By solving the recurrence for the above recursive algorithm, we get the next claim (an instance for the tightness of the claim is given in Fig. 4).

Theorem 3. *There are $\Theta(n \log n)$ subpaths of a cycle C of length n such that no subpath arises as the union of some other two of these subpaths. We may find such a subset in $O(m \log n)$ time.* □

Phase 1. In Phase 1 a natural data structure of the essential set $\mathcal{E}$ is constructed in $O(nm)$ time. First however we contract all consecutive pairs of cycle-edges such that no path starts or ends at their common vertex (a straightforward $O(n+m)$ time procedure); then we sort the input so that the first edges follow clockwise; ties are broken so that the last edges follow anti-clockwise. Our implementation uses Quicksort; however an improved $O(n+m)$ time can be achieved by radix sort [3].

We define the following natural 2D list data structure for $\mathcal{E}$. We visualize the subpath-edge pairs in $\mathcal{E}$ as a table with rows corresponding to subpaths and containing all edges of a fixed subpath while columns corresponding to edges and containing all subpaths containing a fixed edge (see Fig. 5). The fact that this table is drawn on a torus, however, makes this kind of visualization somewhat inappropriate. The actual data structure consists of two collections of doubly linked lists. The first collection contains one list for each subpath I (*row lists*) while the second collection one list for each edge $e \in C$ (*column lists*). For each list we may access the first and last elements. The size of the data structure is (due to possible empty lists) $O(|\mathcal{E}| + m)$.

Now we construct the essential set by parsing each edge $e \in C$ in a cyclic order and considering all subpaths $I_1, \ldots, I_k$ starting at this edge e. Then in an inner loop we select all edges f and consider all subpaths $J_1, \ldots, J_\ell$ starting at edge f. Now we choose all pairs $J_j \subset I_i$; no subset-edge pair (I_i, g) with $g \in J_j$ may be essential. Since for each I_i it suffices to select the longest $J_j \subset I_i$, we may perform this procedure by merging the sequences $I_1, \ldots, I_k$ and $J_1, \ldots, J_\ell$ sorted by the endpoints. For a fixed e and f we use time $O(k+\ell)$; this totals to $O(nm)$.

If we are careful in selecting e and f, it is easy to insert the essential (I_i, f) into the 2D list data structure. We scan f in a linear order starting from e (the first edge of all I_i). For each I_i we maintain a pointer to the first possible edge $g \in I_i$ so that (I_i, g) may be essential; whenever we find $J_j \subset I_i$, we increase this pointer after the last edge of J_j. Under this scenario if the pointer of I_i is equal to f after processing all J_j staring at

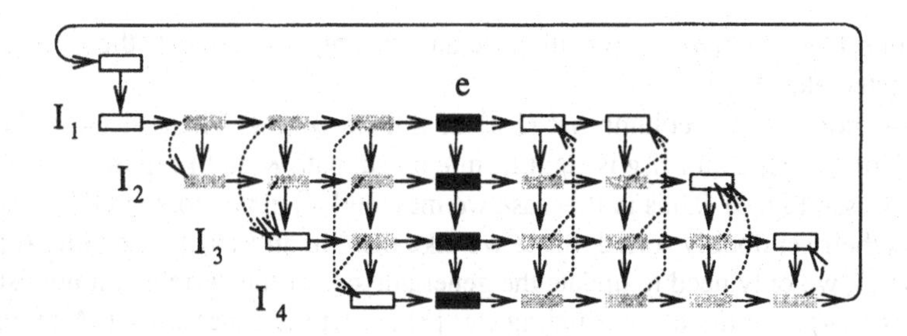

Fig. 5. An instance of a data structure (with solid arrows as pointers drawn only in one direction) for the subset-edge pairs of $\mathcal{E}$ where the last edge over subpath I_4 wraps around. Light shaded elements all cross some of (I_i, e) for $i = 1,2,3,4$ (dark shaded); for these elements dotted arrows indicate one of their marks after deletion.

f, then (I_i, f) is essential: we insert it into the data structure and increase the pointer to the next edge. Insertion takes $O(1)$ time since (I_i, f) is always added to the end of both linked lists (apart from the effect of the wraparound over cycle C that can easily be handled as an exception). Hence Phase 1 takes $O(nm)$ time.

Phase 2. We are able to very efficiently construct the cross-free set $\mathcal{K}$ by choosing a specific order for (I,e) in the inner loop of Phase 2: we select each edge e in a cyclic order and then read the column list of e containing all (I,e). In order to find all (J,f) crossing (I,e), we take advantage of Frank's Theorem 2 that claims (J,e) may not have been deleted prior to (J,f). Hence all (J,f) to be deleted can be found by parsing the row lists of elements (J,e) found on the column list of e.

Our next idea is to process all (I,e) for a fixed I in parallel. We visit the row lists starting at each (J,e) in both directions. It is easy to see that the elements to be deleted form consecutive sequences over these lists; it is also not hard to determine which (I,e) causes the deletion of a certain (J,f). Deletions from the lists take $O(1)$ time; we spend an additional $O(1)$ time to determine that no more elements should be deleted from a row list. Hence we use $O(|\mathcal{E}|+n^2)$ time where an easy bound on $|\mathcal{E}|$ is n^2. Phase 2 thus takes $O(n^2)$ time.

Phase 3. In order to use an algorithm for Dilworth's theorem, we have to construct all cover pairs (the Hasse diagram) of the poset $\mathcal{K}$. We consider row lists for subpaths I in the data structure for $\mathcal{K}$; we proceed with the I so that the first edge of I follow the cyclic order. For two consecutive list members (I,e_1) and (I,e_2) all covers (J,f) for (I,e_1) satisfy that f is clockwise between e_1 and e_2; for such an f we get $(J,f) \succ (I,e_1)$ only if (J,f) is the last in the column list of f with J preceding I. By maintaining the value of these last (J,f) for each f, we are able to find the possible cover (J,f) for a fixed f in $O(1)$ time; thus the total time spent for all subpaths starting at a given edge is proportional to the length of the longest such path. This gives a running time of $O(n^2)$ for this phase, in addition to the time of our Dilworth subroutine of Section 3.

Phase 4. Assume we delete (J,f) because of (I,e) in Phase 2; it gets the labels (J,e) and (I,f). By Lemma 2 if we know the generators for these latter two elements, then we may adjust the generator set for each non-generated (J,f) in $O(1)$ time. Since the generator set keeps changing, we still need an efficient way to update the generators for all generated elements.

Let us consider the column list (I,e) of a fixed edge e. Then all (K,g) keep the endpoint of the generator if g is right from e on subpath K and keep the starting point otherwise (see Fig. 3). Thus in this case we may find all generators in $O(1)$ time if we maintain their set doubly sorted. Since we take the e in anti-clockwise order (opposite of Phase 2), we only need to update the generator set for the current column list (I,e); this takes $O(n)$ time if we use counting sort [3] to sort the generator set. All remaining work can be bounded in the same way as in Phase 2; the running time totals to $O(n^2)$.

3 Algorithms for Dilworth's problem

In this section we give various algorithms for Dilworth's problem where the running time depends on $\tilde{q}$, the size of the cover relation or the Hasse diagram of the poset. For a poset of p elements, the number of comparable pairs typically satisfy $q \gg \tilde{q}$. We give a simple $O(\tilde{q}p)$-time algorithm that we use in our implementation; we give a randomized $O(\tilde{q}\sqrt{p \log p})$-time[1] one that applies to special posets only but uses no elaborate data structures (this algorithm might outperform our implementation in practice); finally we give a general deterministic one with the same time bound but using the Sleator–Tarjan dynamic tree data structure [15] (this algorithm is likely of little practical utility [1]).

The only previous known algorithm for Dilworth's problem is a straightforward reduction to a bipartite matching problem. We make two copies $\mathcal{K}_1$ and $\mathcal{K}_2$ of the ground set $\mathcal{K}$ of the poset and for each pair $x < y$ of $\mathcal{K}$ add an edge between the copy of x in $\mathcal{K}_1$ and the copy of y in $\mathcal{K}_2$. Then the edges of a maximal matching in this graph determine a minimal chain partition of the poset. For a poset of p members and q comparable pairs, we may thus find a chain partition in $O(pq)$ time by the alternating path matching algorithm [3] or in $O(q\sqrt{p})$ time by the Hopcroft–Karp matching algorithm [10].

Our first algorithm modifies the basic alternating path bipartite matching algorithm when one augments a matching M along a path whose edges are alternately inside and outside M. We extend the notion of an alternating path such that between two edges of M we allow arbitrary sequences of covers instead of a single (transitive) pair. This type of alternating paths can be found by breadth-first search (BFS)[3] in the graph with vertices $\mathcal{K}_1$ and $\mathcal{K}_2$ whose edges are (i) all matching edges directed from $\mathcal{K}_2$ towards $\mathcal{K}_1$; (ii) all covers $x \prec y$ with $x \in \mathcal{K}_1$ and $y \in \mathcal{K}_2$; and (iii) in addition all covers $x \prec y$ with both $x, y \in \mathcal{K}_1$. Since the original algorithm also finds augmenting paths by BFS, we simply replace the q by $\tilde{q}$ in the running time bound.

We heuristically improve our matching algorithm by identifying all possible edge-disjoint augmenting paths of the same breadth-first tree. In our experiments we never required significantly more than $\sqrt{p}$ rounds of BFS to identify all augmenting paths, indicating that the experimental running time stays around $\tilde{q}\sqrt{p}$. However for the hard matching instances arising from extreme sparse ($n \approx 2m$) inputs it often took around 10 rounds of BFS to find all augmenting paths of the same length—a task completed in $O(q)$ time by the Hopcroft–Karp algorithm. Our next algorithms hence might outperform this implementation.

[1] Further slight improvements in the log factors are achieved in the full paper [2]

It is not straightforward to modify the Hopcroft–Karp matching algorithm. That algorithm finds shortest alternating paths by a depth-first search [3] of all edges along which the distance to an unmatched vertex decreases. If this distance increases over $\sqrt{p}$, then it switches to the basic alternating path algorithm. The running time $O(q\sqrt{p})$ arises from the following two claims with $d = \sqrt{p}$:

Lemma 4 (Hopcroft and Karp [10]). *At certain stage of the Hopcroft–Karp matching algorithm, let the length of a shortest augmenting path be d. Then (i) the total number of times we backtracked along a given edge in depth-first search is at most d; this totals to $O(qd)$ for all edges; and (ii) the size of the maximum matching may not exceed the size of the current matching by more than p/d.*

When simulating steps over comparable pairs by several steps over covers, the main difficulty arises in (i) of the lemma: the same cover may have to be traversed for several distinct comparable pairs and the bound $O(\tilde{q}d)$ for the number of backtracks will not hold. Instead we will show that the average time spent for simulating a comparable pair is $O(\log p)$; then we may set $d = \sqrt{p/\log p}$ to get a running time of $O(\tilde{q}\sqrt{p \log p})$.

We sketch two algorithms to simulate a single comparable pair by covers in an average $O(\log p)$ time. For details see our full paper [2]. The first idea is to maintain fragments of cover paths so that whenever a new vertex is reached, we may jump to the end of a path passing through. The Sleator–Tarjan dynamic tree data structure [15] is capable of performing all necessary operations in $O(\log p)$ time.

A simpler randomized algorithm can be given for posets with the property that no two elements may be connected by two distinct paths in the Hasse diagram. The posets in Frank's algorithm satisfy this property. In such a poset we select edge xy out of a certain vertex $x \in \mathcal{K}_1$ as follows. Whenever possible, we select $y \in \mathcal{K}_2$ such that there is a matching edge yz; we will then complete the simulation of a comparable pair. Otherwise if we have a unique choice of vertex y, we contract x into y (we use a Union–Find data structure [3]). In all remaining cases we have at least two covers leading out of x; we choose one of them at random. By the special property of the poset, we take $O(\log p)$ such steps before reaching a matching edge, with high probability.

4 Bounds for the size of poset $\mathcal{K}$

The central theorems in our analysis are non-trivial bounds on the vertex and the edge size of poset $\mathcal{K}$ obtained in Phase 3. We show that the poset has $O(n \log n)$ vertices and $O(n^2)$ comparable pairs. These results are tight within small constant multiplicative factors (instances for tightness are in Figs. 4 and 6). By using the Hopcroft–Karp algorithm [10] for Dilworth's problem (with running time $O(q\sqrt{p})$ for a poset of p elements and q comparable pairs), we may hence derive an $O(n^{2.5}\sqrt{\log n})$ running time bound; preprocessing requires $O(m \log n)$ time in addition.

Theorem 5. *Let $T(n)$ denote the maximum number of essential subpath-edge pairs over C that contain no two elements that cross. Then $T(n) = \Theta(n \log n)$.*

Theorem 6. *There are $O(n^2)$ comparable and $\Omega(n^2)$ cover pairs of elements in $\mathcal{K}$. Hence the edge size of the poset of $\mathcal{K}$ is $\Theta(n^2)$.*

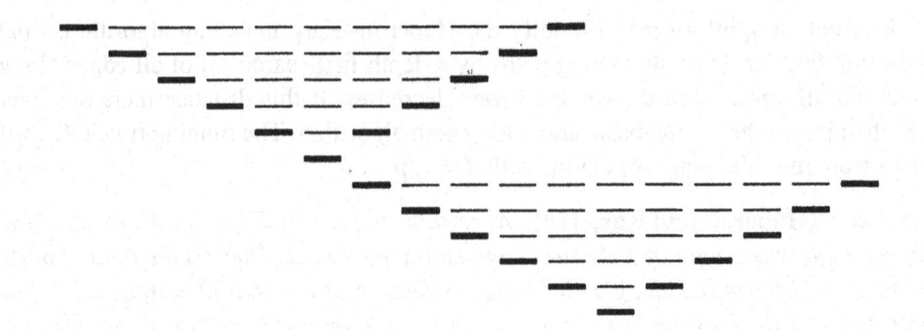

Fig. 6. Any subset-edge pair of the upper shaded area of $n/4$ elements is a cover of any other element of the lower shaded area of another $n/4$ elements. Hence the cover relation of the poset has size $\Omega(n^2)$.

Unlike our experiments indicating that there are significantly less cover pairs than comparable pairs, our theoretic running time bound is not based on our cover edge algorithms of Section 3. We were not able to derive better bounds for cover edges: while there are $O(n^2)$ transitive edges, the instance in Fig. 6 shows a matching $\Omega(n^2)$ bound for the cover edges themselves.

Both proofs apply a divide-and-conquer technique similar to that of the preprocessing step in Section 2. For the sake of simplicity we consider subpaths of a path P; the case of a length n cycle can be reduced to a length $2n$ path by doubling the edges. We subdivide the path P into P_1 and P_2, the first and last $n/2$ edges of P. For Theorem 5 it suffices to prove that there are $O(n)$ elements $(I,e) \in \mathcal{K}$ such that I intersects both P_1 and P_2. Similarly for Theorem 6 it suffices to show that there are $O(n^2)$ pairs of $\mathcal{K}$ with $(I,e) < (J,f)$ and I or J intersecting both P_1 and P_2. The second claim follows relatively easy; we leave the proof for the full paper [2].

For Theorem 5 next we show the $O(n)$ bound. We immediately set aside those $O(n)$ elements $(I,f) \in \mathcal{K}$ where f is the first edge of I. Thus it suffices to show that the cardinality of $\mathcal{K}_1 = \{(I,f) \in \mathcal{K} : I$ contains the last edge of P_1 and f is not the first edge of $I\}$ is $O(n)$.

Let $P_1 = \{e_1, e_2, \dots, e_k\}$. By the definition of essential the following mapping is one-to-one: we map $(I,f) \in \mathcal{K}_1$ to pairs (i,j) with $i < j$ where the first edge of I is e_i and $f = e_j$. We will use (i,j) as shorthand names of subpath-edge pairs (I,f). By the definition of essential and non-crossing we have:

Claim. There are no pairs (i,j) and $(i',j') \in \mathcal{K}_1$ with $i' < i \le j' < j$. □

Notice that if $(i,j) \in \mathcal{K}_1$, then no pair (s,j') for $j' > j$ may be in $\mathcal{K}_1$ by the claim. Hence (i,j) "blocks" all elements (s,j'). Since $j \le |P_1|$, we are done by giving an assignment of the elements $(i,j) \in \mathcal{K}_1$ to values $s \le j$ blocked by (i,j) such that no two elements (i,j) and (i',j') are mapped to the same value.

We achieve an assignment as above by ordering all (i,j) by the increasing value of $j - i$; we break ties by the increasing value of i. Then for all (i,j) we assign the smallest value $s > i$ not assigned to any other element of $\mathcal{K}_1$ preceding (i,j).

To show the correctness of the assignment, by contradiction let (i,j) be the first element in order where the assignment of a value $s \le j$ is not possible. Since the value $i+1$ satisfies $i < i+1 \le j$, in particular $i+1$ is assigned to some (i',j') that precedes (i,j) in the ordering. Then

$$i' < i+1 \le j', \text{i.e.} \quad i' \le i < j' \text{ and} \tag{1}$$
$$j' - i' \le j - i, \tag{2}$$

where (1) is since (i',j') is not a counterexample by the minimal choice of (i,j); and (2) is by the definition of the ordering. Since (i,j) and (i',j') are different and $i' \le i$, from (2) we get $j' < j$. If we assume $i' < i$ (strict inequality holds in (1)), the previous observations are in contradiction with the Claim. Hence we conclude that $i' = i$, i.e. there is an $(i,j') \in \mathcal{K}_1$ with $j' < j$. Let j' be maximum among all elements $(i,j'') \in \mathcal{K}_1$ with $j'' < j$.

Consider now the value $j'+1 \le j$ for j' defined above; by the assumption that $j' < j$ and hence $j'+1 \le j$ the value $j'+1$ is assigned to some $(i'',j'') \in \mathcal{K}_1$ with (again)

$$i'' \le j' < j'' \tag{3}$$
$$j'' - i'' \le j - i. \tag{4}$$

We distinguish three cases by comparing i and i''. If $i'' = i$, then $j'' \le j$ by (4) and $j' < j''$ contradicts with the maximal choice of j'. If $i'' < i$, then $j'' < j$ by (4); $i < j'+1$ since (i,j') is a pair and $x < y$ for all pairs (x,y); finally $j' < j''$ by (3), adding up to $i'' < i \le j''$ and $j'' < j$, contradicting the Claim. In the last case $i < i''$, when $i < i'' \le j'$ and $j' < j''$, both following by (3); the choice of (i,j') and (i'',j'') contradicts the Claim again.

5 Knuth's algorithm vs. ours: performance tests

Our experiments were conducted on a SPARC-10 with 160M memory. We calibrated the machine as designed by the organizers of the First DIMACS Implementation Challenge [11]; we resulted the following user times in seconds (no optimization ... optimization level 4): first test 0.5,0.3,0.3,0.2,0.3; second test 4.6,3.0,2.9,3.0,3.0. In the table n_r and m_r are the new input size after the reductions are applied; p and $\tilde{q}$ are the poset sizes. n, n_r, m, m_r and q are in thousands; $\tilde{q}$ is in millions. 'Knuth" denotes the user time in seconds for [13] with our initial reduction applied; "Frank" denotes the user time of our algorithm. We ran each test twice with two different seeds; we observed no variation (time, n_r, m_r, p or $\tilde{q}$) exceeding 5%.

Our implementation performs well on relative dense inputs (the maximum density is $m = O(n \log n)$ due to the initial reduction); here Knuth's algorithm ran out of memory for the largest instance. The Dilworth instances are easy or completely trivial; matching takes 10% of user time. In contrast our memory consumption is high for sparse inputs and Knuth's algorithm performs much better if the density is extreme low ($n \approx 2n$; the $2m$ first and last edges for the set of subpath are distinct). Here our implementation spends 90% of time for matchings and take slightly more than $\sqrt{p}$ BFS-steps; a better matching algorithm might be needed here.

[1]

$n=$	4	6	9	15	20	4	6	9	15	20	4	6	9	15	20	150				
$m=$	12	18	27	45	60	40	60	90	150	200	4	6	9	15	20	2	3	4.5	7.5	10
$n_r \approx$	3.95	5.9	8.8	14.8	19.6	4	6	9	15	20	3.3	4.9	7.3	12	16	3.8	5.7	8.5	13.5	17.5
$m_r \approx$	9	13.6	21	34	46	16.5	25	40	62	86	3.8	5.6	8.6	14	19	2	3	4.5	7.5	9.9
$p=$	12	18	27	46	60	12.5	18.8	29	48.5	65	8.8	13.5	20	33.5	45	9.3	14	20.6	33.6	43.5
$\bar{q}=$	0.3	0.7	1.5	4	7	0.14	0.3	0.7	2	3.5	0.2	0.5	1.1	3	5.3	0.2	0.5	1	2.4	4
seed	36 157	744 411	70 988	100 37	732 0	948 8	13 901	444 538	450 37	12 7864	14 711	487 466	317 1	99 3153	984 33	105 654	729 84	926 4422	387 917	832 843
Knuth	22	74	257	1000	9000	54	163	501	1500	–	3	11	32	220	450	6	25	25	27	50
Frank	10	23	60	188	440	5.7	14	44	86	250	9	20	55	190	333	18	40	96	290	510

6 Conclusion

We implemented and analyzed a new efficient $O(n^{2.5}\sqrt{\log n} + m \log n)$ time algorithm to find the minimum number of so-called generators of a set of subpaths of a cycle. Previous implementations are only able to handle the case of subpaths of a path instead of a cycle. We compared our implementation (restricted to paths) to Knuth's [13] previous algorithm. While by our results Knuth's algorithm is faster in theory, our experiments with random input indicate the contrary for all but the extreme sparse problems. As of own interest, our experiments resulted in new algorithms for Dilworth's theorem.

References

[1] Ahuja, R.K., T.L. Magnanti and J.B. Orlin, *Network Flows*, Prentice Hall (1993).

[2] Benczúr, A.A., J. Förster and Z. Király, *Dilworth's Theorem and its application for path systems of a cycle—implementation and analysis* (full paper). Available at `http://www.cs.elte.hu/~joergf/`.

[3] Cormen, T.H., C.E. Leiserson and R.L. Rivest, *Introduction to Algorithms*, MIT Press (1990).

[4] Frank, A., Finding minimum generators of path systems, submitted to *JCT* B.

[5] Frank, A., Finding minimum weighted generators of a path system, submitted.

[6] Frank, A. and T. Jordán, Minimal edge-coverings of pairs of sets, *JCT* B **65** (1995), pp. 73–110.

[7] Franzblau, D.S. and D.J. Kleitman, An algorithm for constructing polygons with rectangles, *Information and Control* **63** (1984), pp. 164–189.

[8] Ford, L.R. and D.R. Fulkerson, *Flows in Networks*, Princeton University Press (1962).

[9] Győri, E., A min-max theorem on intervals, *JCT* B **37** (1984), pp. 1–9.

[10] Hopcroft, J.E. and R.M. Karp, An $n^{5/2}$ algorithm for maximum matching in bipartite graphs, *SIAM J. Comp.* **2** (1973), pp. 225–231.

[11] Johnson, D.S. and C.C. McGeoch, eds, *DIMACS Implementation Challenge Workshop: Algorithms for Network Flows and Matching*. AMS and ACM (1993).

[12] Karger, D.R. and M.S. Levine, Finding Maximum Flows in Simple Undirected Graphs is Easier than Bipartite Matching, *30th ACM Symposium on Theory of Computing* (1998).

[13] Knuth, D.E., Irredundant intervals, *ACM Journal of Experimental Algorithmics* **1** (1996)

[14] Lubiw, A., A weighted min-max relation for intervals, *JCT* B **53** (1991), pp. 151–172.

[15] Sleator, D.D. and R.E. Tarjan, A data structure for dynamic trees, *J. Comput. Syst. Sci.* **26** (1983), pp. 362–391.

[16] Tarjan, R.E., *Data Structures and Network Algorithms*, SIAM (1983).

On 2-Coverings and 2-Packings of Laminar Families

Joseph Cheriyan[1]*, Tibor Jordán[2]**, and R. Ravi[3]***

[1] Department of Combinatorics and Optimization,
University of Waterloo, Waterloo ON Canada N2L 3G1,
e-mail: jcheriyan@math.uwaterloo.ca

[2] BRICS†, Department of Computer Science, University of Aarhus,
Ny Munkegade, building 540, DK-8000 Aarhus C, Denmark.
e-mail: jordan@daimi.au.dk

[3] GSIA, Carnegie Mellon University,
Pittsburgh, PA 15213-3890,
e-mail: ravi@cmu.edu

Abstract. Let $\mathcal{H}$ be a laminar family of subsets of a groundset V. A *k-cover* of $\mathcal{H}$ is a multiset C of edges on V such that for every subset S in $\mathcal{H}$, C has at least k edges that have exactly one end in S. A *k-packing* of $\mathcal{H}$ is a multiset P of edges on V such that for every subset S in $\mathcal{H}$, P has at most $k \cdot u(S)$ edges that have exactly one end in S. Here, u assigns an integer capacity to each subset in $\mathcal{H}$.
Our main results are: (a) Given a k-cover C of $\mathcal{H}$, there is an efficient algorithm to find a 1-cover contained in C of size $\leq k|C|/(2k-1)$. For 2-covers, the factor of 2/3 is best possible. (b) Given a 2-packing P of $\mathcal{H}$, there is an efficient algorithm to find a 1-packing contained in P of size $\geq |P|/3$. The factor of 1/3 for 2-packings is best possible.
These results are based on efficient algorithms for finding appropriate colorings of the edges in a k-cover or a 2-packing, respectively, and they extend to the case where the edges have nonnegative weights. Our results imply approximation algorithms for some NP-hard problems in connectivity augmentation and related topics. In particular, we have a 4/3-approximation algorithm for the following problem: Given a tree T and a set of nontree edges E that forms a cycle on the leaves of T, find a minimum-size subset E' of E such that $T + E'$ is 2-edge connected.

1 Introduction

Let $\mathcal{H}$ be a laminar family of subsets of a groundset V. In detail, let V be a groundset, and let $\mathcal{H} = \{S_1, S_2, \dots, S_q\}$ be a set of distinct subsets of V such that for every $1 \leq i, j \leq q$, $S_i \cap S_j$ is exactly one of $\emptyset$, S_i or S_j. A *k-cover* of $\mathcal{H}$ is a multiset of edges, C, such that for every subset S in $\mathcal{H}$, C has at least k edges (counting multiplicities) that have exactly one end in S. A *k-packing* of $\mathcal{H}$ is a multiset of edges, P, such that for every subset S in $\mathcal{H}$, P has at most $k \cdot u(S)$ edges (counting multiplicities) that have

* Supported in part by NSERC research grant OGP0138432.
** Supported in part by the Hungarian Scientific Research Fund no. OTKA T29772 and T30059.
*** Supported in part by NSF career grant CCR–9625297.
† Basic Research in Computer Science, Centre of the Danish National Research Foundation.

exactly one end in S. Here, u assigns an integer capacity to each subset in $\mathcal{H}$. Our main results are:

1. Given a k-cover C of $\mathcal{H}$, there is an efficient algorithm to find a 1-cover contained in C of size $\leq k|C|/(2k-1)$. For 2-covers, the factor of $2/3$ is best possible.
2. Given a 2-packing P of $\mathcal{H}$, there is an efficient algorithm to find a 1-packing contained in P of size $\geq |P|/3$. The factor of $1/3$ is best possible.

All of these results extend to the weighted case, where the edges have nonnegative weights. Also, we show that the following two problems are NP-hard: (1) Given a 2-cover C of $\mathcal{H}$, find a minimum-size 1-cover that is contained in C. (2) Given a 2-packing P of $\mathcal{H}, u$, find a maximum-size 1-packing that is contained in P.

The upper bound of $2/3$ on the ratio of the minimum size of a 1-cover versus the size of a (containing) 2-cover is tight. To see this, consider the complete graph K_3, and the laminar family $\mathcal{H}$ consisting of three singleton sets. Let the 2-cover be $E(K_3)$. A minimum 1-cover has 2 edges from K_3. The same example, with unit capacities for the three singleton sets in $\mathcal{H}$, shows that the ratio of the maximum size of a 1-packing versus the size of a (containing) 2-packing may equal $1/3$. There is an infinite family of similar examples.

An edge is said to *cover* a subset S of V if the edge has exactly one end in S. Our algorithm for finding a small-size 1-cover from a given 2-cover constructs a "good" 3-coloring of (the edges of) the 2-cover. In detail, the 3-coloring is such that for every subset S in the laminar family, at least two different colors appear among the edges covering S. The desired 1-cover is obtained by picking the two smallest (least weight) color classes. Similarly, our algorithm for finding a large-size 1-packing from a given 2-packing constructs a 3-coloring of (the edges of) the 2-packing such that for every subset S in the laminar family, at most $u(S)$ of the edges covering S have the same color. The desired 1-packing is obtained by picking the largest (most weight) color class.

1.1 A Linear Programming Relaxation

Consider the natural integer programming formulation (IP) of our minimum 1-cover problem. Let the given k-cover be denoted by E. There is a (nonnegative) integer variable x_e for each edge $e \in E$. For each subset $S \in \mathcal{H}$, there is a constraint $\sum_{e\in\delta(S)} x_e \geq 1$, where $\delta(S)$ denotes the set of edges covering S. The objective function is to minimize $\sum_e w_e x_e$, where w_e is the weight of edge e. Let (LP) be the following linear program obtained by relaxing all of the integrality constraints on the variables.

$$(LP) \quad z_{LP} = \min \sum_e w_e x_e \quad \text{s.t.} \ \{ \sum_{e\in\delta(S)} x_e \geq 1, \forall S \in \mathcal{H}; \quad x_e \geq 0, \forall e \in E\}.$$

Clearly, (LP) is solvable in polynomial time. The k-cover gives a feasible solution to (LP) by fixing $x_e = 1/k$ for each edge e in the k-cover.

For the minimum 1-cover problem, Theorem 3 below shows that the optimal value of the integer program (IP) is $\leq 4/3$ times the optimal value of a half-integral solution to the LP relaxation (LP). (A feasible solution x to (LP) is called *half-integral* if $x_e \in \{0, \frac{1}{2}, 1\}$, for all edges e.) There are examples where the LP relaxation has a unique optimal solution that is *not* half-integral. For the maximum 1-packing problem, Theorem 6 shows that the optimal value of the integer program is $\geq 2/3$ times the optimal value of a half-integral solution to the LP relaxation.

Recall that a laminar family $\mathcal{H}$ may be represented as a tree $T = T(\mathcal{H})$. (T has a node for V as well as for each set $A_i \in \mathcal{H}$, and T has an edge A_iA_j if $A_j \in \{V\} \cup \mathcal{H}$ is the smallest set containing $A_i \in \mathcal{H}$.)

Two special cases of the minimum 1-cover problem are worth mentioning. (i) If the laminar family $\mathcal{H}$ is such that the tree $T(\mathcal{H})$ is a path, then the LP relaxation has an integral optimal solution. This follows because the constraints matrix of the LP relaxation is essentially a network matrix, see [CCPS 98, Theorem 6.28], and hence the matrix is totally unimodular; consequently, every extreme point solution (basic feasible solution) of the LP relaxation is integral. (ii) If the laminar family $\mathcal{H}$ is such that the tree $T(\mathcal{H})$ is a star (i.e., the tree has one nonleaf node, and that is adjacent to all the leaf nodes) then the LP relaxation has a half-integral optimal solution. This follows because in this case the LP relaxation is essentially the same as the linear program of the fractional matching polytope, which has half-integral extreme point solutions, see [CCPS 98, Theorem 6.13].

1.2 Equivalent Problems

The problem of finding a minimum 1-cover of a laminar family $\mathcal{H}$ from among the multiedges of a k-cover E may be reformulated as a connectivity augmentation problem. Let $T = T(\mathcal{H})$ be the tree representing $\mathcal{H}$; note that $E(T)$ is disjoint from E. Then the problem is to find a minimum weight subset of edges E' contained in E such that $T + E' = (V(T), E(T) \cup E')$ is 2-edge connected; we may assume that E' has no multiedges. Instead of taking T to be a tree, we may take T to be a connected graph. This gives the problem *CBRA* which was initially studied by Eswaran & Tarjan [ET 76], and by Frederickson & Ja'ja' [FJ 81].

Similarly, the problem of finding a maximum 1-packing of a capacitated laminar family $\mathcal{H}, u$ from among the multiedges of a k-packing E may be reformulated as follows. Let $T = T(\mathcal{H})$ be the tree representing $\mathcal{H}$, and let the tree edges have (nonnegative) integer capacities $u : E(T) \to \mathbf{Z}$; the capacity of a set $A_i \in \mathcal{H}$ corresponds to the capacity of the tree edge a_i representing A_i. The k-packing E corresponds to a set of demand edges. The problem is to find a maximum integral multicommodity flow $x : E \to \mathbf{Z}$ where the source-sink pairs (of the commodities) are as specified by E. In more detail, the objective is to maximize the total flow $\sum_{e \in E} x_e$, subject to the capacity constraints, namely, for each tree edge a_i the sum of the x-values over the demand edges in the cut given by $T - a_i$ is $\leq u(a_i)$, and the constraints that x is integral and ≥ 0.

1.3 Approximation Algorithms for NP-hard Problems in Connectivity Augmentation

Our results on 2-covers and 2-packings imply improved approximation algorithms for some NP-hard problems in connectivity augmentation and related topics. Frederickson and Ja'ja' [FJ 81] showed that problem *CBRA* is NP-hard and gave a 2-approximation algorithm. Later, Khuller and Vishkin [KV 94] gave another 2-approximation algorithm for a generalization, namely, find a minimum-weight k-edge connected spanning subgraph of a given weighted graph. Subsequently, Garg et al [GVY 97, Theorem 4.2] showed that problem *CBRA* is max SNP-hard, implying that there is no polynomial-time approximation scheme for *CBRA* modulo the P$\neq$NP conjecture. Currently, the best approximation guarantee known for *CBRA* is 2.

Our work is partly motivated by the question of whether or not the approximation guarantee for problem *CBRA* can be improved to be strictly less than 2 (i.e., to $2-\varepsilon$ for a constant $\varepsilon > 0$). We give a $4/3$-approximation algorithm for an NP-hard problem that is a special case of *CBRA*, namely, the tree plus cycle (*TPC*) problem. See Section 4.

Garg, Vazirani and Yannakakis [GVY 97] show that the above maximum 1-packing problem (equivalently, the above multicommodity flow problem) is NP-hard and they give a 2-approximation algorithm. In fact, they show that the optimal value of an integral 1-packing z_{IP} is $\geq 1/2$ times the optimal value of a fractional 1-packing z_{LP}. We do not know whether the factor $1/2$ here is tight.

It should be noted that the maximum 1-packing problem for the special case of unit capacities (i.e., $u(A_i) = 1$, $\forall A_i \in \mathcal{H}$) *is* polynomial-time solvable. If the capacities are either one or two, and the tree $T(\mathcal{H})$ representing the laminar family $\mathcal{H}$ has height two (i.e., every tree path has length ≤ 4), then the problem may be NP-hard, see [GVY 97, Lemma 4.3].

Further discussion on related topics may be found in the survey papers by Frank [F 94], Hochbaum [Hoc 96], and Khuller [Kh 96]. Jain [J 98] has interesting recent results, including a 2-approximation algorithm for an important generalization of problem *CBRA*.

We close this section by introducing some notation. For a multigraph $G = (V,E)$ and a node set $S \subseteq V$, let $\delta_E(S)$ denote the multiset of edges in E that have exactly one end node in S, and let $d_E(S)$ denote $|\delta_E(S)|$; so $d_E(S)$ is the number of multiedges in the cut $(S, V-S)$.

2 Obtaining a 1-Cover from a k-Cover

This section has our main result on k-covers, namely, there exists a 1-cover whose size (or weight) is at most $k/(2k-1)$ times the size (or weight) of a given k-cover. The main step (Proposition 2) is to show that there exists a "good" $(2k-1)$-coloring of any k-cover. We start with a preliminary lemma.

Lemma 1. *Let V be a set of nodes, and let $\mathcal{H}$ be a laminar family on V. Let E be a minimal k-cover of $\mathcal{H}$. Then there exists a set $X \in \mathcal{H}$ such that $d_E(X) = k$ and no proper subset Y of X is in $\mathcal{H}$.*

Proof. Since E is minimal, there exists at least one set $X \in \mathcal{H}$ with $d_E(X) = k$. We call a node set $X \subseteq V$ a *tight set* if $d_E(X) = k$. Consider an inclusionwise minimal tight set X in $\mathcal{H}$. Suppose there exists a $Y \subset X$ such that $Y \in \mathcal{H}$. If each edge of E that covers Y also covers X, then we have $d_E(Y) = k$. But this contradicts our choice of X. Thus there exists an edge $xy \in E$ covering Y with $x, y \in X$. By the minimality of E, xy must cover a tight set $Z \in \mathcal{H}$. Since $\mathcal{H}$ is a laminar family, Z must be a proper subset of X. This contradiction to our choice of X proves the lemma. □

Proposition 2. *Let V be a set of nodes, and let $\mathcal{H}$ be a laminar family on V. Let E be a minimal k-cover of $\mathcal{H}$. Then there is a $(2k-1)$-coloring of (the edges in) E such that*
(i) each set $X \in \mathcal{H}$ is covered by edges of at least k different colors, and
(ii) for every node v with $d_E(v) \leq k$, all of the edges incident to v have distinct colors.

Proof. The proof is by induction on $|\mathcal{H}|$. For $|\mathcal{H}| = 1$ the results holds since there are k edges in E (since E is minimal) and these can be assigned different colors. (For $|\mathcal{H}| = 0$, $|E| = 0$ so the result holds. However, even if E is nonempty, it is easy to color the edges in an arbitrary order to achieve property (ii).)

Now, suppose that the result holds for laminar families of cardinality $\leq N$. Consider a laminar family $\mathcal{H}$ of cardinality $N+1$, and let E be a minimal k-cover of $\mathcal{H}$. By Lemma 1, there exists a tight set $A \in \mathcal{H}$ (i.e., $d_E(A) = k$) such that no $Y \subset A$ is in $\mathcal{H}$. We contract the set A to one node v_A, and accordingly update the laminar family $\mathcal{H}$. Then we remove the singleton set $\{v_A\}$ from $\mathcal{H}$. Let the resulting laminar family be $\mathcal{H}'$, and note that it has cardinality N. Clearly, E is a k-cover of $\mathcal{H}'$. Let $E' \subseteq E$ be a minimal k-cover of $\mathcal{H}'$. By the induction hypothesis, E' has a $(2k-1)$-coloring that satisfies properties (i) and (ii), i.e., E' has a good $(2k-1)$-coloring.

If the node v_A is incident to $\geq k$ edges of E', then note that E' with its $(2k-1)$-coloring is good with respect to $\mathcal{H}$ (i.e., properties (i) and (ii) hold for $\mathcal{H}$ too). To see this, observe that $k \leq d_{E'}(v_A) \leq d_E(v_A) = k$, so $d_{E'}(v_A) = k$, hence, the k edges of E' incident to v_A get distinct colors by property (ii). Then, for the original node set V, the k edges of E' covering A get k different colors.

Now focus on the case when $d_{E'}(v_A) < k$. Clearly, each edge in $E - E'$ is incident to v_A, since each edge in E not incident to v_A covers some tight set that is in both $\mathcal{H}$ and $\mathcal{H}'$. We claim that the remaining edges of $E - E'$ incident to v_A can be colored and added to E' in such a way that E with its $(2k-1)$-coloring is good with respect to $\mathcal{H}$.

It is easy to assign colors to the edge (or edges) of $E - E'$ such that the k edges of E incident to v_A get different colors. The difficulty is that property (ii) has to be preserved, that is, we must not "create" nodes of degree $\leq k$ that are incident to two edges of the same color. It turns out that this extra condition is easily handled as follows. Let $e \in E - E'$ be an edge incident to v_A, and let $w \in V$ be the other end node of e. If w has degree $\leq k$ for the current subset of E, then e is incident to $\leq (2k-2)$ other edges; since $(2k-1)$ colors are available, we can assign e a color different from the colors of all the edges incident to e. Otherwise (w has degree $> k$ for the current subset of E), the other edges incident to w impose no coloring constraint on e, and we assign e a color different from the colors of the other edges incident to v_A; this is easy since $d_E(v_A) = k$. □

Theorem 3. *Let V be a node set, and let $\mathcal{H}$ be a laminar family on V. Let E be a k-cover of $\mathcal{H}$, and let each edge $e \in E$ have a nonnegative weight $w(e)$. Then there is a 1-cover of $\mathcal{H}$, call it E', such that $E' \subseteq E$ and $w(E') \leq k\,w(E)/(2k-1)$. Moreover, there is an efficient algorithm that given E finds E'; the running time is $O(\min(k|V|^2, k^2|V|))$.*

Proof. We construct a good $(2k-1)$-coloring of the k-cover E by applying Proposition 2 to a minimal k-cover $\tilde{E} \subseteq E$ and then "extending" the good $(2k-1)$-coloring of $\tilde{E}$ to E. That is, we partition E into $(2k-1)$ subsets such that each set X in $\mathcal{H}$ is covered by edges from at least k of these subsets. We take E' to be the union of the cheapest k

of the $(2k-1)$ subsets. Clearly, the weight of E' is at most $k/(2k-1)$ of the weight of E, and (by property (i) of Proposition 2) E' is a 1-cover of $\mathcal{H}$.

Consider the time complexity of the construction in Proposition 2. Let $n = |V|$; then note that $|\mathcal{H}| \leq 2n$ and $|E| \leq 2kn$. The construction is easy to implement in time $O(|\mathcal{H}| \cdot |E|) = O(kn^2)$. Also, for $k < n$, the time complexity can be improved to $O(k^2 \cdot |\mathcal{H}|) = O(k^2 n)$. To see this, note that for each set $A \in \mathcal{H}$ we assign colors to at most k of the edges covering A after we contract A to v_A, and for each such edge e we examine at most $(2k-2)$ edges incident to e. □

3 Obtaining a 1-Packing from a 2-Packing

This section has our main result on 2-packings, namely, there exists a 1-packing whose size (or weight) is at least $1/3$ times the size (or weight) of a given 2-packing. First, we show that there is no loss of generality in assuming that the 2-packing forms an Eulerian multigraph. Then we give a 3-coloring for the edges of the 2-packing such that for each set S in the laminar family at most $u(S)$ edges covering S have the same color. We take the desired 1-packing to be the biggest color class.

Lemma 4. *Let V be a set of nodes, let $\mathcal{H}$ be a laminar family on V, and let $u : \mathcal{H} \to \mathbf{Z}$ assign an integral capacity to each set in $\mathcal{H}$. Let E be a 2-packing of $\mathcal{H}, u$, i.e., for all sets $A_i \in \mathcal{H}$, $d_E(A_i) \leq 2u(A_i)$. If E is a maximal 2-packing, then the multigraph $G = (V,E)$ is Eulerian.*

Proof. If G is not Eulerian, then it has an even number (≥ 2) of nodes of odd degree. Let $A \in \{V\} \cup \mathcal{H}$ be an inclusionwise minimal set that contains ≥ 2 nodes of odd degree. For every proper subset S of A that is in $\mathcal{H}$ and that contains an odd-degree node, note that $d_E(S)$ is odd, hence, this quantity is strictly less than the capacity $2u(S)$. Consequently, we can add an edge (or another copy of the edge) vw where v, w are odd-degree nodes in A to get $E \cup \{vw\}$ and this stays a 2-packing of $\mathcal{H}$, u. This contradicts our choice of E, since E is a maximal 2-packing. Consequently, G has no nodes of odd degree, i.e., G is Eulerian. □

Proposition 5. *Given an Eulerian multigraph $G = (V,E)$, an arbitrary pairing $\mathcal{P}$ of the edges such that for every edge-pair the two edges have a common end node, and a laminar family of node sets $\mathcal{H}$, there is a 3-coloring of E such that*
(i) for each cut $\delta_E(A_i)$, $A_i \in \mathcal{H}$, at most half of the edges have the same color, and
(ii) for each edge-pair e, f in $\mathcal{P}$, the edges e and f have different colors.

Proof. Let $\mathcal{P}$ be a set of triples $[v,e,f]$, where e and f are paired edges incident to the node v. Note that an edge $e = vw$ may occur in two triples $[v,e,f]$ and $[w,e,g]$. W.l.o.g. assume that $\mathcal{P}$ gives, for each node v, a pairing of all the edges incident to v. Then $\mathcal{P}$ partitions E into one or more (edge disjoint) subgraphs $Q_1, Q_2, \ldots$, where each subgraph Q_j is a connected Eulerian multigraph. To see this, focus on the Eulerian tour given by fixing the successor of any edge $e = vw$ to be the other edge in the triple $[w,e,f] \in \mathcal{P}$, assuming e is oriented from v to w; each such Eulerian tour gives a subgraph Q_j.

If $\mathcal{H} = \emptyset$, then we color each subgraph Q_j with 3 colors such that no two edges in the same edge-pair in $\mathcal{P}$ get the same color. This is easy: We traverse the Eulerian tour of Q_j given by $\mathcal{P}$, and alternately assign the colors red and blue to the edges in Q_j, and if necessary, we assign the color green to the last edge of Q_j.

Otherwise, we proceed by induction on the number of sets in $\mathcal{H}$. We take an inclusionwise minimal set $A \in \mathcal{H}$, shrink it to a single node v_A, and update $G = (V,E)$, $\mathcal{H}$ and $\mathcal{P}$ to $G' = (V',E')$, $\mathcal{H}'$ and $\mathcal{P}'$. Here, $\mathcal{H}' = \mathcal{H} - \{A\}$, i.e., the singleton set $\{v_A\}$ is not kept in $\mathcal{H}'$. Also, we add new edge pairs to $\mathcal{P}'$ to ensure that all edges incident to v_A are paired. For a node $v \notin A$, all its triples $[v,e,f] \in \mathcal{P}$ are retained in $\mathcal{P}'$. Consider the pairing of all the edges incident to v_A in G'. For each triple $[v,e,f]$ in $\mathcal{P}$ such that $v \in A$ and each of e,f has one end node in $V - A$ (so e,f are both incident to v_A in G'), we replace the triple by $[v_A,e,f]$. We arbitrarily pair up the remaining edges incident to v_A in G'.

By the induction hypothesis, there exists a good 3-coloring for $G', \mathcal{H}', \mathcal{P}'$. It remains to 3-color the edges with both ends in A. For this, we shrink the nodes in $V - A$ to a single node v_B, and update $G = (V,E), \mathcal{P}, \mathcal{H}$, to $G'' = (V'',E''), \mathcal{P}'', \mathcal{H}''$; note that $\mathcal{H}''$ is the empty family and so may be ignored. We also keep the 3-coloring of $\delta_{E'}(v_A) = \delta_{E''}(v_B)$. Our final goal is to extend this 3-coloring to a good 3-coloring of E'' respecting $\mathcal{P}''$. We must check that this can always be done. Consider the differently-colored edge pairs incident to v_B. Consider any connected Eulerian subgraph Q_j containing one of these edge pairs e_1, e_2; the corresponding triple in $\mathcal{P}''$ is $[v_B, e_1, e_2]$. Let $\tilde{Q}_j$ be a minimal walk of (the Eulerian tour of) Q_j starting with e_2 and ending with an edge f incident to v_B (possibly, $f = e_1$). The number of internal edges in $\tilde{Q}_j$ is $\equiv 0$ or $1 \pmod 2$, and the two terminal edges either have the same color or not. If the number of internal edges in $\tilde{Q}_j$ is nonzero, then it is easy to assign one, two, or three colors to these edges such that every pair of consecutive edges gets two different colors. The remaining case is when $\tilde{Q}_j$ has *no* internal edges, say, $\tilde{Q}_j = v_B, e_2, w, f, v_B$, where w is a node in A. Then edges e_2, f are paired via the common end-node w, i.e., the triple $[w, e_2, f]$ is present in both $\mathcal{P}''$ and $\mathcal{P}$. Then, by our construction of $\mathcal{P}'$ from $\mathcal{P}$, the triple $[v_A, e_2, f]$ is in $\mathcal{P}'$, and so edges e_2 and f (which are paired in $\mathcal{P}'$ and present in $\delta_{E'}(v_A) = \delta_{E''}(v_B)$) must get different colors. Hence, a good 3-coloring of $G', \mathcal{H}', \mathcal{P}'$ can always be extended to give a good 3-coloring of $\tilde{Q}_j$, and the construction may be repeated to give a good 3-coloring of Q_j.

Finally, note that E'' is partitioned by $\mathcal{P}''$ into several connected Eulerian subgraphs $Q_1, Q_2, \ldots$, where some of these subgraphs contain edges of $\delta_{E''}(v_B)$ and others do not. Clearly, the good 3-coloring of $G', \mathcal{H}', \mathcal{P}'$ can always be extended to give a good 3-coloring of each of $Q_1, Q_2, \ldots$, and thus we obtain a good 3-coloring of $G, \mathcal{H}, \mathcal{P}$. □

Theorem 6. *Let V be a node set, let $\mathcal{H}$ be a laminar family on V, and let $u : \mathcal{H} \to \mathbf{Z}$ assign an integer capacity to each set in $\mathcal{H}$. Let E be a 2-packing of $\mathcal{H}$, and let each edge $e \in E$ have a nonnegative weight $w(e)$. Then there is a 1-packing of $\mathcal{H}$, call it E', such that $E' \subseteq E$ and $w(E') \geq w(E)/3$. Moreover, there is an efficient algorithm that given E finds E'; the running time is $O(|V| \cdot |E|)$.*

Proof. If the multigraph (V,E) is not Eulerian, then we use the construction in Lemma 4 to add a set of edges to make the resulting multigraph Eulerian without violating the 2-packing constraints. We assign a weight of zero to each of the new edges. Let us continue to use E to denote the edge set of the resulting multigraph. We construct a good

3-coloring of the 2-packing E by applying Proposition 5. Let F be the most expensive of the three "color classes;" so, the weight of F, $w(F)$, is $\geq w(E)/3$. Note that F is a 1-packing of $\mathcal{H}, u$ by property (i) in the proposition since for every set $A_i \in \mathcal{H}$, we have $d_F(A_i) \leq d_E(A_i)/2 \leq 2u(A_i)$. Finally, we discard any new edges in F (i.e., the edges added by the construction in Lemma 4) to get the desired 1-packing.

Consider the time complexity of the whole construction. It is easy to see that the construction in Proposition 5 for the minimal set $A \in \mathcal{H}$ takes linear time. This construction may have to be repeated $|\mathcal{H}| = O(|V|)$ times. Hence, the overall running time is $O(|V| \cdot |E|)$. □

4 Applications to Connectivity Augmentation and Related Topics

This section applies our covering result (Theorem 3) to the design of approximation algorithms for some NP-hard problems in connectivity augmentation and related topics. The main application is to problem *CBRA*, which is stated below. Problem *CBRA* is equivalent to some other problems in this area, and so we immediately get some more applications.

Recall problem *CBRA*: given a connected graph $T = (V,F)$, and a set of "supply" edges E with nonnegative weights $w : E \to \Re_+$, the goal is to find a minimum-weight subset E' of E such that $T + E' = (V, F \cup E')$ is 2-edge-connected. One application of Theorem 3 is to give a $4/3$-approximation algorithm for the special case of *CBRA* when the LP relaxation has an optimal solution that is half-integral.

Theorem 7. *Given a half-integral solution to the LP relaxation of CBRA of weight z, there is an $O(|V|)$-time algorithm to find an integral solution (i.e., a feasible solution of CBRA) whose weight is $\leq \frac{4}{3}z$.*

Proof. Problem *CBRA* may be restated as the problem of finding a minimum-weight 1-cover of a laminar family $\mathcal{H}$, where the 1-cover must be chosen from the set of supply edges E and each supply edge has a nonnegative weight. To specify $\mathcal{H}$, fix any node $r \in V$ to be the root of T, and focus on the cut edges of T, call them $f_1, f_2, \ldots$. For each of these cut edges $f_1, f_2, \ldots$, let A_i be the (node set of the) component of $T - f_i$ that does not contain r. We take $\mathcal{H} = \{A_1, A_2, \ldots\}$.

Let $x : E \to \{0, \frac{1}{2}, 1\}$ be a half-integral solution to the LP relaxation of *CBRA*, and let $z = \sum_e w_e x_e$. Then x corresponds to a 2-cover C of $\mathcal{H}$, where C has zero, one or two copies of a supply edge e iff $x_e = 0,\ 1$, or 2. By Theorem 3, C contains a 1-cover C' whose weight is $\leq 4z/3$, and moreover, C' can be computed in time $O(|V|)$. □

We have sharper results for the following (NP-complete) special case of problem *CBRA*.

Tree Plus Cycle Problem (TPC):
INSTANCE: A tree $T = (W,F)$ whose set of leaf nodes is $V \subseteq W$, a "supply" cycle $Q = (V,E)$ on the leaves of T (i.e., $d_E(v) = 2$, $\forall v \in V$), and a positive integer N.
QUESTION: Is there a set of edges $E' \subseteq E$ with $|E'| \leq N$ such that $T + E' = (W, F \cup E')$ is 2-edge-connected?

Corollary 8. *There exists a $\frac{4}{3}$-approximation algorithm for TPC. Moreover, there exists a feasible solution $E' \subseteq E(Q)$ of size $\leq 2|V(Q)|/3$.*

Proof. Consider the LP relaxation of problem *TPC*; it is easy to verify that an optimal solution is given by $x_e = 1/2$ for all supply edges $e \in E(Q)$. Now, the result follows directly from Theorem 7. □

Now consider the following problem: given a $(2k-1)$-edge-connected graph $T = (V,F)$ and a set of "supply" edges E with nonnegative weights $w: E \rightarrow \Re_+$, the goal is to find a minimum weight subset E' of E such that $G' = (V, F + E')$ is $(2k)$-edge-connected. Since the edge connectivity of T is odd, this problem is equivalent to problem *CBRA* because all the $(2k-1)$-cuts (minimum cuts) of T can be represented by means of a laminar family. (This follows easily from the fact that the node sets of two minimum cuts do not cross in this case.)

5 NP-completeness Results

First, we show that problem *TPC* (tree plus cycle) is NP-complete. It is convenient to reformulate *TPC* in terms of a laminar family rather than a tree.

Laminar Family Plus Cycle Problem (LPC):
INSTANCE: A laminar family $\mathcal{H}$ on a node set V, a cycle $Q = (V,E)$ on V, and a positive integer N. (Assume $\emptyset, V \notin \mathcal{H}$.)
QUESTION: Is there a 1-cover E' of $\mathcal{H}$ such that $E' \subseteq E$ and $|E'| \leq N$?

We give a polynomial-time reduction from the 3-dimensional matching problem to problem *LPC*. Our reduction is based on the proof of [FJ 81, Theorem 2].

Theorem 9. *Problem LPC is NP-complete.*

Proof. It is easy to see that *LPC* is in NP. Given an instance of *3DM* (that is, three disjoint sets W, X, Y, of cardinality q each, and a set M of 3-edges (triples) $(w_i x_j y_k) \in W \times X \times Y$), construct a connected graph T as follows. First build a star with a "root" r and $3q$ leaves $\{w_1, \ldots, w_q, x_1, \ldots, x_q, y_1, \ldots, y_q\}$ corresponding to the elements of $W \cup X \cup Y$. Then for each 3-edge $(w_i x_j y_k)$ of M add two nodes a_{ijk} and $\bar{a}_{ijk}$ to T and add the edges $w_i a_{ijk}$, $w_i \bar{a}_{ijk}$. Now replace each of the $2q$ nodes corresponding to elements of X and Y by complete graphs (or arbitrary 2-edge-connected graphs) denoted by $X_1, \ldots, X_q, Y_1, \ldots, Y_q$ as follows. Each complete subgraph of this type has $d_M(x_j)8q$ $(d_M(y_k)8q)$ nodes and is partitioned into $d_M(x_j)$ $(d_M(y_k))$ parts (so-called "lanes") of size $8q$ each. (Here, $d_M(x_j)$ and $d_M(y_k)$ denote the number of 3-edges of M containing x_j, respectively, y_k.) The graph constructed is connected and has $2p + 2q$ "leaves" (that is, leaf 2-edge-connected components), where $p := |M|$.

The next step is to define the cycle Q. The nodes of Q are the nodes of the leaves of T. Hence, $|V(Q)| = p(16q+2)$. First, we define p disjoint paths of Q such that each has $16q+2$ nodes (so each of these paths has length $16q+1$). Every 3-edge $(w_i x_j y_k)$ of M defines such a path as follows: take the $8q$ nodes (and edges connecting the consecutive ones) $l_1 l_2 \ldots l_{8q}$ of a lane of X_1 in an arbitrary order, then take the edges $l_{8q} a_{ijk}$, $a_{ijk} \bar{a}_{ijk}$,

$\bar{a}_{ijk}m_{8q}$ for some node m_{8q} of some lane of Y_k, in this order, and then take the other nodes of this lane $m_{8q-1},\dots,m_1$ in an arbitrary order. The lanes are chosen in such a way that these paths are pairwise disjoint. This can be done, since the lanes are pairwise disjoint and each X_j (or Y_k) has $d_M(x_j)$ (or $d_M(y_k)$) lanes.

Now fix a cyclic ordering $e_1,\dots,e_p$ of the 3-edges of M and complete the cycle Q by adding the missing p edges in such a way that the end of the path corresponding to $e_s = (w_i x_j y_k)$ (that is, a node m_1 of a lane in Y_k) is connected to the first node of the path corresponding to $e_{s+1} = w_{i'}x_{j'}y_{k'}$ (that is, to a node l_1 of a lane of $X_{j'}$) for $1 \le s \le p$. Note that each of these edges connects a complete subgraph X_j to a complete subgraph Y_k and all the edges of Q either connect different leaves of T or connect different nodes of some leaf of T. Furthermore, $V(Q)$ equals the union of the nodes of the leaves of T.

The last part of the reduction consists of defining a laminar family $\mathcal{H}$ on $V(Q)$. We define $\mathcal{H}$ by defining two disjoint subfamilies $\mathcal{H}_1$ and $\mathcal{H}_2$. Let $\mathcal{H}_1 := \{S \cap V(Q) : d_T(S) = 1, r \notin S, S \subseteq V(T)\}$ contain intersections of $V(Q)$ and those minimum cuts of T which do not contain the root. It is easy to see that this family is laminar. $\mathcal{H}_2$ consists of $2p$ disjoint collections, each of them defined on the nodes of a lane of a complete subgraph of the form X_j or Y_k of T as follows. Let us fix such a subgraph, say X_1. (The definition is similar for all the $2q$ subgraphs $X_1,\dots,X_q,Y_1,\dots,Y_q$.) Focus on a lane $l_1,\dots,l_{8q}$ of X_1, where the numbering follows the ordering of these nodes in Q. (Hence l_{8q} is connected to some leaf a_{ijk} and l_1 is connected to some Y_k.) This lane adds the following sets to $\mathcal{H}_2$: the singletons $l_1,\dots,l_{8q}$, the sets of nodes of the intervals of Q with end-node pairs (l_{8q-1}, l_{8q-s}) $(2 \le s \le 4q)$ and (l_{4q-r}, l_2) $(1 \le r \le 4q-3)$. Each lane of every complete subgraph X_j, Y_k $(1 \le j,k \le q)$ adds a similar collection to $\mathcal{H}_2$. Clearly, every collection of this type is laminar, and the collections are defined on pairwise disjoint sets of nodes, where each of these sets is included in a minimal element of $\mathcal{H}_1$. Therefore $\mathcal{H}$ is a laminar family on $V(Q)$, where $\mathcal{H} := \mathcal{H}_1 \cup \mathcal{H}_2$. Note that each node of Q belongs to $\mathcal{H}$ as a singleton set.

Observe the following important property, that follows from the structure of these collections and the fact that every node of Q belongs to $\mathcal{H}$. Let $E' \subseteq E(Q)$ be a 1-cover of $\mathcal{H}$. Then

(*) if the edge $l_{8q}l_{8q-1}$ (or similarly l_1l_2, $m_{8q}m_{8q-1}$, m_1m_2) for some lane in an X_j or Y_k is not in E' then $|E'| \ge |V(Q)|/2 + 2q - 1$.

It is easy to see that our reduction is polynomial. We claim that there exists a solution to the given instance of *3DM* (that is, a set of q pairwise disjoint 3-edges of M) if and only if $\mathcal{H}$ has a 1-cover of size at most $p + 8pq + q = |V(Q)|/2 + q$.

First observe that a set E' is a 1-cover if and only if $T + E'$ is 2-edge-connected and E' covers each member of $\mathcal{H}_2$. Moreover, as it was verified in [FJ 81], there is a 3-dimensional matching if and only if there is set E^* of $p+q$ edges in Q^* for which $T^* + E^*$ is 2-edge-connected, where T^* and Q^* arise from T and Q, respectively, by contracting the complete subgraphs (that is, the sets of the form X_j, Y_k, which are 2-edge-connected) to singletons and deleting the edges connecting these complete subgraphs from the cycle.

Suppose that there exists a 3-dimensional matching $M' \subseteq M$. Then there exists a set E^* of size $p+q$ which makes T^* 2-edge-connected and it is easy to see that there exists a set E'' of independent edges in Q which covers $\mathcal{H}_2$. Hence $|E''| = 16qp/2 = 8pq$. Now $E' := E^* \cup E''$ covers $\mathcal{H}$ and $|E'| = 8pq + p + q$, as required.

The proof of the other direction (which relies on $(*)$) is omitted. □

Corollary 10. *The following problem is NP-hard: given a 2-cover C of a laminar family $\mathcal{H}$, find a minimum-size 1-cover that is contained in C.*

Theorem 11. *The following problem is NP-hard: given a 2-packing P of a capacitated laminar family $\mathcal{H}$, u, find a maximum-size 1-packing that is contained in P.*

6 Conclusions

We suspect that our bounds on the ratios for 1-covers versus 2-covers and for 1-packings versus 2-packings hold in general.

1-COVER CONJECTURE: Consider the integer program for a minimum weight 1-cover of a laminar family and its LP relaxation (see Section 1). We conjecture that the ratio of the optimal values is at most $4/3$.

1-PACKING CONJECTURE: Consider the integer program for a maximum weight 1-packing of a capacitated laminar family and its LP relaxation (see Section 1). We conjecture that the ratio of the optimal values is at least $2/3$.

Another interesting question is to find sufficient conditions on the laminar family $\mathcal{H}$ (or, on the tree $T(\mathcal{H})$ representing $\mathcal{H}$) such that the LP relaxation has $\frac{1}{k}$-integral extreme point solutions. As noted in Section 1, the LP relaxation has integral extreme point solutions iff $T(\mathcal{H})$ is a path.

References

[CCPS 98] W. J. Cook, W. H. Cunningham, W. R. Pulleyblank, and A. Schrijver, *Combinatorial Optimization*, John Wiley & Sons, New York, 1998.

[ET 76] K. Eswaran and R.E. Tarjan, "Augmentation problems," *SIAM J. Computing* **5** (1976), 653–665.

[F 94] A. Frank, "Connectivity augmentation problems in network design," in *Mathematical Programming: State of the Art 1994*, (Eds. J. R. Birge and K. G. Murty), The University of Michigan, Ann Arbor, MI, 1994, 34–63.

[FJ 81] G.N.Frederickson and J.Ja'Ja', "Approximation algorithms for several graph augmentation problems," *SIAM J. Comput.* **10** (1981), 270–283.

[GVY 97] N. Garg, M. Yannakakis, and V. Vazirani, "Primal-dual approximation algorithms for integral flow and multicut in trees," *Algorithmica* **18** (1997), 3–20.

[Hoc 96] D. S. Hochbaum, "Approximating covering and packing problems: set cover, vertex cover, independent set, and related problems," in *Approximation algorithms for NP-hard problems*, Ed. D. S. Hochbaum, PWS co., Boston, 1996.

[J 98] K. Jain, "A factor 2 approximation algorithm for the generalized Steiner network problem," Proc. 39th IEEE FOCS, Palo Alto, CA, November 1998.

[Kh 96] S. Khuller, "Approximation algorithms for finding highly connected subgraphs," in *Approximation algorithms for NP-hard problems*, Ed. D. S. Hochbaum, PWS publishing co., Boston, 1996.

[KV 94] S. Khuller and U. Vishkin, "Biconnectivity approximations and graph carvings," *Journal of the ACM* **41** (1994), 214–235.

Random Cayley Graphs with $O(\log|G|)$ Generators Are Expanders

Igor Pak

Department of Mathematics
Yale University
New Haven, CT 06520
pak@math.yale.edu

Abstract. Let G be a finite group. Choose a set S of size k uniformly from G and consider a lazy random walk on the corresponding Cayley graph $\Gamma(G,S)$. We show that for almost all choices of S given $k = 2a\log_2|G|$, $a > 1$, we have $Re\lambda_1 \le 1 - 1/2a$. A similar but weaker result was obtained earlier by Alon and Roichman (see [4]).

1 Introduction

In the past few years there has been a significant progress in analysis of random Cayley graphs. Still for general groups G and small sets of generators, such as of size $O(\log|G|)$, more progress is yet to be made. Our results partially fill this gap.

Here is a general setup of a problem. Let G be a finite group, $n = |G|$. For a given k choose uniformly k random elements $g_1, \ldots, g_k \in G$. Denote by S the set of these elements. By $\Gamma(G,S)$ denote the corresponding oriented Cayley graph. Define *transition matrix* $A = (a_{g,h})$, $g, h \in G$ to be $a_{g,h} = 1/k$ if $g^{-1}h \in S$, and $a_{g,h} = 0$ otherwise. By $1 = |\lambda_0| \ge |\lambda_1| \ge \ldots$ denote the eigenvalues of the A. Note that since A is a real matrix, eigenvalues are either real, or complex numbers which appear in complex conjugate pairs.

Theorem *Let G be a finite group, $n = |G|$. Let $\varepsilon > 0$, $a > 1$ be given. Then*

$$\mathbf{Pr}(Re\lambda_1 > 4/a) \to 0 \quad as\ n \to \infty,$$

where the probability is taken over all choices of $S = \{g_1, \ldots, g_k\}$ of size

$$k > 2a\log_2 n$$

In other words, we get a constant expansion for $k = \Omega(\log|G|)$. In particular, this implies the $O(\log|G|)$ mixing time for random random walks (see [1, 8, 12]). Our technique is based on careful analysis of such random walks and is obtained as an application of the Erdős-Rényi results on random subproducts (see [10]).

Similar result for general groups was first explored by Alon and Roichman in [4], where authors considered *symmetric* Cayley graphs $\Gamma(G,S)$, $S = S^{-1}$. Clearly, transition matrix A has only real eigenvalues. The authors showed that when $k = \Omega(\log n)$ the second largest eigenvalue λ_2 of the random is bounded by a constant. Formally, they showed that given $1 > \delta > 1/e$, $k \geq (1+o(1))2e^4 \ln 2/(\delta e - 1)$ then $E(\lambda_2) < \delta$. Alon and Roichman's analysis relies on the Wigner's semicircle law and seem impossible to generalize to this case. Note also that our technique gives a bound for the case $\delta \leq 1/e$.

Finally, we should mention that there is a number of results on the mixing time on random Cayley graphs, rather than on relaxation times we consider in this paper (see e.g. [8, 12, 13, 14]). In [8, 13] authors consider the case when $k = \Omega(\log^a n)$, where $a > 1$. The analysis in [13] gives also bounds on eigenvalue gap in this case.

2 Random random walks

A *lazy random walk* $\mathcal{W} = \mathcal{W}(\mathcal{G}, \mathcal{S})$ is defined as a finite Markov chain X_t with state space G, and such that $X_0 = e$,

$$X_{t+1} = X_t \cdot g_i^{\varepsilon_i}$$

where $g_i = g_i(t)$ are independent and uniform in $[k] = \{1, \ldots, k\}$; ε_i are independent and uniform in $\{0, 1\}$. By Q^m denote the probability distribution of X_m. If S is a set of generators, then $Q^m(g) \to 1/|G|$, i.e. the walk $\mathcal{W}$ has a uniform stationary distribution U, $U(g) = 1/n$ for all $g \in G$.

Define the *separation distance*

$$\mathbf{s}(m) = |G| \max_{g \in G} \left(\frac{1}{|G|} - Q^m(g) \right)$$

It is easy to see that $0 \leq \mathbf{s}(m) \leq 1$. It is also known (see e.g. [2]) that $\mathbf{s}(m+1) \leq \mathbf{s}(m)$ for all $m > 0$, and $\mathbf{s}(m+l) \leq \mathbf{s}(m) \cdot \mathbf{s}(l)$.

The general problem is to find the smallest m such that $\mathbf{s}(m) \leq \varepsilon$ for almost all choices of S. Clearly, if m is small enough, then almost surely S is not a set of generators and $\mathbf{s}(m) = 1$, $\mathbf{d}(m) \geq 1/2$. The example of $G = \mathbb{Z}_2^r$ shows that if $k < r = \log_2 n$ this is the case. Thus it is reasonable to consider only the case $k \geq \log_2 n$.

It is not hard to see (see [3, 11]) that $\mathbf{s}(m) \sim (Re \lambda_1)^m$ as $m \to \infty$. Thus if we can prove an upper bound for the separation distance with high probability, this will give a desired eigenvalue bound.

3 Random subproducts

Let G be a finite group, $n = |G|$. Throughout the paper we will ignore a small difference between *random subsets* S and *random sequences* J of group elements. The reason is that the two concepts are virtually identical since probability of repetition of elements (having $g_i = g_j$, $1 \leq i < j \leq k$) when $k = O(\log n)$ is exponentially small. Thus in the future we will substitute uniform sets S of size k by the uniform sequences $J \in G^k$, which, of course, can have repeated elements.

Fix a sequence $J = (g_1, \ldots, g_k) \in G^k$. *Random subproducts* are defined as

$$g_1^{\varepsilon_1} \cdot \ldots \cdot g_k^{\varepsilon_k}$$

where $\varepsilon_i \in \{0,1\}$ are given by independent unbiased coin flips. Denote by P_J the probability distribution of the random subproducts on G. Erdős and Rényi showed in [10] that if $g_1, \ldots, g_k$ are chosen uniformly and independently, then :

$$(*)\ \mathbf{Pr}\left(\max_{g \in G}\left|P_J(g) - \frac{1}{n}\right| \le \frac{\varepsilon}{n}\right) > 1 - \delta \text{ for } k \ge 2\log_2 n + 2\log_2 1/\varepsilon + \log_2 1/\delta$$

Proofs of the Theorem is based on $(*)$.

Let $m > 2\log_2 |G|$, and let J be as above. Denote by Q_J the probability distribution Q_J^m of the lazy random walk $\mathcal{W}(\mathcal{G}, \mathcal{S})$ after m steps, where $S = S(J)$ is a set of elements in J. Suppose we can show that with probability $> 1 - \alpha/2$ we have $\mathbf{s}_J(m) = n \max_{g \in G}(1/n - Q_J^m(g)) \le \alpha/2$, where $\alpha \to 0$ as $n \to \infty$. This would imply the theorem. Indeed, we have

$$\begin{aligned} E[\mathbf{s}_J(m)] &\le \mathbf{Pr}(\mathbf{s}_J \le \alpha/2) \cdot \alpha/2 + \mathbf{Pr}(\mathbf{s}_J > \alpha/2) \cdot 1 \\ &< (1 - \alpha/2)\alpha/2 + \alpha/2 < \alpha \to 0 \end{aligned}$$

By definition, Q_J is distributed as random subproducts

$$g_{i_1}^{\varepsilon_1} \cdot \ldots \cdot g_{i_m}^{\varepsilon_m}$$

where $i_1, \ldots, i_m$ are uniform and independent in $[k] = \{1, \ldots, k\}$.

Let $J = (g_1, \ldots, g_k)$ be fixed. For a given $I = (i_1, \ldots, i_m) \in [k]^m$, consider $J(I) = (g_{i_1}, \ldots, g_{i_m})$ and $R_I = P_{J(I)}$. By definition of a lazy random walk we have

$$Q_J = \frac{1}{k^m} \sum_{I \in [k]^m} R_I$$

We will show that for almost all choices of J and I, the probability distribution R_I is almost uniform.

Let $I = (i_1, \ldots, i_m) \in [k]^m$ be a sequence. Define an **L**-*subsequence* $I' = (i_{r_1}, \ldots, i_{r_l})$ to satisfy $1 \le r_1 < \cdots < r_l \le m$, and for all j, $1 \le j \le m$, there exist a unique t, $1 \le t \le m$, such that $r_t \le j$ and $i_{r_t} = i_j$. In other words, we read numbers in I, and whenever we find a new number, we add it to I'. For example, if $I = (2,7,5,1,2,3,2,5,6)$, then $I' = (2,7,5,1,3,6)$ is an **L**-subsequence of length 6. Note that by definition **L**-subsequence is always unique.

Lemma 1. *Let I, J be as above, $n = |G|$. Let I' be a* **L**-*subsequence of I. Then for all $\alpha, \beta > 0$ we have $\max_{g \in G}|R_{I'}(g) - 1/n| \le \alpha/n$ with probability $1 - \beta$ implies $\max_{g \in G}|R_I(g) - 1/n| \le \alpha/n$ with probability $1 - \beta$.*

Lemma 2. *Let $\beta > 0$, $a > 1$, $k = al$. Consider the probability $P(l)$ a sequence $I \in [k]^m$ does not contain an* **L**-*subsequence I' of length l. Then $P(l) \sim 1/a^l$ as $l \to \infty$.*

4 Proof of the Theorem

First we deduce the Theorem from the lemmas and then prove the lemmas.

Proof of Theorem 1. Let I' be a **L**-subsequence of I of length $l > 2\log_2 n + 3\log_2 1/\delta$. Since numbers in I' are all different, for at least $(1-\delta)$ fraction of all $J = \{g_1, \dots, g_k\}$, we have

$$\max_{g \in G} \left| R_{I'}(g) - \frac{1}{n} \right| \leq \frac{\delta}{n}$$

Indeed, this is a restatement of $(*)$ with $\varepsilon = \delta$.

Note here that we do not require the actual group elements g_{i_j}, $i_j \in I'$ be different. By coincidence they can be the same. But we do require that *numbers* in I' are all different, so that the corresponding group elements are independent.

Let $l = \lceil 2\log_2 n + 3\log_2 1/\delta \rceil$, $k > al$, and $m > (1+\varepsilon)\, a l \ln \frac{a}{a-1}$. Denote by $P(l)$ the probability that a uniformly chosen $I \in [k]^m$ does not contain an **L**-subsequence of length l. By Lemma 1, with probability $> (1 - P(I))(1-\delta)$ we have

$$\max_{g \in G} \left(\frac{1}{n} - R_I(g) \right) \leq \frac{\delta}{n}$$

where the the probability is taken over all $I \in [k]^m$ and all $J \in G^k$. Setting $\delta = \delta(\alpha, \varepsilon, n)$ small enough we immediately obtain $\mathrm{s}_J(m) \leq \alpha/2$ with probability $> (1 - \alpha/2)$. where the the probability is taken over all $J \in G^k$. By observations above, this is exactly what we need to prove the theorem.

Now take $\delta = \alpha/4$, $\beta = \varepsilon/2$, $\alpha = 4/a^l$. By Lemma 2, and and since $l > \log_2 n$ we have $P(I) \sim \alpha/4$ for n large enough. We conclude $(1 - P(I))(1-\delta) > (1-\alpha/4)^2 > 1 - \alpha/2$. This finishes proof of Theorem 1. □

5 Proof of Lemmas

Proof of Lemma 1. For any $x, y \in G$ denote by y^x the element $xyx^{-1} \in G$. Clearly, if y is uniform in G and independent of x, then y^x is also uniform in G.

Let Q be a distribution on a group G which depends on $J \in G^m$ and takes values in G. We call Q (α, β)-*good* if with probability $> (1-\beta)$ it satisfies inequality $\max_{g \in G} |Q(g) - 1/n| \leq \alpha/n$.

Consider the following random subproducts:

$$h = g_1^{\varepsilon_1} \cdot \dots \cdot g_r^{\varepsilon_r} \cdot x \cdot g_{r+1}^{\varepsilon_{r+1}} \cdot \dots \cdot g_l^{\varepsilon_l}$$

where x is fixed, while $g_1, \dots, g_l$ are uniform and independent in G, and $\varepsilon_1, \dots, \varepsilon_l$ are uniform and independent in $\{0, 1\}$. We have

$$h = g_1^{\varepsilon_1} \cdot \dots \cdot g_r^{\varepsilon_r} \cdot (g_{r+1}^x)^{\varepsilon_{r+1}} \cdot \dots \cdot (g_l^x)^{\varepsilon_l} \cdot x$$

Thus $h \cdot x^{-1}$ is distributed as R_I, $I = (1, 2, \dots, l)$. Therefore if R_I is (α, β)-good, then distribution of h is also (α, β)-good.

Similarly, let $x, y, \ldots$ be fixed group elements. Then random subproducts

$$h = g_1^{\varepsilon_1} \cdot \ldots \cdot x \cdot g_r^{\varepsilon_r} \cdot \ldots \cdot y \cdot g_l^{\varepsilon_l} \cdot \ldots$$

are distributed as $R_I \cdot f(x, y, \ldots), I = (1, \ldots, r, \ldots, l, \ldots)$. Indeed, pull the rightmost fixed element all the way to the right, then pull the previous one, etc. We conclude that if R_I is (α, β)-good, then distribution of h is also (α, β)-good. Note that in the observation above we can relax a condition that the elements $x, y, \ldots$ are fixed. Since we do not have to change their relative order, it is enough to require that they are independent of the elements g_i to the right of them.

Now let $I = (i_1, \ldots, i_m) \in [k]^m$, and let I' be an **L**-subsequence of I. Define $Q(h)$ to be a distribution of random subproducts

$$h = g_{i_1}^{\varepsilon_1} \cdot \ldots \cdot g_{i_m}^{\varepsilon_m}$$

where all the powers ε_j are fixed except for those of $j \in I'$. We claim that if $R_{I'}$ is (α, β)-good, then $Q(h)$ is also (α, β)-good. Indeed, pull all the elements that are not in I' to the right. By definition of the **L**-subsequence, the elements in I' to the right of those that are not in I' must be different and thus independent of each other. Thus by the observation above $Q(h)$ is also (α, β)-good.

Now, the distribution R_I is defined as an average of the distributions $Q(h)$ over all of the 2^{m-l} choices of values ε_s of elements not in $I' = \left(i_{r_1}, \ldots, i_{r_l}\right)$. Observe that for fixed $g_1, \ldots, g_k$ and different choices the ε_s, $s \neq r_j$ the distributions of subproducts h can be obtained by a shift from each other (i.e. by multiplication on a fixed group element). Therefore each of these distributions has the same separation distance. In other words, each of the J is either "good" altogether or "bad" altogether for all 2^{m-l} choices. Therefore after averaging we obtain an (α, β)-good distribution R_I. This finishes proof of the lemma. □

Proof of Lemma 2. The problem is equivalent to the following question. What is the asymptotic behavior of the probability that in the usual coupon collector's problem with k coupons, after m trials we have at least l different coupons? Indeed, observe that if all m chosen coupons correspond to elements in a sequence $I \in [k]^m$, then *distinct* coupons correspond to **L**-subsequence I' of length l. Note that in our case $k = al$ and $m \to \infty$.

Let τ be the first time we collect l out of k possible coupons. Let us compute the expected time $E(\tau)$. We have

$$E(\tau) = \frac{k}{k} + \frac{k}{k-1} + \frac{k}{k-2} + \cdots + \frac{k}{k-l+1} = k(\ln k - \ln(k-l) + o(1))$$

When $k = al$. We obtain

$$E(\tau) = al\left(\log \frac{a}{a-1} + o(1)\right)$$

Now let $m = (1+\beta)E(\tau)$. The probability $1 - P(l)$ that after m trials we collect l coupons is equal to $\mathbf{Pr}(\tau \leq m)$. Now use a geometric bound on the distribution to obtain the result. We skip the easy details. □

6 Acknowledgements

We would like to thank Gene Cooperman, Martin Hildebrand, László Lovász and Van Vu for helpful remarks.

The author was supported in part by the NSF Postdoctoral Research Fellowship.

References

[1] D. Aldous, P. Diaconis: *Shuffling cards and stopping times*, Amer. Math. Monthly, **93** 1986, pp 333–348.

[2] D. Aldous, P. Diaconis: *Strong uniform times and finite random walks*, Adv. Appl. Math., **8** 1987, pp 69–97.

[3] D. Aldous, J. Fill: *Reversible Markov Chains and Random Walks on Graphs* monograph in preparation 1996.

[4] N. Alon, Y. Roichman: *Random Cayley graphs and expanders*, Rand. Str. Alg., **5** 1994, pp 271–284.

[5] L. Babai: *Local expansion of vertex-transitive graphs and random geneartion in finite groups*, in Proc 23rd ACM STOC 1991, pp 164–174.

[6] L. Babai: *Automorphism groups, isomorphism, reconstruction*, in Handbook of Combinatorics (R. L. Graham et al., eds.), Elsevier 1996.

[7] P. Diaconis: *Group Representations in Probability and Statistics*, IMS, Hayward, California 1988.

[8] C. Dou, M. Hildebrand: *Enumeration and random random walks on finite groups*, Ann. Prob., **24** 1996, pp 987–1000.

[9] P. Erdős, R.R. Hall: *Probabilistic methods in group theory. II*, Houston J. Math., **2** 1976, pp 173–180.

[10] P. Erdős, A. Rényi: *Probabilistic methods in group theory*, Jour. Analyse Mathématique, **14** 1965, pp 127–138.

[11] I. Pak: *Random walks on groups: strong uniform time approach*, Ph.D. Thesis, Harvard U. 1997.

[12] I. Pak: *Random walks on finite groups with few random generators*, Electronic J. Prob., **4** 1999, pp 1–11.

[13] Y. Roichman: *On random random walks*, Ann. Prob.,**24** 1996, pp 1001–1011.

[14] D. Wilson: *Random random walks on* $\mathbb{Z}_2^d$, Prob. Th. Rel. Fields, **108** 1997, pp 441–457.

A Fully Dynamic Algorithm for Recognizing and Representing Proper Interval Graphs

Pavol Hell[1], Ron Shamir[2], and Roded Sharan[2]

[1] School of Computing Science, Simon Fraser University, Burnaby, B.C., Canada V5A1S6.
pavol@cs.sfu.ca.
[2] Department of Computer Science, Tel Aviv University, Tel Aviv, Israel.
{shamir,roded}@math.tau.ac.il.

Abstract. In this paper we study the problem of recognizing and representing dynamically changing proper interval graphs. The input to the problem consists of a series of modifications to be performed on a graph, where a modification can be a deletion or an addition of a vertex or an edge. The objective is to maintain a representation of the graph as long as it remains a proper interval graph, and to detect when it ceases to be so. The representation should enable one to efficiently construct a realization of the graph by an inclusion-free family of intervals. This problem has important applications in physical mapping of DNA.

We give a near-optimal fully dynamic algorithm for this problem. It operates in time $O(\log n)$ per edge insertion or deletion. We prove a close lower bound of $\Omega(\log n/(\log\log n + \log b))$ amortized time per operation in the cell probe model with word-size b. We also construct optimal incremental and decremental algorithms for the problem, which handle each edge operation in $O(1)$ time.

1 Introduction

A graph G is called an *interval graph* if its vertices can be assigned to intervals on the real line so that two vertices are adjacent in G iff their intervals intersect. The set of intervals assigned to the vertices of G is called a *realization* of G. If the set of intervals can be chosen to be inclusion-free, then G is called a *proper interval graph*. Proper interval graphs have been studied extensively in the literature (cf. [7, 13]), and several linear time algorithms are known for their recognition and realization [2, 3].

This paper deals with the problem of recognizing and representing dynamically changing proper interval graphs. The input is a series of operations to be performed on a graph, where an operation is any of the following: Adding a vertex (along with the edges incident to it), deleting a vertex (and the edges incident to it), adding an edge and deleting an edge. The objective is to maintain a representation of the dynamic graph as long as it is a proper interval graph, and to detect when it ceases to be so. The representation should enable one to efficiently construct a realization of the graph. In the *incremental* version of the problem, only addition operations are permitted, i.e., the operations include only the addition of a vertex and the addition of an edge. In the *decremental* version of the problem only deletion operations are allowed.

The motivation for this problem comes from its application to *physical mapping* of DNA [1]. Physical mapping is the process of reconstructing the relative position of DNA fragments, called *clones*, along the target DNA molecule, prior to their sequencing, based on information about their pairwise overlaps. In some biological frameworks the set of clones is virtually inclusion-free - for example when all clones have a similar length (this is the case for instance for cosmid clones). In this case, the physical mapping problem can be modeled using proper interval graphs as follows. A graph G is built according to the biological data. Each clone is represented by a vertex and two vertices are adjacent iff their corresponding clones overlap. The physical mapping problem then translates to the problem of finding a realization of G, or determining that none exists.

Had the overlap information been accurate, the two problems would have been equivalent. However, some biological techniques may occasionally lead to an incorrect conclusion about whether two clones intersect, and additional experiments may change the status of an intersection between two clones. The resulting changes to the corresponding graph are the deletion of an edge, or the addition of an edge. The set of clones is also subject to changes, such as adding new clones or deleting 'bad' clones (such as chimerics [14]). These translate into addition or deletion of vertices in the corresponding graph. Therefore, we would like to be able to dynamically change our graph, so as to reflect the changes in the biological data, as long as they allow us to construct a map, i.e., as long as the graph remains a proper interval graph.

Several authors have studied the problem of dynamically recognizing and representing certain graph families. Hsu [10] has given an $O(m+n\log n)$-time incremental algorithm for recognizing interval graphs. (Throughout, we denote the number of vertices in the graph by n and the number of edges in it by m.) Deng, Hell and Huang [3] have given a linear-time incremental algorithm for recognizing and representing connected proper interval graphs This algorithm requires that the graph will remain connected throughout the modifications. In both algorithms [10, 3] only vertex increments are handled. Recently, Ibarra [11] found a fully dynamic algorithm for recognizing chordal graphs, which handles each edge operation in $O(n)$ time, or alternatively, an edge deletion in $O(n\log n)$ time and an edge insertion in $O(n/\log n)$ time.

Our results are as follows: For the general problem of recognizing and representing proper interval graphs we give a fully dynamic algorithm which handles each operation in time $O(d+\log n)$, where d denotes the number of edges involved in the operation. Thus, in case a vertex is added or deleted, d equals its degree, and in case an edge is added or deleted, $d = 1$. Our algorithm builds on the representation of proper interval graphs given in [3]. We also prove a lower bound for this problem of $\Omega(\log n/(\log\log n+\log b))$ amortized time per edge operation in the cell probe model of computation with word-size b [16]. It follows that our algorithm is nearly optimal (up to a factor of $O(\log\log n)$).

For the incremental and the decremental versions of the problem we give optimal algorithms (up to a constant factor) which handle each operation in time $O(d)$. For the incremental problem this generalizes the result of [3] to arbitrary instances.

As a part of our general algorithm we give a fully dynamic procedure for maintaining connectivity in proper interval graphs. The procedure receives as input a sequence of operations each of which is a vertex addition or deletion, an edge addition or deletion, or a query whether two vertices are in the same connected component. It is assumed that the graph remains proper interval throughout the modifications, since otherwise our main algorithm detects that the graph is no longer a proper interval graph and halts. We show how to implement this procedure in $O(\log n)$ time per operation. In comparison, the best known algorithms for maintaining connectivity in general graphs require $O(\log^2 n)$ amortized time per operation [9], or $O(\sqrt{n})$ worst-case (deterministic) time per operation [4]. We also show that the lower bound of Fredman and Henzinger [5] of $\Omega(\log n/(\log\log n + \log b))$ amortized time per operation (in the cell probe model with word-size b) for maintaining connectivity in general graphs, applies to the problem of maintaining connectivity in proper interval graphs.

The paper is organized as follows: In section 2 we give the basic background and describe our representation of proper interval graphs and the realization it defines. In sections 3 and 4 we present the incremental algorithm. In section 5 we extend the incremental algorithm to a fully dynamic algorithm for proper interval graph recognition and representation. We also derive an optimal decremental algorithm. In section 6 we give a fully dynamic algorithm for maintaining connectivity in proper interval graphs. Finally, in section 7 we prove a lower bound on the amortized time per operation of a fully dynamic algorithm for recognizing proper interval graphs. For lack of space, some of the proofs and some of the algorithmic details are omitted.

2 Preliminaries

Let $G = (V,E)$ be a graph. We denote its set V of vertices also by $V(G)$ and its set E of edges also by $E(G)$. For a vertex $v \in V$ we define $N(v) := \{u \in V : (u,v) \in E\}$ and $N[v] := N(v) \cup \{v\}$. Let R be an equivalence relation on V defined by uRv iff $N[u] = N[v]$. Each equivalence class of R is called a *block* of G. Note that every block of G is a complete subgraph of G. The *size* of a block is the number of vertices in it. Two blocks A and B are *neighbors* in G if some (and hence all) vertices $a \in A, b \in B$, are adjacent in G. A *straight enumeration* of G is a linear ordering Φ of the blocks in G, such that for every block, the block and its neighboring blocks are consecutive in Φ.

Let $\Phi = B_1 < \ldots < B_l$ be an ordering of the blocks of G. For any $1 \le i < j \le l$, we say that B_i is ordered *to the left of* B_j, and that B_j is ordered *to the right of* B_i. A *chordless cycle* is an induced cycle of length greater than 3. A *claw* is an induced $K_{1,3}$. A graph is *claw-free* if it does not contain an induced claw. For basic definitions in graph theory see, e.g., [7].

The following are some useful facts about interval and proper interval graphs.

Theorem 1. *([12]) An interval graph contains no chordless cycle.*

Theorem 2. *([15]) A graph is a proper interval graph iff it is interval and claw-free.*

Theorem 3. *([3]) A graph is a proper interval graph iff it has a straight enumeration.*

Lemma 4 ("The umbrella property"). *Let Φ be a straight enumeration of a connected proper interval graph G. If A, B and C are blocks of G, such that $A < B < C$ in Φ and A is adjacent to C, then B is adjacent to A and to C (see figure 1).*

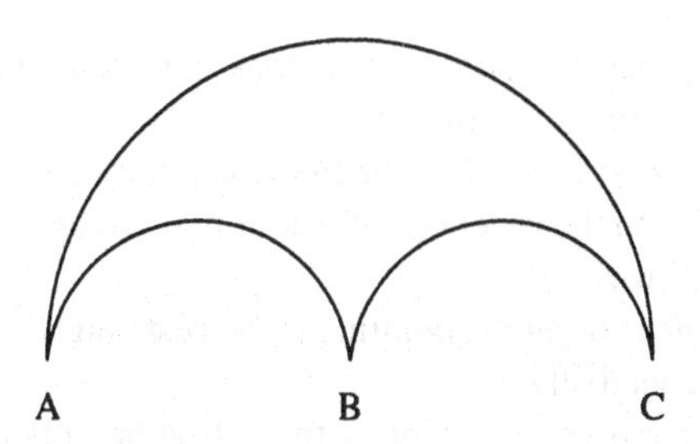

Fig. 1. The umbrella property

Let G be a connected proper interval graph and let Φ be a straight enumeration of G. It is shown in [3] that a connected proper interval graph has a unique straight enumeration up to its full reversal. Define the *out-degree* of a block B w.r.t. Φ, denoted by $o(B)$, as the number of neighbors of B which are ordered to its right in Φ.

We shall use the following representation: For each connected component of the dynamic graph we maintain a straight enumeration (in fact, for technical reasons we shall maintain both the enumeration and its reversal). The details of the data structure containing this information will be described below.

This information implicitly defines a realization of the dynamic graph (cf. [3]) as follows: Assign to each vertex in block B_i the interval $[i, i+o(B_i)+1-\frac{1}{i}]$. The out-degrees and hence the realization of the graph can be computed from our data structure in time $O(n)$.

3 An Incremental Algorithm for Vertex Addition

In the following two sections we describe an optimal incremental algorithm for recognizing and representing proper interval graphs. The algorithm receives as input a series of addition operations to be performed on a graph. Upon each operation the algorithm updates its representation of the graph and halts if the current graph is no longer a proper interval graph. The algorithm handles each operation in time $O(d)$, where d denotes the number of edges involved in the operation. It is assumed that initially the graph is empty, or alternatively, that the representation of the initial graph is known.

A *contig* of a connected proper interval graph G is a straight enumeration of G. The first and the last blocks of a contig are called *end-blocks*. The rest of the blocks are called *inner-blocks*.

As mentioned above, each component of the dynamic graph has exactly two contigs (which are full reversals of each other) and both are maintained by the algorithm. Each operation involves updating the representation. (In the sequel we concentrate on describing only one of the two contigs for each component. The second contig is updated in a similar way.)

3.1 The Data Structure

The following data is kept and updated by the algorithm:

1. For each vertex we keep the name of the block to which it belongs.
2. For each block we keep the following:
   (a) An *end* pointer which is null if the block is not an end-block of its contig, and otherwise points to the other end-block of that contig.
   (b) The *size* of the block.
   (c) Left and right *near* pointers, pointing to nearest neighbor blocks on the left and on the right respectively.
   (d) Left and right *far* pointers, pointing to farthest neighbor blocks on the left and on the right respectively.
   (e) Left and right *self* pointers, pointing to the block.
   (f) A counter.

In the following we shall omit details about the obvious updates to the name of the block of a vertex and to the size of a block.

During the execution of the algorithm we may need to update many far pointers pointing to a certain block, so that they point to another block. In order to be able to do that in $O(1)$ time we use the technique of *nested pointers*: We make the far pointers point to a *location* whose content is the address of the block to which the far pointers should point. The role of this special location will be served by our self-pointers. The value of the left and right self-pointers of B is always the address of B. When we say that a certain left (right) far pointer points to B, we mean that it points to a left (right) self-pointer of B. Let A and B be blocks. In order to change all left (right) far pointers pointing to A so that they point to B, we require that no left (right) far pointer points to B. If this is the case, we simply *exchange* the left (right) self-pointer of A with the left (right) self-pointer of B. This means that: (1) The previous left (right) self-pointer of A is made to point to B, and the algorithm records it as the new left (right) self-pointer of B; (2) The previous left (right) self-pointer of B is made to point to A, and the algorithm records it as the new left (right) self-pointer of A.

We shall use the following notation: For a block B we denote its address in the memory by $\&B$. When we set a far pointer to point to a left or to a right self-pointer of B we will abbreviate and set it to $\&B$. We denote the left and right near pointers of B by $N_l(B)$ and $N_r(B)$ respectively. We denote the left and right far pointers of B by $F_l(B)$ and $F_r(B)$ respectively. We denote its end pointer by $E(B)$. In the sequel we often refer to blocks by their addresses. For example, if A and B are blocks, and $N_r(A) = \&B$, we sometimes refer to B by $N_r(A)$. When it is clear from the context, we also use a name of a block to denote any vertex in that block. Given a contig Φ we denote its reversal by Φ^R. In general when performing an operation, we denote the graph before the operation is carried out by G, and the graph after the operation is carried out by G'.

3.2 The Impact of a New Vertex

In the following we describe the changes made to the representation of the graph in case G' is formed from G by the addition of a new vertex v of degree d. We also give some necessary and some sufficient conditions for deciding whether G' is proper interval.

Let B be a block of G. We say that v is *adjacent* to B if v is adjacent to some vertex in B. We say that v is *fully adjacent* to B if v is adjacent to *every* vertex in B. We say that v is *partially adjacent* to B if v is adjacent to B but not fully adjacent to B.

The following lemmas characterize, assuming that G' is proper interval, the adjacencies of the new vertex.

Lemma 5. *If G' is a proper interval graph then v can have neighbors in at most two connected components of G.*

Lemma 6. *[3] Let C be a connected component of G containing neighbors of v. Let $B_1 < \ldots < B_k$ be a contig of C. Assume that G' is proper interval and let $1 \leq a < b < c \leq k$. Then the following properties are satisfied:*

1. *If v is adjacent to B_a and to B_c, then v is fully adjacent to B_b.*
2. *If v is adjacent to B_b and not fully adjacent to B_a and to B_c, then B_a is not adjacent to B_c.*
3. *If $b = a+1, c = b+1$ and v is adjacent to B_b, then v is fully adjacent to B_a or to B_c.*

One can view a contig Φ of a connected proper interval graph C as a weak linear order $<_\Phi$ on the vertices of C, where $x <_\Phi y$ iff the block containing x is ordered in Φ to the left of the block containing y. We say that Φ' is a *refinement* of Φ if for every $x, y \in V(C)$, $x <_\Phi y$ implies $x <_{\Phi'} y$ (since a contig can be reversed, we also allow complete reversal of Φ).

Lemma 7. *If G is a connected induced subgraph of a proper interval graph G', Φ is a contig of G and Φ' is a straight enumeration of G', then Φ' is a refinement of Φ.*

Note, that whenever v is partially adjacent to a block B in G, then the addition of v will cause B to split into two blocks of G', namely $B \setminus N(v)$ and $B \cap N(v)$. Otherwise, if B is a block of G to which v is either fully adjacent or not adjacent, then B is also a block of G'.

Corollary 8. *If B is a block of G to which v is partially adjacent, then $B \setminus N(v)$ and $B \cap N(v)$ occur consecutively in a straight enumeration of G'.*

Lemma 9. *Let C be a connected component of G containing neighbors of v. Let the set of blocks in C which are adjacent to v be $\{B_1, \ldots, B_k\}$. Assume that in a contig of C, $B_1 < \ldots < B_k$. If G' is proper interval then the following properties are satisfied:*

1. *$B_1, \ldots, B_k$ are consecutive in C.*
2. *If $k \geq 3$ then v is fully adjacent to $B_2, \ldots, B_{k-1}$.*
3. *If v is adjacent to a single block B_1 in C, then B_1 is an end-block.*
4. *If v is adjacent to more than one block in C and has neighbors in another component, then B_1 is adjacent to B_k, and one of B_1 or B_k is an end-block to which v is fully adjacent, while the other is an inner-block.*

Proof. Claims 1 and 2 follow directly from part 1 of Lemma 6. Claim 3 follows from part 3 of Lemma 6. To prove the last part of the lemma let us denote the other component containing neighbors of v by D. Examine the induced connected subgraph H of G whose set of vertices is $V(H) = \{v\} \cup V(C) \cup V(D)$. H is proper interval as an induced subgraph of G. It is composed of three types of blocks: Blocks whose vertices are from $V(C)$, which we will call henceforth C-blocks; blocks whose vertices are from $V(D)$, which we will call henceforth D-blocks; and $\{v\}$ which is a block of H since $H \setminus \{v\}$ is not connected. All blocks of C remain intact in H, except B_1 and B_k which might split into $B_j \setminus N(v)$ and $B_j \cap N(v)$, for $j = 1,k$.

Surely in a contig of H, C-blocks must be ordered completely before or completely after D-blocks. Let Φ denote a contig of H, in which C-blocks are ordered before D-blocks. Let X denote the rightmost C-block in Φ. By the umbrella property, $X < \{v\}$ and moreover, X is adjacent to v. By Lemma 7, Φ is a refinement of a contig of C. Hence, $X \subseteq B_1$ or $X \subseteq B_k$ (more precisely, $X = B_1 \cap N(v)$ or $X = B_k \cap N(v)$). Therefore, one of B_1 or B_k is an end-block.

W.l.o.g. $X \subseteq B_k$. Suppose to the contrary that v is not fully adjacent to B_k. Then by Lemma 7 we have $B_{k-1} \cap N(v) < B_k \setminus N(v) < \{v\}$ in Φ, contradicting the umbrella property. B_1 must be adjacent to B_k, or else G' contains a claw consisting of v, B_1, B_k and a vertex from $V(D) \cap N(v)$. It remains to show that B_1 is an inner-block. Suppose it is an end block. Since B_1 and B_k are adjacent, C contains a single block B_1, a contradiction. Thus, claim 4 is proved.■

3.3 The Algorithm

In our algorithm we rely on the incremental algorithm of Deng, Hell and Huang [3], which we call henceforth the *DHH algorithm*. This algorithm handles the insertion of a new vertex into a graph in $O(d)$ time, provided that all its neighbors are in the same connected component, changing the straight enumeration of this component appropriately. We refer the reader to [3] for more details.

We perform the following upon a request for adding a new vertex v. For each neighbor u of v we add one to the count of the block containing u. We call a block *full* if its counter equals its size, *empty* if its counter equals zero, and *partial* otherwise. In order to find a set of consecutive blocks which contain neighbors of v, we pick arbitrarily a neighbor of v and march down the enumeration of blocks to the left using the left near neighbor pointers. We continue till we hit an empty block or till we reach the end of the contig. We do the same to the right and this way we discover a maximal sequence of nonempty blocks in that component which contain neighbors of v. We call this maximal sequence a *segment*. Only the two extreme blocks of the segment are allowed to be partial or else we fail (by Lemma 9(2)).

If the segment we found contains all neighbors of v then we can use the DHH algorithm in order to insert v into G, updating our internal data structure accordingly. Otherwise, by Lemmas 5 and 9(1) there could be only one more segment which contains neighbors of v. In that case, exactly one extreme block in each segment is an end-block to which v is fully adjacent (if the segment contains more than one block), and the two extreme blocks in each segment are adjacent, or else we fail (by Lemma 9(3,4)).

We proceed as above to find a second segment containing neighbors of v. We can make sure that the two segments are from two different contigs by checking that their end-blocks do not point to each other. We also check that conditions 3 and 4 in Lemma 9 are satisfied. If the two segments do not cover all neighbors of v, we fail.

If v is adjacent to vertices in two distinct components C and D, then we should merge their contigs. Let $\Phi = B_1 < \ldots < B_k$, Φ^R be the two contigs of C. Let $\Psi = B'_1 < \ldots < B'_l$, Ψ^R be the two contigs of D. The way the merge is performed depends on the blocks to which v is adjacent. If v is adjacent to B_k and to B'_1, then by the umbrella property the two new contigs (up to refinements described below) are $\Phi < \{v\} < \Psi$ and $\Psi^R < \{v\} < \Phi^R$. In the following we describe the necessary changes to our data structure in case these are the new contigs. The three other cases are handled similarly.

- Block enumeration: We merge the two enumerations of blocks and put a new block $\{v\}$ in-between the two contigs. Let the leftmost block adjacent to v in the new ordering $\Phi < \{v\} < \Psi$ be B_i and let the rightmost block adjacent to v be B'_j. If B_i is partial we split it into two blocks $\hat{B}_i = B_i \setminus N(v)$ and $B_i = B_i \cap N(v)$ in this order. If B'_j is partial we split it into two blocks $B'_j = B'_j \cap N(v)$ and $\hat{B}'_j = B'_j \setminus N(v)$ in this order.
- End pointers: We set $E(B_1) = E(B'_1)$ and $E(B'_l) = E(B_k)$. We then nullify the end pointers of B_k and B'_1.
- Near pointers: We update $N_l(\{v\}) = \&B_k, N_r(\{v\}) = \&B'_1, N_r(B_k) = \&\{v\}$ and $N_l(B'_1) = \&\{v\}$. Let $B_0 = \emptyset$. In case B_i was split we update $N_r(\hat{B}_i) = \&B_i, N_l(B_i) = \&\hat{B}_i, N_l(\hat{B}_i) = \&B_{i-1}$ and $N_r(B_{i-1}) = \&\hat{B}_i$. Similar updates are made in case B'_j was split to the near pointers of $B'_j, \hat{B}'_j$ and B'_{j+1}.
- Far pointers: If B_i was split we set $F_l(\hat{B}_i) = F_l(B_i), F_r(\hat{B}_i) = \&B_k$ and exchange the left self-pointer of B_i with the left self-pointer of $\hat{B}_i$. If B'_j was split we set $F_r(\hat{B}'_j) = F_r(B'_j), F_l(\hat{B}'_j) = \&B'_1$ and exchange the right self-pointer of B'_j with the right self-pointer of $\hat{B}'_j$. In addition, we set all right far pointers of $B_i, B_{i+1} \dots, B_k$ and all left far pointers of $B'_1, \dots, B'_{j-1}, B'_j$ to $\&\{v\}$ (in $O(d)$ time). Finally, we set $F_l(\{v\}) = \&B_i$ and $F_r(\{v\}) = \&B'_j$.

4 An Incremental Algorithm for Edge Addition

In this section we show how to handle the addition of a new edge (u,v) in $O(1)$ time. We characterize the cases for which $G' = G \cup \{(u,v)\}$ is proper interval and show how to efficiently detect them, and how to update our representation of the graph.

Lemma 10. *If u and v are in distinct components in G, then G' is proper interval iff u and v were in end-blocks of their respective contigs.*

Proof. To prove the 'only if' part let us examine the graph $H = G' \setminus \{u\} = G \setminus \{u\}$. H is proper interval as an induced subgraph of G. If G' is proper interval, then by Lemma 9(3) v must be in an end-block of its contig, since u is not adjacent to any other vertex in the component containing v. The same argument applies to u.

To prove the 'if' part we give a straight enumeration of the new connected component containing u and v in G'. Denote by C and D the components containing u and v respectively. Let $B_1 < \dots < B_k$ be a contig of C, such that $u \in B_k$. Let $B'_1 < \dots < B'_l$ be a contig of D, such that $v \in B'_1$. Then $B_1 < \dots < B_k \setminus \{u\} < \{u\} < \{v\} < B'_1 \setminus \{v\} < \dots < B'_l$ is a straight enumeration of the new component.■

We can check in $O(1)$ time if u and v are in end-blocks of distinct contigs. If this is the case, we update our data structure according to the straight enumeration given in the proof of Lemma 10 in $O(1)$ time.

It remains to handle the case where u and v were in the same connected component C in G. If $N(u) = N(v)$ then by the umbrella property it follows that C contains only three blocks which are merged into a single block in G'. In this case G' is proper interval and updates to the internal data structure are trivial. The following lemma analyses the case where $N(u) \neq N(v)$.

Lemma 11. *Let $B_1 < \ldots < B_k$ be a contig of C, such that $u \in B_i$ and $v \in B_j$ for some $1 \leq i < j \leq k$. Assume that $N(u) \neq N(v)$. Then G' is proper interval iff $F_r(B_i) = B_{j-1}$ and $F_l(B_j) = B_{i+1}$ in G.*

Proof. To prove the 'only if' part assume that G' is proper interval. Since B_i and B_j are not adjacent, $F_r(B_i) \leq B_{j-1}$ and $F_l(B_j) \geq B_{i+1}$. Suppose to the contrary that $F_r(B_i) < B_{j-1}$. Let $z \in B_{j-1}$. If in addition $F_l(B_j) = B_{i+1}$ then $N[v] \supset N[z]$ (this is a strict containment). As v and z are in distinct blocks, there exists a vertex $b \in N[v] \setminus N[z]$. But then, v, b, z, u induce a claw in G', a contradiction. Hence, $F_l(B_j) > B_{i+1}$ and so $F_r(B_{i+1}) < B_j$. Let $x \in B_{i+1}$ and let $y \in F_r(B_{i+1})$. Since u and x are in distinct blocks, either $(u,y) \notin E(G)$ or there is a vertex $a \in N[u] \setminus N[x]$ (or both). In the first case, v, u, x, y and the vertices of the shortest path from y to v induce a chordless cycle in G'. In the second case u, a, x, v induce a claw in G'. Hence, in both cases we arrive at a contradiction. The proof that $F_l(B_j) = B_{i+1}$ is symmetric.

To prove the 'if' part we shall provide a straight enumeration of $C \cup \{u,v\}$. If $B_i = \{u\}$, $F_r(B_{j-1}) = F_r(B_j)$ and $F_l(B_{j-1}) = B_i$ (i.e., $N[v] = N[B_{j-1}]$ in G'), we move v from B_j to B_{j-1}. Similarly, if B_j contained only v, $F_l(B_{i+1}) = F_l(B_i)$ and $F_r(B_{i+1}) = B_j$ (i.e., $N[u] = N[B_{i+1}]$ in G'), we move u from B_i to B_{i+1}. If u was not moved and $B_i \supset \{u\}$, we split B_i into $B_i \setminus \{u\}, \{u\}$ in this order. If v was not moved and $B_j \supset \{v\}$, we split B_j into $\{v\}, B_j \setminus \{v\}$ in this order. It is easy to see that the result is a straight enumeration of $C \cup \{u,v\}$. ∎

We can check in $O(1)$ time if the condition in Lemma 11 holds. If this is the case, we change our data structure so as to reflect the new straight enumeration given in the proof of Lemma 11. This can be done in $O(1)$ time, in a similar fashion to the update technique described in Section 3.3. The details are omitted here. The following theorem summarizes the results of Sections 3 and 4.

Theorem 12. *The incremental proper interval graph representation problem is solvable in $O(1)$ time per added edge.*

5 The Fully Dynamic Algorithm

In this section we give a fully dynamic algorithm for recognizing and representing proper interval graphs. The algorithm performs each operation in $O(d + \log n)$ time, where d denotes the number of edges involved in the operation. It supports four types of operations: Adding a vertex, adding an edge, deleting a vertex and deleting an edge. It is based on the same ideas used in the incremental algorithm. The main difficulty in extending the incremental algorithm to handle all types of operations, is updating the end pointers of blocks when deletions are allowed. To bypass this problem we do not keep end pointers at all. Instead, we maintain the connected components of G, and use this information in our algorithm. In the next section we show how to maintain the connected components of G in $O(\log n)$ time per operation. We describe below how each operation is handled by the algorithm.

5.1 The Addition of a Vertex or an Edge

These operations are handled in essentially the same way as done by the incremental algorithm. However, in order to check if the end-blocks of two segments are in distinct components, we query our data structure of connected components (in $O(\log n)$ time). Similarly, in order to check if the endpoints of an added edge are in distinct components, we check if their corresponding blocks are in distinct components (in $O(\log n)$ time).

5.2 The Deletion of a Vertex

We show next how to update the contigs of G after deleting a vertex v of degree d. Note that G' is proper interval as an induced subgraph of G. Denote by X the block containing v. If $X \supset \{v\}$, then the only change needed is to delete v. We hence concentrate on the case that $X = \{v\}$. We can find in $O(d)$ time the segment of blocks which includes X and all its neighbors. Let the contig containing X be $B_1 < \ldots < B_k$ and let the blocks of the segment be $B_i < \ldots < B_j$, where $X = B_l$ for some $1 \leq i \leq l \leq j \leq k$. Let $B_0 = \emptyset, B_{k+1} = \emptyset$. We make the following updates:

- Block enumeration: If $1 < i < l$, we check whether B_i can be merged with B_{i-1}. If $F_l(B_i) = F_l(B_{i-1}), F_r(B_i) = B_l$ and $F_r(B_{i-1}) = B_{l-1}$, we merge them by moving all vertices from B_i to B_{i-1} (in $O(d)$ time) and deleting B_i. If $l < j < k$ we act similarly w.r.t. B_j and B_{j+1}. Finally, we delete B_l. If $1 < l < k$ and B_{l-1}, B_{l+1} are non-adjacent, then by the umbrella property they are no longer in the same connected component, and the contig should be split into two contigs, one ending at B_{l-1} and one beginning at B_{l+1}.
- Near pointers: If B_i and B_{i-1} were merged, we update $N_r(B_{i-1}) = \&B_{i+1}$ and $N_l(B_{i+1}) = \&B_{i-1}$. Similar updates should be made w.r.t. B_{j-1} and B_{j+1} in case B_j and B_{j+1} were merged. If the contig is split, we nullify $N_r(B_{l-1})$ and $N_l(B_{l+1})$. Otherwise, we update $N_r(B_{l-1}) = \&B_{l+1}$ and $N_l(B_{l+1}) = \&B_{l-1}$.
- Far pointers: If B_i and B_{i-1} were merged, we exchange the right self-pointer of B_i with the right self-pointer of B_{i-1}. Similar changes should be made w.r.t. B_j and B_{j+1}. We also set all right far pointers previously pointing to B_l, to $\&B_{l-1}$; and all left far pointers previously pointing to B_l, to $\&B_{l+1}$ (in $O(d)$ time).

Note that these updates take $O(d)$ time and require no knowledge about the connected components of G.

5.3 The Deletion of an Edge

Let (u,v) be an edge of G to be deleted. Let C denote the connected component of G containing u and v, and let $B_1 < \ldots < B_k$ be a contig of C. If $k = 1$ then B_1 is split into $\{u\}, B_1 \setminus \{u,v\}$ and $\{v\}$, resulting in a straight enumeration of G'. Updates are trivial in this case. If $N[u] = N[v]$ then one can show that $G' = G \setminus \{(u,v)\}$ is a proper interval graph iff C was a clique, so again $k = 1$. We assume henceforth that $k > 1$ and $N(u) \neq N(v)$.

W.l.o.g. $i < j$. If $B_i = \{u\}, B_j = \{v\}, j = i+1$ and B_i, B_j were far neighbors of each other, then we should split the contig into two contigs, one ending at B_i and the other beginning at B_j. Otherwise, updates to the straight enumeration are derived from the following lemma.

Lemma 13. *Let $B_1 < \ldots < B_k$ be a contig of C, such that $u \in B_i$ and $v \in B_j$ for some $1 \leq i < j \leq k$. Assume that $N(u) \neq N(v)$. Then G' is proper interval iff $F_r(B_i) = B_j$ and $F_l(B_j) = B_i$ in G.*

Proof. Assume that G' is proper interval. We will show that $F_r(B_i) = B_j$. The proof that $F_l(B_j) = B_i$ is symmetric. Since B_i and B_j are adjacent in G, $F_r(B_i) \geq B_j$. Suppose to the contrary that $F_r(B_i) > B_j$. Let $x \in F_r(B_i)$. Since x and v are in distinct blocks, either there is a vertex $a \in N[v] \setminus N[x]$ or there is a vertex $b \in N[x] \setminus N[v]$ (or both). In the first case, by the umbrella property $(a,u) \in E(G)$ and therefore u, x, v, a induce a chordless cycle in G'. In the second case, x, b, u, v induce a claw in G'. Hence, in both cases we arrive at a contradiction.

To prove the opposite direction we give a straight enumeration of $C \setminus \{(u,v)\}$. If $B_j = \{v\}, F_l(B_{i-1}) = F_l(B_i)$ and $F_r(B_{i-1}) = B_{j-1}$ (i.e., $N[u] = N[B_{i-1}]$ in G'), we move u into B_{i-1}. If B_i contained only u, $F_r(B_{j+1}) = F_r(B_j)$ and $F_l(B_{j+1}) = B_{i+1}$ (i.e., $N[v] = N[B_{j+1}]$ in G'), we move v into B_{j+1}. If u was not moved and $B_i \supset \{u\}$, then B_i is split into $\{u\}, B_i \setminus \{u\}$ in this order. If v was not moved and $B_j \supset \{v\}$, then B_j is split into $B_j \setminus \{v\}, \{v\}$ in this order. The result is a contig of $C \setminus \{(u,v)\}$.■

If the conditions of Lemma 13 are fulfilled, one has to update the data structure according to its proof. These updates require no knowledge about the connected components of G, and it can be shown that they take $O(1)$ time. Hence, from Sections 5.2 and 5.3 we obtain the following result:

Theorem 14. *The decremental proper interval graph representation problem is solvable in $O(1)$ time per removed edge.*

6 Maintaining the Connected Components

In this section we describe a fully dynamic algorithm for maintaining connectivity in a proper interval graph G in $O(\log n)$ time per operation. The algorithm receives as input a series of operations to be performed on a graph, which can be any of the following: Adding a vertex, adding an edge, deleting a vertex, deleting an edge or querying if two blocks are in the same connected component. The algorithm depends on a data structure which includes the blocks and the contigs of the graph. It hence interacts with the proper interval graph representation algorithm. In response to an update request, changes are made to the representation of the graph based on the structure of its connected components prior to the update. Only then are the connected components of the graph updated.

Let us denote by $B(G)$ the *block graph* of G, that is, a graph in which each vertex corresponds to a block of G and two vertices are adjacent iff their corresponding blocks are adjacent in G. The algorithm maintains a spanning forest F of $B(G)$. In order to decide if two blocks are in the same connected component, the algorithm checks if they belong to the same tree in F.

The key idea is to design F so that it can be efficiently updated upon a modification in G. We define the edges of F as follows: For every two vertices u and v in $B(G)$, $(u,v) \in E(F)$ iff their corresponding blocks are consecutive in a contig of G. Consequently, each tree in F is a path representing a contig. The crucial observation about F is that an addition or a deletion of a vertex or an edge in G induces $O(1)$ modifications to the vertices and edges of F. This can be seen by noting that each modification of G induces $O(1)$ updates to near pointers in our representation of G.

It remains to show how to implement a spanning forest in which trees may be cut when an edge is deleted from F, linked when an edge is inserted to F, and which allows to query for each vertex to which tree does it belong. All these operations are supported by the ET-tree data structure of [8] in $O(\log n)$ time per operation.

We are now ready to state our main result:

Theorem 15. *The fully dynamic proper interval graph representation problem is solvable in $O(d + \log n)$ time per modification involving d edges.*

7 The Lower Bound

In this section we prove a lower bound of $\Omega(\log n / (\log\log n + \log b))$ amortized time per edge operation for fully dynamic proper interval graph recognition in the cell probe model of computation with word-size b [16].

Fredman and Saks [6] proved a lower bound of $\Omega(\log n / (\log\log n + \log b))$ amortized time per operation for the following *parity prefix sum* (PPS) problem: Given an array of integers $A[1], \ldots, A[n]$ with initial value zero, execute an arbitrary sequence of Add(t) and Sum(t) operations, where an Add(t) increases $A[t]$ by 1, and Sum(t) returns $(\sum_{i=1}^{t} A[i])$ *mod* 2. Fredman and Henzinger [5] showed that the same lower bound applies to the problem of maintaining connectivity in general graphs, by showing a reduction from a modified PPS problem, called *helpful parity prefix sum*, for which they proved the same lower bound. A slight change to their reduction yields the same lower bound for the problem of maintaining connectivity in proper interval graphs, as the graph built in the reduction is a union of two paths and therefore proper interval. Using a similar construction we can prove the following result:

Theorem 16. *Fully dynamic proper interval recognition takes $\Omega(\log n / (\log\log n + \log b))$ amortized time per edge operation in the cell probe model with word-size b.*

Acknowledgments

The first author gratefully acknowledges support from NSERC. The second author was supported in part by a grant from the Ministry of Science, Israel. The third author was supported by Eshkol scholarship from the Ministry of Science, Israel.

References

[1] A. V. Carrano. Establishing the order of human chromosome-specific DNA fragments. In A. D. Woodhead and B. J. Barnhart, editors, *Biotechnology and the Human Genome*, pages 37–50. Plenum Press, New York, 1988.

[2] D. Corneil, H. Kim, S. Natarajan, S. Olariu, and A. P. Sprague. Simple linear time recognition of unit interval graphs. *Inf. Proc. Letts.*, 55:99–104, 1995.

[3] X. Deng, P. Hell, and J. Huang. Recognition and representation of proper circular arc graphs. In *Proc. 2nd Integer Programming and Combinatorial Optimization (IPCO)*, pages 114–121, 1992. Journal version: *SIAM Journal on Computing*, 25:390–403, 1996.

[4] D. Eppstein, Z. Galil, G. F. Italiano, and A. Nissenzweig. Sparsification – A technique for speeding up dynamic graph algorithms. In *Proc. 33rd Symp. Foundations of Computer Science*, pages 60–69. IEEE, 1992.

[5] M. Fredman and M. Henzinger. Lower bounds for fully dynamic connectivity problems in graphs. *Algorithmica*. to appear.

[6] M. Fredman and M. Saks. The cell probe complexity of dynamic data structures. In *Proceedings of the 19th Annual ACM Symposium on Theory of Computing*, pages 345–354, 1989.

[7] M. C. Golumbic. *Algorithmic Graph Theory and Perfect Graphs*. Academic Press, New York, 1980.

[8] M. Henzinger and V. King. Randomized dynamic graph algorithms with polylogarithmic time per operation. In *Proceedings of the TwentySeventh Annual ACM Symposium on Theory of Computing*, pages 519–527, 1995.

[9] J. Holm, K. de Lichtenberg, and M. Thorup. Poly-logarithmic deterministic fully-dynamic algorithms for connectivity, minimum spanning tree, 2-edge and biconnectivity. In *Proceedings of the 30th Annual ACM Symposium on Theory of Computing (STOC'98)*, pages 79–89, New York, May 23–26 1998. ACM Press.

[10] W.-L. Hsu. A simple test for interval graphs. In W.-L. Hsu and R. C. T. Lee, editors, *Proc. 18th Int. Workshop (WG '92), Graph-Theoretic Concepts in Computer Science*, pages 11–16. Springer-Verlag, 1992. LNCS 657.

[11] L. Ibarra. Fully dynamic algorithms for chordal graphs. In *Proceedings of the 10th Annual ACM-SIAM Symposium on Discrete Algorithms (SODA'99)*, 1999.

[12] C. G. Lekkerkerker and J. C. Boland. Representation of a finite graph by a set of intervals on the real line. *Fundam. Math.*, 51:45–64, 1962.

[13] F. S. Roberts. Indifference graphs. In F. Harary, editor, *Proof Techniques in Graph Theory*, pages 139–146. Academic Press, New York, 1969.

[14] J. Watson, M. Gilman, J. Witkowski, and M. Zoller. *Recombinant DNA*. W.H. Freeman, New York, 2nd edition, 1992.

[15] G. Wegner. *Eigenschaften der Nerven homologisch einfacher Familien in R^n*. PhD thesis, Göttingen, 1967.

[16] A. Yao. Should tables be sorted. *Assoc. Comput. Mach.*, 28(3):615–628, 1981.

A Fast General Methodology for Information — Theoretically Optimal Encodings of Graphs

Xin He[1] and Ming-Yang Kao[2] and Hsueh-I Lu[3]

[1] Department of Computer Science and Engineering, State University of New York at Buffalo, Buffalo, NY 14260, USA. xinhe@cse.buffalo.edu[†]

[2] Department of Computer Science, Yale University, New Haven, CT 06250, USA. kao-ming-yang@cs.yale.edu[‡]

[3] Department of Computer Science and Information Engineering, National Chung-Cheng University, Chia-Yi 621, Taiwan, ROC. hil@cs.ccu.edu.tw[§]

Abstract. We propose a fast methodology for encoding graphs with information-theoretically minimum numbers of bits. The methodology is applicable to general classes of graphs; this paper focuses on simple planar graphs. Specifically, a graph with property π is called a *π-graph*. If π satisfies certain properties, then an n-node π-graph G can be encoded by a binary string X such that (1) G and X can be obtained from each other in $O(n \log n)$ time, and (2) X has at most $\beta(n) + o(\beta(n))$ bits for any function $\beta(n) = \Omega(n)$ so that there are at most $2^{\beta(n)+o(\beta(n))}$ distinct n-node π-graphs. Examples of such π include all conjunctions of the following sets of properties: (1) G is a planar graph or a plane graph; (2) G is directed or undirected; (3) G is triangulated, triconnected, biconnected, merely connected, or not required to be connected; and (4) G has at most ℓ_1 (respectively, ℓ_2) distinct node (respectively, edge) labels. These examples are novel applications of small cycle separators of planar graphs and settle several problems that have been open since Tutte's census series were published in 1960's.

1 Introduction

Let G be a graph with n nodes and m edges. This paper studies the problem of *encoding* G into a binary string X with the requirement that X can be *decoded* to reconstruct G. We propose a fast methodology for designing a coding scheme such that the bit count of X is information-theoretically optimal. The methodology is applicable to general classes of graphs; this paper focuses on *simple* planar graphs, i.e., planar graphs that are free of self-loops and multiple edges. Specifically, a graph with property π is called a *π-graph*. If π satisfies certain properties, then we can obtain an X such that (1) G and X can be computed from each other in $O(n \log n)$ time and (2) X has at most $\beta(n) + o(\beta(n))$ bits for any function $\beta(n) = \Omega(n)$ so that there are at most $2^{\beta(n)+o(\beta(n))}$ distinct n-node π-graphs.

Examples of suitable π include all conjunctions of the following sets of properties: (1) G is a planar graph or a plane graph; (2) G is directed or undirected; (3) G is

[†] Research supported in part by NSF Grant CCR-9205982.

[‡] Research supported in part by NSF Grant CCR-9531028.

[§] Research supported in part by NSC Grant NSC-88-2213-E-194-022.

triangulated, triconnected, biconnected, merely connected, or not required to be connected; and (4) G has at most ℓ_1 (respectively, ℓ_2) distinct node (respectively, edge) labels. For instance, π can be the property of being a directed unlabeled biconnected plane graph. These examples are novel applications of small cycle separators of planar graphs [12, 11]. They also settle several problems that have been open since Tutte's census series were published in 1960's [18, 20, 17, 19, 21]. Tutte proved that there are $2^{\beta(m)+o(\beta(m))}$ distinct m-edge plane triangulations where $\beta(m) = (\frac{8}{3} - \log_2 3)m + o(m) \approx 1.08m + o(m)$ [18] and that there are $2^{2m+o(n)}$ distinct m-edge n-node triconnected plane graphs that may be non-simple [20]. Note that the rooted trees are the only other non-trivial class of graphs with a known polynomial-time information-theoretically optimal coding scheme, which encodes a tree as nested parentheses using $2(n-1)$ bits.

Previously, Turán [16] used $4m$ bits to encode a plane graph G that may have self-loops. This bit count was improved by Keeler and Westbrook [10] to $3.58m$. They also gave coding schemes for several families of plane graphs. In particular, they used $1.53m$ bits for a triangulated simple G, and $3m$ bits for a connected G free of self-loops and degree-one nodes. For a simple triangulated G, He, Kao, and Lu [5] improved the bit count to $\frac{4}{3}m + O(1)$. For a simple G that is triconnected and thus free of degree-one nodes, they [5] improved the bit count to at most $2.835m$ bits. This bit count was later reduced to at most $\frac{3\log_2 3}{2}m + O(1) \approx 2.378m + O(1)$ by Chuang, Garg, He, Kao, and Lu [2]. These coding schemes all take linear time for encoding and decoding, but their bit counts are not information-theoretically optimal.

For applications that require support of certain queries, Jacobson [7] gave an $\Theta(n)$-bit encoding for a connected and simple planar graph G that supports traversal in $\Theta(\log n)$ time per node visited. Munro and Raman [13] improved this result and gave schemes to encode binary trees, rooted ordered trees and planar graphs. For a general planar G, they used $2m + 8n + o(m+n)$ bits while supporting adjacency and degree queries in $O(1)$ time. Chuang *et al.* [2] reduced this bit count to $2m + (5 + \frac{1}{k})n + o(m + n)$ for any constant $k > 0$ with the same query support. The bit count can be further reduced if only $O(1)$-time adjacency queries are supported, or G is simple, triconnected or triangulated.

Encoding problems with additional query support and for other classes of graphs have also been extensively studied in the literature. For certain graph families, Kannan, Naor and Rudich [8] gave schemes that encode each node with $O(\log n)$ bits and support $O(\log n)$-time testing of adjacency between two nodes. For dense graphs and complement graphs, Kao, Occhiogrosso, and Teng [9] devised two compressed representations from adjacency lists to speed up basic graph search techniques. Galperin and Wigderson [4] and Papadimitriou and Yannakakis [15] investigated complexity issues arising from encoding a graph by a small circuit that computes its adjacency matrix. For labeled planar graphs, Itai and Rodeh [6] gave an encoding of $\frac{3}{2}n\log n + O(n)$ bits. For unlabeled general graphs, Naor [14] gave an encoding of $\frac{1}{2}n^2 - n\log n + O(n)$ bits.

Section 2 discusses the general encoding methodology. Sections 3 and 4 use the methodology to obtain information-theoretically optimal encodings for various classes of planar graphs. Section 5 concludes the paper with some future research directions.

2 The Encoding Methodology

Let $|X|$ be the number of bits in a binary string X. Let $|G|$ be the number of nodes in a graph G. Let $|S|$ be the number of elements, counting multiplicity, in a multiset S.

Fact 1 (see [1, 3]). *Let $X_1, X_2, \ldots, X_k$ be $O(1)$ given binary strings. Let $n = |X_1| + |X_2| + \cdots + |X_k|$. Then there exists an $O(\log n)$-bit string χ, obtainable in $O(n)$ time, such that given the concatenation of $\chi, X_1, X_2, \ldots, X_k$, the index of the first symbol of each X_i in the concatenation can be computed in $O(1)$ time.*

Let $X_1 + X_2 + \cdots + X_k$ denote the concatenation of $\chi, X_1, X_2, \ldots, X_k$ as in Fact 1. We call χ the *auxiliary binary string* for $X_1 + X_2 + \cdots + X_k$.

A graph with property π is called a *π-graph*. Whether two π-graphs are *equivalent* or *distinct* depends on π. For example, let G_1 and G_2 be two distinct embeddings of the same planar graph. If π being a planar graph, then G_1 and G_2 are two equivalent π-graphs. If π being a planar embedding, then G_1 and G_2 are two distinct π-graphs. Let α be the number of distinct n-node π-graphs. Clearly it takes $\lceil \log_2 \alpha \rceil$ bits to differentiate all n-node π-graphs.

Let $B(G)$ be the boundary of the exterior face of a plane graph G. Let M be a table, each of whose entries stores a distinct n-node π-graph. Define $\text{index}_\pi(G, M)$ to be the $\lceil \log_2 \alpha \rceil$-bit index of G in M. Each π-graph is stored in M in a form such that it takes $2^{(n^{O(1)})}$ time to determine whether two π-graphs in this form are equivalent. For example, if π being a directed planar graphs then each n-node π-graph can be stored in M using an adjacency list, in which each node v keeps a list of nodes that have a directed edge leaving v. If π being an undirected plane graph, each n-node π-graph G can be stored in M using an adjacency list of G, $B(G)$, and the counterclockwise order of the neighbors of each node in the plane embedding of G. For brevity, we write $\text{index}_\pi(G, M)$ as $\text{index}_\pi(G)$ if M is clear from the context.

Lemma 2. *Let π be a property such that a table M for n-node π-graphs described as above can be obtained in $2^{(n^{O(1)})}$ time. Then G and $\text{index}_\pi(G, M)$ can be obtained from each other in $2^{(n^{O(1)})}$ time for any n-node π-graph G.*

Proof. Straightforward.

Let G_0 be an input n_0-node π-graph. Let $\lambda = \log\log\log(n_0)$. The encoding algorithm is merely a function call $\text{code}_\pi(G_0)$, where $\text{code}_\pi(G)$ is as follows.

```
Algorithm code_π(G) {
  If |G| ≤ λ then
    return index_π(G)
  else {
    compute π-graphs G1, G2, and a string X, from which G can be recovered;
    return code_π(G1) + code_π(G2) + X;
  }
}
```

Clearly, $\text{code}_\pi(G_0)$ can be decoded to recover G_0.

Algorithm $\text{code}_\pi(G)$ *satisfies the separation property* if there exist two constants c and r, where $0 \leq c < 1$ and $r > 1$, such that the following conditions hold:

P1. $\max(|G_1|, |G_2|) \leq |G|/r$.

P2. $|G_1| + |G_2| = |G| + O(|G|^c)$.

P3. $|X| = O(|G|^c)$.

Let $f(|G|)$ be the time required to obtain $\text{index}_\pi(G)$ and G from each other. Let $g(|G|)$ be the time required to obtain G_1, G_2, X from G, and vice versa.

Theorem 3. *Suppose Algorithm* $\text{code}_\pi(G)$ *satisfies the separation property.*

1. *If the number of distinct n-node π-graphs is at most* $2^{\beta(n)+o(\beta(n))}$ *for some function* $\beta(n) = \Omega(n)$, *then* $|\text{code}_\pi(G_0)| \leq \beta(n_0) + o(\beta(n_0))$ *for any* n_0*-node π-graph* G_0.
2. *If* $f(n) = 2^{(n^{O(1)})}$ *and* $g(n) = O(n)$, *then* G_0 *and* $\text{code}_\pi(G_0)$ *can be obtained from each other in* $O(n_0 \log n_0)$ *time.*

Proof. Statement 1. Many graphs may appear during the execution of Algorithm $\text{code}_\pi(G_0)$. These graphs can be organized as nodes of a binary tree T rooted at G_0, where (i) if G_1 and G_2 are obtained from G by calling $\text{code}_\pi(G)$, then G_1 and G_2 are the children of G in T, and (ii) if $|G| \leq \lambda$, then G has no children in T.

Consider the multiset S consisting of all graphs G that are nodes of T. We partition S into $\ell + 1$ multisets $S(0), S(1), S(2), \ldots, S(\ell)$. $S(0)$ consists of the graphs G with $|G| \leq \lambda$. For $i \geq 1$ $S(i)$ consists of the graphs G with $r^{i-1}\lambda < |G| \leq r^i\lambda$. Let $G_0 \in S(\ell)$. Clearly, $\ell = O(\log \frac{n_0}{\lambda})$.

Define $p = \sum_{H \in S(0)} |H|$. We first show

$$|S(i)| < \frac{p}{r^{i-1}\lambda}, \tag{1}$$

for every $1 \leq i \leq \ell$. Let G be a graph in $S(i)$. Let $S(0,G)$ be the set consisting of the leaf descendants of G in T; for example, $S(0,G_0) = S(0)$. By Condition P2, $|G| \leq \sum_{H \in S(0,G)} |H|$. By Condition P1, no two graphs in $S(i)$ are related in T. Therefore $S(i)$ contains at most one ancestor of H in T for every graph H in $S(0)$. It follows that $\sum_{G \in S(i)} |G| \leq \sum_{G \in S(i)} \sum_{H \in S(0,G)} |H| \leq p$. Since $|G| > r^{i-1}\lambda$ for every G in $S(i)$, Inequality (1) holds.

Suppose the children of G in T are G_1 and G_2. Let $b(G) = |X| + |\chi|$, where χ is the auxiliary binary string for $\text{code}_\pi(G_1) + \text{code}_\pi(G_2) + X$. Let $q = \sum_{i \geq 1} \sum_{G \in S(i)} b(G)$. Clearly $|\text{code}_\pi(G_0)| = \sum_{H \in S(0)} |\text{code}_\pi(H)| + q \leq \sum_{H \in S(0)} (\beta(|H|) + o(\beta(|H|))) + q$. Since $\beta(n) = \Omega(n)$, $|\text{code}_\pi(G_0)| \leq \beta(p) + q + o(\beta(p))$. Therefore Statement 1 can be proved by showing that $p = n_0 + o(n_0)$ and $q = o(n_0)$.

By Condition P3, $|X| = O(|G|^c)$. By Fact 1, $|\chi| = O(\log |G|)$. Thus, $b(G) = O(|G|^c)$, and

$$q = \sum_{i \geq 1} \sum_{G \in S(i)} O(|G|^c). \tag{2}$$

Now we regard the execution of $\text{code}_\pi(G_0)$ as a process of growing T. Let $a(T) = \sum_{H \text{ is a leaf of } T} |H|$. At the beginning of the function call $\text{code}_\pi(G_0)$, T has exactly one node G_0, and thus $a(T) = n_0$. At the end of the function call, T is fully expanded, and thus $a(T) = p$. By Condition P2, during the execution of $\text{code}_\pi(G_0)$, every function call $\text{code}_\pi(G)$ with $|G| > \lambda$ increases $a(T)$ by $O(|G|^c)$. Hence

$$p = n_0 + \sum_{i \geq 1} \sum_{G \in S(i)} O(|G|^c). \tag{3}$$

By Equations (2) and (3), it remains to prove $\sum_{i\geq 1}\sum_{G\in S(i)} |G|^c = o(n_0)$. Note that

$$\sum_{i \geq 1} \sum_{G \in S(i)} |G|^c \leq \sum_{i \geq 1} (r^i\lambda)^c p/(r^{i-1}\lambda) = p\lambda^{c-1} r \sum_{i \geq 1} r^{(c-1)i} = p\lambda^{c-1} O(1) = o(p). \tag{4}$$

By Equations (3) and (4), $p = n_0 + o(p)$, and thus $p = O(n_0)$. Hence

$$\sum_{i \geq 1} \sum_{G \in S(i)} |G|^c = o(n_0).$$

By Equations (2) and (3), $p = n_0 + o(n_0)$ and $q = o(n_0)$, finishing the proof of Statement 1.

Statement 2. Since $|G| \leq r^i\lambda$ and $|S(i)| = O(\frac{n_0}{r^i\lambda})$ for every $0 \leq i \leq \ell$ and for every G in $S(i)$, G_0 and $\text{code}_\pi(G_0)$ can be obtained from each other in time

$$\frac{n_0}{\lambda} O(f(\lambda) + \sum_{1 \leq i \leq \ell} r^{-i} g(r^i\lambda)).$$

Clearly $f(\lambda) = 2^{(\lambda^{O(1)})} = 2^{o(\log\log n_0)} = o(\log n_0)$. Since $\ell = O(\log n_0)$ and $g(n) = O(n)$, $\sum_{1\leq i\leq \ell} r^{-i} g(r^i\lambda)) = \sum_{1\leq i\leq \ell} \lambda = O(\lambda \log n_0)$, and Statement 2 follows immediately.

The next two sections use Theorem 3 to encode various classes of graphs G. Section 3 considers plane triangulations. Section 4 considers planar graphs and plane graphs.

3 Plane triangulation

In this section, a π-graph is a directed or undirected plane triangulation which has at most ℓ_1 (respectively, ℓ_2) distinct node (respectively, edge) labels. The number of distinct n-node π-graphs is $2^{\Omega(n)}$ [18]. Our encoding scheme is based on the next fact.

Fact 4 (See [12]). *Let G be an n-node plane triangulation. We can compute a cycle C of G in $O(n)$ time such that*

- *C has at most $\sqrt{8n}$ nodes; and*
- *the numbers of G's nodes inside and outside C are at most $2n/3$, respectively.*

Let G be a given n-node π-graph. Let C be a cycle of G guaranteed by Fact 4. Let G_{in} (respectively, G_{out}) be the subgraph of G formed by C and the part of G inside (respectively, outside) C. Let x be an arbitrary node on C.

G_1 is obtained by placing a cycle C_1 of three nodes surrounding G_{in} and then triangulating the interior face of C_1 such that a particular node y_1 of C_1 has degree strictly higher than the other two. The labels for the nodes and edges of $G_1 - G_{\text{in}}$ can be assigned arbitrarily.

G_2 is obtained from G_{out} by (a) placing a cycle C_2 of three nodes surrounding G_{out} and then triangulating the interior face of C_2 such that a particular node y_2 of C_2 has degree strictly higher than the other two; and (b) triangulating the interior face of C by placing a new node z inside of C and then connecting it to each node of C by an edge. Similarly, the labels for the nodes and edges of $G_2 - G_{out}$ can be assigned arbitrarily.

Define $\text{dfs}(u,G,v)$ as follows. When $\text{dfs}(u,G,v)$ is called, v is always a node on $B(G)$. Let w be the counterclockwise neighbor of v on $B(G)$. We perform a depth-first search of G starting from v such that (1) the neighbors of each node are visited in the counterclockwise order; and (2) w is the second visited node. Let $\text{dfs}(u,G,v)$ be the binary index of u in the above depth-first search. Let $X = \text{dfs}(x,G_1,y_1)+\text{dfs}(x,G_2,y_2)+\text{dfs}(z,G_2,y_2)$.

Lemma 5. *1. G_1 and G_2 are π-graphs.*
2. *There exists a constant $r > 1$ with $\max(|G_1|,|G_2|) \le n/r$.*
3. *$|G_1|+|G_2| = n+O(\sqrt{n})$.*
4. *$|X| = O(\log n)$.*
5. *G_1, G_2, X can be obtained from G in $O(n)$ time.*
6. *G can be obtained from G_1, G_2, X in $O(n)$ time.*

Proof. Statements 1–5 are straightforward by Fact 4 and the definitions of G_1, G_2 and X. Statement 6 is proved as follows. It takes $O(n)$ time to locate y_1 (respectively, y_2) in G_1 (respectively, G_2) by looking for the node with the highest degree on $B(G_1)$ (respectively, $B(G_2)$). By Fact 1, it takes $O(1)$ time to obtain $\text{dfs}(y_1,G_1,x)$, $\text{dfs}(y_2,G_2,x)$, and $\text{dfs}(y_2,G_2,z)$ from X. Therefore x and z can be located in G_1 and G_2 in $O(n)$ time by depth-first traversal. Now G_{in} can be obtained from G_1 by removing $B(G_1)$ and its incident edges. The cycle C in G_{in} is simply $B(G_{in})$. Also, G_{out} can be obtained from G_2 by removing $B(G_2)$, z, and their incident edges. The C in G_{out} is simply the boundary of the face that encloses z and its incident edges in G_2. Since we know the positions of x in G_{in} and G_{out}, G can be obtained from G_{in} and G_{out} by fitting them together along C by aligning x. The overall time complexity is $O(n)$.

Theorem 6. *Let G_0 be an n_0-node π-graph. Then G_0 and $\text{code}_\pi(G_0)$ can be obtained from each other in $O(n_0 \log n_0)$ time. Moreover, $|\text{code}_\pi(G_0)| \le \beta(n_0)+o(\beta(n_0))$ for any function $\beta(n)$ such that the number of distinct n-node π-graphs is at most $2^{\beta(n)+o(\beta(n))}$.*

Proof. One can easily see that determining whether two n-node π-graphs are equivalent can be done in $2^{(n^{O(1)})}$ time. It is also clear that the class of n-node π-graphs can be enumerated in $2^{(n^{O(1)})}$ time. Therefore the table M for n-node π-graphs can be obtained in $2^{(n^{O(1)})}$ time. By Lemma 2, $\text{index}_\pi(G)$ and G can be obtained from each other in $2^{(|G|^{O(1)})}$ time. Thus the theorem follows from Theorem 3 and Lemma 5.

4 Planar graphs and plane graphs

In this section, π can be any conjunction of the following properties of G:

- G is a planar graph or a plane graph;

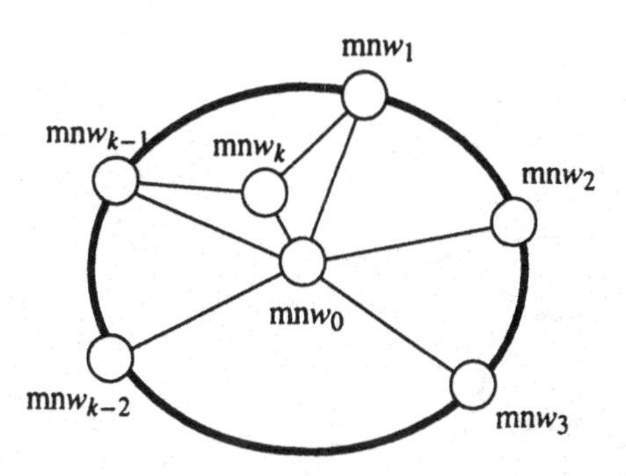

Fig. 1. A k-wheel graph W_k.

- G is directed or undirected;
- G is triconnected, biconnected, connected, or not required to be connected; and
- G has at most ℓ_1 (respectively, ℓ_2) distinct node (respectively, edge) labels.

Clearly there are $2^{\Omega(n)}$ distinct n-node π-graphs.

Let G be an input n-node π-graph. For the cases of π being a planar graph rather than a plane graph, let G be embedded first. Note that this is only for the encoding process to be able to apply Fact 4. At the base level, we still use the table M for π-graphs rather than the much larger table M' for embedded π-graphs. As shown below, the decoding process does not require the π-graphs to be embedded.

Let G' be obtained from triangulating G. Let C be a cycle of G' guaranteed by Fact 4. Let G_C be the union of G and C. Let G_{in} (respectively, G_{out}) be the subgraph of G_C formed by C and the part of G_C inside (respectively, outside) C. Let $C = x_1x_2\cdots x_\ell x_{\ell+1}$, where $x_{\ell+1} = x_1$. By Fact 4, $\ell = O(\sqrt{n})$.

Lemma 7. *Let H be an $O(n)$-node planar graph. There exists an integer k with $n^{0.6} \le k \le n^{0.7}$ such that H does not contain any node of degree k or $k-1$.*

Proof. Assume for a contradiction that such a k does not exist. It follows that the sum of degrees of all nodes in H is at least $(n^{0.6}+n^{0.7})(n^{0.7}-n^{0.6})/4 = \Omega(n^{1.4})$. This contradicts the fact that H is planar.

Let W_k, with $k \ge 3x$, be a *k-wheel graph* defined as follows. As shown in Fig. 1, W_k consists of $k+1$ nodes $w_0, w_1, w_2, \ldots, w_{k-1}, w_k$, where $w_1, w_2, \ldots, w_k, w_1$ form a cycle. w_0 is a degree-k node incident to each node on the cycle. Finally, w_1 is incident to w_{k-1}. Clearly W_k is triconnected. Also, w_1 and w_k are the only degree-four neighbors of w_0 in W_k. Let k_1 (respectively, k_2) be an integer k guaranteed by Lemma 7 for G_{in} (respectively, G_{out}). Now we define G_1, G_2 and X as follows.

G_1 is obtained from G_{in} and a k_1-wheel graph W_{k_1} by adding an edge (w_i, x_i) for every $i = 1, \ldots, \ell$. Clearly for the case of π being a plane graph, G_1 can be embedded such that W_{k_1} is outside G_{in}, as shown in Fig. 2(a). Thus, the original embedding of G_{in} can be obtained from G_1 by removing all nodes of W_{k_1}. The node labels, edge labels, and edge directions of $G_1 - G_{\text{in}}$ can be assigned arbitrarily.

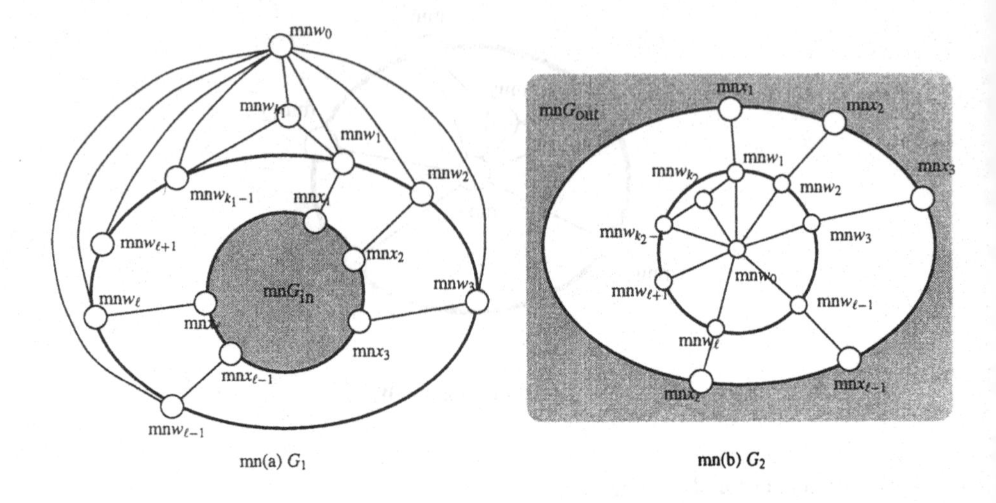

Fig. 2. G_1 and G_2. The gray area of G_1 is G_{in}. The gray area of G_2 is G_{out}.

G_2 is obtained from G_{out} and a k_2-wheel graph W_{k_2} by adding an edge (w_i, x_i) for every $i = 1, \ldots, \ell$. Clearly for the case of π being a plane graph, G_2 can be embedded such that W_{k_2} is inside C, as shown in Fig. 2(b). Thus, the original embedding of G_{out} can be obtained from G_2 by removing all nodes of W_{k_2}. The node labels, edge labels, and edge directions of $G_2 - G_{\text{out}}$ can be assigned arbitrarily.

Let X be an $O(\sqrt{n})$-bit string which encodes k_1, k_2, and whether each edge (x_i, x_{i+1}) is an original edge in G, for $i = 1, \ldots, \ell$.

Lemma 8. *1. G_1 and G_2 are π-graphs.*
2. There exists a constant $r > 1$ with $\max(|G_1|, |G_2|) \le n/r$.
3. $|G_1| + |G_2| = n + O(n^{0.7})$.
4. $|X| = O(\sqrt{n})$.
5. G_1, G_2, X can be obtained from G in $O(n)$ time.
6. G can be obtained from G_1, G_2, X in $O(n)$ time.

Proof. Since W_{k_1} and W_{k_2} are both triconnected, and each node of C has degree at least three in G_1 and G_2, Statement 1 holds for each case of the connectivity of the input π-graph G. Statements 2–5 are straightforward by Fact 4 and the definitions of G_1, G_2 and X. Statement 6 is proved as follows. First of all, we obtain k_1 from X. Since G_{in} does not contain any node of degree k_1 or $k_1 - 1$, w_0 is the only degree-k_1 node in G_1. Therefore it takes $O(n)$ time to identify w_0 in G_1. w_{k_1} is the only degree-3 neighbor of w_0. Since $k_1 > \ell$, w_1 is the only degree-5 neighbor of w_0. w_2 is the common neighbor of w_0 and w_1 that is not adjacent to w_{k_1}. From now on, w_i, for each $i = 3, 4, \ldots, \ell$, is the common neighbor of w_0 and w_{i-1} other than w_{i-2}. Clearly, $w_1, w_2, \ldots, w_\ell$ and thus $x_1, x_2, \ldots, x_\ell$ can be identified in $O(n)$ time. G_{in} can now be obtained from G_1 by removing W_{k_1}. Similarly, G_{out} can be obtained from G_2 and X by deleting W_{k_1} after identifying $x_1, x_2, \ldots, x_\ell$. Finally, G_C can be recovered by fitting G_{in} and G_{out} together by aligning $x_1, x_2, \ldots, x_\ell$. Based on X, G can then be obtained from G_C by removing the edges of C that are not originally in G.

Remark. In the proof for Statement 5 of Lemma 8, identifying the degree-k_1 node (and the k_1-wheel graph W_{k_1}) does not require the embedding for G_1. Therefore the decoding process does not require the π-graphs to be embedded. This is different from the proof of Lemma 5.

Theorem 9. *Let G_0 be an n_0-node π-graph. Then G_0 and $\text{code}_\pi(G_0)$ can be obtained from each other in $O(n_0 \log n_0)$ time. Moreover, $|\text{code}_\pi(G_0)| \leq \beta(n_0) + o(\beta(n_0))$ for any function $\beta(n)$ such that the number of distinct n-node π-graphs is at most $2^{\beta(n)+o(\beta(n))}$.*

Proof. One can easily see that determining whether two n-node π-graphs are equivalent can be done in $2^{(n^{O(1)})}$ time. It is also clear that the class of n-node π-graphs can be enumerated in $2^{(n^{O(1)})}$ time. Therefore the table M for n-node π-graphs can be obtained in $2^{(n^{O(1)})}$ time. By Lemma 2, $\text{index}_\pi(G)$ and G can be obtained from each other in $2^{(|G|^{O(1)})}$ time. The theorem follows from Theorem 3 and Lemma 8.

5 Future directions

The coding schemes presented in this paper require $O(n \log n)$ time for encoding and decoding. An immediate open question is whether one can encode some graphs other than rooted trees in $O(n)$ time using information-theoretically minimum number of bits.

References

[1] T. C. Bell, J. G. Cleary, and I. H. Witten, *Text Compression*, Prentice-Hall, Englewood Cliffs, NJ, 1990.

[2] R. C.-N. Chuang, A. Garg, X. He, M.-Y. Kao, and H.-I. Lu, *Compact encodings of planar graphs via canonical orderings and multiple parentheses*, in Automata, Languages and Programming, 25th Colloquium, K. G. Larsen, S. Skyum, and G. Winskel, eds., vol. 1443 of Lecture Notes in Computer Science, Aalborg, Denmark, 13–17 July 1998, Springer-Verlag, pp. 118–129.

[3] P. Elias, *Universal codeword sets and representations of the integers*, IEEE Transactions on Information Theory, IT-21 (1975), pp. 194–203.

[4] H. Galperin and A. Wigderson, *Succinct representations of graphs*, Information and Control, 56 (1983), pp. 183–198.

[5] X. He, M.-Y. Kao, and H.-I. Lu, *Linear-time succinct encodings of planar graphs via canonical orderings*, SIAM Journal on Discrete Mathematics, (1999). To appear.

[6] A. Itai and M. Rodeh, *Representation of graphs*, Acta Informatica, 17 (1982), pp. 215–219.

[7] G. Jacobson, *Space-efficient static trees and graphs*, in 30th Annual Symposium on Foundations of Computer Science, Research Triangle Park, North Carolina, 30 Oct.–1 Nov. 1989, IEEE, pp. 549–554.

[8] S. Kannan, N. Naor, and S. Rudich, *Implicit representation of graphs*, SIAM Journal on Discrete Mathematics, 5 (1992), pp. 596–603.

[9] M. Y. KAO, N. OCCHIOGROSSO, AND S. H. TENG, *Simple and efficient compression schemes for dense and complement graphs*, Journal of Combinatorial Optimization, 2 (1999), pp. 351–359.

[10] K. KEELER AND J. WESTBROOK, *Short encodings of planar graphs and maps*, Discrete Applied Mathematics, 58 (1995), pp. 239–252.

[11] R. J. LIPTON AND R. E. TARJAN, *A separator theorem for planar graphs*, SIAM Journal on Applied Mathematics, 36 (1979), pp. 177–189.

[12] G. L. MILLER, *Finding small simple cycle separators for 2-connected planar graphs*, Journal of Computer and System Sciences, 32 (1986), pp. 265–279.

[13] J. I. MUNRO AND V. RAMAN, *Succinct representation of balanced parentheses, static trees and planar graphs*, in 38th Annual Symposium on Foundations of Computer Science, Miami Beach, Florida, 20–22 Oct. 1997, IEEE, pp. 118–126.

[14] M. NAOR, *Succinct representation of general unlabeled graphs*, Discrete Applied Mathematics, 28 (1990), pp. 303–307.

[15] C. H. PAPADIMITRIOU AND M. YANNAKAKIS, *A note on succinct representations of graphs*, Information and Control, 71 (1986), pp. 181–185.

[16] G. TURÁN, *On the succinct representation of graphs*, Discrete Applied Mathematics, 8 (1984), pp. 289–294.

[17] W. T. TUTTE, *A census of Hamiltonian polygons*, Canadian Journal of Mathematics, 14 (1962), pp. 402–417.

[18] W. T. TUTTE, *A census of planar triangulations*, Canadian Journal of Mathematics, 14 (1962), pp. 21–38.

[19] W. T. TUTTE, *A census of slicing*, Canadian Journal of Mathematics, 14 (1962), pp. 708–722.

[20] W. T. TUTTE, *A census of planar maps*, Canadian Journal of Mathematics, 15 (1963), pp. 249–271.

[21] W. T. TUTTE, *The enumerative theory of planar maps*, in A survey of combinatorial theory, H. F. Srivastava, C. R. Rao, G. C. Rota, and S. S. Shrikhande, eds., North-Holland Publishing Company, 1973.

Author Index

Springer and the environment

At Springer we firmly believe that an international science publisher has a special obligation to the environment, and our corporate policies consistently reflect this conviction.

We also expect our business partners – paper mills, printers, packaging manufacturers, etc. – to commit themselves to using materials and production processes that do not harm the environment. The paper in this book is made from low- or no-chlorine pulp and is acid free, in conformance with international standards for paper permanency.

Lecture Notes in Computer Science

For information about Vols. 1–1557
please contact your bookseller or Springer-Verlag

Vol. 1558: H. J. van den Herik, H. Iida (Eds.), Computers and Games. Proceedings, 1998. XVIII, 337 pages. 1999.

Vol. 1559: P. Flener (Ed.), Logic-Based Program Synthesis and Transformation. Proceedings, 1998. X, 331 pages. 1999.

Vol. 1560: H. Imai, Y. Zheng (Eds.), Public Key Cryptography. Proceedings, 1999. IX, 327 pages. 1999.

Vol. 1561: I. Damgård (Ed.), Lectures on Data Security. VII, 250 pages. 1999.

Vol. 1562: C.L. Nehaniv (Ed.), Computation for Metaphors, Analogy, and Agents. X, 389 pages. 1999. (Subseries LNAI).

Vol. 1563: Ch. Meinel, S. Tison (Eds.), STACS 99. Proceedings, 1999. XIV, 582 pages. 1999.

Vol. 1565: P. P. Chen, J. Akoka, H. Kangassalo, B. Thalheim (Eds.), Conceptual Modeling. XXIV, 303 pages. 1999.

Vol. 1567: P. Antsaklis, W. Kohn, M. Lemmon, A. Nerode, S. Sastry (Eds.), Hybrid Systems V. X, 445 pages. 1999.

Vol. 1568: G. Bertrand, M. Couprie, L. Perroton (Eds.), Discrete Geometry for Computer Imagery. Proceedings, 1999. XI, 459 pages. 1999.

Vol. 1569: F.W. Vaandrager, J.H. van Schuppen (Eds.), Hybrid Systems: Computation and Control. Proceedings, 1999. X, 271 pages. 1999.

Vol. 1570: F. Puppe (Ed.), XPS-99: Knowledge-Based Systems. VIII, 227 pages. 1999. (Subseries LNAI).

Vol. 1571: P. Noriega, C. Sierra (Eds.), Agent Mediated Electronic Commerce. Proceedings, 1998. IX, 207 pages. 1999. (Subseries LNAI).

Vol. 1573: J.M.L.M. Palma, J. Dongarra, V. Hernández (Eds.), Vector and Parallel Processing – VECPAR'98. Proceedings, 1998. XVI, 706 pages. 1999.

Vol. 1572: P. Fischer, H.U. Simon (Eds.), Computational Learning Theory. Proceedings, 1999. X, 301 pages. 1999. (Subseries LNAI).

Vol. 1574: N. Zhong, L. Zhou (Eds.), Methodologies for Knowledge Discovery and Data Mining. Proceedings, 1999. XV, 533 pages. 1999. (Subseries LNAI).

Vol. 1575: S. Jähnichen (Ed.), Compiler Construction. Proceedings, 1999. X, 301 pages. 1999.

Vol. 1576: S.D. Swierstra (Ed.), Programming Languages and Systems. Proceedings, 1999. X, 307 pages. 1999.

Vol. 1577: J.-P. Finance (Ed.), Fundamental Approaches to Software Engineering. Proceedings, 1999. X, 245 pages. 1999.

Vol. 1578: W. Thomas (Ed.), Foundations of Software Science and Computation Structures. Proceedings, 1999. X, 323 pages. 1999.

Vol. 1579: W.R. Cleaveland (Ed.), Tools and Algorithms for the Construction and Analysis of Systems. Proceedings, 1999. XI, 445 pages. 1999.

Vol. 1580: A. Včkovski, K.E. Brassel, H.-J. Schek (Eds.), Interoperating Geographic Information Systems. Proceedings, 1999. XI, 329 pages. 1999.

Vol. 1581: J.-Y. Girard (Ed.), Typed Lambda Calculi and Applications. Proceedings, 1999. VIII, 397 pages. 1999.

Vol. 1582: A. Lecomte, F. Lamarche, G. Perrier (Eds.), Logical Aspects of Computational Linguistics. Proceedings, 1997. XI, 251 pages. 1999. (Subseries LNAI).

Vol. 1583: D. Scharstein, View Synthesis Using Stereo Vision. XV, 163 pages. 1999.

Vol. 1584: G. Gottlob, E. Grandjean, K. Seyr (Eds.), Computer Science Logic. Proceedings, 1998. X, 431 pages. 1999.

Vol. 1585: B. McKay, X. Yao, C.S. Newton, J.-H. Kim, T. Furuhashi (Eds.), Simulated Evolution and Learning. Proceedings, 1998. XIII, 472 pages. 1999. (Subseries LNAI).

Vol. 1586: J. Rolim et al. (Eds.), Parallel and Distributed Processing. Proceedings, 1999. XVII, 1443 pages. 1999.

Vol. 1587: J. Pieprzyk, R. Safavi-Naini, J. Seberry (Eds.), Information Security and Privacy. Proceedings, 1999. XI, 327 pages. 1999.

Vol. 1590: P. Atzeni, A. Mendelzon, G. Mecca (Eds.), The World Wide Web and Databases. Proceedings, 1998. VIII, 213 pages. 1999.

Vol. 1592: J. Stern (Ed.), Advances in Cryptology – EUROCRYPT '99. Proceedings, 1999. XII, 475 pages. 1999.

Vol. 1593: P. Sloot, M. Bubak, A. Hoekstra, B. Hertzberger (Eds.), High-Performance Computing and Networking. Proceedings, 1999. XXIII, 1318 pages. 1999.

Vol. 1594: P. Ciancarini, A.L. Wolf (Eds.), Coordination Languages and Models. Proceedings, 1999. IX, 420 pages. 1999.

Vol. 1595: K. Hammond, T. Davie, C. Clack (Eds.), Implementation of Functional Languages. Proceedings, 1998. X, 247 pages. 1999.

Vol. 1596: R. Poli, H.-M. Voigt, S. Cagnoni, D. Corne, G.D. Smith, T.C. Fogarty (Eds.), Evolutionary Image Analysis, Signal Processing and Telecommunications. Proceedings, 1999. X, 225 pages. 1999.

Vol. 1597: H. Zuidweg, M. Campolargo, J. Delgado, A. Mullery (Eds.), Intelligence in Services and Networks. Proceedings, 1999. XII, 552 pages. 1999.

Vol. 1598: R. Poli, P. Nordin, W.B. Langdon, T.C. Fogarty (Eds.), Genetic Programming. Proceedings, 1999. X, 283 pages. 1999.

Lecture Notes in Computer Science

For information about Vols. 1–1557
please contact your bookseller or Springer-Verlag

Vol. 1599: T. Ishida (Ed.), Multiagent Platforms. Proceedings, 1998. VIII, 187 pages. 1999. (Subseries LNAI).

Vol. 1601: J.-P. Katoen (Ed.), Formal Methods for Real-Time and Probabilistic Systems. Proceedings, 1999. X, 355 pages. 1999.

Vol. 1602: A. Sivasubramaniam, M. Lauria (Eds.), Network-Based Parallel Computing. Proceedings, 1999. VIII, 225 pages. 1999.

Vol. 1603: J. Vitek, C.D. Jensen (Eds.), Secure Internet Programming. X, 501 pages. 1999.

Vol. 1605: J. Billington, M. Diaz, G. Rozenberg (Eds.), Application of Petri Nets to Communication Networks. IX, 303 pages. 1999.

Vol. 1606: J. Mira, J.V. Sánchez-Andrés (Eds.), Foundations and Tools for Neural Modeling. Proceedings, Vol. I, 1999. XXIII, 865 pages. 1999.

Vol. 1607: J. Mira, J.V. Sánchez-Andrés (Eds.), Engineering Applications of Bio-Inspired Artificial Neural Networks. Proceedings, Vol. II, 1999. XXIII, 907 pages. 1999.

Vol. 1608: S. Doaitse Swierstra, P.R. Henriques, J.N. Oliveira (Eds.), Advanced Functional Programming. Proceedings, 1998. XII, 289 pages. 1999.

Vol. 1609: Z. W. Raś, A. Skowron (Eds.), Foundations of Intelligent Systems. Proceedings, 1999. XII, 676 pages. 1999. (Subseries LNAI).

Vol. 1610: G. Cornuéjols, R.E. Burkard, G.J. Woeginger (Eds.), Integer Programming and Combinatorial Optimization. Proceedings, 1999. IX, 453 pages. 1999.

Vol. 1611: I. Imam, Y. Kodratoff, A. El-Dessouki, M. Ali (Eds.), Multiple Approaches to Intelligent Systems. Proceedings, 1999. XIX, 899 pages. 1999. (Subseries LNAI).

Vol. 1612: R. Bergmann, S. Breen, M. Göker, M. Manago, S. Wess, Developing Industrial Case-Based Reasoning Applications. XX, 188 pages. 1999. (Subseries LNAI).

Vol. 1613: A. Kuba, M. Šámal, A. Todd-Pokropek (Eds.), Information Processing in Medical Imaging. Proceedings, 1999. XVII, 508 pages. 1999.

Vol. 1614: D.P. Huijsmans, A.W.M. Smeulders (Eds.), Visual Information and Information Systems. Proceedings, 1999. XVII, 827 pages. 1999.

Vol. 1615: C. Polychronopoulos, K. Joe, A. Fukuda, S. Tomita (Eds.), High Performance Computing. Proceedings, 1999. XIV, 408 pages. 1999.

Vol. 1617: N.V. Murray (Ed.), Automated Reasoning with Analytic Tableaux and Related Methods. Proceedings, 1999. X, 325 pages. 1999. (Subseries LNAI).

Vol. 1618: J. Bézivin, P.-A. Muller (Eds.), The Unified Modeling Language. Proceedings, 1998. IX, 443 pages. 1999.

Vol. 1619: M.T. Goodrich, C.C. McGeoch (Eds.), Algorithm Engineering and Experimentation. Proceedings, 1999. VIII, 349 pages. 1999.

Vol. 1620: W. Horn, Y. Shahar, G. Lindberg, S. Andreassen, J. Wyatt (Eds.), Artificial Intelligence in Medicine. Proceedings, 1999. XIII, 454 pages. 1999. (Subseries LNAI).

Vol. 1621: D. Fensel, R. Studer (Eds.), Knowledge Acquisition Modeling and Management. Proceedings, 1999. XI, 404 pages. 1999. (Subseries LNAI).

Vol. 1622: M. González Harbour, J.A. de la Puente (Eds.), Reliable Software Technologies – Ada-Europe'99. Proceedings, 1999. XIII, 451 pages. 1999.

Vol. 1625: B. Reusch (Ed.), Computational Intelligence. Proceedings, 1999. XIV, 710 pages. 1999.

Vol. 1626: M. Jarke, A. Oberweis (Eds.), Advanced Information Systems Engineering. Proceedings, 1999. XIV, 478 pages. 1999.

Vol. 1627: T. Asano, H. Imai, D.T. Lee, S.-i. Nakano, T. Tokuyama (Eds.), Computing and Combinatorics. Proceedings, 1999. XIV, 494 pages. 1999.

Col. 1628: R. Guerraoui (Ed.), ECOOP'99 - Object-Oriented Programming. Proceedings, 1999. XIII, 529 pages. 1999.

Vol. 1629: H. Leopold, N. García (Eds.), Multimedia Applications, Services and Techniques - ECMAST'99. Proceedings, 1999. XV, 574 pages. 1999.

Vol. 1631: P. Narendran, M. Rusinowitch (Eds.), Rewriting Techniques and Applications. Proceedings, 1999. XI, 397 pages. 1999.

Vol. 1632: H. Ganzinger (Ed.), Automated Deduction – Cade-16. Proceedings, 1999. XIV, 429 pages. 1999. (Subseries LNAI).

Vol. 1633: N. Halbwachs, D. Peled (Eds.), Computer Aided Verification. Proceedings, 1999. XII, 506 pages. 1999.

Vol. 1634: S. Džeroski, P. Flach (Eds.), Inductive Logic Programming. Proceedings, 1999. VIII, 303 pages. 1999. (Subseries LNAI).

Vol. 1636: L. Knudsen (Ed.), Fast Software Encryption. Proceedings, 1999. VIII, 317 pages. 1999.

Vol. 1638: A. Hunter, S. Parsons (Eds.), Symbolic and Quantitative Approaches to Reasoning and Uncertainty. Proceedings, 1999. IX, 397 pages. 1999. (Subseries LNAI).

Vol. 1639: S. Donatelli, J. Kleijn (Eds.), Application and Theory of Petri Nets 1999. Proceedings, 1999. VIII, 425 pages. 1999.

Vol. 1640: W. Tepfenhart, W. Cyre (Eds.), Conceptual Structures: Standards and Practices. Proceedings, 1999. XII, 515 pages. 1999. (Subseries LNAI).

Vol. 1643: J. Nešetřil (Ed.), Algorithms – ESA '99. Proceedings, 1999. XII, 552 pages. 1999.

Vol. 1644: J. Wiedermann, P. van Emde Boas, M. Nielsen (Eds.), Automata, Languages, and Programming. Proceedings, 1999. XIV, 720 pages. 1999.

Vol. 1647: F.J. Garijo, M. Boman (Eds.), Multi-Agent System Engineering. Proceedings, 1999. X, 233 pages. 1999. (Subseries LNAI).

Vol. 1649: R.Y. Pinter, S. Tsur (Eds.), Next Generation Information Technologies and Systems. Proceedings, 1999. IX, 327 pages. 1999.

Vol. 1650: K.-D. Althoff, R. Bergmann, L.K. Branting (Eds.), Case-Based Reasoning Research and Development. Proceedings, 1999. XII, 598 pages. 1999. (Subseries LNAI).

Vol. 1651: R.H. Güting, D. Papadias, F. Lochovsky (Eds.), Advances in Spatial Databases. Proceedings, 1999. XI, 371 pages. 1999.

Vol. 1653: S. Covaci (Ed.), Active Networks. Proceedings, 1999. XIII, 346 pages. 1999.